Food Science Text Series

The Food Science Text Series provides faculty with the leading teaching tools. The Editorial Board has outlined the most appropriate and complete content for each food science course in a typical food science program and has identified textbooks of the highest quality, written by the leading food science educators.

Harry T. Lawless · Hildegarde Heymann

Sensory Evaluation of Food

Principles and Practices

Second Edition

Harry T. Lawless
Department of Food Science
Cornell University
Stocking Hall, Room 106
14853 Ithaca
NY, USA
htl1@cornell.edu

Hildegarde Heymann
Department of Viticulture and Enology
University of California – Davis
2003 RMI Sensory Building
Davis 95616
CA, USA
hheymann@ucdavis.edu

ISSN 1572-0330
ISBN 978-1-4419-6487-8   e-ISBN 978-1-4419-6488-5
DOI 10.1007/978-1-4419-6488-5
Springer New York Dordrecht Heidelberg London

Library of Congress Control Number: 2010932599

Printed on acid-free paper

Springer is part of Springer Science+Business Media (www.springer.com)

Preface

The field of sensory science has grown exponentially since the publication of the previous version of this work. Fifteen years ago the journal *Food Quality and Preference* was fairly new. Now it holds an eminent position as a venue for research on sensory test methods (among many other topics). Hundreds of articles relevant to sensory testing have appeared in that and in other journals such as the *Journal of Sensory Studies*. Knowledge of the intricate cellular processes in chemoreception, as well as their genetic basis, has undergone nothing less than a revolution, culminating in the award of the Nobel Prize to Buck and Axel in 2004 for their discovery of the olfactory receptor gene super family. Advances in statistical methodology have accelerated as well. Sensometrics meetings are now vigorous and well-attended annual events. Ideas like Thurstonian modeling were not widely embraced 15 years ago, but now seem to be part of the everyday thought process of many sensory scientists.

And yet, some things stay the same. Sensory testing will always involve human participants. Humans are tough measuring instruments to work with. They come with varying degrees of acumen, training, experiences, differing genetic equipment, sensory capabilities, and of course, different preferences. Human foibles and their associated error variance will continue to place a limitation on sensory tests and actionable results. Reducing, controlling, partitioning, and explaining error variance are all at the heart of good test methods and practices. Understanding the product–person interface will always be the goal of sensory science. No amount of elaborate statistical maneuvering will save a bad study or render the results somehow useful and valid. Although methods continue to evolve, appreciation of the core principles of the field is the key to effective application of sensory test methods.

The notion that one can write a book that is both comprehensive and suitable as an introductory text was a daunting challenge for us. Some may say that we missed the mark on this or that topic, that it was either too superficially treated or too in depth for their students. Perhaps we have tried to do the impossible. Nonetheless the demand for a comprehensive text that would serve as a resource for practitioners is demonstrated by the success of the first edition. Its widespread adoption as a university level text shows that many instructors felt that it could be used appropriately for a first course in sensory evaluation.

This book has been expanded somewhat to reflect the advances in methodologies, theory, and analysis that have transpired in the last 15 years. The chapters are now divided into numbered sections. This may be of assistance to educators who may wish to assign only certain critical sections to beginning students. Much of the organization of key chapters has been done with this in mind and in some of the opening sections; instructors will find suggestions about which sections are key for fundamental understanding of that topic or method. In many chapters we have gone out on a limb and specified a "recommended procedure." In cases where there are multiple options for procedure or analysis, we usually chose a simple solution over one that is more complex. Because we are educators, this seemed the appropriate path.

Note that there are two kinds of appendices in this book. The major statistical methods are introduced with worked examples in Appendices A–E, as in the previous edition. Some main chapters also have appended materials that we felt were not critical to understanding the main topic, but might be of interest to advanced students, statisticians, or experienced practitioners. We continue to give reference citations at the end of every chapter, rather than in one big list at the end. Statistical tables have been added, most notably the discrimination tables that may now be found both in the Appendix and in Chapter 4 itself.

One may question whether textbooks themselves are an outdated method for information retrieval. We feel this acutely because we recognize that a textbook is necessarily retrospective and is only one *snapshot in time* of a field that may be evolving rapidly. Students and practitioners alike may find that reference to updated websites, wikis, and such will provide additional information and new and different perspectives. We encourage such investigation. Textbooks, like automobiles, have an element of built-in obsolescence. Also textbooks, like other printed books, are linear in nature, but the mind works by linking ideas. Hyperlinked resources such as websites and wikis will likely continue to prove useful.

We ask your patience and tolerance for materials and citations that we have left out that you might feel are important. We recognize that there are legitimate differences of opinion and philosophy about the entire area of sensory evaluation methods. We have attempted to provide a balanced and impartial view based on our practical experience. Any errors of fact, errors typographical, or errors in citation are our own fault. We beg your understanding and patience and welcome your corrections and comments.

We could not have written this book without the assistance and support of many people. We would like to thank Kathy Dernoga for providing a pre-publication version of the JAR scale ASTM manual as well as the authors of the ASTM JAR manual Lori Rothman and Merry Jo Parker. Additionally, Mary Schraidt of Peryam and Kroll provided updated examples of a consumer test screening questionnaire and field study questionnaires. Thank you Mary. We thank John Hayes, Jeff Kroll, Tom Carr, Danny Ennis, and Jian Bi for supplying additional literature, software, and statistical tables. Gernot Hoffmann graciously provided graphics for Chapter 12. Thank you Dr. Hoffmann. We would like to thank Wendy Parr and James Green for providing some graphics for Chapter 10. Additionally, Greg Hirson provided support with R-Graphics. Thank you, Greg. Additionally, we want to thank the following people for their willingness to discuss the book in progress and for making very useful suggestions: Michael Nestrud, Susan Cuppett, Edan Lev-Ari, Armand Cardello, Marj Albright, David Stevens, Richard Popper, and Greg Hirson. John Horne had also been very helpful in the previous edition, thank you John. Proofreading and editing suggestions were contributed by Kathy Chapman, Gene Lovelace, Mike Nestrud, and Marge Lawless.

Although not directly involved with this edition of the book we would also like to thank our teachers and influential mentors—without them we would be very different scientists, namely Trygg Engen, William S. Cain, Linda Bartoshuk, David

Peryam, David Stevens, Herb Meiselman, Elaine Skinner, Howard Schutz, Howard Moskowitz, Rose Marie Pangborn, Beverley Kroll, W. Frank Shipe, Lawrence E. Marks, Joseph C. Stevens, Arye Dethmers, Barbara Klein, Ann Noble, Harold Hedrick, William C Stringer, Roger Boulton, Kay McMath, Joel van Wyk, and Roger Mitchell.

Ithaca, New York   Harry T. Lawless
Davis, California   Hildegarde Heymann

Contents

Chapter 1

Introduction

Abstract   In this chapter we carefully parse the definition for sensory evaluation, briefly discuss validity of the data collected before outlining the early history of the field. We then describe the three main methods used in sensory evaluation (discrimination tests, descriptive analysis, and hedonic testing) before discussing the differences between analytical and consumer testing. We then briefly discuss why one may want to collect sensory data. In the final sections we highlight the differences and similarities between sensory evaluation and marketing research and between sensory evaluation and commodity grading as used in, for example, the dairy industry.

Sensory evaluation is a child of industry. It was spawned in the late 40's by the rapid growth of the consumer product companies, mainly food companies. . . . Future development in sensory evaluation will depend upon several factors, one of the most important being the people and their preparation and training.

— Elaine Skinner (1989)

Contents

1.1 Introduction and Overview

1.1.1 Definition

The field of sensory evaluation grew rapidly in the second half of the twentieth century, along with the expansion of the processed food and consumer products industries. Sensory evaluation comprises a set of techniques for accurate measurement of human responses to foods and minimizes the potentially biasing effects of brand identity and other information influences on consumer perception. As such, it attempts to isolate the sensory properties of foods themselves and provides important and useful information to product developers, food scientists, and managers about the sensory characteristics of their products. The field was comprehensively reviewed by Amerine, Pangborn, and Roessler in 1965, and more recent texts have been published by Moskowitz et al. (2006), Stone and Sidel (2004), and Meilgaard et al. (2006). These three later sources are practical works aimed at sensory specialists

H.T. Lawless, H. Heymann, *Sensory Evaluation of Food*, Food Science Text Series,
DOI 10.1007/978-1-4419-6488-5_1, © Springer Science+Business Media, LLC 2010

in industry and reflect the philosophies of the consulting groups of the authors. Our goal in this book is to provide a comprehensive overview of the field with a balanced view based on research findings and one that is suited to students and practitioners alike.

Sensory evaluation has been defined as a scientific method used to evoke, measure, analyze, and interpret those responses to products as perceived through the senses of sight, smell, touch, taste, and hearing (Stone and Sidel, 2004). This definition has been accepted and endorsed by sensory evaluation committees within various professional organizations such as the Institute of Food Technologists and the American Society for Testing and Materials. The principles and practices of sensory evaluation involve each of the four activities mentioned in this definition. Consider the words "to evoke." Sensory evaluation gives guidelines for the preparation and serving of samples under controlled conditions so that biasing factors are minimized. For example, people in a sensory test are often placed in individual test booths so that the judgments they give are their own and do not reflect the opinions of those around them. Samples are labeled with random numbers so that people do not form judgments based upon labels, but rather on their sensory experiences. Another example is in how products may be given in different orders to each participant to help measure and counterbalance for the sequential effects of seeing one product after another. Standard procedures may be established for sample temperature, volume, and spacing in time, as needed to control unwanted variation and improve test precision.

Next, consider the words, "to measure." Sensory evaluation is a quantitative science in which numerical data are collected to establish lawful and specific relationships between product characteristics and human perception. Sensory methods draw heavily from the techniques of behavioral research in observing and quantifying human responses. For example, we can assess the proportion of times people are able to discriminate small product changes or the proportion of a group that expresses a preference for one product over another. Another example is having people generate numerical responses reflecting their perception of how strong a product may taste or smell. Techniques of behavioral research and experimental psychology offer guidelines as to how such measurement techniques should be employed and what their potential pitfalls and liabilities may be.

The third process in sensory evaluation is analysis. Proper analysis of the data is a critical part of sensory testing. Data generated from human observers are often highly variable. There are many sources of variation in human responses that cannot be completely controlled in a sensory test. Examples include the mood and motivation of the participants, their innate physiological sensitivity to sensory stimulation, and their past history and familiarity with similar products. While some screening may occur for these factors, they may be only partially controlled, and panels of humans are by their nature heterogeneous instruments for the generation of data. In order to assess whether the relationships observed between product characteristics and sensory responses are likely to be real, and not merely the result of uncontrolled variation in responses, the methods of statistics are used to analyze evaluation data. Hand-in-hand with using appropriate statistical analyses is the concern of using good experimental design, so that the variables of interest are investigated in a way that allows sensible conclusions to be drawn.

The fourth process in sensory evaluation is the interpretation of results. A sensory evaluation exercise is necessarily an experiment. In experiments, data and statistical information are only useful when interpreted in the context of hypotheses, background knowledge, and implications for decisions and actions to be taken. Conclusions must be drawn that are reasoned judgments based upon data, analyses, and results. Conclusions involve consideration of the method, the limitations of the experiment, and the background and contextual framework of the study. The sensory evaluation specialists become more than mere conduits for experimental results, but must contribute interpretations and suggest reasonable courses of action in light of the numbers. They should be full partners with their clients, the end-users of the test results, in guiding further research. The sensory evaluation professional is in the best situation to realize the appropriate interpretation of test results and the implications for the perception of products by the wider group of consumers to whom the results may be generalized. The sensory specialist best understands the limitations of the test procedure and what its risks and liabilities may be.

A sensory scientist who is prepared for a career in research must be trained in all four of the phases mentioned in the definition. They must understand products, people as measuring instruments, statistical analyses, and interpretation of data within the context of research objectives. As suggested in Skinner's quote, the future advancement of the field depends upon the breadth and depth of training of new sensory scientists.

1.1.2 Measurement

Sensory evaluation is a science of measurement. Like other analytical test procedures, sensory evaluation is concerned with precision, accuracy, sensitivity, and avoiding false positive results (Meiselman, 1993). Precision is similar to the concept in the behavioral sciences of reliability. In any test procedure, we would like to be able to get the same result when a test is repeated. There is usually some error variance around an obtained value, so that upon repeat testing, the value will not always be exactly the same. This is especially true of sensory tests in which human perceptions are necessarily part of the generation of data. However, in many sensory test procedures, it is desirable to minimize this error variance as much as possible and to have tests that are low in error associated with repeated measurements. This is achieved by several means. As noted above, we isolate the sensory response to the factors of interest, minimizing extraneous influences, controlling sample preparation and presentation. Additionally, as necessary, sensory scientists screen and train panel participants.

A second concern is the accuracy of a test. In the physical sciences, this is viewed as the ability of a test instrument to produce a value that is close to the "true" value, as defined by independent measurement from another instrument or set of instruments that have been appropriately calibrated. A related idea in the behavioral sciences, this principle is called the validity of a test. This concerns the ability of a test procedure to measure what it was designed and intended to measure. Validity is established in a number of ways. One useful criterion is predictive validity, when a test result is of value in predicting what would occur in another situation or another measurement. In sensory testing, for example, the test results should reflect the perceptions and opinions of consumers that might buy the product. In other words, the results of the sensory test should generalize to the larger population. The test results might correlate with instrumental measures, process or ingredient variables, storage factors, shelf life times,

or other conditions known to affect sensory properties. In considering validity, we have to look at the end use of the information provided by a test. A sensory test method might be valid for some purposes, but not others (Meiselman, 1993). A simple difference test can tell if a product has changed, but not whether people will like the new version.

A good sensory test will minimize errors in measurement and errors in conclusions and decisions. There are different types of errors that may occur in any test procedure. Whether the test result reflects the true state of the world is an important question, especially when error and uncontrolled variability are inherent in the measurement process. Of primary concern in sensory tests is the sensitivity of the test to differences among products. Another way to phrase this is that a test should not often miss important differences that are present. "Missing a difference" implies an insensitive test procedure. To keep sensitivity high, we must minimize error variance wherever possible by careful experimental controls and by selection and training of panelists where appropriate. The test must involve sufficient numbers of measurements to insure a tight and reliable statistical estimate of the values we obtain, such as means or proportions. In statistical language, detecting true differences is avoiding Type II error and the minimization of β-risk. Discussion of the power and sensitivity of tests from a statistical perspective occurs in Chapter 5 and in the Appendix.

The other error that may occur in a test result is that of finding a positive result when none is actually present in the larger population of people and products outside the sensory test. Once again, a positive result usually means detection of a statistically significant difference between test products. It is important to use a test method that avoids false positive results or Type I error in statistical language. Basic statistical training and common statistical tests applied to scientific findings are oriented toward avoiding this kind of error. The effects of random chance deviations must be taken into account in deciding if a test result reflects a real difference or whether our result is likely to be due to chance variation. The common procedures of inferential statistics provide assurance that we have limited our possibility of finding a difference where one does not really exist. Statistical procedures reduce this risk to some comfortable level, usually with a ceiling of 5% of all tests we conduct.

Note that this error of a false positive experimental result is potentially devastating in basic scientific research—whole theories and research programs may develop from spurious experimental implications if results are due to only random chance. Hence we guard against this kind of danger with proper application of statistical tests. However, in product development work, the second kind of statistical error, that of missing a true difference can be equally devastating. It may be that an important ingredient or processing change has made the product better or worse from a sensory point of view, and this change has gone undetected. So sensory testing is equally concerned with not missing true differences and with avoiding false positive results. This places additional statistical burdens on the experimental concerns of sensory specialists, greater than those in many other branches of scientific research.

Finally, most sensory testing is performed in an industrial setting where business concerns and strategic decisions enter the picture. We can view the outcome of sensory testing as a way to reduce risk and uncertainty in decision making. When a product development manager asks for a sensory test, it is usually because there is some uncertainty about exactly how people perceive the product. In order to know whether it is different or equivalent to some standard product, or whether it is preferred to some competitive standard, or whether it has certain desirable attributes, data are needed to answer the question. With data in hand, the end-user can make informed choices under conditions of lower uncertainty or business risk. In most applications, sensory tests function as risk reduction mechanisms for researchers and marketing managers.

In addition to the obvious uses in product development, sensory evaluation may provide information to other corporate departments. Packaging functionality and convenience may require product tests. Sensory criteria for product quality may become an integral part of a quality control program. Results from blind-labeled sensory consumer tests may need to be compared to concept-related marketing research results. Sensory groups may even interact with corporate legal departments over advertising claim substantiation and challenges to claims. Sensory evaluation also functions in situations outside corporate research. Academic research on foods and materials and their properties and processing will often require sensory tests to evaluate the human perception of changes in the products

(Lawless and Klein, 1989). An important function of sensory scientists in an academic setting is to provide consulting and resources to insure that quality tests are conducted by other researchers and students who seek to understand the sensory impact of the variables they are studying. In government services such as food inspection, sensory evaluation plays a key role (York, 1995). Sensory principles and appropriate training can be key in insuring that test methods reflect the current knowledge of sensory function and test design. See Lawless (1993) for an overview of the education and training of sensory scientists—much of this piece still rings true more than 15 years later.

1.2 Historical Landmarks and the Three Classes of Test Methods

The human senses have been used for centuries to evaluate the quality of foods. We all form judgments about foods whenever we eat or drink ("Everyone carries his own inch-rule of taste, and amuses himself by applying it, triumphantly, wherever he travels."—Henry Adams, 1918). This does not mean that all judgments are useful or that anyone is qualified to participate in a sensory test. In the past, production of good quality foods often depended upon the sensory acuity of a single expert who was in charge of production or made decisions about process changes in order to make sure the product would have desirable characteristics. This was the historical tradition of brewmasters, wine tasters, dairy judges, and other food inspectors who acted as the arbiters of quality. Modern sensory evaluation replaced these single authorities with panels of people participating in specific test methods that took the form of planned experiments. This change occurred for several reasons. First, it was recognized that the judgments of a panel would in general be more reliable than the judgments of single individual and it entailed less risk since the single expert could become ill, travel, retire, die, or be otherwise unavailable to make decisions. Replacement of such an individual was a nontrivial problem. Second, the expert might or might not reflect what consumers or segments of the consuming public might want in a product. Thus for issues of product quality and overall appeal, it was safer (although often more time consuming and expensive) to go directly to the target population. Although the tradition of informal, qualitative inspections such as benchtop "cuttings" persists in some industries, they have been gradually replaced by more formal, quantitative, and controlled observations (Stone and Sidel, 2004).

The current sensory evaluation methods comprise a set of measurement techniques with established track records of use in industry and academic research. Much of what we consider standard procedures comes from pitfalls and problems encountered in the practical experience of sensory specialists over the last 70 years of food and consumer product research, and this experience is considerable. The primary concern of any sensory evaluation specialist is to insure that the test method is appropriate to answer the questions being asked about the product in the test. For this reason, tests are usually classified according to their primary purpose and most valid use. Three types of sensory testing are commonly used, each with a different goal and each using participants selected using different criteria. A summary of the three main types of testing is given in Table 1.1.

1.2.1 Difference Testing

The simplest sensory tests merely attempt to answer whether any perceptible difference exists between two types of products. These are the discrimination tests or simple difference testing procedures. Analysis is usually based on the statistics of frequencies and proportions (counting right and wrong answers). From the test results, we infer differences based on the proportions of persons who are able to choose a test product correctly from among a set of similar or control products. A classic example of this test was the triangle procedure, used in the Carlsberg breweries and in the Seagrams distilleries in the 1940s (Helm and Trolle,

1946; Peryam and Swartz, 1950). In this test, two products were from the same batch while a third product was different. Judges would be asked to pick the odd sample from among the three. Ability to discriminate differences would be inferred from consistent correct choices above the level expected by chance. In breweries, this test served primarily as a means to screen judges for beer evaluation, to insure that they possessed sufficient discrimination abilities. Another multiple-choice difference test was developed at about the same time in distilleries for purposes of quality control (Peryam and Swartz, 1950). In the duo–trio procedure, a reference sample was given and then two test samples. One of the test samples matched the reference while the other was from a different product, batch or process. The participant would try to match the correct sample to the reference, with a chance probability of one-half. As in the triangle test, a proportion of correct choices above that expected by chance is considered evidence for a perceivable difference between products. A third popular difference test was the paired comparison, in which participants would be asked to choose which of two products was stronger or more intense in a given attribute. Partly due to the fact that the panelist's attention is directed to a specific attribute, this test is very sensitive to differences. These three common difference tests are shown in Fig. 1.1.

Simple difference tests have proven very useful in application and are in widespread use today. Typically a discrimination test will be conducted with 25–40 participants who have been screened for their sensory acuity to common product differences and who are familiar with the test procedures. This generally provides an adequate sample size for documenting clear sensory differences. Often a replicate test is performed while the respondents are present in the sensory test facility. In part, the popularity of these tests is due to the simplicity of data analysis. Statistical tables derived from the binomial distribution give the minimum number of correct responses needed to conclude statistical

Table 1.1 Classification of test methods in sensory evaluation

Class	Question of interest	Type of test	Panelist characteristics
Discrimination	Are products perceptibly different in any way	"Analytic"	Screened for sensory acuity, oriented to test method, sometimes trained
Descriptive	How do products differ in specific sensory characteristics	"Analytic"	Screened for sensory acuity and motivation, trained or highly trained
Affective	How well are products liked or which products are preferred	"Hedonic"	Screened for products, untrained

Fig. 1.1 Common methods
for discrimination testing
include the triangle, duo–trio,
and paired comparison
procedures.

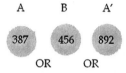

Discrimination Testing Examples

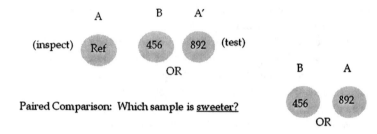

Triangle Test: Choose the sample that is most different

Duo-trio Test: Choose the sample that matches the reference

Paired Comparison: Which sample is <u>sweeter?</u>

significance as a function of the number of participants. Thus a sensory technician merely needs to count answers and refer to a table to give a simple statistical conclusion, and results can be easily and quickly reported.

1.2.2 Descriptive Analyses

The second major class of sensory test methods is those that quantify the perceived intensities of the sensory characteristics of a product. These procedures are known as descriptive analyses. The first method to do this with a panel of trained judges was the Flavor Profile® method developed at the Arthur D. Little consulting group in the late 1940s (Caul, 1957). This group was faced with developing a comprehensive and flexible tool for analysis of flavor to solve problems involving unpleasant off flavors in nutritional capsules and questions about the sensory impact of monosodium glutamate in various processed foods. They formulated a method involving extensive training of panelists that enabled them to characterize all of the flavor notes in a food and the intensities of these notes using a simple category scale and noting their order of appearance. This advance was noteworthy on several grounds. It supplanted the reliance on single expert judges (brewmasters, coffee tasters, and such) with a panel of individuals, under the realization that

the consensus of a panel was likely to be more reliable and accurate than the judgment of a single individual. Second, it provided a means to characterize the individual attributes of flavor and provide a comprehensive analytical description of differences among a group of products under development.

Several variations and refinements in descriptive analysis techniques were forthcoming. A group at the General Foods Technical Center in the early 1960s developed and refined a method to quantify food texture, much as the flavor profile had enabled the quantification of flavor properties (Brandt et al., 1963, Szczesniak et al., 1975). This technique, the Texture Profile method, used a fixed set of force-related and shape-related attributes to characterize the rheological and tactile properties of foods and how these changed over time with mastication. These characteristics have parallels in the physical evaluation of food breakdown or flow. For example, perceived hardness is related to the physical force required to penetrate a sample. Perceived thickness of a fluid or semisolid is related in part to physical viscosity. Texture profile panelists were also trained to recognize specific intensity points along each scale, using standard products or formulated pseudo-foods for calibration.

Other approaches were developed for descriptive analysis problems. At Stanford Research Institute in the early 1970s, a group proposed a method for descriptive analysis that would remedy some of the apparent shortcomings of the Flavor Profile® method and be even more broadly applicable to all sensory properties of a food, and not just taste and texture (Stone et al., 1974). This method was termed Quantitative Descriptive Analysis® or QDA® for short (Stone and Sidel, 2004). QDA® procedures borrowed heavily from the traditions of behavioral research and used experimental designs and statistical analyses such as analysis of variance. This insured independent judgments of panelists and statistical testing, in contrast to the group discussion and consensus procedures of the Flavor Profile® method. Other variations on descriptive procedures were tried and achieved some popularity, such as the Spectrum Method® (Meilgaard et al., 2006) that included a high degree of calibration of panelists for intensity scale points, much like the Texture Profile. Still other researchers have employed hybrid techniques that include some features of the various descriptive approaches (Einstein, 1991). Today many product development groups use hybrid approaches as the advantages of each may apply to the products and resources of a particular company.

Descriptive analysis has proven to be the most comprehensive and informative sensory evaluation tool. It is applicable to the characterization of a wide variety of product changes and research questions in food product development. The information can be related to consumer acceptance information and to instrumental measures by means of statistical techniques such as regression and correlation.

An example of a descriptive ballot for texture assessment of a cookie product is shown in Table 1.2. The product is assessed at different time intervals in

Table 1.2 Descriptive evaluation of cookies–texture attributes

Phase	Attributes	Word anchors
Surface	Roughness	Smooth–rough
	Particles	None–many
	Dryness	Oily–dry
First bite	Fracturability	Crumbly–brittle
	Hardness	Soft–hard
	Particle size	Small–large
First chew	Denseness	Airy–dense
	Uniformity of chew	Even–uneven
Chew down	Moisture absorption	None–much
	Cohesiveness of mass	Loose–cohesive
	Toothpacking	None–much
	Grittiness	None–much
Residual	Oiliness	Dry–oily
	Particles	None–many
	Chalky	Not chalky–very chalky

a uniform and controlled manner, typical of an analytical sensory test procedure. For example, the first bite may be defined as cutting with the incisors. The panel for such an analysis would consist of perhaps 10–12 well-trained individuals, who were oriented to the meanings of the terms and given practice with examples. Intensity references to exemplify scale points are also given in some techniques. Note the amount of detailed information that can be provided in this example and bear in mind that this is only looking at the product's texture—flavor might form an equally detailed sensory analysis, perhaps with a separate trained panel. The relatively small number of panelists (a dozen or so) is justified due to their level of calibration. Since they have been trained to use attribute scales in a similar manner, error variance is lowered and statistical power and test sensitivity are maintained in spite of fewer observations (fewer data points per product). Similar examples of texture, flavor, fragrance, and tactile analyses can be found in Meilgaard et al. (2006).

1.2.3 Affective Testing

The third major class of sensory tests is those that attempt to quantify the degree of liking or disliking of a product, called hedonic or affective test methods. The most straightforward approach to this problem is to offer people a choice among alternative products and see if there is a clear preference from the majority of respondents. The problem with choice tests is that they are not very informative about the magnitude of liking or disliking from respondents. An historical landmark in this class of tests was the hedonic scale developed at the U.S. Army Food and Container Institute in the late 1940s (Jones et al., 1955). This method provided a balanced 9-point scale for liking with a centered neutral category and attempted to produce scale point labels with adverbs that represented psychologically equal steps or changes in hedonic tone. In other words, it was a scale with ruler-like properties whose equal intervals would be amenable to statistical analysis.

An example of the 9-point scale is shown in Fig. 1.2. Typically a hedonic test today would involve a sample of 75–150 consumers who were regular users of the product. The test would involve several alternative versions of the product and be conducted in some central location or sensory test facility. The larger panel size

Quartermaster Corps. 9-point Hedonic Scale

like extremely
like very much
like moderately
like slightly
neither like nor dislike
dislike slightly
dislike moderately
dislike very much
dislike extremely

Scale points chosen to represent equal psychological intervals.

Fig. 1.2 The 9-point hedonic scale used to assess liking and disliking. This scale, originally developed at the U.S. Army Food and Container Institute (Quartermaster Corps), has achieved widespread use in consumer testing of foods.

of an affective test arises due to the high variability of individual preferences and thus a need to compensate with increased numbers of people to insure statistical power and test sensitivity. This also provides an opportunity to look for segments of people who may like different styles of a product, for example, different colors or flavors. It may also provide an opportunity to probe for diagnostic information concerning the reasons for liking or disliking a product.

Workers in the food industry were occasionally in contact with psychologists who studied the senses and had developed techniques for assessing sensory function (Moskowitz, 1983). The development of the 9-point hedonic scale serves as good illustration of what can be realized when there is interaction between experimental psychologists and food scientists. A psychological measurement technique called Thurstonian scaling (see Chapter 5) was used to validate the adverbs for the labels on the 9-point hedonic scale. This interaction is also visible in the authorship of this book—one author is trained in food science and chemistry while the other is an experimental psychologist. It should not surprise us that interactions would occur and perhaps the only puzzle is why the interchanges were not more sustained and productive. Differences in language, goals, and experimental focus probably contributed to some difficulties. Psychologists were focused primarily on the individual person while sensory evaluation specialists were focused primarily on

the food product (the stimulus). However, since a sensory perception involves the necessary interaction of a person with a stimulus, it should be apparent that similar test methods are necessary to characterize this person–product interaction.

1.2.4 The Central Dogma—Analytic Versus Hedonic Tests

The central principle for all sensory evaluation is that the test method should be matched to the objectives of the test. Figure 1.3 shows how the selection of the test procedure flows from questions about the objective of the investigation. To fulfill this goal, it is necessary to have clear communication between the sensory test manager and the client or end-user of the information. A dialogue is often needed. Is the important question whether or not there is any difference at all among the products? If so, a discrimination test is indicated. Is the question one of whether consumers like the new product better than the previous version? A consumer acceptance test is needed. Do we need to know what attributes have changed in the sensory characteristics of the new product? Then a descriptive analysis procedure is called for. Sometimes there are multiple objectives and a sequence of different tests is required (Lawless and Claassen, 1993). This can present problems if all the answers are required at once or under severe time pressure during competitive product development. One of the most important jobs of the sensory specialist in the food industry is to insure a clear understanding and specification of the type of information needed by the end-users. Test design may require a number of conversations, interviews with different people, or even written test requests that specify why the information is to be collected and how the results will be used in making specific decisions and subsequent actions to be taken. The sensory specialist is the best position to understand the uses and limitations of each procedure and what would be considered appropriate versus inappropriate conclusions from the data.

There are two important corollaries to this principle. The sensory test design involves not only the selection of an appropriate method but also the selection of appropriate participants and statistical analyses. The three classes of sensory tests can be divided into two types, analytical sensory tests including discrimination

Fig. 1.3 A flowchart showing methods determination. Based on the major objectives and research questions, different sensory test methods are selected. Similar decision processes are made in panelist selection, setting up response scales, in choosing experimental designs, statistical analysis, and other tasks in designing a sensory test (reprinted with permission from Lawless, 1993).

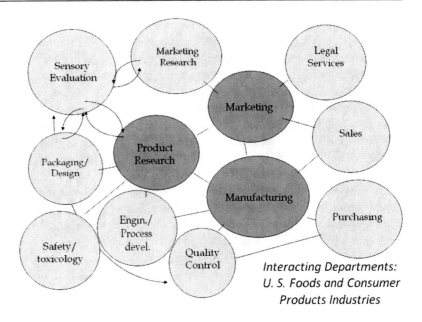

Interacting Departments: U. S. Foods and Consumer Products Industries

and descriptive methods and affective or hedonic tests such as those involved in assessing consumer liking or preferences (Lawless and Claassen, 1993). For the analytical tests, panelists are selected based on having average to good sensory acuity for the critical characteristics (tastes, smells, textures, etc.) of products to be evaluated. They are familiarized with the test procedures and may undergo greater or lesser amounts of training, depending upon the method. In the case of descriptive analysis, they adopt an analytical frame of mind, focusing on specific aspects of the product as directed by the scales on their questionnaires. They are asked to put personal preferences and hedonic reactions aside, as their job is only to specify what attributes are present in the product and at what levels of sensory intensity, extent, amount, or duration.

In contrast to this analytical frame of mind, consumers in an affective test act in a much more integrative fashion. They perceive a product as a whole pattern. Although their attention is sometimes captured by a specific aspect of a product (especially if it is a bad, unexpected, or unpleasant one), their reactions to the product are often immediate and based on the integrated pattern of sensory stimulation from the product and expressed as liking or disliking. This occurs without a great deal of thought or dissection of the product's specific profile. In other words, consumers are effective at rendering impressions based on the integrated pattern of perceptions. In such consumer

tests, participants must be chosen carefully to insure that the results will generalize to the population of interest. Participants should be frequent users of the product, since they are most likely to form the target market and will be familiar with similar products. They possess reasonable expectations and a frame of reference within which they can form an opinion relative to other similar products they have tried.

The analytic/hedonic distinction gives rise to some highly important rules of thumb and some warnings about matching test methods and respondents. It is unwise to ask trained panelists about their preferences or whether they like or dislike a product. They have been asked to assume a different, more analytical frame of mind and to place personal preference aside. Furthermore, they have not necessarily been selected to be frequent users of the product, so they are not part of the target population to which one would like to generalize hedonic test results. A common analogy here is to an analytical instrument. You would not ask a gas chromatograph or a pH meter whether it liked the product, so why ask your analytical descriptive panel (O'Mahony, 1979).

Conversely, problems arise when consumers are asked to furnish very specific information about product attributes. Consumers not only act in a non-analytic frame of mind but also often have very fuzzy concepts about specific attributes, confusing sour and bitter tastes, for example. Individuals often differ markedly in their interpretations of sensory attribute words on a questionnaire. While a trained texture profile panel has no trouble in agreeing how cohesive a product is after chewing, we cannot expect consumers to provide precise information on such a specific and technical attribute. In summary, we avoid using trained panelists for affective information and we avoid asking consumers about specific analytical attributes.

Related to the analytic–hedonic distinction is the question of whether experimental control and precision are to be maximized or whether validity and generalizability to the real world are more important. Often there is a tradeoff between the two and it is difficult to maximize both simultaneously. Analytic tests in the lab with specially screened and trained judges are more reliable and lower in random error than consumer tests. However, we give up a certain amount of generalizability to real-world results by using artificial conditions and a special group of participants. Conversely, in the testing of products by consumers in their own homes we have not only a lot of real-life validity but also a lot of noise in the data. Brinberg and McGrath (1985) have termed this struggle between precision and validity one of "conflicting desiderata." O'Mahony (1988) has made a distinction between sensory evaluation Type I and Type II. In Type I sensory evaluation, reliability and sensitivity are key factors, and the participant is viewed much like an analytical instrument used to detect and measure changes in a food product. In Type II sensory evaluation, participants are chosen to be representative of the consuming population, and they may evaluate food under more naturalistic conditions. Their emphasis here is on prediction of consumer response. Every sensory test falls somewhere along a continuum where reliability versus real-life extrapolation are in a potential tradeoff relationship. This factor must also be discussed with end-users of the data to see where their emphasis lies and what level of tradeoff they find comfortable.

Statistical analyses must also be chosen with an eye to the nature of the data. Discrimination tests involve choices and counting numbers of correct responses. The statistics derived from the binomial distribution or those designed for proportions such as chi-square are appropriate. Conversely, for most scaled data, we can apply the familiar parametric statistics appropriate to normally distributed and continuous data, such as means, standard deviations, t-tests, analysis of variance. The choice of an appropriate statistical test is not always straightforward, so sensory specialists are wise to have thorough training in statistics and to involve statistical and design specialists in a complex project in its earliest stages of planning.

Occasionally, these central principles are violated. They should not be put aside as a matter of mere expediency or cost savings and never without a logical analysis. One common example is the use of a discrimination test before consumer acceptance. Although our ultimate interest may lie in whether consumers will like or dislike a new product variation, we can conduct a simple difference test to see whether any change is perceivable at all. The logic in this sequence is the following: if a screened and experienced discrimination panel cannot tell the difference under carefully controlled conditions in the sensory laboratory, then a more heterogeneous group of consumers is unlikely to see a difference in their less controlled and more variable world. If no difference is perceived, there can logically be no systematic preference. So a more time consuming and costly consumer test can sometimes be avoided by conducting a simpler but more sensitive discrimination test first. The added reliability of the controlled discrimination test provides a "safety net" for conclusions about consumer perception. Of course, this logic is not without its pitfalls—some consumers may interact extensively with the product during a home use test period and may form stable and important opinions that are not captured in a short duration laboratory test, and there is also always the possibility of a false negative result (the error of missing a difference). MacRae and Geelhoed (1992) describe an interesting case of a missed difference in a triangle test where a significant preference was then observed between water samples in a paired comparison. The sensory professional must be aware that these anomalies in experimental results will sometimes arise, and must also be aware of some of the reasons why they occur.

1.3 Applications: Why Collect Sensory Data?

Human perceptions of foods and consumer products are the results of complex sensory and interpretation processes. At this stage in scientific history, perceptions of such multidimensional stimuli as conducted by the parallel processing of the human nervous system are difficult or impossible to predict from instrumental measures. In many cases instruments lack the sensitivity of human sensory systems—smell is a good example. Instruments rarely mimic the mechanical manipulation of foods when tasted nor do they mimic the types of peri-receptor filtering that occur in biological fluids like saliva or mucus that can cause chemical partitioning of flavor materials. Most importantly, instrumental assessments give values that miss an important perceptual process: the interpretation of sensory experience by the human brain prior to responding. The brain lies interposed between sensory input and the generation of responses that form our data. It is a massively parallel-distributed processor and computational engine, capable of rapid feats of pattern recognition. It comes to the sensory evaluation task complete with a personal history and experiential frame of reference. Sensory experience is interpreted, given meaning within the frame of reference, evaluated relative to expectations and can involve integration of multiple simultaneous or sequential inputs. Finally judgments are rendered as our data. Thus there is a "chain of perception" rather than simply stimulus and response (Meilgaard et al., 2006).

Only human sensory data provide the best models for how consumers are likely to perceive and react to food products in real life. We collect, analyze, and interpret sensory data to form predictions about how products have changed during a product development program. In the food and consumer products industries, these changes arise from three important factors: ingredients, processes, and packaging. A fourth consideration is often the way a product ages, in other words its shelf life, but we may consider shelf stability to be one special case of processing, albeit usually a very passive one (but also consider products exposed to temperature fluctuation, light-catalyzed oxidation, microbial contamination, and other "abuses"). Ingredient changes arise for a number of reasons. They may be introduced to improve product quality, to reduce costs of production, or simply because a certain supply of raw materials has become unavailable. Processing changes likewise arise from the attempt to improve quality in terms of sensory, nutritional, microbiological stability factors, to reduce costs or to improve manufacturing productivity. Packaging changes arise from considerations of product stability or other quality factors, e.g., a certain

amount of oxygen permeability may insure that a fresh beef product remains red in color for improved visual appeal to consumers. Packages function as carriers of product information and brand image, so both sensory characteristics and expectations can change as a function of how this information can be carried and displayed by the packaging material and its print overlay. Packaging and print ink may cause changes in flavor or aroma due to flavor transfer out of the product and sometimes transfer of off-flavors into the product. The package also serves as an important barrier to oxidative changes, to the potentially deleterious effects of light-catalyzed reactions, and to microbial infestations and other nuisances.

The sensory test is conducted to study how these product manipulations will create perceived changes to human observers. In this sense, sensory evaluation is in the best traditions of psychophysics, the oldest branch of scientific psychology, that attempts to specify the relationships between different energy levels impinging upon the sensory organs (the physical part of psychophysics) and the human response (the psychological part). Often, one cannot predict exactly what the sensory change will be as a function of ingredients, processes, or packaging, or it is very difficult to do so since foods and consumer products are usually quite complex systems. Flavors and aromas depend upon complex mixtures of many volatile chemicals. Informal tasting in the lab may not bring a reliable or sufficient answer to sensory questions. The benchtop in the development laboratory is a poor place to judge potential sensory impact with distractions, competing odors, nonstandard lighting, and so on. Finally, the nose, eyes, and tongue of the product developer may not be representative of most other people who will buy the product. So there is some uncertainty about how consumers will view a product especially under more natural conditions.

Uncertainty is the key here. If the outcome of a sensory test is perfectly known and predictable, there is no need to conduct the formal evaluation. Unfortunately, useless tests are often requested of a sensory testing group in the industrial setting. The burden of useless routine tests arises from overly entrenched product development sequences, corporate traditions, or merely the desire to protect oneself from blame in the case of unexpected failures. However, the sensory test is only as useful as the amount of reduction in uncertainty that occurs. If there is no uncertainty, there is no need for the sensory test. For example, doing a sensory test to see if there is a perceptible color difference between a commercial red wine and a commercial white wine is a waste of resources, since there is no uncertainty! In the industrial setting, sensory evaluation provides a conduit for information that is useful in management business decisions about directions for product development and product changes. These decisions are based on lower uncertainty and lower risk once the sensory information is provided.

Sensory evaluation also functions for other purposes. It may be quite useful or even necessary to include sensory analyses in quality control (QC) or quality assurance. Modification of traditional sensory practices may be required to accommodate the small panels and rapid assessments often required in on-line QC in the manufacturing environment. Due to the time needed to assemble a panel, prepare samples for testing, analyze and report sensory data, it can be quite challenging to apply sensory techniques to quality control as an on-line assessment. Quality assurance involving sensory assessments of finished products are more readily amenable to sensory testing and may be integrated with routine programs for shelf life assessment or quality monitoring. Often it is desirable to establish correlations between sensory response and instrumental measures. If this is done well, the instrumental measure can sometimes be substituted for the sensory test. This is especially applicable under conditions in which rapid turnaround is needed. Substitution of instrumental measurements for sensory data may also be useful if the evaluations are likely to be fatiguing to the senses, repetitive, involve risk in repeated evaluations (e.g., insecticide fragrances), and are not high in business risk if unexpected sensory problems arise that were missed.

In addition to these product-focused areas of testing, sensory research is valuable in a broader context. A sensory test may help to understand the attributes of a product that consumers view as critical to product acceptance and thus success. While we keep a wary eye on the fuzzy way that consumers use language, consumer sensory tests can provide diagnostic information about a product's points of superiority or shortcomings. Consumer sensory evaluations may suggest hypotheses for further inquiry such as exploration of new product opportunities.

There are recurrent themes and enduring problems in sensory science. In 1989, the ASTM Committee E-18 on Sensory Evaluation of Materials and Products published a retrospective celebration of the origins of sensory methods and the committee itself (ASTM, 1989). In that volume, Joe Kamen, an early sensory worker with the Quartermaster Food and Container Institute, outlined nine areas of sensory research which were active 45 years ago. In considering the status of sensory science in the first decade of the twenty-first century, we find that these areas are still fertile ground for research activity and echo the activities in many sensory labs at the current time. Kamen (1989) identified the following categories:

(1) Sensory methods research. This aimed at increasing reliability and efficiency, including research into procedural details (rinsing, etc.) and the use of different experimental designs. Meiselman (1993), a later sensory scientist at the U.S. Army Food Laboratories, raised a number of methodological issues then and even now still unsettled within the realm of sensory evaluation. Meiselman pointed to the lack of focused methodological research aimed at issues of measurement quality such as reliability, sensitivity, and validity. Many sensory techniques originate from needs for practical problem solving. The methods have matured to the status of standard practice on the basis of their industrial track record, rather than a connection to empirical data that compare different methods. The increased rate of experimental publications devoted to purely methodological comparisons in journals such as the Journal of Sensory Studies and Food Quality and Preference certainly points to improvement in the knowledge base about sensory testing, but much remains to be done.

(2) Problem solving and trouble shooting. Kamen raised the simple example of establishing product equivalence between lots, but there are many such day-to-day product-related issues that arise in industrial practice. Claim substantiation (ASTM E1958, 2008; Gacula, 1991) and legal and advertising challenges are one example. Another common example would be identification of the cause of off-flavors, "taints" or other undesirable sensory characteristics and the detective exercise that goes toward the isolation and identification of the causes of such problems.

(3) Establishing test specifications. This can be important to suppliers and vendors, and also for quality

control in multi-plant manufacturing situations, as well as international product development and the problem of multiple sensory testing sites and panels.

(4) Environmental and biochemical factors. Kamen recognized that preferences may change as a function of the situation (food often tastes better outdoors and when you are hungry). Meiselman (1993) questioned whether sufficient sensory research is being performed in realistic eating situations that may be more predictive of consumer reactions, and recently sensory scientists have started to explore this area of research (for example, Giboreau and Fleury, 2009; Hein et al., 2009; Mielby and Frøst, 2009).

(5) Resolving discrepancies between laboratory and field studies. In the search for reliable, detailed, and precise analytical methods in the sensory laboratory, some accuracy in predicting field test results may be lost. Management must be aware of the potential of false positive or negative results if a full testing sequence is not carried out, i.e., if shortcuts are made in the testing sequence prior to marketing a new product. Sensory evaluation specialists in industry do not always have time to study the level of correlation between laboratory and field tests, but a prudent sensory program would include periodic checks on this issue.

(6) Individual differences. Since Kamen's era, a growing literature has illuminated the fact that human panelists are not identical, interchangeable measuring instruments. Each comes with different physiological equipment, different frames of reference, different abilities to focus and maintain attention, and different motivational resources. As an example of differences in physiology, we have the growing literature on specific anosmias—smell "blindnesses" to specific chemical compounds among persons with otherwise normal senses of smell (Boyle et al., 2006; Plotto et al., 2006; Wysocki and Labows, 1984). It should not be surprising that some olfactory characteristics are difficult for even trained panelists to evaluate and to come to agreement (Bett and Johnson, 1996).

(7) Relating sensory differences to product variables. This is certainly the meat of sensory science in industrial practice. However, many product developers do not sufficiently involve their sensory specialists in the underlying research questions.

They also may fall into the trap of never ending sequences of paired tests, with little or no planned designs and no modeling of how underlying physical variables (ingredients, processes) create a dynamic range of sensory changes. The relation of graded physical changes to sensory response is the essence of psychophysical thinking.

(8) Sensory interactions. Foods and consumer products are multidimensional. The more sensory scientists understand interactions among characteristics such as enhancement and masking effects, the better they can interpret the results of sensory tests and provide informed judgments and reasoned conclusions in addition to reporting just numbers and statistical significance.

(9) Sensory education. End-users of sensory data and people who request sensory tests often expect one tool to answer all questions. Kamen cited the simple dichotomy between analytical and hedonic testing (e.g., discrimination versus preference) and how explaining this difference was a constant task. Due to the lack of widespread training in sensory science, the task of sensory education is still with us today, and a sensory professional must be able to explain the rationale behind test methods and communicate the importance and logic of sensory technology to non-sensory scientists and managers.

1.3.1 Differences from Marketing Research Methods

Another challenge to the effective communication of sensory results concerns the resemblance of sensory data to those generated from other research methods. Problems can arise due to the apparent similarity of some sensory consumer tests to those conducted by marketing research services. However, some important differences exist as shown in Table 1.3. Sensory tests are almost always conducted on a blind-labeled basis. That is, product identity is usually obscured other than the minimal information that allows the product to be evaluated in the proper category (e.g., cold breakfast cereal). In contrast, marketing research tests often deliver explicit concepts about a product—label claims, advertising imagery, nutritional information, or any other information that may enter into the mix

Table 1.3 Contrast of sensory evaluation consumer tests with market research tests

Sensory testing with consumers
    Participants screened to be users of the product category
    Blind-labeled samples—random codes with minimal conceptual information
    Determines if sensory properties and overall appeal met targets
    Expectations based on similar products used in the category
    Not intended to assess response/appeal of product concept
Market research testing (concept and/or product test)
    Participants in product-testing phase selected for positive response to concept
    Conceptual claims, information, and frame of reference are explicit
    Expectations derived from concept/claims and similar product usage
    Unable to measure sensory appeal in isolation from concept and expectations

designed to make the product conceptually appealing (e.g., bringing attention to convenience factors in preparation).

In a sensory test all these potentially biasing factors are stripped away in order to isolate the opinion based on sensory properties only. In the tradition of scientific inquiry, we need to isolate the variables of interest (ingredients, processing, packaging changes) and assess sensory properties as a function of these variables, and not as a function of conceptual influences. This is done to minimize the influence of a larger cognitive load of expectations generated from complex conceptual information. There are many potential response biases and task demands that are entailed in "selling" an idea as well as in selling a product. Participants often like to please the experimenter and give results consistent with what they think the person wants. There is a large literature on the effect of factors such as brand label on consumer response. Product information interacts in complex ways with consumer attitudes and expectancies (Aaron et al., 1994; Barrios and Costell, 2004; Cardello and Sawyer, 1992; Costell et al., 2009; Deliza and MacFie, 1996; Giménez et al., 2008; Kimura et al., 2008; Mielby and Frøst, 2009; Park and Lee, 2003; Shepherd et al., 1991/1992). Expectations can cause assimilation of sensory reactions toward what is expected under some conditions and under other conditions will show contrast effects, enhancing differences when expectations are not met (Siegrist and Cousin, 2009; Lee et al., 2006; Yeomans et al., 2008; Zellner et al., 2004). Packaging and brand information will also affect sensory judgments (Dantas et al., 2004; Deliza et al., 1999; Enneking et al., 2007). So the apparent resemblance of a blind sensory test and a fully concept-loaded market research test are quite illusory. Corporate management needs to be reminded of this important distinction. There continues to be tension between the roles of marketing research and sensory research within companies. The publication by Garber et al. (2003) and the rebuttal to that paper by Cardello (2003) are a relatively recent example of this tension.

Different information is provided by the two test types and both are very important. Sensory evaluation is conducted to inform product developers about whether they have met their sensory and performance targets in terms of perception of product characteristics. This information can only be obtained when the test method is as free as possible from the influences of conceptual positioning. The product developer has a right to know if the product meets its sensory goals just as the marketer needs to know if the product meets its consumer appeal target in the overall conceptual, positioning, and advertising mix. In the case of product failures, strategies for improvement are never clear without both types of information.

Sometimes the two styles of testing will give apparently conflicting results (Oliver, 1986). However, it is almost never the situation that one is "right" and the other is "wrong." They are simply different types of evaluations and are even conducted on different participants. For example, taste testing in market research tests may be conducted only on those persons who previously express a positive reaction to the proposed concept. This seems reasonable, as they are the likely purchasers, but bear in mind that their product evaluations are conducted *after they have already expressed some positive attitudes* and people like to be internally consistent. However, a blind sensory consumer test is conducted on a sample of regular product user, with no prescreening for conceptual interest or attitudes. So they are not necessarily the same sample population in each style of test and differing results should not surprise anyone.

1.3.2 Differences from Traditional Product Grading Systems

A second arena of apparent similarity to sensory evaluation is with the traditional product quality grading systems that use sensory criteria. The grading of agricultural commodities is a historically important influence on the movement to assure consumers of quality standards in the foods they purchase. Such techniques were widely applicable to simple products such as fluid milk and butter (Bodyfelt et al., 1988, 2008), where an ideal product could be largely agreed upon and the defects that could arise in poor handling and processing gave rise to well-known sensory effects. Further impetus came from the fact that competitions could be held to examine whether novice judges-in-training could match the opinions of experts. This is much in the tradition of livestock grading—a young person could judge a cow and receive awards at a state fair for learning to use the same criteria and critical eye as the expert judges. There are noteworthy differences in the ways in which sensory testing and quality judging are performed. Some of these are outlined in Table 1.4.

The commodity grading and the inspection tradition have severe limitations in the current era of highly processed foods and market segmentation. There are fewer and fewer "standard products" relative to the wide variation in flavors, nutrient levels (e.g., low fat), convenience preparations, and other choices that line the supermarket shelves. Also, one person's product defect may be another's marketing bonanza, as in the glue that did not work so well that gave us the ubiquitous post-it notes. Quality judging methods are poorly suited to research support programs. The techniques have been widely criticized on a number of scientific grounds (Claassen and Lawless, 1992; Drake, 2007; O'Mahony, 1979; Pangborn and Dunkley, 1964; Sidel et al., 1981), although they still have their proponents in industry and agriculture (Bodyfelt et al., 1988, 2008).

The defect identification in quality grading emphasizes root causes (e.g., oxidized flavor) whereas the descriptive approach uses more elemental singular terms to describe perceptions rather than to infer causes. In the case of oxidized flavors, the descriptive analysis panel might use a number of terms (oily, painty, and fishy) since oxidation causes a number of qualitatively different sensory effects. Another notable difference from mainstream sensory evaluation is that the quality judgments combine an overall quality scale (presumably reflecting consumer dislikes) with diagnostic information about defects, a kind of descriptive analysis looking only at the negative aspects of products. In mainstream sensory evaluation, the descriptive function and the consumer evaluation would be clearly separate in two distinct tests with different respondents. Whether the opinion of a single expert can effectively represent consumer opinion is highly questionable at this time in history.

Table 1.4 Contrast of sensory evaluation tests with quality inspection

Sensory testing
Separates hedonic (like–dislike) and descriptive information into separate tests
Uses representative consumers for assessment of product appeal (liking/disliking)
Uses trained panelists to specify attributes, but not liking/disliking
Oriented to research support
Flexible for new, engineered, and innovative products
Emphasizes statistical inference for decision making, suitable experimental designs, and sample sizes
Quality inspection
Used for pass–fail online decisions in manufacturing
Provides quality score and diagnostic information concerning *defects* in one test
Uses sensory expertise of highly trained individuals
May use only one or very few trained experts
Product knowledge, potential problems, and causes are stressed
Traditional scales are multi-dimensional and poorly suited to statistical analyses
Decision-making basis may be qualitative
Oriented to standard commodities

1.4 Summary and Conclusions

Sensory evaluation comprises a set of test methods with guidelines and established techniques for product presentation, well-defined response tasks, statistical methods, and guidelines for interpretation of results. Three primary kinds of sensory tests focus on the existence of overall differences among products (discrimination tests), specification of attributes (descriptive analysis), and measuring consumer likes and dislikes (affective or hedonic testing). Correct application of sensory technique involves correct matching of method to the objective of the tests, and this requires good communication between sensory specialists and

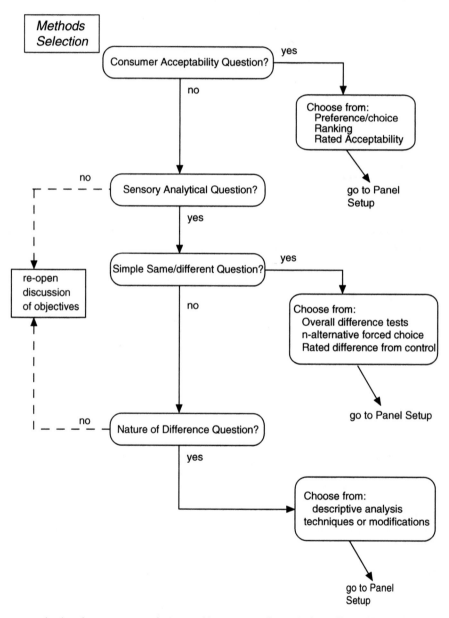

Fig. 1.4 A sensory evaluation department may interact with many other departments in a food or consumer products company. Their primary interaction is in support of product research and development, much as marketing research supports the company's marketing efforts. However, they may also interact with quality control, marketing research, packaging and design groups, and even legal services over issues such as claim substantiation and advertising challenges.

end-users of the test results. Logical choices of test participants and appropriate statistical analyses form part of the methodological mix. Analytic tests such as the discrimination and descriptive procedures require good experimental control and maximization of test precision. Affective tests on the other hand require use of representative consumers of the products and test conditions that enable generalization to how products are experienced by consumers in the real world.

Sensory tests provide useful information about the human perception of product changes due to ingredients, processing, packaging, or shelf life. Sensory evaluation departments not only interact most heavily with new product development groups but may also provide information to quality control, marketing research, packaging, and, indirectly, to other groups throughout a company (Fig. 1.4). Sensory information reduces risk in decisions about product development and strategies for meeting consumer needs. A well-functioning sensory program will be useful to a company in meeting consumer expectations and insuring a greater chance of marketplace success. The utility of the information provided is directly related to the quality of the sensory measurement.

> ..., sensory food science stands at the intersection of many disciplines and research traditions, and the stakeholders are many (Tuorila and Monteleone, 2009).

> Quantities derive from measurement, figures from quantities, comparisons from figures, and victory from comparisons (Sun Tzu – The Art of War (Ch. 4, v.18)).

References

Aaron, J. I., Mela, D. J. and Evans, R. E. 1994. The influence of attitudes, beliefs and label information on perceptions of reduced-fat spread. Appetite, 22(1), 25–38.

Adams, H. 1918. The Education of Henry Adams. The Modern Library, New York.

Amerine, M. A., Pangborn, R. M. and Roessler, E. B. 1965. Principles of Sensory Evaluation of Food. Academic, New York.

ASTM E1958. 2008. Standard guide for sensory claim substantiation. ASTM International, West Conshohocken, PA.

ASTM. 1989. Sensory evaluation. In celebration of our beginnings. Committee E-18 on Sensory Evaluation of Materials and Products. ASTM, Philadelphia.

Barrios, E. X. and Costell, E. 2004. Review: use of methods of research into consumers' opinions and attitudes in food research. Food Science and Technology International, 10, 359–371.

Bett, K. L. and Johnson, P. B. 1996. Challenges of evaluating sensory attributes in the presence of off-flavors. Journal of Sensory Studies, 11, 1–17.

Bodyfelt, F. W., Drake, M. A. and Rankin, S. A. 2008. Developments in dairy foods sensory science and education: from student contests to impact on product quality. International Dairy Journal, 18, 729–734.

Bodyfelt, F. W., Tobias, J. and Trout, G. M. 1988. Sensory Evaluation of Dairy Products. Van Nostrand/AVI Publishing, New York.

Boyle, J. A., Lundström, J. N., Knecht, M., Jones-Gotman, M., Schaal, B. and Hummel, T. 2006. On the trigeminal percept of androstenone and its implications on the rate of specific anosmia. Journal of Neurobiology, 66, 1501–1510.

Brandt, M. A., Skinner, E. Z. and Coleman, J. A. 1963. Texture profile method. Journal of Food Science, 28, 404–409.

Brinberg, D. and McGrath, J. E. 1985. Validity and the Research Process. Sage Publications, Beverly Hills, CA.

Cardello, A. V. 2003. Ideographic sensory testing vs. nomothetic sensory research for marketing guidance: comments on Garber et al. Food Quality and Preference, 14, 27–30.

Cardello, A. V. and Sawyer, F. M. 1992. Effects of disconfirmed consumer expectations on food acceptability. Journal of Sensory Studies, 7, 253–277.

Caul, J. F. 1957. The profile method of flavor analysis. Advances in Food Research, 7, 1–40.

Claassen, M. and Lawless, H. T. 1992. Comparison of descriptive terminology systems for sensory evaluation of fluid milk. Journal of Food Science, 57, 596–600, 621.

Costell, E., Tárrega, A. and Bayarri, S. 2009. Food acceptance: the role of consumer perception and attitudes. Chemosensory Perception. doi:10.1007/s12078-009-9057-1.

Dantas, M. I. S., Minim, V. P. R., Deliza, R. and Puschmann, R. 2004. The effect of packaging on the perception of minimally processed products. Journal of International Food and Agribusiness Marketing, 2, 71–83.

Deliza, R., Rosenthal, A., Hedderley, D., MacFie, H. J. H. and Frewer, L. J. 1999. The importance of brand, product information and manufacturing process in the development of novel environmentally friendly vegetable oils. Journal of International Food and Agribusiness Marketing, 3, 67–77.

Deliza, R. and MacFie, H. J. H. 1996. The generation of sensory expectation by external cues and its effect on sensory perception and hedonic ratings: A review. Journal of Sensory Studies, 11, 103–128.

Drake, M. A. 2007. Invited Review: sensory analysis of dairy foods. Journal of Dairy Science, 90, 4925–4937.

Einstein, M. A. 1991. Descriptive techniques and their hybridization. In: H. T. Lawless and B. P. Klein (eds.), Sensory Science Theory and Applications in Foods. Marcel Dekker, New York, pp. 317–338.

Enneking, U., Neumann, C. and Henneberg, S. 2007. How important intrinsic and extrinsic product attributes affect purchase decision. Food Quality and Preference, 18, 133–138.

Gacula, M. C., Jr. 1991. Claim substantiation for sensory equivalence and superiority. In: H. T. Lawless and B. P. Klein (eds.), Sensory Science Theory and Applications in Foods. Marcel Dekker, New York, pp. 413–436.

Garber, L. L., Hyatt, E. M. and Starr, R. G. 2003. Measuring consumer response to food products. Food Quality and Preference, 14, 3–15.

Giboreau, A. and Fleury, H. 2009. A new research platform to contribute to the pleasure of eating and healthy food behaviors through academic and applied food and hospitality research. Food Quality and Preference, 20, 533–536

Giménez, A., Ares, G. and Gámbaro, A. 2008. Consumer attitude toward shelf-life labeling: does it influence acceptance? Journal of Sensory Studies, 23, 871–883.

Hein, K. A., Hamid, N., Jaeger, S. R. and Delahunty, C. M. 2009. Application of a written scenario to evoke a consumption context in a laboratory setting: effects on hedonic ratings. Food Quality and Preference. doi:10.1016/j.foodqual.2009.10.003

Helm, E. and Trolle, B. 1946. Selection of a taste panel. Wallerstein Laboratory Communications, 9, 181–194.

Jones, L. V., Peryam, D. R. and Thurstone, L. L. 1955. Development of a scale for measuring soldier's food preferences. Food Research, 20, 512–520.

Kamen, J. 1989. Observations, reminiscences and chatter. In: Sensory Evaluation. In celebration of our Beginnings. Committee E-18 on Sensory Evaluation of Materials and Products. ASTM, Philadelphia, pp. 118–122.

Kimura, A., Wada, Y., Tsuzuki, D., Goto, S., Cai, D. and Dan, I. 2008. Consumer valuation of packaged foods. Interactive effects of amount and accessibility of information. Appetite, 51, 628–634.

Lawless, H. T. 1993. The education and training of sensory scientists. Food Quality and Preference, 4, 51–63.

Lawless, H. T. and Claassen, M. R. 1993. The central dogma in sensory evaluation. Food Technology, 47(6), 139–146.

Lawless, H. T. and Klein, B. P. 1989. Academic vs. industrial perspectives on sensory evaluation. Journal of Sensory Studies, 3, 205–216.

Lee, L., Frederick, S. and Ariely, D. 2006. Try it, you'll like it. Psychological Science, 17, 1054–1058.

MacRae, R. W. and Geelhoed, E. N. 1992. Preference can be more powerful than detection of oddity as a test of discriminability. Perception and Psychophysics, 51, 179–181.

Meilgaard, M., Civille, G. V. and Carr, B. T. 2006. Sensory Evaluation Techniques. Fourth Second edition. CRC, Boca Raton.

Meiselman, H. L. 1993. Critical evaluation of sensory techniques. Food Quality and Preference, 4, 33–40.

Mielby, L. H. and Frøst, M. B. 2009. Expectations and surprise in a molecular gastronomic meal. Food Quality and Preference. doi:10.1016/j.foodqual.2009.09.005

Moskowitz, H. R., Beckley, J. H. and Resurreccion, A. V. A. 2006. Sensory and Consumer Research in Food Product Design and Development. Wiley-Blackwell, New York.

Moskowitz, H. R. 1983. Product Testing and Sensory Evaluation of Foods. Food and Nutrition, Westport, CT.

Oliver, T. 1986. The Real Coke, The Real Story. Random House, New York.

O'Mahony, M. 1988. Sensory difference and preference testing: The use of signal detection measures. Chpater 8 In: H. R. Moskowitz (ed.), Applied Sensory Analysis of Foods. CRC, Boca Raton, FL, pp. 145–175.

O'Mahony, M. 1979. Psychophysical aspects of sensory analysis of dairy products: a critique. Journal of Dairy Science, 62, 1954–1962.

Pangborn, R. M. and Dunkley, W. L. 1964. Laboratory procedures for evaluating the sensory properties of milk. Dairy Science Abstracts, 26, 55–121.

Park, H. S. and Lee, S. Y. 2003. Genetically engineered food labels, information or warning to consumers? Journal of Food Products Marketing, 9, 49–61.

Peryam, D. R. and Swartz, V. W. 1950. Measurement of sensory differences. Food Technology, 4, 390–395.

Plotto, A., Barnes, K. W. and Goodner, K. L. 2006. Specific anosmia observed for β-ionone, but not for α-ionone: Significance for flavor research. Journal of Food Science, 71, S401–S406.

Shepherd, R., Sparks, P., Belleir, S. and Raats, M. M. 1991/1992. The effects of information on sensory ratings and preferences: The importance of attitudes. Food Quality and Preference, 3, 1–9.

Sidel, J. L., Stone, H. and Bloomquist, J. 1981. Use and misuse of sensory evaluation in research and quality control. Journal of Dairy Science, 61, 2296–2302.

Siegrist, M. and Cousin, M-E. 2009. Expectations influence sensory experience in a wine tasting. Appetite, 52, 762–765.

Skinner, E. Z. 1989. (Commentary). Sensory evaluation. In celebration of our beginnings. Committee E-18 on Sensory Evaluation of Materials and Products. ASTM, Philadelphia, pp. 58–65.

Stone, H. and Sidel, J. L. 2004. Sensory Evaluation Practices, Third Edition. Academic, San Deigo.

Stone, H., Sidel, J., Oliver, S., Woolsey, A. and Singleton, R. C. 1974. Sensory evaluation by quantitative descriptive analysis. Food Technology 28(1), 24, 26, 28, 29, 32, 34.

Sun Tzu (Sun Wu) 1963 (trans.), orig. circa 350 B.C.E. The Art of War. S.B. Griffith, trans. Oxford University.

Szczesniak, A. S., Loew, B. J. and Skinner, E. Z. 1975. Consumer texture profile technique. Journal of Food Science, 40, 1253–1257.

Tuorila, H. and Monteleone, E. 2009. Sensory food science in the changing society: opportunities, needs and challenges. Trends in Food Science and Technology, 20, 54–62.

Wysocki, C. J. and Labows, J. 1984. Individual differences in odor perception. Perfumer and Flavorist, 9, 21–24.

Yeomans, M. R., Chambers, L., Blumenthal, H. and Blake, A. 2008. The role of expectation in sensory and hedonic evaluation: The case of salmon smoked ice-cream. Food Quality and Preference, 19, 565–573.

York, R. K. 1995. Quality assessment in a regulatory environment. Food Quality and Preference, 6, 137–141.

Zellner, D. A., Strickhouser, D. and Tornow, C. E. 2004. Disconfirmed hedonic expectations produce perceptual contrast, not assimilation. The American Journal of Psychology, 117, 363–387.

Chapter 2

Physiological and Psychological Foundations of Sensory Function

Abstract   This chapter reviews background material underpinning sensory science and sensory evaluation methodologies. Basic and historical psychophysical methods are reviewed as well as the anatomy, physiology, and function of the chemical senses. The chapter concludes with a discussion of multi-modal sensory interactions.

There is no conception in man's mind which hath not at first, totally or in parts, been begotten upon by the organs of sense.

—Thomas Hobbes, Leviathan (1651)

Contents

2.1 Introduction

In order to design effective sensory tests and provide insightful interpretation of the results, a sensory professional must understand the functional properties of the sensory systems that are responsible for the data. By a functional property, we mean a phenomenon like mixture interactions such as masking or suppression. Another example is sensory adaptation, a commonly observed decrease in responsiveness to conditions of more or less constant stimulation. In addition, it is useful to understand the anatomy and physiology of the senses involved as well as their functional limitations. A good example of a functional limitation is the threshold or minimal amount of a stimulus needed for perception. Knowing about the anatomy of the senses

H.T. Lawless, H. Heymann, *Sensory Evaluation of Food*, Food Science Text Series,
DOI 10.1007/978-1-4419-6488-5_2, © Springer Science+Business Media, LLC 2010

can help us understand how consumers and panelists interact with the products to stimulate their senses and by what routes. Different routes of smelling, for example, are the orthonasal or sniffing route, when odor molecules enter the nose from the front (nostrils), versus retronasal smell, when odor molecules pass into the nose from the mouth or from breathing out, and thus have a reversed airflow pathway from that of external sniffing.

Another basic area that the sensory professional should have as background knowledge involves the sensory testing methods and human measurement procedures that are the historical antecedents to the tests we do today. This is part of the science of psychophysics, the quantification and measurement of sensory experiences. Psychophysics is a very old discipline that formed the basis for the early studies in experimental psychology. Parallels exist between psychophysics and sensory evaluation. For example, the difference test using paired comparisons is a version of the method used for measuring difference thresholds called the method of constant stimuli. In descriptive analysis with trained panels, we work very hard to insure that panelists use singular uni-dimensional scales. These numerical systems usually refer to a single sensory continuum like sweetness or odor strength and are thus based on changes in perceived intensity. They do not consider multiple attributes and fold them into a single score like the old-quality grading methods. Thus there is a clear psychophysical basis for the attribute scales used in descriptive analysis.

This chapter is designed to provide the reader some background in the sensory methods of psychophysics. A second objective is to give an overview of the structure and function of the chemical senses of taste, smell, and the chemesthetic sense. Chemesthesis refers to chemically induced sensations that seem to be at least partly tactile in nature, such as pepper heat, astringency, and chemical cooling. These three senses together comprise what we loosely call flavor and are the critical senses for appreciating foods, along with the tactile, force, and motion-related experiences that are part of food texture and mouthfeel. Texture is dealt with in Chapter 11 and color and appearance evaluations in Chapter 12. The auditory sense is not a large part of food perception, although many sounds can be perceived when we eat or manipulate foods. These provide another sense modality to accompany and reinforce our texture perceptions, as in the case of crisp or crunchy foods, or the audible hissing sound we get from carbonated beverages (Vickers, 1991).

One growing area of interest in the senses concerns our human biodiversity, differences among people in sensory function. These differences can be due to genetic, dietary/nutritional, physiological (e.g., aging), or environmental factors. The research into the genetics of the chemical senses, for example, has experienced a period of enormous expansion since the first edition of this book. The topic is too large and too rapidly changing to receive a comprehensive treatment here. We will limit our discussion of individual differences and genetic factors to those areas that are well understood, such as bitter sensitivity, smell blindness, and color vision anomalies. The sensory practitioner should be mindful that people exist in somewhat different sensory worlds. These differences contribute to the diversity of consumer preferences. They also limit the degree to which a trained panel can be "calibrated" into uniform ways of responding. Individual differences can impact sensory evaluations in many ways.

2.2 Classical Sensory Testing and Psychophysical Methods

2.2.1 Early Psychophysics

The oldest branch of experimental psychology is that of psychophysics, the study of relationships between physical stimuli and sensory experience. The first true psychophysical theorist was the nineteenth century German physiologist, E. H. Weber. Building on earlier observations by Bernoulli and others, Weber noted that the amount that a physical stimulus needed to be increased to be just perceivably different was a constant ratio. Thus 14.5 and 15 ounces could be told apart, but with great difficulty, and the same could be said of 29 and 30 ounces or 14.5 and 15 drams (Boring, 1942). This led to the formulation of Weber's law, generally written nowadays as

$$\Delta I/I = k \qquad (2.1)$$

where ΔI is the increase in the physical stimulus that was required to be just discriminably different from some starting level, I. The fraction, $\Delta I/I$, is sometimes called the "Weber fraction" and is an index of how well the sensory system detects changes. This relationship proved generally useful and provided the first quantitative operating characteristic of a sensory system. Methods for determining the difference threshold or just-noticeable-difference (j.n.d.) values became the stock in trade of early psychological researchers.

These methods were codified by G. T. Fechner in a book called *Elemente der Psychophysik* (*Elements of Psychophysics*) in 1860. Fechner was a philosopher as well as a scientist and developed an interest in Eastern religions, in the nature of the soul, and in the Cartesian mind–body dichotomy. Fechner's broader philosophical interests have been largely overlooked, but his little book on sensory methods was to become a classic text for the psychology laboratory. Fechner also had a valuable insight. He realized that the j.n.d. might be used as a unit of measurement and that by adding up j.n.d.s one could construct a psychophysical relationship between physical stimulus intensity and sensory intensity. This relationship approximated a log function, since the integral of $1/x\, dx$ is proportional to the natural log of x. So a logarithmic relationship appeared useful as a general psychophysical "law:"

$$S = k \log I \qquad (2.2)$$

where S is sensation intensity and I is once again the physical stimulus intensity. This relationship known as Fechner's law was to prove a useful rule of thumb for nearly 75 years, until it was questioned by acoustical researchers who supplanted it with a power law (see Section 2.2.3).

2.2.2 The Classical Psychophysical Methods

Fechner's enduring contribution was to assemble and publish the details of sensory test methods and how several important operating characteristics of sensory systems could be measured. Three important methods were the method of limits, the method of constant stimuli (called the method of right and wrong cases in those days), and the method of adjustment or average error (Boring, 1942). The methods are still used today in some research situations and variations on these methods form part of the toolbox of applied sensory evaluation. Each of the three methods was associated

with a particular type of measured response of sensory systems. The method of limits was well suited to determine absolute or detection thresholds. The method of constant stimuli could be used to determine difference thresholds and the method of adjustment to establish sensory equivalence.

In the method of limits the physical stimulus is changed by successive discrete steps until a change in response is noted. For example, when the stimulus is increasing in intensity, the response will change from "no sensation" to "I detect something." When the stimulus is decreasing in intensity, at some step the response will change back to "no sensation." Over many trials, the average point of change can be taken as the person's absolute threshold (see Fig. 2.1). This is the minimum intensity required for detection of the stimulus. Modern variations on this method often use only an ascending series and force the participants to choose a target sample among alternative "blank" samples at each step. Each concentration must be discriminated from a background level such as plain water in the case of taste thresholds. Forced-choice methods for determining thresholds are discussed in detail in Chapter 6.

In the method of constant stimuli, the test stimulus is always compared against a constant reference level (a standard), usually the middle point on a series of physical intensity levels. The subject's job is to respond to each test item as "greater than" or "less

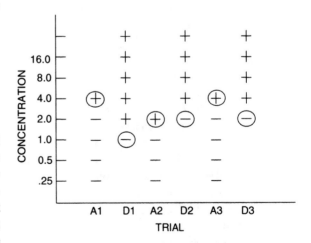

Fig. 2.1 An example of the method of limits. The *circled reversal points* would be averaged to obtain the person's threshold. *A*: ascending series. *D*: descending series. In taste and smell, only ascending series are commonly used to prevent fatigue, adaptation or carry-over of persistent sensations.

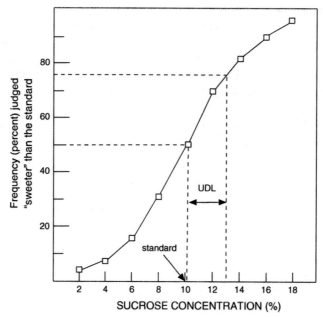

Fig. 2.2 A psychometric function derived from the Method of Constant Stimuli, a repeated series of paired comparisons against a constant (standard) stimulus, in this case 10% sucrose. Frequency of judgments in which the comparison stimulus is judged sweeter than the standard are plotted against concentration. The difference threshold is determined by the concentration difference between the standard and the interpolated 75% (or 25%) point. UDL: Upper difference limen (threshold).

than" the standard. Many replications of each intensity level are presented. The percentage of times the response is "greater than" can be plotted as in Fig. 2.2. This S-shaped curve is called a psychometric function (Boring, 1942). The difference threshold was taken as the difference between the 50 and 75% points interpolated on the function. The method of constant stimuli bears a strong resemblance to current techniques of paired comparison, with two exceptions. One point of difference is that the method was geared toward interval estimation, rather than testing for statistically significant differences. That is, the technique estimated points on the psychometric function (25, 50, and 75%) and researchers were not concerned with statistical significance of difference tests. Also, a range of comparison stimuli were tested against the standard and not just a single paired comparison of products.

The third major method in classical psychophysics was the method of adjustment or average error. The subject was given control over a variable stimulus like a light or a tone and asked to match a standard in brightness or loudness. The method could be used to determine difference thresholds based on the variability of the subject over many attempts at matching, for example, using the standard deviation as a measure of difference threshold. A modern application is in measuring sensory tradeoff relationships. In this type of experiment the duration of a very brief tone could be balanced against a varying sound pressure level to yield a constant perception of loudness. Similarly, the duration of a flash of light could be traded off against its photometric intensity to create a constant perceived brightness. For very brief tones or brief flashes, there is summation of the intensity over time in the nervous system, so that increasing duration can be balanced against decreasing physical intensity to create a constant perception. These methods have proven useful in understanding the physiological response of different senses to the temporal properties of stimuli, for example, how the auditory and visual systems integrate energy over time.

Adjustment methods have not proven so useful for assessing sensory equivalence in applied food testing, although adjustment is one way of trying to optimize an ingredient level (Hernandez and Lawless, 1999; Mattes and Lawless, 1985). Pangborn and co-workers employed an adjustment method to study individual preferences (Pangborn, 1988; Pangborn and Braddock, 1989). Adding flavors or ingredients "to taste" at the benchtop is a common way of initially formulating products. It is also fairly common to make formula changes to produce approximate sensory matches to some target, either a standard formula or perhaps some competitor's successful product. However, the method as applied in the psychophysics laboratory is an unwieldy technique for the senses of taste and smell where elaborate equipment is needed to provide adjustable stimulus control. So methods of equivalency adjustment are somewhat rare with food testing.

2.2.3 Scaling and Magnitude Estimation

A very useful technique for sensory measurement has been the direct application of rating scales to measure the intensity of sensations. Historically known as the "method of single stimuli," the procedure is highly cost efficient since one stimulus presentation yields one data point. This is in contrast to a procedure like the method of constant stimuli, where the presentation of many pairs is necessary to give a frequency count of the number of times each level is judged stronger than a standard. Rating scales have many uses. One of the most common is to specify a psychophysical function, a quantitative relationship between the perceived intensity of a sensation and the physical intensity of the stimulus. This is another way of describing a dose–response curve or in other words, capturing the input–output function of a sensory system over its dynamic range.

The technique of magnitude estimation grew out of earlier procedures in which subjects would be asked to fractionate an adjustable stimulus. For example, a subject would be asked to adjust a light or tone until it seemed half as bright as a comparison stimulus. The technique was modified so that the experimenter controlled the stimulus and the subject responded using (unrestricted) numbers to indicate the proportions or ratios of the perceived intensities. Thus if the test stimulus was twice as bright as the standard, it would be assigned a number twice as large as the rating for the standard and if one-third as bright, a number one-third as large. An important observation in S. S. Stevens' laboratory at Harvard was that the loudness of sounds was not exactly proportional to the decibel scale. If Fechner's log relationship was correct, rated loudness should grow in a linear fashion with decibels, since

they are a log scale of sound pressure relative to a reference ($db = 20 \log (P/P_0)$ where P is the sound pressure and P_0 is the reference sound pressure, usually a value for absolute threshold). However, discrepancies were observed between decibels and loudness proportions.

Instead, Stevens found with the direct magnitude estimation procedure that loudness was a power function of stimulus intensity, with an exponent of about 0.6. Scaling of other sensory continua also gave power functions, each with its characteristic exponent (Stevens, 1957, 1962). Thus the following relationship held:

$$S = kI^n \text{ or } \log S = n \log I + \log k \qquad (2.3)$$

where n was the characteristic exponent and k was a proportionality constant determined by the units of measurement. In other words, the function formed a straight line in a log–log plot with the exponent equal to the slope of the linear function. This was in contrast to the Fechnerian log function which was a straight line in a semilog plot (response versus log physical intensity).

One of the more important characteristics of a power function is that it can accommodate relationships that are expanding or positively accelerated while the log function does not. The power function with an exponent less than one fits a law of diminishing returns, i.e., larger and larger physical increases are required to maintain a constant proportional increase in the sensation level. Other continua such as response to electric shocks and some tastes were found to have a power function exponent greater than one (Meiselman, 1971; Moskowitz, 1971; Stevens, 1957). A comparison of power functions with different exponents is shown in Fig. 2.3.

Many sensory systems show an exponent less that one. This shows a compressive energy relationship that may have adaptive value for an organism responding to a wide range of energy in the environment. The range from the loudest sound one can tolerate to the faintest audible tone is over 100 dB. This represents over 10 log units of sound energy, a ratio of 10 billion to one. The dynamic range for the visual response of the eyes to different levels of light energy is equally broad. Thus exponents less than one have ecological significance for sensory systems that are tuned to a broad range of physical energy levels.

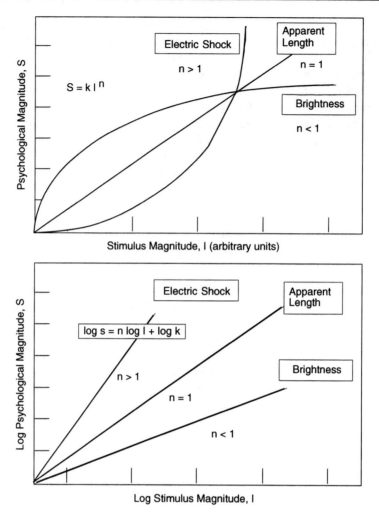

Fig. 2.3 Power function exponents less than, equal to, or greater than one generate different curves. In a log–log plot the exponent becomes the slope of a *straight line*.

Magnitude estimation as a test method and the resulting form of the power function formed an interlocking and valid system in Stevens' thinking. Power function exponents were predictable from various experiments. For example, in a cross-modality matching experiment, separate scaling functions were derived for two continua (e.g., brightness and loudness). One continuum was then scaled as a function of the other without using numbers. For example, a subject would be told to adjust the brightness of a light so it matched the loudness of a tone (fixed by the experimenter). The exponent in the matching experiment could be accurately predicted from the ratios of the exponents in the two separate scaling experiments.

When setting the sensations equal, the following relationships should hold:

$$\text{loudness} = \text{brightness} = k \log I^{n_1} = k \log I^{n_2} \quad (2.4)$$

and

$$n_1 \log(I_{\text{sound}}) + (a\,\text{constant}) = n_2 \log(I_{\text{light}}) \\ + (a\,\text{constant}) \quad (2.5)$$

and

$$\log(I_{\text{sound}}) = n_2/n_2 \log(I_{\text{light}}) + (a\,\text{constant}) \quad (2.6)$$

so that plotting a function of log sound intensity as a function of log light intensity would give a straight line with slope equal to n_2/n_1. This technique was very reliable (Stevens, 1959) and was often used as an undergraduate laboratory demonstration.

2.2.4 Critiques of Stevens

Other researchers were not so willing to accept the simple idea that the numbers applied to stimuli were in fact a direct reflection of the perceived sensation intensity. After all, the sensation was a subjective experience and the person had to decide what numbers to apply to the experience. So the simple stimulus–response idea was replaced by the notion that there were at least two separate processes: a psychophysical relationship translating stimulus intensity into subjective experience and response output function by which the subject applied numbers or some other response categories to the stimulus. Obviously, different scaling techniques could produce different response matching functions, so it was not surprising that an open-ended scaling task like magnitude estimation and a fixed-range scaling task like category ratings produced different psychophysical functions (a power function and a log function, respectively).

An extended argument ensued between the proponents of magnitude estimation and proponents of other scaling techniques like simple category scales (Anderson, 1974). The magnitude estimation camp claimed that the technique was capable of true ratio scale measurement, like measurements of physical quantities in the natural sciences (length, mass, heat, etc.). This was a preferable level of measurement than other techniques that merely rank ordered stimuli or measured them on interval scales (see Chapter 7). Opponents of these assertions remained unconvinced. They pointed out that the interlocking theory of the power law and the method that generated it were consistent, but self-justifying or circular reasoning (Birnbaum, 1982).

One problem was that category scales gave data consistent with Fechner's log function. Indirect scales did as well, so these two methods produced a consistent system (McBride, 1983). Category scales already had widespread use in applied sensory testing at about the time Stevens was spreading the doctrines of ratio-level scaling and magnitude estimation (Caul, 1957). Given the argument that only one kind of scale could

be a true or valid representation of sensations and the fact that they were nonlinearly related (Stevens and Galanter, 1957) an "either/or" mentality soon developed. This is an unfortunate distraction for applied sensory workers. For many practical purposes, the category and magnitude scaling data are very similar, especially over the small ranges of intensities encountered in most sensory tests (Lawless and Malone 1986).

2.2.5 Empirical Versus Theory-Driven Functions

Both the log function and the power function are merely empirical observations. There are an unlimited number of mathematical relationships that could be fit to the data and many functions will appear nearly linear in log plots. An alternative psychophysical relationship has been proposed that is based on physiological principles. This is a semi-hyperbolic function derived from the law of mass action and is mathematically equivalent to the function used to describe the kinetics of enzyme–substrate relationships. The Michaelis–Menten kinetic equation states the velocity of an enzyme–substrate reaction as a function of the substrate concentration, dissociation constant, and the maximum rate (Lehninger 1975; Stryer, 1995). Another version of this equation was proposed by Beidler, a pioneering physiologist, for description of the electrical responses of taste nerves and receptor cells (Beidler, 1961). The relationship is given by

$$R = (R_{max}C)/(k + C) \qquad (2.7)$$

where R is response, R_{max} is the maximal response, and k is the concentration at which response is half-maximal. In enzyme kinetics, k is a quantity proportional to the dissociation constant of the enzyme–substrate complex. Since taste involves the binding of a molecule to a protein receptor, it is perhaps not surprising that there is a parallel between taste response and an enzyme–substrate binding relationship. So this relationship has stirred some interest among researchers in the chemical senses (Curtis et al., 1984; McBride, 1987). In a plot of log concentration, the function forms an S-shaped curve, with an initial flat portion, a steep rise and then another flat zone representing saturation of response at high levels (see Fig. 2.4).

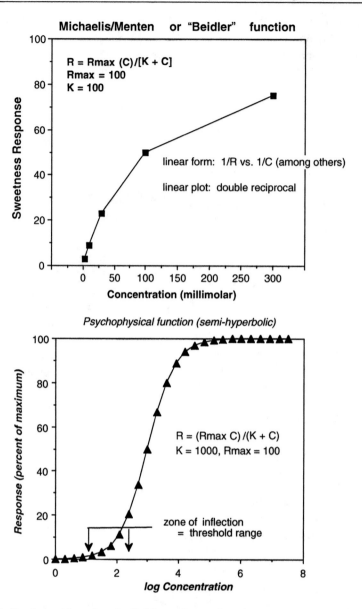

Fig. 2.4 The semi-hyperbolic relationship attributed to Beidler, cast in the same form as the Michaelis–Menten enzyme kinetic equation. This type of function is a common form of a dose–response curve for chemical stimuli. In the *lower* panel the semi-log plot takes the form of an S-curve.

This is intuitively appealing. The response at levels below threshold should hover around some baseline and then grow faster as threshold is surpassed (Marin et al., 1991). The function should eventually flatten out as it approaches a maximum response as all receptor sites are filled and/or as the maximum number of taste nerves respond at their maximum rate. In other words the system must saturate at some point.

2.2.6 Parallels of Psychophysics and Sensory Evaluation

Each of the psychophysical techniques mentioned in this section has its parallel or application in applied sensory evaluation. The emphasis of sensory psychology is on studying the person as the research object of interest, while applied sensory evaluation uses people to understand the sensory properties of products. Because any sensory event is an interaction of person and stimulus, the parallels in techniques should not surprise us. The major psychophysical research questions and methods and their sensory evaluation parallels are shown in Table 2.1. Threshold measurement has its applications in determining the minimum levels for impact of flavor compounds and the concentration ranges in which taints or off-flavors are likely to cause problems. Difference thresholds are similar in many ways to difference testing, with both scenarios making use of forced-choice or comparison procedures. Scaling is done in the psychophysics laboratory to determine psychophysical functions, but can also be used to describe sensory changes in product characteristics as a function of ingredient levels. So there are many points of similarity.

The remainder of this chapter is devoted to basic information on the structure and function of the flavor senses, since they have strong influence on the acceptability of foods. The visual and tactile senses are discussed only briefly, as separate chapters are devoted to color and visual perception generally (Chapter 12), and to texture evaluation (Chapter 11). For further information on sensory function the reader should go to basic texts on the senses such as Goldstein (1999) or the comprehensive *Handbook of Perception* (Goldstein, 2001).

2.3 Anatomy and Physiology and Functions of Taste

2.3.1 Anatomy and Physiology

Specialized sense organs on the tongue and soft palate contain the receptors for our sense of taste. Taste receptors are in the cell membranes of groups of about 30–50 cells clustered in a layered ball called a taste bud. These cells are modified epithelial cells (skin-like cells) rather than neurons (nerve cells) and they have a lifespan of about a week. New cells differentiate from the surrounding epithelium, migrate into the taste bud structure and make contact with sensory nerves. A pore at the top of the taste bud makes contact with the outside fluid environment in the mouth and taste molecules are believed to bind to the hair-like cilia at or near the opening. An illustration of this structure is shown in Fig. 2.5. Taste cells in a bud are not independently operating receptors, but make contact with each other and share junctions between cells for common signaling functions. The taste receptor cells make contact with the primary taste nerves over a gap or synaptic connection. Packets of neurotransmitter molecules are released into this gap to stimulate the taste nerves and send the taste signals on to the higher processing centers of the brain.

Table 2.1 Questions and methods in psychophysics and sensory evaluation

Question	Psychophysical study	Sensory evaluation examples
At what level is the stimulus detected?	Detection or absolute threshold measurement	Thresholds, taint investigation, flavor impact studies, dilution methods
At what level can a change be perceived?	Difference thresholds, just-noticeable-difference	Difference testing
What is the relationship between physical intensity and sensory response?	Scaling via direct numerical responses or indirect scales from difference thresholds	Scaling attribute intensity as in descriptive analysis
What is the matching relationship between two stimuli?	Adjustment procedures, trade-off relationships	Adjusting ingredients to match or optimize

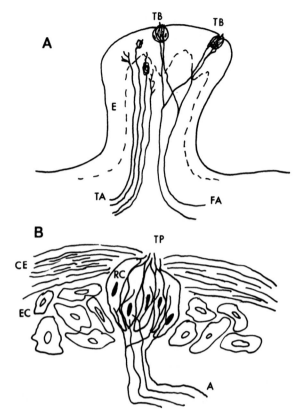

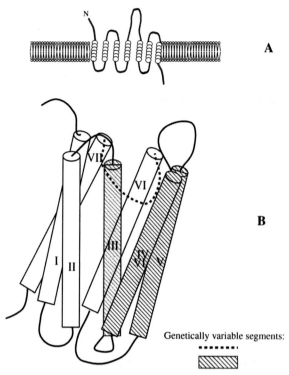

Fig. 2.5 (**a**) Cross-sectional drawing of a fungiform papilla. E, epithelium; TB, taste buds; TA, trigeminal afferent nerves terminating in various branched endings or encapsulated receptor structures; FA—facial nerve (chorda tympani) taste afferents terminating in taste buds. (**b**) Cross-sectional drawing of a taste bud. CE, cornified epithelium; EC, epithelial cells that may differentiate into taste receptor cells; RC, taste receptor cells, TP taste pore; A, axons from primary taste nerves making synaptic contact with receptor cells.

Fig. 2.6 (**a**) Planar schematic of a 7-transmembrane chemoreceptor protein. 7TMs have helical segments inside the membrane and several intracellular and extracellular peptide loops. The T1Rs for sweet and umami reception are associated as dimers with a long N-terminal; The T2Rs for bitter reception do not. (**b**) Schematic of the 7-transmembrane olfactory receptor, modeled after the structure of rhodopsin, the visual receptor protein. Transmembrane segments are symbolized by the cylinders and extracellular and intracellular loops by the heavy lines connecting them. Genetically variable segments include the barrels of segments II, IV, and V and the extracellular loop connecting segments VI and VII, making these sections candidates for a receptor pocket.

Through genetic research, the nature and types of taste receptor proteins have now been characterized. For sweet, bitter, and umami tastes, two families of receptor proteins are functional, the T1Rs for sweet and umami and the T2Rs for bitter tastes. These receptor proteins have seven transmembrane segments connected by intracellular and extracellular loops (hence "7TMs"). Figure 2.6 shows the arrangement of a 7TM with its genetically variable segments, which is also the structure of the family of odor receptors and the visual receptor, rhodopsin. The T1R proteins have about 850 amino acids and a large extracellular N-terminus, sometimes referred to as a "venus flytrap domain" after the hypothetical pockets formed by the paired (dimer) forms of these receptors. The T2Rs

have about 300–330 amino acids and a short extracellular N-terminus (Bachmanov and Beauchamp, 2007). The two families can exist side by side in taste buds, but are expressed in different cells (Sugita, 2006). The family of T2Rs contains about 40 active human variants with 38 intact genes currently known (Bachmanov and Beauchamp, 2007). Different T2Rs may be coexpressed in the same cells. This may explain why most bitter taste substances are similar in quality and difficult to differentiate. The number and variability of this family may be responsible for the ability of mammals to react to a wide range of molecular structures among the various bitter substances. The hT2R38 variant has been identified as the receptor for molecules such as PTC (phenylthiocarbamide or phenylthiourea) and PROP (6-*n*-propylthiouracil) to which there is a genetically based taste "blindness." The mutations in hT2R38 responsible for this inherited insensitivity have been identified (Bufe et al., 2005; Kim et al., 2003).

The T1Rs comprise only three peptide chains in two combinations, forming heterodimers. One dimer is the T1R1/T1R2 combination that is sensitive to glutamate and thus functions as an umami taste receptor. The other dimer is a T1R2/T1R3 combination that functions as the sweet receptor. The umami and sweet receptors are expressed in different taste receptor cells. Both the T1Rs and the T2Rs are G-protein coupled receptors (GPCRs) as are olfactory and visual receptors. The G-protein is an intracellular messenger consisting of three subunits, associated to the receptor inside the cell membrane. Stimulation of the taste receptor (i.e., binding to the 7-TM) leads to separation of the G-protein subunits, which can then activate other enzyme systems within the cell, causing a cascade of amplified events. Notably, G-protein subunits may activate adenylate cyclase, leading to production of cyclic AMP and/or phospholipase C, producing inositol triphosphate (IP3) (Sugita, 2006). Both cAMP and IP3 cause further activation of intracellular mechanisms such as activation or inactivation of ion channels in the cell membrane. These events lead to calcium influx or release, which is required for binding of neurotransmitter vesicles (packets) to the cell membrane and release of neurotransmitter molecules into the synapse to stimulate the associated taste nerve.

Salt and sour taste mechanisms appear to work more directly on ion channels, rather than via GPCRs. Sodium entering the cell is responsible for a cell membrane potential change (an ionic/electrical gradient) associated with calcium influx. Various ion channels have been proposed for mediating salty taste. Protons for sour taste may enter taste receptor cells and then stimulate ion channels such as the family of acid-sensitive ion channels (ASICs) or potassium conductance channels (Bachmanov and Beauchamp, 2007; Da Conceicao Neta et al., 2007; Sugita, 2006). Evidence points to the involvement of members of the transient receptor potential family in sour transduction, specifically members of the polycystic kidney disease family of receptors (PKD, so named from the syndromes in which they were first identified) (Ishimaru et al., 2006). Recent work has also suggested a taste sensitivity to

free fatty acids, due to the presence of a fatty acid transporter, CD36, in taste receptor cells (Bachmanov and Beauchamp, 2007). This could serve as a supplement to the textural cues which are usually thought of as the main signal for fat in the oral cavity.

The taste buds themselves are contained in specialized structures consisting of bumps and grooves on the tongue. The tongue is not a smooth uniform surface. The upper surface is covered with small cone-shaped filiform papillae. These serve a tactile function but do not contain taste buds. Interspersed among the filiform papillae, especially on the front and edges of the tongue are slightly larger mushroom-shaped fungiform papillae, often more reddish in color. These small button-shaped structures contain from two to four taste buds each, on the average (Arvidson, 1979). There are over a hundred on each side of the anterior tongue, suggesting an average of several hundred taste buds in the normal adult fungiform papillae (Miller and Bartoshuk, 1991). Along the sides of the tongue there are several parallel grooves about two-thirds of the way back from the tip to the root, called the foliate papillae. Each groove contains several hundred taste buds. Other specialized structures are about seven large button-shaped bumps arranged in an inverted-V on the back of the tongue, the circumvallate papillae. They contain several hundred taste buds in the outer grooves or moat-like fissures that surround them. Taste buds are also located on the soft palate just behind where the hard or bony part of the palate stops, an important but often overlooked area for sensing taste. The root of the tongue and upper part of the throat are also sensitive to tastes. Frequency counts of taste buds show that people with higher taste sensitivity tend to possess more taste buds (Bartoshuk et al., 1994).

Four different pairs of nerves innervate the tongue to make contact with these structures. This may explain in part why the sense of taste is resistant to disease, trauma, and aging, in contrast to the sense of smell (Weiffenbach, 1991). The fungiform papillae are innervated by the chorda tympani branches of the facial nerves (cranial nerve VII), which as its name suggests, crosses the eardrum. This circuitous route has actually permitted monitoring of human taste nerve impulses during surgery on the middle ear (Diamant et al., 1965). The glossopharyngeal nerves (cranial nerve IX) send branches to the rear of the tongue and the vagus nerve (cranial X) to the far posterior areas on the tongue root. The greater superficial petrosal branch of the facial nerve goes to the palatal taste area (Miller and Spangler, 1982; Nejad, 1986). Any one of the four classical taste qualities can be perceived on any area of the tongue, so the old-fashioned map of the tongue with different tastes in different areas is not accurate. For example, thresholds for quinine are lower on the front of the tongue than the circumvallate area (Collings, 1974).

Saliva plays an important part in taste function, both as a carrier of sapid molecules to the receptors and because it contains substances capable of modulating taste response. Saliva contains sodium and other cations, bicarbonate capable of buffering acids, and a range of proteins and mucopolysaccharides that give it its slippery and coating properties. There are recent suggestions that salivary glutamate may be capable of altering food flavor perception (Yamaguchi and Kobori, 1994). Whether saliva is actually necessary for taste response is a matter of historical controversy. At least in short time spans it does not seem to be required, as extensive rinsing of the tongue with deionized water through a flow system does not inhibit the taste response, but can actually sharpen it (McBurney, 1966).

2.3.2 Taste Perception: Qualities

Various perceptual qualities have been proposed as taste categories throughout history (Bartoshuk, 1978) but the consistent theme was that four qualities suffice for most purposes. These are the classical taste qualities of sweet, salty, sour, and bitter. Various others have been proposed to join the group of fundamental categories, most notably metallic, astringent, and umami. Umami is an oral sensation stimulated by salts of glutamic or aspartic acids. Astringency is a chemically induced complex of tactile sensations. These are discussed below. The metallic taste is occasionally used to describe the side tastes of sweeteners such as acesulfame-K and is a sensation experienced in certain taste disorders (Grushka and Sessle, 1991; Lawless and Zwillinberg, 1983). The classical four taste qualities are probably not sufficient to describe all taste sensations (O'Mahony and Ishii, 1986). However, they describe many taste experiences and have common reference materials, making them quite useful for practical sensory evaluation.

The umami sensation, roughly translated from Japanese as "delicious taste," is attributed to the taste of monosodium glutamate (MSG) and ribosides such as salts of $5'$ inosine monophosphate (IMP) and $5'$ guanine monophosphate (GMP) (Kawamura and Kare, 1987). The sensation is distinguishable from that of saltiness, as direct comparison with equally intense NaCl solutions demonstrates. The sensation is sometimes rendered in English by the term "brothy" due to its resemblance to the sensations from bouillon or soup stocks. "Savory" or "meaty" are alternatives (Nagodawithana, 1995). The taste properties of glutamate and aspartate salts form the building blocks of flavor principles in some ethnic (notably Asian) cuisines, and so perhaps it is not surprising that Japanese, for example, have no difficulty in using this taste term (O'Mahony and Ishii, 1986). Occidental subjects, on the other hand, seem to be able to fractionate the taste into the traditional four categories (Bartoshuk et al., 1974). Many animals including humans possess receptors for glutamate (Scott and Plata-Salaman, 1991; Sugita, 2006).

2.3.3 Taste Perception: Adaptation and Mixture Interactions

The sense of taste has two important functional properties that also have parallels in the sense of smell, sensory adaptation, and mixture interactions. Adaptation can be defined as a decrease in responsiveness under conditions of constant stimulation. It is a property of sensory systems that act to alert an organism to changes; the status quo is rarely of interest. We become largely adjusted to the ambient level of stimulation, especially in the chemical, tactile, and thermal senses. Placing your foot in a hot bath can be alarming at first, but the skin senses adapt. Our eyes constantly adapt to ambient levels of light, as we notice upon entering a dark movie theater. We are generally unaware of the sodium in our saliva, but rinsing the tongue with deionized water and representing that concentration of NaCl will produce a sensation above threshold. Adaptation is easily demonstrated in taste if the stimulus can be maintained on a controlled area of the tongue, for example, when a solution is flowed over the extended tongue or through a chamber (Kroeze, 1979;

McBurney, 1966). Under these conditions, most taste sensations will disappear in a minute or two. However, when the stimulus is not so neatly controlled, as in eating or in pulsatile stimulation, the adaptation is less robust and in some cases disappears (Meiselman and Halpern, 1973).

One other important discovery accompanied experiments on taste adaptation. Concentrations of NaCl or any other tastant below the adapting level—of which pure water was the extreme example—would take on other taste qualities. Thus water after salt adaptation can taste sour and/or bitter. Water tastes sweet after quinine or acid and tastes bitter after sucrose (McBurney and Shick, 1971). Figure 2.7 shows the response to concentrations of NaCl after different adaptation conditions. Above the adapting concentration, there is a salty taste. At the adapting concentration, there is little or no taste. Below the adapting concentration there is a sour–bitter taste that is strongest when water itself is presented. Water can take on any one of the four qualities, depending upon what has preceded it. This should alert sensory evaluation workers to the need for controlling or at least considering the effects of taste adaptation. Both the solvent and the taste molecules themselves can elicit sensory responses.

A second feature of taste function is the tendency for mixtures of different tastes to show partially inhibitory or masking interactions. Thus a solution of quinine and sucrose is less sweet than an equal concentration of sucrose tasted alone (i.e., when the sucrose in the two solutions is in equimolar concentration). Similarly the mixture is less bitter than equimolar quinine tasted alone. The general pattern is that all four classical taste qualities show this inhibitory pattern, commonly called mixture suppression (McBurney and Bartoshuk, 1973). In many foods these interactions are important in determining the overall appeal of the flavors and how they are balanced. For example, in fruit beverages and wines, the sourness of acids can be partially masked by sweetness from sugar . The sugar thus serves a dual role—adding its own pleasant taste while decreasing the intensity of what could be an objectionable level of sourness (Lawless, 1977). Some of these mixture inhibition effects, like the inhibition of bitterness by sweetness, appear to reside in the central nervous system (Lawless, 1979) while others, such as the inhibition of bitterness by salt, are more likely due to peripheral mechanisms at the receptors themselves (Kroeze and Bartoshuk, 1985).

There are a few exceptions to the pattern of inhibition where hyperadditive relationships, sometimes

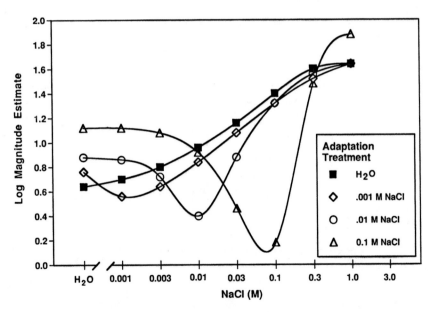

Fig. 2.7 Taste and water taste of NaCl following different adapting concentrations. The key shows the adapting (pretreatment) concentrations. Taste intensity reaches a minimum at each adapting level. Above the adapting levels, increased salty taste is reported. Below the adapting level, sour–bitter taste is reported, reaching a maximum with water. (from McBurney (1966), copyright 1966, by the American Psychological Association, reprinted with permission).

called enhancement or synergism occur. Hyperadditive effects imply that there is a higher taste intensity in the mixture than would be predicted on the basis of simple addition of component effects. However, how this outcome is predicted is controversial (Ayya and Lawless, 1992; Frank et al., 1989b). The most well-known claim of synergy is the interaction of MSG with the ribosides mentioned above. These are clearly hyperadditive by any definition. Addition of even small subthreshold amounts in mixtures will produce strong taste sensations (Yamaguchi, 1967) and there is strongly interactive binding enhancement at taste receptors that could be the physiological reason for this effect (Cagan, 1981). A second area of enhancement is seen with sweetness from salt in low concentrations added to sugar. NaCl has an intrinsic sweet taste seen at low levels that is normally masked by the saltiness at higher levels (Bartoshuk et al., 1978; Murphy et al., 1977). This may explain some of the beneficial effects of small amounts of salt in foods. A third case of hyperadditivity appears in the sweetener mixtures (Ayya and Lawless, 1992; Frank et al., 1989b). The search for synergistic mixtures of sweeteners and of other flavors is ongoing, due to the potential cost savings in this food ingredient category.

Finally, one can ask what happens to mixture suppression when one or more of the components has reduced impact? Figure 2.8 shows a release from inhibition that follows adaptation to one component of a mixture. Both the sweetness of sucrose and the bitterness of quinine are partially suppressed when present in a mixture. After adaptation to sucrose, the bitterness of a quinine/sucrose mixture rebounds to the level it would be perceived at in an equimolar unmixed quinine solution (Lawless, 1979). Likewise the sweetness rebounds after the bitterness is reduced by adaptation to quinine. These interactions are quite common in everyday eating. They can be easily demonstrated during a meal with tasting wines, since many wines contain sugar/acid (sweet/sour) taste mixtures. A wine will seem too sour after eating a very sweet dessert. Similarly, tasting a wine after eating a salad dressed with vinegar makes the wine seem too sweet and lacking in acid ("flabby"). These are simply the adapting effects upon the components of the wine, decreasing some tastes and enhancing others through release from inhibition. A similar effect can be seen in mixtures of three components, especially with salt. In a bitter–sweet mixture of urea and sucrose, for example, the usually suppression of bitterness and sweetness will be observed. But when a sodium salt is added to the mixture, there is a disproportionate effect of the salt inhibiting the bitter taste and consequently the sweet taste is enhanced (Breslin and Beauchamp, 1997). This effect is another explanation of the reported flavor enhancement in various foods when salt is added.

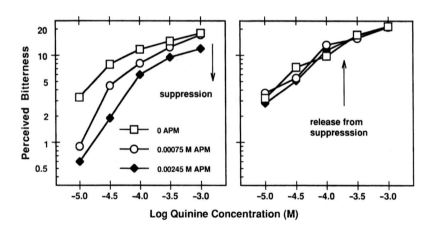

Fig. 2.8 Mixture suppression and release. The *left panel* shows perceived bitterness of quinine (*filled circles*) and mixtures with 0.00075 M aspartame (*squares*) and 0.00245 M aspartame (*open circles*) following adaptation to water. Mixture suppression is shown by reduced bitterness when the sweet taste is present in the mixtures. The *right panel* shows the same items after adaptation to sucrose, reducing the sweetness and returning the bitterness to its unsuppressed level (from Lawless (1979), copyright 1979, by the American Psychological Association, reprinted with permission).

2.3.4 Individual Differences and Taste Genetics

Wide individual differences in taste sensitivity exist, particularly for bitter compounds. The best example of this is the genetically inherited insensitivity to compounds containing the functional group –N–C $=$ S typified by certain aromatic thiourea compounds. This taste "blindness" was primarily studied using the compound phenylthiourea, originally called phenylthiocarbamide or PTC (Blakeslee, 1932; Fox, 1932). Due to the potential toxicity of PTC as well as its tendency to give off an odor, more recent studies have used the compound 6-*n*-propylthiouracil (PROP) which is highly correlated with PTC response (Lawless, 1980). Their structures are shown in Fig. 2.9. The minimum detectable concentrations (thresholds) of these compounds, PTC and PROP, follow a bimodal distribution, with about 1/3 of Caucasian persons unable to detect the substance at the concentration detected by most people. Thresholds tests as well as ratings for bitterness above threshold both allow differentiation into "taster" (sensitive) and "nontaster" (insensitive) groups (Lawless, 1980). Nontasters have a modification in the TAS2R38 taste receptor and show a simple Mendelian pattern of inheritance. Many other bitter substances such as quinine also show wide variation (Yokomukai et al., 1993), but none so dramatic as PTC and PROP.

Recent studies have identified hypersensitive groups of "supertasters" and counts of papillae and taste buds are correlated with taste sensitivity and responsiveness (Miller and Bartoshuk, 1991). Due to the enhanced trigeminal innervation in such individuals with a higher papillae density, it is perhaps not surprising that a relationship between PROP sensitivity and some lingual tactile sensations such as the sensitivity

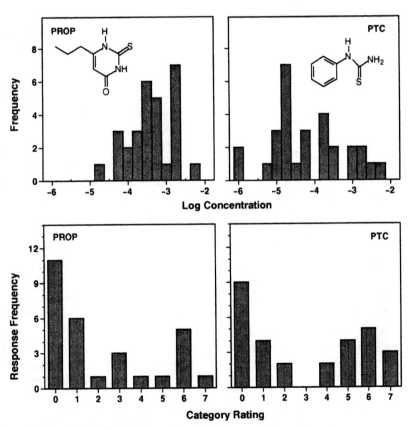

Fig. 2.9 PTC and PROP detection thresholds (*left panels*) and perceived intensity ratings (*right panels*) of 0.0001 M PTC and 0.00056 M PROP (from Lawless (1980), by permission of Information Retrieval Limited (IRL) and Oxford University Press). Note that PTC gives a better separation of taster and nontaster groups, especially with the perceived intensity ratings.

to fat have been found. A large number of other correlates to PROP sensitivity have been observed including sensitivity to the bitterness of caffeine, saccharin, and responses to capsaicin (Bartoshuk, 1979; Hall et al., 1975; Karrer and Bartoshuk, 1995). However, many of these correlations are low and some lower than the correlations among traditional tastants (Green et al., 2005, see also Schifferstein and Frijters, 1991). The current view, then, is that taste and chemesthesis are mostly independent systems and the sensory professional should be cautious in trying to use any general marker like PROP sensitivity as a predictor of individual response (Green et al., 2005). A potentially important finding is that persons who are insensitive to a bitter compound such as PTC will not show some mixture suppression effects (since they perceive no bitterness, there is no inhibition) on other flavors (Lawless, 1979). This illustrates a more general principle, that depending upon what we *do not* sense in a product, the other flavors may be enhanced for us, in a similar fashion to the effect of release from suppression.

2.4 Anatomy and Physiology and Functions of Smell

2.4.1 Anatomy and Cellular Function

The olfactory receptors are located in two small portions of epithelium very high in the nasal cavity. This remote location may serve some protective function against damage, but it also means that only a small percentage of the airborne substances flowing through the nose actually reach the vicinity of the sensory organs. In order to counter this factor, the olfactory sense has several attributes that enhance its sensitivity. There are several million receptors on each side of the nose and they have a terminal knob protruding into the mucus with about 20–30 very fine cilia which "float" in the mucus layer (Fig. 2.10). One function of these cilia is to increase the surface area of the cell, exposing the receptors to chemical stimuli. The main body of the olfactory receptor cells lies inside the epithelium and they each send a thin axon into the olfactory bulbs.

Another anatomical amplification factor is that the millions of receptors send nerve fibers into a much smaller number (perhaps 1,000) of glomerular structures in the olfactory bulb, after passing through a bony plate in the top of the nose. The glomeruli are dense areas of branching and synaptic contact of the olfactory receptors onto the next neurons in the olfactory pathway. Several thousand olfactory sensory neurons converge onto only 5–25 mitral cells in each glomerulus (Firestein, 2001). The mitral cells in turn send axons onto more central brain structures. The olfactory nerves project to many different sites in the brain, some of them closely associated with emotion, affect, and memory (Greer, 1991).

Unlike the taste receptors that are modified epithelial cells, the olfactory receptors are true nerve cells. They are unusual neurons in that they have a limited life span—they are replaced in about a month. The ability of the olfactory system to maintain its functional connections in the face of this turnover and replacement is a great puzzle of neural science. Other parts of the nervous system do not readily regenerate when damaged, so unlocking the mystery of olfactory replacement may provide benefits to those suffering from nervous system damage. The olfactory system is not immune from damage, however. A common injury occurs when a blow to the head severs the nerve fibers from the olfactory receptors as they pass through small passages in the bony cribriform plate on their way into the olfactory bulbs. This is sometimes self-repairing but often is not, leaving the individual without a functioning sense of smell, and therefore deprived of most food flavor perception for life. Sensory panel leaders need to be aware of the condition of total loss of smell, called anosmia, and screen panelists for sensory analysis duties with tests of olfaction such as a smell identification tests (Doty, 1991).

The mechanisms of odor reception are now well understood, starting with the discovery of a family of about 1,000 genes for olfaction in mammals, a discovery that earned Buck and Axel the Nobel Prize in 2004 (Buck and Axel, 1991). This may be the single largest gene family in the human genome. About 350 of these receptor types are active in humans. The receptors are G-protein coupled receptors, like the bitter receptors and visual receptor molecules. They have a sequence indicating seven transmembrane segments connected by intracellular and extracellular loops and have short N-terminals, like the bitter family of T2R receptors. Within the peptide sequences, there are from 10 to 60% variability (Firestein, 2001) with strong divergence in

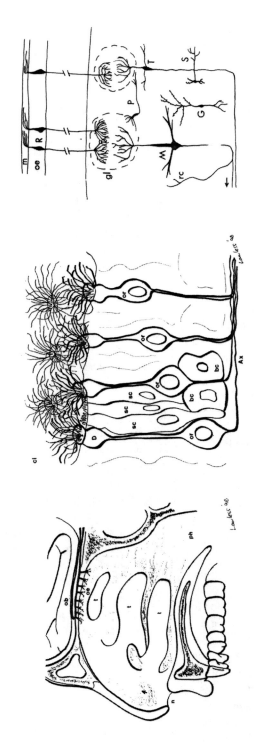

Fig. 2.10 *Left panel* gross nasal anatomy, n, external nares; t, turbinate bones or nasal conchae; ph, nasopharynx; ob, olfactory bulb at the base of the anterior cerebrum; oe, olfactory epithelium. Axon tracts leaving the olfactory epithelium pass into the bulb through small openings the cribriform plate. *Center panel* Diagram of olfactory epithelium. ci, cilia; D, dendritic termination of olfactory receptor cell; or, olfactory receptor cells; sc, supporting cells; bc, basal cells; Ax axon bundles. *Right Panel* basic cells types in the olfactory bulb. Olfactory receptors (R) send axons to glomeruli (gl) to make contact with apical dendrites of mitral (M) and tufted (T) cells, the output neurons from the bulb to higher structures. Cross-connections (some inhibitory) within layers are made by periglomerular cells (P), short axon cells (S), and granule cells (G) as well as recurrent collateral axons (rc). m, mucuous layer containing olfactory receptor cell cilia; oe, olfactory epithelium.

the third, fourth, and fifth transmembrane regions (see Fig. 2.6). These three "barrels" face one another and may form a receptor pocket about 1/3 of the way into the membrane. Identifying the kinds of molecules (ligands) that bind in these pockets has proven difficult due to the difficulty in expressing olfactory receptors in model systems. In the one case in which this was successful, the receptor was found to be specifically tuned to octanal and very similar molecules (Zhao et al., 1998).

The intracellular mechanisms for stimulation are similar to those of the G-coupled receptors in taste. Binding to the receptor results in activation of the G-protein subunits, which in turn activate enzymes such as adenylyl cyclase. This turns ATP into cyclic AMP which in turn activates various ion channels. An influx of $Na+$ and $Ca++$ ions causes the inside of the cell to become less negatively charged and when this membrane potential reaches a 20 mV threshold, an action potential is generated that travels down the nerve axon and results in neurotransmitter release. This is an amplification process, as the enzyme cascade can create about a thousand molecules of cAMP per second and hundreds of thousands of ions can cross through each open channel (Firestein, 2001). The calcium ions also open an outward flowing chloride ion channel, which serves as a kind of intracellular battery to reinforce the membrane potential change.

Different odor qualities are seen in spatial patterns (Kauer, 1987). Each odor receptor cell expresses only one type of receptor protein. Receptor cells with the same protein project to the same set of glomeruli. Similar odors also tend to map onto overlapping regions (Firestein, 2001). So different odors are represented by activation of different segments of the olfactory bulb. However, the matter is somewhat complicated by the fact that receptors are tuned to multiple odor molecules, and conversely, many odor molecules can stimulate a wide array of receptors. This has led to the combinatorial code for odor quality (Malnic et al., 1999). The brain recognizes the pattern of response across the array of neurons in order to "decide" on the odor quality or type. Viewed this way, olfaction appears to be the prototypical pattern recognition mechanism. Such a code can explain why some odorants change in their quality when the concentration increases. Additional receptors with higher thresholds for that compound are recruited as concentrations increase, altering the patterned array.

2.4.2 Retronasal Smell

Arguably, the largest contribution to the diversity of flavors comes from the volatile airborne molecules sensed by the olfactory receptors. Whether sniffed through the external nares in the direction of normal inspiration or arising from odors present in the mouth, the vast diversity of what we come to know as food flavors is mediated by smell. Due to the tendency to localize aromatics from foods in the mouth, many people do not realize that the olfactory sense is responsible for sensing most flavors other than the simple five tastes described above. Much of what we normally speak of as taste is really smell (Murphy et al., 1977; Murphy and Cain, 1980). The lemon character of a lemon, for example, is derived not from lemon taste (which is only sour, sweet, and bitter) but from the terpene aroma compounds that arise in the mouth and pass up into the nasal cavity from the rear direction (retronasally), opposite to that from sniffing. This highlights the dual role of olfaction as both an external sensory system and an internal sensory system (Rozin, 1982).

A simple demonstration can convince anyone of the importance of this internal smelling or retronasal smell. Take a sip of a simple fruit beverage or juice while holding the nose pinched shut. Take care to note the sensations present in the mouth, primarily the sweet and sour tastes. Now swallow the sample and while keeping the mouth shut, release the nostrils and exhale. In about a second or so, the fruit flavor will appear. Pinching the nose shut effectively blocks the retronasal passage of flavor volatiles up to the olfactory receptors (Murphy and Cain, 1980). When that route is facilitated by swallowing and exhaling, the contribution of smell becomes clear. The tendency of people to label internal smells as "tastes" probably contributes to the claims of sweetness enhancement by volatile flavors such as vanilla and maltol. This is a simple mislocation and mislabeling of the sensation (see Chapter 9). Learning to distinguish aromatics from true tastes is one of the first tasks in panel training for any sensory analysis of food flavor. Note that volatiles in the oral cavity may also have stimulatory effects there, but these seem to be limited to trigeminal stimuli such as menthol (Halpern, 2008). In most respects, orthonasal and retronasal smells are qualitatively similar.

It has been claimed that most (or even all) of retronasal smell arises from a kind of pumping action of air into the nose when people swallow, a so-called swallow breath (Buettner et al., 2002). However, a simple demonstration of exhalation without swallowing shows that this is not the only mechanism for retronasal smell: Take a small volume of liquid into the mouth and swirl it around. Expectorate. Do not swallow! Breathe in while holding the nose pinched shut. Release the nasal pinch and breathe out. There will be a clear impression of the volatile flavors that are perceived by retronasal smell. This kind of exhalation-induced flavor perception (a matter of smell) is commonly practiced by judges such as wine tasters when they differentiate aroma in the glass from aroma in the mouth. So the swallow breath is not absolutely required for retronasal smell. The swallow breath may be an important part of perception during normal eating, but it may be supplemented by other mechanisms in both eating and formal sensory evaluations.

2.4.3 Olfactory Sensitivity and Specific Anosmia

The olfactory sensitivity of humans and other animals is remarkable. Our ability to detect many potent odorants at very low concentrations still surpasses the sensitivity of nearly all instrumental means of chemical analysis. Many important flavor compounds are detectable in the parts per billion range, such as sulfur-containing compounds like ethyl mercaptan, a cabbage or skunk-like compound, so potent that it is employed as a gas odorization agent. Some food flavors are even more potent, like the methoxy pyrazine compounds that occur in bell peppers. Other small organic molecules are not so effective at stimulating the olfactory sense. The vast array of terpene aroma compounds responsible for citrus, herbal, mint, and pine-like aromas are usually potent in the parts-per million range. In contrast, alcohol compounds like ethanol are only sensed when their concentrations reach parts per thousand, so although we may think of alcohol as "smelly," in contrast to potent chemicals such as the pyrazines, it is not a very effective odor molecule.

A danger in flavor research is to assume that since a chemical has been identified in a product, and that chemical has an odor when smelled from a bottle that resembles the natural flavor, it will necessarily contribute to the flavor in the natural product. For example, limonene has been often used as a marker compound for orange juice aroma, but analysis of orange samples shows that it is often present well below threshold (Marin et al., 1987). It has the status of a "red herring" or a misleading compound. The critical question is whether the concentration in the product exceeds the threshold or minimum detectable concentration. Compounds present below their thresholds are unlikely to contribute to the perceived flavor, although some summation of the effects of similar compounds is always a possibility. This kind of threshold analysis for estimating flavor impact is discussed further in Chapter 6. The approach uses "odor units"—multiples of threshold—as evidence of a potential sensory contribution.

Thresholds are highly variable both within and across individuals (Lawless et al., 1995; Stevens et al., 1988). Some individuals with an otherwise normal sense of smell are unable to detect some families of similar smelling compounds. This is a condition called specific anosmia, as opposed to general anosmia or a total inability to smell. Specific anosmia is operationally defined as a condition in which an individual has a smell threshold more than two standard deviations above the population mean concentration (Amoore et al., 1968; Amoore, 1971). Common specific anosmias include an insensitivity to the following compounds of potential importance in foods: androstenone, a component of boar taint (Wysocki and Beauchamp, 1988); cineole, a common terpene component in many herbs (Pelosi and Pisanelli, 1981); several small branched-chain fatty acids important in dairy flavors (Amoore et al., 1968; Brennand et al., 1989); diacetyl, a lactic bacteria by-product (Lawless et al., 1994); trimethyl amine, a fish spoilage taint (Amoore and Forrester, 1976); isobutyraldehyde, responsible for malty flavors (Amoore et al., 1976); and carvone, a terpene in mint and other herbs (Pelosi and Viti, 1978, but see also Lawless et al., 1995). A sensory panel leader must be aware that each panel member has somewhat different olfactory equipment and that it may not be possible to force a panel into total agreement on all flavors. Also, a panelist with one specific anosmia may be a poor judge of that particular odor, but may function perfectly well on most other flavors. It makes little sense to exclude this panelist from participation unless the odor in question is a key component of all the foods being evaluated. This diversity presents a challenge in panel screening and detection of outliers in data analysis.

The sense of smell has a rather poor ability to discriminate intensity levels. This is observed in several ways. Measured difference thresholds for smell are often quite large compared to other sense modalities (Cain, 1977) and the power function exponents are often quite low (Cain and Engen, 1969). Early experiments on the ability of untrained subjects to identify or consistently label odor categories showed that people could reliably identify only about three levels of odor intensity (Engen and Pfaffmann, 1959). However, not all of the problem may be in the nose. In reviewing the historical literature on differential sensitivity, Cain (1977) reported that the Weber fraction (Section 2.2.1) falls in the range of about 25–45% for many odorants. This is about three times the size of the change needed to discriminate between levels of auditory or visual stimuli. Much of the problem was due to variation in the physical stimulus as confirmed by gas chromatography. The sniff bottles' concentration variation was highly correlated with discrimination performance, with stimulus variation accounting for 75% of the variance in discrimination. Thus historical estimates of odor difference thresholds may be too high.

2.4.4 Odor Qualities: Practical Systems

In contrast to its limited ability to distinguish intensity changes, the sense of smell provides us with a remarkably wide range of odor qualities. Experiments on odor identification show that the number of familiar odors people can label is quite large, seemingly with no upper bound (Desor and Beauchamp, 1974). However, the process of labeling odors itself is not easy. Often we know a smell but cannot conjure up the name, called a tip-of-the-nose phenomenon, in an analogy to saying a word is "on the tip of your tongue" (Lawless, 1977). This difficulty in verbal connection is one reason why many clinical tests of smell use a multiple choice format (Cain, 1979; Doty, 1991) to separate true problems in smelling from problems in verbal labeling. Our sense of smell is also limited in the ability to analytically recognize many components in complex odor mixtures (Laing et al., 1991;

Laska and Hudson, 1992). We tend to perceive odors as whole patterns rather than as collections of individual features (Engen and Ross, 1973; Engen, 1982). This tendency makes odor profiling and flavor description a difficult task for sensory panelists (Lawless, 1999). It seems more natural to react to odors as pleasant or unpleasant. The analytical frame of mind for odor and flavor perception demanded in sensory analysis is more difficult.

In spite of the common adage in psychology texts that there is no accepted scheme for classifying primary odors, there is quite strong agreement among flavor and fragrance professionals about categories for smells (Brud, 1986). Perfumers share a common language, developed in part on the basis of perceptual similarities within categories (Chastrette et al., 1988) and upon the sources of their ingredients. However, these schemes are generally unfamiliar to those outside these professions and may seem laden with technical jargon. Odor classification poses several challenges and problems. First, the number of differentiable categories is large. Early attempts at odor classification erred on the side of oversimplification. An example is Linnaeus's seven categories: aromatic, fragrant, musky, garlicky, goaty, repulsive, and nauseating, to which Zwaardemaker added ethereal and burned. A second impediment to the understanding of odor classification outside the flavor and fragrance world is that many of the original categories derive from the source materials of vendors of such ingredients. Thus they have a class for aldehydic (from aldehydes used as perfume fixatives, later an important ingredient in perfumes such as Chanel No. 5) and a class for balsamic fragrances. This nomenclature can seem a bit mysterious to the outsider. Balsamic fragrances include pine-woody sorts of smells combined with sweeter smells like vanilla. This example raises the question whether the perfumery categories can be broken down into more basic elements. Another approach to the problem proposed that odor categories be based on specific anosmias, since they may represent lack of a specific receptor type for a related group of compounds (Amoore, 1971). However, such attempts so far reduce to systems that are too small.

Nonetheless, there is considerable agreement among workers in different fields about quality categories for smells. For example, Table 2.2 shows a practical descriptive system for fragrances in consumer products derived solely from the experience

Table 2.2 Odor category systems

Functional odor categories[a]	Factor analysis groups[b]
Spicy	Spicy
Sweet (vanilla, maltol)	Brown (vanilla, molasses)
Fruity, non-citrus	Fruity, non-citrus
Woody, nutty	Woody
	Nutty
Green	Green
Floral	Floral
Minty	Cool, minty
Herbal, camphoraceous	Caraway, anise
(other)	Animal
	Burnt
	Sulfidic
	Rubber

[a]Descriptive attributes derived via principles of non-overlap and applicability to consumer products
[b]Factor analysis groups derived from ratings of aroma compounds on 146 attribute list

and intuition of the panel leaders during training. The second system is based on a categorization of tobacco flavors derived from a factor analysis of hundreds of odor terms and aromatic compounds (Civille and Lawless, 1986; Jeltema and Southwick, 1986). Given the different approaches and product areas, the agreement is surprisingly parallel. The terms for the tobacco work were derived from the ASTM list of odor character notes that contain 146 descriptors. This list provides a useful starting point for odor description (Dravnieks, 1982) but it is far from exhaustive and contains both general and specific terms. Other multivariate analyses of fragrance materials have yielded systems with similar categories (<20) (Zarzo and Stanton, 2006).

Other terminology systems for aromatic flavors have been developed for specific industries. This narrows the problem somewhat and makes the task of developing and odor classification system more manageable. One popular system is shown in Fig. 2.11 for wine aroma, arranged in a wheel format with hierarchical structure (Noble et al., 1987). A similar approach was taken with a circular arrangement of beer flavor terms (Meilgaard et al., 1982). The outer terms represent fairly specific aroma notes. Each outer term has an associated recipe for a flavor standard to act as a prototype/standard for training wine panelists. The system has embedded category structure that makes it easy to use. Interior terms act as more general categories subsuming the more specific outer terms. The

more general terms have practical value. Sometimes a wine may have some fruity character, but this will not be sufficiently distinct or specific to enable the panelist to classify the aroma as a specific berry, citrus, or other fruit. In that case there is some utility in having panelists simply estimate the general (overall) fruity intensity. Different parts of the wheel may apply more or less to different varietal wines and slightly different versions may evolve for different wine types, e.g., for sparkling wines.

2.4.5 Functional Properties: Adaptation, Mixture Suppression, and Release

An important operating characteristic of the flavor senses is their tendency to adapt or to become unresponsive to stimuli which are stable in space and time. This is perhaps most obvious for olfaction in everyday life. When one enters the home of a friend, we often notice the characteristic aroma of the house—the residual smells of their cooking and cleaning, personal care products, of babies or smokers, of pets or perfumes. These odors seem to characterize and permeate a house in its carpets and draperies. After several minutes, these aromas go largely unnoticed by a visitor. The sense of smell has adapted. There is no new information coming in, so attention and sensory function turn in other directions. In smell, like taste and the thermal senses, adaptation can be profound (Cain and Engen, 1969).

The sense of smell also shows mixture interactions. Odors of different qualities tend to mask or suppress one another, much like mixture suppression in taste. This is how most air fresheners work, by a process of odor counteraction via intensity suppression. The effect can easily be seen in two component mixtures where the odors are very different and easily separated perceptually, like lavender oil and pyridine (Cain and Drexler, 1974). Figure 2.12 shows pyridine/lavender mixtures, estimates of the intensity of the pyridine component at different levels of lavender, and estimates of the lavender intensity at different levels of pyridine (from Lawless, 1977). Odor intensity decreases as a function of the concentration of the other component. Such intensity interactions are most likely common in all complex food flavors.

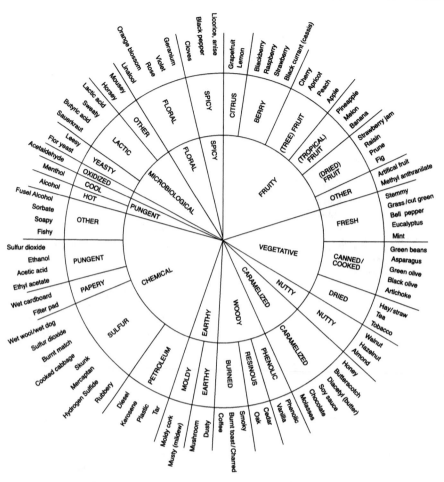

Fig. 2.11 The wine aroma "wheel" a system for arranging common wine aroma characteristics in a three-tiered categorical system. Inner terms are more comprehensive while the outer terms are more specific. Reference materials for the outer terms are given in the original paper. From Noble et al. (1987), courtesy of Ann Noble.

The contrast produced by release from mixture suppression also occurs in olfaction. Figure 2.13 shows a two-component odor mixture of vanillin and cinnamaldehyde. These odor components are distinguishable, i.e., they do not seem to blend into an new or inseparable mixture. Adapting the nose to one component makes the other one stand out (Lawless, 1987). This is an old analytical strategy used by some perfumers. When trying to analyze a competitor's fragrance, some components may be readily distinguished in the complex mixture and others may be obscured. If the nose is fatigued to one of the known components, the other components may seem to emerge, allowing them to be more readily identified. Patterns of adaptation to the strongest component of a flavor over time may explain in part why some

complex foods or beverages like wine seem to change in character over several minutes of repeated tastings.

The phenomena of adaptation and release present important considerations for sensory testing and a good reason why sensory tests should be done in an odor-free environment. Testing against the background of ambient odors will alter the quality and intensity profile of whatever is being tested. After a short period the olfactory system becomes immune to whatever is ambient in the building, less responsive to those aromatics if they occur in the test product, and more responsive to other flavors or aromas present due to the release from suppression effect. This makes testing in a factory, for example, potentially troublesome unless care is taken to insure that the test area is odor free or at least neutral in its background smell.

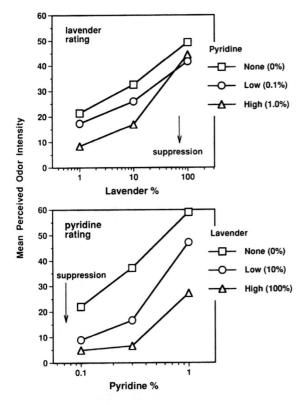

Fig. 2.12 Odor mixture inhibition in mixtures of lavender oil and pyridine. Decreased intensity of lavender is seen as a function of mixture with pyridine odor (*upper panel*) and decreased pyridine odor as a function of mixture with lavender (*lower panel*) (from Lawless, 1977 with permission).

Odor quality interactions are less predictable. Some odors seem to blend while others remain distinct. In general, odor mixtures bear a resemblance in their character to the quality characteristics of the individual components. For example, Laing and Wilcox (1983) showed that in binary mixtures, the odor profiles were generally similar to or predictable from the profiles of the components, although any intensity mismatch tended to favor the dominant component at the expense of the weaker item. This would suggest that emergent qualities or deriving a completely new odor as a function of mixing is rare. However, anecdotes exist about multicomponent mixtures in which the odor of the emergent pattern is not clearly present in any single component. For example, a mixture of ten or so medium chain aldehydes (C6–C16) produces a smell reminiscent of old wax crayons (Lawless, 1996). Furthermore, natural flavors consist of mixtures of many chemical components and no single

chemical may possess the odor quality characteristic of the blend. The odor of cocoa is a distinctive smell, but it is difficult to find any single chemical component which produces this impression. In an analysis of cheese aroma by gas chromatographic sniffing, the components had no cheese aromas in their individual characteristics (Moio et al., 1993). Burgard and Kuznicki (1990) noted that such synthesis may be the rule: "Coffee aroma is contributed to by several hundreds of compounds, a great many of which do not smell anything like coffee" (p. 65).

2.5 Chemesthesis

2.5.1 Qualities of Chemesthetic Experience

A variety of chemically induced sensations can be perceived in the oral and nasal cavities as well as the external skin. These chemically induced sensations do not fit neatly into the traditional classes of tastes and smells. They are called chemesthetic sensations in an analogy to "somesthesis" or the tactile and thermal sensations perceived over the body surface (Green and Lawless, 1991; Lawless and Lee, 1994). Many of these sensations are perceived through stimulation of the trigeminal nerve endings in the mouth, nose, or eyes. They include the heat-related irritative sensations from chili pepper and other spices, the non-heat related irritations from horseradish, mustard, and wasabi, the lachrymatory (tear-inducing) stimuli from onions, the cooling sensations from menthol and other cooling agents, and irritation from carbon dioxide. Other classes of sensations that are sometimes grouped with these are astringency, which is a chemically induced tactile sensation and the so-called metallic taste. Others could be added, but they are beyond the scope of this text. The ones discussed here are the common and major types of experiences found in foods and consumer products.

The importance of chemesthesis is evident from anatomical and also economic considerations. Much of the chemesthetic flavor sensations are mediated by the trigeminal nerves and the size of the trigeminal tracts relative to the other chemical sense nerves is impressive. One study found three times as many trigeminal fibers in the fungiform papillae of the rat than the facial

Fig. 2.13 Odor mixture inhibition and release following adaptation in mixtures of vanillin and cinnamaldehyde. *Open bars*, perceived intensity of vanilla odor; *Hatched bars*, perceived intensity of cinnamon odor. After adaptation to vanillin, the cinnamon odor returns to its unmixed level. After adaptation to cinnamaldehyde, the vanilla odor returns to its unmixed level (from Lawless (1984) with permission of the Psychonomic Society).

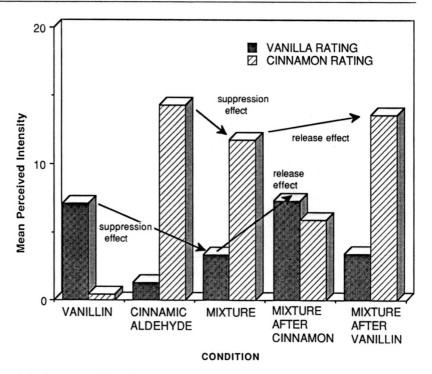

(taste) nerve fibers innervating taste buds (Farbman and Hellekant, 1978). So these papillae are not just taste sensory organs, but might be more accurately classified as organs for the perception of chili pepper burn (Lawless and Stevens, 1988). Even the taste bud itself seems organized to provide trigeminal access to the oral milieu. Trigeminal fibers ascend around the taste bud forming a chalice-like structure (Whitehead et al., 1985), possibly enhancing their access to the external environment.

The economic impact of trigeminal flavors on the food and flavor industry is growing. Carbon dioxide is a trigeminal stimulus and the carbonated beverage business—soda, beer, sparkling wines, etc.—amounts to huge sales worldwide. Putting aside CO_2, we can ask about the economic impact of individual spices or their use in various products. In the United States, so-called ethnic foods are experiencing a period of rapid growth due to a continuing influx of immigration of peoples from cultures with hot spicy cuisines and a growing trend toward less neophobic and more adventurous dining on the part of many Americans. Sales of salsa have surpassed the sales of ketchup since 1992. New programs of research have added whole new categories of chemesthetic flavorants, such as "tingle" compounds.

2.5.2 Physiological Mechanisms of Chemesthesis

A variety of specialized nerve endings from the tactile somatosensory systems can be observed histologically in skin and other epithelial tissues. For purposes of nociception, especially those induced by chemicals, it has long been thought that free nerve endings are the likely sensors. Generally, the nerve fibers involved in nociception are small diameter and slowly conducting c-class nerves. Many of the chemesthetic sensations are mediated by a special family of receptor proteins known as Transient Receptor Potential (TRP) channels (Silver et al., 2008). These proteins form cation channels and consist of four associated subunits. Each subunit contains a long peptide with six sections that cross the cell membrane and each contains a single pore region. Originally discovered in Drosophila photoreceptors, a wide variety of these functional channels have been found in various organs and many different cells (Patapoutian et al., 2003; Venkatachalam and Montell, 2007). The first chemoreceptive TRP to be characterized was the TRPV1, a so-called vanilloid receptor that is sensitive to capsaicin as well as acidic pH, heat, and mechanical stimulation. One member of the TRPM family, TRPM8, is sensitive to menthol and other cooling compounds. TRPP3 channels have been implicated in sour taste transduction as they are responsive to acids, and may form a functional sour receptor. A type of TRP channel which is responsive to a very wide range of chemical stimuli including irritants and pungent stimuli such as wasabi and horseradish is the TRPA channel (Tai et al., 2008). TRP channels may also act in concert with the GPRC's to affect taste cell transduction for sweet, bitter, and umami tastes (TRPM5). The capsaicin-sensitive TRPV1 channel and the TRPM5 channel found in some taste receptor cells may participate in the sensing of some aspects of complex tasting divalent salts (iron, zinc, copper, etc.) (Riera et al., 2009). Because some TRPs are sensitive to both temperature and chemical stimulation, simultaneous or sequential combinations cause enhancements. For example, capsaicin can enhance heat pain from thermal stimulation, probably through a common action on TRP1V channels and menthol can enhance cold-induced pain, probably through common action on TRPM8 channels (Albin et al. 2008). For a review of these important chemoreceptive mechanisms, see Calixto et al. (2005), Silver et al. (2008), and Venkatachalam and Montell (2007).

2.5.3 Chemical "Heat"

An actively studied category of chemesthetic sensations are those that arise from pepper compounds such as capsaicin from chili peppers, piperine from black pepper, and the ginger compounds such as zingerone. The potency of capsaicin is noteworthy, with thresholds below 1 ppm. This is about 100 times as potent as piperine and other irritants, based on dilution to threshold measures such as the Scoville procedure (discussed in Chapter 6). In pure form, capsaicin causes a warm or burning type of irritation with little or no apparent taste or smell (Green and Lawless, 1991; Lawless, 1984). The most obvious sensory characteristic of stimulation with the pepper compounds is their long-lasting nature. Stimulation with capsaicin, piperine, or ginger oleoresin at concentrations above threshold may last 10 min or longer (Lawless, 1984). So these flavor types are well suited to the application of time–intensity profiling (see Chapter 8). Other irritants such as ethanol and salt produce less persistent effects over time.

The temporal properties of capsaicin are complex. When stimulation is followed by a short rest period, a type of desensitization or numbing of the oral tissues sets in (Green, 1989). Application of the red pepper compound, capsaicin, to the skin or oral epithelium has profound desensitizing effects (Jansco, 1960; Lawless and Gillette, 1985; Szolscanyi, 1977). This nicely parallels the animal experimentation showing a generalized desensitization after injection with capsaicin (Burks et al., 1985; Szolcsanyi, 1977), which is believed to result from the depletion of substance P, a neurotransmitter in the somatic pain system. Since effects of substance P have also been linked to the functioning of endorphins (Andersen et al., 1978), there is a suggestion that the kind of craving or addiction that occurs for spicy foods may be endorphin-related. High dietary levels of capsaicin also result in a chronic desensitization, as shown in psychophysical tests (Lawless et al., 1985). Figure 2.14 shows a desensitization effect seen in sequences during a psychophysical study, and also the apparent chronic desensitization that occurs in people who consume chili peppers or spices derived from red pepper on a regular basis (Prescott and Stevenson, 1996). Sensitization is also observed when the rest period is omitted and stimulation proceeds in rapid sequences; the irritation continues to build to higher levels (Stevens and Lawless, 1987; Green, 1989). These tendencies to sensitize and desensitize make sensory evaluations of pepper heat somewhat difficult if more than one trial per session is required. A calibrated descriptive panel may be useful, one whose abilities can help bridge the time delays required between repeated observations.

In addition to their numbing and sensitizing effects, irritant stimulation in the oral or nasal cavity evokes strong defensive reflexes in the body, including sweating, tearing, and salivary flow. There is a strong correspondence between sensory ratings of pepper heat intensity and the evoked salivary flow from the same subjects taken simultaneously with ratings (Lawless, 1984). This provides a nice demonstration that sensory ratings should not be dismissed as merely "subjective" in that they have obvious correlates in "objectively" measurable physiological reflexes.

An unresolved question in the realm of chemical irritation is the degree to which different sensory qualities are evoked (Green and Lawless, 1991). This is difficult to study due to a lack of vocabulary to

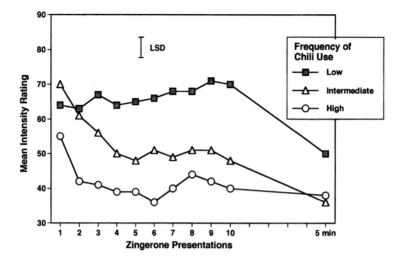

Fig. 2.14 Zingerone desensitization as a function of dietary use, numbers of exposures and a break in stimulation. The differences in the height of the curves demonstrate the chronic desensitization that is correlated with high dietary intake of pungent spices. The symbols at the far right demonstrate the within-session desensitization that occurs during a hiatus in stimulation, as commonly seen with capsaicin, the irritant component of red (chili) peppers. The latter effect is more pronounced for those with low dietary intake. From Prescott and Stevenson (1996) with permission.

describe, at least in English, the experiences from pepper burn, CO_2, mustard, and so on. Experience with spices suggests that there are a variety of irritative flavor experiences and not all irritations are the same. Studies of synergistic interaction in mixtures and potentiation with different irritants in rapid sequences are suggestive of the possibility of multiple receptor mechanisms for oral chemical irritation (Lawless and Stevens, 1989, 1990). Direct measurement of qualitative differences was attempted in a descriptive study by Cliff and Heymann (1992) using a variety of irritant flavor materials. They found evidence for differences in lag time (short versus long onset) and burning versus tingling sensations among the irritants tested. A lexicon for carbonation was developed by Harper and McDaniel (1993) and involved descriptors for cooling, taste, trigeminal (bite, burn, numbing), and tactile/mechanoreception properties.

2.5.4 Other Irritative Sensations and Chemical Cooling

The trigeminal flavor senses also affect food flavor in other ways. Even such benign stimuli as NaCl can be irritative at high concentrations (Green and Gelhard, 1989). Carbon dioxide is a potent irritant in the nasal cavity, as are many organic compounds (Cain

and Murphy, 1980; Cometto-Muñiz and Cain, 1984; Commetto-Muñiz and Hernandez, 1990). Completely anosmic individuals can detect many odor compounds, presumably from the ability of odorants to stimulate the trigeminal nerve branches in the nasal cavity (Doty et al., 1978). There is an irritative component to many common odorants and flavor compounds. A variety of highly reactive sulfur compounds have been identified in other irritative spices and food flavors, such as compounds from horseradish, mustard, and the lacrimatory (tear-inducing) factor from onions and related vegetables (Renneccius, 2006). Ethanol and cinnamaldehyde are other examples of other common flavors that are irritative (Prescott and Swain-Campbell, 2000).

Carbonation, or the perception of dissolved CO_2, involves a truly multimodal stimulus. In addition to the tactile stimulation of mechanoreceptors, CO_2 acts on both trigeminal receptors (Dessirier et al., 2000) and gustatory receptors (Chandrashekar et al., 2009). Both of these chemical sensations involve the enzyme carbonic anhydrase, which can convert CO_2 to carbonic acid. For the sense of taste, the stimulation with CO_2 appears to involve the extracellular anhydrase enzyme and the transient receptor potential (TRP) mechanism (PDK2L1) of sour receptor cells (Chandrashekar et al., 2009). This is consistent with the enhancement of sour taste by CO_2 and suppression of sweetness (Cowart, 1998; Hewson et al., 2009). The role of nociceptors in

CO_2 perception is further substantiated by its desensitization by capsaicin (Dessirier et al., 2000).

Using the method of magnitude estimation, Yau and McDaniel (1990) examined the power function exponent (see Section 2.2.4) for carbonation. Over a range of approximately one to four volumes CO_2 per volume of H_2O, sensation intensity grew as a power function with an exponent of about 2.4, a much higher value than in most other modalities. The exponent is consistent with high sensitivity to changes in carbonation levels. Given the involvement of TRP mechanisms in both nociception and temperature sensing, interactions between carbonation and temperature might be expected. An enhancement of irritation, tactile sensations, cooling, and cold pain have all been observed with carbonation of solutions served at low temperatures (Green, 1992; Harper and McDaniel, 1993; Yau and McDaniel, 1991). Yau and McDaniel (1991) noted a small increase in tactile intensity at low temperatures (3–10°C). This may be an example of a phenomenon called Weber's illusion, in which Weber noted that a cold coin seemed heavier than a warm one, an early clue to the overlap in tactile and thermal sensing mechanisms.

Menthol, a compound that has both odor properties and is capable of causing cool sensations, is a trigeminal stimulus with obvious commercial significance in confections, oral health care, and tobacco products (Patel et al., 2007). Menthol has been found to interact with thermal stimulation in complex ways . Menthol enhances cool stimuli as would be expected, but can either enhance or inhibit warm stimuli depending upon the conditions of stimulation (Green, 1985, 1986). The sensory properties of menthol itself are complex, inducing a number of cooling, warming, aromatic, and other sensory effects depending upon the isomer, concentration, and temporal parameters (Gwartney and Heymann, 1995, 1996). A large number of hyperpotent cooling compounds have been patented, many of which can produce cooling without the odor sensations of menthol (Leffingwell, 2009; Renneccius, 2006).

2.5.5 Astringency

Tannins in foods are chemical stimuli and yet the astringent sensations they produce are largely tactile.

They make the mouthfeel rough and dry and cause a drawing, puckery, or tightening sensation in the cheeks and muscles of the face (Bate Smith, 1954). There are two approaches to defining astringency. The first is to emphasize the causes of astringent sensations, i.e., those chemicals which readily induce astringency. For example, ASTM (1989) defines astringency as "the complex of sensations due to shrinking, drawing or puckering of the epithelium as a result of exposure to substances such as alums or tannins." A more perceptually based definition is that of Lee and Lawless (1991): "A complex sensation combining three distinct aspects: drying of the mouth, roughing of oral tissues, and puckery or drawing sensations felt in the cheeks and muscles of the face." Principal component analysis has shown these sub-qualities to be independent factors and furthermore, distinctly separate from taste sensations such as sourness (Lawless and Corrigan, 1994). The fact that astringent sensations can be sensed from areas of the mouth such as the lips, that are lacking in taste receptors, further substantiates their classification as tactile rather than a gustatory sensations (Breslin et al., 1993).

The mechanisms for astringency involve the binding of tannins to salivary proteins and mucins (slippery constituents of saliva), causing them to aggregate or precipitate, thus robbing saliva of its ability to coat and lubricate oral tissues (Clifford, 1986; McManus et al., 1981). We feel this result as rough and dry sensations on oral tissues. Other mechanisms may also contribute to astringency in addition to the binding of tannins to salivary proteins (Murray et al., 1994). Acids commonly used foods also induce astringency in addition to their sour taste (Rubico and McDaniel, 1992; Thomas and Lawless, 1995). The astringent impact of acids is pH dependent (Lawless et al., 1996; Sowalski and Noble, 1998) suggesting that a direct attack on epithelial tissues or a pH-dependent denaturation of the lubricating salivary proteins may also occur.

The interaction of mucins and proline-rich proteins (PRPs) in saliva with tannins may be a key part of astringency mechanisms as protein content is a correlate of sensory response (Kallikathraka et al., 2001). Binding of polyphenols to PRPs is well known in the beer and fruit juice industries as it can give rise to turbidity known as chill-haze (Siebert, 1991). A similar visible haze generation reaction has been shown to occur with tannic acid mixed with saliva (Horne et al., 2002). Haze development of saliva is an in vitro measure, a correlate of predicting individual responses to astringency in products such as wine and a potential measure for screening and selecting panelists for astringency evaluation (Condelli et al., 2006) as well as analysis of wine samples. The individual differences show an inverse relationship: panelists with high haze development and higher salivary flow rates are less reactive (have lower ratings). Their enhanced mucin or protein content may provide greater "protection" of oral surfaces against astringent compounds. Another important individual difference in astringent reactions is salivary flow rate (Fisher et al., 1994). Individuals with a higher flow rate tend to "clean up" faster after astringent stimulation. Repeated stimulation with astringent substances tends to cause a buildup of tactile effects rather than a decrement as one might see in taste adaptation or capsaicin desensitization. Figure 2.15 shows an increase in astringency upon repeated stimulation, as might happen with multiple sips of a beverage such as wine. Note that the pattern changes as function of tannin concentration, interstimulus interval between sips and to a small extent, as a function of the volume tasted (Guinard et al., 1986).

2.5.6 Metallic Taste

Another quality of chemical sensations that is sometimes referred to as a taste are the metallic sensations that arise from placing different metals in the mouth or from contact with iron or copper salts. Two common reference standards for metallic taste in descriptive analysis training are (1) rinses with ferrous sulfate and (2) a clean copper penny (Civille and Lyon, 1996). Research now shows that these are quite different sensations in terms of their mechanisms, although they both are described as "metallic" perhaps because they may occur at the same time.

The so-called metallic taste after rinses with ferrous sulfate solutions is actually a case of retronasal smell. The sensation is virtually abolished if the nose is pinched shut during tasting (Epke et al., 2008; Lawless et al., 2004, 2005). Because metal salts are not volatile, this olfactory sensation probably arises from the ferrous ions catalyzing a rapid lipid oxidation in the mouth, creating well-known potent odor compounds such as 1-octen-3-one (Lubran et al., 2005).

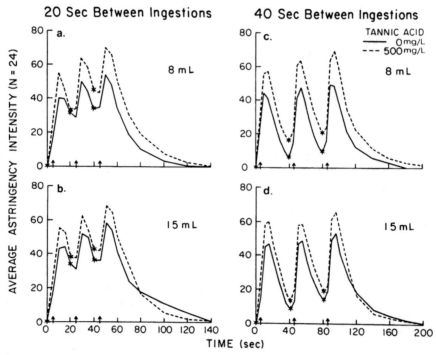

Fig. 2.15 Average time–intensity curves for astringency in wine with 0 or 500 mg/l of added tannic acid upon three successive ingestions. Sample uptake and swallowing are indicated by a *star* and *arrow*, respectively. From Guinard et al. (1986) by permission of the American Society for Enology and Viticulture.

A second kind of metallic sensation is the one that arises from the "clean copper penny." If one scratches the copper off part of the surface of a US penny, exposing the zinc core, the metallic sensation increases dramatically (Lawless et al., 2005). Due to the different electrical potentials of the different metals, a small current is created, making this a case of electrical taste stimulation (McClure and Lawless, 2007; Stevens et al., 2008). In the clinical literature on electrogustometry, in which electrical taste stimulation is used for diagnostic tests, the term "metallic" is often reported. The sensory analyst should be careful to distinguish between these two kinds of sensations. If the sensation is abolished or dramatically diminished by nasal occlusion, then it is a case of retronasal olfactory sensations, possibly due to potent lipid oxidation products. If not, there may be metals in the system leading to small electrical potentials. There is also the possibility of a third kind of metallic sensation that may be a true taste, but this is still controversial.

2.6 Multi-modal Sensory Interactions

Food is a multi-modal experience, so it should come as no surprise that the sensations from one sensory modality may influence judgments and perceptions from another. Through our experience, we learn about the pairings of colors and tastes, colors, and odors and come to have expectations about what sensations may accompany one another. Through repeated pairings or through natural co-occurrence of different tastes and flavors, an association can be built up leading to integration of those experiences (Stevenson et al., 1999). Brain imaging of regions of the frontal cortex supports that notion that the merging of these sensations into coherent percepts are "real" perceptions and not just some kind of response bias (Small et al., 2004). Interactions between sensory modalities and their possible neural substrates have been reviewed by Delwiche (2004), Small and Prescott (2004), and Verhagen and Engelen (2006). The discussions that follow will focus on those interactions that have been most heavily studied and are most relevant to foods: taste/odor, flavor/irritation (chemesthesis), and color/flavor. Other inter-modality interactions are discussed in the review papers mentioned above.

2.6.1 Taste and Odor Interactions

An reliable observation from the psychophysical literature is that sensation intensities of tastes and odors are additive or slightly hypo-additive (Hornung and Enns, 1984, 1986; Murphy et al., 1977; Murphy and Cain, 1980). The pattern of results is that intensity ratings show about 90% additivity. That is, when framed as a simple question about the summation of gustatory and olfactory intensity ratings in producing overall ratings of flavor strength, there is little evidence for interactions between the two modalities.

However, there have been many other studies showing enhancement of specific taste qualities, notably sweetness, in the presence of odors. An important tendency, especially among untrained consumers, is to misattribute some volatile olfactory sensations to "taste," particularly retronasally perceived odors. Retronasal smell is poorly localized and often perceived as a taste from the oral cavity. Murphy and coworkers (1977, 1980) noted that the odorous compounds, ethyl butyrate and citral, contributed to judgments of "taste" magnitude. This illusion is eliminated by pinching the nostrils shut during tasting, which prohibits the retronasal passage of volatile materials and effectively cuts off the volatile flavor impressions.

Another observation is that harsh tastes can suppress and pleasant tastes can enhance ratings of volatile flavor intensity. Von Sydow et al. (1974) examined ratings for taste and odor attributes in fruit juices that varied in added sucrose. Ratings for pleasant odor attributes increased and those for unpleasant odor attributes decreased as sucrose concentration increased. No changes in headspace concentrations of volatiles were detected. Von Sydow et al. interpreted this as evidence for a psychological effect as opposed to a physical interaction. A similar effect was found for blackberry juice flavor at varying levels of sucrose and acidity (Perng and McDaniel, 1989). Sucrose-enhanced fruit flavor ratings while juices with high acid level showed lower fruit ratings.

When retronasal smell is permitted, a common finding is that sweetness is enhanced (Delwiche, 2004) and odors are enhanced as well. The effect depends upon the specific odor/taste pairings. Aspartame enhanced fruitiness of orange and strawberry solutions (sucrose showed no effect) and a somewhat greater enhancement occurred for orange than for strawberry (Wiseman and McDaniel, 1989). Sweetness was enhanced by strawberry odor, but not by peanut butter odor (Frank and Byram, 1988). Some authors have argued that the sweetness enhancement depends upon the congruence and/or similarity of the taste and odor. This makes sense because many odors are referred to as smelling like tastes, such as the sweet smell of honey or the sour smell of vinegar (Small and Prescott, 2005). The spatial and temporal contiguity of odors and tastes when foods are consumed may also be important in facilitating this effect.

The degree of cultural experience panelists have with particular combinations seems important. There is an influence of learned expectancies (Stevenson et al., 1995). The pattern of learned correlations may determine how and when effects such as sweet taste enhancement are seen. Common experience with the co-occurrence of sweet tastes and carmelization odors, for example, may drive some sweetness enhancement effects. The influence of associative learning is shown by the fact that sweetness enhancement is predicted by initial sweetness ratings of odors and that pairings of formerly neutral odors with a sweet taste will induce this enhancement effect (Prescott, 1999; Stevenson et al., 1998).

Is this effect a true enhancement or simply an inflation of sweetness ratings due to taste/smell confusion? Evidence for the "reality" of the effect comes from the observation that a sweet smelling odor can suppress the rated sourness of a citric acid solution, just like a sweet taste would (Stevenson et al., 1999). A number of brain imaging studies have identified multi-modal neural activity in brain regions such as the orbitofrontal cortex (see Small and Prescott, 2005; Verhagen and Engelen, 2006). This has led to the interesting speculation that sniffing a sweet odor might evoke the entire experience of a taste/odor pairing (i.e., a flavor) that has been encoded in memory (Small and Prescott, 2005). Dalton et al. (2000) showed that detection thresholds for an odorant were reduced when subjects held a taste in the mouth, but only when the taste was congruent. However, in another study, sweetness enhancement by subthreshold odors was not observed (Labbe and Martin, 2009).

These interactions change with instructions and with training. In one study, citral–sucrose mixtures were evaluated using both direct scaling and "indirect" scale values derived from triangle test performance

(Lawless and Schlegel, 1984). A pair which was barely discriminable according to triangle tests received significantly different sweetness ratings when separate taste and odor attributes were scaled. Focused attention produces different results than appreciation of the product as a unitary whole. Sweetness enhancement by ethyl maltol decreased when panelists were trained to distinguish tastes from smells (Bingham et al., 1990). In another study, sensory profile training did not seem to promote the associative learning needed for odor/sweetness enhancement (Labbe and Martin, 2009). Along these lines, having subjects take an analytic (rather than synthetic) approach to odor/taste mixtures negates the odor-enhanced sweetness (Prescott et al., 2004). Taken together, these results show that attentional mechanisms or modality-specific training can alter the effect substantially.

A further consideration is that the responses that subjects are instructed to make also influence the apparent taste–odor interactions (van der Klaauw and Frank, 1996). Strawberry odor enhances the sweetness of sucrose–strawberry solutions (Frank et al., 1989a), an effect reminiscent of the enhancement reported by Wiseman and McDaniel (1989) and also the mislabeling of volatile sensations as taste intensity estimates seen by Murphy et al. (1977). However, when subjects are instructed to make total intensity ratings and then partition them into their components, no significant enhancement of sweetness is seen (Frank et al., 1990, 1993; Lawless and Clark, 1992). Odor–taste enhancement, then could in many cases merely be a case of response shifting, and not a truly increased sensation of sweetness at all.

This finding has broad implications for the ways in which sensory evaluations, particularly descriptive analyses in which multiple attributes of complex foods are rated, should be conducted. It also suggests some caution in substantiating claims for various synergies or enhancement effects in which ratings are restricted to too few attributes. Respondents may choose to "dump" some of their impressions into the most suitable category or the only allowable response if the attribute they perceive is otherwise unavailable on the ballot (Lawless and Clark, 1992). Alleged enhancements such as the effect of maltol on sweetness should be viewed with caution unless the response biases inherent in mislabeling smells as tastes can be ruled out. These effects are discussed at length in Chapter 9.

2.6.2 Irritation and Flavor

Two other groups of interactions between modalities are important in foods. One is the interaction of chemical irritation with flavors and the second are effects in flavor ratings caused by changes in visual appearance. Anyone who has compared flat soda to carbonated soda will recognize that the tingle imparted by carbon dioxide will alter the flavor balance in a product, usually to its detriment when the carbonation is not present. Flat soda is usually too sweet. Decarbonated champagne is usually very poor wine.

Several psychophysical studies have examined interactions of trigeminal irritation from chemicals with taste and with odor perception. As in most laboratory psychophysics, these studies have focused on perceived intensity changes in single chemicals simple mixtures. The first workers to examine effects of chemical irritation on olfaction found mutual inhibition of smell by carbon dioxide in the nose (Cain and Murphy, 1980). This occurs even though the onset of the sting from carbon dioxide is delayed somewhat compared to the onset of smell sensations. Since many smells also have an irritative component (Doty et al., 1978; Tucker, 1971), it is probable that some of this inhibition is a common event in everyday flavor perception. If a person had decreased sensitivity to nasal irritation the balance of aromatic flavor perception might be shifted in favor of the olfactory components. If irritation is reduced, then the inhibitory effects of irritation would also be reduced.

Does chili burn mask tastes in the mouth, the way that carbon dioxide sting masks smell in the nose? Partial inhibition of taste responses has been found following pretreatment of oral tissues with capsaicin, particularly inhibition of sour and bitter tastes (Karrer and Bartoshuk, 1995; Lawless and Stevens, 1984; Lawless et al., 1985; Prescott et al., 1993; Prescott and Stevenson, 1995, but see also Cowart, 1987). Note that capsaicin desensitization takes several minutes to develop, i.e., it depends upon a delay between treatment and test stimuli (Green, 1989). Such a temporal gap would have occurred to varying degrees in pretreatment experiments with tastants. Also, since capsaicin inhibition is most reliably observed for substances sometimes reported as partially irritative, the inhibitory effect seen in pretreatment studies may be due to desensitization to an irritative component

of the "tastants," rather than a direct effect on gustatory intensity per se (e.g., Karrer and Bartoshuk, 1995).

Tastes can modulate or ameliorate chili burn. There are folk remedies in various cultures, such as starchy corn, ghee, pineapple, sugar, and beer. Systematic studies of trying to wash out chili burn with different tasting rinses have shown some effect for sweet (most pronounced), sour, and perhaps salt (Sizer and Harris, 1985; Stevens and Lawless, 1986). Cold stimuli provide a temporary but potent inhibition of pepper burn, as known to many habitués of ethnic restaurants. Since capsaicin is lipid soluble, the Indian remedy of ghee (clarified butter) has some merit. Sour things stimulate salivary flow, which may provide some relief to abused oral tissues. The combination of fatty, sour, cold, and sweet suggests chilled yogurt as a good choice. A culinary practice of alternating cool, sweet chutneys with hot curries would seem to facilitate these interactions.

2.6.3 Color–Flavor Interactions

Finally, let us consider the effects of appearance on flavor perception. The literature concerning color–flavor interactions is quite extensive and interested researchers are cautioned that it is complex and at times contradictory (e.g., Lavin and Lawless, 1998). We make no attempt here to provide a comprehensive review.

Humans are a visually driven species. In many societies with mature culinary arts, the visual presentation of a food is as important as its flavor and texture characteristics. A common finding is that when foods are more deeply colored, they will obtain higher ratings for flavor intensity (e.g., Dubose et al., 1980; Zellner and Kautz, 1990). Effects of colored foods on flavor intensity and flavor identification are discussed in Stillman (1993). Miscolored foods or flavors are less effectively identified (Dubose et al., 1980). However, the pattern of results is mixed and inconsistent in this literature (see Delwiche, 2004). Once again, learned associations may drive the patterns of influence. Morrot et al. (2001) found that more red wine descriptors were used by a panel when a white wine was intentionally miscolored red.

An example of visual influences on food perception can be found in the literature on perception of milks of varying fat content. Most people believe that skim milk is easily differentiated from whole milk or even from 2% low fat milk by appearance, flavor, and texture (mouthfeel). However, most of their perception of fat content is driven by appearance (Pangborn and Dunkley, 1964; Tuorila, 1986). Trained descriptive panelists readily differentiate skim milk from 2% on the basis of appearance (color) ratings, mouthfeel, and flavor. However, when visual cues are removed, discrimination is markedly impaired (Philips et al., 1995). When tested in the dark with cold milk, discrimination of skim milk from 2% milk drops almost to chance performance, a result that many skim milk drinkers find difficult to swallow. This research emphasizes that humans react to the ensemble of sensory stimulation available from a food. Even "objective" descriptive panelists may be subject to visual bias.

2.7 Conclusions

An important knowledge base for any sensory professional is an appreciation of the function of the senses through which we obtain our data. Understanding the physiological processes of the senses helps us take into account the limits of sensory function and how sensations interact. The historical underpinnings of sensory methods lie in the discipline of psychophysics, the systematic study of relationships between stimulus and response. Psychophysical thinking, then, is not just about methods for sensory testing, but a view of sensory function that looks at relationships among *variables*. This is a valuable point of view that can enhance the contribution of a sensory group to their product development clients. One of our industrial colleagues used to ask his product developers not to send him products to test. At first glance such a statement seems outrageous. But the key was in his next request: "Send me *variables* to test." This approach is advantageous as it brings a deeper understanding of the relationships between ingredient or process variables and sensory response. It moves the sensory specialist beyond simple hypothesis testing and into the realm of theory building and modeling, in other words more like engineering than the all too common pattern of simple yes/no hypothesis testing.

References

Albin, K. C., Carstens, M. I. and Carstens, E. 2008. Modulation of oral heat and cold pain by irritant chemicals. Chemical Senses, 33, 3–15.

Amoore, J. E. 1971. Olfactory genetics and anosmia. In: L. M. Beidler (ed.), Handbook of Sensory Physiology. Springer, Berlin, pp. 245–256.

Amoore, J. E. and Forrester, L. J. 1976. The specific anosmia to trimethylamine: The fishy primary odor. Journal of Chemical Ecology, 2, 49–56.

Amoore, J. E., Forrester, L. J. and Pelosi, P. 1976. Specific anosmia to isobutyraldehyde: The malty primary odor. Chemical Senses, 2, 17–25.

Amoore, J. E., Venstrom, D. and Davis, A. R. 1968. Measurement of specific anosmia. Perceptual and Motor Skills, 26, 143–164.

Andersen, R. K., Lund, J. P. and Puil, E. 1978. Enkephalin and substance P effects related to trigeminal pain. Canadian Journal of Physiology and Pharmacology, 56, 216–222.

Anderson, N. 1974. Algebraic models in perception. In: E. C. Carterette and M. P. Friedman (eds.), Handbook of Perception. II. Psychophysical Judgment and Measurement. Academic, New York, pp. 215–298.

Arvidson, K. 1979. Location and variation in number of taste buds in human fungiform papillae. Scand. Journal of Dental Research, 87, 435–442.

ASTM 1989. Standard definitions of terms relating to sensory evaluation of materials and products. In Annual Book of ASTM Standards. American Society for Testing and Materials, Philadelphia, p. 2.

Ayya, N. and Lawless, H. T. 1992. Qualitative and quantitative evaluation of high-intensity sweeteners and sweetener mixtures. Chemical Senses, 17, 245–259.

Bachmanov, A. A. and Beauchamp, G. K. 2007. Taste receptor genes. Annual Review of Nutrition, 27, 389–414.

Bartoshuk, L. M. 1978. History of taste research. In: E. C. Carterette and M. P. Friedman (eds.), Handbook of Perception. IVA, Tasting and Smelling. Academic, New York, pp. 2–18.

Bartoshuk, L. M. 1979. Bitter taste of saccharin related to the genetic ability to taste the bitter substance 6-N-Propylthiouracil. Science, 205, 934–935.

Bartoshuk, L. M., Cain, W. S., Cleveland, C. T., Grossman, L. S., Marks, L. E., Stevens, J. C. and Stolwijk, J. A. 1974. Saltiness of monosodium glutamate and sodium intake. Journal of the American Medical Association, 230, 670.

Bartoshuk, L. M., Duffy, V. B. and Miller, I. J. 1994. PTC/PROP tasting: Anatomy, psychophysics and sex effects. Physiology and Behavior, 56, 1165–1171.

Bartoshuk, L. M., Murphy, C. L. and Cleveland, C. T. 1978. Sweet taste of dilute NaCl. Physiology and Behavior, 21, 609–613.

Bate Smith, E. C. 1954. Astringency in foods. Food Processing and Packaging, 23, 124–127.

Beidler, L. M. 1961. Biophysical approaches to taste. American Scientist, 49, 421–431.

Bingham, A. F., Birch, G. G., de Graaf, C., Behan, J. M. and Perring, K. D. 1990. Sensory studies with sucrose-maltol mixtures. Chemical Senses, 15, 447–456.

Birnbaum, M. H. 1982. Problems with so called "direct" scaling. In: J. T. Kuznicki, A. F. Rutkiewic and R. A. Johnson (eds.), Problems and Approaches to Measuring Hedonics (ASTM STP 773). American Society for Testing and Materials, Philadelphia, pp. 34–48.

Blakeslee, A. F. 1932. Genetics of sensory thresholds: Taste for phenylthiocarbamide. Proceedings of the National Academy of Science USA, 18, 120–130.

Boring, E. G. 1942. Sensation and Perception in the History of Experimental Psychology. Appleton-Century-Crofts, New York.

Brennand, C. P., Ha, J. K. and Lindsay, R. C. 1989. Aroma properties and thresholds of some branched-chain and other minor volatile fatty acids occurring in milkfat and meat lipids. Journal of Sensory Studies, 4, 105–120.

Breslin, P. A. S. and Beauchamp, G. K. 1997. Salt enhances flavour by suppressing bitterness. Nature, 387, 563.

Breslin, P. A. S., Gilmore, M. M., Beauchamp, G. K. and Green, B. G. 1993. Psychophysical evidence that oral astringency is a tactile sensation. Chemical Senses, 18, 405–417.

Brud, W. S. 1986. Words versus odors: How perfumers communicate. Perfumer and Flavorist, 11, 27–44.

Buck, L. and Axel, R. 1991. A novel multigene family may encode odorant receptors: A molecular basis for odor recognition. Cell, 65, 175–187.

Buettner, A., Beer, A., Hannig, C., Settles, M. and Schieberle, P. 2002. Physiological and analytical studies on flavor perception dynamics as induced by the eating and swallowing process. Food Quality and Preference, 13, 497–504.

Bufe, B., Breslin, P. A. S., Kuhn, C., Reed, D. R., Tharp, C. D., Slack, J. P., Kim, U.-K., Drayna, D. and Meyerhof, W. 2005. The molecular basis of individual differences in phenylthiocarbamide and propylthiouracil bitterness perception. Current Biology, 15, 322–327.

Burgard, D. R. and Kuznicki, J. T. 1990. Chemometrics: Chemical and Sensory Data. CRC, Boca Raton.

Burks, T. F., Buck, S. H. and Miller, M. S. 1985. Mechanisms of depletion of substance P by capsaicin. Federation Proceedings, 44, 2531–2534.

Cagan, R. H. 1981. Recognition of taste stimuli at the initial binding interaction. In: R. H. Cagan and M. R. Kare (eds.), Biochemistry of Taste and Olfaction. Academic, New York, pp. 175–204.

Cain, W. S. 1977. Differential sensitivity for smell: "Noise" at the nose. Science, 195 (25 February), 796–798.

Cain, W. S. 1979. To know with the nose: Keys to odor identification. Science, 203, 467–470.

Cain, W. S. and Drexler, M. 1974. Scope and evaluation of odor counteraction and masking. Annals of the New York Academy of Sciences, 237, 427–439.

Cain, W. S. and Engen, T. 1969. Olfactory adaptation and the scaling of odor intensity. In: C. Pfaffmann (ed.), Olfaction and Taste III. Rockefeller University, New York, pp. 127–141.

Cain, W. S. and Murphy, C. L. 1980. Interaction between chemoreceptive modalities of odor and irritation. Nature, 284, 255–257.

Calixto, J. B., Kassuya, C. A., Andre, E. and Ferreira, J. 2005. Contribution of natural products to the discovery of the transient receptor potential (TRP) channels family and their functions. Pharmacology & Therapeutics, 106, 179–208.

Caul, J. F. 1957. The profile method of flavor analysis. Advances in Food Research, 7, 1–40.

Chandrashekar, J., Yarmolinksy, D., von Buchholtz, L., Oka, Y., Sly, W., Ryba, N. J. P. and Zuker, C. S. 2009. The taste of carbonation. Science, 326, 443–445.

Chastrette, M., Elmouaffek, E. and Sauvegrain, P. 1988. A multi-dimensional statistical study of similarities between 74 notes used in perfumery. Chemical Senses, 13, 295–305.

Civille, G. L. and Lawless, H. T. 1986. The importance of language in describing perceptions. Journal of Sensory Studies, 1, 203–215.

Civille, G. V. and Lyon, B. G. 1996. Aroma and Flavor Lexicon for Sensory Evaluation. ASTM DS 66. American Society for Testing and Materials, West Coshohocken, PA.

Cliff, M. and Heymann, H. 1992. Descriptive analysis of oral pungency. Journal of Sensory Studies, 7, 279–290.

Clifford, M. N. 1986. Phenol-protein interactions and their possible significance for astringency. In: Birch, G. G. and M. G. Lindley (eds.), Interactions of Food Components. Elsevier, London, pp. 143–163.

Collings, V. B. 1974. Human taste response as a function of locus on the tongue and soft palate. Perception & Psychophysics, 16, 169–174.

Cometto-Muñiz, J. E. and Cain, W. S. 1984. Temporal integration of pungency. Chemical Senses, 8, 315–327.

Commetto-Muñiz, J. E. and Hernandez, S. M. 1990. Odorous and pungent attributes of mixed and unmixed odorants. Perception & Psychophysics, 47, 391–399.

Condelli, N., Dinnella, C., Cerone, A., Monteleone, E. and Bertucciolo, M. 2006. Prediction of perceived astringency induced by phenolic compounds II: Criteria for panel selection and preliminary application on wine samples. Food Quality and Preference, 17, 96–107.

Cowart, B. J. 1987. Oral chemical irritation: Does it reduce perceived taste intensity? Chemical Senses, 12, 467–479.

Cowart, B. J. 1998. The addition of CO_2 to traditional taste solutions alters taste quality. Chemical Senses, 23, 397–402.

Curtis, D. W., Stevens, D. A. and Lawless, H. T. 1984. Perceived intensity of the taste of sugar mixtures and acid mixtures. Chemical Senses, 9, 107–120.

Da Conceicao Neta, E. R., Johanningsmeier, S. D. and McFeeters, R. F. 2007. The chemistry and physiology of sour taste – A review. Journal of Food Science, 72, R33–R38.

Dalton, P., Doolittle, N., Nagata, H. and Breslin, P. A. S. 2000. The merging of the senses: Integration of subthreshold taste and smell. Nature Neuroscience, 3, 431–432.

Delwiche, J. 2004. The impact of perceptual interactions on perceived flavor. Food Quality and Preference, 15, 137–146.

Dessirier, J.-M., Simons, C. T., Carstens, M. I., O'Mahony, M. and Carstens, E. 2000. Psychophysical and neurobiological evidence that the oral sensation elicited by carbonated water is of chemogenic origin. Chemical Senses, 25, 277–284.

Desor, J. A. and Beauchamp, G. K. 1974. The human capacity to transmit olfactory information. Perception & Psychophysics, 16, 551–556.

Diamant, H., Oakley, B., Strom, L. and Zotterman, Y. 1965. A comparison of neural and psychophysical responses to taste stimuli in man. Acta Physiologica Scandinavica, 64, 67–74.

Doty, R. L. 1991. Psychophysical measurement of odor perception in humans. In: D. G. Laing, R. L. Doty and W. Breipohl (eds.), The Human Sense of Smell. Springer, Berlin, pp. 95–143.

Doty, R. L., Brugger, W. E., Jurs, P. C., Orndorff, M. A., Snyder, P. J. and Lowry, L. D. 1978. Intranasal trigeminal stimulation from odorous volatiles: Psychometric responses from anosmic and normal humans. Physiology and Behavior, 20, 175–185.

Dravnieks, A. 1982. Odor quality: Semantically generated multidimensional profiles are stable. Science, 218, 799–801.

Dubose, C. N., Cardello, A. V. and Maller, O. 1980. Effects of colorants and flavorants on identification, perceived flavor intensity and hedonic quality of fruit-flavored beverages and cake. Journal of Food Science, 45, 1393–1399, 1415.

Engen, T. 1982. The Perception of Odors. Academic, New York.

Engen, T. and Pfaffmann, C. 1959. Absolute Judgments of Odor Intensity. Journal of Experimental Psychology, 58, 23–26.

Engen, T. and Ross, B. 1973. Long term memory for odors with and without verbal descriptors. Journal of Experimental Psychology, 100, 221–227.

Epke, E., McClure, S. T. and Lawless, H. T. 2008. Effects of nasal occlusion and oral contact on perception of metallic taste from metal salts. Food Quality and Preference, 20, 133–137.

Farbman, A. I. and Hellekant, G. 1978. Quantitative analyses of fiber population in rat chorda tympani nerves and fungiform papillae. American Journal of Anatomy, 153, 509–521.

Fechner, G. T. 1966 (translation, orig. 1860). Elements of Psychophysics. E. H. Adler (trans.). D. H. Howes and E. G. Boring (eds.), Holt, Rinehart and Winston, New York.

Firestein, S. 2001. How the olfactory system makes senses of scents. Nature, 413, 211–218.

Fisher, U., Boulton, R. B. and Noble, A. C. 1994. Physiological factors contributing to the variability of sensory assessments: Relationship between salivary flow rate and temporal perception of gustatory stimuli. Food Quality and Preference, 5, 55–64.

Fox, A. L. 1932. The relationship between chemical constitution and taste. Proceedings of the National Academy of Sciences USA, 18, 115–120.

Frank, R. A. and Byram, J. 1988. Taste-smell interactions are tastant and odorant dependent. Chemical Senses, 13, 445.

Frank, R. A., Ducheny, K. and Mize, S. J. S. 1989a. Strawberry odor, but not red color enhances the sweetness of sucrose solutions. Chemical Senses, 14, 371.

Frank, R. A., Mize, S. J. and Carter, R. 1989b. An Assessment of binary mixture interactions for nine sweeteners. Chemical Senses, 14, 621–632.

Frank, R. A., van der Klaauw, N. J. and Schifferstein, H. N. J. 1993. Both perceptual and conceptual factors influence taste-odor and taste-taste interactions. Perception and Psychophysics, 54, 343–354.

Frank, R. A., Wessel, N. and Shaffer, G. 1990. The enhancement of sweetness by strawberry odor is instruction dependent. Chemical Senses, 15, 576–577.

Goldstein, E. B. 1999. Sensation & Perception, Fifth Edition. Brooks/Cole Publishing, Pacific Grove, CA.

Goldstein, E. B. (ed.). 2001. Handbook of Perception. Blackwell Publishers, Inc., Malden, MA.

Green, B. G. 1985. Menthol modulates oral sensations of warmth and cold. Physiology and Behavior, 35, 427–434.

Green, B. G. 1986. Menthol inhibits the perception of warmth. Physiology and Behavior, 38, 833–838.

Green, B. G. 1989. Capsaicin sensitization and desensitization on the tongue produced by brief exposures to a low concentration. Neuroscience Letters, 107, 173–178.

Green, B. G. 1992. The effects of temperature and concentration on the perceived intensity and quality of carbonation. Chemical Senses, 17, 435–450.

Green, B. G. and Gelhard, B. 1989. Salt as an oral irritant. Chemical Senses, 14, 259–271.

Green, B. G. and Lawless, H. T. 1991. The psychophysics of somatosensory chemoreception in the nose and mouth. In: T. V. Getchell, L. M. Bartoshuk, R. L. Doty and J. B. Snow (eds.), Smell and Taste in Health and Disease. Raven, New York, NY, pp. 235–253.

Green, B. G., Alvarez-Reeves, M., Pravin, G. and Akirav, C. 2005. Chemesthesis and taste: Evidence of independent processing of sensation intensity. Physiology and Behavior, 86, 526–537.

Greer, C. A. 1991. Structural organization of the olfactory system. In: T. V. Getchell, R. L. Doty, L. M. Bartoshuk and J. B. Snow (eds.), Smell and Taste in Health and Disease. Raven, New York, pp. 65–81.

Grushka, M. and Sessle, B. J. 1991. Burning mouth syndrome. In: T. V. Getchell, R. L. Doty, L. M. Bartoshuk and J. B. Snow (eds.), Smell and Taste in Health and Disease. Raven, New York, NY, pp. 665–682.

Guinard, J.-X., Pangborn, R. M. and Lewis, M. J. 1986. The time course of astringency in wine upon repeated ingestion. American Journal of Enology and Viticulture, 37, 184–189.

Gwartney, E. and Heymann, H. 1995. The temporal perception of menthol. Journal of Sensory Studies, 10, 393–400.

Gwartney, E. and Heymann, H. 1996. Profiling to describe the sensory characteristics of a simple model menthol solution. Journal of Sensory Studies, 11, 39–48.

Hall, M. L., Bartoshuk, L. M., Cain, W. S. and Stevens, J. C. 1975. PTC taste blindness and the taste of caffeine. Nature, 253, 442–443.

Halpern, B. P. 2008. Mechanisms and consequences of retronasal smelling: Computational fluid dynamic observations and psychophysical observations. Chemosense, 10, 1–8.

Harper, S. J. and McDaniel, M. R. 1993. Carbonated water lexicon: Temperature and CO_2 level influence on descriptive ratings. Journal of Food Science, 58, 893–898.

Hernandez, S. V. and Lawless, H. T. 1999. A method of adjustment for preference levels of capsaicin in liquid and solid food systems among panelists of two ethnic groups. Food Quality and Preference 10, 41–49.

Hewson, L., Hollowood, T., Chandra, S. and Hort, J. 2009. Gustatory, olfactory and trigeminal interactions in a model carbonated beverage. Chemosensory Perception, 2, 94–107.

Hobbes, T. 1651. M. Oakeshott (ed.), Leviathan: Or the Matter, Forme and Power of a Commonwealth Ecclesiastical and Civil. Collier Books, New York, NY, 1962 edition.

Horne, J., Hayes, J. and Lawless, H. T. 2002. Turbidity as a measure of salivary protein reactions with astringent substances. Chemical Senses, 27, 653–659.

Hornung, D. E. and Enns, M.P. 1984. The independence and integration of olfaction and taste. Chemical Senses, 9, 97–106.

Hornung, D. E. and Enns, M. P. 1986. The contributions of smell and taste to overall intensity: A model. Perception and Psychophysics, 39, 385–391.

Ishimaru, Y., Inada, H., Kubota, M., Zhuang, H., Tominaga, M. and Matsunami, H. 2006. Transient receptor potential family members PDK1L3 and PKD2L1 form a candidate sour taste receptor. Proceedings of the National Academy of Science, 103, 12569–12574.

Jansco, N. 1960. Role of the nerve terminals in the mechanism of inflammatory reactions. Bulletin of Millard Fillmore Hospital, Buffalo, 7, 53–77.

Jeltema, M. A. and Southwick, E. W. 1986. Evaluations and application of odor profiling. Journal of Sensory Studies, 1, 123–136.

Kallikathraka, S., Bakker, J., Clifford, M. N. and Vallid, L. 2001. Correlations between saliva composition and some T-I parameters of astringency. Food Quality and Preference, 12, 145–152.

Karrer, T. and Bartoshuk, L. M. 1995. Effects of capsaicin desensitization on taste in humans. Physiology & Behavior, 57, 421–429.

Kauer, J. S. 1987. Coding in the olfactory system. In: T. E. Finger and W. L. Silver (eds.), Neurobiology of Taste and Smell, Wiley, New York, NY, pp. 205–231.

Kawamura, Y. and Kare, M. R. 1987. Umami: A Basic Taste. Marcel Dekker, New York.

Kim, U.-K., Jorgenson, E., Coon, H. Leppert, M.. Risch, N. and Drayna, D. 2003. Positional cloning of the human quantitative trait locus underlying taste sensitivity to phenylthiocarbamide. Science, 299, 1221–1225.

Kroeze, J. H. A. 1979. Masking and adaptation of sugar sweetness intensity. Physiology and Behavior, 22, 347–351.

Kroeze, J. H. A. and Bartoshuk, L. M. 1985. Bitterness suppression as revealed by split-tongue taste stimulation in humans. Physiology and Behavior, 35, 779–783.

Labbe, D. and Martin, N. 2009. Impact of novel olfactory stimuli and subthreshold concentrations on the perceived sweetness of sucrose after associative learning. Chemical Senses, 34, 645–651.

Laing, D. G., Livermore, B. A. and Francis, G. W. 1991. The human sense of smell has a limited capacity for identifying odors in mixtures. Chemical Senses, 16, 392.

Laing, D. G. and Willcox, M. E. 1983. perception of components in binary odor mixtures. Chemical Senses, 7, 249–264.

Laska, M. and Hudson, R. 1992. Ability to discriminate between related odor mixtures. Chemical Senses, 17, 403–415.

Lavin, J. and Lawless, H. T. 1998. Effects of color and odor on judgments of sweetness among children and adults, Food Quality and Preference, 9, 283–289.

Lawless, H. 1977. The pleasantness of mixtures in taste and olfaction. Sensory Processes, 1, 227–237.

Lawless, H. T. 1979. Evidence for neural inhibition in bittersweet taste mixtures. Journal of Comparative and Physiological Psychology, 93, 538–547.

Lawless, H. T. 1980. A comparison of different methods for assessing sensitivity to the taste of phenylthiocarbamide PTC. Chemical Senses, 5, 247–256.

Lawless, H. T. 1984. Oral chemical irritation: Psychophysical properties. Chemical Senses, 9, 143–155.

Lawless, H. T. 1987. An olfactory analogy to release from mixture suppression in taste. Bulletin of the Psychonomic Society, 25, 266–268.

Lawless, H. T. 1996. Flavor. In: E. C. Carterrette and M. P. Friedman (eds.), Cognitive Ecology. Academic, San Diego, pp. 325–380.

Lawless, H. T. 1999. Descriptive analysis of complex odors: Reality, model or illusion? Food Quality and Preference, 10, 325–332.

Lawless, H. T. and Clark, C. C. 1992. Psychological biases in time intensity scaling. Food Technology, 46(11), 81, 84–86, 90.

Lawless, H. T. and Corrigan, C. J. 1994. Semantics of astringency. In: K. Kurihara (ed.), Olfaction and Taste XI. Proceedings of the 11th International Symposium on Olfaction and Taste and 27th Meeting, Japanese Association for Smell and Taste Sciences. Springer, Tokyo, pp. 288–292.

Lawless, H. T. and Gillette, M. 1985. Sensory responses to oral chemical heat. In: D. D. Bills and C. J. Mussinan (eds.), Characterization and Measurement of Flavor Compounds. American Chemical Society, Washington, DC, pp. 27–42.

Lawless, H. T. and Lee, C. B. 1994. The common chemical sense in food flavor. In: T. E. Acree and R. Teranishi (eds.), Flavor Science, Sensible Principles and Techniques. American Chemical Society, Washington, pp. 23–66.

Lawless, H. T. and Malone, G. J. 1986. Comparisons of rating scales: Sensitivity, replicates and relative measurement. Journal of Sensory Studies, 1, 155–174.

Lawless, H. T. and Schlegel, M. P. 1984. Direct and indirect scaling of taste – odor mixtures. Journal of Food Science, 49, 44–46.

Lawless, H. T. and Stevens, D. A. 1984. Effects of oral chemical irritation on taste. Physiology and Behavior, 32, 995–998.

Lawless, H. T. and Stevens, D. A. 1988. Responses by humans to oral chemical irritants as a function of locus of stimulation. Perception & Psychophysics, 43, 72–78.

Lawless, H. T. and Stevens, D. A. 1989. Mixtures of oral chemical irritants. In: D. G. Laing, W. S. Cain, R. L. McBride and B. W. Ache (eds.), Perception of Complex Smells and Tastes. Academic Press Australia, Sydney, pp. 297–309.

Lawless, H. T. and Stevens, D. A. 1990. Differences between and interactions of oral irritants: Neurophysiological and perceptual implications. In: B. G. Green and J. R. Mason (eds.), Chemical Irritation in the Nose and Mouth. Marcel Dekker, New York, NY, pp. 197–216.

Lawless, H. T. and Zwillenberg, D. 1983. Clinical methods for testing taste and olfaction. Transactions of the Pennsylvania Academy of Ophthalmology and Otolaryngology, Fall, 1983, 190–196.

Lawless, H. T., Horne, J. and Giasi, P. 1996. Astringency of acids is related to pH. Chemical Senses, 21, 397–403.

Lawless, H. T., Rozin, P. and Shenker, J. 1985. Effects of oral capsaicin on gustatory, olfactory and irritant sensations and flavor identification in humans who regularly or rarely consumer chili pepper. Chemical Senses, 10, 579–589.

Lawless, H. T., Thomas, C. J. C. and Johnston, M. 1995. Variation in odor thresholds for l-carvone and cineole and correlations with suprathreshold intensity ratings. Chemical Senses, 20, 9–17.

Lawless, H. T., Antinone, M. J., Ledford, R. A. and Johnston, M. 1994. Olfactory responsiveness to diacetyl. Journal of Sensory Studies, 9, 47–56.

Lawless, H. T., Stevens, D. A., Chapman, K. W. and Kurtz, A. 2005. Metallic taste from ferrous sulfate and from electrical stimulation. Chemical Senses, 30, 185–194.

Lawless, H. T., Schlake, S., Smythe, J., Lim, J., Yang, H., Chapman, K. and Bolton, B. 2004. Metallic taste and retronasal smell. Chemical Senses 29, 25–33.

Lee, C. B. and Lawless, H. T. 1991. Time-course of astringent sensations. Chemical Senses, 16, 225–238.

Leffingwell, J. C. 2009. Cool without menthol and cooler than menthol. Leffingwell and Associates. http://www.leffingwell.com/cooler_than_menthol.htmv

Lehninger, A. L. 1975. Biochemistry, Second Edition. Worth Publishers, New York.

Lubran, M. B., Lawless, H. T., Lavin, E. and Acree, T. E. 2005. Identification of metallic-smelling 1 octen-3-one and 1-nonen-3-one from solutions of ferrous sulfate. Journal of Agricultural and Food Chemistry, 53, 8325–8327.

Malnic, B., Hirono, J., Sato, T. and Buck, L. B. 1999. Combinatorial receptor codes for odors. Cell, 96, 713–723.

Marin, A. B., Acree, T. E. and Hotchkiss, J. 1987. Effects of orange juice packaging on the aroma of orange juice. Paper presented at the 194th ACS National Meeting, New Orleans, LA, 9/87.

Marin, A. B., Barnard, J., Darlington, R. B. and Acree, T. E. 1991. Sensory thresholds: Estimation from dose-response curves. Journal of Sensory Studies, 6(4), 205–225.

Mattes, R. D. and Lawless, H. T. 1985. An adjustment error in optimization of taste intensity. Appetite, 6, 103–114.

McBride, R. L. 1983. A JND-scale/category scale convergence in taste. Perception & Psychophysics, 34, 77–83.

McBride, R. L. 1987. Taste psychophysics and the Beidler equation. Chemical Senses, 12, 323–332.

McBurney, D. H. 1966. Magnitude estimation of the taste of sodium chloride after adaptation to sodium chloride. Journal of Experimental Psychology, 72, 869–873.

McBurney, D. H. and Bartoshuk, L. M. 1973. Interactions between stimuli with different taste qualities. Physiology and Behavior, 10, 1101–1106.

McBurney, D. H. and Shick, T. R. 1971. Taste and water taste of 26 compounds for man. Perception & Psychophysics, 11, 228–232.

McClure, S. T. and Lawless, H. T. 2007. A comparison of two electric taste stimulation devices, metallic taste responses and lateralization of taste. Physiology and Behavior, 92, 658–664.

McManus, J. P., Davis, K. G., Lilley, T. H. and Halsam, E. 1981. Polyphenol interactions. Journal of the Chemical Society, Chemical Communications, 309–311.

Meilgaard, M. C., Reid, D. S. and Wyborski, K. A. 1982. Reference standards for beer flavor terminology system. Journal of the American Society of Brewing Chemists, 40, 119–128.

Meiselman, H. L. 1971. Effect of presentation procedure on taste intensity functions. Perception and Psychophysics, 10, 15–18.

Meiselman, H. L. and Halpern, B. P. 1973. Enhancement of taste intensity through pulsatile stimulation. Physiology and Behavior, 11, 713–716.

Miller, I. J. and Bartoshuk, L. M. 1991. Taste perception, taste bud distribution and spatial relationships. In: T. V. Getchell, R. L. Doty, L. M. Bartoshuk and J. B. Snow (eds.), Smell and Taste in Health and Disease. Raven, New York, pp. 205–233.

Miller, I. J. and Spangler, K. M. 1982. Taste bud distribution and innervation on the palate of the rat. Chemical Senses, 7, 99–108.

Moio, L., Langlois, D., Etievant, P. X. and Addeo, F. 1993. Powerful odorants in water buffalo and bovine mozzarella cheese by use of extract dilution sniffing analysis. Italian Journal of Food Science, 3, 227–237.

Morrot, G., Brochet, F. and Dubourdieu, D. 2001. The color of odors. Brain & Language, 79, 309–320.

Moskowitz, H. R. 1971. The sweetness and pleasantness of sugars. American Journal of Psychology, 84, 387–405.

Murray, N. J., Williamson, M. P., Lilley, T. H. and Haslam, E. 1994. Study of the interaction between salivary proline-rich proteins and a polyphenol by 1H-NMR spectroscopy. European Journal of Biochemistry, 219, 923–935.

Murphy, C. and Cain, W. S. 1980. Taste and olfaction: Independence vs. interaction. Physiology and Behavior, 24, 601–605.

Murphy, C., Cain, W. S. and Bartoshuk, L. M. 1977. Mutual action of taste and olfaction. Sensory Processes, 1, 204–211.

Nagodawithana, T. W. 1995. Savory Flavors. Esteekay Associates, Milwaukee, WI.

Nejad, M. S. 1986. The neural activities of the greater superficial petrosal nerve of the rat in response to chemical stimulation of the palate. Chemical Senses, 11, 283–293.

Noble, A. C., Arnold, R. A., Buechsenstein, J., Leach, E. J., Schmidt, J. O. and Stern, P. M. 1987. Modification of a standardized system of wine aroma terminology. American Journal of Enology and Viticulture, 38(2), 143–146.

O'Mahony, M. and Ishii, R. 1986. Umami taste concept: Implications for the dogma of four basic tastes. In: Y. Kawamura and M. R. Kare (eds.), Umami: A Basic Taste. Marcel Dekker, New York, NY, pp. 75–93.

Pangborn, R. M. 1988. Relationship of personal traits and attitudes to acceptance of food attributes. In: J. Solms, D. A. Booth, R. M. Pangborn and O. Rainhardt (eds.), Food Acceptance and Nutrition. Academic, New York, NY, pp. 353–370.

Pangborn, R. M. and Braddock, K. S. 1989. Ad libitum preferences for salt in chicken broth. Food Quality and Preference, 1, 47–52.

Pangborn, R. M. and Dunkley, W. L. 1964. Laboratory procedures for evaluating the sensory properties of milk. Dairy Science Abstracts, 26, 55–62.

Patapoutian, A., Peier, A. M., Story, G. M. and Viswanath, V. 2003. TermoTRP channels and beyond: Mechanisms of temperature sensation. Nature Reviews/Neuroscience, 4, 529–539.

Patel, T., Ishiuji, Y. and Yosipovitch, G. 2007. Menthol: A refreshing look at this ancient compound. Journal of the American Academy of Dermatology, 57, 873–878.

Pelosi, P. and Pisanelli, A. M. 1981. Specific anosmia to 1,8-cineole: The camphor primary odor. Chemical Senses, 6, 87–93.

Pelosi, P. and Viti, R. 1978. Specific anosima to I-carvone: The minty primary odour. Chemical Senses and Flavour, 3, 331–337.

Perng, C. M. and McDaniel, M. R. 1989. Optimization of a blackberry juice drink using response surface methodology. In: Institute of Food Technologists, Program and Abstracts, Annual Meeting. Institute of Food Technologists, Chicago, IL, p. 216.

Philips, L. G., McGiff, M. L., Barbano, D. M. and Lawless, H. T. 1995. The influence of nonfat dry milk on the sensory properties, viscosity and color of lowfat milks. Journal of Dairy Science, 78, 2113–2118.

Prescott, J. 1999. Flavour as a psychological construct: Implications for perceiving and measuring the sensory qualities of foods. Food Quality and Preference, 10, 349–356.

Prescott, J. and Stevenson, R. J. 1995. Effects of oral chemical irritation on tastes and flavors in frequent and infrequent users of chili. Physiology and Behavior, 58, 1117–1127.

Prescott, J. and Stevenson, R. J. 1996. Psychophysical responses to single and multiple presentations of the oral irritant zingerone: Relationship to frequency of chili consumption. Physiology and Behavior, 60, 617–624.

Prescott, J. and Swain-Campbell, N. 2000. Reponses to repeated oral irritation by capsaicin, cinnamaldehyde and ethanol in PROP tasters and non-tasters. Chemical Senses, 25, 239–246.

Prescott, J., Allen, S. and Stephens, L. 1993. Interactions between oral chemical irritation, taset and temperature. Chemical Senses, 18, 389–404.

Prescott, J., Johnstone, V. and Francis, J. 2004. Odor/taste interactions: Effects of different attentional strategies during exposure. Chemical Senses, 29, 331–340.

Renneccius, G. 2006. Flavor Chemistry and Technology. Taylor and Francis, Boca Raton.

Riera, C. E., Vogel, H., Simon, S. A., Damak, S. and le Coutre, J. 2009. Sensory attributes of complex tasting divalent salts are mediated by TRPM5 and TRPV1 channels. Journal of Neuroscience, 29, 2654–2662.

Rozin, P. 1982. "Taste-smell confusions" and the duality of the olfactory sense. Perception & Psychophysics, 31, 397–401.

Rubico, S. M. and McDaniel, M. R. 1992. Sensory evaluation of acids by free-choice profiling. Chemical Senses, 17, 273–289.

Schifferstein, H. N. J. and Frijters, J. E. R. 1991. The perception of the taste of KCl, NaCl and quinine HCl is not related to PROP sensitivity. Chemical Senses, 16(4), 303–317.

Scott, T. R. and Plata-Salaman, C. R. 1991. Coding of taste quality. In: T. V. Getchell, R. L. Doty, L. M. Bartoshuk and J. B. Snow (eds.), Smell and Taste in Health and Disease. Raven, New York, NY, pp. 345–368.

Siebert, K. 1991. Effects of protein-polyphenol interactions on beverage haze, stabilization and analysis. Journal of Agricultural and Food Chemistry, 47, 353–362.

Silver, W. L., Roe, P., Atukorale, V., Li, W. and Xiang, B.-S. 2008. TRP channels and chemosensation. Chemosense, 10, 1, 3–6.

Sizer, F. and Harris, N. 1985. The influence of common food additives and temperature on threshold perception of capsaicin. Chemical Senses, 10, 279–286.

Small, D. M. and Prescott, J. 2005. Odor/taste integration and the perception of flavor. Experimental Brain Research, 166, 345–357.

Small, D. M., Voss, J., Mak, Y. E., Simmons, K. B., Parrish, T. R. and Gitelman, D. R. 2004. Experience-dependent neural integration of taste and smell in the human brain. Journal of Neurophysiology, 92, 1892–1903.

Sowalski, R. A. and Noble, A. C. 1998. Comparison of the effects of concentration, pH and anion species on astringency and sourness of organic acids. Chemical Senses, 23, 343–349.

Stevens, D. A. and Lawless, H. T. 1986. Putting out the fire: Effects of tastants on oral chemical irritation. Perception & Psychophysics, 39, 346–350.

Stevens, D. A. and Lawless, H. T. 1987. Enhancement of responses to sequential presentation of oral chemical irritants. Physiology and Behavior, 39, 63–65.

Stevens, D.A., Baker, D., Cutroni, E., Frey, A., Pugh, D. and Lawless, H. T. 2008. A direct comparison of the taste of electrical and chemical stimuli. Chemical Senses, 33, 405–413.

Stevens, J. C., Cain, W. S. and Burke, R. J. 1988. Variability of olfactory thresholds. Chemical Senses, 13, 643–653.

Stevens, S. S. 1957. On the psychophysical law. Psychological Review, 64, 153–181.

Stevens, S. S. 1959. Cross-modality validation of subjective scales for loudness, vibration and electric shock. Journal of Experimental Psychology, 57, 201–209.

Stevens, S. S. 1962. The surprising simplicity of sensory metrics. American Psychologist, 17, 29–39.

Stevens, S. S. and Galanter, E. H. 1957. Ratio scales and category scales for a dozen perceptual continua. Journal of Experimental Psychology, 54, 377–411.

Stevenson, R. J., Prescott, J. and Boakes, R. A. 1995. The acquisition of taste properties by odors. Learning and Motivation 26, 433–455.

Stevenson, R. J., Boakes, R. A. and Prescott, J. 1998. Changes in odor sweetness resulting from implicit learning of a simultaneous odor-sweetness association: An example of learned synesthesia. Learning and Motivation, 29, 113–132.

Stevenson, R. J., Prescott, J. and Boakes, R. A. 1999. Confusing tastes and smells: How odors can influence the perception of sweet and sour tastes. Chemical Senses, 24, 627–635.

Stillman, J. A. 1993. Color influence flavor identification in fruit-flavored beverages. Journal of Food Science, 58, 810–812.

Stryer, L., 1995. Biochemistry, Fourth Edition. W. H. Freeman, New York, NY.

Sugita, M. 2006. Review. Taste perception and coding in the periphery. Cellular and Molecular Life Sciences, 63, 2000–2015.

Szolscanyi, J. 1977. A pharmacological approach to elucidation of the role of different nerve fibers and receptor endings in mediation of pain. Journal of Physiology (Paris), 73, 251–259.

Tai, C, Zhu, C. and Zhou, N. 2008. TRPA1: The central molecule for chemical sensing in pain pathway? The Journal of Neuroscience, 28(5):1019–1021

Thomas, C. J. C. and Lawless, H. T. 1995. Astringent subqualities in acids. Chemical Senses, 20, 593–600.

Tucker, D. 1971. Nonolfactory responses from the nasal cavity: Jacobson's Organ and the trigeminal system. In:

L. M. Beidler (eds.), Handbook of Sensory Physiology IV(I). Springer, Berlin, pp. 151–181.

Tuorila, H. 1986. Sensory profiles of milks with varying fat contents. Lebensmitter Wissenschaft und Technologie, 19, 344–345.

van der Klaauw, N. J. and Frank, R. A. 1996. Scaling component intensities of complex stimuli: The influence of response alternatives. Environment International, 22, 21–31.

Venkatachalam, K. and Montell, C. 2007. TRP channels. Annual Reviews in Biochemistry, 76, 387–414.

Verhagen, J. V. and Engelen, L. 2006. The neurocognitive bases of human food perception: Sensory integration. Neuroscience and Biobehavioral Reviews, 30, 613–650.

von Sydow, E., Moskowitz, H., Jacobs, H. and Meiselman, H. 1974. Odor-taste interactions in fruit juices. Lebensmittel Wissenschaft und Technologie, 7, 18–20.

Weiffenbach, J. M. 1991. Chemical senses in aging. In: T. V. Getchell, R. L. Doty, L. M. Bartoshuk and J. B. Snow (eds.), Smell and Taste in Health and Disease. Raven, New York, pp. 369–378.

Whitehead, M. C., Beeman, C. S. and Kinsella, B. A. 1985. Distribution of taste and general sensory nerve endings in fungiform papillae of the hamster. American Journal of Anatomy, 173, 185–201.

Wiseman, J. J. and McDaniel, M. R. 1989. Modification of fruit flavors by aspartame and sucrose. Institute of Food Technologists, Annual Meeting Abstracts, Chicago, IL.

Vickers, Z. 1991. Sound perception and food quality. Journal of Food Quality, 14, 87–96.

Wysocki, C. J. and Beauchamp, G. K. 1988. Ability to smell androstenone is genetically determined. Proceedings of the National Academy of Sciences USA, 81, 4899–4902.

Yamaguchi, S. 1967. The synergistic taste effect of monosodium glutamate and disodium 5'inosinate. Journal of Food Science, 32, 473–475.

Yamaguchi, S. and Kobori, I. 1994. Humans and appreciation of umami taste. Olfaction and Taste XI. In: K. Kurihara (ed.), Proceedings of the 11th International Symposium on Olfaction and Taste and 27th Meeting, Japanese Association for Smell and Taste Sciences. Springer, Tokyo, pp. 353–356.

Yau, N. J. N. and McDaniel, M. R. 1990. The power function of carbonation. Journal of Sensory Studies, 5, 117–128.

Yau, N. J. N. and McDaniel, M. R. 1991. The effect of temperature on carbonation perception. Chemical Senses, 16, 337–348.

Yokomukai, Y., Cowart, B. J. and Beauchamp, G. K. 1993. Individual differences in sensitivity to bitter-tasting substances. Chemical Senses, 18, 669–681.

Zarzo, M. and Stanton, D. T. 2006. Identification of latent variables in a semantic odor profile database using principal component analysis. Chemical Senses, 31, 713–724.

Zellner, D. A. and Kautz, M. A. 1990. Color affects perceived odor intensity. Journal of Experimental Psychology: Human Perception and Performance, 16, 391–397.

Zhao, H., Ivic, L., Otaki, J. M., Hashimoto, M., Mikoshiba, K. and Firestein, S. 1998. Functional expression of a mammalian odorant receptor. Science, 279, 237–242.

Chapter 3

Principles of Good Practice

Abstract This chapter outlines the standards of good practice in performing sensory evaluation studies. It briefly covers the sensory testing environment and its requirements, serving samples to panelists, and creating serving procedures, planning. There is a short section on designing experiments including design and treatment structures. Subsequently, it then covers general panelist screening, selecting, and training as well as an overview of panelist incentives. The legal ramifications and requirements of using humans as subjects of sensory tests are also described. Lastly the chapter discusses data collection and tabulation.

Some of the reasons some experimenters advance in trying to resist a (scientific approach) to their work are that: (a). there is no reason to suppose that there will be a bias; (b). it means much more work; (c). things might get mixed up.

There is no reason to suppose that there will not be a bias. As regards (b), one may ask, "more than what?" for that a valid experiment takes more work than an invalid experiment is irrelevant to a man who is wanting to make valid inferences. As regards (c), one feels sympathy, but if an experimenter isn't willing to do a decent job why doesn't he choose some other easier way of earning a living.

—Brownlee (1957, p. 1)

Contents

3.1 Introduction

In later chapters of this textbook we will often state that a particular method should be performed using standard sensory practices. This chapter will describe what we mean by "standard sensory practices." Table 3.1 provides a checklist of many of the good practice guidelines discussed in this chapter; this table can be used by sensory specialists to ensure that the study has

H.T. Lawless, H. Heymann, *Sensory Evaluation of Food*, Food Science Text Series,
DOI 10.1007/978-1-4419-6488-5_3, © Springer Science+Business Media, LLC 2010

Table 3.1 Sensory checklist[a]

Test objective
Test type
Panelist
    Recruitment
        Method of contact
        Supervisory approval
    Screening
        Informed consent
        Incentives
    Training
Sample
        Size and shape
        Volume
        Carrier
        Serving temperature
        Maximum holding time
Test setup
        Panelist check-in
        Palate cleansers
        Instructions
            To technicians
            To panelists
        Score sheets
            Instructions
            Type of scales
            Attribute words
            Anchor words
        Coding
        Randomization/counterbalancing
        Booth items
            Pencils
            Napkins
            Spit cups
        Clean up
        Disposal arrangements (important if security risk)
        Receipts if incentive is monetary
        Panelist debriefing
Test area
        Separation of panelists
        Temperature
        Humidity
        Light conditions
        Noise (auditory)
        Background odor/clean air handling/positive pressure
        Accessibility
        Security

[a]This checklist is a quick way of making sure that the sensory specialist has thought of many of the good practice guidelines discussed in this chapter

been thought through. It should be remembered that a good sensory specialist will always follow the standard practices because that would help ensure that he/she will obtain consistent, actionable data. However, an experienced sensory scientist will occasionally break the standard practice guidelines. When one breaks these rules one always has to be fully aware of the consequences, the risks entailed, and whether one still can get valid data from the study.

3.2 The Sensory Testing Environment

Much of the information in this section comes from our experiences in visiting, designing, and operating sensory facilities both in industrial and in university settings. The section was also written with reference to Amerine, Pangborn, and Roessler (1965), Jellinek (1985), Eggert and Zook (2008), Stone and Sidel (2004), and Meilgaard et al. (2006). We feel that anyone planning on constructing or renovating a sensory facility should read Eggert and Zook (2008) and view the accompanying CD, this is an extremely valuable resource. The sensory facility should be located close to potential judges but not in the middle of areas with extraneous odors and/or noise. This means that in a meat-processing plant the sensory area should not be near the smokehouse and in a winery the facility should be out of earshot from the noise of the bottling line. The sensory booth area must be easily accessible to the panelists and if the facility will be used by consumer panelists or panelists that will be traveling some distance then there should be ample, easy parking available. This frequently means that the sensory facility should be on the ground floor of a building and that the area should be near the entrance to the complex. In companies with security concerns, the sensory preparation facility should be within the secure area but the panelist waiting room and possibly the sensory booth area should be in an area that is easily accessible and possibly not secure.

When designing the sensory testing area, the traffic pattern of the panelists should be kept in mind. Panelists should enter and exit the facility without passing through the preparation area or the office areas of the facility. This is to prevent panelists from having physical or visual access to information that may bias their responses. For example, if panelists happen to see some empty jars of a specific brand in a trash can it may bias their responses if they expect to evaluate that brand as one of their coded samples. Additionally, for security reasons it is not a good idea to have panelists wandering through the sensory area where they may pick up information about projects or other panelists.

3.2.1 Evaluation Area

In its simplest form the facility would need an evaluation area. This may be as simple as a large room that could be used with tables or temporary booths placed on tables. It is always important to remember that if the evaluation occurs in a quiet, uninterrupted manner the likelihood of success is increased. It is especially important that the panelists not influence each other. If temporary booths are not available the sensory specialist should at the very least arrange the tables in room so that the participants do not face each other. Kimmel et al. (1994) arranged a room with tables in such a way that the panelists (in their case children) could not influence each other. If at all possible, separate the panelists with portable plywood booths (see Fig. 3.1 for manufacturing instructions). These can be made inexpensively and will allow panelists to be separated during testing.

Some consumer testing companies use a classroom style where each consumer is seated at a small table with space for a computer screen and the samples. The advantage of this situation is that it is portable (the evaluation area can be set up in hotels, conference rooms, church basements, etc.) and the whole group can receive any verbal instructions simultaneously. If color or appearance is important make sure that the testing area is well lit with balanced daylight-type fluorescent bulbs, however, see Chapter 12 for further information on color evaluation.

In a situation where sensory evaluation is an integral part of the product development and quality assurance cycle of the product a more permanent evaluation area should be constructed. In most sensory facilities, the evaluation area should encompass a discussion area, a booth area and, frequently, a waiting room area for the panelists (Fig. 3.2).

The waiting area should have comfortable seating, be well lit, and clean. This area is often the panelists' first introduction to the facility and should make them feel that the operation is professional and well organized. This area should be modeled on the waiting room of a medical practitioner. The sensory specialist

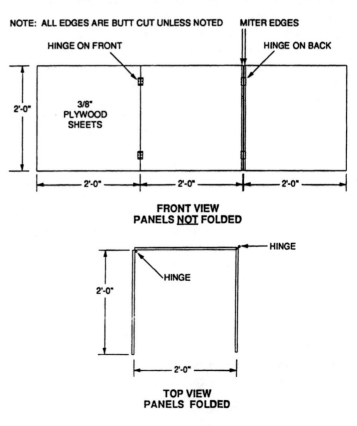

Fig. 3.1 Construction information for portable sensory booths.

Fig. 3.2 Floor plan of the three sensory facilities in the Robert Mondavi Sensory Building on the University of California at Davis Campus. A1, is a preparation area that includes a range and four ovens; A2 are preparation areas without cooking facilities; B1 is a sensory booth area with 24 individual booths; B2 are two separate sensory booths areas, each with six booths; C is the sensory waiting area with chairs, tables, and sofas; D indicates the work spaces for individual employees and students; E is a focus room with a two-way mirror. (Used with permission from ZGF Architects LLP, Portland, Oregon).

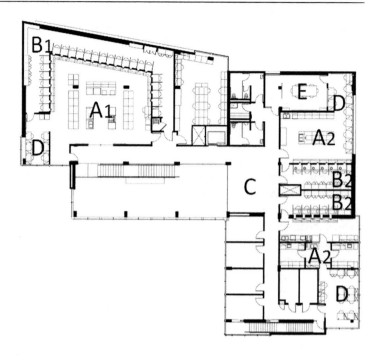

should always try to minimize the waiting time of panelists but sometimes this is unavoidable. To relieve the tedium of waiting the area should be equipped with some light reading. In some facilities, a child care area for panelists' children may also be included. In this case, care must be taken to prevent the noise and distraction from this area from interfering with the panelists' concentration during product evaluation.

In some consumer testing facilities, a briefing area may be adjacent to the waiting room or orientations may be done in the waiting room itself. The orientation area is very useful if chairs in the room are arranged in rows or a semi-circle. Then instructions, as to procedures, can be given to a whole group at once before they enter the test booths or discussion room. Questions can be fielded and volunteer panelists having difficulties can be further instructed or weeded out.

The discussion room would usually be arranged similarly to a conference room but the decor and the furnishings should be simple and in colors that would not affect the panelists' concentration. The area should be easily accessible to the panelists and to the preparation area. However, the panelists should not have visual or physical access to the preparation area. The sections on climate control, lighting, etc., of the booth area are equally applicable to the discussion area.

In many sensory facilities the booth area is the heart of the operation. This area should be isolated from the preparation area, be comfortable but not too casual in appearance. The area should always be clean and professional looking. Once again, neutral or non-distracting colors are advisable. The room should be kept quiet to facilitate panelist concentration. There are probably as many versions of booth areas as there are sensory facilities. Some of the variations are cosmetic and others affect the functionality of the space. In this section we will describe some variations, highlighting advantages and disadvantages of each. We will concentrate on booths used for food evaluations; however, specialized booths are often required for the evaluation of personal care products such as shaving creams, soaps, deodorants and home care products like insecticides, floor waxes, and detergents. An example of a purpose-built sensory booth is Renault motor company's poly-sensorial booth described by Eterradossi et al. (2009).

The number of booths in facilities we have seen ranges from as low as 3 to as high as 25. The number is usually constrained by the space available. However, the sensory scientist should attempt to have the maximum possible number of booths constructed, since the booth availability is frequently a bottleneck in test volume causing undue delay for panelists or decreasing the number of panelists that can be accommodated. Booth sizes vary greatly in different facilities but the ideal booth is about 1 m by 1 m in size. Smaller booths may make the panelist feel more "cramped" and this could potentially affect concentration. On the other hand, excessively large booths waste space. The booths should be separated from each other by opaque dividers that extend about 50 cm beyond the front edge of the counter top and 1 m above the counter top. This is to prevent panelists in adjacent booths from affecting each other's concentration. The corridor behind the booths should be wide enough for the panelists to comfortably move into and out of the booth area. Additionally, in the United States, if the booths are to be used by disabled persons the guidelines on corridor widths, seating configurations, and counter top heights of the Americans with Disabilities Act of 1990 (42 USC 126 § 12101–12213) should be followed.

The booth counter height is usually either desk or table height (76 cm) or kitchen counter height (92 cm). The height of the booth counter is constrained by the height of the serving counter on the other side of the booth pass-through hatch. We have seen booths where the serving counter was at kitchen counter height and the booth counter was at table height. The potential mess when samples were passed from the higher kitchen counter to the lower booth counter should discourage anyone else from constructing this type of booth. In general, either counter height is used. The table height counters allow the panelists to sit in comfortable chairs but demands that the sensory specialists bend to pass samples through the serving hatches. The amount of bending is minimized when the counter is at kitchen counter height but then the panelists should be supplied stools of adjustable height.

Serving hatches should be large enough to accommodate sample trays, score sheets, and yet small enough to minimize panel observation of the preparation/serving area. The hatches are often about 45 cm wide and 40 cm high; however, the exact size is dependent on the size of the sample serving trays that would be used in the facility. The most popular serving hatches are either the sliding door style or the bread box style. The sliding door style has a door that either slides up or to the side. These doors have the advantage that they do not occupy space either in the booth or on the serving counter. The major disadvantage of these doors is that the panelists can see through the open space into the preparation area. The amount of visual

information gleaned by the panelist can be minimized if the sensory specialist stands in front of the open space when serving samples. The bread box design has a metal hatch that is either open to the booth area or to the serving area, but not to both simultaneously. The advantage is that the bread box visually separates the panelist from the serving/preparation area but the disadvantage is that the hatch takes up counter space both in the booth and on the serving counter. The serving hatch should be mounted flush with the counter top, allowing the sensory specialist to easily slide the sample trays into and out of the booth.

The booth should be equipped with electrical outlets for use with computerized data entry systems as well as for electrical appliances that may be needed in the evaluation of a specific product. Data entry systems will be discussed later in this chapter. The installation of sinks in the booth should be discouraged. These sinks are frequently a major source of odor contamination in the booth and are very difficult to maintain in a completely sanitary fashion. It is better to use disposable spittoons and water glasses rather than sinks. When the serving hatch is closed, the panelist should have some means of communicating with the person serving the samples. The ideal communication link is a lighted two-way signal system. In some instances, the communication link can also be an intercom between the booths and the preparation area. In other cases, a card or a simple piece of colored plastic is used, the panelist pushing the signal through a small slot under the serving hatch to gain the attention of the person serving the samples.

Preparation areas will differ based on the product lines evaluated in the particular facility. For example, a facility designed to be used exclusively for frozen desserts would have no need for ovens, but would need ample freezer space. On the other hand, a facility designed for meat evaluations would need freezer and refrigerator space as well as ovens, stove tops, and other appliances used to cook meat. For these reasons it is somewhat difficult to give many rules as to the appliances needed in the preparation area, but there are some appliances and features that would be required in nearly all preparation areas.

The area needs a great deal of storage space. Refrigerated storage is necessary for samples, reference standards, and food treats (incentives) for panelists. Frozen storage space is needed for samples that require freezing. Additionally, cabinet storage space is required for utensils, serving dishes, serving trays, spittoons, paper ballots, computer printouts of data and statistical analyses, reports, photocopies of literature, etc. Many preparation areas have a lack of adequate storage space. If you as the sensory specialist have any input into the design of a sensory facility insist on ample storage space.

The other area that is often inadequate is the horizontal space required for setting up sensory tests. The counter space should be large enough to allow the specialists to set up one or two sessions' worth of serving dishes at the same time. The space can be re-used if food service trays and vertical food service carts are used as a holding space prior to serving samples. The entire area should be constructed with materials that are easy to clean and to maintain. Dishwashers, sinks with garbage disposals and trash cans should be installed in the preparation area. There should also be an adequate clean water supply for cleaning purposes as well as a supply of tasteless and odorless water to be used by the panelists for rinsing between samples. Double distilled water or bottled water from a reputable dealer are generally preferred. Additionally, depending on the types of products to be tested, other appliances such as electrical or gas cook tops and ovens, microwave ovens, deep-fat fryers may also be required. If oven and cook tops are installed, the area then requires hoods with charcoal filters or outside venting to control odors from the cooking area(s). The list of possible appliances is almost endless. In some facilities, flexibility has been designed into the preparation area with moveable case goods, flexible electrical and water hookups, and the potential to roll in new appliances and remove appliances that are not needed for a specific test. Again, space for storage of specialized equipment such as rice cookers or tea pots must be considered when designing a facility. In addition, local restaurant building codes should be consulted to make sure that sprinklers (used for fire safety), water quality, sewer, and all other utilities are adequate in the preparation area.

3.2.2 Climate Control

The booth and discussion areas should be climate controlled and odor free. The use of replaceable active carbon filters in the ventilation system ducts supplying these areas is encouraged. These areas should have excellent ventilation. A slight positive air pressure in these areas can minimize odor transfer from the preparation area. The sensory scientist should make sure that any cleaning supplies used in the booth and discussion areas do not add extraneous odors. These areas should be as noise free and distraction free as possible. Signs requiring silence in the hallways around these areas during testing times are helpful. Additionally, the noise added by nearby mechanical systems, e.g., freezers, air conditioners, processing equipment should be minimized.

The temperature and relative humidity for the booth and discussion areas should be 20–22°C and 50–55% relative humidity. These conditions would make the environment comfortable for the panelists and will prevent them from being distracted by the temperature or the humidity.

Illumination in these areas should be at least 300–500 lx at the table surface. Ideally it should be controllable with a dimmer switch to a maximum of 700–800 lx, the usual illumination intensity in offices. Incandescent lights are modifiable, by changing light bulbs, and versatile by allowing one to control both the light intensity and the light color. However, heat buildup can be a problem and should be accounted for when designing the booth area. The lighting should be even and shadow free on the counter surface. There are special lighting requirements for color evaluations and these will be discussed in Chapter 12.

The above discussion was for an average sensory facility used for food testing. However, for some product ranges more specialized facilities should be constructed. For example, if the facility is to be used to test ambient odor, for odor thresholds, room air deodorants, odors associated with household cleaners, etc., then either an odor room or a dynamic olfactory testing area should be created.

A dynamic olfactory test area would contain an olfactometer. In an olfactometer a gaseous sample flows continuously through tubes and the sample is diluted by mixing with odor-free air. Panelists would evaluate the samples at an exit port using a face mask or specially designed sniffing ports (Takagi, 1989). An odor room can be used by more than one panelist simultaneously. The odor evaluation area consists of an anteroom and a test room. The anteroom shields the test chamber from the external environment. The odor area should be constructed of odor free, easy to clean, non-absorptive materials. Stainless steel, porcelain, glass, or epoxy paints may be appropriate. The test room should have a ventilation system that can completely remove odorized air and introduce a controllable odor-free background.

3.3 Test Protocol Considerations

3.3.1 Sample Serving Procedures

The sensory specialist should be very careful to standardize all serving procedures and sample preparation techniques except the variable(s) under evaluation. For example, in a study to evaluate the effect of accelerated ripening on cheddar cheese flavor we decided to do a triangle test. Two technicians were assigned to the project and they divided cutting the cheese samples into 1 cm^3 cubes, each technician cutting either cheese A or cheese B. One technician was very precise and all the cubes she cut were exactly 1 cm^3. The other technician was less precise and her cubes varied slightly in size. Once the cubes were placed in serving containers it became obvious that the panelists could identify the odd sample by visual inspection only. These cheese samples were thrown away and the more precise technician was assigned to do all the cubing. However, she could not cut the cheese and serve it immediately. The cubes had to be stored overnight in a refrigerator. The technician decided to store the cubes from cheese A in one refrigerator and those from cheese B in another. The two refrigerators varied slightly in their temperature settings. The next morning, the samples were served to the panelists, who by simply touching the samples could identify the odd sample. The samples then had to be stored again, but in the same refrigerator to equilibrate the temperature difference. In a different study, involving threshold determination by discrimination testing, the layout of the samples prior to serving along a benchtop allowed a temperature decrease in samples set nearer to a room air conditioner. Some panelists could pick out the samples that were different based on this small temperature difference.

If carriers or combinations of products are required the timing of this process must be standardized. For example, if milk is poured on a breakfast cereal, the amount of time between pouring and tasting must be the same for all samples. It may not be wise to simply pass the milk in a container into the test booth for the panelists to add without instructions. They may pour the milk on all the samples at the outset, with the result that the last one evaluated has a much different texture than the first.

As can be seen from these examples, the sensory specialist should pay careful attention to the following areas when writing the test protocol and when performing the study: the visual appearance of the sample, sample size and shape, and sample serving temperature. Additionally, the sensory specialist should decide which serving containers should be used, whether the sample should be served with a carrier, how many samples should be served in a session, whether the panelists should rinse their mouths between samples, whether samples are to be expectorated or swallowed and how many samples should be served in a session. In the following sections each of these issues will be discussed, many of the suggestions made in these sections are based on our own experiences in a variety of sensory settings.

3.3.2 Sample Size

If the samples are evaluated in a discrimination test and the appearance of the sample is not the variable under evaluation then the samples should appear identical. If it is not possible to standardize the appearance exactly, a sequential monadic serving order may be used (Stone and Sidel, 2004). However, if there is a possibility that the panelists may remember that the samples were not identical in appearance then a discrimination test is not appropriate.

Cardello and Segars (1989) found that sample size affected the intensity scores assigned to textural attributes by panelists, even when the panelists were unaware of the sample size differential. De Wijk et al. (2003) confirmed these results with a different product. These results make it very important that the sensory specialist specifies the sample size and shape used in study, since the possibility exists that a different sample size may have led to different results. Therefore, when deciding on the sample size to serve the sensory specialist should keep a few questions in mind, namely What is the purpose of this study? How large is the normal portion size for this product? How large is a normal mouthful of this product? How many attributes does the panelist have to evaluate on this product? Is it possible to easily manipulate the size of the product? The answers to these questions should lead the sensory specialist to a reasoned decision in determining the size of the sample to be served. Keep in mind that it is better to err slightly on the side of a more generous portion size than a stingier one. In some cases a minimum amount to be eaten may be specified. This is potentially important in consumer tests where some participants may be timid about tasting novel products. However, a reasonable balance between cost associated with the product, storage, and preparation in relation to the sample size should be maintained.

3.3.3 Sample Serving Temperatures

The serving temperature of the product must be specified in the test protocol. Serving temperatures and holding time can present difficulties with some products such as meats. One approach to this is to serve the items in containers that are themselves warmed. In our laboratories and others, sand baths heated in an oven to a fixed temperature (usually 50°C) are used. Small glass beakers or ceramic crucibles used as holding dishes sit embedded in the sand baths and these in turn hold the samples to be tested. Even with this arrangement it is important to minimize the time samples are held or at the very least maintain this time as a constant across panelists.

In dairy products such as fluid milk, sensory characteristics may be accentuated if the product is warmed to a temperature above those of storage. In some tests where sensitivity and discrimination are the primary concerns, this is less realistic but a serving temperature allowing better discrimination is warranted. Thus fluid milk can be served at 15°C instead of the more usual 4°C to enhance the perception of volatile flavors. Ice cream should be tempered at −15°C to −13°C for at least 12 h before serving since scooping is difficult if the ice cream is colder. At higher temperatures the ice cream would melt. It is also usually best to scoop ice cream directly from the freezer immediately before serving rather than to scoop the portions and store these in a freezer. In this latter case the surface of the sample portion is inclined to become icier than the outer portion of a freshly scooped sample.

When samples are served at ambient temperatures the sensory specialist should measure and record the ambient temperature during each session. For samples served at non-ambient the serving temperature should be specified as well as the method of maintaining that temperature, whether it is sand baths, thermos flasks, water baths, warming tables, refrigerators, freezers, etc. The temperatures of samples that are served at non-ambient temperatures should be checked at the time of serving to ensure that the specified temperatures were achieved. Additionally, the specialist should specify the sample holding time at the specified temperature.

If samples are to be held for an extended period, the test protocol should include a discrimination test, with sufficient power (see Chapter 4) to determine if the holding period leads to changes in the sensory attributes of the product. If no changes occur then the samples could be held for an extended period. However, if products are to be held at elevated temperatures for any period the sensory specialist should also monitor potential microbial growth that could compromise the safety of the panelists.

3.3.4 Serving Containers

It is difficult to give rigid rules as to the choice of containers since different conditions exist in different sensory facilities. In some facilities, it is expensive and time consuming to wash many dishes, thus specialists in these cases would tend toward using disposable containers. In other facilities, there may be financial or environmental constraints that preclude the use of disposable dishes. The best advice is to use common sense when deciding which containers to use. The sensory specialist should choose the container that is most convenient, yet the choice of container should not negatively affect the sensory attributes of the product. For example, Styrofoam cups are very convenient to use since they are disposable and can easily be labeled using either a permanent ink marker or a stick-on label, yet we have found that these containers can adversely affect the flavor characteristics of hot beverages. If three-digit codes are applied via marking pens, care must be taken to insure that the ink does not impart an aroma.

3.3.5 Carriers

The issue of whether or not to use carriers poses some problems for the sensory specialist and deserves careful consideration. "Carriers" usually refer to materials that form a base or vehicle for the food being tested, but may more broadly be considered as any other food that accompanies the one being tested so that they are ingested (and tasted) together. Examples are cream fillings in pastries, butter on bread, spices in a sauce, and salad dressing on lettuce leaves.

In discrimination testing, the goal is often to make a test that will be very sensitive to product differences. A carrier can mask or disguise differences or minimize panelists' abilities to perceive the difference due to the addition of other flavors and modifications to the texture and mouthfeel characteristics. In some cases the carrier may simply increase the overall complexity of the sensory impressions to the point where the panelists are overwhelmed. In these cases the use of a carrier might not be desirable since it will decrease the effective sensitivity of the test for detecting sensory differences. If there are serious consequences from missing a difference (Type II error, see Chapter 4) then the use of a carrier that could potentially mask differences is not recommended.

If on the other hand, a false alarm or detection of a false positive difference (Type I error, see Chapter 4) poses serious problems, then the obscuring of a difference by the carrier is less detrimental. The degree of realism added by the carrier may complicate the situation, but it could prevent the detection of a difference that might be meaningless to consumers. The sensory specialist should discuss with the client whether the degree of realism in the test is a concern. For a food product that is rarely consumed alone and almost always involves a carrier, the "artificiality" of the situation where the carrier is omitted may be a major psychological problem to the panelists, especially in consumer testing. An example would be cherry pie fillings which are rarely eaten without pie crust.

So, there are two guidelines for consideration in determining whether a carrier should be used: the relative consequences of missing a difference versus a false positive test result and the degree of realism that is deemed necessary. Often the complications created by use of the carrier will lower the degree of sample control and uniformity that is possible, so this must be considered. Careful discussion of these issues with the client can help clarify the best approach. In some cases it may be advisable to do the test both with and without the carrier if time and resources permit. This can be very informative about the size of the perceivable difference as well as the nature of the interactions between the carrier and the food to be tested.

Stone and Sidel (2004) give the following interesting example of a compromise in the use of a carrier, where the food product (a pizza sauce) is influenced by the carrier (crust) in such important ways that the preparation, but not the testing, had to involve the substrate: "... it was agreed that flavor interactions with crust components resulted in a sauce flavor that could not be achieved by heating the sauce alone. However, owing to variability in pizza crust within a brand, it was determined that the pizza would be cooked and the sauce scraped from the crust and tested by itself. Thus the chemical reactions were allowed to occur and the subjects' responses were not influenced by crust thickness or other non test variables."

3.3.6 Palate Cleansing

The goal of palate cleansers should be to aid in the removal of residual materials from previous samples. An anecdote frequently told at wine tasting events says that serving rare roast beef slices will help undo the effects of high tannin in red wine samples. This makes some sense chemically. The proteins of the meat and its juices could form a complex removing tannins from solution—reducing the "pucker" of the wine. There have been numerous studies on palate cleansers to remove red wine astringency (see Ross et al., 2007 for a relatively recent example). However, it would seem that no true consensus has been reached on the ideal palate cleanser to use in these conditions.

Lucak and Delwiche (2009) evaluated the effects of a range of palate cleansers (chocolate, pectin solution, table water crackers, warm water, water, and whole milk) on foods representing various tastes and mouthfeel effects such as jelly beans (sweet), coffee (bitter), smoked sausage (fatty), tea (astringent), spicy tortilla chip (pungent), mint (cooling), and applesauce (non-lingering). They found that table water crackers were the only palate cleanser effective across all representative foods.

A study of off-flavors in fish examined the difficulties panelists have when cleaning the palate of methyl isoborneol, a compound associated with earthy, muddy, or musty aroma (Bett and Johnson, 1996). They suggested the use of untainted fish itself as the cleanser to use between test samples. This would make sense in that the fish flesh is an effective binder of the odor compound in question. However, these authors did raise the concern that this would involve time and expense in using additional fish samples as a palate cleanser.

3.3.7 Swallowing and Expectoration

In most analytical sensory tests, swallowing is avoided and samples are expectorated. This is assumed to provide less carry-over or unwanted influence of one product to the next. Also swallowing high-fat products can add unnecessary calories to panelists' diets. Of course, in consumer testing where acceptability is being measured, swallowing and post-ingestion effects can affect consumers' opinions on the products. Also in consumer testing generalizing to the natural consumption is a concern and here having respondents swallowing the products is acceptable. Kelly and Heymann (1989) studied the effect of swallowing versus expectoration on thresholds and fatigue effects in paired comparison and triangle tests using added salt in kidney beans and added milk fat to skim milk. They found no significant effects. However, it should be noted that the power of the test was low and thus the likelihood of finding a difference was slight. A time–intensity evaluation of Yerba mate infusions by Calviño et al. (2004) found that swallowing versus expectoration did not affect the perceived bitterness intensity of the infusion but that spitting did increase the rate of decay of the sensation.

One advantage of swallowing in analytical sensory testing is the stimulation of sensory receptors in the throat. This can be important in some products and flavor systems. For example, throat burn is important in pepper samples and "throat catch" (another type of chemical irritation) is characteristic of chocolate.

3.3.8 Instructions to Panelists

These should be very clear and concise. It is frequently desirable to give the instructions on how to perform the sensory evaluation both verbally, before the panelists enter the booth area, and in written form on the score sheet. These instructions should be pre-tested by having someone unfamiliar with sensory testing and the project attempts to follow them. We have frequently been amazed at how easily panelists misread or misunderstand what seemed to us to be simple, clear instructions. This usually occurs because we are too familiar with the testing methodology and thus read more into the instructions than is really there. The sensory specialist should always be aware of this potential problem.

The instructions to technicians and staff should also be very clear and preferably should be written. It is useful to have the technicians repeat the explanation of the procedure to the sensory specialist. This will assure that there were no communication gaps between the sensory specialist and the people performing the study. Additionally, for many tests it is useful to develop a standard operating procedure and to keep this available in laboratory notebooks.

3.3.9 Randomization and Blind Labeling

Samples should be blind labeled with random three-digit codes to avoid bias and sample order should be randomized to avoid artifacts due to order of presentation. Table 3.2 gives step-by-step instructions to set up discrimination and preference tests and Table 3.3 does the same for rating, ranking, and hedonic tests. Figures 3.3, 3.4, 3.5, 3.6, and 3.7 show master sheets prepared according to the instructions in Tables 3.2 and 3.3.

3.4 Experimental Design

This chapter is not designed to be a comprehensive discussion on experimental design. Excellent books and chapters have been written on experimental design and we would refer the reader to Cochran and Cox (1957), Gacula and Singh (1984), Milliken and Johnson (1984), MacFie (1986), Petersen (1985), Hunter (1996), and Gacula (1997).

3.4.1 Designing a Study

In this section we want to highlight some major issues that should be kept in mind by the sensory specialist when designing an experiment. At the beginning

Table 3.2 Step-by-step instructions for setting up discrimination and preference tests

1. Prepare master sheet (see the completed master sheets in Figs. 3.3, 3.4 3.5, and 3.6).
    a. Fill in the sample identification at top. For paired difference or paired preference tests two columns should be filled in (A, B). For constant reference duo-trio tests three columns should be indicated (Reference A, A, and B). For balanced reference duo-trio tests four columns are needed (Reference A, Reference B, A, and B). Triangle tests also need four columns filled in (A, A, B, and B). Only the researchers should know the identity of the A and B.
    b. Fill in judge numbers (i.e., 1, 2, 3. . .). Assign each judge a number and make sure that a key to these numbers is included in the notebook associated with study. It is simpler if a specific judge retains that number throughout the study.
    c. Create permutations of sample presentation. For paired difference or paired preference tests there are two possible permutations (AB, BA). For the balanced reference duo-trio tests there are four possible permutations (R_A AB, R_A BA, R_B AB, R_B BA).and the constant reference duo-trio tests has two possible permutations (R_A AB, R_A BA). For triangle tests there are six possible permutations (AAB, ABA, BAA, BBA, BAB, ABB). Each serving order should be assigned a number.
    d. Determine order of sample presentation. Using a table of random permutations, numbers are read from top to bottom within a column. Use only numbers corresponding to the number of serving orders in the test. Write the number (with a red pen, indicated in **bold** on Figs. 3.3, 3.4, 3.5, 3.6 and 3.7) in a blank column and then write the order that the samples will appear on the serving tray in the upper right hand corner of each square on the master sheet. This indicates the order in which each sample is presented to each judge.
    e. Assign three-digit random code numbers to each sample for each judge. Start from any point on the table of random numbers and use three digits for each number. Do not use numbers that may have meaning to the judges (i.e., 13, 666, 911). Write the random numbers on the master sheet, one for each sample for each judge (use blue or black pen, indicated in *italics* in Figs. 3.3, 3.4, 3.5, 3.6 and 3.7). An occasional duplicate of a number may be found on a random number table, if so, skip the duplicate number.
2. Write random number codes on the sample containers. Use the random code numbers which were written on the master sheet. Code numbers on sample containers should match the appropriate code numbers on the master sheet. The sample containers to be filled with the reference samples should not be coded R_A or R_B but should be coded only with an R.
3. Prepare score sheet. Fill in the date, the judge number, and the random code numbers in the sequence in which the samples are to be evaluated (as indicated by the random permutations).
4. Prepare samples.
    a. Prepare an organized arrangement for portioning samples. A simple method is to make a master sheet template with sufficient space for the sample containers to be placed in the squares. This template may be made out of any large paper or available substitute. Allowing a 3 in. square for each sample is suggested, however, this will vary depending on the sample container itself.
    b. Assemble sample containers on template. Once all the containers are placed on the template it should be identical in appearance to the master sheet.
    c. Portion samples into containers.
5. Assemble samples for each judge on a tray in the sequence that they are to be evaluated. Also, place the score sheet on the tray and water for rinsing the palate. Double check serving order.
6. Serve samples to judges for evaluations.
7. Decode score sheet on the master sheet. Circle the code that the judge circled. (Use a pen, *never*use pencil on master sheets or score sheets). In this way, decoding is simple and orderly. In order to analyze the data, it must be represented numerically. This may be according to the number of correct judgments (paired difference test, triangle test, and duo-trio test) or number of judges preferring sample A or B (paired preference). Make sure that a column is left for this purpose.
8. Analyze the data.

of any project the sensory specialist and all the parties that are requesting the study should define the objective of the study. To ensure that all parties are clearly communicating, the sensory specialist should rephrase all the objectives as questions. These should be circulated among all parties, who should provide feedback to the sensory specialist. The sensory specialist, in consultation with the client(s), should identify the tests required to answer the questions. At this point it is usually instructive for the specialist to design the perfect experiment without any cost constraints. This exercise is instructive because the process allows the specialist to clearly indicate the "ideal." Then when time and cost constraints are added and the specialist has to redesign the study to a scaled down version there is a clear picture of what is "given up" in this process. In some situations the scaled down version may not be capable of answering the test objectives. When this happens, the specialist and the client(s) must renegotiate the cost and time constraints and/or the test objectives. It is usually better to decrease the number of test objectives to those with the highest

Table 3.3 Step-by-step instructions for setting up ranking, rating and hedonic tests

1. Prepare master sheet (see Fig. 3.7 for completed master sheet).
    a. Fill in sample identification at top. In the example, in Fig. 3.7 for a study of fish, this may be Scrod, Cod, Tuna, Hake. Only the researchers should know the identity of the products or samples.
    b. Fill in judge numbers (i.e., 1, 2, 3...). Assign each judge a number and make sure that a key to these numbers is placed in the study notebook. It is simpler if a specific judge retains that number throughout the study.
    c. Assign three-digit random code numbers to each sample for each judge. Start from any point on the table of random numbers and use three digits for each number. Never use numbers that may have meaning to the judges (i.e., 13, 666, 911). Write the random numbers on the master sheet, one for each sample for each judge (use blue or black pen, indicated in *italics* in Fig. 3.8). An occasional duplicate of a number may be found on a random number table. If so, skip the duplicate number.
    d. Determine order of sample presentation. Using a table of random permutations, numbers are read from top to bottom within a column. Use only numbers corresponding to the number of samples being tested (i.e., for four samples: use only numbers 1, 2, 3, and 4; read the numbers in the order they appear). Write the number (with a red pen, indicated in bold in Fig. 3.7) in the upper right-hand corner of each square on the master sheet. This indicates the order in which each sample is presented to each judge. In the example, the first sample is served fourth, the second sample is served first, etc. for judge 1.
2. Write the random codes on the sample containers. Use the random code numbers which are written on the master sheet. Code numbers on sample containers should match the appropriate code number on the master sheet. If there are enough people working together, this can be done as random numbers are recorded on the master sheet.
3. Prepare score sheet. Fill in the date, the judge number, and the random code numbers in the sequence in which the samples are to be evaluated (as indicated by random permutations).
4. Prepare samples.
    a. Prepare an organized arrangement for portioning samples. A simple method is to make a master sheet template with sufficient space for the sample containers to be placed in the squares. This template may be made out of any large paper or available substitute. Allowing a 3 in. square for each sample is suggested, however, this will vary depending on the sample container itself.
    b. Assemble sample containers on template. Once all the containers are placed on the template it should be identical to the master sheet.
    c. Portion samples into containers.
5. Assemble samples for each judge on a tray in the sequence that they are to be evaluated. Also, place the score sheet on the tray and water for rinsing the palate. Double check serving order.
6. Serve samples to judges for evaluations.
7. Decode score sheet on the master sheet. When judges are asked to rate only one attribute a blank column is left between columns of random code numbers. When asked to rate more than one term more blank columns (one column for each terms rated) should be left. These columns provide space for recording judge scores after completion of the test. (Use a pen, *never* use pencil on master sheets or score sheets). In this way, decoding is simple and orderly.
8. Analyze the data.

priorities rather than cutting the power of the test (see Appendix E for power issues). There is no point in performing a study that is inadequate in answering the major test objectives. If it is not possible to design an adequate study, the specialist must ask for more resources.

Next, the sensory specialist should meticulously scrutinize the study step by step. The idea is to ask questions at each point about the worst possible scenario and how the study could be improved to minimize these contingencies. Sensory studies are more complex than they appear at first glance and the potential for complications and mistakes is always present. Samples may be lost, contaminated, or otherwise mishandled. Panelists may drop out before completing the test sequence. Participants may not correctly follow the test protocol or they may misunderstand instructions. Technical personnel can make mistakes in serving order sequences. Unwanted fluctuations in sample temperature or other conditions may enter the picture. Most of these problems can be eliminated or minimized in a well-designed test.

Once the study has been redesigned it is a good idea to write down a "skeleton" statistical analysis. This will give the specialist a good idea about the degrees of freedom associated with significance tests. It is also helpful to sketch out potential figures and tables that will be used in the final report.

Permutation numbers (Perm #) AB = 1
                              BA = 2

Judge	Perm. #	A	B		
1	2	169^2	507^1		
2	1	212^1	194^2		
3	1	962^1	644^2		
4	2	273^2	693^1		
Etc.					

Fig. 3.3 Example of a master sheet for a paired preference test.

Permutation numbers (Perm #) R_A AB = 1
                              R_A BA = 2

Judge	Perm. #	R_A	A	B	
1	1	R	557^1	485^2	
2	2	R	636^2	684^1	
3	1	R	325^1	238^2	
4	2	R	401^2	159^1	
etc..					

Fig. 3.4 Example of a master sheet for a constant reference duo–trio test.

3.4.2 Design and Treatment Structures

We like to use description of the experimental design elucidated by Milliken and Johnson (1984). These authors divided experimental design into two basic structures, namely treatment structure and design structure. They describe the treatment structure as set of samples or treatments that the client(s) selected to study in the specific project. The design structure

Permutation numbers (Perm #) R_A AB = 1
                              R_A BA = 2
                              R_B AB = 3
                              R_B BA = 4

Judge	Perm. #	R_A	R_B	A	B
1	4		R^1	557^3	485^2
2	1	R^1		636^2	684^3
3	2	R^1		325^3	238^2
4	3		R^1	401^2	159^3
etc..					

Fig. 3.5 Example of a master sheet for a balanced reference duo–trio test.

Permutation numbers (Perm #) BAA = 1   BBA = 4
                              ABA = 2   BAB = 5
                              AAB = 3   ABB = 6

Judge	Perm. #	A	A	B	B
1	5		495^2	926^1	183^3
2	4	292^3		899^1	854^2
3	2	797^1	630^3	315^2	
4	3	888^1	566^2	981^3	
5	1	267^2	531^3	469^1	
6	6		201^1	239^2	827^3
etc..					

Fig. 3.6 Example of a master sheet for a triangle test.

is defined by sensory specialists when they group experimental units into blocks. These two structures are linked by the randomization performed by the

Judge	Scrod	Cod	Tuna	Hake	
1	909^4	623^3	703^2	903^1	
2	690^1	558^2	578^3	383^4	
3	694^3	373^1	693^4	290^2	
4	890^2	763^4	787^1	661^3	
etc..					

Fig. 3.7 Example of a master sheet for a rating, ranking of hedonic test.

sensory specialist prior to the study and together they make up the experimental design of the study. The sensory specialist should let treatment structure dictate neither a poor design structure nor a favorite or frequently used design structure affect the selection of treatments.

3.4.2.1 Design Structures

Completely Randomized Design (CRD)

In this design all the samples are randomly assigned to all the panelists. Most of the experimental designs associated with sensory studies are performed to avoid or minimize artifacts due to order of sample presentation. The simplest solution to this problem is to make sure that the sample presentation order is completely randomized across all panelists. This technique works quite well in situations where the number of samples is small and all samples can be evaluated by all panelists in a single session. CRD is the ideal design for a central location consumer test where each panelist evaluates each sample. For example, in a mall intercept, test panelists are asked to express their degree of liking for each of four cola products. Each panelist receives the four colas in a randomly assigned sequence.

CRD designs also include random assignment of products to people where each individual only sees one product. These so-called consumer monadic tests are common in consumer field studies. These are also called between-groups comparisons since there are

different groups of people evaluating each product. The product group forms a block. An example would be a study with three versions of a product. The total consumer group is divided into three subgroups with people randomly assigned to a group. Each group tests one product, then fills out a questionnaire. Justification for monadic designs arises when (1) the test would be too time consuming or lengthy to have all people evaluate all products; (2) the use of one product would be likely to influence opinion of another; or (3) the use of the product changes the environment, person, or substrate. The last effect is common with consumer products (e.g., floor wax, insecticide) and personal care products (e.g., skin cream, hair conditioner). Time pressure to complete a test might also dictate a monadic design in consumer field work.

With trained analytical panels, the samples should be evaluated in replicate (often triplicate) to ascertain judge to judge variation. If the number of samples is sufficiently small it may be possible to have each panelist evaluate all samples in replicate in a single session using CRD. However, this is often not possible and then the sensory scientist would use a randomized complete block design (RCBD).

Randomized Complete Block Design (RCBD)

In a randomized complete block design each treatment (usually samples) is randomly assigned to each unit (usually panelists) within each block (often sessions). This design is frequently used when trained analytical panelists cannot evaluate all samples in replicate in a single session. In this case the best solution is to have each panelist evaluate all samples in a single session and then have them return for a subsequent session to re-evaluate all the samples. An example is a descriptive analysis study of six ice creams made with fat replacers. In a single session the panelists can only evaluate six samples. However, the samples should be evaluated in triplicate. The panelists must attend three sessions to evaluate all the samples in triplicate. In this study the blocks are the sessions and the six samples are randomized across those panelists within each block.

Incomplete Block Design

Incomplete block designs are used when there are too many treatments in the experiment for the panelists to judge all samples in a single session (block). In this case the panelists evaluate subsets of samples in individual sessions. The objective may be for each panelist to ultimately evaluate all samples often in replicate or it may be that panelists only see a subset of samples. An example of the first type of incomplete block design is the descriptive analysis of 13 vanilla samples performed by one of the authors (Heymann, 1994). The panelists could not evaluate all 13 samples in a single session. We chose an incomplete block design with four samples per block (session) and 13 blocks (sessions) (plan 11.22, Cochran and Cox, 1957). At the end of the study all the panelists had evaluated each of the 13 vanillas four times. The second type of incomplete block design is often used in consumer studies where the purpose is to screen flavor or fragrance candidates from a large pool of potential flavors or fragrances. For example, there may be twenty eight possible fragrances for a new floor wax, but due to the fatiguing nature of the fragrances consumers cannot rate their liking for more than four fragrances in a session. By choosing the appropriate incomplete block design (plan 11.38, Cochran and Cox, 1957) 63 groups of nine consumers would evaluate four fragrances in a screening test to pick the most liked fragrances.

3.4.2.2 Treatment Structures

One-Way Treatment Structure

In this case a set of treatments (samples) are chosen without assuming a relationship among the treatments. In sensory studies this occurs when a product set is chosen from among the brands on the market. In these cases there is no assumption the product made by one company is related to that from another company. For example, in a study of the sensory characteristics of black tea, the sensory specialist may choose four black teas, one made by each of four nationally known companies. The one-way treatment structure for this study then has four samples that are not related to each other in any way except that they are national brands of tea.

Two-Way Treatment Structure

For two-way treatment structures a set of samples are created by combining levels of two different types of treatments. In the sensory setting one may choose a product set from among the brands on the market and then each of these products is prepared in two different ways prior to sensory evaluation. For example, to return to the tea example used above, the sensory specialist decides to evaluate the teas as a hot beverage and as an iced tea. The treatment structure for this study is then two way with a total of eight treatments (four teas at two temperatures). Two-way treatment structures are also known as factorial treatment structures.

Other Treatment Structures

Many other possible treatment structures exist and are used in sensory studies. Examples include fractional factorial structures, a one-way structure of controls combined with a two-way factorial arrangement treatment structure. Split-plot and repeated measures experimental designs are created from incomplete block design structures and factorial arrangement treatment structures with two or more types of treatments. In a simple split-plot experimental design there are two sizes of experimental units and the treatments can be applied to differently sized experimental units by randomization. An example would be the following: each of six varieties of potatoes is grown in three rows, randomly assigned, in a field; the potatoes are harvested and the potatoes from each row are kept in separate containers. The potatoes are the cooked, using three cooking techniques. Each container is split into three batches and randomly assigned a cooking procedure. The cooked potatoes are evaluated by a descriptive analysis panel for texture. In this case the experimental unit to evaluate variety is the row and the experimental unit to evaluate the cooking procedure is the batch (Milliken and Johnson (2004).

A simple repeated measures designs is similar to a simple split-plot designs in terms of the two sizes of experimental units but the levels of at least one treatment (usually time) cannot be randomly assigned. For example, broccoli is harvested and randomly assigned to be packaged in four different packaging materials. The packages are stored and a sensory descriptive panel evaluates the samples daily in triplicate over a 2-week period. In this case the one experimental unit is the packaging type and the other is time (daily).

The following are some of the textbooks with numerous examples of more complex treatment structures: Milliken and Johnson (1984), Cochran and Cox (1957), Petersen (1985).

3.4.2.3 Randomization

The setting up instructions in Tables 3.2 and 3.3 only indicate how sample order may be randomized. It is usually better to also ensure that sample order is counterbalanced, as far as possible. When sample order is counterbalanced each serving sequence occurs an equal number of times. To determine if a specific master sheet is counterbalanced, one must determine the number of times each serving sequence appears. In a fully counterbalanced design all potential serving sequences will occur an equal number of times. It is possible to use specially designed serving sequences allowing the sensory scientist to not only have completely counterbalanced designs but also have serving sequences that are completely balanced. In other words, every sample is preceded by every other sample an equal number of times (MacFie et al., 1989; Wakeling and MacFie, 1995). These designs are especially helpful when the possibility of carry-over effects between samples exist (Muir and Hunter, 1991/1992; Schlich 1993; Williams and Arnold, 1991/1992). They are also helpful as "insurance" against carry-over effects, since their use allows one to determine carry-over effects post hoc.

Randomization of presentation orders is required for statistical validity but it is also important due to presentation order effects, specifically, first-position order effects. Position order effects occur when the perception of the sample is affected by the position in the presentation sequence that the sample is assessed at. In other words, the first sample is perceived differently than subsequent samples, solely due to its position in the line-up. This so-called first-position effect is quite strong (especially in consumer studies) and the sensory scientist should attempt to mitigate the effect. Randomization with each sample in the first position an equal number of times decreases the effect by spreading it across all samples. A better solution is to serve a dummy sample first followed by the true samples—in this case the panelists are told that they would be served say five samples, but unbeknownst to them the first sample is a dummy and samples two

through five are the actual samples. There also seems to be a small but persistent final sample effect.

3.5 Panelist Considerations

3.5.1 Incentives

Some incentive to participate in a sensory study is usually necessary in order to motivate people to volunteer. Sensory specialists should not expect automatic agreement of a person when they are asked to be on a panel and should be realistic about the benefits for that person. "What is in it for me?" is a reasonable question that sensory panel leaders should be ready to answer. In academic settings the days of ordering graduate students to participate are long gone. Likewise, in industry, sensory panel participation should be a volunteer activity. If it is required as a condition of employment (this is not recommended, except in the case noted below), the nature of the participation and the testing must be spelled out at length during the interviews and hiring process, otherwise the voluntary nature of participation is violated.

A guideline for motivating participation is the concept of the token incentive. By "token" we mean that the incentive is just enough to get the person to participate in the evaluation, but not so much that it is the sole reason for the participation. Obviously, if people are paid a great deal they will do just about anything, but an overpaid individual may have little or no motivation to concentrate and work during the evaluation sessions. In other words, they are just "in it for the money." The importance of the token incentive, payment or reward is different in different testing situations. In consumer work, where there is little or no loyalty, commitment or long-term interest in the testing program, the payment is of primary concern. For employees, students, or academic staff who participate in a sensory test, there are other reasons to become involved, such as positive feelings from helping out in the testing program. In some cases and in some cultures, the sense of social responsibility or support for the group effort may be strong enough so that the tokens may be quite minimal.

Common token incentives include snacks or "treats." This can serve as a social or coffee-break time for employees or staff and the opportunity for social interaction may itself become a motivating factor.

Small gifts for repeated testing and free company products are also common incentives. For very high levels of participation, larger gifts or social activities such as a luncheon or a holiday party can recognize the contributions of regular participants. One of us and at least one company that we know of uses a raffle system to entice panelists to attend. An entry is made after each test session. The more tests a person attends the better the chance of winning a prize. This system works well as long as the winners are rotated (you cannot win 2 months in a row) and the sensory professionals themselves are not eligible.

One of the most important incentives for participation is management recognition. When management acknowledges participation in sensory panels as an important contribution to the research effort, recruiting panelists become a great deal easier. Support for sensory evaluation must extend through all levels of management. If only top management supports sensory panel participation, the support quickly becomes "lip service." Supervisors may resent the time employee panelists may spend away from their main job. Thus, it is important to get the cooperation and support of the panelist's direct supervisor as well as all those higher up. An enlightened management will recognize that sensory panel participation enhances job skills, provides a broader motivation for project success, and can serve as a welcome break from routine activities that may enhance overall job performance. It is the job of the sensory professionals to communicate these benefits and to secure management support and to make sure that the supporting attitudes are made known to all potential panelists.

In some companies descriptive panel members are actually additional part-time employees. In this case these employees' only job description is to be panel members. If a sensory specialist decides to employ such descriptive panelists, there must be enough work to keep the panel busy on an ongoing basis. During slow times the panelists may work on re-training exercises or they may be laid off (not a good way to keep the panelists motivated).

3.5.2 Use of Human Subjects

Sensory specialists should be very aware of the health and safety of their panelists. These panelists are human subjects and the specialist should know and follow the guidelines that constrain the use of human subjects. The basis for the guidelines associated with the use of human subjects is the Nuremberg Code of Ethics in Medical Research (United States v. Karl Brandt et al., 1949) and the declaration of Helsinki (Morris, 1966). These guidelines principally state the following:

(1) It is essential the subjects give voluntary consent to participation.
(2) The subject should have the legal capacity to give the consent.
(3) The subject should be able to exercise free power of choice about participating in the study.
(4) The study should yield fruitful results for the good of society.
(5) The researchers should protect the rights and welfare of all the subjects.
(6) The researchers must ensure that the risks to the subjects associated with the study do not outweigh either the potential benefits to the participants or the expected value of the knowledge sought to society.
(7) Above all, the researchers must ensure that each person participating in the study had the right of adequate and informed consent without undue duress.

In legal language, most sensory studies pose no risk "above the ordinary risks of daily life." This includes any inherent risks associated with an individual's chosen occupation (the risks of being an astronaut are greater than those of a college professor). In general, in the United States, sensory testing of foods are often exempt from human subjects oversight scrutiny under the Federal Register (CFR 56:117 § A7 28102). The subjects may be at increased risk if the research plan deviates from the application of accepted and established methods. Physical risks may sometimes be present. For example, food ingredients and additives are sometimes tested during the product development cycle before these ingredients have achieved government approval such as the "generally recognized as safe" (GRAS) list in the United States. Employees who are asked to participate in such tests should be told of all possible risks and as always, participation should be voluntary. Additionally, sensory specialists should be sensitive to psychological risks such as embarrassment when mistakes are made. If panel results are published, shown or otherwise made available, as in some panelist training and monitoring situations, care must be taken to protect the feelings and, if possible, the identity of the outliers in the data.

In academic settings in the United States, all studies involving human subjects must be reviewed and approved by the particular institution's Human Subjects' Institutional Review Board (Belmont Report, 1979; Edgar and Rothman, 1995). In industrial settings this is not required. However, the ethical sensory specialist will adhere to the principles associated with responsible research involving human subjects (Sieber, 1992).

3.5.3 Panelist Recruitment

The sensory specialist must make sure that the people who are recruited know what is expected of them during the study. It is best to view their participation in the study as a contractual relationship. As much information as possible about time commitment and the product categories should be available to the potential panelists before they commit to the project. Panelists must also be told clearly what they may expect to gain from the study, such as daily treats, money. In most settings, the sensory specialist must be sure that panelists have approval from their supervisors to participate. Additionally, in academic settings in the United States, depending on the specific institution's Human Subjects Institutional Review Board, the sensory specialist may also need to make sure that each panelist voluntarily signs an informed consent form prior to participation in the study.

3.5.4 Panelist Selection and Screening

For certain product categories it may be necessary to have the panelists undergo a medical screening prior to participating in the sensory study. Additionally, the sensory specialist may need to screen the sensory acuity of the potential panelists. However, the specialist should allow some leeway in the sensory deficiencies of the potential panelists. Some people may be very discriminating in general, but have one or two problem areas. Also, many average panelists will improve

markedly with training. Therefore it is not necessary to have only the most highly discriminating panelists at the outset of training.

To screen for panelists the sensory scientist should create a battery of tests that are appropriate to the products to be evaluated and the general tasks required of the panelists. If the panelists are only going to be doing discrimination testing then the screening tests should only involve discrimination tasks. On the other hand, if the panelists are going to do scaling tasks then the screening tests may involve both discrimination and scaling tasks. The key to screening, however, is not to over-test panelists before performing true product evaluations. Too many screening tests could decrease the panelists' enthusiasm and motivation when it comes time to do "real" products. Judicious decisions related to the amount of screening needed for a specific study are important.

3.5.4.1 Examples of Screening Tests

The sensory scientist can create a series of discrimination tests differing in difficulty. In other words, the sensory scientist creates a series of product formulations that are more and more difficult to tell apart. Jellinek (1985) discusses how to select panelists using an extensive training course. She required the panelist to meet a stringent series of minimum requirements, before the panelist will be allowed to participate in sensory evaluation studies. These are generally applicable to a broad range of food testing. If the sensory program is more limited in scope, a series of tests may focus on the specific attributes to be encountered in the food products to be tested.

Additionally, it is helpful during panel screening to determine whether the panelist can discriminate the key ingredient flavors and the possible taints (off-flavors) in the product. It is possible to ask the panelists to rank order the intensity of the key ingredient flavors in the product or to rank order increasing levels of taints in the product. Panelists could also be asked to use multiple choice tests to describe aroma, flavor, and mouthfeel characteristics of the products. The sensory specialist can use these data to determine the extent of panel training. Such testing may illuminate areas needing additional work or identify panelists requiring special consideration and training.

If possible, the sensory specialist should recruit two to three times as many persons as needed for the panel. Then rank the panelists' performance on several screening test measures and invite the top performers to participate in the actual panel. A sensory specialist must be very tactful when potential panelists are told that their services will not be needed. When the panelists are informed of their performance on the screening tests, the specialist should use general labels, all of which should be positive in connotation. For example, the group could be divided into good smellers, very good smellers, and excellent smellers, rather than describing groups by using adjectives like "bad" or "poor." It is necessary to be very diplomatic and very careful not to insult people. All potential panelists must be made to feel appreciated, even if they were not invited to participate at this time since they may be recruited for a different study at a later time.

Records of all screening tests should be kept to compare future performance of these and new panelists. It is very possible that the performance of some panelists will improve over time and that of others will get worse. The specialist should plan from the beginning what to do about panel attrition because it will occur. The decision must be made whether panelists will be trained and added over time or whether the final panel size will be smaller than originally planned.

The most important fact to remember is that good panelists are not born but they can be created through the hard work of the panelist and the sensory specialist. Most individuals of average sensory activity can be trained to a level of very high, reliable, and accurate sensory evaluation performance.

3.5.5 Training of Panelists

The amount of training required is dependent on the task and the level of sensory acuity desired. For most descriptive tests extensive and in-depth training is necessary (see Chapter 10). For many discrimination tests, only minimal training is involved. In these cases the panelists are oriented to the task and that is the extent of the training (see Chapter 4).

During the training phase, especially for descriptive panels, the sensory specialist must make the panelists realize sensory work is difficult and requires attention and concentration. During extensive training sessions it is helpful if the panel develops an esprit de corps and this can be facilitated during training by having the panelists work as a team. As mentioned earlier, panelists are easier to train and likely to remain more motivated during the entire study if they feel that the sensory work done by them is valued by management. Attrition and turnover on panels are a factor in all settings. The sensory specialist must plan for this from the first day of recruitment. It is sometimes possible for new trainees to work with experienced panelists, such as people who had been trained for another product category.

3.5.6 Panelist Performance Assessment

The performance of trained panelists used over long periods of time may fluctuate, as the panelists become more or less motivated to participate and to concentrate on the task at hand during evaluation sessions. Also if people do not participate for awhile due to transfers, vacations, leaves-of-absence, etc., their performance may deteriorate and require re-training. Many companies have panelist assessment and reporting programs in place. These can be as simple as plotting the scores given by each panelist against the mean scores for the panel or as elaborate as using multiple assessment programs like those described by Sinesio et al. (1990), Naes and Solheim (1991), Mangan (1992), and Schlich (1996). The Panel Check program (available for free from the European Sensory Network (ESN) website) incorporates most of the above assessment programs in a single simple-to-use software package. We will revisit this issue in Chapter 10.

3.6 Tabulation and Analysis

3.6.1 Data Entry Systems

With the decrease in costs associated with personal computers a number of data entry systems have become very readily available. In this section we are not going to compare the systems currently available, since these would be obsolete within a few years. We are, however, going to list some principles that should be kept in mind when sensory specialists explore the feasibility of different data entry systems in their facility.

1. The limitations of the computer system should not dictate the form of the test. Before purchasing a system the sensory specialists should be sure that all the tests used in their situation can be programmed with the specific software system.
2. Purchasing online computerized systems requires a careful evaluation of cost-savings in terms of technician time and data entry time "by hand" and the pay-back time versus the overall cost of the system as well as the time needed to become comfortable with using the system.
3. In most situations the testing volume is the primary determinant of the need for automation or direct online entry. In situations where small volumes of many different types of tests are performed, a computerized system may also be useful.
4. The sensory specialist should be aware that less expensive alternatives to online data entry exist: the use of digitizing to enter data or the use of optical scanning.
5. The advantages of computerization of the sensory booth include

   (a) the speed of receiving test results
   (b) a ready interface between the data entry system and statistical and graphing programs
   (c) a reduction in the errors involved in data entry ("key punching")

6. Disadvantages include

   (a) consumers may be unfamiliar with computers and ill-at-ease with using the system. Their concentration may shift to the response system rather than the products
   (b) errors in use may go undetected if data are analyzed "automatically," e.g., without inspection
   (c) computer programs may not be flexible enough to handle variation in experimental designs or requirements for specific scale types

3.7 Conclusion

In many ways the good practice techniques associated with sensory testing are based on common sense. Many of the coding and setting up practices seem very cumbersome at first glance, but the goal is to insure that the specialist always, at all times, knows which sample is in which coded container, because inevitably at some point in a study a sample will be spilled. Sensory specialists should continually ask themselves whether a specific serving container, serving procedure, panelists' recruitment method seem logical and sensible.

Additionally, the use of good practice techniques improves the quality of the tests performed and this in turn will instill client confidence which ultimately leads to increased management respect for the results of sensory tests.

References

Amerine, M. A., Pangborn, R. M. and Roessler, E. R. 1965. Principles of Sensory Evaluation of Foods, Academic, New York, Ch. 6.

Belmont Report. 1979. Ethical Principles and Guidelines for the Protection of Human Subjects Research. The National Commission for the Protection of Human Subjects of Biomedical and Behavioral Research. National Institutes of Health. Office for the Protection from Risks Research. Washington, DC.

Bett, K. L. and Johnson, P. B. 1996. Challenges of evaluating sensory attributes in the presence of off-flavors. Journal of Sensory Studies, 11, 1–17.

Brownlee, K. A. 1957. The principles of experimental design. Industrial Quality Control, 13, 1–9.

Calviño, A., González Fraga, S. and Garrido, D. 2004. Effects of sampling conditions on temporal perception of bitterness in Yerba mate (*Ilex paraguariensis*) infusions. Journal of Sensory Studies, 19, 193–210.

Cardello, A. V. and Segars, R. A. 1989. Effects of sample size and prior mastication on texture judgments. Journal of Sensory Studies, 4, 1–18.

Cochran, W. G. and Cox, G. M. 1957. Experimental Designs. Wiley, New York.

de Wijk, R., Engelen, L. Prinz, J. F. and Weenen, H. 2003. The influence of bite size and multiple bites on oral texture sensations. Journal of Sensory Studies, 18, 423–435.

Edgar, H. and Rothman, D. J. 1995. The institutional review board and beyond: Future challenges to the ethics of human experimentation. The Milbank Quarterly, 73, 489–506

Eggert, J. and Zook K. 2008. Physical Requirement Guidelines for Sensory Evaluation Laboratories, Second Edition. ASTM Special Technical Publication 913. American Society for Testing and Materials, West Conshohocken, PA.

Eterradossi, O., Perquis, S. and Mikec, V. 2009. Using appearance maps drawn from goniocolorimetric profiles to predict sensory appreciation of red and blue paints. Color Research and Appreciation, 34, 68–74.

Gacula, M. C. and Singh, J. 1984. Statistical methods in food and consumer research. Academic, Orlando, FL.

Gacula, M. C. 1997. Descriptive sensory analysis in practice. Food and Nutrition, Trumbull, CT.

Heymann, H. 1994. A comparison of descriptive analysis of vanilla by two independently trained panels. Journal of Sensory Studies, 9, 21–32.

Hunter, E. A. 1996. Experimental design. In: Naes, T. and Risvik, E. (eds.), Multivariate Analysis of Data in Sensory Science. Elsevier Science, Amsterdam, The Netherlands, pp. 37–69.

Jellinek, G. 1985. Sensory Evaluation of Food: Theory and Practice. Ellis Horwood Series in Food Science and Technology, Chichester, England.

Kelly, F. B. and Heymann, H. 1989. Contrasting the effects of ingestion and expectoration in sensory difference tests. Journal of Sensory Studies, 3, 249–255.

Kimmel, S. A., Sigman-Grant, M. and Guinard, J-X. 1994. Sensory testing with young children. Food Technology 48, 92–99

Lucak, C. L. and Delwiche, J. F. 2009. Efficacy of various palate cleansers with representative foods. Chemosensory Perception, 2, 32–39.

MacFie, H. J. H. 1986. Aspects of experimental design. In: Piggott, J. R. (ed.), Statistical Procedures in Food Research. Elsevier Applied Science, London, pp. 1–18.

MacFie, H. J. H., Greenhoff, K., Bratchell, N. and Vallis, L. 1989. Designs to balance the effect of order of presentation and first-order carry-over effects in hall tests. Journal of Sensory Studies, 4, 129–148.

Mangan P. A. P. 1992. Performance assessment of sensory panelists. Journal of Sensory Studies, 7, 229–252.

Meilgaard, M., Civille, G. V. and Carr, B. T. 2006. Sensory Evaluation Techniques, Fourth Edition. CRC, Taylor & Francis Group, Boca Raton, FL.

Milliken, G. A. and Johnson, D. E. 1984. Analysis of Messy Data: Vol. 1. Van Nostrand Reinhold, New York.

Milliken, G. A. and Johnson, D. E. 2004. Analysis of Messy Data: Vol. 1, Second Edition. Chapman & Hall/CRC, New York.

Morris, C. 1966. Human Testing and the Court Room. Use of Human Subjects in Safety Evaluation of Food Chemicals. Publication 1491. National Academy of Sciences. National Research Council. Washington, DC, pp. 144–146.

Muir, D. D. and Hunter, E. A. 1991/1992. Sensory evaluation of cheddar cheese: Order of tasting and carry-over effects. Food Quality and Preference, 3, 141–145.

Naes, T. and Solheim, S. 1991. Detection and interpretation of variation within and between assessors in sensory profiling. Journal of Sensory Studies, 6, 159–177.

Petersen, R. G. 1985. Design and Analysis of Experiments. Marcel Dekker, New York.

Ross, C. F., Hinken, C. and Weller, K. 2007. Efficacy of palate cleansers for reduction of astringency carryover during repeated ingestions of red wine. Journal of Sensory Studies, 22, 293–312.

Schlich, P. 1993. Use of change-over designs and repeated measurements in sensory and consumer studies. Food Quality and Preference, 4, 2223–235.

Schlich, P. 1996. Defining and validating assessor compromises about product distances and attribute correlations. In: Naes, T. and Risvik, E. (eds.), Multivariate Analysis If Data in Sensory Science. Elsevier, B. V. Amsterdam, The Netherlands.

Sieber, J. E. 1992. Planning Ethically Responsible Research: A Guide for Students and Internal Review Boards. Applied Social Research Methods Series, Vol. 31. Sage Publications, Inc., Newbury Park, CA.

Sinesio, F., Risvik, E. and Rødbotten, M. 1990. Evaluation of panelist performance in descriptive profiling of rancid sausages: A multivariate study. Journal of Sensory Studies, 5, 33–52.

Stone, H. and Sidel, J. L. 2004. Sensory Evaluation Practices, Third Edition. Elsevier Academic, San Diego, CA.

Takagi, S. F. 1989. Standardization olfactometries in Japan—a review over ten years. Chemical Senses, 14, 24–46.

United States v. Karl Brandt, et al. 1949. The Medical Case: Trials of War Criminals before the Nuremberg Military Tribunals under Control Council Law No. 10. Vol. 2. U.S. Government Printing Office, Washington, DC, pp. 181–183

Wakeling, I. N. and MacFie, H. J. H. 1995. Designing consumer trials for first and higher orders of carry-over effect when only a subset of k samples from p may be tested. Food Quality and Preference, 6, 299–308.

Williams, A. A. and Arnold, G. M. 1991/1992. The influence of presentation factors on the sensory assessment of beverages. Food Quality and Preference, 3, 101–107.

Chapter 4

Discrimination Testing

Abstract Discrimination tests in most situations will only allow the sensory specialist to determine that two products perceptibly differ from one another or not. In this chapter we describe the more familiar discrimination tests such as paired comparison, duo–trio, triangle, dual standard, and A-not-A, as well as less used tests such as ABX and sorting tests. Data analysis techniques for these tests are described in detail (binomial, chi-square, z-, and beta-binomial distributions). Additionally, we begin the discussion of the effect of statistical power in sensory tests—this is further discussed in Chapter 5 and the Appendix of the book. The need for replication in sensory discrimination tests and the analysis of these data are discussed. Lastly, we discuss the need for warm-up samples in certain situations and well as some common issues arising from the interpretation of the results of sensory discrimination tests.

> *Chance favors only those who knows how to court her.*
> —Charles Nicolle

Contents

4.1 Discrimination Testing

Discrimination tests should be used when the sensory specialist wants to determine whether two samples are perceptibly different (Amerine et al., 1965; Meilgaard et al., 2006; Peryam 1958; Stone and Sidel, 2004). It is possible for two samples to be chemically different in formulation but for humans not to perceive this difference. Product developers exploit this possibility when they reformulate a product by using different ingredients while simultaneously not wanting the consumer to detect a difference. For example, an ice cream manufacturer may want to substitute the expensive vanilla flavor used in their premium vanilla ice cream with a cheaper vanilla flavor. However, they also may

H.T. Lawless, H. Heymann, *Sensory Evaluation of Food*, Food Science Text Series,
DOI 10.1007/978-1-4419-6488-5_4, © Springer Science+Business Media, LLC 2010

not want the consumer to perceive a difference in the product. A properly executed discrimination test with sufficient power indicating that the two ice cream formulations are not perceptibly different would allow the company to make the substitution with lowered risk. This is an ideal use of sensory discrimination testing. Discrimination testing may also be used when a processing change is made which the processor hopes would not affect the sensory characteristics of the product. In both of these cases the objective of the discrimination test is not to reject the null hypothesis, this is also known as similarity testing.

However, when a company reformulates a product to make a "new, improved" version then the discrimination test could be used to indicate that the two formulations are perceived to be different. In this case the objective of the discrimination is to reject the null hypothesis. If the data indicate that the two formulations are perceptibly different then the sensory scientist has to do a test that would indicate that the "new" formulation is perceived to be an improvement by the targeted consumer (see Chapters 13–15).

If the difference between the samples is very large and thus obvious, discrimination tests are not useful. If preliminary bench testing indicates that the two samples will be perceptibly different to all panelists then these discrimination tests should not be used. In such cases it may be useful to do scaling techniques to indicate the exact magnitude of the difference between the samples (see Chapter 7). In other words, discrimination testing is most useful when the differences between the samples are subtle. However, these subtle differences make the risk of Type II errors more likely (see later in this chapter and Appendix E).

Discrimination tests are usually performed when there are only two samples. It is possible to do multiple difference tests to compare more than two products but this is not efficient or statistically defensible. Usually ranking or scaling techniques will prove to be more effective (see Chapter 7).

There are a number of different discrimination tests available including triangle tests, duo-trio tests, paired comparison tests, *n*-alternative forced choice tests, tetrad tests (Frijters, 1984), polygonal and polyhedral tests (Basker, 1980). In Chapter 1, we briefly outlined the history associated with the triangle, duo–trio, and paired comparison tests. In the following section the more usual discrimination tests and their uses are described in more detail.

4.2 Types of Discrimination Tests

See Table 4.1 for a summary of the types of available discrimination tests and Table 4.2 for the process of doing a discrimination test.

4.2.1 Paired Comparison Tests

There are two analytical sensory forms of this test, namely the directional paired comparison (also known as the two-alternative forced choice) test and the difference paired comparison (also known as the simple difference or the same/different) test. The decision to use one or the other form is dependent on the objective of the study. If the sensory scientist knows that the two samples differ only in a specific sensory attribute then the two-alternative forced choice (2-AFC) method is used. In fact, as we will discuss in Chapter 5, it is always more efficient and powerful to use a directional paired comparison test specifying the sensory attribute in which the samples differ (if known) than to ask the panelists to identify the different sample. On the other hand, if the sensory scientist does not know in which sensory attribute(s) the samples differ than other techniques, such as the difference paired comparison must be employed, despite the subsequent loss of power.

For both paired comparison methods the probability of selection of a specific product, by chance alone (guessing), is one chance in two. However, as explained in Chapter 5 the situation is a little fuzzier for the same/different test where the probability is affected by the individual panelist's decision criterion. In both cases the null hypothesis states that in the long run (across all possible replications and samples of people) when the underlying population cannot discriminate between the samples they will pick each product an equal number of times. Thus the probability of the null hypothesis is $P_{pc} = 0.5$. Remember that P_{pc}, the proportion that we are making an inference (a conclusion) about, refers to the proportion we would see correct in the underlying population (and not the proportion correct in our sample or the data). That is why statistical hypothesis testing is part of inferential statistics. What the null hypothesis states in mathematical terms can also be verbalized as follows: If the

Table 4.1 Types of available discrimination tests

Class of test	Test	Samples: inspection phase	Samples: test phase	Task/instructions	Chance probability
Oddity	Triangle	(None)	A, A′, B (or A, B, B′)	Choose the most different sample	1/3
Matching	Constant reference duo–trio	Ref-A	A, B	Match sample to reference	1/2
	Balanced reference duo–trio	Ref-A, Ref-B	A, B	Match sample to reference	1/2
	ABX	Ref-A, Ref-B	A (or B)	Match sample to reference	1/2
	Dual standard	Ref-A, Ref-B	A, B	Match both pairs	1/2
Forced choice	Paired comparison	(None)	A, B	Choose sample with most of specified attribute	1/2
	3-AFC	(None)	A, A′,B	(Same)	1/3
	n-AFC	(None)	$A_1 - A_{n-1}$, B	(Same)	$1/n$
	Dual pair	(None)	A, B and A, A′	Choose A, B (different pair)	1/2
Sorting	Two out of five	(None)	A, A′, B, B′, B″	Sort into two groups	1/10
	4/8 "Harris–Kalmus"	(None)	$A_1 - A_4$, $B_1 - B_4$	Sort into two groups	1/70
Yes/no	Same–different	(None)	Pairs: A, A′ or A, B	Choose response: "Same" or "different"	N/A[a]
(Response choice)	A, not-A	Ref-A	A or B	Choose response: "A" or "not-A"	N/A[a]

[a]For the yes/no tests, a criterion may be set by each individual and therefore the probability may not be equal to 1/2. See Chapter 5 for further discussion of criterion in yes/no tasks

Table 4.2 Steps in conducting a difference test

1. Obtain samples and confirm test purpose, details, timetable, and panelists' training (e.g., training with the process) with client.
2. Decide testing conditions (sample size, volume, temperature, etc.) and clear with client.
3. Write instructions to the panelists and construct ballot.
4. Recruit potential panelists.
5. Screen panelists for acuity.
6. Train to do specific difference test (can use colors or shapes or spiked samples).
7. Set up counterbalanced orders.
8. Assign random three-digit codes and label sample cups/plates.
9. Conduct test.
10. Analyze results.
11. Communicate results to client or end user.

underlying population cannot discriminate between the samples then the probability of choosing sample A (that is the P_A) is equal to the probability of choosing sample B (P_B). Mathematically, this may be written as

$$H_0 : P_A = P_B = \frac{1}{2} \qquad (4.1)$$

However, as we will see the verbal forms of the alternate hypotheses for the two paired comparison tests differ.

4.2.1.1 Directional Paired Comparison Method (or the Two-Alternative Forced-Choice Method)

In this case, the experimenter wants to determine whether the two samples differ in a specified dimension, such as sweetness, yellowness, crispness. The two samples are presented to the panelist simultaneously and the panelist is asked to identify the sample that is higher in the specified sensory attribute. Figure 4.1 shows a sample score sheet. The panelist must clearly understand what the sensory specialist

Fig. 4.1 Example of a directional paired comparison (2-AFC) score sheet.

Please rinse your mouth with water before starting. There are two samples in each of the two paired comparison sets for you to evaluate. Taste each of the coded samples in the set in the sequence presented, from left to right, beginning with Set 1. Take the entire sample in your mouth. NO RETASTING. Within each pair, circle the number of the sweeter sample. Rinse with water between samples and expectorate all samples and water. Then proceed to the next set and repeat the tasting sequence.

Set

1   _____   _____

2   _____   _____

means by the specified dimension and the panelist should therefore be trained to identify the specified sensory attribute. The panelist should also be trained to perform the task as described by the score sheet. The directional paired comparison test has two possible serving sequences (AB, BA). These sequences should be randomized across panelists with an equal number of panelists receiving either sample A or sample B first.

The test is one tailed since the experimenter knows which sample is supposed to be higher in the specified dimension. The alternative hypothesis for the directional paired comparison test is that if the underlying population can discriminate between the samples based on the specified sensory attribute then the sample higher in the specified dimension (say A) will be chosen more often as higher in intensity of the specified dimension than the other sample (say B), this is P_{pc}. Mathematically this may be written as Eq. (4.2)

$$H_A : P_{pc} > \frac{1}{2} \qquad (4.2)$$

The results of the paired directional (2-AFC) test indicate the direction of the specified difference between the two samples. The sensory specialist must be sure that the two samples *only* differ in the single specified sensory dimension. This is often a problem with sensory discrimination testing of foods because changing one parameter frequently affects many other sensory attributes of the products. For example, removing some of the sugar from a sponge cake will likely make the cake less sweet but it would also affect the texture and the browning of the cake. In this case the directional paired comparison would not be an appropriate discrimination test to use.

4.2.1.2 Difference Paired Comparison (or the Simple Difference Test or the Same/Different Test)

This technique is similar to the triangle and duo–trio tests but it is not often used. It is best used, instead of the triangle or duo–trio test, when the product has a lingering effect or is in short supply and the presentation of three samples simultaneously would not be feasible (Meilgaard et al., 2006). In this case, the experimenter wants to determine whether the two samples differ without specifying the dimension(s) of the potential difference. An example would be if the study involves two sponge cakes, identical in formulation, except for the amount of sugar used. It is likely that the two cakes will differ in sweetness but probably also in texture and crust color.

The panelists are presented simultaneously with the two samples and are asked whether they perceive the samples to be the same or different. See Fig. 4.2 for a sample score sheet. The panelists only need to compare the two samples and decide whether they are similar or different. Humans easily make these types of comparisons and thus the task is relatively easy for the panelists. Thus, the panelists must be trained to understand the task as described by the score sheet but they need not be trained to evaluate specified sensory dimensions. The difference paired comparison method has four possible serving sequences (AA, BB, AB, BA). These sequences should be randomized across panelists with each sequence appearing an equal number of times.

The test is one tailed since the experimenter knows the correct answer to the question asked of each of the panelists, i.e., whether the two samples served to a

Fig. 4.2 Example of a difference paired comparison score sheet.

Date _____

Name _____

Please rinse your mouth with water before starting. There are two samples in each of the two paired comparison sets for you to evaluate. Taste each of the coded samples in the set in the sequence presented, from left to right, beginning with Set 1. Take the entire sample in your mouth. NO RETASTING. Are the samples within each set the same or different? Circle the corresponding word. Rinse with water between samples and expectorate all samples and water. Then proceed to the next set and repeat the tasting sequence.

Set

1 _____   _____   SAME   DIFFERENT

2 _____   _____   SAME   DIFFERENT

specific panelists were the same or different. The alternative hypothesis for the difference paired comparison test states that the samples are perceptibly different and that the population will correctly indicate that the samples are the same or different more frequently than 50% of the time. The mathematical form is

$$H_A : P_{pc} > \frac{1}{2} \tag{4.3}$$

The verbal form of the alternative hypothesis is that the population would be correct (saying that AB and BA pairs are different and that AA and BB pairs are the same) more than half the time. The results of the paired difference test will only indicate whether the panelists could significantly discriminate between the samples. Unlike the paired directional test, no specification or direction of difference is indicated. In other words, the sensory scientist will only know that the samples are perceptibly different but not in which attribute(s) the samples differed. An alternative analysis is presented in the Appendix to this chapter, where each panelist sees an identical pair (AA or BB) and one test pair (AB or BA) in randomized sequence.

4.2.2 Triangle Tests

In the triangle test, three samples are presented simultaneously to the panelists, two samples are from the same formulation and one is from the different formulation. Each panelist has to indicate either which sample is the odd sample or which two samples are most similar. The usual form of the score sheet asks the panelist to indicate the odd sample. However, some sensory specialists will ask the panelist to indicate the pair of similar samples. It probably does not matter which question is asked. However, the sensory specialist should not change the format when re-using panelists since they will get confused. See Fig. 4.3 for a sample score sheet. Similarly to the paired difference test the panelist must be trained to understand the task as described by the score sheet.

The null hypothesis for the triangle test states that the long-run probability (P_t) of making a correct selection when there is no perceptible difference between the samples is one in three ($H_0:P_t = 1/3$). The alternative hypothesis states that the probability that the underlying population will make the correct decision

Date _____

Name _____

Set _____

Rinse your mouth with water before beginning. Expectorate the water into the container provided. You received three coded samples. Two of these samples are the same and one is different. Please taste the samples in the order presented, from left to right. Circle the number of the sample that is different (odd). Rinse your mouth with water between samples and expectorate all samples and the water.

Fig. 4.3 Example of a triangle score sheet.

_____   _____   _____

when they perceive a difference between the samples will be larger than one in three.

$$H_A : P_t > \frac{1}{3} \qquad (4.4)$$

This is a one-sided alternative hypothesis and the test is one tailed. In this case there are six possible serving orders (AAB, ABA, BAA, BBA, BAB, ABB) which should be counterbalanced across all panelists. As with the difference paired comparison, the triangle test allows the sensory specialist to determine if two samples are perceptibly different but the direction of the difference is not indicated by the triangle test. Again, the sensory scientist will only know that the samples are perceptibly different but not in which attribute(s) the samples differed.

4.2.3 Duo–Trio Tests

In the duo–trio tests, the panelists also receive the three samples simultaneously. One sample is marked reference and this sample is the same formulation as one of the two coded samples. The panelists have to pick the coded sample that is most similar to reference. The null hypothesis states that the long-run probability (P_{dt}) of the population making a correct selection when there is no perceptible difference between the samples is one in two (H_0: $P_{dt} = 1/2$). The alternate hypothesis is that if there is a perceptible difference between the samples the population would match the reference and the sample correctly more frequently than one in two times.

$$H_A : P_{dt} > \frac{1}{2} \qquad (4.5)$$

Again, the panelists should be trained to perform the task as described by the score sheet correctly. Duo–trio tests allow the sensory specialist to determine if two samples are perceptibly different but the direction of the difference is not indicated by the duo–trio test. In other words, the sensory scientist will only know that the samples are perceptibly different but not in which attribute(s) the samples differed.

There are two formats to the duo–trio test, namely the constant reference duo–trio test and the balanced reference duo–trio test. From the point of view of the panelists the two formats of the duo–trio test are identical (see Figs. 4.4a and b), but to the sensory specialist the two formats differ in the sample(s) used as the reference.

4.2.3.1 Constant Reference Duo–Trio Test

In this case, all panelists receive the same sample formulation as the reference. The constant reference duo–trio test has two possible serving orders (R_A BA, R_A AB) which should be counterbalanced across all panelists. The constant reference duo–trio test seems to be more sensitive especially if the panelists have had prior experience with the product (Mitchell, 1956). For example, if product X is the current formulation (familiar to the panelists) and product Z is a new reformulation then a constant reference duo–trio test with product X as reference would be the method of choice.

4.2.3.2 Balanced Reference Duo–Trio Test

With the balanced reference duo–trio test half of the panelists receive one sample formulation as the

Date _____
Name _____

Before starting please rinse your mouth with water and expectorate. There are three samples in each of the two duo–trio sets for you to evaluate. In each set, one of the coded pairs is the same as the reference. For each set taste the reference first. Then taste each of the coded samples in the sequence presented, from left to right. Take the entire sample in your mouth. NO RETASTING. Circle the number of the sample which is most similar to the reference. Do not swallow any of the sample or the water. Expectorate into the container provided. Rinse your mouth with water between sets 1 and 2.

Fig. 4.4a Example of a constant reference duo–trio score sheet.

Set
1   Reference   _____   _____
2   Reference   _____   _____

Fig. 4.4b Example of a
balanced reference duo–trio
score sheet.

Date _____

Name _____

Before starting please rinse your mouth with water and expectorate. There are three samples in
each of the two duo–trio sets for you to evaluate. In each set, one of the coded pairs is the same
as the reference. For each set taste the reference first. Then taste each of the coded samples in
the sequence presented, from left to right. Take the entire sample in your mouth. NO
RETASTING. Circle the number of the sample which is most similar to the reference. Do not
swallow any of the sample or the water. Expectorate into the container provided. Rinse your
mouth with water between sets 1 and 2.

Set
1   Reference   _____   _____

2   Reference   _____   _____

reference and the other half of the panelists receive
the other sample formulation as the reference. In this
case, there are four possible serving orders (R_A BA,
R_A AB, R_B AB, R_B BA) which should be counterbalanced across all panelists. This method is used when
both products are prototypes (unfamiliar to the panelists) or when there is not a sufficient quantity of the
more familiar product to perform a constant reference
duo–trio test.

4.2.4 n-Alternative Forced Choice (n-AFC) Methods

The statistical advantages and hypotheses associated
with and the uses of the n-AFC tests will be discussed
in detail in Chapter 5. As we have seen the 2-AFC
method is the familiar directional paired comparison
method. The three-alternative forced choice (3-AFC)
method is similar to a "directional" triangle method
where the panelists receive three samples simultaneously and are asked to indicate the sample(s) that
are higher or lower in a specified sensory dimension
(Frijters, 1979). In any specific 3-AFC study there are
only three possible serving orders (AAB, ABA, BAA
or BBA, BAB, ABB) that should be counterbalanced
across all panelists. As with the 2-AFC the specified
sensory dimension must be the only perceptible dimension in which the two samples may differ. The panelists
must be trained to identify the sensory dimension evaluated. They must also be trained to perform the task as
described by the score sheet (Fig. 4.5).

The three-alternative forced choice test will allow
the sensory scientist to determine if the two samples

differ in the specified dimension and which sample is
higher in perceived intensity of the specified attribute.
The danger is that other sensory changes will occur in
a food when one attribute is modified and these may
obscure the attribute in question. Another version of
the n-AFC asks panelists to pick out the weakest or
strongest in overall intensity, rather than in a specific
attribute. This is a very difficult task for panelists when
they are confronted with a complex food system.

4.2.5 A-Not-A tests

There are two types of A-not-A tests referenced in
the literature. The first and the more commonly used
version has a training phase with the two products
followed by monadic evaluation phase (Bi and Ennis,
2001a, b), we will call this the standard A-not-A test.
The second version is essentially a sequential paired
difference test or simple difference test (Stone and
Sidel, 2004), which we will call the alternate A-not-A test. The alternate A-not-A test is not frequently
used. In the next section we will discuss the alternate
A-not-A test first since the statistical analysis for this
version is similar to that of the paired comparison discrimination test. The statistical analyses for the various
standard A-not-A tests are based on a different theory and somewhat more complex and will be discussed
later.

4.2.5.1 Alternate A-Not-A test

This is a sequential same/difference paired difference
test where the panelist receives and evaluates the first

Fig. 4.5 Example of a
three-alternative forced choice
score sheet.

Date _____
Name _____

Please rinse your mouth with water before starting. There are three samples in the set for you to
valuate. Taste each of the coded samples in the set in the sequence presented, from left to right.
Take the entire sample in your mouth. NO RETASTING. Within the group of three, circle the
number of the sweeter sample. Rinse with water between samples and expectorate all samples
and water.

_____   _____   _____

sample, that sample is then removed. Subsequently, the panelist receives and evaluates the second sample. The panelist is then asked to indicate whether the two samples were perceived to be the same or different. Since the panelists do not have the samples available simultaneously they must mentally compare the two samples and decide whether they are similar or different. Thus, the panelists must be trained to understand the task as described by the score sheet but they need not be trained to evaluate specified sensory dimensions. The alternate A-not-A test, like the difference paired comparison method, has four serving sequences (AA, BB, AB, BA). These sequences should be randomized across panelists with each sequence appearing an equal number of times. The test is one tailed since the experimenter knows the correct answer to the question asked of the panelists namely whether the two samples are the same or different. The null hypothesis of the alternate A-not-A test is the same as the difference paired comparison null hypothesis (H_0: $P_{pc} = 0.5$). The alternative hypothesis for this form of the A-not-A test is that if the samples are perceptibly different the population will correctly indicate that the samples are the same or different more frequently than one in two times. This alternative hypothesis is also the same as that of the difference paired comparison test (H_A: $P_{pc} > 1/2$).

The results of the A-not-A test only indicate whether the panelists could significantly discriminate between the samples when they are not presented simultaneously. Like the paired difference test, no direction of difference is indicated. In other words, the sensory scientist will only know that the samples are perceptibly different but not in which attribute(s) the samples differed.

This version of the A-not-A test is frequently used when the experimenter cannot make the two formulations have exactly the same color or shape or size, yet the color or shape or size of the samples are not

relevant to the objective of the study. However, the differences in color or shape or size have to be very subtle and only obvious when the samples are presented simultaneously. If the differences are not subtle the panelists are likely to remember these and they will make their decision based on these extraneous differences.

4.2.5.2 Standard A-Not-A Test

Panelists inspect multiple examples of products that are labeled "A" and usually also products that are labeled "not-A." Thus there is a learning period. Then once the training period has been completed the panelists receive samples one at a time and are asked whether each one is either A or not-A. As discussed by Bi and Ennis (2001a) the standard A-not-A test potentially has four different designs. For the monadic A-not-A test the panelist, after the training phase, is presented with a single sample (either A or not-A). In the paired A-not-A version the panelist, after completion of the training phase, is presented with a pair of samples, sequentially (one A and one not-A, counter balanced across panelists). In the replicated monadic A-not-A version the panelist, after completion of training, receives a series of samples of either A or not-A but not both. This version is rarely used in practice. Lastly, in the replicated mixed A-not-A version the panelist, after completion of training, receives a series of A and not-A samples. Each of these different formats requires different statistical models and using an inappropriate model could lead to a misleading conclusion. As described by Bi and Ennis (2001a) "The statistical models for the A-Not A method are different from that of other discrimination methods such as the m-AFC, the triangle, and the duo–trio methods."

"Pearson's and McNemar's chi-square statistics with one degree of freedom can be used for the standard A-Not A method while binomial tests based on the proportion of correct responses can be used for the m-AFC, the triangle, and the duo–trio methods. The basic difference between the two types of difference tests is that the former involves a comparison of two proportions (i.e., the proportion of "A" responses for the A sample versus that for the Not A sample) or testing independence of two variables (sample and response) while the latter is a comparison of one proportion with a fixed value (i.e., the proportion of correct responses versus the guessing probability)". Articles by Bi and Ennis (2001a, b) clearly describe data analysis methods for these tests.. Additionally, the article by Brockhoff and Christensen (2009) describes a R-package called SensR (http://www.cran.r-project.org/package=sensR/) that may be used for the data analyses of some Standard A-not-A tests. The data analyses associated with the standard A-not-A tests are beyond the scope of this textbook, but see the Appendix of this chapter which shows the application of the McNemar chi-square for a simple A-not-A test where each panelist received one standard product (a "true" example of A) and one test product. Each is presented separately and a judgment is collected for both products.

4.2.6 Sorting Methods

In sorting tests the panelists are given a series of samples and they are asked to sort them into two groups. The sorting tests can be extremely fatiguing and are not frequently used for taste and aroma sensory evaluation but they are used when sensory specialists want to determine if two samples are perceptibly different in tactile or visual dimensions. The sorting tests are statistically very efficient since the long-run probability of the null hypotheses of the sorting tests can be very small. For example, the null hypothesis of the two-out-of-five test is 1 in 10 ($P_{2/5} = 0.1$) and for the Harris–Kalmus test the null hypothesis is 1 in 70 ($P_{4/8} = 0.0143$). These tests are discussed below.

4.2.6.1 The Two-Out-of-Five Test

The panelists receive five samples and are asked to sort the samples into two groups, one group should contain the two samples that are different from the other

three samples (Amoore et al., 1968). Historically, this test was used for odor threshold work where the samples were very weak and therefore not very fatiguing (Amoore, 1979). The probability of correctly choosing the correct two samples from five by chance alone is equal to 0.1. This low probability of choosing the correct pair by chance is the main advantage of the method. However, major disadvantage of this method is the possibility of sensory fatigue. The panelists would have to make a number of repeat evaluations and this could be extremely fatiguing for samples that have to be smelled and tasted. This technique works well when the samples are compared visually or by tactile methods but it is usually not appropriate for samples that must be smelled or tasted. Recently Whiting et al. (2004) compared the two-out-of-five and the triangle test in determining perceptible differences in the color of liquid foundation cosmetics. They found that the triangle test results gave weak correlations with the instrumental color-differences but that the results of the two-out-of-five test were well correlated with the instrumental values.

4.2.6.2 The Harris–Kalmus Test

The Harris–Kalmus test was used to determine individual thresholds for phenyl thiocarbamide (PTC, a.k.a. phenyl thiourea, PTU). In this test panelists are exposed to increasing concentration levels of PTC in groups of eight (four samples containing water and four samples containing the current concentration of PTC). The panelists are asked to sort the eight samples into two groups of four. If the panelist does the sorting task incorrectly he/she is then exposed to the next higher concentration of PTC. The sorting task continues until the panelist correctly sorts the two groups of four samples. That concentration level of PTC is then identified as the threshold level for that panelist (Harris and Kalmus, 1949–1950). The method has the same disadvantage as the two-out-of-five test, in that it could be fatiguing. However, as soon as the panelist correctly sorts the samples the researcher concludes that the panelist is sensitive to PTC. Panelists insensitive to PTC only "taste" water in the solutions and are thus not fatigued. A shortened version of this test using three-out-of-six was used by Lawless (1980) for PTC and PROP (6-n-propyl thiouracil) thresholds.

4.2.7 The ABX Discrimination Task

The ABX discrimination task, as its name intends to suggest, is a matching-to-sample task. The panelist receives two samples, representing a control sample and a treatment sample. As in other discrimination tasks, the "treatment" in food research is generally an ingredient change, a processing change or a variable having to do with packaging or shelf life. The "X" sample represents a match to one of the two inspected samples and the panelist is asked to indicate which one is the correct match. The chance probability level is 50% and the test is one tailed, as the alternative hypothesis is performance in the population above 50% (but not below). In essence, this task is a duo–trio test in reverse (Huang and Lawless, 1998). Instead of having only one reference, two are given, as in the dual standard discrimination test. In theory, this allows the panelists to inspect the two samples and to discover for themselves the nature of the sensory difference between the samples, if any. As the differences are completely "demonstrated" to the panelists, the task should enjoy the same advantage as the dual standard test (O'Mahony et al., 1986) in that the participants should be able to focus on one or more attributes of difference and use these cues to match the test item to the correct sample. The inspection process of the two labeled samples may also function as a warm-up period. The test may also have some advantage over the dual standard test since only one item, rather than two are presented, thus inducing less sensory fatigue, adaptation, or carry-over effects. On the other hand, giving only one test sample provides less evidence as to the correct match, so it is unknown whether this test would be superior to the dual standard. As in other general tests of overall difference (triangle, duo–trio) the nature of the difference is not specified and this presents a challenge to the panelists to discover relevant dimensions of sensory difference and not be swayed by apparent but random differences. As foods are multi-dimensional, random variation in irrelevant dimensions can act as a false signal to the panelists and draw their attention to sensory features that are not consistent sources of difference (Ennis and Mullen, 1986).

This test has been widely used as a forced choice measure of discrimination in psychological studies, for example, in discrimination of speech sounds and

in measuring auditory thresholds (Macmillan et al., 1977; Pierce and Gilbert, 1958). Several signal detection models (see Chapter 5) are available to predict performance using this test (Macmillan and Creelman, 1991). The method has been rarely if ever applied to food testing, although some sensory scientists have been aware of it (Frijters et al., 1980). Huang and Lawless (1998) did not see any advantages to the use of this test over more standard discrimination tests.

4.2.8 Dual-Standard Test

The dual standard was first used by Peryam and Swartz (1950) with odor samples. It is essentially a duo–trio test with two reference standards—the control and the variant. The two standards allow the panelists to create a more stable criterion as to the potential difference between the samples. The potential serving orders for this test are $R_{(A)}$ $R_{(B)}$, AB, $R_{(A)}$ $R_{(B)}$ BA, $R_{(B)}$ $R_{(A)}$ AB, $R_{(B)}$ $R_{(A)}$ BA. The probability of guessing the correct answer by chance is 0.5 and the data analyses for this test are identical to that of the duo–trio test. Peryam and Swartz felt quite strongly that the technique would work best with odor samples due to the relatively quick recovery and that the longer recovery associated with taste samples would preclude the use of the test. The test was used by Pangborn and Dunkley (1966) to detect additions of lactose, algin gum, milk salts, and proteins to milk. O'Mahony et al. (1986) working with lemonade found that the dual-standard test elicited superior performance over the duo–trio test. But O'Mahony (personal communication, 2009) feels that this result is in error, since the panelists were not instructed to evaluate the standards prior to each pair evaluation and therefore the panelists were probably reverting to a 2-AFC methodology. This would be in agreement with Huang and Lawless (1998) who studied sucrose additions to orange juice and they did not find superiority in performance between the dual standard and the duo–trio or the ABX tests.

4.3 Reputed Strengths and Weaknesses of Discrimination Tests

If the batch-to-batch variation within a sample formulation is as large as the variation between formulations then the sensory specialist should not use triangle or duo–trio tests (Gacula and Singh, 1984). In this case the paired comparison difference test could be used but the first question that the sensory specialist should ask is whether the batch-to-batch variation should not be studied and improved prior to any study of new or different formulations.

The major weakness of all discrimination tests is that they do not indicate the magnitude of the sensory difference(s) between the sample formulations. As the simple discrimination tests are aimed at a yes/no decision about the existence of a sensory difference, they are not designed to give information on the magnitude of a sensory difference, only whether one is likely to be perceived or not. The sensory specialist should not be tempted to conclude that a difference is large or small based on the significance level or the probability (p-value) from the statistical analysis. The significance and p-value depend in part upon the number of panelists in the test as well as the inherent difficulty of the particular type of discrimination test method. So these are no acceptable indices of the size of the perceivable difference. However, it is sensible that a comparison in which 95% of the judges answered correctly has a larger sensory difference between control and test samples than a comparison in which performance was only at 50% correct. This kind of reasoning works only if a sufficient number of judges were tested, the methods were the same, and all test conditions were constant. Methods for interval level scaling of sensory differences based on proportions of correct discriminations in forced choice tests are discussed further in Chapter 5 as Thurstonian scaling methods. These methods are indirect measures of small differences. They are also methodologically and mathematically complex and require certain assumptions to be met in order to be used effectively. Therefore we feel that the sensory specialist is wiser to base conclusions about the degree of difference between samples on scaled (direct) comparisons, rather than indirect estimates from choice performance in discrimination tests. However, there are alternative opinions in the sensory community and we suggest that interested parties read Lee and O'Mahony (2007).

With the exception of the 2-AFC and 3-AFC tests the other discrimination tests also do not indicate the nature of the sensory difference between the samples. The major strength of the discrimination tests is that the task that the panelists perform is quite simple and

intuitively grasped by the panelists. However, it is frequently the very simplicity of these tests that lead to the generation of garbage data. Sensory specialists must be very aware of the power, replication, and counterbalancing issues associated with discrimination tests. These issues are discussed later in this chapter.

4.4 Data Analyses

The data from discrimination tests may be analyzed by any of the following statistical methods. The three data analyses are based on the binomial, chi-square, or normal distributions, respectively. All these analyses assume that the panelists were forced to make a choice. Thus they had to choose one sample or another and could not say that they did not know the answer. In other words, each panelist either made a correct or incorrect decision, but they all made a decision.

4.4.1 Binomial Distributions and Tables

The binomial distribution allows the sensory specialist to determine whether the result of the study was due to chance alone or whether the panelists actually perceived a difference between the samples. The following formula allows the sensory scientists to calculate the probability of success (of making a correct decision; p) or the probability of failure (of making an incorrect decision; q) using the following formula.

$$P(y) = \frac{n!}{y!(n-y)!} p^y p^{n-y} \qquad (4.6)$$

where

n = total number of judgments
y = total number of correct judgments
p = probability of making the correct judgment by chance

In this formula, $n!$ describes the mathematical factorial function which is calculated as $n \times (n-1) \times (n-2) \ldots \times 2 \times 1$. Before the widespread availability of calculators and computers, calculation of the binomial formula was quite complicated, and even now it remains somewhat tedious. Roessler et al. (1978) published a series of tables that use the binomial formula to calculate the number of correct judgments and their probability of occurrence. These tables make it very easy to determine if a statistical difference were detected between two samples in discrimination tests. However, the sensory scientist may not have these tables easily available thus he/she should also know how to analyze discrimination data using statistical tables that are more readily available. We abridged the tables from Roessler et al. (1978) into Table 4.3. Using this table is very simple. For example, in a duo–trio test using 45 panelists, 21 panelists correctly matched the sample to the reference. In Table 4.3, in the section for duo–trio tests, we find that the table value for 45 panelists at 5% probability is 29. This value is larger than 21 and therefore the panelists could not detect a difference between the samples. In a different study, using a triangle test, 21 of 45 panelists correctly identified the odd sample. In Table 4.3, in the section for triangle tests, we find that the table value for 45 panelists at 5% probability is 21. This value is equal to 21 and therefore the panelists could detect a significant difference between the samples at the alpha probability of 5%.

O_2 = observed number incorrect choices

E_1 = expected number of correct choices

E_1 is equal to total number of observations (n) times probability (p) of a correct choice, by chance alone in a single judgment where

$p = 0.100$ for the two-out-of-five test

$p = 0.500$ for duo–trio, paired difference, paired directional, alternate A-not-A tests

$p = 0.333$ for triangle tests

E_2 = expected number of incorrect choices

E_2 is equal to total number of observations (n) times probability (q) of an incorrect choice, by chance alone in a single judgment where $q = 1-p$

$q = 0.900$ for the two-out-of-five test

$q = 0.500$ for duo–trio, paired difference, paired directional, alternate A-not-A tests, ABX tests

$q = 0.667$ for triangle tests

The use of discrimination tests allows the sensory scientist to determine whether two products are statistically perceived to be different, therefore the degrees of freedom equal one (1). Therefore, a χ^2 table using df $= 1$ should be consulted, for alpha (α) at 5% the critical χ^2 value is 3.84. For other alpha levels consult the chi-square table in the Appendix.

4.4.2 The Adjusted Chi-Square (χ^2) Test

The chi-square distribution allows the sensory scientist to compare a set of observed frequencies with a matching set of expected (hypothesized) frequencies. The chi-square statistic can be calculated from the following formula (Amerine and Roessler, 1983), which includes the number –0.5 as a continuity correction. The continuity correction is needed because the χ^2 distribution is continuous and the observed frequencies from discrimination tests are integers. It is not possible for one-half of a person to get the right answer and so the statistical approximation can be off by as much as $\frac{1}{2}$, maximally.

$$\chi^2 = \left[\frac{(|O_2 - E_2| - 0.5)^2}{E_1} \right] + \left[\frac{(|O_2 - E_2| - 0.5)^2}{E_2} \right]$$

(4.7)

where

O_1 = observed number of correct choices

4.4.3 The Normal Distribution and the Z-Test on Proportion

The sensory specialist can also use the areas under the normal probability curve to estimate the probability of chance in the results of discrimination tests. The tables associated with the normal curve specify areas under the curve (probabilities) associated with specified values of the normal deviate (z). The following two formulae (Eqs. (4.8) and (4.9)) can be used to calculate the z-value associated with the results of a specific discrimination test (Stone and Sidel, 1978):

$$z = \frac{[P_{obs} - P_{chance}] - \frac{1}{2N}}{\sqrt{pq/N}}$$

(4.8)

where

$P_{obs} = X/N$

P_{chance} = probability of correct decision by chance

For triangle test: $P_{chance} = 1/3$

Table 4.3   Minimum numbers of correct judgments[a] to establish significance at probability levels of 5 and 1% for paired difference and duo–trio tests (one tailed, $p = 1/2$) and the triangle test (one tailed, $p = 1/3$)

Paired difference and duo–trio tests			Triangle test		
Number of trials (n)	Probability levels		Number of trials (n)	Probability levels	
	0.05	0.01		0.05	0.01
5	5	–	3	3	–
6	6	–	4	4	–
7	7	7	5	4	5
8	7	8	6	5	6
9	8	9	7	5	6
10	9	10	8	6	7
11	9	10	9	6	7
12	10	11	10	7	8
13	10	12	11	7	8
14	11	12	12	8	9
15	12	13	13	8	9
16	12	14	14	9	10
17	13	14	15	9	10
18	13	15	16	9	11
19	14	15	17	10	11
20	15	16	18	10	12
21	15	17	19	11	12
22	16	17	20	11	13
23	16	18	21	12	13
24	17	19	22	12	14
25	18	19	23	12	14
26	18	20	24	13	15
27	19	20	25	13	15
28	19	21	26	14	15
29	20	22	27	14	16
30	20	22	28	15	16
31	21	23	29	15	17
32	22	24	30	15	17
33	22	24	31	16	18
34	23	25	32	16	18
35	23	25	33	17	18
36	24	26	34	17	19
37	24	26	35	17	19
38	25	27	36	18	20
39	26	28	37	18	20
40	26	28	38	19	21
41	27	29	39	19	21
42	27	29	40	19	21
43	28	30	41	20	22
44	28	31	42	20	22
45	29	31	43	20	23
46	30	32	44	21	23
47	30	32	45	21	24
48	31	33	46	22	24
49	31	34	47	22	24
50	32	34	48	22	25
60	37	40	49	23	25
70	43	46	50	23	26

Table 4.3 (continued)

Paired difference and duo–trio tests			Triangle test		
Number of trials (n)	Probability levels		Number of trials (n)	Probability levels	
80	48	51	60	27	30
90	54	57	70	31	34
100	59	63	80	35	38
110	65	68	90	38	42
120	70	74	100	42	45
130	75	79	110	46	49
140	81	85	120	50	53
150	86	90	130	53	57
160	91	96	140	57	61
170	97	101	150	61	65
180	102	107	160	64	68
190	107	112	170	68	72
200	113	117	180	71	76
			190	75	80
			200	79	83

[a]Created in EXCEL 2007 using B. T. Carr's Discrimination Test Analysis Tool EXCEL program (used with permission)

For duo–trio and paired comparison tests:
$P_{chance} = \frac{1}{2}$
X = number of correct judgments
N = total number of judgments.

Alternately one can use the following equation:

$$z = \frac{X - np - 0.5}{\sqrt{npq}} \qquad (4.9)$$

where

X = number of correct responses
n = total number of responses
p = probability of correct decision by chance
For triangle test: $p = 1/3$
For duo–trio and paired comparison tests: $p = \frac{1}{2}$
  and in both cases $q = 1 - p$

As with the χ^2 calculation a continuity correction of –0.5 has to be made. Consult a Z-table (area-under-normal-probability curve) to determine the probability of this choice being made by chance. The critical Z-value for a one-tailed test at alpha (α) of 5% is 1.645. See the Z-table in the Appendix for other values.

4.5 Issues

4.5.1 The Power of the Statistical Test

Statistically, there are two types of errors that the sensory scientist of any sensory method can make when testing the null hypothesis (H_0). These are the Type I (α or alpha) and Type II (β or beta) errors (see Appendix E for a more extensive discussion). A Type I error occurs when the sensory scientist rejects the null hypothesis (H_0) when it is actually true. When making a Type I error in a discrimination test we would conclude that the two products are perceived to be different when they are actually not perceptibly different. The Type I error is controlled by the sensory scientist choice of the size of alpha. Traditionally, alpha is chosen to be very low (0.05, 0.01, or 0.001) which means that there is a 1 in 20, 1 in 100, and 1 in 1,000 chance, respectively, of making a Type I error. A Type II error occurs when the sensory scientist accepts the null hypothesis (H_0) when it is actually false. The Type II error is based on the size of and it is the risk of not finding a difference when one actually exists and it is defined as 1–beta. In other words, the power of a test could be defined as the probability of finding a difference if one actually exists or it is the probability of making the correct decision that the two samples are perceptibly different.

The power of the test is dependent on the magnitude of the difference between the samples, the size of alpha, and the number of judges performing the test.

4.5.1.1 Why Is Power Important When Performing Discrimination Tests?

A candy manufacturer wants to show that the new formulation of their peanut brittle is crunchier than the old formulation. Prior to the study the sensory scientist had decided which probability of making a Type I error (alpha) would be acceptable. If the sensory scientist had decided that alpha should be 0.01, then she/he had a 1 in 100 chance of committing a Type I error. Consider than that this candy maker performs a two-alternative forced choice test and the data indicate that the null hypothesis should be rejected. The sensory scientist is confronted with two possibilities. In the first case, the null hypothesis is actually false and should be rejected; in this case the new formulation is actually crunchier than the old formulation. In the second case, the null hypothesis is actually true and the sensory scientist has made a Type I error. In this type of study the Type I error is usually minimized because the sensory scientist wants to be quite certain that the new formulation is different from the old formulation. In this case the power of the test is only of passing interest.

Consider a second scenario. An ice cream manufacturer wants to substitute the expensive vanilla flavor used in their premium vanilla ice cream with a cheaper vanilla flavor. However, they do not want the consumer to perceive a difference in the product. They perform a triangle test to determine if a panel could tell a difference between the samples. The data indicate that the null hypothesis should not be rejected. Again the sensory scientist is confronted with two possibilities. In the first case, the null hypothesis is true and the two formulations are not perceptibly different. In the second case the samples are perceptibly different but the sensory scientist was making a Type II error. In this type of study the Type II error should be minimized (power of the test should be maximized) so that the sensory scientist can state with some confidence that the samples are not perceptibly different.

In many published discrimination studies the authors claim that a discrimination test indicated that two samples were not significantly different. Frequently, the power of these tests is not reported, but it can often be calculated post hoc. It is unfortunately the case that the power of these tests is often very low, suggesting that the research would not have revealed a difference even if a difference existed. Or in statistical jargon, the probability of a Type II error was high.

4.5.1.2 Power Calculations

Discrimination test power calculations are not simple. However, this does not absolve the sensory scientist from attempting to determine the power associated with a specific study, especially when the objective of the study is to make an equivalent ingredient substitution and therefore the objective of the study is not to reject the null hypothesis. In general, the sensory specialist should consider using a large sample size when power needs to be high ($N = 50$ or greater). This is essential in any situation where there are serious consequences to missing a true difference.

The sensory scientist will frequently make a post hoc power calculation. In this calculation the power of the test is calculated after the completion of the study. The sensory scientist can also make an a priori calculation of the number of judgments needed for a specific power. In both cases the sensory scientist must make a series of assumptions and all power calculations will only be as good as these assumptions. The scientists must be as extremely careful when making the required assumptions for the power calculations. A number of authors (Amerine et al., 1965; Ennis, 1993; Gacula and Singh, 1984; Kraemer and Thiemann, 1987; Macrae, 1995; Schlich, 1993) have studied power calculations for discrimination tests and have prepared tables that can be used to determine either the power of a specified discrimination test or the number of judgments needed for a specific level of power. The different tables (Ennis, 1993; Schlich, 1993) would lead one to slightly different conclusions as to the power of the test. The reason for these differences is that these calculations are based on a series of assumptions and slight differences in assumptions can lead to differences in power calculations. Power calculations will be discussed in more detail in Chapter 5 and Appendix E. Additionally the R-package SensR (http://www.cran.r-project.org/package=sensR/) written by Brockhoff and Christensen (2009) allows one to calculate the power associated with most discrimination tests.

4.5.2 Replications

As seen in the previous section and in the power section of Appendix E the number of judgments made in a discrimination test is very important. The number of judgments can be increased by using more panelists or by having a smaller number of panelists perform more tests. These two methods of increasing the number of judgments are clearly not equivalent. The ideal way to increase the number of judgments is to use more panelists. This is the only way in which the sensory specialist can be assured that all judgments were made independently. All the data analysis methods discussed above assume that all judgments were made entirely independently of one another (Roessler et al., 1978). Frequently and perhaps unfortunately, the industrial sensory scientist has only a limited number of panelists available. In this case the number of judgments may be increased by having each panelist evaluate the samples more than once in a session. In practice this is rather simply done. The panelist receives a set of samples which he/she evaluates. The samples and the score sheet are returned and then the panelist would receive a second set of samples. In some cases the panelist may even receive additional replications. It should be remembered that if the same panelists repeat their judgments on the same samples, there is a possibility these judgments are not totally independent. In other words, the replicate evaluations made by a specific individual may be related to each other. The use of replication increases the power of the difference test by increasing the number of judgments; however, depending on the assumptions relating to effect size (see Appendix E) and the type of difference test used the increase in power may be similar or less than when one uses the same number of independent judgments (Brockhoff, 2003). As can be seen in Table 4.4, extracted from Tables 3 and 4 in Brockhoff (2003) assuming an alpha of 5% and a medium effect size (37.5% above chance discriminating) for a triangle test the power for the independent judgments is more than for the replicated judgments. On the other hand for an alpha of 5% and a small effect size (25% above chance discriminator) for a duo–trio test the values are quite similar. Therefore replication of discrimination tests and the effect of this on power are not simple.

4.5.2.1 Analyzing Replicate Discrimination Tests

The important caution is that sensory scientists should not simply combine the data from replications and count the number of observations as the number of replicates multiplied by the number of panelists. They are not independent judgments and the situation needs to be examined closely before any such combination can be justified. There are a few simpler options that are available and a more complex option, the beta-binomial model.

Table 4.4 Limits of power (%) based on Monte Carlo simulations (extracted from Tables 3 and 4 of Brockhoff, 2003)

n[a]	k[b]$=1$	k[b]$=2$	k[b]$=3$	k[b]$=4$	k[b]$=5$
(a) Triangle test with alpha=5% and medium-effect size (37.5% above chance discriminator)					
12	40	70	81	90	91
24	74				
36	88				
48	97				
(b) Duo–trio test with alpha=5% and small-effect size (25% above chance discriminator)					
12	11	28	39	46	58
24	27				
36	37				
48	44				

[a]Number of panelists; [b]Number of replications

Simpler Options

First, the replications can be analyzed separately as independent tests. This can provide information as to whether there is a practice, warm-up, or learning effect if the proportion correct increases over trials. This could be useful information because a consumer in the real world will have multiple opportunities to interact with a product, and usually not just a single tasting. If later replications are statistically significant (and the first is not), that is usually grounds for concluding that the samples are in fact perceivably different. It is also possible, of course, that fatigue or adaptation or carry-over could have an effect on later replications, so the sensory scientist needs to consider the nature of each specific product and situation and make a reasoned judgment. If the replications lead to different results, further investigation or analysis may be necessary.

A second approach for duplicate tests is to simply tabulate the proportion of panelists that got both tests correct. Now the chance probability for a three sample test is 1/9 and for a test like the duo–trio or paired comparison, it becomes $1/4$. The same z-score formula applies for the binomial test on proportions (Eq. (4.10)) as

$$z = \frac{(P_{obs} - p) - 1/2n}{\sqrt{pq/n}} = \frac{(X - np) - 1/2}{\sqrt{npq}} \quad (4.10)$$

where P_{obs} is the proportion correct ($=X/n$), X is the actual number correct, n is the number of judges, p is the chance probability, and $q = 1-p$.

Solving for X as a function of n, and using $p = 1/9$ for the triangle or 3-AFC tests, we get the following for $Z = 1.645$ ($p < 0.05$, one tailed):

$$X = n/9 + 0.517\sqrt{n} + 0.5 \quad (4.11)$$

and for $p = 1/4$ for the duplicated paired or duo–trio tests:

$$X = n/4 + 0.712\sqrt{n} + 0.5 \quad (4.12)$$

These can easily be programmed on a spreadsheet, but do not forget to change the value of z if you wish to calculate the critical value of X for other probability levels. Of course, you must round up the value of X to the next highest integer since you are counting individuals. This approach is somewhat conservative, as it only considers those people who answered correctly on both tests, and it is possible that a person might miss the first test but get the replicate correct as a true discrimination. The solution to this issue (considering that some people might have partially correct discriminations) is to use a chi-square test to compare the observed frequencies against what one would expect by chance for zero, one or two correct judgments (e.g., 4/9, 4/9, and 1/9 for the replicated triangle or 3/8, 3/8, and $1/4$ for the replicated duo–trio).

4.5.2.2 Are Replications Statistically Independent?

Another approach to replication is to test whether the two replicates are in some way independent, seem to be varying randomly or whether there are systematic patterns in the data. One such approach is the test of Smith (1981), which can be used for two replications (i.e., a duplicate test). This test essentially determines whether there are significantly more correct choices on one replication, or whether they are not significantly different. It uses a binomial test on proportions with $p = 1/2$, so the binomial tables for the duo–trio test are applicable or for triangles it uses $p=1/3$, and thus the binomial tables for the triangle test would be applicable.

The total number of correct responses in each replication ($C1$, $C2$) are added together ($M = C1 + C2$). M represents the total number of trials (n) and either $C1$ or $C2$ (whichever is larger) is used to represent the number of correct responses in the study. If $C1$ or $C2$ is larger than the minimum number of judgments required for significance then the difference in proportions of correct responses between the two replications is significant and the replication data cannot be pooled. Each replication must then be analyzed independently. If the larger of $C1$ and $C2$ is less than the minimum number of judgments required for significance then the difference in proportions of correct responses between the two replications is not significant and the replication data can be pooled. The combined data can then be analyzed as if each judgment was made by a different panelist.

For example, a sensory specialist was asked to determine if two chocolate chip cookie formulations (one with sucrose as the sweetener and the other with a non-caloric sweetener) were perceptibly different from each other. The sensory specialist decided to use a constant reference duo–trio test to determine if the two formulations differed. A panel of 35 judges did the study in duplicate. In the first replication 28 judges correctly matched the reference and in the second replication 20 judges correctly matched the reference. The sensory scientist has to determine if the data can be pooled and if there is a significant perceived difference between the two cookie formulations.

Using Smith's test he found that $M = 28 + 20 = 48$, and that $C1 = 28$ is the larger of $C1$ and $C2$. Table 4.1 for duo–trio tests indicated that for $n = 48$ the minimum number of correct judgments for an alpha (α) of 5% is 31. Thus, 28 is less than 31 and the data from the two replications were pooled. The combined data were therefore 48 correct decisions out of a potential 70 (2×35). The sensory scientist decided to use the z-calculation to determine the exact probability of finding this result. Using Eq. (4.2), with 48 as the number of correct responses, with 70 as the total number of responses, with p equal to 1/2, $z = 2.9881$. The z-table showed that the exact probability of this value occurring by chance was 0.0028. The panelists could therefore perceive a difference between the two formulations.

The Beta-Binomial Model

The test devised by Smith does not address the issue of whether some panelists have different patterns across replicates than others. If people had systematic trends (e.g., some people getting both correct, easily discriminating and others missing consistently) you could still get a non-significant result by Smith's test, yet the data would hardly be independent from trial to trial. This issue is addressed in the beta-binomial model, which looks at patterns of consistency (versus. randomness) among the panelists. Although Smith's test is appropriate when detecting differences between replicates, it is not an airtight proof that replications are independent should the test not meet the significance level. The beta-binomial model allows us to pool replicates, but makes some adjustments in the binomial calculations to make the criteria more conservative when the data are not fully independent.

The beta-binomial model assumes that the performance of panelists is distributed like a beta distribution. This distribution has two parameters, but they can be summarized in a simple statistic called gamma.

Gamma, which varies from zero to one, is a measure of the degree to which there are systematic behaviors of panelists (like always answering correctly or incorrectly), in which case gamma approaches one, or whether people seem to behave independently from trial to trial (gamma approaches zero). You can think of this as a kind of test of the independence of the observations, but from the perspective of individual performance, rather than group data as in Smith's test. Gamma is basically an estimate of the degree to which people's total number of correct choices varies from the panel mean. It is given by the following formula (Bi, 2006, p. 110):

$$\gamma = \frac{1}{r-1} \left[\frac{rS}{\mu(1-\mu)n} - 1 \right] \qquad (4.13)$$

where

$r =$ the number of replicates
$S =$ a measure of dispersion
$\mu =$ mean proportion correct for the group (looking at each person's individual proportions as shown below)
$n =$ the number of judges. S and μ are defined as follows:

$$\mu = \frac{\sum_{i=1}^{n} x_i / r}{n} \qquad (4.14)$$

where $x_i =$ the number of correct judgments summed across replicates for that panelist.

So μ is the mean of the number of correct replicates. S is defined as

$$S = \sum_{i=1}^{n} ((x_i / r) - \mu)^2 \qquad (4.15)$$

Once gamma is found we have two choices. We can test to see whether the beta-binomial fits better than the simple binomial. This is essentially testing whether gamma is different from zero. The alternative is to go directly to beta-binomial tables for different levels of gamma. In these tables (see Table O), we combine replicates to get the total number of correct judgments, and compare that to the critical number required, for the total number of judgments (number of panelists times number of replicates). The tables adjust the binomial model requirements to be more conservative as gamma increases (i.e., as the panelists look less random and more systematic).

To test whether the beta-binomial is a better fit, we use the following Z-test (Bi, 2006, p. 114):

$$Z = \frac{E - nr}{\sqrt{2nr(r - 1)}} \qquad (4.16)$$

where E is another measure of dispersion defined as

$$E = \sum_{i=1}^{n} \frac{(x_1 - rm)^2}{m(1 - m)} \qquad (4.17)$$

and m is the mean proportion correct defined as

$$m = \sum_{i=1}^{n} x_i / nr \qquad (4.18)$$

The advantage of doing the Z-test is that should you find a significant Z, then there is evidence that the panelists are not random, but that there are probably groups of consistent discriminators and also perhaps some people who are consistently not discriminating. In other words, a non-zero gamma is evidence for a consistently perceived difference for at least some of the panel! If the Z-test is non-significant, then one option is to pool replicates and just use the simple binomial table. Note that now you have effectively increased the sample size and given more power to the test. A good example of this approach can be found in Liggett and Delwiche (2005).

4.5.3 Warm-Up Effects

Much has been written concerning the potential advantages of giving warm-up samples before discrimination tests (e.g., Barbary et al., 1993; Mata-Garcia et al., 2007). A warm-up strategy involves repeated alternate tasting of the two different versions of the product, with the panelist's knowledge that they are different. Often (but not always) they are encouraged to try to figure out the way or ways in which the products differ. This is similar to giving a familiarization sample or dummy sample before the actual test, but with "warm up" it involves a more extended period of tasting. A single warm-up sample was in fact part of the original version of the duo–trio test (Peryam and Swartz, 1950).

However, evidence for the advantage of this added procedure is not very strong. In two early reports, a larger number of significant triangle tests were seen with warm-up for a wine sample and a fruit beverage (O'Mahony and Goldstein, 1986) and for NaCl solutions and for orange juices (O'Mahony et al., 1988). In the latter study it was unclear whether naming the difference gave any additional advantage. Later studies showed mixed results.

Thieme and O'Mahony (1990) found good discrimination for A, not-A, and paired comparison tests with warm-up but a direct comparison of the same kind of test, with and without warm-up, was lacking, so it is difficult to draw conclusions from that study. Dacremont et al. (2000) showed no effect of warm-up for the first trial of repeated triangles with naïve assessors nor with highly experienced judges. Panelists who were of intermediate experience did show some benefit of the warm-up. Kim et al. (2006) reported increased discrimination for the triangle, duo–trio, and same–different tests, but a 2-AFC test in which panelists were told to identify the NaCl sample (versus water) was also conducted before the warmed-up tests, so the cause of the increased discrimination in this study is not clear. Angulo et al. (2007) reported a small but nonsignificant increase in discriminability with a relatively less sensitive group in a 2-AFC test. Rousseau et al. (1999) looked at effects of a primer (single example) food and a familiarization with mustard samples before discrimination tests. The primer had no effect and the familiarization appeared to cause small increases in discriminability.

Taken together, these studies suggest that a warm-up protocol might have some benefits. The sensory practitioner should weigh the possible benefits against the extra burden on the panelist if the kind of extensive warm-up (three to ten pairs) that are usually done in the laboratory studies is adopted.

4.5.4 Common Mistakes Made in the Interpretation of Discrimination Tests

If a discrimination test had been performed properly, with adequate power and the sensory scientist finds that the two samples were not perceptibly different then there is no point in performing a subsequent consumer preference test with these samples. Logically, if two samples are sensorially perceived to be the same then one sample cannot be preferred over another. However, if a subsequent consumer preference test indicates that the one of the samples is preferred over the other, then the scientist must carefully review the adequacy of the discrimination test, especially the power of the test. Of course, any test is a sampling experiment, so there is always some probability of a wrong decision. Therefore a discrimination test does not "prove" that there is no perceivable difference and follow-up preference tests will sometimes be significant.

Sometimes, novice sensory scientists will do a preference tests and find that there was no significant preference for one sample over the other. This means that the two samples are liked or disliked equally. However, it does not mean that the samples are not different from one another. It is very possible for two samples to be perceptibly very different and yet to be preferred equally. For example, American consumers may prefer apples and bananas equally, yet that does not mean that apples are indistinguishable from bananas.

Appendix: A Simple Approach to Handling the A, Not-A, and Same/Different Tests

Both of these tests have response choices, rather than a sample choice. The choice of either response is affected by the criterion used by each panelist. For example, as a panelist one may ask oneself: Do I want to be really strict and be sure these products are different, or can I call them different if I think there is just an inkling of a difference? These criteria are clearly quite different from one another and will dramatically affect the outcome of the test. However, the sensory scientist does not know (not can he find out) which criterion each panelist used (sometimes even the panelists do not know since they do not explicitly decided on a criterion).

In order to get around this problem, we can give a control sample of true A in the A, not-A test or an identical pair ("same") in the same/different test. The question then becomes whether the percent of choice of "not-A" for the test sample was greater than the

choice of "not-A" for the control (i.e., true A) sample. Similarly, we can ask if the proportion of "different" responses was higher for the test pair than it was for the identical control pair. So we are comparing against a sensible baseline.

So far so good. A simple binomial test on proportions or a simple chi-square would seem to do it. But in most situations, we give both the true A and the test sample to the same person. In the same/different test, we would have given both a control pair (identical samples) and the test pair (samples that are physically different and might in fact be called "different"). The binomial test and the chi-square assume independent observations, but now we have two measurements on the same person (clearly NOT independent). So the appropriate statistic is provided by the McNemar test. Let us look at the A, not-A situation. We cast the data in a two-way table as follows, with everyone counted into one of the four cells (1, 2, 3, and 4 are the actual frequency counts, not percents):

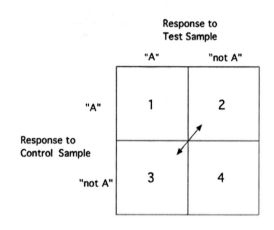

Now, the people who are giving the same answer on both trials are not very interesting. They are not transmitting any information as to whether the products are different. They are counted in cells #1 and #4 above. The critical question is whether there are significantly more people in cell #2 (who call the test sample "not A" and the control sample "A") than there are people who do the reverse in cell #3. If the cells have about the same counts, then let us face it, there is not much evidence for a difference. But if a lot of people call the test sample "not-A" and they recognize the control sample as a good example of what A should be like, then we have evidence that something important has in fact changed. The difference is perceivable.

So we need to compare the size of cell 2 to cell 3. The McNemar test does just this for us. Let C_2 be the count in cell #2, and C_3 be the count in cell #3. Here is the formula:

$$\chi^2 \frac{(|C_2 - C_3| - 1)^2}{C_2 + C_3}$$

This χ^2 test has one degree of freedom, so the critical chi-square value that must be exceeded for significance is 3.84, if we use the standard alpha at 5%.

A worked example:

During an A-not-A test a group of 50 panelists were received (in randomized order: a control sample (the true A) and a test sample). The results are displayed in the figure below:

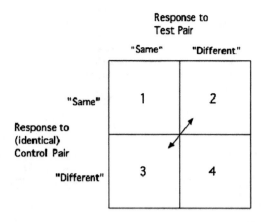

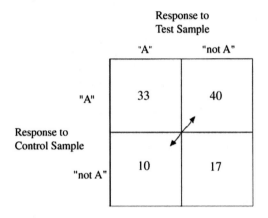

$$\chi^2 \frac{(|40 - 10| - 1)^2}{40 + 10} = 16.82 \text{ which is larger than } 3.84.$$

The panelist therefore found a significant difference between the control and the test samples.

The same kind of chart can be drawn for the same–different test and the same comparison can be made:

If there are just a few more people in cell 3 than cell 2, it is probably random variation and there is no difference. If there is a LOT more people in cell 3 and you get a significant chi-square but in the "wrong direction," there is something wrong with your study (maybe you switched the codes, for example). Also, if there are a lot of people in cells 1 and 4, that is a concern because those folks are not distinguishing much, or maybe they have some "lazy" or lax criteria.

References

Amerine, M. A., Pangborn, R. M. and Roessler, E. B. 1965. Principles of sensory evaluation. Academic, New York, NY.

Amerine, M. A. and Roessler, E. B. 1983. Wines, their sensory evaluation, Second Edition. W.H. Freeman, San Francisco, CA.

Amoore, J. E., Venstrom, D. and Davis, A. R. 1968. Measurement of specific anosmia. Perceptual Motor Skills, 26, 143–164.

Amoore, J. 1979. Directions for preparing aqueous solutions of primary odorants to diagnose eight types of specific anosmias. Chemical Senses and Flavour, 4, 153–161.

Angulo, O., Lee, H.-S. and O'Mahony, M. 2007. Sensory difference tests, over dispersion and warm-up. Food Quality and Preference, 18, 190–195.

Barbary, O., Nonaka, R., Delwiche, J., Chan, J. and O'Mahony, M. 1993. Focused difference testing for the assessment of differences between orange juice made from orange concentrate. Journal of Sensory Studies, 8, 43–67.

Basker, D. 1980. Polygonal and polyhedral taste testing. Journal of Food Quality, 3, 1–10.

Bi, J. and Ennis, D. M. 2001a. Statistical methods for the A-Not A method. Journal of Sensory Studies, 16, 215–237.

Bi, J. and Ennis, D. M. 2001b. The power of the A-Not A method. Journal of Sensory Studies, 16, 343–359.

Bi, J. 2006. Sensory Discrimination Tests and Measurements: Statistical Principles, Procedures and Tables. Blackwell Publishing Professional, Ames, IA.

Brockhoff, P. B. 2003. The statistical power in difference tests. Food Quality and Preference, 14, 405–417.

Brockhoff, P. B. and Christensen, R. H.B. 2009. Thurstonian models for sensory discrimination tests as generalized linear models. Journal of Food Quality and Preference, doi:10.1016/j.foodqual.2009.04.003.

Dacremont, C., Sauvageot, F. and Ha Duyen, T. 2000. Effect of assessors expertise level on efficiency of warm-up for triangle tests. Journal of Sensory Studies, 15, 151–162.

Ennis, D. M. and Mullen, K. 1986. Theoretical aspects of sensory discrimination. Chemical Senses, 11, 513–522.

Ennis, D. M. 1993. The power of sensory discrimination methods. Journal of Sensory Studies, 8, 353–370.

Frijters, J. E. R. 1979. Variations of the triangular method and the relationship of its unidimensional probabilistic model to three-alternative forced choice signal detection theories. British Journal of Mathematical and Statistical Psychology, 32, 229–241.

Frijters, J. E. R. 1984. Sensory difference testing and the measurement of sensory discriminability. In: J. R. Piggott (ed.), Sensory Analysis of Food. Elsevier Applied Science Publications, London, pp.117–140.

Frijters, J. E. R., Kooistra, A. and Vereijken, P. F.G. 1980. Tables of d' for the triangular method and the 3-AFC signal detection procedure. Perception and Psychophysics, 27, 176–178.

Gacula, M. C. and Singh, J. 1984. Statistical methods in food and consumer research. Academic, Orlando, FL.

Harris, H. and Kalmus, H. 1949. The measurement of taste sensitivity to phenylthiourea. Annals of Eugenics, 15, 24–31.

Huang, Y.-T., and Lawless, H. T. 1998. Sensitivity of the ABX discrimination test. Journal of Sensory Studies, 13, 229–239; 8, 229–239.

Kim, H.-J., Jeon, S. Y., Kim, K.-O. and O'Mahony, M. 2006. Thurstonian models and variance I: Experimental confirmation of cognitive strategies for difference tests and effects of perceptual variance. Journal of Sensory Studies, 21, 465–484.

Kraemer, H. C. and Thiemann, S. 1987. How many subjects: Statistical power analysis in research. Sage, Newbury Park, CA.

Lawless, H. T. 1980. A comparison of different methods used to assess sensitivity to the taste of phenylthiocarbamide (PTC).Chemical Senses, 5, 247–256.

Lee, H-S., and O'Mahony, M. 2007. The evolution of a model: A review of Thurstonian and conditional stimulus effects on difference testing. Food Quality and Preference, 18, 369–383.

Liggett, R. A. and Delwiche, J. F. 2005. The beta-binomial model: Variability in over- dispersion across methods and over time. Journal of Sensory Studies, 20, 48–61.

Macmillan, N. A., Kaplan, H. L. and Creelman, C. D. 1977. The psychophysics of categorical perception. Psychological Review, 452–471.

Macmillan, N. A. and Creelman, C. D. 1991. Detection Theory: A User's Guide. University Press, Cambridge, UK.

Macrae, A. W. 1995. Confidence intervals for the triangle test can give reassurance that products are similar. Food Quality and Preference, 6, 61–67.

Mata-Garcia, M., Angulo, O. and O'Mahony, M. 2007. On warm-up. Journal of Sensory Studies, 22, 187–193.

Meilgaard, M., Civille, C. V., and Carr, B. T. 2006. Sensory Evaluation Techniques, Fourth Edition. CRC, Boca Raton, FL.

Mitchell, J. W. 1956. The effect of assignment of testing materials to the paired and odd position in the duo-trio taste difference test. Journal of Food Technology, 10, 169–171.

Nicolle, C. 1932. Biologie de l"Invention Alcan Paris, quoted in Beveridge, W. I.B. 1957. The Art of Scientific Investigation, Third Edition. Vintage Books, New York. p. 37.

O'Mahony, M. and Goldstein, L. R. 1986. Effectives of sensory difference tests: Sequential sensitivity analysis for liquid food stimuli. Journal of Food Science, 51, 1550–1553.

O'Mahony, M., Wong, S. Y. and Odbert, N. 1986. Sensory difference tests: Some rethinking concerning the general rule that more sensitive tests use fewer stimuli. Lebensmittel Wissenschaft und Technologie, 19, 93–95.

O'Mahony, M., Thieme, U. and Goldstein, L. R. 1988. The warm-up effect as a means of increasing the discriminability of sensory difference tests. Journal of Food Science, 53, 1848–1850.

Pangborn, R. M. and Dunkley, W. L. 1966. Sensory discrimination of milk salts, nondialyzable constituents and algin gum in milk. Journal of Dairy Science, 49, 1–6.

Peryam, D. R. 1958. Sensory difference tests. Journal of Food Technology, 12, 231–236.

Peryam, D. R. and Swartz, V. W. 1950. Measurement of sensory differences. Food Technology, 4, 390–395.

Pierce, J. R. and Gilbert, E. N. 1958. On AX and ABX limens. Journal of the Acoustical Society of America, 30, 593–595.

Roessler, E. B., Pangborn, R. M., Sidel. J. L. and Stone, H. 1978. Expanded statistical tables for estimating significance in paired-preference, paired difference, duo-trio and triangle tests. Journal of Food Science, 43, 940–941.

Rousseau, B., Rogeaux, M. and O'Mahony, M. 1999. Mustard discrimination by same-different and triangle tests: aspects of irritation, memory and tau criteria. Food Quality and Preference, 10, 173–184.

Schlich, P. 1993. Risk tables for discrimination tests. Journal of Food Quality and Preference, 4, 141–151.

Smith, G. L. 1981. Statistical properties of simple sensory difference tests: Confidence limits and significance tests. Journal of the Science of Food and Agriculture, 32, 513–520.

Stone, H. and Sidel, J. L. 1978. Computing exact probabilities in sensory discrimination tests. Journal of Food Science, 43, 1028–1029.

Stone, H. and Sidel, J. L. 2004. Sensory Evaluation Practices, Third Edition. Academic, Elsevier, New York.

Thieme, U. and O'Mahony, M. 1990. Modifications to sensory difference test protocols: The warmed-up paired comparison, the single standard duo-trio and the A, not-A test modified for response bias. Journal of Sensory Studies, 5, 159–176.

Whiting, R., Murray, S., Ciantic, Z. and Ellison, K. 2004. The use of sensory difference tests to investigate perceptible colour-difference in a cosmetic product. Color Research and Application, 29, 299–304.

Chapter 5

Similarity, Equivalence Testing, and Discrimination Theory

Abstract   This chapter discusses equivalence testing and how difference tests are modified in their analyses to guard against Type II error (missing a true difference). Concepts of test power and required sample sizes are discussed and illustrated. An alternative approach to equivalence, namely interval testing is introduced along with the concept of paired one-sided tests. Two theoretical approaches to the measurement of the size of a difference are introduced: discriminator theory (also called guessing models) and the signal detection or Thurstonian models.

Difference testing method constitute a major foundation for sensory evaluation and consumer testing. These methods attempt to answer fundamental questions about stimulus and product similarity before descriptive or hedonic evaluations are even relevant. In many applications involving product or process changes, difference testing is the most appropriate mechanism for answering questions concerning product substitutability.

—D. M. Ennis (1993)

Contents

5.1 Introduction

Discrimination, or the ability to differentiate two stimuli, is one of the fundamental processes underlying other sensory-based responses. As suggested by Ennis above, if two items cannot be discriminated, there is no basis for description of differences between them, nor for consumer preference. The previous chapter discussed simple discrimination tests that are used to gather evidence that a product has changed from some previous version. We might make an ingredient change, a cost reduction, a processing or packaging change, do a shelf life test against a fresh control, or a quality control test against some standard product. Questions arise as to whether the difference in

H.T. Lawless, H. Heymann, *Sensory Evaluation of Food*, Food Science Text Series,
DOI 10.1007/978-1-4419-6488-5_5, © Springer Science+Business Media, LLC 2010

the products is perceivable. Discrimination or simple difference tests are appropriate for these questions. When we find evidence of a difference, the methods are straightforward and the interpretation is usually clear-cut as well. However, a great deal of sensory testing is done in situations where the critical finding is one of equivalence or similarity. That is, a no-difference result has important implications for producing, shipping, and selling our product. Shelf life and quality control tests are two examples. Cost reductions and ingredient substitutions are others.

This is a much trickier situation. It is often said that "science cannot prove a negative" and the statistical version of this is that you cannot really prove that the null hypothesis is correct. But in a way, this is exactly what we are trying to do when we amass evidence that two products are equivalent or sufficiently similar that we can substitute one for the other without any negative consequences.

The issue is not so easy as just finding "no significant difference." A failure to reject the null is always ambiguous. Just because two products are "not significantly different" does not necessarily mean that they are equivalent, sensorially. There are a many reasons why we may have failed to find a statistically significant difference. We may not have tested enough people relative to the amount of error variability in our measurements. The error variability may be high for any number of reasons, such as lack of sample control, poor testing environment, unqualified judges, poor instructions, and/or a bad test methodology. It is easy to do a sloppy experiment and get a non-significant result.

Students and sensory scientists should recall that there are two kinds of statistical errors that can be made in any test. These are shown in Fig. 5.1. We can reject the null and conclude that products are different when they are not. This is our familiar Type I error that we try to keep to a long-term minimum called our alpha level. This is commonly set at 5%, and why we try to use probability levels of 0.05 as our cutoffs in statistical significance testing. This kind of error is dangerous in normal experimental science and so it is the first kind of error that we worry about. If a graduate student is studying a particular effect, but that effect was a spurious false-positive result from some earlier test, then he or she is wasting time and resources. If a product developer is trying to make an improved product, but that hunch is based on an earlier false result, the effort is doomed. Some people refer to Type I error as a "false alarm." The second kind of error occurs when we miss a difference that is really there. This is a Type II error, when we do not reject the null, but the alternative hypothesis is really the true state of the situation.

		Test Result: "Difference"	Test Result: "No difference"
Actual Situation	Difference exists	(correct conclusion)	"Miss" Type II error beta-risk
	No difference	False rejection Type I error alpha-risk	(correct conclusion)

Fig. 5.1 The statistical decision matrix showing the two types of error, Type I, when the null is rejected but there is really no difference, and Type II, when there is a difference but none is detected by the test (null false but accepted). The long-term risk of Type I under a true null is the alpha risk. Beta risk is managed by the choices made in the experiment of *N*, alpha, and the size of the difference one is trying to detect.

Type II error has important business ramifications, including lost opportunities and franchise risk. We can miss an opportunity to improve our product if we do not find a significant effect of some ingredient or processing change. We can risk losing market share or "franchise risk" if we make a poorer product and put it into the marketplace because we did not detect a negative change. So this kind of error is of critical importance to sensory workers and managers in the foods and consumer products industries.

The first sections of this chapter will deal with ways to gather evidence that two products are similar or equivalent from a sensory perspective. The first section will illustrate some commonsense approaches and the question of test power. Then some formal tests for similarity or equivalence will be considered, after we look at a model for estimating the size of a difference based on the proportion of people discriminating.

After discussing basic approaches to similarity and equivalence, this chapter will examine more sophisticated models for measuring sensory differences from discrimination results. Sensory professionals need to do more than simply "turn the crank" on routine tests and produce binary yes/no decisions about statistical significance. They are also required to understand the relative sensitivity of different test methods, the decision processes and foibles of sensory panelists, and the potential pitfalls of superficial decisions. For these reasons, we have included in this chapter sections on the theory of signal detection and its related model, Thurstonian scaling. Many questions can arise from apparently simple discrimination tests in applied research, as the following examples show:

(1) Clients may ask, "OK, you found a difference, but was it a big difference?"
(2) When is it acceptable to conclude that two products are sensorially equivalent when the test says simply that I did not get enough correct answers to "reject the null?"
(3) What can I do to insure that the test is as sensitive as possible and does not miss an important difference (i.e., protect against Type II error)?
(4) Why are some test methods more stringent or more difficult than others? Can this be measured?
(5) What are the behaviors and decision processes that influence sensory-based responses?

Each of these questions raise difficult issues without simple answers. This chapter is designed to provide the sensory professional with approaches to these questions and a better understanding of methods, panelists, and some enhanced interpretations of discrimination results. For further detail, the books by Bi (2006a) on discrimination testing, by Welleck (2003) on equivalence testing, and the general sensory statistics book by Gacula et al. (2009) are recommended.

5.2 Common Sense Approaches to Equivalence

Historically, many decisions about product equivalence were based on a finding of no significant difference in a simple discrimination test. This is a risky enterprise. If the difference is subtle, it is easy to miss it. The test might have included too few panelists for the size of the effect. We may have let unexpected sources of variability creep into the test situation, ones that could easily overwhelm the difference. We may have used unqualified panelists because too many of our regular testers were on vacation or out sick that week. Perhaps our sample-handling equipment like

heat lamps were not working that day. Any number of reasons could contribute to a sloppy test that would create situation in which a difference could be missed. So why was this approach so prevalent during the early history of sensory testing?

There are some common sense situations in which it may make sense to consider a non-significant result as important. The first requirement is that the test instrument must be proven to detect differences in previous tests. By "test instrument" we mean the entire scenario – a particular method, a known population of panelists, specific test conditions, these type of products, etc. In a company with an ongoing test program this kind of repeated scenario may provide a reasonable insurance policy that when a significant difference is not found, it was in fact not due to a faulty instrument, as the instrument has a track record. For example, a major coffee company may have ongoing tests to insure that the blend and roasting conditions produce a reliable, reproducible flavor that the loyal customers will recognize as typical and accept. Other controls may be introduced to further demonstrate the effectiveness of the test method. Known defective or different samples may be given in calibration tests to demonstrate that the method can pick up differences when they are in fact expected. Conversely, known equivalents or near duplicates may be given to illustrate that the method will not result in an unacceptable rate of false alarms. Finally, the panel may consist of known discriminators who have been screened to be able to find differences and who have a track record of detecting differences that are confirmed by other tests on the same samples, such as consumer tests or descriptive analysis panels.

These kinds of controls are illustrated in a paper on cross-adaptation of sweet taste materials (Lawless and Stevens, 1983). In cross-adaptation studies, exposure to one taste substance may have an adapting effect, i.e., cause a decrease in intensity of a second substance. To claim that there is full cross-adaptation, the decrease must be about the same as with self-exposure, i.e., the decrement should be the same as when the substance follows itself. To claim no cross-adaptation, the test substance must have an intensity equivalent to its taste after plain water adaptation. Both of these are essentially equivalence tests. In order to accept these results it is necessary to prove that the second or test item is capable of being adapted (for example, it can adapt itself) and that the first-presented substance in fact can cause an adaptation-related decrease (i.e., it also has an effect on itself). Given these two control situations, the claim of cross-adaptation (or lack thereof) as an equivalence statement becomes trustworthy.

Simple logical controls can be persuasive in an industrial situation in which there is an ongoing testing program. Such an ongoing testing program may be in place for quality control, shelf-life testing, or situations in which a supplier change or ingredient substitution is common due to variable sources of raw materials. This kind of logic is most applicable to situations in which the conditions of testing do not vary. If there is a sudden decrease in the panel size during a vacation period, for example, it becomes more difficult to claim that "we have a good instrument" and therefore a non-significant difference is trustworthy. All the testing conditions, including the panel size, must remain fairly constant to make such a claim.

5.3 Estimation of Sample Size and Test Power

A more statistical approach to equivalence is to manage the sample size and the test power to minimize the probability of a Type II error, i.e., the chance of missing a true difference. There are commonly accepted formulas for calculating the required sample size for a certain test power. At first glance, this seems rather straightforward. However, there is one important part of the logic that managers (and often students) find troubling. One must specify not only the acceptable amount of alpha- and beta-risks in these calculations, but also *the size of the difference one is trying to detect.* Conversely, how much of a difference would one allow, and still call the two products "equivalent" on a sensory basis? Managers, when faced with this question, may reply that they want no difference at all, but this is unrealistic and not possible within the constraints of statistical testing. Some degree of difference, no matter how minor, must be specified as a lower tolerable limit.

The common equation for calculating the necessary sample size is given as follows (from Amerine et al., 1965):

$$N = \left[\frac{Z_\alpha \sqrt{p_0 q_0} + Z_\beta \sqrt{p_a q_a}}{p_0 - p_a} \right]^2 \quad (5.1)$$

where Z_α and Z_β are the Z-scores associated with the chosen levels of alpha- and beta-risk, p_0 is the chance probability in the test and p_a is the probability chosen for the alternative hypothesis (as always, $q = 1-p$). This is the equation for determining the required panel size, N, as a function of the alpha-risk, beta-risk, the chance probability level, and the effect size one does not want to miss. A similar equation (see Appendix at the end of this chapter) is used for scaled data, in which the degree of difference can be specified as a difference on a scale or number of standard deviations.

The effect size or size of the allowable difference is given in the denominator. It is this quantity that management must choose in order to determine what is sufficiently "equivalent." Strategically, management may not want to go out on a limb and may delegate this choice to the statisticians or sensory personnel involved in the program, so the sensory professional must be prepared to make recommendations based on prior experience with the product. Knowledge of the degree of consumer tolerance for differences is key in making any such recommendation. In a vacuum of consumer information, this choice can be exceedingly difficult.

In this case the size of the difference is given by stating some percentage of correct choices that is higher than the chance level, noted as p_a in Eq. (5.1). You can think of this as a percent correct associated with an alternative hypothesis, or simply as a percent correct that one would not want to miss detecting. Above this level, there are too many people detecting the difference for our management level of comfort with the product change. In the next section we will introduce a useful way to think about this level, in terms of the proportion of people detecting the difference. This proportion is different than the actual observed percent correct, because we have to apply a correction for chance, i.e., the possibility that some people get the correct answer just by guessing correctly. More on this below.

For now, let us examine two worked examples for a triangle test. In the first example, we will set alpha and beta at 5%, so our one-tailed z-values will both be 1.645. Let us allow a fairly liberal alternative hypothesis percentage of 2/3 correct or 66.7%. This might be considered a fairly large difference, as the proportion of people truly detecting, after the correction for chance, is 50%. In other words, we might expect half the population to detect this change. On the other hand,

50% detection is considered one definition of a threshold (see Chapter 6), so from that perspective this might be an acceptable difference.

Working through the math, we get the following equation and result:

$$\left[\frac{1.645\sqrt{(0.33)(0.67)} + 1.645\sqrt{(0.67)(0.33)}}{0.33 - 0.67} \right]^2 = 21.6$$

So for this kind of test, looking for what some managers might call a "gross difference" we need about 22 panelists. Now let us see what happens when we make the difference smaller. In this example we can only allow a correct performance level of 50% (which corresponds to 25% true detection of the difference after correction for chance or one person in four actually seeing the difference). The new equation is

$$\left[\frac{1.645\sqrt{(0.33)(0.67)} + 1.645\sqrt{(0.50)(0.50)}}{0.33 - 0.50} \right]^2 = 91.9$$

So now that we have lowered the size of the allowable difference a little, the required panel size has expanded to 92 judges. This would be a fairly large triangle test panel by most common industrial standards. Unfortunately if you are trying to get evidence for sensory equivalence and you can permit only a small difference, you are going to need a lot of panelists! There is just no way around it, unless one goes to replicated measures and a beta-binomial approach as discussed in the previous chapter. The power of difference tests with small panels can be alarmingly low, as discussed in Appendix E. Further calculations and tables for triangle and duo–trio tests are found there as well. The important factor to note in our two examples is that it is the size of the difference as specified in the denominator of Eq. (5.1) that has the biggest influence on the panel size requirements. In the next section, we will examine a simple traditional way of choosing our acceptable size of a difference based on the estimated proportion of panelists discriminating.

5.4 How Big of a Difference Is Important? Discriminator Theory

How can we calculate some true measure of discrimination after adjustment for the chance level of performance? That is, a certain percent correct is expected by guessing alone in the face of no discrimination at all. An old historical approach to this was to provide a correction for the guessing level, i.e., the level of performance expected in the face of no discrimination whatsoever. The formula for the corrected proportion is known as Abbot's formula (Finney, 1971), and is given by

$$UpperC.I.95\% =$$
$$\left[1.5(X/N) - 0.5\right] + 1.645(1.5)\sqrt{\frac{(X/N)(1-X/N)}{N}} \quad (5.2)$$

where $P_{observed}$ is the observed proportion correct and P_{chance} is the proportion expected by chance guessing.

This formula has been widely applied since the 1920's in fields such as pharmacology, toxicology, and even educational testing. In pharmacology, it is used to adjust for the size of the placebo effect, i.e., those test subjects that improve without the drug. In toxicology it is used to adjust for the baseline fatality rate in the control group (i.e., those not exposed to the toxin but who die anyway). This formula has also been employed in educational testing where multiple-choice tests are common, but adjustment for guessing is desired. Another version of this formula appears in publications discussing the issue of estimating true discriminators in the sample and separating them from an estimated proportion of people who are merely guessing correctly (e.g., Morrison, 1978). The formula will also be used in the next chapter when forced-choice methods are used in threshold determinations. Chance-adjusted discrimination was unfortunately discussed initially as "recognition" in the early sensory literature (Ferdinandus et al., 1970) but we will stick with the terms discrimination and discriminators here. "Recognition" in the psychological literature implies a match to something stored in memory and that is not really the issue in discriminating a difference among samples.

The model is simple but it embraces two assumptions. The first assumption states that there are two kinds of panelists during a particular test—discriminators, who see the true difference and select the correct item and non-discriminators who see no difference and guess. The second assumption contains the logical notion that non-discriminators include people who guess correctly and those who guess incorrectly. The best estimate of the proportion guessing correctly is the chance performance level. Thus the total number of correct judgments comes from two sources: People who see the difference and answer correctly and those who guess correctly.

In forced choice threshold measures (see Chapter 6) 50% correct performance *after adjustment for chance* is taken as a working definition (Antinone et al., 1994; Lawless, 2010; Morrison, 1978; Viswanathan et al., 1983). For example, in the triangle test or a three-alternative forced choice test, the chance level is 1/3, so 50% above chance or 50% adjusted for chance is 66.7% correct. If a paired test or duo–trio was employed, the 50% chance level now requires a 75% correct discrimination to be at threshold, when threshold is defined as a 50% correct proportion after adjustment. Another approach is to work backward, i.e., try to find the percent correct that one would expect given a targeted proportion of discriminators in the test. This is given by the re-arrangement of Abbott's formula as follows:

$$P_{observed} = P_{adjusted} + P_{chance}(1 - P_{adjusted}) \quad (5.3)$$

So for our threshold example, if we had a 3-AFC test and we required 50% discriminators, we would expect one-third of the remaining (i.e., $1 - P_{adjusted}$) non-discriminators to guess correctly, thus adding 1/6 (or 1/3 of 0.5) to the 50% who see the difference and giving us 66.7% correct.

This discrimination ability should be viewed as momentary. It is not necessarily a stable trait for a given judge. That is, a given judge does not have to "always" be a discriminator or a non-discriminator. Furthermore, we are not attempting to determine who is a discriminator, we are only estimating the likely proportion of such people given our results. This is an important point that is sometimes misinterpreted in the sensory literature. A sensory panel leader who is accustomed to screening panelists to determine if they have good discriminating ability may view "discriminators" as people with a more or less stable ability and classify them as such. This is not the point here. In the guessing models, the term "discriminator" is not used to single out any individuals, in fact there is no way of knowing who was a discriminator, nor is there any need to know in order to apply the model. The model simply estimates the most likely proportion of people who are momentarily in a discriminating state and thus answer correctly as opposed to those who might be answering correctly by chance. In other words, the issue is how often people were likely to have noticed the difference.

If we choose to think about numbers correct rather than proportions, we can use a simple translation of Abbott's formula. How are the numbers of discriminators and non-discriminators estimated? The best estimate of the number of non-discriminators who guess correctly is based on the chance performance level. Once again, the total number of correct choices by the panel reflects the sum of the discriminators plus the fraction of the non-discriminators who guess correctly. This leads to the following equations: Let N = number of panelists, C = the number of correct answers, and D = the number of discriminators. For a triangle test, the following relationships should hold:

$$C = D + \frac{1}{3}(N - D) \quad (5.4)$$

This is simply a transformation of Abbott's formula (as in Eq. (5.3)), expressed in numbers observed rather than proportions.

Here is an example: Suppose we do a triangle test with 45 judges and 21 choose the odd sample correctly. We conclude that there is a significant difference at $p < 0.05$. But how many people were actually discriminators? In this example, $N = 45$ and $C = 21$. Solving for D in Eq. (5.4):

$$21 = D + \frac{1}{3}(45 - D) = \frac{2}{3}D + 15$$

and thus $D = 9$.

In other words, our best estimate is that 9 people out of 45 (21% of the sample) were the most likely number to have actually seen the difference. Note that this is very different from the percentage correct or 21/45 (=47%). Framed this way, a client may view the sensory result from quite a different perspective and one that is potentially more useful than the raw percent correct.

Table 5.1 shows how the number of required discriminators for various tests increases as a function of sample size. Of course, as the number of judges increases, it takes a smaller and smaller proportion of correct responses to exceed our minimum level above chance for statistical significance. This is due to the fact that our confidence intervals around the observed proportions shrink as sample size increases. We are simply more likely to have estimated a point nearer to the true population proportion correct. The number of judges getting a correct answer will also need to increase, as shown in the table. However, the number of discriminators increases at a slower rate.

Table 5.1 Number correct versus estimated discriminators

N	Minimum correct, $p = 1/2$	Estimated number discriminating	Minimum correct, $p = 1/3$	Estimated number discriminating
10	9	6	7	4
15	12	7	9	5
20	15	8	11	6
25	18	9	13	7
30	20	10	15	7
35	23	11	17	8
40	26	11	19	8
45	29	12	21	9
50	32	13	23	9
55	35	13	25	9
60	37	14	27	10
65	40	14	29	10
70	43	15	31	10
75	46	15	33	11
80	49	16	35	11
85	51	16	36	11
90	54	17	39	12
95	57	17	40	12
100	59	17	42	12

Minimum correct gives the required number for significance at $p < 0.05$, one tailed

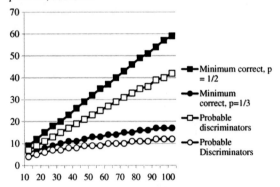

For large sample sizes, we only need a small proportion of discriminators to push us over the statistically critical proportion for significance. Another important message for clients and management here is that although we may have found a significant difference, *not everyone can discern it.*

This way of looking at difference tests has several benefits but one serious shortcoming. One advantage is that this concept of "the proportion of discriminators" gives management an additional point of reference for interpretation of the meaning of difference tests. Statistical significance is a poor reference point, in that the significance of a given proportion correct also depends upon the number of judges. As *N*, the number of judges increases, the minimum proportion correct required for statistical significance gets smaller and smaller, approaching a level nearer to chance. So statistical significance, while providing necessary evidence against a chance result, is a poor yardstick for business decisions and is only a binary choice. The estimated proportion of discriminators is not dependent upon sample size (although confidence intervals around it are).

Another advantage to this model is that it provides a yardstick for setting panel size and a point of reference for the test of significant similarity, as outlined below. In determining a desired panel size, the experimenter must make a decision about the size of the alternative hypothesis that must be detected if it is true. That is, how much of a difference do we want to be sure to detect if it in fact exists? The correction for guessing provides one such benchmark for these calculations. Once we have decided upon a critical proportion of discriminators, we can calculate what proportion correct would be expected from the addition of (chance) guessers. This proportion correct becomes the alternative hypothesis proportion in Eq. (5.1). In other words, the choice of the alternative hypothesis can now be based on the observed proportion required to give a certain level of discriminators. We merely have to apply Abbott's formula to see what percent correct is required.

The choice should consider a strategy based on the level of normal product variability and what consumers will expect. In one situation, there might be strong brand loyalty and consumers who demand high product consistency. In that case a low proportion of discriminators might be desired in order to make a process change or an ingredient substitution. Another product might have commonplace variation that is tolerated by consumers, like some fruit or vegetable commodities, or wines from different vintages. In this case a higher proportion of discriminators might be tolerated in a difference test. As we will discuss next, in the statistical test for significant similarity, the proportion of discriminators must be decided upon in order to test for performance below that critical level, as evidence for a difference too small to be practically significant.

Let us consider a triangle test with 90 judges and 42 choose the odd sample correctly. According to Table L, this is a significant difference at $p < 0.01$. Working with Eq. (5.2), we find that the proportion of discriminators is about 20% $(42/90 - 1/3)/(1 - 1/3) = 0.20$. So one-fifth of the test group is our best estimate here of the proportion discriminating. For a product with an extremely brand-loyal user group, this could be considered very risky. On the other hand, for a product in which there is some degree of expected variability, the proportion might not be a practical concern for worry, in spite of the statistical significance.

5.5 Tests for Significant Similarity

Another approach to the problem of demonstrating product similarity or equivalence was presented by Meilgaard et al. (2006). It is based on the fact that we do not have to test against the chance performance level in applying the binomial test on proportions. Rather, we can test against some higher level of expected proportion correct, and see whether we are significantly *below* that level in order to make a decision about two products being sensorially similar enough. This is shown graphically in Fig. 5.2. Our usual tests for discrimination involve a statistical test against the chance performance level and we look for a point at which the confidence interval around our observed proportion no longer overlaps the chance

level. This is just another way of thinking about the minimum level required for a significant difference. When the error bar no longer overlaps the chance level, we put that minimum number correct (for a given N) in our tables for the triangle test, duo–trio, etc. A higher proportion correct will be less likely to overlap and a larger sample size will shrink the error bars. As the "N" increases, the standard error of the proportion gets smaller. Thus higher proportions correct and larger panel sizes lead to less likely overlap with the chance level and thus a significant difference.

However, we can also test to see whether we are below some other level. The binomial test on proportions can be applied to other benchmarks as well. How can we determine this other benchmark? Once again, our proportion of allowable discriminators will give us a value to test against. We may have a very conservative situation, in which we can allow no more than 10% discriminators, or we might have a less critical or discerning public and be able to allow 30% or 50% discriminators or more. From the proportion of discriminators, it becomes a simple matter to calculate the other proportion we should test against, and see

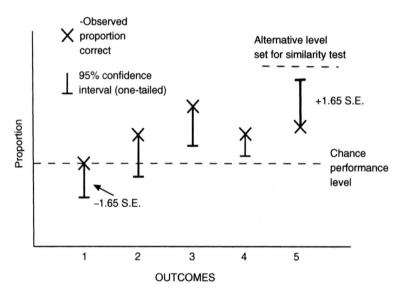

Fig. 5.2 Difference testing and similarity testing outcomes. In outcome one, the performance is at the chance level, so there is obviously no difference. In the outcome, two performances are above chance, but the 95% one-tailed confidence interval overlaps the chance level, so no statistically significant difference is found. This level would be below the tabulated critical number of correct answers for that number of judges. In outcome three, the level correct and the associated confidence level are now above

the chance level so a statistically significant difference is found. In outcome four, the level correct is lower than the third example, but the standard error has become smaller, due to an increase of N, so the outcome is also significant. In the fifth example, there is a significantly significant similarity, because the outcome and its associated one-tailed confidence interval are *below* the acceptable level based on the maximum allowable proportion of discriminators.

whether we are below that level. This is simply using Abbott's formula in reverse (Eq. (5.3)).

Tables H1 and H2 show critical values for significant similarity in the triangle test and for the duo–trio. Other tables are given in Meilgaard et al (2006). Here is a worked example. Suppose we conduct a triangle test with 72 panelists and 28 choose the correct (odd) sample. Do we have evidence for significant similarity? From Table H1 we see that for a criterion of no more than 30% discriminators, the critical value for a beta-risk of 10% is 32. Because we are *below* this value, we can conclude that the products are significantly similar.

If you examine these tables closely, you will note that there is a very narrow window for some proportions of discriminators and for a low number of panelists. There are empty cells in the table since we need a large panel size and low standard errors (small confidence intervals) in order to squeeze our result and the confidence interval between our test proportion and the chance proportion. The chance proportion forms, of course, a lower bound because it is not expected to see performance below chance. Once again, as in our power calculations, having a large sample size may be necessary to protect against Type II error, and probably a larger number of judges than we employ in most difference testing.

Let us look at this approach in some detail. The method of similarity testing is based upon the comparison of a maximum allowable proportion of discriminators to a confidence interval around the observed proportion correct. You can think of this test as involving three steps. First, we set an upper limit on the proportion of acceptable discriminators. Note that this involves professional judgment, knowledge of the products, and the business situation regarding consumer expectations about product consistency. It is not found in any statistics book. This is the same process as we discussed in Section 5.3 for choosing the size of the difference we need to detect. Second, we translate this into an expected percent correct by working through Eq. (5.3). The test then compares the inferred proportion plus its upper confidence interval to the maximum allowable proportion you set first. If the calculated value is less than the acceptable limit, we can conclude statistically significant similarity.

Here is the derivation of the formula. There are two items we need to know, the proportion correct we would expect, based on our proportion of

discriminators, and the confidence interval boundaries around observed proportion correct given the number of judges. The proportion of discriminators and proportion expected correct are calculated just as in Eq. (5.3). The confidence interval of a proportion is given by $\pm Z$ (standard error), where $Z =$ normal deviate for our desired level of confidence. For an upper one-tailed confidence interval at 95%, $Z = 1.65$. Equation (5.5) shows the standard error of the proportion, SE_p and Eq. (5.6) the confidence interval:

$$SE_p = \sqrt{\frac{(X/N)(1 - X/N)}{N}} = \sqrt{pq/N} \qquad (5.5)$$

where X is the number correct, N is the total number of judges, $p = X/N$ and $q = 1-p$.

$$CI_{95\%} = (X/N) \pm Z(SE_p) \qquad (5.6)$$

where Z is the Z-score for 95% confidence. The remaining step is to recast our confidence interval to include the fact that we are working with a limit on the number of discriminators, not the simple percent correct. For the triangle test, for example, the proportion of discriminators, D/N, is $1.5(X/N)–0.5$. Note that the standard error associated with the proportion of discriminators also is 1.5 times as large as the standard error associated with our observed proportion. So, our upper confidence interval boundary for discriminators now becomes

$$\text{Upper } CI_{95\%} = [1.5(X/N) - 0.5] + 1.645(1.5)$$
$$\sqrt{\frac{(X/N)(1-X/N)}{N}}$$
$$(5.7)$$

Here is a worked example. Suppose we do a triangle test with 60 panelists, and 30 get the correct answer. We can ask the following questions: What is the best estimate of the number of discriminators? The proportion of discriminators? What is the upper 95% confidence interval on the number of discriminators? What is the confidence interval on the proportion of discriminators? Finally, could we conclude that there is significant similarity, based on a maximum allowable proportion of discriminators of 50%?

The solution is as follows: Let $X =$ number correct, $D =$ number of discriminators, so $X = 30$ and $N = 60$. We have $1.5(30)–0.5(60) = D$ or 15 discriminators, or 25% of our judges detecting the difference, as our best estimate.

The standard error is given by

$$1.5\sqrt{\frac{(30/60)[1 - (30/60)]}{60}} = 0.097 = 9.7\%$$

and the upper bound on the confidence interval is given by Eq. (5.7), or $z(\text{SE})$ + proportion of discriminators = $1.65\ (0.097) + 0.25 = 0.41$ (or 41%).

So if our maximum allowable proportion of discriminators was 50%, we would have evidence that 95% of the time we would fall below this acceptable level. In fact, we would have 41% discriminating or less, given our observed percent correct of 50% which gives us our calculated best estimate of discriminators at 25%. This worked example is given to illustrate the approach. For practical purposes, the tables shown in Meilgaard et al. (2006) can be used as a simple reference without performing these calculations.

A similar approach was taken by Schlich (1993). He combined the questions of difference and similarity by calculating simultaneous alpha- and beta-risks for different panel sizes at the critical number correct for significant differences. Some of these values are shown in Table N. The table has two entries, one for the required number of judges and a second for the critical number correct at the crossover point. If the observed number correct is equal to or higher than the tabulated value, you can conclude that there is a significant difference. If the number correct in your test is lower, you can conclude significant similarity, based on the allowable proportion of discriminators you have chosen as an upper limit and the beta-risk. These tables can be very useful, but they required that you adopt the specified panel size that is stipulated for the conditions you have chosen for beta and proportion of discriminators.

5.6 The Two One-Sided Test Approach (TOST) and Interval Testing

The notion that equivalence can be concluded from a non-significant test, even with high test power, has largely been rejected by the scientific community concerned with bioequivalence (Bi, 2005, 2007). For example, the FDA has published guidelines for bioequivalence testing based on an interval testing approach (USFDA, 2001). This kind of test demands that the value of interest falls inside some interval and thus is sometimes referred to as interval testing. In general, this kind of testing is done on some scaled variable, like the amount of a drug delivered to the bloodstream in a certain specific period. Such a scaled variable is different from most discrimination tests, which are based on proportions, not some measured quantity that varies continuously. However, some scaled sensory data may fall under this umbrella, such as descriptive data or consumer acceptability ratings on a hedonic scale. Discrimination tests and preference tests are also amenable to this approach (Bi, 2006a; MacRae, 1995).

Logically, an interval test can be broken down into two parts, one test against an upper acceptable limit and one test against a lower acceptable limit. This is similar to finding some confidence intervals for an acceptable range of variation. In the case of many discrimination tests only the upper limit is of interest. The situation can then be a single one-tailed test. For paired comparison tests, the TOST method is described in detail by Bi (2007). In this article he shows some differences between the TOST estimates and the conventional two-sided confidence interval approach. Some authors recommend comparing the interval testing approach at $100(1-\alpha/2)$ to TOST because the interval testing approach at $100(1-\alpha)$ is too conservative and lacks statistical power (Gacula et al., 2009).

For scaled data, we may wish to "prove" that our test product and control product have mean values within some acceptable range. This approach can be taken with descriptive data, for instance, or acceptability ratings or overall degree-of-difference ratings. Bi (2005) described a non-central t-test for this situation. This is similar to a combination of t-tests in which we are testing that the observed difference between the means falls within some acceptable limit. An example of this approach is shown in Appendix at the end of this chapter. For purposes of using these models for equivalence, the sensory professional is advised to work closely with a statistical consultant. Further information on formal equivalence testing can be found in Welleck (2003) and Gacula et al. (2009).

Alternatives to the TOST method have been given by Ennis and Ennis (2010) and have some advantages to TOST from a statistical perspective. An alternative to TOST that is applicable to a non-directional 2-AFC (e.g., a two-tailed 2-AFC, much like a paired preference) has been proposed (Ennis and Ennis, 2010). Note that under this approach, establishing an equivalence or parity situation usually requires a much larger sample size (N) than simple tests for difference. Useful tables derived from this theory are given in Ennis (2008). The theory states that a probability value for the equivalence test can be obtained from an exact binomial or more simply from a normal approximation as follows:

$$p = \phi \left(\frac{|x| - \theta}{\sigma} \right) - \phi \left(\frac{|-x| + \theta}{\sigma} \right) \qquad (5.8)$$

where phi (Φ) symbolizes the cumulative normal distribution area (converting a z-score back to a p-value), theta (θ) is the half-interval for parity such as ± 0.05, and sigma (σ) is the estimated standard error of the proportion (square root of pq/N). x in this case is the difference between the observed proportion and the null (for 2-AFC, subtract 0.5). A worked example is given in Chapter 13, Section 13.3.5. Note that the tables given in Ennis (2008) are specific to the 2-AFC and may not be used for other tests such as the duo–trio.

5.7 Claim Substantiation

A special case of equivalence testing arises when a food or consumer product manufacturer wishes to make a claim of equivalence or parity against a competitor. Such a claim might take the form of a statement such as "our product is just as sweet as product X." Because of the legal ramifications of this kind of test, and the need to prove that the result lies within certain limits, large numbers of consumers are typically required for such a test, with recommended sample sizes from 300 to 400 as a minimum (ASTM, 2008a). This is a different scenario than most simple discrimination tests that use laboratory panels of 50–75 judges. The special case of proving preference equality (products chosen equally often in a preference test) is discussed further in Chapter 19 on strategic research.

The simple case of a paired comparison test (2-AFC) is amenable to this kind of analysis. As noted above, Bi (2007) discussed the TOST approach to equivalence for 2-AFC with worked examples. There are two different statistical scenarios: In one case we wish to make an equality claim, and in the second, we want to make a claim that our product is "unsurpassed." The equality claim involves two tests, because neither product can have more of the stated attribute

than another. The unsurpassed claim is a simple one-tailed alternative and just requires showing that our product is not significantly lower than the other product. For both of these tests, we have to choose some lower bound that we cannot cross. For example, a common criterion for the equality claim is that the true population percentage in the paired test lies between 45 and 55% of choices. Thus a 5% difference is considered "close enough" or of no appreciable practical significance.

The equality claim requires that neither one of the products cross over the lower bound and can be viewed as two one-tailed tests. Tables for the minimum number allowable in such tests can be found in ASTM (2008a). For the unsurpassed claim we are stating that our product is not inferior or not lower in the attribute in question. For this purpose the test takes the following form of a simple binomial-approximation Z-score:

$$Z = \frac{(P_{obs} - 0.45) - (1/2N)}{\sqrt{\frac{(0.45)(0.55)}{N}}} \qquad (5.9)$$

where P_{obs} is the proportion observed for your test product. In the case of large sample sizes ($N>200$), the value of the continuity correction, $1/2N$, becomes negligible. Note that this is a one-tailed test, so the obtained Z must be greater than or equal to 1.645. If the obtained Z is greater than that value, you have support for a claim that your product is not lower than the competitor. In the case of our sweetness claim, we can be justified in saying our product is "just as sweet."

5.8 Models for Discrimination: Signal Detection Theory

In the preceding sections, we looked at the size of the sensory difference in terms of the proportion of panelists who could be discriminating in any given test. The calculations are based on an adjustment for chance correct guesses. However, the chance probability level does not tell the whole story, because some tests are harder or involve more variability than others, even at the same chance probability level. As an example, the triangle test is "more difficult" than the 3-AFC test, because it is a more difficult cognitive task to find the odd sample (one that entails higher variability), as opposed to simply choosing the strongest or weakest. In this section, we will look at a more sophisticated model in which this issue can be taken into account. From this theory, we can derive a universal index of sensory similarity or difference and one that takes into account the difficulty or variability inherent in different discrimination tests.

One of the most influential theories in psychophysics and experimental psychology is signal detection theory (SDT). The approach grew from the efforts of engineers and psychologists who were concerned with the decisions of human observers under conditions that were less than perfect (Green and Swets, 1966). An example is the detection of a weak signal on a radar screen, when there is additional visual "noise" present in the background. Mathematically, this theory is closely related to an earlier body of theory developed by Thurstone (1927). Although they worked in different experimental paradigms, credit should be given to Thurstone for the original insight that a scaled difference could be measured based on performance and error variability. Although signal detection usually deals with threshold-level sensations, any question of perceived differences (when such differences are small) can be addressed by SDT. For a good introduction to SDT, the book on psychophysics by Baird and Noma (1978) is useful and for a more detailed look, Macmillan and Creelman (1991) is recommended.

is clearly perceivable before they respond. An example in industry might be in quality control, where the mistaken rejection of a batch of product could incur a large cost if the batch has to be reworked or discarded. On the other hand, a high-margin upscale product with a brand-loyal and knowledgeable consumer base might require narrower or more stringent criteria for product acceptance. Any sensory problem at all might cause rejection or retesting of a batch. The criterion for rejection would cast a wider net to be sure to catch potential problem items before they can offend the brand-loyal purchasers.

So a decision process is layered on top of the actual sensory experience. A person may set a criterion that is either conservative or lax in terms of how much evidence they need to respond positively. Here is a simple example of a decision process involved in perception: Suppose you have just purchased a new pair of stereo headphones. It is the end of a long work week and you have gone home to enjoy your favorite music in a comfortable chair. As you settle in and turn the music up very loud, you think the phone might be ringing. Let us consider two scenarios: In one case, you are expecting a job offer following a successful interview. Do you get up and check the phone? In another case you are expecting a relative to call who wants to borrow money. Do you get up? What are the relative payoffs and penalties associated with deciding that the phone has actually rung? What does it take to get you up out of that comfortable chair?

5.8.1 The Problem

In traditional threshold experiments, the physical strength of a weak stimulus is raised until the level is found where people change their responses from "No, I do not taste (or smell, see, hear, feel) anything" to "Yes, now I do." This is the original procedure for the method of limits (see Chapter 2). The difficulty in such an experiment is that there are many biases and influences that can affect a person's response, in addition to the actual sensation they experience. For example, they may expect a change in sensation and anticipate the level where something will be noticeable. Conversely, a person might adopt a very conservative stance and want to be very sure that something

5.8.2 Experimental Setup

In a classic signal detection experiment, two levels of a stimulus are to be evaluated, for example, the background or blank stimulus called the "noise" and some weak but higher level of stimulus intensity near threshold called the "signal" (Baird and Noma, 1978; Macmillan and Creelman, 1991) Stimuli are presented one at a time and the observer would have to respond "Yes, I think that is a case of the signal," or "No, I think that is the case of the noise." So far this resembles the A, not-A test in sensory evaluation. Both signal and noise would be presented many times and the data would be tabulated in a two-by-two response matrix as shown in Fig. 5.3. Over many presentations, some correct decisions would be made when a signal is in

Response

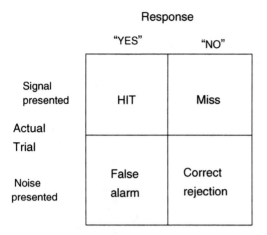

Fig. 5.3 The stimulus–response matrix for a signal detection experiment. The hit rate is the proportion of times the subject responds "yes" or "signal" when in fact the signal was presented. The false alarm rate is the proportion of noise trials when the subject also responds "yes" or "signal" (noise presented). These two outcomes define the response space since the misses are the total of the signal trials (which the experimenter has designed into the study) minus hits, and the correct rejections are likewise the number of noise trials minus false alarms (there is only one degree of freedom per row).

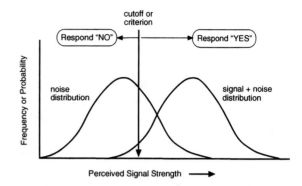

Fig. 5.4 Signal detection assumptions include normally distributed experiences from signal and noise trials, with equal variance and the establishment of a stable criterion or cutoff level, above which the subject responds "yes" and below which the subject responds "no".

fact presented and these are called "hits" in signal detection terminology. Since the stimuli are confusable, sometimes the observer would respond positively on noise trials as well, mislabeling them as signal. These are called "false alarms." There are also situations in which the signal is presented and the observer fails to call it a signal (a "miss") and cases in which the noise trial is correctly labeled a noise trial. However, since we have set up the experimental design and know how many signal and noise trials we have presented, the total number of signal trials is equal to the hits plus misses and the total number of noise trials equals false alarms plus correct rejections. In other words, there is only one degree of freedom in each row and we can define the observer's behavior by examining only hit and false alarm rates.

5.8.3 Assumptions and Theory

The theory makes a few sensible assumptions (Baird and Noma, 1978; Green and Swets, 1966). Over many trials, the sensations from signal and noise are normally distributed with equal variance. That is,

sometimes a more intense sensation will be felt from the signal, and sometimes a weaker sensation, and over many trials these experiences will be normally distributed around some mean. Similarly the noise trials will sometimes be perceived as strong enough so that they are mistakenly called a "signal." Once the observer is familiar with the level of sensations evoked, he or she will put a stable criterion in place. When the panelist decides that the sensation is stronger than a certain level, the response will be "signal" and if weaker than a certain amount a "noise" response will be given. This situation is shown in Fig. 5.4. Remember that the panelist does not know if it is a signal or noise trial, they just respond based on how strong the sensation is to them.

Variabilities in the signal and noise are reasonable assumptions. There is spontaneous variation in the background level of activity in sensory nerves, as well as variance associated with the adaptation state of the observer, variation in the stimulus itself, and perhaps in the testing environment. The greater the overlap in the signal and noise distributions, the more difficult the two stimuli are to tell apart. This shows up in the data as more false alarms relative to the occurrence of hits. Of course, in some situations, the observer will be very cautious and require a very strong sensation before responding "yes, signal." This will not only minimize the false alarm rate but also lower the hit rate. In other situations, the observer might be very lax, and respond "yes" a lot, producing a lot of hits, but at the cost of increased false alarms. The hit and false alarm rates will co-vary as the criterion changes.

Now we need to connect the performance of the observer (Fig. 5.3) to the underlying scheme in Fig. 5.4 to come up with a scaled estimate of performance and one that is independent of where observers set their particular criteria for responding. The separation of the two distributions can be specified as the difference between their means and the unit of measurement as the standard deviations of the distributions (a convenient yardstick). Here is the key idea: The proportion of signal trials that are hits corresponds to the area underneath the signal distribution to the right of the criterion, i.e., the sensation stronger than our cutoff, so response is "yes" to a signal presentation. Similarly, the proportion of false alarm trials represents the tail of the noise distribution to the right of the cutoff, i.e., sensations stronger than criterion but drawn from the noise experiences. This scheme is shown in Fig. 5.5.

All we need to estimate, then, is the distance from the criterion level or cutoff to the mean of each distribution. These can be found from the z-scores relating the proportion to the distance in standard deviation units. Since we know the relationship between proportions and z-scores, the two distances

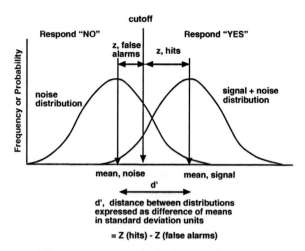

Fig. 5.6 How d' is calculated based on the signal detection scheme. Using the conversion of proportions (areas) to z-scores, the overall difference (d', pronounced "d prime") is given by the z-value for hits minus the z-value for false alarms.

can be estimated, and then summed, as shown in Fig. 5.6. Due to the way that z-scores are usually tabulated, this turns out to be a process of subtraction and the value of sensory difference called d' ("d-prime") is equal to the z-score for the proportion of hits minus the z-score for the proportion of false alarms.

5.8.4 An Example

The great advantage of this approach is that we can estimate this sensory difference independently of where the observer sets the criterion for responding. Whether the criterion is very lax or very conservative, the hit and false alarm z-scores will change to keep d' the same. The criterion can slide around, but for a given set of products for the same panelist, the difference between the two distributions remains the same. When the criterion is moved to the right, fewer false alarms result and also fewer hits. (Note that the z-score will change sign when the criterion passes the mean of the signal distribution). If the criterion becomes very lax, the hit and false alarm rates will both go up and the Z-score for false alarms will change sign if the proportion of false alarms is over 50% of the noise trials. Table 5.2 may be used for conversion of proportions of hits and false alarms to z-scores. Figure 5.7 shows a criterion shift for two approximately equal levels of discriminability. The upper panel shows a conservative criterion with only 22% hits and 5% false

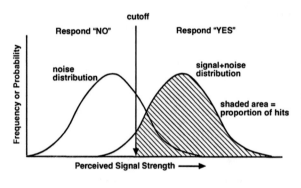

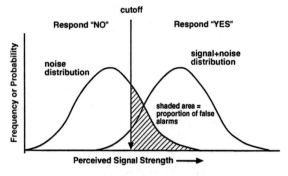

Fig. 5.5 Signal and noise distributions are shaded to show the area corresponding to the proportions of hits and false alarms, respectively. These proportions can then be converted to z-scores.

Table 5.2 Proportions and Z-scores for calculation of d'

Proportion	Z-score	Proportion	Z-score	Proportion	Z-score	Proportion	Z-score
0.01	−2.33	0.26	−0.64	0.51	0.03	0.76	0.71
0.02	−2.05	0.27	−0.61	0.52	0.05	0.77	0.74
0.03	−1.88	0.28	−0.58	0.53	0.08	0.78	0.77
0.04	−1.75	0.29	−0.55	0.54	0.10	0.79	0.81
0.05	−1.64	0.30	−0.52	0.55	0.13	0.80	0.84
0.06	−1.55	0.31	−0.50	0.56	0.15	0.81	0.88
0.07	−1.48	0.32	−0.47	0.57	0.18	0.82	0.92
0.08	−1.41	0.33	−0.44	0.58	0.20	0.83	0.95
0.09	−1.34	0.34	−0.41	0.59	0.23	0.84	0.99
0.10	−1.28	0.35	−0.39	0.60	0.25	0.85	1.04
0.11	−1.23	0.36	−0.36	0.61	0.28	0.86	1.08
0.12	−1.18	0.37	−0.33	0.62	0.31	0.87	1.13
0.13	−1.13	0.38	−0.31	0.63	0.33	0.88	1.18
0.14	−1.08	0.39	−0.28	0.64	0.36	0.89	1.23
0.15	−1.04	0.40	−0.25	0.65	0.39	0.90	1.28
0.16	−0.99	0.41	−0.23	0.66	0.41	0.91	1.34
0.17	−0.95	0.42	−0.20	0.67	0.44	0.92	1.41
0.18	−0.92	0.43	−0.18	0.68	0.47	0.03	1.48
0.19	−0.88	0.44	−0.15	0.69	0.50	0.94	1.55
0.20	−0.84	0.45	−0.13	0.70	0.52	0.95	1.64
0.21	−0.81	0.46	−0.10	0.71	0.55	0.96	1.75
0.22	−0.77	0.47	−0.08	0.72	0.58	0.97	1.88
0.23	−0.74	0.48	−0.05	0.73	0.61	0.98	2.05
0.24	−0.71	0.49	−0.03	0.74	0.64	0.99	2.33
0.25	−0.67	0.50	0.00	0.75	0.67	0.995	2.58

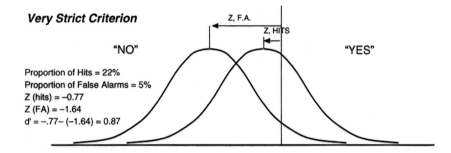

Very Strict Criterion

Z, F.A.
Z, HITS
"NO"   "YES"

Proportion of Hits = 22%
Proportion of False Alarms = 5%
Z (hits) = −0.77
Z (FA) = −1.64
d' = −.77− (−1.64) = 0.87

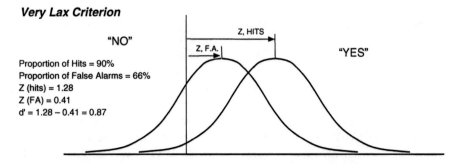

Very Lax Criterion

Z, HITS
"NO"
Z, F.A.   "YES"

Proportion of Hits = 90%
Proportion of False Alarms = 66%
Z (hits) = 1.28
Z (FA) = 0.41
d' = 1.28 − 0.41 = 0.87

Fig. 5.7 The measure of sensory difference, d', will remain constant for the same observer and the same stimuli, even though the criterion shifts. In the upper case, the criterion is very strict and the observer needs to be sure before responding "yes" so there is a low proportion of hits but also a low proportion of false alarms. In the lower panel the subject sets a lax criterion, responding "yes" most of the time, catching a lot of hits, but at the expense of many false alarms.

alarms. Referring to Table 5.2, the Z-scores for these proportions are –0.77 and –1.64, respectively, giving a d' value of –0.77–(–1.64) or +0.87. The lower panel illustrates a less conservative mode of responding, with 90% hits and 66% false alarms. Table 5.2 shows the Z-scores to be 1.28 and 0.41, again giving a d' of 0.87.

In other words, the d' does not change, even though the criterion has shifted. This theory permits a determination of the degree of sensory discriminability, regardless of the bias or criterion of the observer. In the next section, we will examine how the theory can be extended to include just about any discrimination test.

5.8.5 A Connection to Paired Comparisons Results Through the ROC Curve

How can we connect the SDT approach to the kinds of discrimination tests used in sensory evaluation? One way to see the connection is to look at the receiver operating characteristic or ROC curve. This curve defines a person's detection ability across different settings of the criterion. In the ROC curve, hit rate in different situations is plotted as a function of false alarm rate. Figure 5.8 shows how two observers would behave in several experiments with the same levels of the stimulus and noise. Payoffs and penalties could be varied to produce more conservative or more lax behaviors, as they often were in the early psychophysical studies. As criterion shifts, the performance moves along the characteristic curve for that observer and for those particular stimuli. If the hit rate and false alarm rates were equal, there is no discrimination of the two levels, and d' is zero. This is shown by the dotted diagonal line in the figure. Higher levels of discrimination (higher levels of d') are shown by curves that bow more toward the upper left of the figure. Observer 2 has a higher level of discrimination, since there are more hits at any given false alarm rate or fewer false alarms at a given hit rate. The level of discrimination in this figure, then is proportional to the area under the ROC curve (to the right and below), a measure that is related to d'. Note that the dotted diagonal cuts off one-half of the area of the figure. One-half (50%) is the performance you would expect in a paired comparison test if there were no difference between the products. From this you can see that there should be a correspondence

between the area under the ROC curve (which is proportional to d') and the performance we would expect in a 2-AFC or paired comparison test. Results from other kinds of discrimination tests such as the triangle, duo–trio, and 3-AFC can be mathematically converted to d' measures (Ennis, 1993).

5.9 Thurstonian Scaling

Thurstone was dealing with the kinds of studies done in traditional psychophysics, like the method of constant stimuli (see Chapter 2). This method is basically just a series of paired comparisons against a constant or standard stimulus. Thurstone realized that if you got 95% correct in a paired test, the sensory difference ought to be bigger than if you only got 55% correct. So he set out to come up with a method to measure the degree of difference, working from the percent correct in a paired test. In doing this he formulated a law of comparative judgment (Thurstone, 1927).

5.9.1 The Theory and Formulae

Thurstone's law of comparative judgment can be paraphrased as follows: Let us assume that the panelist will compare two similar products, A and B, over several trials and we will record the number of times A is judged stronger than B. Thurstone proposed that the sensory events produced by A and B would be normally distributed. Thurstone called these perceptual distributions "discriminal dispersions," but they are analogous to the distributions of signal and noise events in the yes/no task. The proportion of times A is judged stronger than B (the datum) comes from a process of mental comparison of the difference between the two stimuli. Evaluating a difference is analogous to a process of subtraction. Sometimes the difference will be positive (A stronger than B) and sometimes the difference will be negative (B stronger than A) since the two items are confusable. One remaining question is how the sampling distribution for these differences would arise from the two underlying discriminal dispersions as shown in Fig. 5.9. The laws of statistics can help here, since it is possible to relate the difference sampling distribution (lower curve) to

ROC Curve - Differing Sensitivities

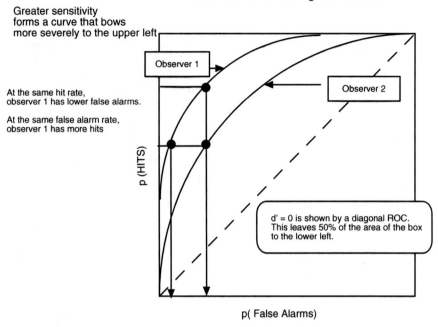

Greater sensitivity forms a curve that bows more severely to the upper left

Observer 1

At the same hit rate, observer 1 has lower false alarms.

At the same false alarm rate, observer 1 has more hits

Observer 2

p (HITS)

d' = 0 is shown by a diagonal ROC. This leaves 50% of the area of the box to the lower left.

p(False Alarms)

The area under the ROC curve is another measure of discrimination. Also, there is a relationship between this measure and forced-choice data. (Area under the curve is approximately what would happen in 2 -AFC).

Fig. 5.8 The ROC curve, or receiver operating characteristic, shows the behavior of a single individual under various criterion shifts, plotting the proportion of hits against the proportion of false alarms. Better discrimination (higher d') is shown by a curve that bows more toward the upper left corner. Thus observer 1 has better performance and better discrimination ability than observer 2. The area under the ROC curve (to the right and below) is another measure of discrimination and can be converted to d' values. Note that the diagonal describes no discrimination, when hits always equal the rate of false alarms. Also note that the area of the box below and to the right is 50% which would be the performance in a paired test or 2-AFC when there is no difference. Thus the area under the ROC curve is expected to be proportional to the performance one would observe with a given observer and a given pair of stimuli in a 2-AFC test.

the sensory events depicted by the upper distributions. This statistical result is given by the following equation:

$$S_{\text{diff}} = \sqrt{S_a^2 + S_b^2 + 2rS_aS_b} \qquad (5.10)$$

$$M_{\text{difference}} = Z\sqrt{S_a^2 + S_b^2 + 2rS_aS_b} \qquad (5.11)$$

where M is the difference scale value, Z is the z-score corresponding to the proportion of times A is judged stronger than B, and S_a and S_b are the standard deviations of the original discriminal dispersions. The "r" represents the correlation between sensations from A and B, which might be negative in the case of contrast or positive in the case of assimilation. If we make the assumptions that S_a and S_b are equal and $r = 0$ (no sequential dependency of the two stimuli) then the equation simplifies to

$$M = Z\sqrt{2}S \qquad (5.12)$$

where S is the common standard deviation. These simplified assumptions are referred to as "Thurstone's Case V" in the statistical literature (Baird and Noma, 1978). The mean of the difference scores is statistically the same as the difference of the two means. So to get to d', which is the mean difference divided by the original standard deviation, we have to multiply our z-score (from the percent correct) by the square

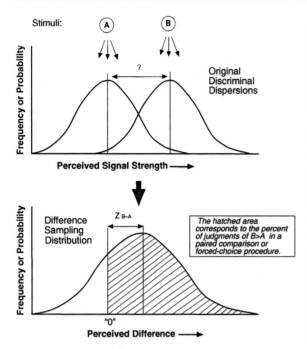

Fig. 5.9 The Thurstone model proposes that the proportion of times one stimulus is judged greater than another is predicted by a difference sampling distribution, which in turn arises from the sensory variability and degree of overlap in the original stimuli.

root of 2. In other words, the z-score value is smaller than what would be estimated from the d' of the yes/no signal detection experiment by the square root of two (Macmillan and Creelman, 1991). The distance of the mean from an arbitrary zero point can be determined by a z-score transformation of the proportion of times A is judged stronger than B. We can conveniently work with the zero point as the mean of distribution for the weaker of the two stimuli (Baird and Noma, 1978). Like d', this gives us a measure that is expressed in standard deviation units.

5.9.2 Extending Thurstone's Model to Other Choice Tests

We can extend this kind of scale value to any kind of choice test and tables have been published for various conversions of percent correct to d' or delta values (Bi, 2006a; Ennis, 1993; Frijters et al., 1980; Ura, 1960). Delta is sometimes used to refer to a population variable (rather than d' which is a sample statistic) but the meaning is the same in terms of the sensory difference

it describes. Other theorists saw the applicability of a signal detection model to forced-choice data. Ura (1960) and Frijters et al. (1980) published the mathematical relationships to relate triangle test performance to d' as well as other test procedures commonly used in food science.

Ennis (1990, 1993) has examined Thurstonian models as one way of showing the relative power of difference tests. Like Frijters, he showed that for a given level of scaled difference a lower level of percent correct is expected in the triangle test, as opposed to the 3-AFC test. On the basis of the variability in judging differences versus intensities, one expects higher performance in the 3-AFC test. This has become famous as the "paradox of the discriminatory non-discriminators" (Byer and Abrams, 1953; Frijters, 1979). In the original paper, Byer and Abrams noted that it was possible for many panelists to answer incorrectly under the triangle test instructions, but still accurately choose the strongest sample when it was the odd one (or the weakest when it was the odd one). An example of how this could occur is shown in Fig. 5.10. That is, for the 3-AFC instructions, there was a higher percent correct. Frijters (1979) was able to show that the different percents correct for the triangle and 3-AFC in Byer and Abram's data actually yielded the same d' value, hence resolving the apparent

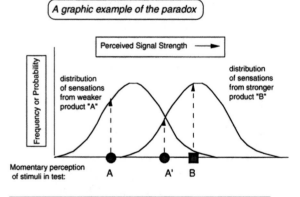

Fig. 5.10 An example of the paradox of the discriminatory non-discriminators. In a hypothetical triad of products, the odd sample, B, is chosen as the strongest sample, a correct answer for the 3-AFC test. However, one of the duplicate samples, A, is momentarily perceived as the outlier, leading to an incorrect choice as the odd sample under the triangle test instructions.

paradox. Delwiche and O'Mahony (1996) demonstrated that similar predictions can be made for tetradic (four-stimulus procedures).

The ability to reconcile the results from different experimental methods and convert them to a common scale is a major advantage of the signal detection and Thurstonian approaches (Ennis, 1993). Studies continue to show the constancy of d' estimates across tasks and methods (Delwiche and O'Mahony, 1996; Stillman and Irwin, 1995), although there are occasional exceptions where other factors come into play to complicate the situation and require further model development (Antinone et al., 1994; Lawless and Schlegel, 1984). Applying the correct Thurstonian model requires that you understand the cognitive strategy of the panelist (O'Mahony et al., 1994). For example, am I looking for the smallest of three pairs of differences (a triangle strategy for finding the odd sample) or am I trying to discern the strongest of three intensities (a 3-AFC "skimming" strategy)? If one has a different strategy for a given test method or task, the resulting d' value will not reflect what people are actually doing in the test. For example, if a number of panelists are "skimming" for the strongest sample, but have been given triangle instructions, the d' will not make sense if taken from the triangle test tables.

Another complicating factor concerns sequential effects in groups of products that are presented at the same time. The discriminability of two items depends not only on the relative strength of signal versus noise sensations but also on the sequence in which items are presented. Thus a strong stimulus following a weak one (signal after noise trial) may give a stronger sensation of difference than a noise trial following a signal trial. O'Mahony and Odbert (1985) have shown how this leads to better performance for some discrimination tests over others in a theory called "sequential sensitivity analysis." Ennis and O'Mahony (1995) showed how sequential effects can be incorporated into a Thurstonian model. Another factor concerns the fact that most foods are multi-dimensional and the simple SDT and Thurstone models are usually formalized as a one-dimensional variation. Ennis and Mullen (1986) using a multivariate model showed how variation in irrelevant dimensions could degrade performance.

The important conclusion for sensory professionals to draw from this theory is that the common tests for overall difference, e.g., the triangle and duo–trio, are

not very sensitive tests. That is, for a given d' value, a much larger panel size needs to be tested to be sure that the difference is detected by the test. This is in comparison to the forced-choice procedures such as the paired comparison and 3-AFC tests which will detect a statistically significant difference for a given d' at much smaller panel sizes (Ennis, 1993). Put a different way, for a given panel size, the triangle test could easily miss a difference that the 3-AFC test would detect, as seen in the Byer and Abrams "paradox." Unfortunately, when an ingredient or processing change is made in a complex food, one cannot always predict any simple singular attribute to use for the AFC tests, nor perhaps even overall strength of flavor, taste, etc. So the sensory professional is stuck using the less sensitive type of test. In the face of a statistically significant result, this is not a problem. But if the equivalence decision is based on a non-significant test outcome, the decision to conclude equivalence can be very risky.

5.10 Extensions of the Thurstonian Methods, *R*-Index

5.10.1 Short Cut Signal Detection Methods

One additional method deserves mention in the application of signal detection models to the discrimination testing situation. A practical impediment to the application of signal detection theory to foods has been the large number of trials necessary in the traditional yes/no signal detection experiment. With foods, and especially in applied difference testing, it has rarely been possible to give the large numbers of trials to each subject needed to accurately estimate an individual's d' value.

O'Mahony (1979) saw the theoretical advantage to signal detection measures, and proposed short-cut rating scale methods to facilitate the application of signal detection in food evaluations. The *R*-index is one example of an alternative measure developed to provide an index of discrimination ability but without the stringent assumptions entailed by d', namely equal and normally distributed variances from signal and noise distributions. The area under the ROC curve is another measure of discrimination that does not depend upon the exact forms of the signal and noise distributions (see Fig. 5.8). The R-index is one such measure and converts rating scale performance to an index related to the percentage of area under the ROC curve, a measure of discrimination. It also gives an indication of what we would expect as performance in two-alternative forced-choice task, which is of course, mathematically related to d'.

5.10.2 An Example

Here is an example of the R-index approach. In a typical experiment, 10 signal and 10 noise trials are given. Subjects are pre-familiarized with signal stimuli (called "A") and noise stimuli (called "B" in this example). Subjects are asked to assign a rating scale value to each presentation of A and B, using labels such as "A, definitely," "A, maybe", "B, maybe" and "B, definitely."

For a single subject, performance on the 20 trials might look like this:

	Ratings			
	A, definitely	A, maybe	B, maybe	B, definitely
Signal presented	5	2	2	1
Noise presented	1	2	3	4

Obviously, there is some overlap in these distributions and the stimuli are somewhat confusable.

R is calculated as follows: Pair every rating of the signal with every rating of the noise as if there were paired comparisons. In this example, there are 10×10 or 100 hypothetical pairings. R calculates how many times would the signal "A" be identified correctly or called the stronger of the pair. First, we consider the five cases in which signal (A) was rated "A, definitely." When paired against the $(2 + 3 + 4 =)$ 9 cases in which the noise trial (B) received a lower rating (i.e., noise was judged less like "A" than signal), 45 correct judgments would have been made if there were actually paired tests. For the five cases in which the signal was "A, definitely" were paired with noise rated "A, definitely," there are five ties (5×1), so we presume that half the trials (2.5) would be correct and half incorrect if a choice were forced. We then to continue

to make these hypothetical pairings of each rating of A with ratings of B, based upon the frequencies in each cell of our matrix. Thus the ratings of signal as "A, maybe" give $2 \times 7 = 14$ "correct" pairings (i.e., A rated higher than B) and the ratings of signal as "B, maybe" give $2 \times 4 = 8$ correct pairings. There are 17 total ties (counted as 8.5 correct pairings). The R-index, then is $45 + 14 + 8 + 8.5$ (for ties) $= 75.5$. In other words, our best guess about the total percent correct in a two-alternative forced-choice task like a paired comparison test would be about 75.5%.

This result indicates a slight degree of difference and that this pair of items is sometimes confusable. Obviously, as the two stimuli have less overlapping response patterns, there is better discrimination and a higher R-value. Remember that this would correspond to 75.5% of the area below the ROC curve for this person. Taking a z-score and multiplying by the square root of 2 gives us a d' of 0.957 (Bi, 2006b). This value, close to one, also suggests that the difference is above threshold but not very clear. Statistical tests for the R-index, including confidence intervals for similarity testing are given in Bi (2006b).

As in other signal detection methods, the R-index allows us to separate discrimination performance from the bias or criterion a person sets for responding. For example, we might have an observer who is very conservative and calls all stimuli noise or labels them "B" in our example. If the observer assigned all A-trials (signals) to "B, maybe" and all B-trials (noise) to "B, definitely" then the R-index would equal 100, in keeping with perfect discrimination. The fact that all stimuli were considered examples of "B" or noise shows a strong response bias, but this does not matter. We have evidence for perfect discrimination due to the assignment of the two stimuli to different response classes, even though the observer was very biased to use only one part of the rating scale. Another advantage of R-index methods is that far fewer trials need be given as compared with the yes/no procedure.

5.11 Conclusions

A common issue in applied sensory testing is whether the experimental sample is close enough to a control or standard sample to justify a decision to substitute it for the standard. This is an exceedingly difficult question from a scientific perspective, since it seems to depend upon proving a negative, or in statistical terms, on proving the null hypothesis. However, failure to reject the null can occur for many reasons. There may be truly no difference, there may be insufficient sample size (N too low), or there may be too much variability or error obscuring the difference (standard deviations too high). Since this situation is so ambiguous, in experimental science we are usually justified in withholding a decision if we find no significant effect. Statisticians often say "there is insufficient evidence to reject the null" rather than "there is no significant difference." However, in industrial testing, the non-significant difference can be practically meaningful in helping us decide that a sample is like some control, as long as we have some knowledge of the sensitivity and power of our test. For example, if we know the track record of a given panel and test method for our particular product lines, we can sometimes make reasonable decisions based on a non-significant test result.

An alternative approach is to choose some acceptable interval of the degree of difference and see whether we are inside that interval or below some acceptable limit. This chapter has approached the degree-of-difference issue from two perspectives. The first was to convert our percent correct to an adjusted percent correct based on the traditional correction for guessing given by Abbott's formula. This allows us to estimate the percent of people actually discriminating, assuming a simple two-category model (either you see the difference or you guess). The second approach is to look at the degree of difference, or conversely the power of the test to detect that difference, as a function of a Thurstonian scaled value such as delta or d'. This value provides a more universal yardstick for sensory differences, as it takes into account the difficulty or variability inherent in the test, and also the cognitive strategy of the panelist in different tasks.

Note that there is an important limitation to the correction-for-guessing models. The guessing model and the Thurstonian model have different implications regarding the difficulty of the triangle and 3-AFC test. The guessing model only considers the proportion correct and the chance performance rate in estimating the proportion of discriminators. For the same proportion correct in the triangle and the 3-AFC test, there will be the same estimated proportion of discriminators since they have the same chance probability level (1/3).

However, the Thurstonian/SDT model tells us that the triangle test is harder. For the same proportion correct in an experiment, there must be much better discriminability of the items to achieve that level in the triangle test. In other words, it was necessary for the products to be more dissimilar in the triangle test—since the triangle test is harder, it took a bigger difference to get to the observed proportion, as opposed to the 3-AFC. Obviously, being a "discriminator" in a triangle test requires a larger perceptual difference than being a discriminatory in 3-AFC. So the notion of discriminators is specific to the method employed. However, in spite of this logical limitation, the correction-for-guessing approach has some value in helping to make decisions about sample size, beta-risk, and power estimation. As long as one is always using the same test method, the problem of different d'-values need not come into play.

The use of d' as a criterion has an important limiting factor as well. The variance of a d' value is given as the value of a B-factor divided by N, the number of judges or observations (ASTM, 2008b) (see Table O for values of B). Unfortunately, the B-factor passes through a minimum near a d' or 2.0 and starts to increase again as d' approaches zero. This makes it difficult, from any practical perspective, to find a significant difference between some d' that you might choose as an acceptable upper limit and a low level of d' that you may find in the test you perform. For all practical purposes, testing an obtained d' against a d' limit less than 1.5 is not very efficient and demonstrating that a d' is significantly lower than 1.0 is very difficult given the size of most discrimination testing panels ($N = 50$–100). For this reason, conclusions about similarity using d' need to be based on simple rules-of-thumb, for example, by comparing the level of d' to those that have previously been found to be acceptable (see ASTM (2008b) for further discussion).

In conclusion, we offer the following guidelines for those seeking evidence of sensory equivalence or similarity: First, apply the common sense principles discussed at the beginning of this chapter. Make sure you have a sensitive test instrument that is capable of detecting differences. If possible, include a control test to show that the method works or be prepared to illustrate the track record of the panel with previous results. Second, do power and sample size calculations to be sure you have an adequate panel size and adequate appreciation of what the test is likely to detect or miss. Third, get management to specify how much of a difference is acceptable. A company with a long history of difference or equivalence testing may have a benchmark d', a proportion of discriminators or some other benchmark or degree of difference that is acceptable. Fourth, adopt one of the statistical approaches such as a similarity test, interval testing (see Ennis and Ennis, 2010, for a new approach), or TOST to prove that you are below (or within) some acceptable limit of variation. Finally, be aware of the power of your test to detect a given degree of difference. The best measures of degree of difference from choice tests are given by the Thurstonian delta or d' values which are independent of the particular test method.

Appendix: Non-Central t-Test for Equivalence of Scaled Data

Bi (2007) described a similarity test for two means, as might come from some scaled data such as acceptability ratings, descriptive panel data, or quality control panel data. The critical test statistic is T_{AH} after the original authors of the test, Anderson and Hauck. If we have two means, M_1 and M_2, from two groups of panelists with N panelists per group and a variance estimate, S, the test proceeds as follows:

$$T_{AH} = \frac{M_1 - M_2}{s\sqrt{2/N}} \qquad (5.13)$$

The variance estimate, S, can be based on the two samples, where

$$S^2 = \frac{S_1^2 + S_2^2}{N} \qquad (5.14)$$

and we must also estimate a non-centrality parameter, δ,

$$\delta = \frac{\Delta_0}{s\sqrt{2/N}} \qquad (5.15)$$

where Δ_0 is the allowable difference interval.

The calculated p-value is then

$$p = t_\nu(|T_{AH}| - \delta) - t_\nu(-|T_{AH}| - \delta) \qquad (5.16)$$

and t_ν is the p-value from the common central t-distribution value for $\nu = 2(N-1)$ degrees of freedom. If p is less than our cutoff, usually 0.05, then we can conclude that our difference is within the acceptable interval and we have equivalence.

For paired data, the situation is even simpler, but in order to calculate your critical value, you need a calculator for critical points of the non-central F-distribution, as found in various statistical packages.

To apply this, perform a simple dependent samples (paired data) t-test. Determine the maximum allowable difference in terms of the scale difference and normalize this by stating it in standard deviation units. The obtained value of t is then compared to the critical value as follows:

$$C = \sqrt{F} \qquad (5.17)$$

where the F value corresponds to a value for the non-central F-distribution for 1, $N-1$ degrees of freedom, and a non-centrality parameter, given by $N(\varepsilon)$, and (ε) is the size of the critical difference in standard deviation units. If you do not have easy access to a calculator for the critical values of a non-central F, a very useful table is given in Gacula et al. (2009) where the value of T may be directly compared to the critical value based on an alpha level of 0.05 and various levels of (ε) (Appendix Table A.30, pp. 812–813 in Gacula et al., 2009). The absolute value of the obtained t-value must be less than the critical C value to fall in the range of significant similarity or equivalence.

Worked examples can be found in Bi (2005) and Gacula et al. (2009).

References

ASTM. 2008a. Standard guide for sensory claim substantiation. Designation E-1958-07. Annual Book of Standards, Vol. 15.08. ASTM International, West Conshohocken, PA, pp. 186–212.

ASTM. 2008b. Standard practice for estimating Thurstonian discriminal differences. Designation E-2262-03. Annual Book of Standards, Vol. 15.08. ASTM International, West Conshohocken, PA, pp. 253–299.

Amerine, M. A., Pangborn, R. M. and Roessler, E. B. 1965. Principles of Sensory Evaluation of Food, Academic Press, New York, pp. 437–440.

Antinone, M. A., Lawless, H. T., Ledford, R. A. and Johnston, M. 1994. The importance of diacetyl as a flavor component in full fat cottage cheese. Journal of Food Science, 59, 38–42.

Baird, J. C. and Noma, E. 1978. Fundamentals of Scaling and Psychophysics. Wiley, New York.

Bi, J. 2005. Similarity testing in sensory and consumer research. Food Quality and Preference, 16, 139–149.

Bi, J. 2006a. Sensory Discrimination Tests and Measurements. Blackwell, Ames, IA.

Bi, J. 2006b. Statistical analyses for R-index. Journal of Sensory Studies, 21, 584–600.

Bi, J. 2007. Similarity testing using paired comparison method. Food Quality and Preference, 18, 500–507.

Byer, A. J. and Abrams, D. 1953. A comparison of the triangle and two-sample taste test methods. Food Technology, 7, 183–187.

Delwiche, J. and O'Mahony, M. 1996. Flavour discrimination: An extension of the Thurstonian "paradoxes" to the tetrad method. Food Quality and Preference, 7, 1–5.

Ennis, D. M. 1990. Relative power of difference testing methods in sensory evaluation. Food Technology, 44(4), 114, 116–117.

Ennis, D. M. 1993. The power of sensory discrimination methods. Journal of Sensory Studies, 8, 353–370.

Ennis, D. M. 2008. Tables for parity testing. Journal of Sensory Studies, 32, 80–91.

Ennis, D. M. and Ennis J. M. 2010. Equivalence hypothesis testing. Food Quality and Preference, 21, 253–256.

Ennis, D.M. and Mullen, K. 1986. Theoretical aspects of sensory discrimination. Chemical Senses, 11, 513–522.

Ennis, D. M. and O'Mahony, M. 1995. Probabilistic models for sequential taste effects in triadic choice. Journal of Experimental Psychology: Human Perception and Performance, 21, 1–10.

Ferdinandus, A., Oosterom-Kleijngeld, I. and Runneboom, A. J. M. 1970. Taste testing. MBAA Technical Quarterly, 7(4), 210–227.

Finney, D. J. 1971. Probit Analysis, Third Edition. Cambridge University, New York.

Frijters, J. E. R. 1979. The paradox of the discriminatory nondiscriminators resolved. Chemical Senses, 4, 355–358.

Frijters, J. E. R., Kooistra, A. and Vereijken, P. F. G. 1980. Tables of d' for the triangular method and the 3-AFC signal detection procedure. Perception and Psychophysics, 27(2), 176–178.

Gacula, M. C., Singh, J., Altan, S. and Bi, J. 2009. Statistical Methods in Food and Consumer Research. Academic and Elsevier, Burlington, MA.

Green, D.M. and Swets, J. A. 1966. Signal Detection Theory and Psychophysics. Wiley, New York.

Lawless, H. T. 2010. A simple alternative analysis for threshold data determined by ascending forced-choice method of limits. Journal of Sensory Studies, 25, 332–346.

Lawless, H. T. and Schlegel, M. P. 1984. Direct and indirect scaling of taste—odor mixtures. Journal of Food Science, 49, 44–46.

Lawless, H. T. and Stevens, D. A. 1983. Cross-adaptation of sucrose and intensive sweeteners. Chemical Senses, 7, 309–315.

Macmillan, N. A. and Creelman, C. D. 1991. Detection Theory: A User's Guide. University Press, Cambridge.

MacRae, A. W. 1995. Confidence intervals for the triangle test can give reassurance that products are similar. Food Quality and Preference, 6, 61–67.

Meilgaard, M., Civille, G. V. and Carr, B. T. 2006. Sensory Evaluation Techniques, Fourth Edition. CRC, Boca Raton.

Morrison, D. G. 1978. A probability model for forced binary choices. American Statistician, 32, 23–25.

O'Mahony, M. A. 1979. Short-cut signal detection measures for sensory analysis. Journal of Food Science, 44(1), 302–303.

O'Mahony, M. and Odbert, N. 1985. A comparison of sensory difference testing procedures: Sequential sensitivity analysis and aspects of taste adaptation. Journal of Food Science, 50, 1055.

O'Mahony, M., Masuoka, S. and Ishii, R. 1994. A theoretical note on difference tests: Models, paradoxes and cognitive strategies. Journal of Sensory Studies, 9, 247–272.

Schlich, P. 1993. Risk tables for discrimination tests. Food Quality and Preference, 4, 141–151.

Stillman, J. A. and Irwin, R. J. 1995. Advantages of the same-different method over the triangular method for the measurement of taste discrimination. Journal of Sensory Studies, 10, 261–272.

Thurstone, L. L. 1927. A law of comparative judgment. Psychological Review, 34, 273–286.

Ura, S. 1960. Pair, triangle and duo-trio test. Reports of Statistical Application Research. Japanese Union of Scientists and Engineers, 7, 107–119.

USFDA. 2001. Guidance for Industry. Statistical Approaches to Bioequivalence. U.S. Department of Health and Human Services, Food and Drug Administration, Center for Drug Evaluation and Research (CDER). http://www.fda.gov/cder/guidance/index.htm

Viswanathan, S., Mathur, G. P., Gnyp, A. W. and St. Peirre, C. C. 1983. Application of probability models to threshold determination. Atmospheric Environment, 17, 139–143.

Welleck, S. 2003. Testing Statistical Hypotheses of Equivalence. CRC (Chapman and Hall), Boca Raton, FL.

Chapter 6

Measurement of Sensory Thresholds

Abstract This chapter discusses the concept of threshold and contrasts the conceptual notion with the idea of threshold as a statistically derived quantity. A simple method for determining detection thresholds based on ASTM method E-679 is illustrated with a worked example. Other methods for determining thresholds are discussed as well as alternative analyses.

A light may be so weak as not sensibly to dispel the darkness, a sound so low as not to be heard, a contact so faint that we fail to notice it. In other words, a finite amount of the outward stimulus is required to produce any sensation of its presence at all. This is called by Fechner the law of the threshold—something must be stepped over before the object can gain entrance to the mind.
—(William James, 1913, p. 16)

Contents

6.1 Introduction: The Threshold Concept

One of the earliest characteristics of human sensory function to be measured was the absolute threshold. The absolute or detection threshold was seen as an energy level below which no sensation would be produced by a stimulus and above which a sensation would reach consciousness. The concept of threshold was central to Fechner's psychophysics. His integration of Weber's law produced the first psychophysical relationship. It depended upon the physical intensity being measured with the threshold for sensing changes as the unit (Boring, 1942). Early physiologists like Weber and Fechner would use the classical method of limits to measure this point of discontinuity, the beginning of the psychophysical function. In the method of limits, the energy level would be raised and lowered and the average point at which the observer changed response from "no sensation"

H.T. Lawless, H. Heymann, *Sensory Evaluation of Food*, Food Science Text Series,
DOI 10.1007/978-1-4419-6488-5_6, © Springer Science+Business Media, LLC 2010

to "yes, I perceive something" would be taken as the threshold. This specification of the minimum energy level required for perception was one of the first operating characteristics of sensory function to be quantified. Historically, the other common property to be measured was the difference threshold or minimal increase in energy needed to produce a noticeable increase in sensation. Together, these two measures were used to specify the psychophysical function, which to Fechner was a process of adding up difference thresholds once the absolute (minimum) threshold had been surpassed.

In practice, some complications arise in trying to apply the threshold idea. First, anyone who attempts to measure a threshold finds that there is variability in the point at which observers change their response. Over multiple measurements there is variability even within a single individual. In a sequence of trials, even within the same experimental session, the point at which a person changes his or her responses will differ. An old story has it that S.S. Stevens, one of the pioneers of twentieth century psychophysics, used the following classroom demonstration at Harvard: Students were asked to take off their wristwatches and hold them at about arm's length, then count the number of ticks they heard in 30 s (back in the day when spring-wound watches still made ticking sounds). Assuming the watch of one of these Harvard gentlemen made uniform ticking sounds, the common result that one would hear some but not all of the ticks illustrated the moment-to-moment variation in auditory sensitivity. Of course, there are also differences among individuals, especially in taste and smell sensitivity. This led to the establishment of common rules of thumb for defining a threshold, such as the level at which detection occurs 50% of the time.

The empirical threshold (i.e., what is actually measured) remains an appealing and useful concept to many workers involved in sensory assessments. One example is in the determination of flavor chemicals that may contribute to the aromatic properties of a natural product. Given a product like apple juice, many hundreds of chemical compounds can be measured through chemical analysis. Which ones are likely to contribute to the perceived aroma? A popular approach in flavor analysis assumes that only those compounds that are present in concentrations above their thresholds will contribute. A second example of the

usefulness of a threshold is in defining a threshold for taints or off-flavors in a product. Such a value has immediate practical implications for what may be acceptable versus unacceptable levels of undesired flavor components. Turning from the product to the sensory panelists, a third application of thresholds is as one means of screening individuals for their sensitivity to key flavor components. The measurement of a person's sensitivity has a long history in clinical testing. Common vision and hearing examinations include some measurements of thresholds. In the chemical senses, threshold measurements can be especially useful, due to individual differences in taste and smell acuity. Conditions such as specific anosmia, a selective olfactory deficit, can be important in determining who is qualified for sensory test panel participation (Amoore, 1971).

Another appealing aspect of the threshold concept is that the values for threshold are specified in physical intensity units, e.g., moles per liter of a given compound in a product. Thus many researchers feel comfortable with threshold specification since it appears to be free from the subjective units of rating scales or sensory scores. However, threshold measurements are no more reliable or accurate than other sensory techniques and are usually very labor intensive to measure. Perhaps most importantly, thresholds represent only one point on a dose–response curve or psychophysical function, so they tell us little about the dynamic characteristics of sensory response as a function of changes in physical concentration. How the sensory system behaves above threshold requires other kinds of measurements.

In this chapter, we will look at some threshold definitions and approaches and their associated problems. Next, we will examine some practical techniques for threshold measurement and discuss a few applications. Throughout, we will pay special attention to the problems of variability in measurement and the challenges that this poses for researchers who would use thresholds as practical measures of peoples' sensitivities to a given stimulus, or conversely, of the potency or biological activity of that stimulus in activating sensory perceptions. Most of the examples chosen come from olfaction and taste, as the chemical senses are especially variable and are prone to difficulties due to factors such as sensory adaptation.

6.2 Types of Thresholds: Definitions

What is a threshold? The American Society for Testing and Materials (ASTM) provides the following definition that captures the essence of the threshold concept for the chemical senses: "A concentration range exists below which the odor or taste of a substance will not be detectable under any practical circumstances, and above which individuals with a normal sense of smell or taste would readily detect the presence of the substance."—ASTM method E-679-79 (2008a, p. 36).

Conceptually, the absolute or detection threshold is the lowest physical energy level of a stimulus or lowest concentration in the case of a chemical stimulus that is perceivable. This contrasts with empirical definitions of threshold. When we try to measure this quantity, we end up establishing some practical rule to find an arbitrary value on a range of physical intensity levels that describes a probability function for detection. In 1908, the psychologist Urban recognized the probabilistic nature of detection and called such a function a psychometric function, as shown in Fig. 6.1 (Boring, 1942). We portray this function as a smooth curve in order to show how the original concept of a fixed threshold boundary was impossible to measure in practice. That is, there is no one energy level below which detection never occurs and above which detection always occurs.

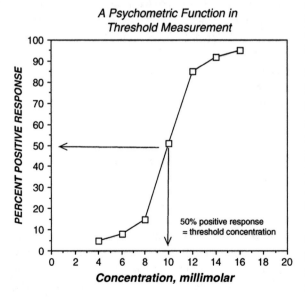

A Psychometric Function in Threshold Measurement

Fig. 6.1 A psychometric function.

It is not a sudden step function. There is a probability function determined by an empirical method of measurement on which we define some arbitrary point as the threshold.

Recognition thresholds are also sometimes measured. These are the minimum levels that take on the characteristic taste or smell of the stimulus and are often a bit higher than detection thresholds. For example, dilute NaCl is not always salty, but at low concentrations just above the detection threshold is perceived as sweet (Bartoshuk et al., 1978). The concentration at which a salty taste is apparent from NaCl is somewhat higher. In food research, it is obvious that the recognition threshold for a given flavor in a food would be a useful thing to know, and perhaps more useful than detection thresholds, since both the percept and the appropriate label have been made consciously available and actionable to the taster. In the case of off-flavors or taints, recognition may have strong hedonic correlates in predicting consumer rejection.

To be recognized and identified, discrimination from the diluent is only one requirement. In addition, the observer must assign the appropriate descriptor word to the stimulus. However, it is difficult to set up a forced-choice experiment for identification in some modalities. In taste, for example, you can have the observer pick from the four (or five) taste qualities, but there is no assurance that these labels are sufficient to describe all sapid substances (O'Mahony and Ishii, 1986). Furthermore, one does not know if there is an equal response bias across all four alternatives. Thus the expected frequencies or null hypothesis for statistical testing or difference from chance responding is unclear. In an experiment on bitter tastes, Lawless (1980) attempted to control for this bias by embedding the to-be-recognized bitter substances in a series that also included salt, acid, and sugar. However, the success of such a procedure in controlling response biases is unclear and at this time there are no established methods for recognition thresholds that have adequately addressed this problem.

The difference threshold has long been part of classical psychophysics (see Chapter 2). It represents the minimum physical change necessary in order for a person to sense the change 50% of the time. Traditionally, it was measured by the method of constant stimuli (a method of comparison to a constant reference) in which a series of products were raised and lowered around the level of the reference. The subject would be asked to say which of member of the pair was stronger and the point at which the "stronger" judgment occurred 75% (or 25%) of the time was taken as the difference threshold or "just-noticeable-difference" (JND).

One can think of sensory discrimination tests (triangles and such) as a kind of difference threshold measurement. The main difference between a psychophysical threshold test and a sensory discrimination test is that the psychophysical procedure uses a series of carefully controlled and usually simple stimuli of known composition. The sensory product test is more likely to have only two products, and the pair is either deemed different or not, based on a criterion of statistical significance. But clearly the two kinds of tests are related. Along these lines, one can think of the absolute threshold as a special case of a difference threshold, when the standard happens to be some blank or baseline stimulus (such as pure air or pure water).

In addition to detection, recognition, and difference thresholds, a fourth category is the terminal threshold or region in which no further increase in response is noted from increasing physical stimulus intensity (Brown et al., 1978). In other words, the sensory response has reached some saturation level, beyond which no further stimulation is possible due to maximal responding of receptors or nerves or some physical process limiting access of the stimulus to receptors. This makes sense in terms of neurophysiology as well. There are only a limited number of receptors and nerves and these have a maximal response rate. This idea fits well with the notion of a threshold as a discontinuity or inflection point in the psychophysical function (Marin et al., 1991).

However, in practice, this level is rarely approached. There are few foods or products in which the saturation level is a common level of sensation, although some very sweet confections and some very hot pepper sauces may be exceptions. For many continua, the saturation level is obscured by the addition of new sensations such as pain or irritation (James, 1913). For example, some odors have a down-turn in the psychophysical function at high levels, as trigeminal irritation begins to take place, that may in turn have an inhibiting effect on odor intensity (Cain, 1976; Cain and Murphy, 1980). Another example is in the bitter side taste of saccharin. At high levels, the bitterness will overtake the sweet sensation for some individuals. This makes it difficult to find a sweetness match of saccharin to other sweeteners at high levels (Ayya and Lawless, 1992). Further increases in concentration only increase bitterness and this additional sensation has an inhibiting effect on sweet perception. So although saturation of response seems physiologically reasonable, the complex sensations evoked by very strong stimuli mediate against any measurement of this effect in isolation.

Recently, a new type of threshold has been proposed for consumer rejection of a taint or off-flavor. Prescott et al. (2005) examined the levels at which consumers would show an aversion to cork taint from trichloroanisole in wines. Using a paired preference tests with increasing levels of trichloroanisole, they defined the rejection threshold as the concentration at which there was a statistically significant preference for an untainted sample. This novel idea may find wide application in flavor science and in the study of specific commodities (like water) in which the chemistry and origins of taints are fairly well understood (for another example, see Saliba et al., 2009). The method in its original form requires some refinement as to the criterion for threshold because statistical significance is a poor choice. As they noted in their paper, the level of statistical significance depends upon the number of judges, a more or less arbitrary choice of the experimenter (not a function of the sensory response of the participants). A better choice would be something akin to the difference threshold, i.e., the concentration at which 75% preference was reached. Of course, confidence intervals can be drawn around any such level for those that need statistical assurance.

6.3 Practical Methods: Ascending Forced Choice

In the early days of psychophysics, the method of limits was the most common approach to measuring thresholds. In this procedure, stimulus intensity would be raised in an ascending series and then lowered in a descending series to find points at which the observer's response changed from a negative to a positive response or from positive to negative. Over several

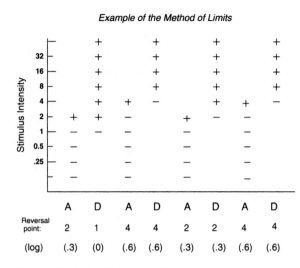

Fig. 6.2 Method of limits example.

ascending and descending runs, an average changing point could be taken as the best estimate of threshold (McBurney and Collings, 1977). This method is illustrated in Fig. 6.2.

Although this procedure seems straightforward, it has several problems. First, the descending series may cause such fatigue or sensory adaptation that the observer fails to detect stimulus presentations that would be clearly perceived if they were presented in isolation. To avoid the adaptation or fatigue problem that is common in the taste and smell senses, the method is usually performed only in an ascending series. A second difficulty is that different persons may set different criteria for how much of a sensation they require before changing their response. Some people might be very conservative and have to be positively sure before they respond, while others might take any inkling at all as a reason to report a sensation. Thus the classical method of limits is contaminated by the panelist's individual bias or criterion, which is not a function of their sensitivity, i.e., what the test is actually trying to measure. This is a central issue in the theory of signal detection (see Chapter 5). To address the problem of uncontrolled individual criterion, later workers introduced a forced choice element to the trials at each intensity level or concentration step (e.g., Dravnieks and Prokop, 1975). This combines the method of limits with a discrimination test. The task requires that the observer gives objective proof of detection by discriminating the target stimulus from

the background level. A forced choice technique is compatible with signal detection principles and is bias free, since the observer does not choose whether or not to respond—response is required on each trial.

6.4 Suggested Method for Taste/Odor/Flavor Detection Thresholds

6.4.1 Ascending Forced-Choice Method of Limits

This procedure is based on a standard method designated ASTM E-679 (ASTM, 2008a). It follows the classical method of limits, in which the stimulus intensity, in this case concentration of a taste or odor chemical, is raised in specified steps until the substance is detected. The procedure adds a forced choice task in which the substance to be detected is embedded in a set of stimuli or products that includes other samples that do not contain any of the added substance. The stimulus or product with the taste or odor chemical is called a "target" and the other items with no added chemical are often referred to as "blanks." One can use various combinations of targets and blanks, but it is common to have one target and in the case of E-679, two additional blanks. So the task is a three-alternative forced choice task (3-AFC), because the person being tested is forced to choose the one different sample in the set of three. That is if they are uncertain, they are told to guess.

6.4.2 Purpose of the Test

This method is designed to find the minimum level (minimum concentration) of a substance that is detected by 50% of the sample group. In practice, this is calculated as the geometric mean of the individual threshold estimates. The geometric mean is a reasonable choice because it is often very close to the median (50th percentile) of a positively skewed distribution. Threshold data tend to show high outliers, i.e., some insensitive individuals cause positive skew.

6.4.3 Preliminary Steps

Before the test is conducted, there are several tasks and some choices that must be made, as shown in Table 6.2. First, a sample of the substance of known purity must be obtained. Second, the diluent (solvent, base) or carrier must be chosen. For the detection threshold for flavors, for example, it is common to use some kind of pure water such as deionized or distilled. Third, the size of the concentration steps must be chosen. It is common to use factors of two or three. In other words, the concentrations will be made up in a geometric progression, which are equal steps on a log scale. Fourth, some sample concentrations should be set up for preliminary or "benchtop screening" to estimate the range in which the threshold is likely to occur. This can be done by successive dilutions using factors of five or ten, but beware the effects of adaptation on reducing one's sensitivity to subsequent test items. Exposure to a strong sample early in this series may cause subsequent samples to seem odorless or tasteless, when they might in fact be perceived when tasted alone. The outcome of the preliminary test should bracket the likely concentration range, so that most, if not all, of the people who participate in the formal test will find an individual threshold estimate somewhere within the test series. It is common to use about eight to ten steps in this procedure.

Next, the panel should be recruited or selected. A sample group should have at least 25 participants. If the goal is to generalize the result to some larger population, then the panel should be representative of that population with respect to age, gender, and so on and a larger panel of 100 or more is recommended. It is common practice to exclude people with known health problems that could affect their sense of taste or smell and individuals with obvious sensory deficits in the modality being tested. Of course, all the appropriate setup work must be done that is associated with conducting any sensory test, such as securing a test room that is free from odors and distractions, scheduling the panelists, setting up the questionnaire or answer sheet, writing instructions for the participants. See Chapter 3 for further details on good practices in sensory testing. For threshold work it is especially important to have clean odor-free glassware or plastic cups that are absolutely free of any odor that would contaminate the test samples. In odor testing the sample vessels are usually covered to preserve the equilibrium in the headspace above the liquid. The covers are removed by each panelist at the moment of sniffing and then replaced. Finally, external sources of odor must be minimized or eliminated, such as use of perfumes or fragrances by participants, hand lotions, or other fragranced cosmetics that could contaminate the sample vessels or the general area. Avoid using any markers or writing instruments that might have an odor when marking the samples. As always, sample cups or vessels should be marked with blind codes, such as randomly chosen three digit numbers. The experimenter must set up random orders for the three items at each step and use a different randomization for each test subject. This

Table 6.1 Types of thresholds

Detection (absolute) threshold:	Point at which the substance is differentiated from the background
Recognition threshold:	Point at which the substance is correctly named
Difference threshold:	(just-noticeable-difference, JND) Point at which the change in concentration is noted
Terminal threshold:	Point at which no further intensity increase is found with increasing concentration
Consumer rejection threshold:	Point at which a consumer preference occurs for a sample *not* containing the substance

Table 6.2 Preliminary tasks before threshold testing

1. Obtain test compound of known purity (note source and lot number)
2. Choose and obtain the solvent, carrier, or food/beverage system
3. Set concentration/dilution steps, e.g., 1/3, 1/9, 1/27
4. Begin benchtop screening to bracket/approximate threshold range
5. Choose number of dilution steps
6. Recruit/screen panelists. $N \geq 25$ is desirable
7. Establish procedure and pilot test if possible
8. Write verbatim instructions for panelists

should be recorded on a master coding sheet showing the randomized three-digit codes and which sample is the correct choice or target item.

6.4.4 Procedure

The steps in the test are shown in Table 6.3. The participant or test subject is typically seated before a sample tray containing the eight or so rows of three samples. Each row contains one target sample and two blank samples, randomized. The instructions, according to E-679-04 (ASTM, 2008a) are the same as in the triangle test, that is to pick out the sample which is different from the other two. The subject is told to evaluate the three samples in each row once, working from left to right. The test proceeds through all the steps of the concentration series and the answers from the subject are recorded, with a forced guess if the person is uncertain. According to E-679, if a person misses at the highest level, that level will be repeated. If a person answers correctly through the entire series, the lowest level will also be repeated for confirmation. If the response changes in either case, it is the repeated trial that is counted.

6.4.5 Data Analysis

Figure 6.3 shows an example of how the data are analyzed and the threshold value is determined. First, an individual estimated threshold is determined for each person. This is defined as the concentration that is the geometric mean of two values (the square root of the product of the two values). One value is the concentration at which they first answered correctly and all higher concentrations were also correct. The other value is the concentration just below that, i.e., the last incorrect judgment. This interpolation provides some protection against the fact that the forced-choice procedure will tend to slightly overestimate the individual's threshold (i.e., the concentration at which they have a 0.5 probability of sensing that something is different from the blanks). If the subject gets to the top of the series with an incorrect judgment, or starts at the bottom with all judgments correct, then a value is extrapolated beyond the test series. At the top, it is the geometric mean of the highest concentration tested and the next concentration that would have been used in the series if the series had been continued. At the bottom, it is the geometric mean of the lowest concentration tested and the next lower concentration that would have been used had the series been continued lower. This is an arbitrary rule, but it is not unreasonable. Once these individual best estimates are tabulated, the group threshold is the geometric mean of the individual values. The geometric mean is easily calculated by taking the log of each of the individual concentration values, finding the average of the logs, and then taking the antilog of this value (equivalent to taking the Nth root of the product of N observations).

6.4.6 Alternative Graphical Solution

An alternative analysis is also appropriate for this kind of data set. Suppose that 3-AFC tests had been conducted and the group percent correct calculated at each step. For examples, see Antinone et al. (1994)

Table 6.3 Ascending forced-choice testing steps

1. Obtain randomized or counterbalanced orders via software program or random number generator.
2. Setup trays or other staging arrangements for each participant, based on random orders.
3. Instruct participants in procedure per verbatim script developed earlier.
4. Show suprathreshold example (optional).
5. Present samples and record results. Force a choice if participant is unsure.
6. Tally results for panel as series of correct/incorrect answers.
7. Calculate estimated individual thresholds: Geometric mean of first correct answer with all higher concentrations correct and last incorrect step.
8. Take geometric mean of all individual threshold estimates to get group threshold value.
9. Plot graphic results of proportion correct against log concentration. Interpolate 66.6% correct point and drop line to concentration axis to get another estimate of threshold (optional).
10. Plot upper and lower confidence interval envelopes based on $\pm 1.96(p\,(1-p)/N)$. Drop lines from the upper and lower envelopes at 66.6% to concentration axis to convert envelope to concentration interval.

Panelist	Concentration ug/L								BET	log(BET)
	2	3.5	6	10	18	30	60	100		
1	+	o	+	+	+	+	+	+	2.6	0.415
2	o	o	o	+	+	+	+	+	7.7	0.886
3	o	+	o	o	o	o	+	+	42	1.623
.	.	.	.	.	.	.	.	.	.	.
.	.	.	.	.	.	.	.	.	.	.
.	.	.	.	.	.	.	.	.	.	.
N	o	o	+	+	o	+	o	+	77	1.886
Prop. Corr.	0.44	0.49	0.61	0.58	0.65	0.77	0.89	0.86	Mean (log(BET))	1.149
									10^1.149 =	14.093

Fig. 6.3 Sample data analysis from ascending 3-AFC method Notes: Correct choices indicated by + and incorrect by o. BET, Best estimate of individual threshold, defined as the geometric mean of the first correct trial with all subsequent trials correct and the previous (incorrect) trial. The group threshold is calculated from the geometric mean of the BET values. In practice, this is done by taking the logs of the BET values, finding the mean of the logs (x), then taking the antilog of that value (or 10^x).

and Tuorila et al. (1981). We can take the marginal count of the number of correct choices from the bottom row in Fig. 6.3 and expressing it as the proportion correct. As the concentration increases, this proportion should go from near the chance level (1/3) to nearly 100% correct. Often this curve will form an S-curve similar to the cumulative normal distribution. The threshold can then be defined as the level at which performance is 50% correct, once we have adjusted the data for chance, i.e., the probability that a person could guess correctly (Morrison, 1978; Tuorila et al., 1981; Viswanathan et al., 1983). This is done by Abbott's formula, a well-known correction for guessing as shown in Eqs. (6.1) and (6.2):

$$P_{corr} = (P_{obs} - P_{chance})/(1 - P_{chance}) \qquad (6.1)$$

where P_{corr} is the chance-corrected proportion, P_{obs} is the observed proportion correct in the data, and P_{chance} is the chance probability, e.g., 1/3 for the 3-AFC. Another form is

$$P_{req} = (P_{chance} - P_{corr})/(1 - P_{chance}) \qquad (6.2)$$

where P_{req} is the observed proportion that is required in order to achieve a certain chance corrected level of performance. So if one needed to get a chance corrected proportion of 0.5 (i.e., a threshold, 50% detection) in a 3-AFC test, you would need to see 1/3 + 0.5 (1–1/3) or 2/3 (= 66.7%) correct.

Once a line or curve is fitted to the data, the concentration at which the group would achieve 66.6% correct can be solved (or simply interpolated by eye

if the data are fairly linear and a curve is fit by eye). A useful equation that can be fit to many data sets is based on logistic regression, shown in Eq. (6.3) (e.g., Walker et al., 2003).

$$\ln\left(\frac{p}{1-p}\right) = b_0 + b_1 \log C \qquad (6.3)$$

where p is the proportion correct at concentration C and b_0 and b_1 are the intercept and slope. The quantity, $p/1-p$, is sometimes referred to as the odds ratio. The interpolation is shown in Fig. 6.4. Note that this also

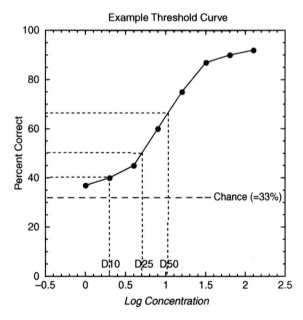

Fig. 6.4 Sample threshold curve and interpolation. D10, D25 and D50 show interpolated detection levels for 10%, 25% and 50% of persons, respectively.

allows one to estimate percentages of the population that would detect with other probabilities and not just the arbitrary 50% detection that we use as the threshold value. That is, one could interpolate at 10 or 90% detection if that was of interest. A lower percentage of detection might be of interest, for example, in setting level to protect consumers from an off-flavor or taint.

This graphical method has certain assumptions and limitations that the user should be aware of. First, it assumes that persons are either detecting or guessing (Morrison, 1978). In reality, every person has an individual threshold gradient or gradually increasing probability of detection around their own threshold. Second, the model does not specify what *percent of the time* that a given percentage of the group will detect. In the data set examined below, the ASTM method and the graphical solution provide a good estimate of when 50% of the group will detect 50% of the time. More extensive statistical models have been developed for this kind of data and an extensive paper on alternative statistical analyses is given in USEPA (2001), again using the data set we have chosen as an example below.

6.4.7 Procedural Choices

Note that although the instructions are the same as in the triangle test, all the possible combinations of the three samples are not used, i.e., the test is not a fully counterbalanced triangle. Only the three possible orders that are given by combinations of two blanks and one target are used. In a fully counterbalanced triangle, the additional three combinations of two targets and one blank would have been used (thus a total of six possible), but this is not done according to E-679. For taste or flavor, there is generally no rinsing between samples, although testers may be instructed to rinse between rows (triads). If possible, it is wise to give the subject a preliminary sample at a detectable level, in order to show them the target item that they will be trying to sense in the test. Of course, one must be careful when using such an above-threshold sample so that it does not adapt or fatigue the senses. An appropriate waiting time and/or rinsing requirement should be used to prevent any effect on the subsequent test samples in the formal test. The experimenters should also decide whether they will allow re-tasting or not.

Re-tasting could either confuse the subjects or it might help them get a better idea of which item is the target. We would generally argue against re-tasting, because that will introduce a variable that is left up to the individual subject and will thus differ among people. Some will choose to re-taste and others will not. So, on the basis of maintaining a consistent test procedure across all participants, re-tasting is generally not recommended.

Another important choice is that of a "stopping rule." In the published version of E-679, every subject must continue to the top of the series. There are some pitfalls in this, because of the possibility that the senses will become fatigued or adapted by the high levels at the top of the series, especially for an individual with a low personal threshold. For this reason, some threshold procedures introduce a "stopping rule." For example, the panelist may be allowed to stop tasting after giving three correct answers at adjacent levels (Dravnieks and Prokop, 1975). This prevents the problem of exposing a sensitive individual to an overwhelming stimulus at high levels. Such an experience, if unpleasant (such as a bitter taste), might even cause them to quit the test. On the downside, the introduction of a stopping rule can raise the false positive rate. We can think of a false positive as finding a threshold value for an individual that is due to guessing only. In the most extreme case, it would be a person who is completely insensitive (e.g., anosmic to that compound if it is an odor threshold) finding a threshold somewhere in the series. With an eight-step series, for the ASTM standard rule (everyone completes the series), the probability of finding a threshold somewhere in steps one through eight, for a completely anosmic person who is always guessing is 33.3%. For the three-in-a-row stopping rule, the chances of the anosmic person making three lucky guesses in a row somewhere rise above 50%. The sensory professional must weigh the possible negatives from exposing the participant to strong stimuli against the increased possibility of false positives creating a low-threshold estimate when using a stopping rule.

6.5 Case Study/Worked Example

For the ascending forced choice method of limits (ASTM E-679), we can use a published data set for odor thresholds. The actual data set is reproduced in the Appendix at the end of this chapter. The data are from a study conducted to find the odor detection threshold for methyl tertiary butyl ether (MTBE), a gasoline additive that can contaminate ground water, rendering some well waters unpotable (Stocking et al., 2001; USEPA, 2001). The ASTM procedure was followed closely, including the triangle test instructions (choose the sample different from the other two), using the 3-AFC in eight concentration steps differing by a factor of about 1.8. Individual best estimates were taken as the geometric mean of the last step missed and the first step answered correctly, with all higher steps also correct. Individuals who got the first and all subsequent steps correct (there were 10/57 or 17.5% of the group) had their estimated threshold assigned as the geometric mean of the first concentration and the hypothetical concentration one step below that which would have been used had the series been extended down. A similar extrapolation/estimation was performed at the high end for persons that missed the target on the eighth (highest) level.

The geometric mean of the individual threshold estimates across a panel of 57 individuals, balanced for gender and representing a range of ages, was 14 μg/l (14 ppb). Figure 6.5 shows the graphical solution, which gives a threshold of about 14 ppb, in good agreement with the geometric mean calculation. This is the interpolated value for 66.7% correct, the chance-adjusted level for 50% probability of detection in the group. Confidence intervals (CI) for this level can be found by constructing upper and lower curves form an envelope of uncertainty around the fitted curve. The standard error is given by the square root of $(p(1-p)/N)$ or in this case 0.062 for $p = 1/3$ and $N = 57$.

The 95% CI is found by multiplying the z-score for 0.95 ($= 1.96$) times the standard error, in this case equal to $\pm 0.062(1.96)$ or ± 0.122. Constructing curves higher and lower than the observed proportions by this amount will then permit interpolation at the 66.7% level to find concentrations for the upper and lower CI bounds. This method is simple, but it provides conservative (wider) estimate of the confidence intervals that found with some other statistical methods such as bootstrap analysis (USEPA, 2001). Another method for error estimation based on the standard error of the regression line is given in Lawless (2010).

Note that by the graphical method, the interpolated value for 10% detection (= 40% correct by Abbott's formula) will be at about 1–2 ppb. Similarly the interpolated value for 25% detection (50% correct by Abbott's formula), will be between 3 and 4 ppb. These values are practically useful to a water company who wanted to set lower limits on the amount of MTBE that could be detected by proportions of the population below the arbitrary threshold value of 50% (Dale et al., 1997).

6.6 Other Forced Choice Methods

Ascending forced-choice procedures are widely used techniques for threshold measurement in the experimental literature on taste and smell. One early example of this approach is in the method for determining sensitivity to the bitter compound phenylthiourea, formerly called phenylthiocarbamide or PTC, and the related compounds 6-*n*-propylthiouracil or PROP. Approximately one-third of Caucasian peoples are insensitive to the bitterness of these compounds, as a function of several mutations in a bitter receptor that usually manifests as a simple homozygous recessive status for this trait (Blakeslee, 1932; Bufe et al., 2005). Early researchers felt the need to have a very stringent test of threshold, so they intermingled four blank samples (often tap water) with four target samples at each

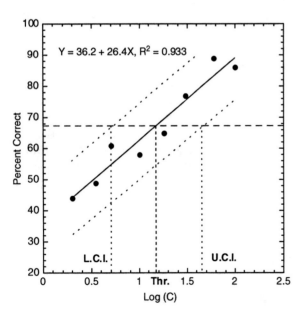

Fig. 6.5 Interpolation of threshold from the data of Stocking et al. (2001).

concentration step (Harris and Kalmus, 1949). The chance probability of sorting correctly is only 0.014, so this is a fairly difficult test. In general, the formula for the chance probability of sorting any one level of X target samples among N total samples is given by Eq. (6.4):

$$p = X!/[N!/(N - X)!]   \qquad (6.4)$$

Obviously, the larger the number of target and blank samples, the more stringent the test and the higher the eventual threshold estimate. However, arbitrarily increasing X and N may make the task tedious and may lead to other problems such as fatigue and flagging motivation among the participants. The rigor of the test estimate must be weighed against undue complexity that could lead to failures to complete the series or poor quality data.

Another example of a threshold test for olfaction is Amoore's technique for assessing specific anosmia (Amoore, 1979; Amoore et al., 1968). Specific anosmia describes a deficit in the ability to smell a compound or closely related family of compounds among people with otherwise normal olfactory acuity. Being classified as anosmic was operationally defined by Amoore as having olfactory detection thresholds more than two standard deviations above the population mean (Amoore et al., 1968). The test is sometimes called a "two-out-of-five" test because at each concentration level there are two target stimuli containing the odorant to be tested and three diluent or blank control samples. The tester must sort the samples correctly in this two-out-of-five test, and the chance probability of obtaining correct sorting by merely guessing is one in ten. Performance is normally confirmed by testing the next highest concentration (an example of a "stopping rule"). The chance occurrence of sorting correctly on two adjacent levels is then 1 in 100. This makes the test somewhat difficult but provides a good deal of insurance against a correct answer by guessing.

Another way to reduce the chance performance on any one level is to require multiple correct answers at any given concentration. This is part of the rationale behind the Guadagni multiple pairs test (Brown et al., 1978) in which up to four pairs may be given for a two-alternative forced choice test in quadruplicate at any one concentration. Brown et al. commented upon the user-friendliness of this technique, i.e., how simple it was to understand and administer to participants. A

variation was used by Stevens et al. (1988) in a landmark paper on the individual variability in olfactory thresholds. In this case, five correct pairs were required to score the concentration as correctly detected, and this performance was confirmed at the next highest concentration level. The most striking finding of this study was that among the three individuals tested 20 times, their individual thresholds for butanol, pyridine, and phenylethylmethylethyl carbinol (a rose odorant) varied over 2,000- to 10,000-fold in concentration. Variation within an individual was as wide as the variation typically seen across a population of test subjects. This surprising result suggests that *day-to-day variation in olfactory sensitivity is large and that thresholds for an individual are not very stable* (for an example, see Lawless et al., 1995). More recent work using extensive testing of individuals at each concentration step suggests that these estimates of variability may be high. Walker et al. (2003) used a simple yes/no procedure (like the A, not-A test, or signal detection test) with 15 trials of targets and 15 trials of blanks at each concentration level. Using a model for statistical significant differences between blank and target trials, they were able to get sharp gradients for the individual threshold estimates.

In summary, an ascending forced-choice method is a reasonably useful compromise between the need to precisely define a threshold level and the problems encountered in sensory adaptation and observer fatigue when extensive measurements are made. However, the user of an ascending forced-choice procedure should be aware of the procedural choices that can affect the obtained threshold value. The following choices will affect the measured value: the number of alternatives (both targets and blanks), the stopping rule, or the number of correct steps in a row required to establish a threshold, the number of replicated correct trials required at any one step, and the rule to determine at what level of concentration steps the threshold value is assigned. For example, the individual threshold might be assigned at the lowest level correct, the geometric mean between the lowest level correct and highest level incorrect. Other specific factors include the chosen step size of concentration units (factors of two or three are common in taste and smell), the method of averaging or combining replicated ascending runs on the same individual and finally the method of averaging or combining group data. Geometric means are commonly used for the last two purposes.

6.7 Probit Analysis

It is often useful to apply some kind of transformation or graphing method to the group data to linearize the curve used to find the 50% point in a group. Both the psychometric curve that represents the behavior of an individual in multiple trials of a threshold test and the cumulative distribution of a group will resemble an S-shaped function similar to the cumulative normal distribution. A number of methods for graphing such data are shown in the ASTM standard E-1432 (ASTM, 2008b). One simple way to graph the data is simply to plot the cumulative percentages on "probability paper." This pre-printed solution provides a graph in which equal standard deviations are marked off along the ordinate, effectively stretching the percentile intervals at the ends and compressing them in the midrange to conform to the density of the normal distribution. Another way to achieve the straightening of the S-shaped response curve is to transform the data by taking z-scores. Statistical packages for data analysis often provide options for transformation of the data.

A related method was once widely used in threshold measurement, called Probit analysis (ASTM, 2008b; Dravnieks and Prokop, 1975; Finney, 1971). In this approach, the individual points are transformed relative to the mean value, divided by the standard deviation and then a constant value of +5 is added to translate all the numbers to positive values for convenience. A linear fitted function can now be interpolated at the value of 5 to estimate the threshold as in Fig. 6.6. The conversion (to a z-score +5) tends to make an S-shaped curve more linear. An example of this can be found in the paper by Brown et al. (1978), using data from a multiple paired test. First the percent correct is adjusted for chance. Then the data are transformed from the percent correct (across the group) at each concentration level by conversion to z-scores and a constant of 5 is added. The mean value or Probit equal to 5 can be found by interpolation or curve fitting. An example of this technique for estimating threshold from a group of 20 panelists is shown in Meilgaard et al. (1991) and in ASTM (2008b). Probit plots can be used for any cumulative proportions, as well as ranked data and analysis of individuals who are more extensively tested than in the 3-AFC method example shown earlier.

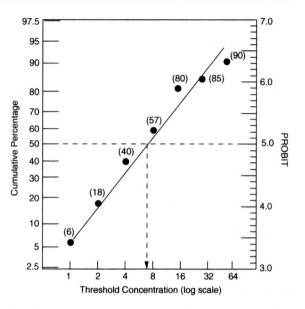

Fig. 6.6 An example of Probit analysis. Numbers in parentheses are the cumulative percentages of panelists reaching threshold at each concentration step. Note the uneven scale on the left axis. Probits mark off equal standard deviations and are based on the z-score for any proportion plus the constant, 5. Interpolation at the 50% or Probit 5.0 gives the threshold value.

6.8 Sensory Adaptation, Sequential Effects, and Variability

Individual variability, both among a group of people and within an individual over repeated measurements presents a challenge to the idea that the threshold is anything like a fixed value. For example, stable olfactory thresholds of an individual are difficult to measure. The test–retest correlation for individual's olfactory threshold is often low (Punter, 1983). Even within an individual, threshold values will generally decrease with practice (Engen, 1960; Mojet et al., 2001; Rabin and Cain, 1986), and superimposed upon this practice effect is a high level of seemingly random variation (Stevens et al., 1988). Individuals may become sensitive to odorants to which they were formerly anosmic, apparently through simple exposure (Wysocki et al., 1989). Increased sensitivity as a function of exposure may be a common phenomenon among women of childbearing age (Dalton et al., 2002; Diamond et al., 2005).

Sensory adaptation and momentary changes in sensitivity due to sequences may have occurred in the experiments of Stevens et al. (1988) and could have contributed to some instability in the measurements. As predicted by sequential sensitivity analysis (Masuoka et al., 1995; O'Mahony and Odbert, 1985) the specific stimulus sequence will render discrimination more or less difficult. After the stronger of two stimuli, an additional second strong stimulus presented next may be partially adapted and seem weaker than normal. Stevens et al. remarked that sometimes subjects would get all five pairs correct at one level with some certainty that they "got the scent" but lost the signal at the next level before getting it back. This report and the reversals of performance in threshold data are consistent with adaptation effects temporarily lessening sensitivity. The sensory impression will sometimes "fade in and out" at levels near threshold.

In attempts to avoid adaptation effects, other researchers have gone to fewer presentations of the target stimulus. For example, Lawless et al. (1995) used one target among three blank stimuli, a 4-AFC test that has appeared in previous studies (e.g., Engen, 1960; Punter, 1983). This lowers the chance performance level and lessens the potential adaptation at any one concentration step. To guard against the effects of correct guessing, threshold was taken as the lowest concentration step with a correct choice when all higher concentrations were also correct. Thresholds were measured in duplicate ascending runs in a test session, and a duplicate session of two more ascending runs was run on a second day. Correlations across the four ascending series ranged from 0.75 to 0.92 for cineole and from 0.51 to 0.92 for carvone. For carvone, thresholds were better duplicated within a day ($r = 0.91$ and 0.88) than across days (r from 0.51 to 0.70). This latter result suggests some drift over time in odor thresholds, in keeping with the variability seen by Stevens et al. (1988). However, results with this ascending method may not be this reliable for all compounds. Using the ascending 4-AFC test and a sophisticated olfactometer, Punter (1983) found median retest correlations for 11 compounds to be only 0.40. The sense of taste may fare somewhat better. In a study of electrogustometric thresholds with ascending paired tests requiring five correct responses, retest correlations for an elderly population were 0.95 (Murphy et al., 1995).

In many forced-choice studies, high variability in smell thresholds is also noted across the testing pool.

Brown et al. (1978) stated that for any test compound, a number of insensitive individuals would likely be seen in the data set, when 25 or more persons were tested to determine an average threshold. Among any given pool of participants, a few people with otherwise normal smell acuity will have high thresholds. This is potentially important for sensory professionals who need to screen panelists for detection of specific flavor or odor notes such as defects or taints. In an extensive survey of thresholds for branched-chain fatty acids, Brennand et al. (1989) remarked that "some judges were unable to identify the correct samples in the pairs even in the highest concentrations provided" and that " panelists who were sensitive to most fatty acids found some acids difficult to perceive" (p. 109). Wide variation in sensitivity was also observed to the common flavor compound, diacetyl, a buttery-smelling by-product of lactic bacteria fermentation (Lawless et al., 1994). Also, simple exposure to some chemicals can modify specific anosmia and increase sensitivity (Stevens and O'Connell, 1995).

6.9 Alternative Methods: Rated Difference, Adaptive Procedures, Scaling

6.9.1 Rated Difference from Control

Another practical procedure for estimating threshold has involved the use of ratings on degree-of-difference scales, where a sample containing the to-be-recognized stimulus is compared to some control or blank stimulus (Brown et al., 1978; Lundahl et al., 1986). Rated difference may use a line scale or a category scale, ranging from no difference or "exact same" to a large difference, as discussed in Chapter 4. In these procedures ratings for the sensory difference from the control sample will increase as the intensity of the target gets stronger. A point on the plot of ratings versus concentration is assigned as threshold. In some variations on this method, a blind control sample is also rated. This provides the opportunity to estimate a baseline or false alarm rate based on the ratings (often nonzero) of the control against itself. Identical samples will often get nonzero difference estimates due to the moment-to-moment variability in sensations.

In one application of this technique for taste and smell thresholds, a substance was added in various levels to estimate the threshold in a food or beverage. In each individual trial, three samples would be compared to the control sample with no added flavor—two adjacent concentration steps of the target compound and one blind control sample (Lundahl et al., 1986). Samples were rated on a simple 9-point scale, from zero (no difference) to eight (extreme difference). This provided a comparison of the control to itself and a cost-effective way of making three comparisons in one set of samples. Since sample concentrations within the three rated test samples were randomized, the procedure was not a true ascending series and was dubbed the "semi-ascending paired difference method."

How is the threshold defined in these procedures? One approach is to compare the difference ratings for a given level with the difference ratings given to the control sample. Then the threshold can be based on some measure of when these difference ratings diverge, such as when they become significantly different by a t-test (see Brown et al., 1978). Another approach is simply to subtract the difference score given to the blind control from the difference score give to each test sample and treat these adjusted scores as a new data set. In the original paper of Lundahl et al. (1986), this latter method was used. In the analysis, they performed a series of t-tests. Two values were taken to bracket the range of the threshold. The upper level was the first level yielding a significant t-test versus zero, and the lower level was the nearest lower concentration yielding a significant t-test versus the first. This provided an interval in which the threshold (as defined by this method) lies between the two bracketing concentrations.

One problem with this approach is that when the threshold is based on the statistical significance of t-statistics (or any such significance test), the value of threshold will depend upon the number of observations in the test. This creates a nonsensical situation where the threshold value will decrease as a function of the number of panelists used in the test. This is an irrelevant variable, a choice of the experimenter, and has nothing to do with the physiological sensitivity of the panelist or the biological potency of the substance being tested, a problem recognized by Brown et al. (1978) and later by Marin et al. (1991). Marin et al. also pointed out that a group threshold, based on a larger number of observations than an individual threshold, would be lower than the mean

of the individual thresholds, due to the larger number of observations, another oddity of using statistical significance to determine the threshold.

Instead of using statistical significance as a criterion, Marin et al. determined the point of maximum curvature on the dose–response curve as the threshold.

Such an approach makes sense from consideration of the general form of the dose–response (psychophysical) curve for most tastes and odors. Figure 6.7 shows a semi-log plot for the Beidler taste equation, a widely applied dose–response relationship in studies of the chemical senses (see Chapter 2). This function has two sections of curvature (change in slope, i.e., acceleration) when plotted as a function of log concentration. There is a point at which the response appears to be slowly increasing out from the background noise and then rises steeply to enter the middle of the dynamic range of response. The point of maximum curvature can be estimated graphically or determined from curve fitting and finding the maximum rate of change (i.e., maximum of the second derivative) (Marin et al., 1991).

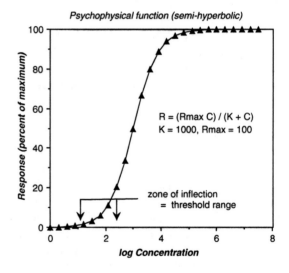

Fig. 6.7 Beidler curve.

6.9.2 Adaptive Procedures

Popular methods for threshold measurement for visual and auditory stimuli have been procedures in which the next stimulus intensity level to be tested depends upon detection or non-detection at the previous interval. In these procedures, the subject will track around the threshold level, ascending in intensity when performance is incorrect (or non-detection is the response) and descending in physical intensity when performance is correct (or detection is indicated). A common example of this procedure is in some automated hearing tests, in which the person being tested pushes a button as long as the signal is not audible. When the button is depressed, the tone will increase in intensity and when the button is released, the tone will decrease in intensity. This automated tracking procedure leads to a series of up and down records, and an average of reversal points is usually taken to determine the threshold. Adaptive procedures may be more efficient than a traditional procedure like the method of limits. They focus on the critical range around threshold and do not waste time testing intensity levels very much higher or very much lower than the threshold (McBurney and Collings, 1977). Further information on these methods can be found in Harvey (1986).

With discrete stimuli, rather than those that are played constantly as in the example of the hearing test, the procedure can be used for the taste and smell modalities as well. This procedure is sometimes called a staircase method, since the record of ascending and descending trials can be connected on graph paper to produce a series of step intervals that visually resemble a staircase. An example is shown in Fig. 6.8. The procedure creates a dependence of each trial on previous trials that may lead to some expectations and bias on the part of the respondent. Psychophysical researchers have found ways to undo this sequential dependence to counteract observer expectancies. One example is the double random staircase procedure (Cornsweet, 1962) in which trials from two staircase sequences are randomly intermixed. One staircase starts above the threshold and descends, while the other starts below the threshold and ascends. On any given trial, the observer is unaware which of the two sequences the stimulus is chosen from. As in the simple staircase procedure, the level chosen for a trial depends upon detection or discrimination in the previous trial, but of that particular sequence. Further refinements of the procedure involve the introduction of forced-choice (Jesteadt, 1980) to eliminate response bias factors involved in simple yes/no detection.

Another modification to the adaptive methods has been to adjust the ascending and descending rules so that some number of correct or incorrect judgments is required before changing intensity levels, rather than the one trial as in the simple staircase (Jesteadt, 1980). An example is the "up down transformed response" rule or UDTR (Wetherill and Levitt, 1965). Wetherill and Levitt gave an example where two positive judgments were required before moving down, and only one negative judgment at a given level before moving up. An example is shown in Fig. 6.9. Rather than estimating the 50% point on a traditional psychometric function, this more stringent requirement now tends to converge on the 71% mark as an average of the peaks and valleys in the series. A forced choice can be added to an adaptive procedure. Sometimes

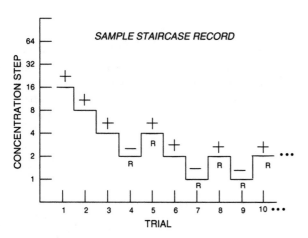

Fig. 6.8 Staircase example.

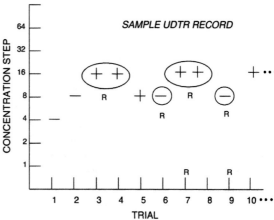

Fig. 6.9 Staircase example.

the initial part of the test sequence is discarded in analysis as it is unrepresentative of the final threshold and derives from a time when the observer may still be warming up to the test procedure. Examples of the up-down procedure can be found in the literature on PTC/PROP tasting (e.g., Reed et al., 1995). Recent advances in adaptive methods have shown that thresholds may be estimated in very few trials using these procedures, a potential advantage for taste and smell measurement (Harvey, 1986; Linschoten et al., 1996).

6.9.3 Scaling as an Alternative Measure of Sensitivity

Threshold measurements are not the only way to screen individuals for insensitivity to specific compounds like PTC or to screen for specific anosmia. Do thresholds bear any relation to suprathreshold responding? While it has been widely held that there is no necessary relationship between threshold sensitivity and suprathreshold responding (Frijters, 1978; Pangborn, 1981), this assertion somewhat overstates the case. Counter-examples of good correlations can be seen in tests involving compounds like PTC where there are insensitive groups. For example, there is a –0.8 correlation between simple category taste intensity ratings for PTC and the threshold, when the rated concentration is near the antimode or center between the modes of a bimodal threshold frequency distribution (Lawless, 1980). Thus ratings of a carefully chosen level can be used for a rapid screening method for PTC taster status (e.g., Mela, 1989).

Similar results have been noted for smell. Berglund and Högman (1992) reported better reliability of suprathreshold ratings than threshold determinations in screening for olfactory sensitivity. Stevens and O'Connell (1991) used category ratings of perceived intensity as well as qualitative descriptors as a screening tool before threshold testing for specific anosmia. Threshold versus rating correlations were in the range of –0.6 for cineole, –0.3 for carvone, and –0.5 for diacetyl (Lawless et al., 1994, 1995). The correlations were obtained after subtraction of ratings to a blank stimulus, in order to correct for differences in scale usage. Thus there is a moderate negative correlation of sensitivity and rated intensity when one

examines the data across a highly variable group as is the case with specific anosmia or tasting PTC bitterness. The correlation is negative since higher thresholds indicate lower sensitivity and thus lower rated intensity.

6.10 Dilution to Threshold Measures

6.10.1 Odor Units and Gas-Chromatography Olfactometry (GCO)

In this section, several applied methods will be described that make use of the threshold concept in trying to determine the sensory impact of various flavors and odor materials. The first group of methods concerns the olfactory potency of volatile aroma compounds as they are found in foods or food components. The issue here becomes one of not just threshold determination, but determination of both the threshold and the actual concentration present in a food sample. The ratio of these concentrations (actual concentration to threshold concentration) can help indicate whether or not a given flavor substance is likely to contribute to the overall sensory impression in a food. These ratios are commonly called "odor units." The second much older method is similar in logic and was developed to determine the point at which the irritative or heat sensations from pepper compounds would be first detectable when diluted to a given extent, the Scoville procedure. Both of these techniques then use dilution-to-threshold as a measure of sensory impact.

When a complex natural product like a fruit extract is analyzed for its chemical components, hundreds or even thousands of chemicals may be identified, many of which have odor properties. The number of potential flavor compounds identified in any product seems only to be limited by the resolution and sensitivity of the current methods in analytical chemistry. These methods are always improving leading to longer and longer lists of possible contributing flavor materials (Piggott, 1990). Flavor scientists need to find a way to narrow the list or to separate those compounds which are most likely contributing to the overall flavor from those compounds that are present in such low concentrations that they are probably not important. Obviously, a sensory-based method is needed in conjunction with the analytical chemistry to provide a bioassay for possible sensory impact (Acree, 1993).

Thresholds can be useful in addressing this kind of problem. The reasoning goes that only those compounds that are present in the product in concentrations above their threshold are likely to be contributors to the flavor of the product. There are a number of potential flaws in this thinking discussed below, but for now let us see how this can be put to use. Given a concentration C present in a natural product, a dimensionless quantity can be derived by dividing that concentration by the threshold concentration C_t, and the ratio C/C_t defines the number of odor units (or flavor units) for compounds assessed by smell. According to this logic, only those compounds with odor units greater than one will contribute to the aroma of the product. This reasoning is extended sometimes to include the idea that the greater the number of odor units, the greater the potential contribution. However, it is now widely recognized that the odor unit is a concentration multiple and not a measure of subjective magnitude. Only direct scaling methods can assess the actual magnitude of sensation above threshold and the psychophysical relationship between concentration and odor intensity (Frijters, 1978). Furthermore, this idea ignores the possibility of subthreshold additivity or synergy (Day et al., 1963). A closely related group of chemical compounds might all be present below their individual thresholds, but together could stimulate common receptors so as to produce an above-threshold sensation. Such additivity is not predicted by the odor unit approach and such a group of compounds could be missed in dilution analysis.

Nonetheless, thresholds provide at least one iso-intense reference point on the dose response curve, so they have some utility as a measure of potency used to compare different odor compounds. In analyzing a food, one could look up literature values for all the identified compounds in the product in one of the published compendia of thresholds (e.g., ASTM, 1978; van Gemert, 2003). If the concentration in the product is determined, then the odor unit value can be calculated by simply dividing by threshold. However, it is important to remember that the literature values for thresholds depend upon the method and the medium

of testing. Unless the same techniques are used and the same medium was used as the carrier (rarely the case) the values may not be necessarily comparable for different compounds.

A second approach is to actually measure the dilutions necessary to reach threshold for each compound, starting with the product itself. This necessitates the use of a separatory procedure, so that each compound may be individually perceived. The combination of gas chromatography with odor port sniffing of a dilution series is a popular technique (Acree, 1993). Various catchy names have been applied to such techniques in the flavor literature, including Aroma Extract Dilution Analysis (for examples, see Guth and Grosch, 1994; Milo and Grosch, 1993; Schieberle and Grosch, 1988), CHARM analysis (Acree et al., 1984) or more generically, gas chromatography olfactometry or GCO. The basis of these techniques is to have subjects respond when an odor is perceived when sniffing the exit port during a GC run. In recent years, the effluent has been embedded in a cooled, humidified air stream to improve the comfort of the observer and to increase sensory resolution of the eluting compounds. Over several dilutions, the response will eventually drop out, and the index of smell potency is related to the reciprocal of the dilution factor. The sniffer's responses occur on a time base that can be cross-referenced to a retention index and then the identity of the compound can be determined by a combination of retention index, mass spectrometry, and aroma character. In practice, these techniques considerably shorten the list of potential aroma compounds contributing to flavor in a natural product (e.g., Cunningham et al., 1986).

The method has also been used as an assessment technique for measuring the sensitivity of human panelists, as opposed to being a tool to determine the sensory impact of flavor compounds (Marin et al., 1988). In this approach, the gas chromatograph once again serves as an olfactometer. Known compounds can be presented as a dilution series of mixtures. Variation in individual thresholds can readily be assessed for a variety of compounds since they can be combined in a GC run as long as they have different retention times, i.e., do not co-elute on the column of choice. The potential use of GCO for screening panelists, for assessing odor responses in general, and for assessing specific anosmias has also been attempted (Friedrich and Acree, 2000; Kittel and Acree, 2008).

6.10.2 Scoville Units

Another example of a dilution method is the traditional Scoville procedure for scoring pepper heat in the spice trade. This procedure was named for W. Scoville who worked in the pharmaceutical industry in the early twentieth century. He was interested in the topical application of spice compounds like pepper extracts as counterirritants, and he needed to establish units that could be used to measure their potency. His procedure consisted of finding the number of dilutions necessary for sensations to disappear and then using this number of dilutions as an estimate of potency. In other words, potency was defined as a reciprocal threshold. A variation of this procedure was adopted by the Essential Oil Association, British Standards Institution, International Standards Organization, American Spice Trade Association (ASTA), and adopted as an Indian standard method (for a review, see Govindarajan, 1987).

The procedure defined units of pungency as the highest dilution at which a definite "bite" would be perceived and thus contains instructions consistent with a recognition threshold. Scoville units were dilution factors, now commonly given as mL/g. ASTA Method 21 (ASTA, 1968) is widely used and contains some modifications in an attempt to overcome problems with the original Scoville procedure. In brief, the method proceeds as follows: Panelists are screened for acuity relative to experienced persons. Dilution schedules are provided which simplify calculations of the eventual potency. Solutions are tested in 5% sucrose and negligible amounts of alcohol. Five panelists participate and concentrations are given in ascending order around the estimated threshold. Threshold is defined as the concentration at which three out of five judges respond positively.

This method is difficult in practice and a number of additional variations have been tried to improve on the accuracy and precision of the method (Govindarajan, 1987). Examples include the following: (1) substitution of other rules for 3/5, e.g., mean + SD of 20–30 judgments, (2) use of a triangle test, rather than simple yes/no at each concentration, (3) requiring recognition of pungency (Todd et al., 1977), (4) reduction of sucrose concentration in the carrier solution to 3%, and (5) use of a rating scale from 1 (definitely not detectable) to 6 (definitely detectable). This latter

modification defined a detection threshold at mean scale value of 3.5. Almost all these methods specify mandatory rest periods between samples due to the long lasting nature of these sensations. The measurements are still difficult. One problem is that capsaicin, the active heat principle in red pepper, is prone to desensitize observers within a session and also regular consumers of hot spices also become less sensitive, leading to wide individual differences in sensitivity among panelists (Green, 1989; Lawless et al., 1985). Alternative procedures have been developed based on rating scales with fixed physical references (Gillette et al., 1984) and these have been endorsed by ASTM (2008c) as standard test methods. These rating scale procedures show good correlations with instrumental measures of capsaicin content in pepper samples and can be cross-referenced to Scoville units for those who prefer to do business in the traditional units (Gillette et al., 1984).

6.11 Conclusions

Threshold measurements find three common uses in sensory analysis and flavor research. First, they can be used to compare the sensitivities of different panelists. Second, they can be used as an index of the biological potency of a flavor compound. Third, they can provide useful information regarding the maximum tolerable levels of an off-flavor or taint. A variety of different techniques have been used to find thresholds or have employed the threshold concept in practical flavor work. Examples of different threshold methods are given in Table 6.4. In spite of their practical applications, the usefulness of threshold measures is often questioned in sensory evaluation. One criticism is that thresholds are only one point on an intensity function and thus they do not tell us anything about above-threshold responding. There are some good examples in which thresholds do not predict or do not correlate very well with suprathreshold responses. For example, patients irradiated for cancer may lose their sense of taste temporarily and thresholds return to normal long before suprathreshold responsiveness is recovered (Bartoshuk, 1987). However, as we have seen both in the case of PTC tasting and in specific anosmias, insensitive individuals (as determined by their threshold) will also tend to be less responsive above

Table 6.4 Threshold procedures

Method	Citations/examples	Response	Threshold
Ascending forced choice	ASTM E-679-79	3-AFC	Geometric mean of individual threshold
Ascending forced choice	Stevens et al. (1988)	2-AFC, 5 replicates	Lowest correct set with confirmation at next concentration
Semi-ascending paired difference	Lundahl et al. (1986)	Rated difference from control	t-test for significant difference from zero with all blank trials (blind control scores) subtracted
Adaptive, up-down transformed response rule	Reed et al. (1995)	2-AFC	Average of responses ascending after one incorrect, descending after two correct at each level
Double Random Staircase	Cornsweet (1962)	Yes/No	Average of reversal points
CHARM analysis	Acree et al. (1986)	Yes/No	Nonresponse on descending concentration runs (implied)

threshold. These correlations are strong when comparing individuals of very different sensitivities, but the threshold–suprathreshold parallel may not extend to all flavor compounds. A more complete understanding of the whole dynamic range of dose–response, as found in scaling studies, would be more informative.

Other shortcomings need to be kept in mind by sensory workers who would use threshold measurements for making product decisions. First, thresholds are statistical constructs only. Signal detection theory warns us that the signal and noise diverge in a continuous fashion and discontinuities in perception may be an idealized construction that is comfortable but unrealistic. There is no sudden transition from nondetection to 100% detection. Any modern concept of threshold must take into account that a range of values, rather than a single point is involved in the specification. Thresholds depend upon the conditions of measurement. For example, as the purity of the diluent increases, the threshold for taste will go down. So a measure of the true absolute threshold for taste (if such a thing existed) would require water of infinite purity. The threshold exists not in this abstract sense, but only

as a potentially useful construct of our methods and informational requirements.

Finally, because of the problems mentioned above, sensory professionals need to keep the following principles firmly in mind when working with threshold procedures: First, changes in method will change the obtained values. Literature values cannot be trusted to extend to a new product or a new medium or if changes are made in test procedure. Next, threshold distributions do not always follow a normal bell curve. There are often high outliers and possibly cases of insensitivity due to inherited deficits like specific anosmia (Amoore, 1971; Brown et al., 1978; Lawless et al., 1994). Threshold values for an individual are prone to high variability and low test–retest reliability. An individual's threshold measure on a given day is not necessarily a stable characteristic of that person (Lawless et al., 1995; Stevens et al., 1988). Practice effects can be profound, and thresholds may stabilize over a period of time (Engen, 1960; Rabin and Cain, 1986). However, group averaged thresholds are reliable (Brown et al., 1978; Punter, 1983) and provide a useful index of the biological activity of a stimulus.

Appendix: MTBE Threshold Data for Worked Example

	Concentration (μg/L)									
Panelist	2	3.5	6	10	18	30	60	100	BET	log(BET)
1	+	o	+	+	+	+	+	+	4.6	0.663
2	o	o	o	+	+	+	+	+	7.7	0.886
3	o	+	o	o	o	o	+	+	42	1.623
4	o	o	o	o	o	o	+	+	42	1.623
5	+	o	+	o	+	+	o	+	77	1.886
6	o	o	+	+	+	+	+	+	4.6	0.663

Panelist	Concentration (μg/L)								BET	log(BET)
	2	3.5	6	10	18	30	60	100		
7	o	+	+	o	+	+	+	+	13	1.114
8	+	+	o	+	+	+	+	+	7.7	0.886
9	o	o	+	o	+	+	+	+	13	1.114
10	o	o	o	o	o	o	o	+	77	1.886
11	+	o	+	+	+	+	+	+	4.6	0.663
12	o	o	o	o	+	o	+	o	132	2.121
13	+	+	+	+	+	+	+	+	1.4	0.146
14	+	+	+	+	+	+	+	+	1.4	0.146
15	o	+	+	o	+	+	+	+	13	1.114
16	o	o	+	o	o	o	+	o	132	2.121
17	+	o	+	+	+	+	+	+	4.6	0.663
18	o	o	+	+	+	+	+	+	4.6	0.663
19	+	+	+	+	+	+	+	+	1.4	0.146
20	+	o	+	+	o	+	o	o	132	2.121
21	+	+	+	+	+	+	+	+	1.4	0.146
22	+	+	+	+	+	+	+	+	1.4	0.146
23	+	+	o	o	+	+	+	+	13	1.114
24	+	+	+	+	+	+	+	+	1.4	0.146
25	o	+	+	+	o	+	+	+	23	1.362
26	o	+	+	+	o	+	+	+	23	1.362
27	+	o	o	o	+	o	o	+	77	1.886
28	o	o	+	+	+	+	+	o	132	2.121
29	o	+	o	o	+	+	+	+	13	1.114
30	o	+	+	o	+	+	+	+	13	1.114
31	+	o	o	+	o	o	+	o	132	2.121
32	+	+	+	+	o	o	o	+	77	1.886
33	+	+	+	o	+	o	+	+	42	1.623
34	o	o	o	o	o	o	+	o	132	2.121
35	o	o	o	+	o	+	+	o	132	2.121
36	o	o	o	+	o	+	+	+	23	1.362
37	+	+	+	+	+	+	+	+	1.4	0.146
38	+	o	+	+	+	+	+	+	1.4	0.146
39	+	+	o	o	+	+	+	+	13	1.114
40	o	o	+	o	+	+	+	+	13	1.114
41	+	+	+	+	+	+	+	+	1.4	0.146
42	o	+	o	+	+	+	+	+	7.7	0.886
43	o	o	o	o	o	o	+	+	42	1.623
44	o	+	+	o	+	o	+	+	42	1.623
45	o	+	+	o	+	+	+	+	13	1.114
46	+	+	+	+	+	+	o	+	77	1.886
47	o	+	o	o	o	+	+	o	132	2.121
48	o	+	o	+	o	+	+	+	23	1.362
49	o	o	o	+	o	+	+	+	23	1.362
50	o	o	+	+	+	+	+	+	4.6	0.663
51	o	o	o	+	+	+	+	+	7.7	0.886
52	+	+	+	+	+	+	+	+	1.4	0.146
53	+	+	+	+	+	+	+	+	1.4	0.146
54	+	o	o	o	o	+	+	+	23	1.362
55	o	o	+	+	+	+	+	+	4.6	0.663
56	o	o	o	o	o	+	+	+	23	1.362
57	o	o	+	+	o	+	o	+	77	1.886
Prop. Corr.	0.44	0.49	0.61	0.58	0.65	0.77	0.89	0.86	Mean (log(BET))	1.154
									10^1.154=	14.24

References

Acree, T. E. 1993. Bioassays for flavor. In: T. E. Acree and R. Teranishi (eds.), Flavor Science, Sensible Principles and Techniques. American Chemical Society Books, Washington, pp. 1–20.

Acree, T. E., Barnard, J. and Cunningham, D. G. 1984. A procedure for the sensory analysis for gas chromatographic effluents. Food Chemistry, 14, 273–286.

American Spice Trade Association 1968. Pungency of capsicum spices and oleoresins. American Spice Trade Association Official Analytical Methods, 21.0, 43–47.

Amoore, J. E. 1971. Olfactory genetics and anosmia. In: L. M. Beidler (ed.), Handbook of Sensory Physiology. Springer, Berlin, pp. 245–256.

Amoore, J. E. 1979. Directions for preparing aqueous solutions of primary odorants to diagnose eight types of specific anosmia. Chemical Senses and Flavor, 4, 153–161.

Amoore, J. E., Venstrom, D. and Davis, A. R. 1968. Measurement of specific anosmia. Perceptual and Motor Skills, 26, 143–164.

Antinone, M. A., Lawless, H. T., Ledford, R. A. and Johnston, M. 1994. The importance of diacetyl as a flavor component in full fat cottage cheese. Journal of Food Science, 59, 38–42.

ASTM. 1978. Compilation of Odor and Taste Threshold Values Data. American Society for Testing and Materials, Philadelphia.

ASTM. 2008a. Standard practice for determining odor and taste thresholds by a forced-choice ascending concentration series method of limits, E-679-04. Annual Book of Standards, Vol. 15.08. ASTM International, Conshocken, PA, pp. 36–42.

ASTM. 2008b. Standard practice for defining and calculating individual and group sensory thresholds from forced-choice data sets of intermediate size, E-1432-04. ASTM International Book of Standards, Vol. 15.08. ASTM International, Conshocken, PA, pp.82–89.

ASTM. 2008c. Standard test method for sensory evaluation of red pepper heat, E-1083-00. ASTM International Book of Standards, Vol. 15.08. ASTM International, Conshocken, PA, pp. 49–53.

Ayya, N. and Lawless, H. T. 1992. Qualitative and quantitative evaluation of high-intensity sweeteners and sweetener mixtures. Chemical Senses, 17, 245–259.

Bartoshuk, L. M. 1987. Psychophysics of taste. American Journal of Clinical Nutrition, 31, 1068–1077.

Bartoshuk, L. M., Murphy, C. and Cleveland, C. T. 1978. Sweet taste of dilute NaCl: Psychophysical evidence for a sweet stimulus. Physiology and Behavior, 21, 609–613.

Berglund, B. and Högman, L. 1992. Reliability and validity of odor measurements near the detection threshold. Chemical Senses, 17, 823–824.

Blakeslee, A. F. 1932. Genetics of sensory thresholds: Taste for phenylthiocarbamide. Proceedings of the National Academy of Sciences, 18, 120–130.

Brennand, C. P., Ha, J. K. and Lindsay, R. C. 1989. Aroma properties and thresholds of some branched chain and other minor volatile fatty acids occurring in milkfat and meat lipids. Journal of Sensory Studies, 4, 105–120.

Boring, E. G. 1942. Sensation and Perception in the History of Experimental Psychology. Appleton-Century-Crofts, New York, pp. 41–42.

Brown, D. G. W., Clapperton, J. F., Meilgaard, M. C. and Moll, M. 1978. Flavor thresholds of added substances. American Society of Brewing Chemists Journal, 36, 73–80.

Bufe, B., Breslin, P. A. S., Kuhn, C., Reed, D., Tharp, C. D., Slack, J. P., Kim, U.-K., Drayna, D. and Meyerhof, W. 2005. The molecular basis of individual differences in phenylthiocarbamide and propylthiouracil bitterness perception. Current Biology, 15, 322–327.

Cain, W. S. 1976. Olfaction and the common chemical sense: Some psychophysical contrasts. Sensory Processes, 1, 57–67.

Cain, W. S. and Murphy, C. L. 1980. Interaction between chemoreceptive modalities of odor and irritation. Nature, 284, 255–257.

Cornsweet, T. M. 1962. The staircase method in psychophysics. American Journal of Psychology, 75, 485–491.

Cunningham, D. G., Acree, T. E., Barnard, J., Butts, R. M. and Braell, P. A. 1986. CHARM analysis of apple volatiles. Food Chemistry, 19, 137–147.

Dale, M. S., Moylan, M. S., Koch, B. and Davis, M. K. 1997. Taste and odor threshold determinations using the flavor profile method. Proceedings of the American Water Works Association Water-Quality Technology Conference, Denver, CO.

Dalton, P., Doolittle, N. and Breslin, P. A. S. 2002. Gender-specific induction of enhanced sensitivity to odors. Nature Neuroscience, 5, 199–200.

Day, E. A., Lillard, D. A. and Montgomery, M. W. 1963. Autooxidation of milk lipids.III. Effect on flavor of the additive interactions of carbonyl compounds at subthreshold concentrations. Journal of Dairy Science, 46, 291–294.

Diamond, J., Dalton, P., Doolittle, N. and Breslin, P. A. S. 2005. Gender-specific olfactory sensitization: Hormonal and cognitive influences. Chemical Senses, 30(suppl 1), i224–i225.

Dravnieks, A. and Prokop, W. H. 1975. Source emission odor measurement by a dynamic forced-choice triangle olfactometer. Journal of the Air Pollution Control Association, 25, 28–35.

Engen, T. E. 1960. Effects of practice and instruction on olfactory thresholds. Perceptual and Motor Skills, 10, 195–198.

Finney, D. J. 1971. Probit Analysis, Third Edition. Cambridge University, London.

Friedrich, J. E. and Acree, T. E. 2000. Design of a standard set of odorants to test individuals for specific anosmia. Frontiers in Flavour Science [Proc. 9th Weurman Flavour Res. Symp.], 230–234.

Frijters, J. E. R. 1978. A critical analysis of the odour unit number and its use. Chemical Senses and Flavour, 3, 227–233.

Gillette, M., Appel, C. E. and Lego, M. C. 1984. A new method for sensory evaluation of red pepper heat. Journal of Food Science, 49, 1028–1033.

Green, B. G. 1989. Capsaicin sensitization and desensitization on the tongue produced by brief exposures to a low concentration. Neuroscience Letters, 107, 173–178.

Govindarajan, V. S. 1987. Capsicum – production, technology, chemistry and quality. Part III. Chemistry of the color, aroma and pungency stimuli. CRC Critical Reviews in Food Science and Nutrition, 24, 245–311.

Guth, H. and Grosch, W. 1994. Identification of the character impact odorants of stewed beef juice by instrumental analysis and sensory studies. Journal of Agricultural and Food Chemistry, 42, 2862–2866.

Harris, H. and Kalmus, H. 1949. The measurement of taste sensitivity to phenylthiourea (P.T.C.). Annals of Eugenics, 15, 24–31.

Harvey, L. O. 1986. Efficient estimation of sensory thresholds. Behavior Research Methods, Instruments and Computers, 18, 623–632.

James, W. 1913. Psychology. Henry Holt and Co., New York.

Jesteadt, W. 1980. An adaptive procedure for subjective judgments. Perception and Psychophysics, 28(1), 85–88.

Kittel, K. M and Acree, T. E. 2008. Investigation of olfactory deficits using gas-chromatography olfactometry. Manuscript submitted, available from the authors.

Lawless, H. T. 1980. A comparison of different methods used to assess sensitivity to the taste of phenylthiocarbamide PTC. Chemical Senses, 5, 247–256.

Lawless, H. T. 2010. A simple alternative analysis for threshold data determined by ascending forced-choice method of limits. Journal of Sensory Studies, 25, 332–346.

Lawless, H. T., Antinone, M. J., Ledford, R. A. and Johnston, M. 1994. Olfactory responsiveness to diacetyl. Journal of Sensory Studies, 9(1), 47–56.

Lawless, H. T., Rozin, P. and Shenker, J. 1985. Effects of oral capsaicin on gustatory, olfactory and irritant sensations and flavor identification in humans who regularly or rarely consumer chili pepper. Chemical Senses, 10, 579–589.

Lawless, H. T., Thomas, C. J. C. and Johnston, M. 1995. Variation in odor thresholds for l-carvone and cineole and correlations with suprathreshold intensity ratings. Chemical Senses, 20, 9–17.

Linschoten, M. R., Harvey, L. O., Eller, P. A. and Jafek, B. W. 1996. Rapid and accurate measurement of taste and smell thresholds using an adaptive maximum-likelihood staircase procedure. Chemical Senses, 21, 633–634.

Lundahl, D. S., Lukes, B. K., McDaniel, M. R. and Henderson, L. A. 1986. A semi-ascending paired difference method for determining the threshold of added substances to background media. Journal of Sensory Studies, 1, 291–306.

Masuoka, S., Hatjopolous, D. and O'Mahony, M. 1995. Beer bitterness detection: Testing Thurstonian and sequential sensitivity analysis models for triad and tetrad methods. Journal of Sensory Studies, 10, 295–306.

Marin, A. B., Acree, T. E. and Barnard, J. 1988. Variation in odor detection thresholds determined by charm analysis. Chemical Senses, 13, 435–444.

Marin, A. B., Barnard, J., Darlington, R. B. and Acree, T. E. 1991. Sensory thresholds: Estimation from dose-response curves. Journal of Sensory Studies, 6(4), 205–225.

McBurney, D. H. and Collings, V. B. 1977. Introduction to Sensation and Perception. Prentice-Hall, Englewood Cliffs, NJ.

Meilgaard, M., Civille, G. V. and Carr, B. T. 1991. Sensory Evaluation Techniques, Second Edition. CRC Press, Boca Raton, FL.

Mela, D. 1989. Bitter taste intensity: The effect of tastant and thiourea taster status. Chemical Senses, 14, 131–135.

Milo, C. and Grosch, W. 1993. Changes in the odorants of boiled trout (Salmo fario) as affected by the storage of the raw material. Journal of Agricultural and Food Chemistry, 41, 2076–2081.

Mojet, J., Christ-Hazelhof, E. and Heidema, J. 2001. Taste perception with age: Generic or specific losses in threshold sensitivity to the five basic tastes. Chemical Senses 26, 854–860.

Morrison, D. G. 1978. A probability model for forced binary choices. American Statistician, 32, 23–25.

Murphy, C., Quiñonez, C. and Nordin, S. 1995. Reliability and validity of electrogustometry and its application to young and elderly persons. Chemical Senses, 20(5), 499–515.

O'Mahony, M. and Ishii, R. 1986. Umami taste concept: Implications for the dogma of four basic tastes. In: Y. Kawamura and M. R. Kare (eds.), Umami: A Basic Taste. Marcel Dekker, New York, pp. 75–93..

O'Mahony, M. and Odbert, N. 1985. A comparison of sensory difference testing procedures: Sequential sensitivity analysis and aspects of taste adaptation. Journal of Food Science, 50, 1055–1060.

Pangborn, R. M. 1981. A critical review of threshold, intensity and descriptive analyses in flavor research. In: Flavor '81. Walter de Gruyter, Berlin, pp. 3–32.

Piggott, J. R. 1990. Relating sensory and chemical data to understand flavor. Journal of Sensory Studies, 4, 261–272.

Prescott, J., Norris, L., Kunst, M. and Kim, S. 2005. Estimating a "consumer rejection threshold" for cork taint in white wine. Food Quality and Preference, 18, 345–349.

Punter, P. H. 1983. Measurement of human olfactory thresholds for several groups of structurally related compounds. Chemical Senses, 7, 215–235.

Rabin, M. D. and Cain, W. S. 1986. Determinants of measured olfactory sensitivity. Perception and Psychophysics, 39(4), 281–286.

Reed, D. R., Bartoshuk, L. M., Duffy, V., Marino, S. and Price, R. A. 1995. Propylthiouracil tasting: Determination of underlying threshold distributions using maximum likelihood. Chemical Senses, 20, 529–533.

Saliba, A. J., Bullock, J. and Hardie, W. J. 2009. Consumer rejection threshold for 1,8 cineole (eucalyptol) in Australian red wine. Food Qualithy and Preference, 20, 500–504.

Schieberle, P. and Grosch, W. 1988. Identification of potent flavor compounds formed in an aqueous lemon oil/citric acid emulsion. Journal of Agricultural and Food Chemistry, 36, 797–800.

Stevens, D. A. and O'Connell, R. J. 1991. Individual differences in threshold and quality reports of subjects to various odors. Chemical Senses, 16, 57–67.

Stevens, D. A. and O'Connell, R. J. 1995. Enhanced sensitivity to adrostenone following regular exposure to pemenone. Chemical Senses, 20, 413–419.

Stevens, J. C., Cain, W. S. and Burke, R. J. 1988. Variability of olfactory thresholds. Chemical Senses, 13, 643–653.

Stocking, A. J., Suffet, I. H., McGuire, M. J. and Kavanaugh, M. C. 2001. Implications of an MTBE odor study for setting drinking water standards. Journal AWWA, March 2001, 95–105.

Todd, P. H., Bensinger, M. G. and Biftu, T. 1977. Determination of pungency due to capsicum by gas-liquid chromatography. Journal of Food Science, 42, 660–665.

Tuorila, H., Kurkela, R., Suihko, M. and Suhonen. 1981. The influence of tests and panelists on odour detection thresholds. Lebensmittel Wissenschaft und Technologie, 15, 97–101.

USEPA (U.S. Environmental Protection Agency). 2001. Statistical analysis of MTBE odor detection thresholds in drinking water. National Service Center for Environmental Publications (NSCEP) # 815R01024, available from http://nepis.epa.gov.

Van Gemert, L. 2003. Odour Thresholds. Oliemans, Punter & Partners, Utrecht, The Netherlands.

Viswanathan, S., Mathur, G. P., Gnyp, A. W. and St. Peirre, C. C. 1983. Application of probability models to threshold determination. Atmospheric Environment, 17, 139–143.

Walker, J. C., Hall, S. B., Walker, D. B., Kendall-Reed, M. S., Hood, A. F. and Nio, X.-F. 2003. Human odor detectability: New methodology used to determine threshold and variation. Chemical Senses, 28, 817–826.

Wetherill, G. B. and Levitt, H. 1965. Sequential estimation of points on a psychometric function. British Journal of Mathematical and Statistical Psychology, 18(1), 1–10.

Wysocki, C. J., Dorries, K. M. and Beauchamp, G. K. 1989. Ability to perceive androstenone can be acquired by ostensibly anosmic people. Proceedings of the National Academy of Science, USA, 86, 7976–7989.

Chapter 7

Scaling

Abstract Scaling describes the application of numbers, or judgments that are converted to numerical values, to describe the perceived intensity of a sensory experience or the degree of liking or disliking for some experience or product. Scaling forms the basis for the sensory method of descriptive analysis. A variety of methods have been used for this purpose and with some caution, all work well in differentiating products. This chapter discusses theoretical issues as well as practical considerations in scaling.

The vital importance of knowing the properties and limitations of a measuring instrument can hardly be denied by most natural scientists. However, the use of many different scales for sensory measurement is common within food science; but very few of these have ever been validated....
—(Land and Shepard, 1984, pp. 144–145)

Contents

7.1 Introduction

People make changes in their behavior all the time based on sensory experience and very often this involves a judgment of how strong or weak something feels. We add more sugar to our coffee if it is not sweet enough. We adjust the thermostat in our home if it is too cold or too hot. If a closet is too dark to find your shoes you turn the light on. We apply more force to

H.T. Lawless, H. Heymann, *Sensory Evaluation of Food*, Food Science Text Series,
DOI 10.1007/978-1-4419-6488-5_7, © Springer Science+Business Media, LLC 2010

chew a tough piece of meat if it will not disintegrate to allow swallowing. These behavioral decisions seem automatic and do not require a numerical response. But the same kinds of experiences can be evaluated with a response that indicates the strength of the sensation. What was subjective and private becomes public data. The data are quantitative. This is the basis of scaling.

The methods of scaling involve the application of numbers to quantify sensory experiences. It is through this process of numerification that sensory evaluation becomes a quantitative science subject to statistical analysis, modeling, prediction, and hard theory. However, as noted in the quote above, in the practical application of sensory test methods, the nature of this process of assigning numbers to experiences is rarely questioned and deserves scrutiny. Clearly numbers can be assigned to sensations by a panelist in a variety of ways, some by mere categorization, or by ranking or in ways that attempt to reflect the intensity of sensory experience. This chapter will illustrate these techniques and discuss the arguments that have been raised to substantiate the use of different quantification procedures.

Scaling involves sensing a product or stimulus and then generating a response that reflects how the person perceives the intensity or strength of one or more of the sensations generated by that product. This process is based on a psychophysical model (see Chapter 2). The psychophysical model states that as the physical strength of the stimulus increases (e.g., the energy of a light or sound or the concentration of a chemical stimulus) the sensation will increase in some orderly way. Furthermore, panelists are capable of generating different responses to indicate these changes in what they experience. Thus a systematic relationship can be modeled of how physical changes in the real world result in changing sensations.

Scaling is a tool used for showing differences and degrees of difference among products. These differences are usually above the threshold level or just-noticeable difference. If the products are very similar and there is a question of whether there is any difference at all, the discrimination testing methods are more suitable (Chambers and Wolf, 1996). Scaling is usually done in one of the two scenarios. In the first, untrained observers are asked to give responses to reflect changes in intensity and it is presumed that (1) they understand the attribute they are asked to scale, e.g., sweetness and (2) there is no need to train or calibrate them to use the scale. This is the kind of scaling done to study a dose–response curve or psychophysical function. Such a study would perhaps be done on a student sample or a consumer population. A second kind of scaling is done when trained panelists are used as if they were measuring instruments, as in descriptive analysis. In this case they may be trained to insure a uniform understanding of the attribute involved (e.g., astringency) and often they are calibrated with reference standards to illustrate what is a little and what is a lot of this attribute. In this case the focus is in on the products being tested and not the more basic process of specifying a psychophysical function that has general application.

Note that these are "cheap data" (a term used by my advisor in graduate school, but perhaps "cost-effective" sounds a little less negative). One stimulus gives at least one data point. This is in contrast to indirect methods like forced choice tests. Many responses on a triangle test are needed to give one data point, i.e., the percent correct. Fechner and others referred to scaling as "the method of single stimuli" and considered it less reliable than the choice methods that were used to generate difference thresholds. However, direct scaling came into its own with the advent of magnitude estimation, an open-ended numerical response method. One or another type of scaling forms the basis for virtually all descriptive analysis techniques. In descriptive analysis, panelists generate scaled responses for various sensory attributes to reflect their subjective intensity.

There are two processes involved in scaling as shown in Fig. 7.1. The first is the psychophysical chain of events in which some energy or matter impinges upon receptors and the receptors send signals to the

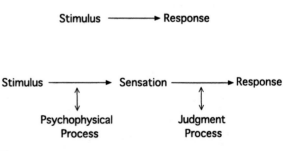

Fig. 7.1 The two processes involved in scaling. The first, physiological process is the psychophysical translation of energy in the outside world into sensation, i.e., conscious experience. The second is the translation of that experience into some response. The psychophysical process can be modified by physiological processes such as adaptation and masking. The judgment function can be modified by cognitive processes such as contextual effects, number usage, and other response biases.

brain. These signals are interpreted in conscious perception as a sensation with some intensity or strength. The translation can be modified (i.e., the experience will change) by processes like adaptation or masking from another stimulus. The second process is the translation of that experience into an overt response (the data). This judgment function is influenced by the nature of the scaling task that the panelist is asked to perform. Factors such as contextual effects, the choice of comparison products, and response biases of that particular person can modify the process. The better the data reflect the experience, the more valid is the scaling method. The sensory professional must be careful to avoid response methods that introduce biases or non-sensory influences on the response output. For example, I might be asked to generate some open-ended numerical response to reflect my perception, but I might have some "favorite" numbers I find easy to use (e.g., integers or multiples of 2, 5, and 10) so this number bias interferes to some degree with the translation of my experience into a truly accurate response.

This chapter will focus on various methods that have been used in sensory evaluation and in psychophysics for scaling. Theory, principles, and issues will be discussed to provide depth of understanding. For the student who wishes to learn just the basic practices, Sections 7.3 and 7.4 are the most practically relevant sections. Section 7.5 illustrates some alternative methods that have appeared in the sensory evaluation literature, but have not at this time enjoyed widespread adoption in industrial practice. The sensory scientist should be aware of these additional methods for the potential advantages they may provide.

7.2 Some Theory

Measurement theory tells us that numbers can be assigned to items in different ways. This distinction was popularized by S. S. Stevens, the major proponent of magnitude estimation (1951). At least four ways of assigning numbers to events exist in common usage. These are referred to as nominal scaling, ordinal scaling, interval scaling, and ratio scaling.

In nominal scaling, numbers are assigned to events merely as labels. Thus gender may be coded as a "dummy variable" in statistical analysis by assigning a zero to males and a one to females; no assumption is made that these numbers reflect any ordered property

of the sexes. They merely serve as convenient labels. The meals at which a food might be eaten could be coded with numbers as categories—one for breakfast, two for lunch, three for supper, and four for snacks. The assignment of a number for analysis is merely a label, a category or pigeonhole. The appropriate analysis of such data is to make frequency counts. The mode, the most frequent response, is used as a summary statistic for nominal data. Different frequencies of response for different products or circumstances can be compared by chi-square analysis or other nonparametric statistical methods (Siegel, 1956; see Appendix B). The only valid comparisons between individual items with this scale is to say whether they belong to the same category or to different ones (an equal versus not equal decision).

In ordinal scaling, numbers are assigned to recognize the rank order of products with regard to some sensory property, attitude, or opinion (such as preference). In this case increasing numbers assigned to the products represent increasing amounts or intensities of sensory experience. So a number of wines might be rank ordered for perceived sweetness or a number of fragrances rank ordered from most preferred to least preferred. In this case the numbers do not tell us anything about the relative differences among the products. We cannot draw conclusions about the degree of difference perceived nor the ratio or magnitude of difference. In an analogy to the order of runners finishing in a race, we know who placed first, second, third, etc. But this order does indicate neither the finishing distances between contestants nor the differences in their elapsed times. In general, analyses of ranked data can report medians as the summary statistic for central tendency or other percentiles to give added information. As with nominal data, nonparametric statistical analyses (see Appendix B) are appropriate when ranking is done (Siegel, 1956).

The next level of scaling occurs when the subjective spacing of responses is equal, so the numbers represent equal degrees of difference. This is called interval-level measurement. Examples in the physical sciences would be the centigrade and Fahrenheit scales of temperature. These scales have arbitrary zero points but equal divisions between values. The scales are inter-convertible through a linear transformation, for example, $°C = 5/9 \ (°F–32)$. Few scales used in sensory science have been subjected to tests that would help establish whether they achieved an interval level of measurement and yet this level is often assumed.

One scale with approximately equal subjective spacing is the 9-point category scale used for like–dislike judgments, the 9-point hedonic scale (Peryam and Girardot, 1952). The phrases are shown below:

Like extremely
Like very much
Like moderately
Like slightly
Neither like nor dislike
Dislike slightly
Dislike moderately
Dislike very much
Dislike extremely

These response choices are commonly entered as data by assignment of the numbers one through nine. Extensive research was conducted to find the apparent spacing of various adjective labels for the scale points (Jones and Thurstone, 1955; Jones et al., 1955). The technique for deciding on the subjective spacing was the use of one kind of Thurstonian scaling method. The original data do not fully support the notion of equality of spacing, as discussed below. However, the scale worked well in practice, so this tradition of integer assignment has persisted. The method is illustrated in Appendix 1 at the end of this chapter. Thurstonian theory is discussed further in Chapter 5.

The advantage of interval-level measurement is that the data allow added interpretation. In a horse-racing example, we know the order the horses finished in and about how many "lengths" separated each horse. A second advantage is that more powerful statistical methods may be brought to bear the parametric methods. Computation of means, t-tests, linear regression, and analysis of variance are appropriate analyses.

Another even more desirable level of measurement is ratio measurement. In this case the zero level is fixed and not arbitrary and numbers will reflect relative proportions. This is the level of measurement commonly achieved in the physical sciences for quantities like mass, length, and temperature (on the absolute or Kelvin scale). Statements can be made that this item has twice as much length or mass than that item. Establishing whether a sensory scaling method actually assigns numbers to represent the relative proportions of different sensation intensities is a difficult matter. It has been widely assumed that the method of magnitude estimation is a priori a ratio scaling procedure. In magnitude estimation, subjects are instructed to assign numbers in relative proportions that reflect the strength of their sensations (Stevens, 1956). However, ratio instructions are easy to give, but whether the scale has ratio properties in reflecting a person's actual subjective experiences is difficult to determine, if not impossible.

Because of these different measurement types with different properties, the sensory professional must be careful about two things. First, statements about differences or ratios in comparing the scores for two products should not be made when the measurement level is only nominal or ordinal. Second, it is risky to use parametric statistics for measurements that reflect only frequency counts or rankings (Gaito, 1980; Townsend and Ashby, 1980). Nonparametric methods are available for statistical analyses of such data.

7.3 Common Methods of Scaling

Several different scaling methods have been used to apply numbers to sensory experience. Some, like magnitude estimation, are adapted from psychophysical research, and others, like category scaling have become popular through practical application and dissemination in a wide variety of situations. This section illustrates the common techniques of category scales, line marking, and magnitude estimation. The next section discusses the less frequently used techniques of hybrid category–ratio scales, indirect scales, and ranking as alternatives. Two other methods are illustrated. Intensity matching across sensory modalities, called cross-modality matching, was an important psychophysical technique and a precedent to some of the category–ratio scales. Finally, adjustable rating techniques in which panelists make relative placements and are able to alter their ratings are also discussed.

7.3.1 Category Scales

Perhaps the oldest method of scaling involves the choice of discrete response alternatives to signify increasing sensation intensity or degrees of liking and/or preference. The alternatives may be presented in a horizontal or vertical line and may offer choices of integer numbers, simple check boxes, or word phrases. Examples of simple category scales are shown in

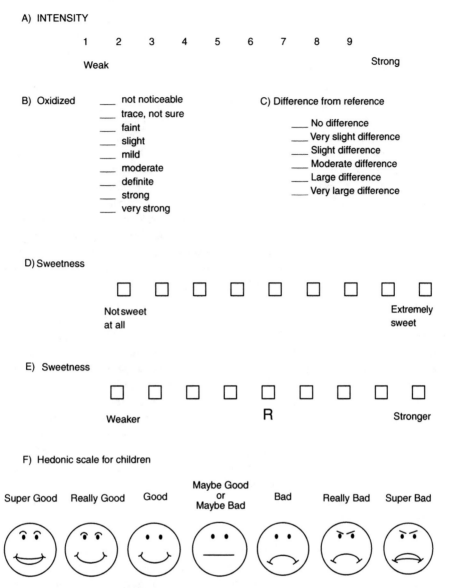

Fig. 7.2 Examples of category scales. (**a**) a simple integer scale for sensation strength (after Lawless and Malone, 1986b); (**b**) a verbal scale for degree of oxidized flavor (after Mecredy et al., 1974; (**c**) a verbal scale for degree of difference from some reference or control sample (after Aust et al., 1985), (**d**) a simple check-box scale for perceived intensity; (**e**) a simple check-box scale for difference in intensity from some reference sample, marked R (after Stoer and Lawless, 1993); (**f**) a facial scale suitable for use with children, after Chen et al. (1996).

Fig. 7.2. The job of the consumer or panelist is to choose the alternative that best represents their reaction or sensation. In a category scale the number of alternative responses is limited. Seven to 15 categories are commonly used for intensity scaling depending upon the application and the number of gradations that the panelists are able to distinguish in the products. As panel training progresses, perceptual discrimination of intensity levels will often improve and more scale points may be added to allow the panel to make finer distinctions. A key idea is to present an easily understandable word like "sweetness" and ask the participant to evaluate the perceived intensity of that attribute. A second important factor concerns the verbal labels that appear along the alternatives. At the very least, the low and high ends of the scale must be labeled with words that make sense, e.g., "not sweet at all" to "extremely sweet."

A wide variety of these scales have been used. A common version is to allow integer numerical responses of approximately nine points (e.g., Lawless and Malone, 1986a, b). Further gradations may be allowed. For example, Winakor et al. (1980) allowed options from 1 to 99 in rating attributes of fabric hand-feel. In the Spectrum method (Meilgaard et al., 2006) a 15-point category scale is used, but allows intermediate points in tenths, rendering it (at least in theory) a 150-point scale. In hedonic or affective testing, a bipolar scale is common, with a zero or neutral point of opinion at the center (Peryam and Girardot, 1952). These are often shorter than the intensity scales. For example, in the "smiling face" scale used with children, only three options may be used for very young respondents (Birch et al., 1980, 1982), although with older children as many as nine points may be used (Chen et al., 1996; Kroll, 1990). Lately there has been a move away from using labels or integers, in case these may be biasing to subjects. People seem to have favorite numbers or tendencies to use some numbers more than others (e.g., Giovanni and Pangborn, 1983). A solution to this problem is to use an unlabeled check-box scale as shown in Fig. 7.2.

In early applications of category scaling, the procedure specifically instructed subjects to use the categories to represent equal spacing. They might also be instructed to distribute their judgments over the available scale range, so the strongest stimulus was rated at the highest category and the weakest stimulus at the lowest category. This use of such explicit instructions surfaces from time to time. An example is in Anderson's (1974) recommendation to show the subject specific examples of bracketing stimuli that are above and below the anticipated range of items in the set to be judged. A related method is the relative scaling procedure of Gay and Mead (1992) in which subjects place the highest and lowest stimuli at the scale endpoints (discussed below). The fact that there is an upper boundary to the allowable numbers in a category scaling task may facilitate the achievement of a linear interval scale (Banks and Coleman, 1981).

A related issue concerns what kind of experience the high end anchor refers to. Muñoz and Civille (1998) pointed out that for descriptive analysis panels, the high end-anchor phrase could refer to different situations. For example, is the term "extremely sweet" referring to all products (a so-called universal scale)? Or is the scale anchored in the panelists' minds only to products in this category? In that case, extremely sweet for a salt cracker refers to something different than extremely sweet for a confectionary product or ice cream. Or is the high end of the scale the most extreme attribute for this product? That would yield a product-specific scale in which comparisons between different attributes could be made, e.g., this cracker is much sweeter than it is salty, but not comparisons to another type of product. These are important concerns for a descriptive panel leader.

However, some experimenters nowadays avoid any extra instructions, allowing the subject or panelist to distribute their ratings along the scale as they see fit. In fact, most people will have a tendency to distribute their judgments along most of the scale range, although some avoid the end categories, reserving them in case extreme examples show up. However, panelists do not like to overuse one part of the scale and will tend to move these judgments into adjoining response categories (Parducci, 1965). These tendencies are discussed in Chapter 9.

In practice, simple category scales are about as sensitive to product differences as other scaling techniques, including line marking and magnitude estimation (Lawless and Malone, 1986a, b). Due to their simplicity, they are well suited to consumer work. In addition, they offer some advantages in data coding and tabulation for speed and accuracy as they are easier to tabulate than measuring lines or recording the more variable magnitude estimates that may include fractions. This presumes, of course, that the data are being tabulated manually. If the data are recorded online using a computer-assisted data collection system, this advantage vanishes. A wide variety of scales with fixed alternatives are in current use. These include Likert-type scales used for opinions and attitudes, which are based on the degree to which a person agrees or disagrees with a statement about the product. Examples of such scales used in consumer tests are found in Chapter 14. Lately the term "Likert scale" has been used to refer to any category type of scale, but we prefer to reserve the name of Likert to the agree/disagree attitude scale in keeping with his original method (Likert, 1932). The flexibility of categorical alternatives for many different situations is thus one important aspect of the appeal of this kind of response measurement.

7.3.2 Line Scaling

A second widely used technique for intensity scaling involves making a mark or slash on a line to indicate the intensity of some attribute. Marking a line is also referred to as using a graphic-rating scale or a visual analog scale. The response is recorded as the distance of the mark from one end of the scale, usually whatever end is considered "lower." Line marking differs from category scales in the sense that the person's choices seem more continuous and less limited. In fact, the data are limited to the discrete choices measurable by the data-encoding instrument, such as the resolution of a digitizer or the number of pixels resolvable on a computer screen. The fundamental idea is that the panelist makes a mark on a line to indicate the intensity or amount of some sensory characteristic. Usually only the endpoints are labeled and marked with short line segments at right angles to the main line. The end-anchor lines may be indented to help avoid end effects associated with the reluctance of subjects to use the ends of scale. Other intermediate points may be labeled. One variation uses a central reference point representing the value of a standard or baseline product on the scale. Test products are scaled relative to that reference. Some of these variations are shown in Fig. 7.3. These techniques are very popular in descriptive analysis in which multiple attributes are evaluated by trained panels.

The first use of line scales for sensory evaluation appears in an experiment from the Michigan State Agricultural Experiment Station conducted during World War II (Baten, 1946). Various storage temperatures of apples were tested. A simple category scale for fruit appeal was used (ratings from very desirable to very undesirable with seven alternatives) and then a 6 in. line scale was used, with the words "very poor" over the left end and "excellent" over the right end. Responses on the line were measured in inches. A poll of the participants revealed a strong preference for the category scale over the line marking scale. However, Baten reported that the *t*-values comparing apples were about twice as large using the line-marking technique, implying a greater ability to statistically differentiate the items using the line scale. Unfortunately, Baten did not report any numerical values for the *t*-statistics, so it is difficult to evaluate the size of the advantage he saw.

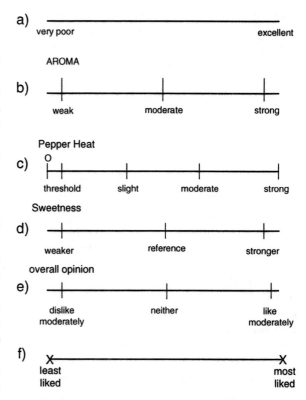

Fig. 7.3 Examples of line-marking scales: (a) with endpoints labeled (after Baten, 1946); (b) with indented "goal posts" (after Mecredy et al., 1974); (c) with additional points labeled as in ASTM procedure E-1083 (ASTM, 2008b); (d) a line for ratings relative to a reference as in Stoer and Lawless (1993); (e) hedonic scaling using a line; (f) the adjustable scale of Gay and Mead (1992) as pictured by Villanueva and D Silva (2009).

An important historical trend to use lines in descriptive analysis was instrumental in the popularization of line marking. Stone et al. (1974) recommended the use of line marking for Quantitative Descriptive Analysis (QDA), then a relatively new approach to specifying the intensities of all important sensory attributes. It was important to have a scaling method in QDA that approximated an interval scale, as analysis of variance was to become the standard statistical technique for comparing products in descriptive analysis. The justification for the application in QDA appears to rest on the previous findings of Baten regarding the sensitivity of the method and the writings of Norman Anderson on functional measurement theory (Anderson, 1974). In his approach, Anderson used indented end anchors (see Weiss, 1972, for another example). Anderson also showed his subjects examples of the high and low stimuli that one might encounter in the experiment, and sometimes even more extreme examples, in order to orient and stabilize their scale usage. Whether such examples make sense for sensory evaluation is questionable.

Since the advent of QDA, the line-marking techniques have been applied in many different situations requiring sensory response. In an early sensory application, Einstein (1976) successfully used a line-marking scale with consumers to evaluate beer attributes of flavor intensity, body, bitterness, and aftertaste. By "successful" we mean that statistically significant differences were obtained among test samples. The use of line marking is not limited to foods and consumer products. Measurement of pain and pain relief has employed line marking in clinical settings, using both vertical and horizontal lines (Huskisson, 1983; Sriwatanakul et al., 1983). Lawless (1977) used a line-marking technique along with ratio instructions for both intensity and hedonic judgments in taste and odor mixture studies. This was a hybrid procedure in which subjects were instructed to mark lines as if performing magnitude estimation. For example, if one product were twice as sweet as a previous item, a mark would be made twice as far down the line (which could be extended if the panelist ran out of room). Villanueva and colleagues used a scale with equally spaced dots along the line and obtained good results for acceptability scaling (Villanueva and Da Silva, 2009; Villanueva et al., 2005). In comparisons of category ratings, line marking, and magnitude estimation, the line-marking method is about as sensitive to product differences as other scaling techniques (Lawless and Malone, 1986a, b).

Marking a point on a line has also been used widely in time–intensity scaling methods. The simplest version of this is to move a pointer along a scale while a moving roll of paper is marked to see the continuous changes in sensation over time. Originally this could be done with a simple marking pen held by the participant (e.g., Moore and Shoemaker, 1981). The record of the pen track would usually be obscured from the person's view, so as not to exert any influence on their response. The pen-tracking method has also been used with ratio instructions (Lawless and Skinner, 1979). In some cases, the participant has turned a dial or other response device while observing a linear display (Lawless and Clark, 1992). Often the time–intensity scale will look much like a vertical "thermometer"

with a cursor that moves up and down via the computer mouse. Time–intensity methods are reviewed more fully in Chapter 8.

7.3.3 Magnitude Estimation

7.3.3.1 The Basic Techniques

A popular technique for scaling in psychophysical studies has been the method of magnitude estimation. In this procedure, the respondent is instructed to assign numbers to sensations in proportion to how strong the sensation seems. Specifically, the ratios between the numbers are supposed to reflect the ratios of sensation magnitudes that have been experienced. For example, if product A is given the value of 20 for sweetness intensity and product B seems twice as sweet, B is given a magnitude estimate of 40. The two critical parts of the technique are the instructions given to the participant and the techniques for data analysis. Two primary variations of magnitude estimation have been used. In one method, a standard stimulus is given to the subject as a reference and that standard is assigned a fixed value such as 10. All subsequent stimuli are rated relative to this standard, sometimes called a "modulus." It is often easier for panelists if the reference (i.e., the item used as the modulus) is chosen from somewhere near the middle of the intensity range.

In the other variation of magnitude estimation, no standard stimulus is given and the participant is free to choose any number he or she wishes for the first sample. All samples are then rated relative to this first intensity, although in practice people probably "chain" their ratings to the most recent items in the series. Because people can choose different ranges of numbers in this "non-modulus" magnitude estimation, the data have to be treated to bring all judgments into the same range, an extra step in the analysis. Variations on magnitude estimation and guidelines for data analysis are found in ASTM Standard Test Method E 1697–05 (ASTM, 2008a).

In the psychophysical laboratory, where magnitude estimation has found its primary usage, generally only one attribute is rated at a time. However, rating multiple attributes or profiling has been used in taste studies (McBurney and Bartoshuk, 1973; McBurney and Shick, 1971; McBurney et al., 1972) and this can naturally be extended to foods with multiple taste and aromatic attributes. Magnitude estimation has not been used very often for descriptive analysis, but in principle there is no reason why it could not be used for that purpose.

Participants should be cautioned to avoid falling into previous habits of using bounded category scales, e.g., limited ranges of numbers from zero to ten. This may be a difficult problem with previously trained panels that have used a different scaling method, as people like to stick with a method they know and feel comfortable with. Panelists who show such behavior may not understand the ratio nature of the instructions. It is sometimes useful with a new panelist to have the participant perform a warm-up task to make sure they understand the scaling instructions. The warm-up task can involve estimation of the size or area of different geometric figures (Meilgaard et al., 2006) or the length of lines (McBurney et al., 1972). Sometimes it is desired to have panelists rate multiple attributes at the same time or to break down overall intensity into specific qualities. If this kind of "profiling" is needed, the geometric figures can include areas with different shading or the lines can be differently colored. A practice task is highly recommended so that the sensory scientist can check on whether the participant understands the task.

Values of zero are allowed in this method as some of the products may in fact have or no sensation for a given attribute (like no sweetness in our example). Of course, the rating of zero should not be used for the reference material. While the value of zero is consistent with common sense for products with no sensation of some attributes, it does complicate the data analysis as discussed below.

7.3.3.2 Instructions

The visual appearance of the ballot in magnitude estimation is not critical; it is the instructions and the participant's comprehension of the ratio nature of the judgments that are important. Some ballots even allow the subject/participant to view all previous ratings. Here are sample instructions for the use of magnitude estimation with a reference sample or modulus with a fixed number assigned to it:

> Please taste the first sample and note its sweetness. This sample is given the value of "10" for its sweetness

intensity. Please rate all other samples in proportion to this reference. For example, if the next sample is twice as sweet, assign it a value of "20", if half as sweet, assign it a value of "5" and if 3.5 times as sweet, assign in a value of 35. In other words, rate the sweetness intensity so that your numbers represent the ratios among the intensities of sweetness. You may use any positive numbers including fractions and decimals.

The other major variation on this method uses no reference. In this case the instructions may read as follows:

> Please taste the first sample and note its sweetness. Please rate all other samples relative to this reference, applying numbers to the samples to represent the ratios of sweetness intensity among the samples. For example, if the next sample was twice as sweet, you would give it a number twice as big as the rating assigned to the first sample, if half as sweet, assign it a number half as big and if 3.5 times as sweet, assign it a number 3.5 times as big. You may use any positive numbers including fractions and decimals.

7.3.3.3 Data Treatment

In non-modulus methods, participants will generally choose some range of numbers they feel comfortable with. The ASTM procedure suggests having them pick a value between 30 and 100 for the first sample, and avoiding any number that seems small. If participants are allowed to choose their own number range, it becomes necessary to re-scale each individual's data to bring them into a common range before statistical analysis (Lane et al., 1961). This will prevent subjects who choose very large numbers from having undue influence on measures of central tendency (means) and in statistical tests. This rescaling process has been referred to as "normalizing" (ASTM, 2008a) although it has nothing to do with the normal distribution or Z-scores. A common method for rescaling proceeds as follows:

(1) Calculate the geometric mean of each individual's ratings across their data set.
(2) Calculate the geometric mean of the entire data set (of all subjects combined).
(3) For each subject, construct a ratio of the grand geometric mean of the entire data set to each person's geometric mean. The value of this ratio provides a post hoc individual rescaling factor for each subject. In place of the grand geometric mean,

any positive numerator may also be chosen in constructing this factor, e.g., a value of 100.

(4) Multiply each data point for a given person by their individual rescaling factor. Do this for all participants using their own individual re-scaling factors.

These re-scaled data are then analyzed. Note that due to the extra data treatment step in this method, it is simpler to use the modulus-based variation with a standard reference item.

Magnitude estimation data are often transformed to logs before data analysis (Butler et al., 1987; Lawless, 1989). This is done primarily because the data tend to be positively skewed or log-normally distributed. There tends to be some high outlying values for any given sample. Perhaps this is not surprising because the scale is open-ended at the top, and bounded by zero at the bottom. Transformation into log data and/or taking geometric means presents some problems, however, when the data contain zeros. The log of zero is undefined. Any attempt to take a geometric mean by calculating the product of N items will yield a zero on multiplying. Several approaches have been taken to this problem. One is to assign a small positive value to any zeros in the data, perhaps one-half of the smallest rating given by a subject (ASTM, 2008a). The resulting analysis, however, will be influenced by this choice. Another approach is to use the median judgments in constructing the normalization factor for non-modulus methods. The median is less influenced by the high outliers in the data than the arithmetic mean.

7.3.3.4 Applications

For practical purposes, the method of magnitude estimation may be used with trained panels, consumers, and even children (Collins and Gescheider, 1989). However, the data do tend to be a bit more variable than other bounded scaling methods, especially in the hands of untrained consumers (Lawless and Malone, 1986b). The unbounded nature of the scale may make it especially well suited to sensory attributes where an upper boundary might impose restrictions on the panelists' ability to differentiate very intense sensory experiences in their ratings. For example, irritative or painful sensations such as chili pepper intensity might all be rated near the upper bound of a category scale

of intensity, but an open-ended magnitude estimation procedure would allow panelists more freedom to differentiate and report variations among very intense sensations.

With hedonic scaling of likes and dislikes, there is an additional decision in using magnitude estimation scaling. Two options have been adopted in using this technique, one employing a single continuum or unipolar scale for amount of liking and the other applying a bipolar scale with positive and negative numbers plus a neutral point (Pearce et al., 1986). In bipolar magnitude scaling of likes and dislikes, positive and negative numbers are allowed in order to signify ratios or proportions of both liking and disliking (e.g., Vickers, 1983). An alternative to positives and negatives is to have the respondent merely indicate whether the number represents liking or disliking (Pearce et al., 1986). In unipolar magnitude estimation only positive numbers (and sometimes zeros) are allowed, with the lower end of the scale representing no liking and higher numbers given to represent increasing proportions of liking (Giovanni and Pangborn, 1983; Moskowitz and Sidel, 1971). It is questionable whether a unipolar scale is a sensible response task for the participant, as it does not recognize the fact that a neutral hedonic response may occur, and that there are clearly two modes of reaction, one for liking and one for disliking. If one assumes that all items are on one side of the hedonic continuum—either all liked to varying degrees or all disliked to varying degrees then the one-sided scale makes sense. However, it is a rare situation with foods or consumer product testing in which at least some indifference or change of opinion was not visible in at least some respondents. So a bipolar scale fits common sense.

7.4 Recommended Practice and Practical Guidelines

Both line scales and category scales may be used effectively in sensory testing and consumer work. So we will not expend much effort in recommending one of these two common techniques over another. Some practical concerns are given below to help the student or practitioner avoid some potential problems. The category–ratio or labeled magnitude scales may facilitate comparisons among different groups, and this issue is discussed below in Section 7.5.2.

7.4.1 Rule 1: Provide Sufficient Alternatives

One major concern is to provide sufficient alternatives to represent the distinctions that are possible by panelists (Cox, 1980). In other words, a simple 3-point scale may not suffice if the panel is highly trained and capable of distinguishing among many levels of the stimuli. This is illustrated in the case of the flavor profile scale, which began with five points to represent no sensation, threshold sensation, weak, moderate, and strong (Caul, 1957). It was soon discovered that additional intermediate points were desirable for panelists, especially in the middle range of the scale where many products would be found. However, there is a law of diminishing returns in allowing too many scale points—further elaboration allows better differentiation of products up to a point and then the gains diminish as the additional response choices merely capture random error variation (Bendig and Hughes, 1953).

A related concern is the tendency, especially in consumer work, to simplify the scale by eliminating options or truncating endpoints. This brings in the danger caused by end-use avoidance. Some respondents may be reluctant to use the end categories, just in case a stronger or weaker item may be presented later in the test. So there is some natural human tendency to avoid the end categories. Truncating a 9-point scale to a 7-point scale may leave the evaluator with what is functionally only a 5-point scale for all practical purposes. So it is best to avoid this tendency to truncate scales in experimental planning.

7.4.2 Rule 2: The Attribute Must Be Understood

Intensity ratings must be collected on an attribute which the participants understand and about which they have a general consensus and agreement as to its meaning. Terms like sweetness are almost universal but a term like "green aroma" might be interpreted in different ways. In the case of a descriptive panel, a good deal of effort may be directed at using reference standards to illustrate what is meant by a specific term. In the case of consumer work, such training is not done, so if any intensity ratings are collected, they must be about simple terms about which people generally

agree. Bear in mind that most early psychophysics was done on simple attributes like the loudness of a sound or the heaviness of a weight. In the chemical senses, with their diverse types of sensory qualities and fuzzy consumer vocabulary, this is not so straightforward.

Other problems to avoid include mixing sensation intensity (strength) and hedonics (liking), except in the just-right scale where this is explicit. An example of a hedonically loaded sensory term is the adjective "fresh." Whatever this means to consumers, it is a poor choice for a descriptive scale because it is both vague and connotes some degree of liking or goodness to most people. Vague terms are simply not actionable when it comes to giving feedback to product developers about what needs to be fixed. Another such vague term that is popular in consumer studies is "natural." Even though consumers might be able to score products on some unknown basis using this word, the information is not useful as it does not tell formulators what to change if a product scores low. A similar problem arises with attempting to scale "overall quality." Unless quality has been very carefully defined, it cannot be scaled.

7.4.3 Rule 3: The Anchor Words Should Make Sense

In setting up the scales for descriptive analysis or for a consumer test, the panel leader should carefully consider the nature of the verbal end anchors for each scale as well as any intermediate anchors that may be needed. Should the scale be anchored from "very weak" to "very strong" or will there be cases in which the sensory attribute is simply not present? If so, it makes sense to verbally anchor the bottom of the scale with "not at all" or "none." For example, a sweetness scale could be anchored with "not at all sweet" and a scale for "degree of oral irritation" could be anchored using "none."

7.4.4 To Calibrate or Not to Calibrate

If a high degree of calibration among the panelists is desired, then physical standards can be given for intensity. Often this is done with end examples as discussed above, but it may be advantageous to give examples of intermediate points on the scales as well. An example of this kind of calibration is found in the ASTM procedures for evaluating pepper heat, where three points on the scale are illustrated (weak, moderate, and approaching strong) (ASTM, 2008b). The traditional texture profile technique (Brandt et al., 1963) used nine scale points for most texture attributes like hardness and would give examples of common products representing each point on the scale. In the Spectrum descriptive method, the scales for intensity are intended to be comparable across all attributes and all products so scale examples are given from various sensory domains representing points on the 15-point scale for intensity (Meilgaard et al., 2006). Whether or not this degree of calibration is required for a specific project should be considered. There may also be a limit to the ability to stabilize the scale usage of respondents. There are limits on the abilities of humans to be calibrated as measuring instruments (Olabi and Lawless, 2008) and in spite of decades of research in scaling, this is not well understood. People differ in their sensitivities to various tastes and odors and thus may honestly differ in their sensory responses.

Another decision of the test designer will be whether to assign physical examples to intermediate scale points. Although reference items are commonly shown for the end categories, this is less often done with intermediate categories. The apparent advantage is to achieve a higher level of calibration, a desirable feature for trained descriptive panelists. A potential disadvantage is the restriction of the subject's use of the scale. What appears to be equal spacing to the experimenter may not appear so to the participant. In that case, it would seem wiser to allow the respondent to distribute his or her judgments among the scale alternatives, without presuming that the examples of intermediate scale points are in fact equally spaced. This choice is up to the experimenter. The decision reflects one's concerns as to whether it is more desirable to work toward calibration or whether there is more concern with potential biasing or restriction of responses.

7.4.5 A Warning: Grading and Scoring are Not Scaling

In some cases pseudo-numerical scales have been set up to resemble category scales, but the examples cut

across different sensory experiences, mixing qualities. An example is the pseudo-scale used for baked products, where the number 10 is assigned for perfect texture, 8 for slight dryness, 6 for gumminess, and 4 if very dry (AACC, 1986). Gumminess and dryness are two separate attributes and should be scaled as such. This is also an example of quality grading, which is not a true scaling procedure. When the numbers shift among different sensory qualities, this violates the psychophysical model for scaling the intensity of a single attribute. Although numbers may be applied to the grades, they cannot be treated statistically, as the average of "very dry " (4) and slightly dry (8) is not gummy (6) (see Pangborn and Dunkley, 1964, for a critique of this in the dairy grading arena). The numbers in a quality-grading scheme do not represent any kind of unitary psychophysical continuum.

7.5 Variations—Other Scaling Techniques

An important idea in scaling theory is the notion that people may have a general idea of how weak or strong sensations are, and that they can compare different attributes of a product for their relative strength, even across different sensory modalities. So, for example, someone could legitimately say that this product tastes much more salty than it is sweet. Or that the trumpets are twice as loud as the flutes in a certain passage in a symphony. Given that this notion is correct, people would seem to have a general internal scale for the strength of sensations. This idea forms the basis for several scaling methods. It permits the comparison of different sensations cross-referenced by their numerical ratings, and even can be used to compare word responses. Methods derived from this idea are discussed next.

7.5.1 Cross-Modal Matches and Variations on Magnitude Estimation

The method of magnitude estimation has a basis in earlier work such as fractionation methods and the method of sense ratios in the older literature (Boring,

1942) where people would be asked to set one stimulus in a given sensation ratio to another. The notion of allowing any numbers to be generated in response to stimuli, rather than adjusting stimuli to represent fixed numbers appeared somewhat later (Moskowitz, 1971; Richardson and Ross, 1930; Stevens, 1956). An important outcome of these studies was the finding that the resulting psychophysical function generally conformed to a power law of the following form:

$$R = kI^n \qquad (7.1)$$

or after log transformation:

$$\log(R) = n\log(I) + \log(k) \qquad (7.2)$$

where R was the response, e.g., perceived loudness (mean or geometric mean of the data) and I was the physical stimulus intensity, e.g., sound pressure, and k was a constant of proportionality that depends upon the units of measurement. The important characteristic value of any sensory system was the value for n, the exponent of the power function or slope of the straight line in a log–log plot (Stevens, 1957). The validity of magnitude estimation then came to hang on the validity of the power law—the methods and resulting functions formed an internally consistent theoretical system. Stevens also viewed the method as providing a direct window into sensation magnitude and did not question the idea that these numbers generated by subjects might be biased in some way. However, the generation of responses is properly viewed as combining at least two processes, the psychophysical transformation of energy into conscious sensation and the application of numbers to those sensations. This response process was not given due consideration in the early magnitude estimation work. Responses as numbers can exhibit nonlinear transformations of sensation (Banks and Coleman, 1981; Curtis et al., 1968) so the notion of a direct translation from sensation to ratings is certainly a dangerous oversimplification.

Ratio-type instructions have been applied to other techniques as well as to magnitude estimation. A historically important psychophysical technique was that of cross-modality matching, in which the sensation levels or ratios would be matched in two sensory continua such as loudness and brightness. One continuum would be adjusted by the experimenter and the other by the subject. For example, one would try to make the brightness of the lights about in the same proportions as the loudness of the sounds. Stevens (1969) proposed that these experiments could validate the power law, since the exponents of the cross-modality matching function could be predicted from the exponents derived from separate scaling experiments. Consider the following example:

For one sensory attribute (using the log transform, Eq. (7.2)),

$$\log R_1 = n_1 \log I_1 + \log k_1 \qquad (7.3)$$

and for a second sensory attribute

$$\log R_2 = n_2 \log I_2 + \log k_2 \qquad (7.4)$$

Setting $R_1 = R_2$ in cross-modality matching gives

$$n_1 \log I_1 + \log k_1 = n_2 \log I_2 + \log k_2 \qquad (7.5)$$

and rearranging,

$$\log I_1 = (n_2/n_1)\log I_2 + \text{a constant} \qquad (7.6)$$

If one plots $\log I_1$ against $\log I_2$ from a cross-modality matching task, the slope of the function can be predicted from the ratio of the slopes of the individual exponents (i.e., n_2/n_1, which you can derive from two separate magnitude estimation tasks). This prediction holds rather well for a large number of compared sensory continua (Steven, 1969). However, whether it actually provides a validation for the power law or for magnitude estimation has been questioned (e.g., Ekman, 1964).

For practical purposes, it is instructive that people can actually take disparate sensory continua and compare them using some generalized notion of sensory intensity. This is one of the underpinnings of the use of a universal scale in the Spectrum descriptive procedure (Meilgaard et al., 2006). In that method, different attributes are rated on 15-point scales that can (in theory) *be meaningfully compared*. In other words, a 12 in sweetness is twice as intense a sensation as a 6 in saltiness. Such comparisons seem to makes sense for tastes and flavors but may not cut across all other modalities. For example, it might seem less sensible to compare the rating given for the amount of chocolate chips in a cookie to the rating given for the cookie's hardness—these seem like quite different experiences to quantify.

Cross-modality matches have been performed successfully with children even down to age 4, using line length compared to loudness (Teghtsoonian, 1980). This may have some advantage for those with limited verbal skills or trouble understanding the numerical concepts needed for category scales or magnitude estimation. The use of line length points out that the line-marking technique might be considered one form of a cross-modality matching scale. Some continua may seem simpler, easier, or more "natural" to be matched (for example, hand grip force to perceived strength of tooth pain). King (1986) matched the pitches of tones to concentrations of benzaldehyde and Ward (1986) used duration as a matching continuum for loudness and brightness. One of the advantages of the cross-modality matching procedure is that it is possible to specify the intensity of a sensation in physical units, i.e., as a physical level on the other continuum. So sweetness, for example, could be represented in a decibel (sound pressure) equivalent. In one amusing variation on this idea, Lindvall and Svensson (1974) used hedonic matching to specify the unpleasantness of combustion toilet fumes to different levels of H_2S gas that were sniffed from an olfactometer. Thus the lucky participant could dial up a concentration that was perceived as being equally as offensive as the test samples.

If line marking can be considered a kind of cross-modality match, then why not the numbers themselves? It should be possible to cross-reference one continuum to another simply through instructions to use a common response scale. Stevens and Marks (1980) developed the technique of magnitude matching to do just that (see also Marks et al., 1992). Subjects were instructed to judge loudness and brightness on a common scale of intensity so that if a sound had the same sensory impact as the brightness of a light, then the stimuli would be given the same number. This type of cross-referencing should facilitate comparisons among people. For example, if it can be assumed that two people have the same response to salt taste or to loudness of tones, then differences in some other continuum like bitter taste, hot chili pepper intensity, or a potential olfactory loss can be cross-referenced through salty taste or through loudness of tones they have rated in magnitude matching (e.g., Gent and Bartoshuk, 1983). Furthermore, if numbers can provide a cross-referencing continuum, then why not scale the word phrases used as anchor points on a category

or line scale? The idea of scaling word phrases takes shape in the labeled magnitude scales discussed next.

7.5.2 Category–Ratio (Labeled Magnitude) Scales

A group of hybrid techniques for scaling has recently enjoyed some popularity in the study of taste and smell, for hedonic measurement and other applications. One of the problems with magnitude estimation data is that it does not tell in any absolute sense whether sensations are weak or strong, only giving the ratios among them. This group of scales attempts to provide ratio information, but combines it with common verbal descriptors along a line scale to provide a simple frame of reference. They are referred to as category–ratio scales, or more recently, labeled magnitude scales. They all involve a horizontal or vertical line with deliberately spaced labels and the panelists' task is to make a mark somewhere along the line to indicate the strength of their perception or strength of their likes or dislikes. In general, these labeled line scales give data that are consistent with those from magnitude estimation (Green et al., 1993). An unusual characteristic of these scales is the verbal high end-anchor phrase, which often refers to the "strongest imaginable."

The technique is based on early work by Borg and colleagues, primarily in the realm of perceived physical exertion (Borg, 1982, 1990; see Green et al., 1993). In developing this scale, Borg assumed that the semantic descriptors could be placed on a ratio scale and that they defined the level of perceptual intensity and that all individuals experienced the same perceptual range. Borg suggested that for perceived exertion, the maximal sensation is roughly equivalent across people for this sensory "modality" (Marks et al., 1983). For example, it is conceivable that riding a bicycle to the point of physical exhaustion produces a similar sensory experience for most people. So the scale came to have the highest label referring to the strongest sensation imaginable.

This led to the development of the labeled magnitude scale (LMS) shown in Fig. 7.4. It is truly a hybrid method since the response is a vertical line-marking task but verbal anchors are spaced according to calibration using ratio-scaling instructions (Green

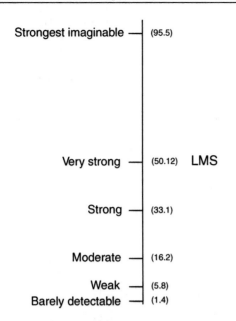

Strongest imaginable — (95.5)

Very strong — (50.12) LMS

Strong — (33.1)

Moderate — (16.2)

Weak — (5.8)
Barely detectable — (1.4)

Fig. 7.4 The labeled magnitude scale (LMS) of Green et al. (1993).

et al., 1993). In setting up the scale, Green and colleagues had subjects provide magnitude estimates of different verbal descriptors after giving magnitude estimates of familiar oral sensations (e.g., the bitterness of celery, the burn of cinnamon gum). These results were generally in line with previous scaling of verbal descriptors, so-called semantic scaling. Other researchers have developed scales with only direct scaling of the verbal descriptors and have not always included a list of everyday or common experiences (Cardello et al., 2003; Gracely et al., 1978a, b).

After the introduction of the LMS, a number of researchers tried to extend this approach into the realm of hedonics (measuring food acceptability). A widely used scale is the labeled affective magnitude (LAM) scale developed by Schutz and Cardello (2001). They used direct ratio scaling of the verbal descriptors of the 9-point hedonic scale and included Borg's type of high (and low) end anchor ("greatest imaginable like/dislike"). The scale is shown in Fig. 7.5. This shows some advantages in differentiating well-liked items (El Dine and Olabi, 2009; Greene et al., 2006; Schutz and Cardello, 2001), although that finding is not universal (Lawless et al., 2010a). The LAM scale or similar versions of it have been applied in a variety of studies with different foods (Chung and Vickers, 2007a, b; El Dine and Olabi, 2009; Forde and Delahunty, 2004; Hein et al., 2008; Keskitalo et al.,

2007; Lawless et al., 2010a, b, c). A growing number of similar scales have been developed for various applications including oral pleasantness/unpleasantness (the "OPUS" scale, Guest et al., 2007), perceived satiety (the "SLIM" scale, Cardello et al., 2005), clothing fabric comfort (the "CALM" scale, Cardello et al., 2003), and odor dissimilarity (Kurtz et al., 2000). All of these scales depend upon a ratio scaling task to determine the spacing of the verbal descriptors and almost all use a Borg-type high end-anchor phrase. Others will surely be developed.

Instructions to participants have differed in the use of these scales. In the first application of the LMS, Green et al. (1993) instructed subjects to first choose the most appropriate verbal descriptor, and then to "fine tune" their judgment by placing a mark on the line between that descriptor and the next most appropriate one. In current practice less emphasis may be placed on the consideration of the verbal labels and instructions may be given to simply make a mark "anywhere" on the line. A common observation with the hedonic versions of the scale is that some panelists will mark at or very near a verbal descriptor, seeming to use it as a category scale (Cardello et al., 2008; Lawless et al., 2010a). The proportion of people displaying this behavior may depend upon the physical length of the line (and not the instructions or examples that may be shown) (Lawless et al., 2010b).

Results may depend in part on the nature of the high end-anchor example and the frame of reference of the subject in terms of the sensory modality they are thinking about. Green et al. (1996) studied the application of the LMS to taste and odor, using descriptors for the upper bound as "strongest imaginable" taste, smell, sweetness, etc. Steeper functions (a smaller response range) were obtained when mentioning individual taste qualities. This appears to be due to the omission of painful experiences (e.g., the "burn of hot peppers") from the frame of reference when sensations were scaled relative to only taste. The steepening of the functions for the truncated frame of reference is consistent with the idea that subjects expanded their range of numbers as seen in other scaling experiments (e.g., Lawless and Malone, 1986b). The fact that subjects appear to adjust their perceptual range depending on instructions or frame of reference suggests that the scales have relative and not absolute properties, like most other scaling methods. Cardello et al. (2008) showed that the hedonic version of the scale (the LAM

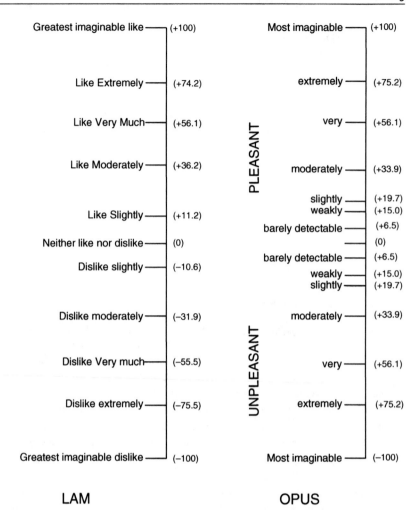

Fig. 7.5 Affective labeled magnitude scales, including the LAM scale (Cardello and Schutz, 2004) and the OPUS scale (Guest et al., 2007).

scale) will also show such range effects. A compressed range of responses is obtained when the frame of reference is greatest imaginable like (dislike) for an "experience of any kind" rather than something more delimited like "foods and beverages." Apparently the compression is not very detrimental to the ability of the LAM scale to differentiate products (Cardello et al., 2008, but see also Lawless et al., 2010a).

Is this a suitable method for cross-subject comparisons? To the extent that Borg's assumptions of common perceptual range and the similarity of the high end-anchor experience among people are true, the method might provide one approach to valid comparisons of the ratings among different respondents. This would facilitate comparisons of clinical groups or patients with sensory disorders or genetically different individuals such as anosmics, PTC/PROP taster groups. Bartoshuk and colleagues (1999, 2003,

2004a, b, 2006) have argued that the labeled magnitude scales should anchor their endpoints to "sensations of any kind" as such a reference experience would allow different individuals to use the scale in similar ways/and thus facilitate inter-individual comparisons. Scales with this kind of high end anchor have been termed "generalized" labeled magnitude scales (or gLMS). However, the sensory evaluation practitioner should be aware of the compression effects that can occur with this kind of scale, which could potentially lead to lessened differentiation among products.

7.5.3 Adjustable Rating Techniques: Relative Scaling

A few methods have been tried that allow consumers or panelists to change their ratings. An example is the "rank-rating" technique developed by O'Mahony, Kim, and colleagues (Kim and O'Mahony, 1998; Lee et al., 2001; O'Mahony et al., 2004; Park et al., 2004). In this method, the consumer has a visual scale in front of him or her on the table and after tasting a sample is instructed to physically place the sample on the scale. Subsequent samples are also tasted and placed and the important feature is that the consumer can change the position of any previous item based on their perception of the new sample(s). This procedure has relatively little to do with rank ordering per se (in fact ties can be allowed). Cordinnier and Delwiche (2008) chose the more descriptive name of "positional relative rating" for this technique.

O'Mahony and Kim examined the efficiency of this technique, mostly in simple salt solutions, on the basis of people's ordering of salt solutions in increasing concentrations (Kim and O'Mahony, 1998; Park et al., 2004). The important data were "reversals" in which a higher concentration was rated lower than a lower one, and vice versa. A good scale would minimize the amount of reversals. Given this criterion, the rank-rating has fewer errors than non-adjustable ratings. At this point it is not clear whether this apparent advantage arises because people are allowed to re-taste, or that they are allowed to re-position previously tasted items. Both factors may be important. There is some evidence that adjustable ratings can produce statistically significant ratings with fewer subjects, but the procedure can take up to twice as long as normal ratings. A limitation of this technique is that only one attribute can be evaluated at a time. If a second or third attribute is needed, the procedure starts over. This may be acceptable for consumer like/dislike ratings (O'Mahony et al., 2004), but would not be suitable for a descriptive analysis task. The option to change previous ratings is an interesting notion and is allowed by some of the sensory data-collection software packages at this time. Whether the option is advantageous should be the subject of further study. Re-tasting (Koo et al., 2002; Lee et al., 2001) is a potentially important feature of this method, and should be evaluated separately, and in consideration of the extra adaptation, fatigue, or carryover that could occur with some products.

A completely relative rating procedure is the method of Gay and Mead (Gay and Mead, 1992; Mead and Gay, 1995). In this task, a panelist inspects the entire set of samples and places the most intense (or most liked) at the top end of the scale and the least intense (or least liked) at the bottom. All other samples are distributed along the full scale. This can provide good differentiation of the products, but obviously any absolute information about what is weak or strong, liked or disliked, is lost (as is the case with magnitude estimation). Because all the ratings are truly relative, contextual effects such as contrast might be expected to be larger with such a technique, but this is not known.

Another relative scaling method is when panelists purposely rate each sample relative to a reference item, which is usually marked on the center of the response scale. Relative-to-reference scaling was studied by Larson-Powers and Pangborn (1978) and compared to traditional scaling for the descriptive profiling of beverages and gelatins. Significant differences were found with the relative scale ("anchored" in their terms) in 22.8% of all possible comparisons as compared with 19.5% of comparisons using the unanchored scale. However, panelists were given the relative scale first, and more practice using that scale. In an extensive study of both trained and untrained respondents, Stoer and Lawless (1993) found a similar advantage, with 33% of all possible comparisons were significant for the relative scaling versus 27% for traditional scaling. However, this was not a statistically significant increase based on meta-analytic comparisons (Rosenthal, 1987). The relative-to-reference scale was also discussed by Land and Shepard (1984), who noted that it facilitates comparisons across occasions that would be otherwise difficult to make. They also warn that the choice of standard may have an effect on the scaling functions that result. The task is certainly easy for subjects and was touted by Land and Shepard as showing "good reproducibility" (see also Mahoney et al., 1957). Whether the method offers any consistent advantage over traditional scaling is questionable. It may be useful in situations where comparison to a reference is a natural feature or explicit objective of the experiment at hand, for example in quality control or shelf life studies where an identified control sample is used as a baseline for comparison.

7.5.4 Ranking

Another alternative to traditional scaling is the use of ranking procedures. Ranking is simply ordering the products from weakest to strongest on the stated attribute or from least liked to most liked for consumer acceptance testing. Ranking has the advantages of simplicity in instructions to subjects, simplicity in data handling, and minimal assumptions about level of measurement since the data are treated as ordinal. Although ranking tests are most often applied to hedonic data, they are also applicable to questions of sensory intensity. When asked to rank items for the intensity of a specific attribute, e.g., the sourness of taste of several fruit juices, for example, the ranking test is merely an extension of a paired comparison procedure into more than two products. Due to its simplicity, ranking may be an appropriate choice in situations where participants would have difficulty understanding scaling instructions. In working with non-literates, young children, across cultural boundaries or in linguistically challenging situations, ranking is worth considering (Coetzee and Taylor, 1996). This is especially true if decisions are likely to be very simple, e.g., whether or not two juice samples differ in perceived sourness. Ranking may allow differentiation of products that are all similar in acceptability. With medications, for example, all formulas may be to some degree unpalatable. It might be useful then to use ranking in choosing alternative flavorings in order to find the least offensive.

Analysis of ranked data is straightforward. Simple rank-sum statistics can be found in the tables published by Basker (1988) and Newell and MacFarlane (1987, see also Table J). Another very sensitive test of differences in ranked data is the Friedman test, also known as the "analysis of variance on ranks." These are discussed in Appendix B. The tests are rapid, straightforward, and easy to perform. It is also possible to convert other data to rankings. This is a conservative approach if the interval nature of the data is in question or when violations of statistical assumptions such as the normality of the data are suspect. For example, Pokorný et al. (1986) used ranking analysis of line-marking data to compare the profiles of different raspberry beverages sweetened with aspartame.

7.5.5 Indirect Scales

A conservative approach to scaling is to use the variance in the data as units of measurement, rather than the numbers taken at face value. For example, we could ask how many standard deviations apart are the mean values for two products. This is a different approach to measurement than simply asking how many scale units separate the means on the response scale. On a 9-point scale, one product may receive mean rating of seven, and another nine, making the difference two scale units. If the pooled standard deviation is two units, however, they would only be one unit apart on a variability-based scale. As one example, Conner and Booth (1988) used both the slope and the variance of functions from just-right scales to derive a "tolerance discrimination ratio." This ratio represents a measure of the degree of difference along a physical continuum (such as concentration of sugar in lime drink) that observers find to make a meaningful change in their ratings of difference-from-ideal (or just-right). This is analogous to finding the size of a just-noticeable difference, but translated into the realm of hedonic scaling. Their insight was that it is not only the slope of the line that is important in determining this tolerance or liking-discrimination function, but also the variance around that function.

Variability-based scales are the basis for scaling in Thurstone's models for comparative judgment (Thurstone, 1927) and its extension into determining distances between category boundaries (Edwards, 1952). Since the scale values can be found from choice experiments as well as rating experiments, the technique is quite flexible. How this type of scaling can be applied to rating data is discussed below in the derivation of the 9-point hedonic scale words. When the scale values are derived from a choice method like the triangle test or paired comparison method, this is sometimes called "indirect scaling" (Baird and Noma, 1978; Jones, 1974). The basic operation in Thurstonian scaling of choice data is to convert the proportion correct in a choice experiment (or simply the proportions in a two-tailed test like paired preference) to Z-scores. The exact derivation depends upon the type of test (e.g., triangle versus 3-AFC) and the cognitive strategy used by the subject. Tables for Thurstonian scale values from various tests such as the triangle test were given by Frijters et al. (1980) and Bi (2006) and some tables are given in the Appendix. Mathematical details of Thurstonian scaling are discussed in Chapter 5.

Deriving measures of sensory differences in such indirect ways presents several problems in applied sensory evaluation so the method has not been widely used. The first problem is one of economy in data collection. Each difference score is derived from a separate discrimination experiment such as a paired comparison test. Thus many subjects must be tested to get a good estimate of the proportion of choice, and this yields just one scale value. In direct scaling, each participant gives at least one data point for each item tasted. Direct scaling allows for easy comparisons among multiple products, while the discrimination test must be done on one pair at a time. Thus the methods of indirect scaling are not cost-efficient.

A second problem can occur if the products are too clearly different on the attribute in question, because then the proportion correct will approach 100%. At that point the scale value is undefined as they are some unknown number of standard deviations apart. So the method only works when there are small differences and some confusability of the items. In a study of many products, however, it is sometimes possible to compare only adjacent or similar items, e.g., products that differ in small degrees of some ingredient or process variable. This approach was taken by Yamaguchi (1967) in examining the synergistic taste combination of monosodium glutamate and disodium 5′ inosinate. Many different levels of the two ingredients were tasted, but because the differences between some levels were quite apparent, an incomplete design was used in which only three adjacent levels were compared.

Other applications have also used this notion of variability as a yardstick for sensory difference or sensation intensity. The original approach of Fechner in constructing a psychophysical function was to accumulate the difference thresholds or just-noticeable differences (JNDs) in order to construct the log function of psychophysical sensory intensity (Boring, 1942; Jones, 1974). McBride (1983a, b) examined whether JND-based scales might give similar results to category scales for taste intensity. Both types of scaling yielded similar results, perhaps not surprising since both tend to conform to log functions. In a study of children's preferences for different odors Engen (1974) used a paired preference paradigm, which was well suited to the abilities of young children to respond in a judgment task. He then converted the paired preference proportions to Thurstonian scale values via Z-scores and was able to show that the hedonic range of children was smaller than that of adults.

Another example of choice data that can be converted to scale values is best–worst scaling, in which a consumer is asked to choose the best liked and least liked samples from a set of three or more items (Jaeger et al., 2008). With three products, it can be considered a form of a ranking task. When applied to sensory intensity, this is sometimes known as maximum-difference or "max-diff." Best–worst scaling is also discussed in Section 13.7. Simple difference scores may be calculated based on the number of times an item is called best versus worst and these scores are supposed to have interval properties. If a multinomial logistic regression is performed on the data, they are theorized to have true ratio properties (Finn and Louviere, 1992). A practical problem with the method, however, is that so many products must be tasted and compared, rendering it difficult to perform with foods (Jaeger and Cardello, 2009).

The sensory professional should bear in mind that in spite of their theoretical sophistication, the indirect methods are based on variability as the main determinant of degree of difference. Thus any influence, which increases variability, will tend to decrease the measured differences among products. In the well-controlled psychophysical experiment under constant standard conditions across sessions and days, this may not be important—the primary variability lies in the resolving power of the participant (and secondarily in the sample products). But in drawing conclusions across different days, batches, panels, factories, and such, one has a less pure situation to consider. Whether one considers the Thurstonian-type indirect measures comparable across different conditions depends upon the control of extraneous variation.

7.6 Comparing Methods: What is a Good Scale?

A large number of empirical studies have been conducted comparing the results using different scaling methods (e.g., Birnbaum, 1982; Giovanni and Pangborn, 1983; Hein et al., 2008; Jaeger and Cardello, 2009; Lawless and Malone, 1986a, b; Lawless et al., 2010a; Marks et al., 1983; Moskowitz and Sidel, 1971; Pearce et al., 1986; Piggot and Harper, 1975; Shand et al., 1985; Vickers, 1983; Villanueva and Da Silva, 2009; Villanueva et al., 2005). Because scaling data are often used to identify differences between products, the ability to detect differences is one important practical criterion for how useful a scaling method can be (Moskowitz and Sidel, 1971). A related criterion is the degree of error variance or similar measures such as size of standard deviations or coefficients of variation. Obviously, a scaling method with low interindividual variability will result in more sensitive tests, more significant differences, and lower risk of Type II error (missing a true difference). A related issue is the reliability of the procedure. Similar results should be obtained upon repeated experimentation.

Other practical considerations are important as well. The task should be user friendly and easy to understand for all participants. Ideally, the method should be applicable to a wide range of products and questions, so that the respondent is not confused by changes in response type over a long ballot or questionnaire. If panelists are familiar with one scale type and are using it effectively, there may be some liability in trying to introduce a new or unfamiliar method. Some methods, like category scales, line scales, and magnitude estimation, can be applied to both intensity and hedonic (like–dislike) responses. The amount of time required to code, tabulate and process the information may be a concern, depending upon computer-assisted data collection and other resources available to the experimenters.

As in any method, validity or accuracy are also issues. Validity can only be judged by reference to some external criterion. For hedonic scaling, one might want the method to correspond to other behaviors such as choice or consumption (Lawless et al., 2010a). A related criterion is the ability of the scale to identify or uncover consumer segments with different preferences (Villanueva and Da Silva, 2009).

Given these practical considerations, we may then ask how the different scaling methods fare. Most published studies have found about equal sensitivity for the different scaling methods, provided that the methods are applied in a reasonable manner. For example, Lawless and Malone (1986a, b) performed an extensive series of studies (over 20,000 judgments) with consumers in central location tests using different sensory continua including olfaction, tactile, and visual modalities. They compared line scales, magnitude estimation, and category scales. Using the degree of statistical differentiation among products as the criterion for utility of the methods, the scales performed about equally well. A similar conclusion was reached by Shand et al. (1985) for trained panelists. There

was some small tendency for magnitude estimation to be marginally more variable in the hands of consumers as opposed to college students (Lawless and Malone, 1986b). Statistical differentiation increased over replicates, as would be expected as people came to understand the range of items to be judged (see Hein et al., 2008, for another example of improvement over replication in hedonic scaling). Similar findings for magnitude estimation and category scales in terms of product differentiation were found by Moskowitz and Sidel (1971), Pearce et al. (1986), Shand et al. (1985), and Vickers (1983) although the forms of the mathematical relations to underlying physical variables was often different (Piggot and Harper, 1975). In other words, as found by Stevens and Galanter (1957) there is often a curvilinear relationship between the data from the two methods. However, this has not been universally observed and sometimes simple linear relationships have been found (Vickers, 1983). Similar results for category scales and line scales were found by Mattes and Lawless (1985).

Taken together, these empirical studies paint a picture of much more parity among methods than one might suppose given the number of arguments over the validity of scaling methods in the psychological literature. With reasonable spacing of the products and some familiarization with the range to be expected, respondents will distribute their judgments across the available scale range and use the scale appropriately to differentiate the products. A reasonable summary of the literature comparing scale types is that they work about equally well to differentiate products, given a few sensible precautions.

7.7 Issues

7.7.1 "Do People Make Relative Judgments" Should They See Their Previous Ratings?

Baten's (1946) report of an advantage for the line scale is illustrative of how observant many researchers were in the older literature. He noted that a category scale with labeled alternatives might help some judges, but could hinder others by limiting the alternative responses (i.e., judgments that might fall in-between categories). The line scale offers a continuously graded choice of alternatives, limited only by the measurement abilities in data tabulation. Baten also noted that the line scale seemed to facilitate a relative comparison among the products. This was probably due to his placement of the scales one above the other on the ballot, so judges could see both markings at the same time. In order to minimize such contextual effects it is now more common to remove the prior ratings for products to achieve a more independent judgment of the products. However, whether that is ever achieved in practice is open to question—humans are naturally comparative when asked to evaluate items, as discussed in Chapter 9. Furthermore, there may be potential for increased discrimination in methods like the relative positioning technique. The naturally comparative nature of human judgment may be something we could benefit from rather than trying to fight this tendency by over-calibration.

7.7.2 Should Category Rating Scales Be Assigned Integer Numbers in Data Tabulation? Are They Interval Scales?

There is also a strong suspicion that many numerical scaling methods may produce only ordinal data, because the spacing between alternatives is not subjectively equal. A good example is the common marketing research scale of "excellent—very good—good—fair—poor." The subjective spacing between these adjectives is quite uneven. The difference between two products rated good and very good is a much smaller difference than that between products rated fair and poor. However, in analysis we are often tempted to assigned numbers one through five to these categories and take means and perform statistics as if the assigned numbers reflected equal spacing. This is a pretense at best. A reasonable analysis of the 5-point excellent to poor scale is simply to count the number of respondents in each category and to compare frequencies. Sensory scientists should not assume that any scale has interval properties in spite of how easy it is to tabulate data as an integer series.

7.7.3 Is Magnitude Estimation a Ratio Scale or Simply a Scale with Ratio Instructions?

In magnitude estimation, subjects are instructed to use numbers to reflect the relative proportions between the intensities experienced from different stimuli. A beverage that is twice as sweet as another should be given a response number that is twice as large. S. S. Stevens felt that these numbers were accurate reflections of the experience, and so the scale had ratio properties. This assumed a linear relationship between the subjective stimulus intensities (sensations or percepts) and the numerical responses. However, there is a wealth of information showing that the process of numerical assignment is prone to a series of contextual and number usage biases that strongly question whether this process is linear (Poulton, 1989). So Stevens' original view of accepting the numerical reporting as having face validity seems misplaced. Although it would be advantageous to achieve a level of measurement that did allow conclusions about proportions and ratios ("this is liked twice as much as that"), this seems not fully justified at this time. It is important to differentiate between a method that has ratio-type instructions, and one that yields a true ratio scale of sensation magnitude, where the numbers actually reflect proportions between intensities of sensory experiences.

7.7.4 What is a "Valid" Scale?

An ongoing issue in psychophysics is what kind of scale is a true reflection of the subject's actual sensations? From this perspective, a scale is valid when the numbers generated reflect a linear translation of subjective intensity (the private experience). It is well established that category scales and magnitude estimates, when given to the same stimuli, will form a curve when plotted against one another (Stevens and Galanter, 1957). Because this is not a linear relationship, one method or the other must result from a non-linear translation of the subjective intensities of the stimuli. Therefore, by this criterion, at least one scale must be "invalid."

Anderson (1974, 1977) proposed a functional measurement theory to address this issue. In a typical experiment, he would ask subjects to do some kind of combination task, like judging the total combined intensity of two separately presented items (or the average lightness of two gray swatches). He would set up a factorial design in which every level of one stimulus was combined with every level of the other (i.e., a complete block). When plotting the response, a family of lines would be seen when the first stimulus continuum formed the X-axis and the second formed a family of lines. Anderson argued that only when the response combination rule was additive, *and* the response output function was linear, would a parallel plot be obtained (i.e., there would be no significant interaction term in ANOVA). This argument is illustrated in Fig. 7.6. In his studies using simple line and category scales, parallelism was obtained in a number of studies, and thus he reasoned that magnitude estimation was invalid by this criterion. If magnitude estimation is invalid, then its derivatives such as the LMS and LAM scales are similarly suspect.

Others have found support for the validity of magnitude estimation in studies of binaural loudness summation (Marks, 1978). This argument continues and is difficult to resolve. A review of the matter was published by Gescheider (1988). For the purposes of sensory evaluation, the issue is not terribly important for two reasons. First, any scale that produces statistically significant differentiation of products is a useful scale. Second, the physical ranges over which category scaling and magnitude estimation produce different results is usually quite large in any psychophysical study. In most product tests, the differences are much more subtle and generally do not span such a wide dynamic range. The issue dissolves from any practical perspective.

7.8 Conclusions

Much sound and fury has been generated over the years in the psychophysical literature concerning what methods yield valid scales. For the sensory practitioner, these issues are less relevant because the scale values do not generally have any absolute meaning. They are only convenient indices of the relative intensities or appeal of different products. The degree of difference may be a useful piece of information, but often we are simply interested in which product is stronger or weaker in some attribute, and whether the difference is both statistically significant and practically meaningful.

Scaling provides a quick and useful way to get intensity or liking information. In the case of descriptive analysis, scaling allows collection of quantitative data on multiple attributes. The degree of variability or noise in the system is, to a large part, determined by whether the panelists have a common frame of reference. Thus reference standards for both the attribute terms and for the intensity anchors are useful. Of course, with consumer evaluations or a psychophysical study such calibration is not possible and usually not desired. The variability of consumer responses should offer a note of caution in the interpretation of consumer scaling data.

Students and sensory practitioners should examine their scaling methods with a critical eye. Not every task that assigns numbers will have useful scale properties like equal intervals. Bad examples abound

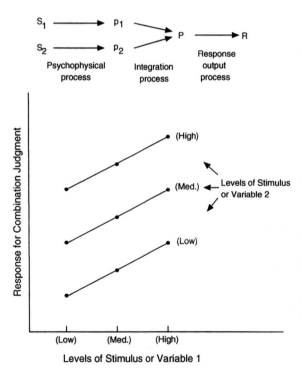

Fig. 7.6 The functional measurement scheme of Anderson (1974).

in the commodity grading (quality scoring) literature (Pangborn and Dunkley, 1964). For example, different numbers may be assigned to different levels of oxidation, but that is scoring a physical condition based on inferences from sensory experience. It is not a report of the intensity of some experience itself. It is not tracking changes along a single perceptual continuum in the psychophysical sense. Scoring is not scaling.

> All hedonic scales seem to measure what they are intended to measure rather effectively, as long as no gross mistakes are made (Peryam, 1989, p. 23).

Appendix 1: Derivation of Thurstonian-Scale Values for the 9-Point Scale

The choice of adjective words for the 9-point hedonic scale is a good example of how carefully a scale can be constructed. The long-standing track record of this tool demonstrates its utility and wide applicability in consumer testing. However, few sensory practitioners actually know how the adjectives were found and what criteria were brought to bear in selecting these descriptors (slightly, moderately, very much, and extremely like/dislike) from a larger pool of possible words. The goal of this section is to provide a shorthand description of the criteria and mathematical method used to select the words for this scale.

One concern was the degree to which the term had consensual meaning in the population. The most serious concern was when a candidate word had an ambiguous or double meaning across the population. For example, the word "average" suggests an intermediate response to some people, but in the original study by Jones and Thurstone (1955) there were a group of people who equated it with "like moderately" perhaps since an average product in those days was one that people would like. These days, one can think of negative connotations to the word "average" as in "he was only an average student." Other ambiguous or bimodal terms were "like not so much" and "like not so well." Ideally, a term should have low variability in meaning, i.e., a low standard deviation, no bimodality, and little skew. Part of this concern with the normality of the distribution of psychological reactions to a word was the fact that the developers used Thurstone's model

for categorical judgment as a means of measuring the psychological-scale values for the words. This model is at its most simple form when the items to be scaled show normal distributions of equal variance.

Which leads us to the numerical method. Jones and Thurstone modified a procedure used earlier by Edwards (1952). A description of the process and results can be found in the paper "Development of a scale for measuring soldiers' food preferences" by Jones et al. (1955). Fifty-one words and phrases formed the candidate list based on a pilot study with 900 soldiers chosen to be a representative sample of enlisted personnel. Each phrase was presented on a form with a rating scale from –4 to +4 with a check off format. In other words, each person read each phrase and assigned in an integer value from –4 to +4 (including zero as an option). This method would seem to presume that these integers were themselves an interval scale of psychological magnitude, an assumption that to our knowledge has never been questioned.

Of course, the mean scale values could now be assigned on a simple and direct basis, but the Thurstonian methods do not use the raw numbers as the scale, but transform them to use standard deviations as the units of measurement. So the scale needs to be converted to Z-score values. The exact steps are as follows:

1. Accumulate frequency counts for all the tested words across the –4 to +4 scale. Think of these categories as little "buckets" into which judgments have been tossed.
2. Find the marginal proportions each value from –4 to +4 (summed across all test items). Add up the proportions from lowest to highest to get a cumulative proportion for each bucket.
3. Convert these proportions to z-scores in order to re-scale the boundaries for the original –4 to +4 cutoffs. Let us call these the "category z-values" for each of the "buckets." The top bucket will have a value of 100%, so it will have no z-score (undefined/infinite).
4. Next examine each individual item. Sum its individual proportions across the categories, from where it is first used until 100% of the responses are accumulated.
5. Convert the proportions for the item to Z-scores. Alternatively, you can plot these proportions on "cumulative probability paper," a graphing format

that marks the ordinate in equal standard deviations units according to the cumulative normal distribution. Either of these methods will tend to make the cumulative S-shaped curve for the item into a straight line. The X-axis value for each point is the "category z-value" for that bucket.

6. Fit a line to the data and interpolate the 50% point on the X-axis (the re-scaled category boundary estimates). These interpolated values for the median for each item now form the new scale values for the items.

An example of this interpolation is shown in Fig. 7.7. Three of the phrases used in the original scaling study of Jones and Thurstone (1955) are pictured, three that were not actually chosen but for which we have approximate proportions and z-scores from their figures. The small vertical arrows on the X-axis show the scale values for the original categories of −4 to +3 (+4 has cumulative proportion of 100% and thus the z-score is infinite). Table 7.1 gives the values and

proportions for each phrase and the original categories. The dashed vertical lines dropped from the intersection at the zero z-score (50% point) show the approximate mean values interpolated on the X-axis (i.e., about −1.1 for "do not care for it" and about +2.1 for "preferred."). Note that "preferred" and "don't care for it" have a linear fit and steep slope, suggesting a normal distribution and low standard deviation. In contrast, "highly unfavorable" has a lower slope and some curvilinearity, indicative of higher variability, skew, and/or pockets of disagreement about the connotation of this term.

The actual scale values for the original adjectives are shown in Table 7.2, as found with a soldier population circa 1950 (Jones et al., 1955). You may note that the words are not equally spaced, and that the "slightly" values are closer to the neutral point than some of the other intervals, and the extreme points are a little farther out. This bears a good deal of similarity to the intervals found with the LAM scale as shown in the column where the LAM values are re-scaled to the same range as the 9-point Thurstonian Values.

Appendix 2: Construction of Labeled Magnitude Scales

There are two primary methods for constructing labeled magnitude scales and they are very similar. Both require magnitude estimates from the participants to scale the word phrases used on the lines. In one case, just the word phrases are scaled, and in the second method, the word phrases are scaled among a list of common everyday experiences or sensations that most people are familiar with. The values obtained by the simple scaling of just the words will depend upon the words that are chosen, and extremely high examples (e.g., greatest imaginable liking for any experience) will tend to compress the values of the interior phrases (Cardello et al., 2008). Whether this kind of context effect will occur for the more general method of scaling amongst common experiences is not known. But the use of a broad frame of reference could be a stabilizing factor.

Here is an example of the instructions given to subjects in construction of a labeled affective magnitude scale. Note that for hedonics, which are a bipolar continuum with a neutral point, it is necessary to collect a tone or valence (plus or minus) value as well as the overall "intensity" rating.

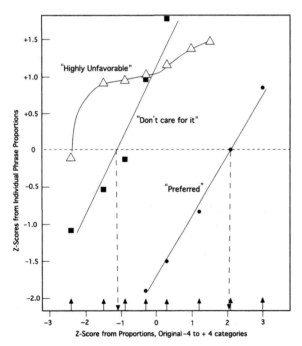

Fig. 7.7 An illustration of the method used to establish spacings and scale values for the 9-point hedonic scale using Thurstonian theory. *Arrows* on the X-axis show the scale points for the z-scores based on the complete distribution of the original −4 to +4 ratings. The Y-axis shows the actual z-scores based on the proportion of respondents using that category for each specific term. Re-plotted from data provided in Jones et al. (1955).

Table 7.1 Examples of scaled phrases used in Fig. 7.7

Original category	Proportion	Z-score	"Preferred" Proportion	Z-score	"Do not care for it" Proportion	Z-score	"Highly unfavorable" Proportion	Z-score
4	1.000	(undef.)	0.80	0.84				
3	0.999	3.0	0.50	0.00			0.96	1.75
2	0.983	2.1	0.20	−0.84			0.93	1.48
1	0.882	1.2	0.07	−1.48			0.92	1.41
0	0.616	0.3	0.03	−1.88	0.96	1.75	0.90	1.28
−1	0.383	−0.3			0.83	0.95	0.86	1.08
−2	0.185	−0.9			0.55	0.13	0.84	0.99
−3	0.068	−1.5			0.30	−0.52	0.82	0.92
−4	0.008	−2.4			0.14	−1.08	0.46	−0.10

Table 7.2 Actual 9-point scale phrase values and comparison to the LAM values

Descriptor	Scale value (9-point)	Interval	LAM value	LAM rescaled	Interval
Like extremely	4.16		74.2	4.20	
Like very much	2.91	1.26	56.1	3.18	1.02
Like moderately	1.12	1.79	36.2	2.05	1.13
Like slightly	0.69	0.43	11.2	0.63	1.52
Neither like nor dislike	0.00	0.69	0.0	0.00	0.63
Dislike slightly	−0.59	0.59	−10.6	−0.60	0.60
Dislike moderately	−1.20	0.61	−31.9	−1.81	1.21
Dislike very much	−2.49	1.29	−55.5	−3.14	1.33
Dislike extremely	−4.32	1.83	−75.5	−4.28	1.14

Next to each word label a response area appeared similar to this:

Phrase: Tone: + − 0   How much:
Like extremely   _____   _____

Words or phrases are presented in random order. After reading a word they must decide whether the word is positive, negative or neutral and place the corresponding symbol on the first line. If the hedonic tone was not a neutral one (zero value), they are instructed to give a numerical estimate using modulus-free magnitude estimation. The following is a sample of the instructions taken from Cardello et al. (2008):

> After having determined whether the phrase is positive or negative or neutral and writing the appropriate symbol (+, −, 0) on the first line, you will then assess the strength or magnitude of the liking or disliking reflected by the phrase. You will do this by placing a number on the second blank line (under "How Much"). For the first phrase that you rate, you can write any number you want on the line. We suggest you do not use a small number for this word/phrase. The reason for this is that subsequent words/phrases may reflect much lower levels of liking or disliking. Aside from this restriction you can use any numbers you want. For each subsequent
>
> word/phrase your numerical judgment should be made proportionally and in comparison to the first number. That is, if you assigned the number 800 to index the strength of the liking/disliking denoted by the first word/phrase and the strength of liking/disliking denoted by the second word/phrase were twice as great, you would assign the number 1,600. If it were three times as great you would assign the number 2,400, etc. Similarly, if the second word/phrase denoted only 1/10 the magnitude of liking as the first, you would assign it the number 80 and so forth. If any word/phrase is judged to be "neutral" (zero (0) on the first line) it should also be given a zero for its magnitude rating.

In the cased of Cardello et al. (2008), positive and negative word labels were analyzed separately. Raw magnitude estimates were equalized for scale range using the procedure of Lane et al. (1961). All positive and negative magnitude estimates for a given subject were multiplied by an individual scaling factor. This factor was equal to the ratio of the grand geometric mean (of the absolute value of all nonzero ratings) across all subjects divided by the geometric mean for that subject. The geometric mean magnitude estimates for each phrase were then calculated based on this range-equated data. These means became the distance from the zero point for placement of the phrases along the scale, usually accompanied by a short cross-hatch mark at that point.

References

AACC (American Association of Cereal Chemists). 1986. Approved Methods of the AACC, Eighth Edition. Method 90–10. Baking quality of cake flour, rev. Oct. 1982. The American Association of Cereal Chemists, St. Paul, MN, pp. 1–4.

Anderson, N. H. 1974. Algebraic models in perception. In: E. C. Carterette and M. P. Friedman (eds.), Handbook of Perception. Psychophysical Judgment and Measurement, Vol. 2. Academic, New York, pp. 215–298.

Anderson, N. H. 1977. Note on functional measurement and data analysis. Perception and Psychophysics, 21, 201–215.

ASTM. 2008a. Standard test method for unipolar magnitude estimation of sensory attributes. Designation E 1697-05. In: Annual Book of ASTM Standards, Vol. 15.08, End Use Products. American Society for Testing and Materials, Conshohocken, PA, pp. 122–131.

ASTM. 2008b. Standard test method for sensory evaluation of red pepper heat. Designation E 1083-00. In: Annual Book of ASTM Standards, Vol. 15.08, End Use Products. American Society for Testing and Materials, Conshohocken, PA, pp. 49–53.

Aust, L. B., Gacula, M. C., Beard, S. A. and Washam, R. W., II. 1985. Degree of difference test method in sensory evaluation of heterogeneous product types. Journal of Food Science, 50, 511–513.

Baird, J. C. and Noma, E. 1978. Fundamentals of Scaling and Psychophysics. Wiley, New York.

Banks, W. P. and Coleman, M. J. 1981. Two subjective scales of number. Perception and Psychophysics, 29, 95–105.

Bartoshuk, L. M., Snyder, D. J. and Duffy, V. B. 2006. Hedonic gLMS: Valid comparisons for food liking/disliking across obesity, age, sex and PROP status. Paper presented at the 2006 Annual Meeting, Association for Chemoreception Sciences.

Bartoshuk, L. M., Duffy, V. B., Fast, K., Green, B. G., Prutkin, J. and Snyder, D. J. 2003. Labeled scales (e.g. category, Likert, VAS) and invalid across-group comparisons: What we have learned from genetic variation in taste. Food Quality and Preference, 14, 125–138.

Bartoshuk, L. M., Duffy, V. B., Green, B. G., Hoffman, H. J., Ko, C.-W., Lucchina, L. A., Marks, L. E., Snyder, D. J. and Weiffenbach, J. M. 2004a. Valid across-group comparisons with labeled scales: the gLMS versus magnitude matching. Physiology and Behavior, 82, 109–114.

Bartoshuk, L. M., Duffy, V. B., Chapo, A. K., Fast, K., Yiee, J. H., Hoffman, H. J., Ko, C.-W. and Snyder, D. J. 2004b. From psychophysics to the clinic: Missteps and advances. Food Quality and Preference, 14, 617–632.

Bartoshuk, L. M., Duffy, V. B., Fast, K., Green, B. Kveton, J., Lucchina, L. A., Prutkin, J. M., Snyder, D. J. and Tie, K. 1999. Sensory variability, food preferences and BMI in non-medium and supertasters of PROP. Appetite, 33, 228–229.

Basker, D. 1988. Critical values of differences among rank sums for multiple comparisons. Food Technology, 42(2), 79, 80–84.

Baten, W. D. 1946. Organoleptic tests pertaining to apples and pears. Food Research, 11, 84–94.

Bendig, A. W. and Hughes, J. B. 1953. Effect of number of verbal anchoring and number of rating scale categories upon transmitted information. Journal of Experimental Psychology, 46(2), 87–90.

Bi, J. 2006. Sensory Discrimination Tests and Measurement. Blackwell, Ames, IA.

Birch, L. L., Zimmerman, S. I. and Hind, H. 1980. The influence of social-affective context on the formation of children's food preferences. Child Development, 51, 865–861.

Birch, L. L., Birch, D., Marlin, D. W. and Kramer, L. 1982. Effects of instrumental consumption on children's food preferences. Appetite, 3, 125–143.

Birnbaum, M. H. 1982. Problems with so-called "direct" scaling. In: J. T. Kuznicki, R. A. Johnson and A. F. Rutkiewic (eds.), Selected Sensory Methods: Problems and Approaches to Hedonics. American Society for Testing and Materials, Philadelphia, pp. 34–48.

Borg, G. 1982. A category scale with ratio properties for intermodal and interindividual comparisons. In: H.-G. Geissler and P. Pextod (Eds.), Psychophysical Judgment and the Process of Perception. VEB Deutscher Verlag der Wissenschaften, Berlin, pp. 25–34.

Borg, G. 1990. Psychophysical scaling with applications in physical work and the perception of exertion. Scandinavian Journal of Work and Environmental Health, 16, 55–58.

Boring, E. G. 1942. Sensation and Perception in the History of Experimental Psychology. Appleton-Century-Crofts, New York.

Brandt, M. A., Skinner, E. Z. and Coleman, J. A. 1963. The texture profile method. Journal of Food Science, 28, 404–409.

Butler, G., Poste, L. M., Wolynetz, M. S., Agar, V. E. and Larmond, E. 1987. Alternative analyses of magnitude estimation data. Journal of Sensory Studies, 2, 243–257.

Cardello, A. V. and Schutz, H. G. 2004. Research note. Numerical scale-point locations for constructing the LAM (Labeled affective magnitude) scale. Journal of Sensory Studies, 19, 341–346.

Cardello, A. V., Lawless, H. T. and Schutz, H. G. 2008. Effects of extreme anchors and interior label spacing on labeled magnitude scales. Food Quality and Preference, 21, 323–334.

Cardello, A. V., Winterhaler, C. and Schutz, H. G. 2003. Predicting the handle and comfort of military clothing fabrics from sensory and instrumental data: Development and application of new psychophysical methods. Textile Research Journal, 73, 221–237.

Cardello, A. V., Schutz, H. G., Lesher, L. L. and Merrill, E. 2005. Development and testing of a labeled magnitude scale of perceived satiety. Appetite, 44, 1–13.

Caul, J. F. 1957. The profile method of flavor analysis. Advances in Food Research, 7, 1–40.

Chambers, E. C. and Wolf, M. B. 1996. Sensory Testing Methods. ASTM Manual Series, MNL 26. ASTM International, West Conshohocken, PA.

Chen, A. W., Resurreccion, A. V. A. and Paguio, L. P. 1996. Age appropriate hedonic scales to measure the food preferences of young children. Journal of Sensory Studies, 11, 141–163.

Chung, S.-J. and Vickers, 2007a. Long-term acceptability and choice of teas differing in sweetness. Food Quality and Preference 18, 963–974.

Chung, S.-J. and Vickers, 2007b. Influence of sweetness on the sensory-specific satiety and long-term acceptability of tea. Food Quality and Preference, 18, 256–267.

Coetzee, H. and Taylor, J. R. N. 1996. The use and adaptation of the paired comparison method in the sensory evaluation of hamburger-type patties by illiterate/semi-literate consumers. Food Quality and Preference, 7, 81–85.

Collins, A. A. and Gescheider, G. A. 1989. The measurement of loudness in individual children and adults by absolute magnitude estimation and cross modality matching. Journal of the Acoustical Society of America, 85, 2012–2021.

Conner, M. T. and Booth, D. A. 1988. Preferred sweetness of a lime drink and preference for sweet over non-sweet foods. Related to sex and reported age and body weight. Appetite, 10, 25–35.

Cordinnier, S. M. and Delwiche, J. F. 2008. An alternative method for assessing liking: Positional relative rating versus the 9-point hedonic scale. Journal of Sensory Studies, 23, 284–292.

Cox, E. P. 1980. The optimal number of response alternatives for a scale: A review. Journal of Marketing Research, 18, 407–422.

Curtis, D. W., Attneave, F. and Harrington, T. L. 1968. A test of a two-stage model of magnitude estimation. Perception and Psychophysics, 3, 25–31.

Edwards, A. L. 1952. The scaling of stimuli by the method of successive intervals. Journal of Applied Psychology, 36, 118–122.

Ekman, G. 1964. Is the power law a special case of Fechner's law? Perceptual and Motor Skills, 19, 730.

Einstein, M. A. 1976. Use of linear rating scales for the evaluation of beer flavor by consumers. Journal of Food Science, 41, 383–385.

El Dine, A. N. and Olabi, A. 2009. Effect of reference foods in repeated acceptability tests: Testing familiar and novel foods using 2 acceptability scales. Journal of Food Science, 74, S97–S105.

Engen, T. 1974. Method and theory in the study of odor preferences. In: A. Turk, J. W. Johnson and D. G. Moulton (Eds.), Human Responses to Environmental Odors. Academic, New York.

Finn, A. and Louviere, J. J. 1992. Determining the appropriate response to evidence of public concern: The case of food safety. Journal of Public Policy and Marketing, 11, 12–25.

Forde, C. G. and Delahunty, C. M. 2004. Understanding the role cross-modal sensory interactions play in food acceptability in younger and older consumers. Food Quality and Preference, 15, 715–727.

Frijters, J. E. R., Kooistra, A. and Vereijken, P. F. G. 1980. Tables of d' for the triangular method and the 3-AFC signal detection procedure. Perception and Psychophysics, 27, 176–178.

Gaito, J. 1980. Measurement scales and statistics: Resurgence of an old misconception. Psychological Bulletin, 87, 564–587.

Gay, C., and Mead, R. 1992 A statistical appraisal of the problem of sensory measurement. Journal of Sensory Studies, 7, 205–228.

Gent, J. F. and Bartoshuk, L. M. 1983. Sweetness of sucrose, neohesperidin dihydrochalcone and sacchar in is related to genetic ability to taste the bitter substance 6-n-propylthiouracil. Chemical Senses, 7, 265–272.

Gescheider, G. A. 1988. Psychophysical scaling. Annual Review of Psychology, 39, 169–200.

Giovanni, M. E. and Pangborn, R. M. 1983. Measurement of taste intensity and degree of liking of beverages by graphic scaling and magnitude estimation. Journal of Food Science, 48, 1175–1182.

Gracely, R. H., McGrath, P. and Dubner, R. 1978a. Ratio scales of sensory and affective verbal-pain descriptors. Pain, 5, 5–18.

Gracely, R. H., McGrath, P. and Dubner, R. 1978b. Validity and sensitivity of ratio scales of sensory and affective verbal-pain descriptors: Manipulation of affect by Diazepam. Pain, 5, 19–29.

Green, B. G., Shaffer, G. S. and Gilmore, M. M. 1993. Derivation and evaluation of a semantic scale of oral sensation magnitude with apparent ratio properties. Chemical Senses, 18, 683–702.

Green, B. G., Dalton, P., Cowart, B., Shaffer, G., Rankin, K. and Higgins, J. 1996. Evaluating the "Labeled Magnitude Scale" for measuring sensations of taste and smell. Chemical Senses, 21, 323–334.

Greene, J. L., Bratka, K. J., Drake, M. A. and Sanders, T. H. 2006. Effective of category and line scales to characterize consumer perception of fruity fermented flavors in peanuts. Journal of Sensory Studies, 21, 146–154.

Guest, S., Essick, G., Patel, A., Prajpati, R. and McGlone, F. 2007. Labeled magnitude scales for oral sensations of wetness, dryness, pleasantness and unpleasantness. Food Quality and Preference, 18, 342–352.

Hein, K. A., Jaeger, S. R., Carr, B. T. and Delahunty, C. M. 2008. Comparison of five common acceptance and preference methods. Food Quality and Preference, 19, 651–661.

Huskisson, E. C. 1983. Visual analogue scales. In: R. Melzack (Ed.), Pain Measurement and Assessment. Raven, New York, pp. 34–37.

Jaeger, S. R.; Jørgensen, A. S., AAslying, M. D. and Bredie, W. L. P. 2008. Best-worst scaling: An introduction and initial comparison with monadic rating for preference elicitation with food products. Food Quality and Preference, 19, 579–588.

Jaeger, S. R. and Cardello, A. V. 2009. Direct and indirect hedonic scaling methods: A comparison of the labeled affective magnitude (LAM) scale and best-worst scaling. Food Quality and Preference, 20, 249–258.

Jones, F. N. 1974. History of psychophysics and judgment. In: E. C. Carterette and M. P. Friedman (Eds.), Handbook of Perception. Psychophysical Judgment and Measurement, Vol. 2. Academic, New York, pp. 11–22.

Jones, L. V. and Thurstone, L. L. 1955. The psychophysics of semantics: An experimental investigation. Journal of Applied Psychology, 39, 31–36.

Jones, L. V., Peryam, D. R. and Thurstone, L. L. 1955. Development of a scale for measuring soldier's food preferences. Food Research, 20, 512–520.

Keskitalo, K. Knaapila, A., Kallela, M., Palotie, A., Wessman, M., Sammalisto, S., Peltonen, L., Tuorila, H. and Perola, M. 2007. Sweet taste preference are partly genetically determined: Identification of a trait locus on Chromosome 16[1-3]. American Journal of Clinical Nutrition, 86, 55–63.

Kim, K.-O. and O'Mahony, M. 1998. A new approach to category scales of intensity I: Traditional versus rank-rating. Journal of Sensory Studies, 13, 241–249.

King, B. M. 1986. Odor intensity measured by an audio method. Journal of Food Science, 51, 1340–1344.

Koo, T.-Y., Kim, K.-O., and O'Mahony, M. 2002. Effects of forgetting on performance on various intensity scaling protocols: Magnitude estimation and labeled magnitude scale (Green scale). Journal of Sensory Studies, 17, 177–192.

Kroll, B. J. 1990. Evaluating rating scales for sensory testing with children. Food Technology, 44(11), 78–80, 82, 84, 86.

Kurtz, D. B., White, T. L. and Hayes, M. 2000. The labeled dissimilarity scale: A metric of perceptual dissimilarity. Perception and Psychophysics, 62, 152–161.

Land, D. G. and Shepard, R. 1984. Scaling and ranking methods. In: J. R. Piggott (ed.), Sensory Analysis of Foods. Elsevier Applied Science, London, pp. 141–177.

Lane, H. L., Catania, A. C. and Stevens, S. S. 1961. Voice level: Autophonic scale, perceived loudness and effect of side tone. Journal of the Acoustical Society of America, 33, 160–167.

Larson-Powers, N. and Pangborn, R. M. 1978. Descriptive analysis of the sensory properties of beverages and gelatins containing sucrose or synthetic sweeteners. Journal of Food Science, 43, 47–51.

Lawless, H. T. 1977. The pleasantness of mixtures in taste and olfaction. Sensory Processes, 1, 227–237.

Lawless, H. T. 1989. Logarithmic transformation of magnitude estimation data and comparisons of scaling methods. Journal of Sensory Studies, 4, 75–86.

Lawless, H. T. and Clark, C. C. 1992. Psychological biases in time intensity scaling. Food Technology, 46, 81, 84–86, 90.

Lawless, H. T. and Malone, J. G. 1986a. The discriminative efficiency of common scaling methods. Journal of Sensory Studies, 1, 85–96.

Lawless, H. T. and Malone, G. J. 1986b. A comparison of scaling methods: Sensitivity, replicates and relative measurement. Journal of Sensory Studies, 1, 155–174.

Lawless, H. T. and Skinner, E. Z. 1979. The duration and perceived intensity of sucrose taste. Perception and Psychophysics, 25, 249–258.

Lawless, H. T., Popper, R. and Kroll, B. J. 2010a. Comparison of the labeled affective magnitude (LAM) scale, an 11-point category scale and the traditional nine-point hedonic scale. Food Quality and Preference, 21, 4–12.

Lawless, H. T., Sinopoli, D. and Chapman, K. W. 2010b. A comparison of the labeled affective magnitude scale and the nine point hedonic scale and examination of categorical behavior. Journal of Sensory Studies, 25, S1, 54–66.

Lawless, H. T., Cardello, A. V., Chapman, K. W., Lesher, L. L., Given, Z. and Schutz, H. G. 2010c. A comparison of the effectiveness of hedonic scales and end-anchor compression effects. Journal of Sensory Studies, 28, S1, 18–34.

Lee, H.-J., Kim, K.-O., and O'Mahony, M. 2001. Effects of forgetting on various protocols for category and line scales of intensity. Journal of Sensory Studies, 327–342.

Likert, R. 1932. Technique for the measurement of attitudes. Archives of Psychology, 140, 1–55.

Lindvall, T. and Svensson, L. T. 1974. Equal unpleasantness matching of malodourous substances in the community. Journal of Applied Psychology, 59, 264–269.

Mahoney, C. H., Stier, H. L. and Crosby, E. A. 1957. Evaluating flavor differences in canned foods. II. Fundamentals of the simplified procedure. Food Technology 11, Supplemental Symposium Proceedings, 37–42.

Marks, L. E. 1978. Binaural summation of the loudness of pure tones. Journal of the Acoustical Society of America, 64, 107–113.

Marks, L. E., Borg, G. and Ljunggren, G. 1983. Individual differences in perceived exertion assessed by two new methods. Perception and Psychophysic, 34, 280–288.

Marks, L. E., Borg, G. and Westerlund, J. 1992. Differences in taste perception assessed by magnitude matching and by category-ratio scaling. Chemical Senses, 17, 493–506.

Mattes, R. D. and Lawless, H. T. 1985. An adjustment error in optimization of taste intensity. Appetite, 6, 103–114.

McBride, R. L. 1983a. A JND-scale/category scale convergence in taste. Perception and Psychophysics, 34, 77–83.

McBride, R. L. 1983b. Taste intensity and the case of exponents greater than 1. Australian Journal of Psychology, 35, 175–184.

McBurney, D. H. and Shick, T. R. 1971. Taste and water taste for 26 compounds in man. Perception and Psychophysics, 10, 249–252.

McBurney, D. H. and Bartoshuk, L. M. 1973. Interactions between stimuli with different taste qualities. Physiology and Behavior, 10, 1101–1106.

McBurney, D. H., Smith, D. V. and Shick, T. R. 1972. Gustatory cross-adaptation: Sourness and bitterness. Perception and Psychophysics, 11, 228–232.

Mead, R. and Gay, C. 1995. Sequential design of sensory trials. Food Quality and Preference, 6, 271–280.

Mecredy, J. M. Sonnemann, J. C. and Lehmann, S. J. 1974. Sensory profiling of beer by a modified QDA method. Food Technology, 28, 36–41.

Meilgaard, M., Civille, G. V. and Carr, B. T. 2006. Sensory Evaluation Techniques, Fourth Edition. CRC, Boca Raton, FL.

Moore, L. J. and Shoemaker, C. F. 1981. Sensory textural properties of stabilized ice cream. Journal of Food Science, 46, 399–402.

Moskowitz, H. R. 1971. The sweetness and pleasantness of sugars. American Journal of Psychology, 84, 387–405.

Moskowitz, H. R. and Sidel, J. L. 1971. Magnitude and hedonic scales of food acceptability. Journal of Food Science, 36, 677–680.

Muñoz, A. M. and Civille, G. V. 1998. Universal, product and attribute specific scaling and the development of common lexicons in descriptive analysis. Journal of Sensory Studies, 13, 57–75.

Newell, G. J. and MacFarlane, J. D. 1987. Expanded tables for multiple comparison procedures in the analysis of ranked data. Journal of Food Science, 52, 1721–1725.

Olabi, A. and Lawless, H. T. 2008. Persistence of context effects with training and reference standards. Journal of Food Science, 73, S185–S189.

O'Mahony, M., Park, H., Park, J. Y. and Kim, K.-O. 2004. Comparison of the statistical analysis of hedonic data using analysis of variance and multiple comparisons versus and R-index analysis of the ranked data. Journal of Sensory Studies, 19, 519–529.

Pangborn, R. M. and Dunkley, W. L. 1964. Laboratory procedures for evaluating the sensory properties of milk. Dairy Science Abstracts, 26–55–62.

Parducci, A. 1965. Category judgment: A range-frequency model. Psychological Review, 72, 407–418.

Park, J.-Y., Jeon, S.-Y., O'Mahony, M. and Kim, K.-O. 2004. Induction of scaling errors. Journal of Sensory Studies, 19, 261–271.

Pearce, J. H., Korth, B. and Warren, C. B. 1986. Evaluation of three scaling methods for hedonics. Journal of Sensory Studies, 1, 27–46.

Peryam. D. 1989. Reflections. In: Sensory Evaluation. In Celebration of our Beginnings. American Society for Testing and Materials, Philadelphia, pp. 21–30.

Peryam, D. R. and Girardot, N. F. 1952. Advanced taste-test method. Food Engineering, 24, 58–61, 194.

Piggot, J. R. and Harper, R. 1975. Ratio scales and category scales for odour intensity. Chemical Senses and Flavour, 1, 307–316.

Pokorný, J., Davídek, J., Prnka, V. and Davídková, E. 1986. Nonparametric evaluation of graphical sensory profiles for the analysis of carbonated beverages. Die Nahrung, 30, 131–139.

Poulton, E. C. 1989. Bias in Quantifying Judgments. Lawrence Erlbaum, Hillsdale, NJ.

Richardson, L. F. and Ross, J. S. 1930. Loudness and telephone current. Journal of General Psychology, 3, 288–306.

Rosenthal, R. 1987. Judgment Studies: Design, Analysis and Meta-Analysis. University Press, Cambridge.

Shand, P. J., Hawrysh, Z. J., Hardin, R. T. and Jeremiah, L. E. 1985. Descriptive sensory analysis of beef steaks by category scaling, line scaling and magnitude estimation. Journal of Food Science, 50, 495–499.

Schutz, H. G. and Cardello, A. V. 2001.. A labeled affective magnitude (LAM) scale for assessing food liking/disliking. Journal of Sensory Studies, 16, 117–159.

Siegel, S. 1956. Nonparametric Statistics for the Behavioral Sciences. McGraw-Hill, New York.

Sriwatanakul, K., Kelvie, W., Lasagna, L., Calimlim, J. F., Wels, O. F. and Mehta, G. 1983. Studies with different types of visual analog scales for measurement of pain. Clinical Pharmacology and Therapeutics, 34, 234–239.

Stevens, J. C. and Marks, L. M. 1980. Cross-modality matching functions generated by magnitude estimation. Perception and Psychophysics, 27, 379–389.

Stevens, S. S. 1951. Mathematics, measurement and psychophysics. In: S. S. Stevens (ed.), Handbook of Experimental Psychology. Wiley, New York, pp. 1–49.

Stevens, S. S. 1956. The direct estimation of sensory magnitudes—loudness. American Journal of Psychology, 69, 1–25.

Stevens, S. S. 1957. On the psychophysical law. Psychological Review, 64, 153–181.

Stevens, S. S. 1969. On predicting exponents for cross-modality matches. Perception and Psychophysics, 6, 251–256.

Stevens, S. S. and Galanter, E. H. 1957. Ratio scales and category scales for a dozen perceptual continua. Journal of Experimental Psychology, 54, 377–411.

Stoer, N. L. and Lawless, H. T. 1993. Comparison of single product scaling and relative-to-reference scaling in sensory evaluation of dairy products. Journal of Sensory Studies, 8, 257–270.

Stone, H., Sidel, J., Oliver, S., Woolsey, A. and Singleton, R. C. 1974. Sensory Evaluation by quantitative descriptive analysis. Food Technology, 28, 24–29, 32, 34.

Teghtsoonian, M. 1980. Children's scales of length and loudness: A developmental application of cross-modal matching. Journal of Experimental Child Psychology, 30, 290–307.

Thurstone, L. L. 1927. A law of comparative judgment. Psychological Review, 34, 273–286.

Townsend, J. T. and Ashby, F. G. 1980. Measurement scales and statistics: The misconception misconceived. Psychological Bulletin, 96, 394–401.

Vickers, Z. M. 1983. Magnitude estimation vs. category scaling of the hedonic quality of food sounds. Journal of Food Science, 48, 1183–1186.

Villanueva, N. D. M. and Da Silva, M. A. A. P. 2009. Performance of the nine-point hedonic, hybrid and self-adjusting scales in the generation of internal preference maps. Food Quality and Preference, 20, 1–12.

Villanueva, N. D. M., Petenate, A. J., and Da Silva, M. A. A. P. 2005. Comparative performance of the hybrid hedonic scale as compared to the traditional hedonic, self-adjusting and ranking scales. Food Quality and Preference, 16, 691–703.

Ward, L. M. 1986. Mixed-modality psychophysical scaling: Double cross-modality matching for "difficult" continua. Perception and Psychophysics, 39, 407–417.

Weiss, D. J. 1972. Averaging: an empirical validity criterion for magnitude estimation. Perception and Psychophysics, 12, 385–388.

Winakor, G., Kim, C. J. and Wolins, L. 1980. Fabric hand: Tactile sensory assessment. Textile Research Journal, 50, 601–610.

Yamaguchi, S. 1967. The synergistic effect of monosodium glutamate and disodium 5′ inosinate. Journal of Food Science 32, 473–477.

Chapter 8

Time–Intensity Methods

Abstract   Time–intensity methods represent a special form of intensity scaling that is either repeated at short intervals or continuous. It offers some advantages over a single intensity estimate, giving more detailed information on changes in flavor and texture over time. This chapter reviews the history of these methods, various current techniques, issues, and approaches to data analysis and provides examples of various applications.

In general, humans perceived tastes as changing experiences originating in the mouth, which normally existed for a limited time and then either subsided or transformed into qualitatively different gustatory perceptions. Taste experiences did not begin at the moment of stimulus arrival in the mouth, did not suddenly appear at full intensity, were influenced by the pattern of taste stimulation, and often continued well beyond stimulus removal.

—(Halpern, 1991, p. 95)

Does your chewing gum lose its flavor (on the bedpost overnight)?
—Bloom and Brever, lyrics (recorded by Lonnie Donegan, May 1961, Mills Music, Inc./AA Music)

Contents

8.1 Introduction

Perception of aroma, taste, flavor, and texture in foods is a dynamic not a static phenomenon. In other words, the perceived intensity of the sensory attributes change from moment to moment. The dynamic nature of food sensations arises from processes of chewing, breathing, salivation, tongue movements, and swallowing (Dijksterhuis, 1996). In the texture profile method for instance, different phases of food breakdown were recognized early on as evidenced by the separation of characteristics into first bite, mastication, and residual phases (Brandt et al., 1963). Wine tasters often discuss

H.T. Lawless, H. Heymann, *Sensory Evaluation of Food*, Food Science Text Series,
DOI 10.1007/978-1-4419-6488-5_8, © Springer Science+Business Media, LLC 2010

how a wine "opens in the glass," recognizing that the flavor will vary as a function of time after opening the bottle and exposing the wine to air. It is widely believed that the consumer acceptability of different intensive sweeteners depends on the similarity of their time profile to that of sucrose. Intensive sweeteners with too long a duration in the mouth may be less pleasant to consumers. Conversely, a chewing gum with long-lasting flavor or a wine with a "long finish" may be desirable. These examples demonstrate how the time profile of a food or beverage can be an important aspect of its sensory appeal.

The common methods of sensory scaling ask the panelists to rate the perceived intensity of the sensation by giving a single (uni-point) rating. This task requires that the panelists must "time-average" or integrate any changing sensations or to estimate only the peak intensity in order to provide the single intensity value that is required. Such a single value may miss some important information. It is possible, for example, for two products to have the same or similar time-averaged profiles or descriptive specifications, but differ in the order in which different flavors occur or when they reach their peak intensities.

Time–intensity (TI) methods provide panelists with the opportunity to scale their perceived sensations over time. When multiple attributes are tracked, the profile of a complex food flavor or texture may show differences between products that change across time after a product is first tasted, smelled, or felt. For most sensations the perceived intensity increases and eventually decreases but for some, like perceived toughness of meat, the sensations may only decrease as a function of time. For perceived melting, the sensation may only increase until a completely melted state is reached.

When performing a TI study the sensory specialist can obtain a wealth of detailed information such as the following for each sample: the maximum intensity perceived, the time to maximum intensity, the rate and shape of the increase in intensity to the maximum point, the rate and shape of the decrease in intensity to half-maximal intensity and to the extinction point, and the total duration of the sensation. Some of the common time–intensity parameters are illustrated in Fig. 8.1. The additional information derived from time–intensity methods is especially useful when studying sweeteners or products like chewing gums and hand lotions that have a distinctive time profile.

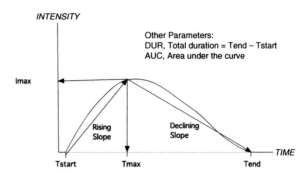

Fig. 8.1 Example of a time–intensity curve and common curve parameters extracted from the record.

The remainder of this chapter will be devoted to an overview of the history and applications of this method, as well as recommended procedures and analyses. For the student who wants only the basic information, the following sections are key: variations on the method (Section 8.3), steps and recommended procedures (Section 8.4), data analysis options (Section 8.5), and conclusions (8.8).

8.2 A Brief History

Holway and Hurvich (1937) published an early report of tracking taste intensity over time. They had their subjects trace a curve to represent the sensations from a drop of either 0.5 or 1.0 M NaCl placed on the anterior tongue surface over 10 s. They noted several general effects that were later confirmed as common trends in other studies. The higher concentration led to a higher peak intensity, but the peak occurred later, in spite of a steeper rising slope. Most importantly they noted that taste intensity was not strictly a function of concentration: "while the concentration originally placed on the tongue is 'fixed,' the intensity varies in a definite manner from moment to moment. Saline intensity depends on time as well as concentration." A review of studies of temporal factors in taste is found in Halpern (1991). A review of TI studies of the 1980s and early 1990s can be found in Cliff and Heymann (1993b).

Sjostrom (1954) and Jellinek (1964) also made early attempts to quantify the temporal response to perceived sensory intensities. These authors asked their panelists to indicate their perceived bitterness of beer at 1 s intervals on a ballot, using a clock to indicate time.

They then constructed TI curves by plotting the x–y coordinates (time on the x-axis and perceived intensity on the y-axis) on graph paper. Once panelists had some experience with the method it was possible to ask them simultaneously to rate the perceived intensities of two different attributes at 1 s intervals. Neilson (1957) in an attempt to make the production of the TI curves easier, asked panelists to indicate perceived bitterness directly on graph paper at 2 s timed intervals. The clock could be distracting to the panelists and thus Meiselman (1968), studying taste adaptation and McNulty and Moskowitz (1974), evaluating oil-in-water emulsions, improved the TI methodology by eliminating the clock. These authors used audible cues to tell the panelists when to enter perceived intensities on a ballot, placing the timekeeping demands on the experimenter rather than the participant.

Larson-Powers and Pangborn (1978), in another attempt to eliminate the distractions of the clock or audible cues, employed a moving strip-chart recorder equipped with a foot pedal to start and stop the movement of the chart. Panelists recorded their responses to the perceived sweetness in beverages and gelatins sweetened with sucrose or synthetic sweeteners, by moving a pen along the cutter bar of the strip-chart recorder. The cutter bar was labeled with an unstructured line scale, from none to extreme. A cardboard cover was placed over the moving chart paper to prevent the panelists from watching the evolving curves and thus preventing them from using any visual cues to bias their responses. A similar setup was independently developed and used at the General Foods Technical Center in 1977 to track sweetness intensity (Lawless and Skinner, 1979). In this apparatus, the actual pen carriage of the chart recorder was grasped by the subject, eliminating the need for them to position a pen; also the moving chart was obscured by a line scale with a pointer attached to the pen carriage. In yet another laboratory at about the same time, Birch and Munton (1981) developed the "SMURF" version of TI scaling (short for "Sensory Measuring Unit for Recording Flux"). In the SMURF apparatus, the subject turned a knob graded from 1 to 10, and this potentiometer fed a variable signal to a strip-chart recorder out of sight of the panelist. The use of strip-chart recorders provided the first continuous TI data-collection methods and freed the panelists from any distractions caused by a clock or auditory signal. However, the methods required a fair degree of mental and physical coordination by the participants. For example, in the Larson-Powers setup, the strip-chart recorder required the panelists to use a foot pedal to run the chart recorder, to place the sample in the mouth, and to move the pen to indicate the perceived intensity. Not all panelists were suitably coordinated and some could not do the evaluation. Although the strip-chart recorder made continuous evaluation of perceived intensities possible, the TI curves had to be digitized manually, which was extremely time consuming.

The opportunity to use computers to time sample an analog voltage signal quite naturally led to online data collection to escape the problem of manual measurement of TI curves. To the best of our knowledge, the first computerized system was developed at the US Army Natick Food Laboratories in 1979 to measure bitter taste adaptation. It employed an electric sensor in the spout just above the subject's tongue in order to determine the actual onset of stimulus arrival. Subthreshold amounts of NaCl were added to the stimulus and thus created a conductance change as the flow changed from the preliminary water rinse to the stimulus interface in the tube. A special circuit was designed to detect the conductance change and to connect the response knob to the visual line scale. Like the SMURF apparatus developed by Birch and Munton, the subject turned a knob controlling a variable resistor. The output of this potentiometer moved a pointer on a line scale for visual feedback while a parallel analog signal was sent to an analog-to-digital converter and then to the computer. The programming was done using FORTRAN subroutines on a popular lab computer of that era. The entire system is shown in Fig. 8.2.

The appearance of desktop computers led to an explosion in the use of TI methodology in the 1980s and 1990s. A number of thesis research projects from U.C. Davis served as useful demonstrations of the method (Cliff, 1987; Dacanay, 1990; Rine, 1987) and the method was championed by Pangborn and coworkers (e.g., Lee and Pangborn, 1986). Several scientists (Barylko-Pikielna et al., 1990; Cliff, 1987; Guinard et al., 1985; Janusz et al., 1991; Lawless, 1980; Lee, 1985; Rine, 1987; Yoshida, 1986) developed computerized TI systems using a variety of hardware and software products. Computerized TI systems are now commercially available as part of data-collection software ensembles, greatly enhancing the ease and availability of TI data collection and data processing.

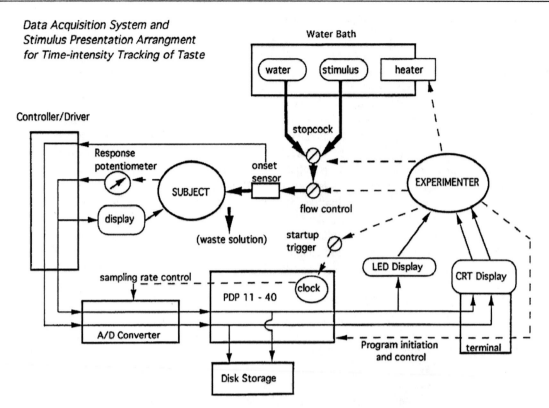

Data Acquisition System and Stimulus Presentation Arrangment for Time-intensity Tracking of Taste

Fig. 8.2 An early computerized system for time–intensity scaling used for tracking bitterness adaptation in a flow system for stimulating the anterior tongue. *Heavy lines* indicate the flow of stimulus solution, *solid lines* the flow of information, and *dashed lines* the experimenter-driven process control. Stimulus arrival at the tongue was monitored by conductivity sensors fitted into the glass tube just above the subject's tongue. The subjects' responses change on a line scale while the experimenter could view the conductivity and response voltage outputs on the computer's display screen, which simultaneously were output to a data file. The system was programmed in FORTRAN subroutines controlling the clock sampling rate and analog-to-digital conversion routine. From Lawless and Clark (1992), reprinted with permission.

However, despite the availability of computerized systems, some research was still conducted using the simple time cueing at discrete intervals (e.g., Lee and Lawless, 1991; Pionnier et al., 2004), and the semi-manual strip-chart recorder method (Ott et al., 1991; Robichaud and Noble, 1990). A discussion of some common applications of TI methods is given in Section 8.6.

8.3 Variations on the Method

8.3.1 Discrete or Discontinuous Sampling

The sensory scientist has several options for collecting time-dependent sensory data. The methods for time-related scaling can be divided into four groups. The oldest approach is simply to ask the panelists to rate the intensity of sensation during different phases of consuming a food. This is particularly applicable to texture which may be evaluated in phases such as initial bite, first chew, mastication, and residual. An example of time division during the texture evaluation of a baked product is shown in Table 8.1.

When using a descriptive panel, it may be useful to have residual flavor or mouthfeel sensations rated at a few small intervals, e.g., every 30 s for 2 min, or immediately after tasting and then again after expectoration. For an example of this approach used with hot pepper "burn" see Stevens and Lawless (1986). Each measurement is then treated like a separate descriptive attribute and analyzed as a separate variable, with little or no attempt to reconstruct a time-connected record like the TI curve shown in Fig. 8.1. For researchers

Table 8.1 Texture attributes at different phases of descriptive analysis

Phase	Attributes	Word anchors
Surface	Roughness	Smooth–rough
	Particles	None–many
	Dryness	Oily–dry
First bite	Fracturability	Crumbly–brittle
	Hardness	Soft–hard
	Particle size	Small–large
First chew	Denseness	Airy–dense
	Uniformity of chew	Even–uneven
Chew down	Moisture absorption	None–much
	Cohesiveness of mass	Loose–cohesive
	Toothpacking	None–much
	Grittiness	None–much
Residual	Oiliness	Dry–oily
	Particles	None–much
	Chalky	Not chalky–very chalky

interested in some simple aspect like "strength of bitter aftertaste" this method may suffice.

Another related approach is to ask for repeated ratings of a single or just a few attributes at repeated smaller time intervals, usually cued by the panel leader or experimenter. These ratings are then connected and graphed on time axis. This is a simple procedure that can be used to track changes in the intensity of a flavor or texture attribute and requires no special equipment other than a stopwatch or other timing device. The panel must be trained to rate their sensations upon the time cue and to move rapidly through the list of attributes. The cue may be given verbally or on a computer screen. It is not known how many attributes can be rated in this way, but with faster time cueing and shorter intervals, obviously fewer attributes may be rated. This method also requires some faith in the assumption that the attributes are being rated close to the actual time when the cue is given. The accuracy with which panelists can do this is unknown, but given that there is a reaction time delay in any perceptual judgment, there must be some inherent error or delay-related variance built into the procedure. An example of the repeated, discrete time interval method with verbal cueing and multiple attributes can be found in studies of sweetener mixtures (e.g., Ayya and Lawless, 1992) and astringency (Lee and Lawless, 1991). The time record is treated as a connected series and time is analyzed as one factor (i.e., one independent variable) in the statistical analysis.

8.3.2 "Continuous" Tracking

A third and widely used method for TI scaling is continuous tracking of flavor or texture using an analog response device such as a lever, a rotating knob, a joystick, or computer mouse. The response device may change a variable resistance and the resulting current is fed through an analog-to-digital conversion. The signal is time sampled at whatever rate is programmed into the recording device. As noted above, continuous records may also be produced by using a chart recorder but digitizing the records may be quite laborious. The advantage of continuous tracking is the detail in the flavor or texture experience that is captured in the record (Lee, 1989; Lee and Pangborn, 1986). It is difficult to capture the rising phase of a flavor with a verbal due or discrete point methods, as the onset of many tastes and odors is quite rapid. The continuous tracking methods are very widely used and are discussed further in Section 8.4. Although the records are continuous, the jagged nature of these records indicates that panelists are not moving the response device in a continuous manner.

Two-dimensional response tasks have been developed so that two attributes can be tracked simultaneously (Duizer et al., 1995). In an experiment on sweetness and mint flavor of chewing gum, it was possible for panelists to track both flavor perceptions simultaneously (Duizer et al., 1996). Panelists were trained to move a mouse diagonally and a visual scale included pointers on both horizontal and vertical scales to represent the intensity of the individual attributes. With a slowly changing product like chewing gum, with a sampling time that is not too frequent (every 9–15 s in this case), the technique would seem to be within the capabilities of human observers to either rapidly shift their attention or to respond to the overall pattern of the combined flavors.

However, as currently used, most TI tracking methods must repeat the evaluation in order to track additional attributes. Ideally, this could lead to a composite profile of all the dynamic flavor and texture attributes in a product and how they changed across time. Such an approach was proposed by DeRovira (1996), who showed how the descriptive analysis spider-web plot of multiple attributes could be extended into the time dimension to produce a set of TI curves and thus to characterize an entire profile.

8.3.3 Temporal Dominance Techniques

A fourth method for gathering time-dependent changes has been to limit the reported profile to a subset of key sensations, called the Temporal Dominance of Sensations (TDS) method (Pineau et al., 2009). This method is still evolving, and descriptions of the procedure and analysis vary somewhat. The basic idea is to present a set of predetermined attributes together on the computer screen for the panelist's choice and scales for rating the intensity of each. The important choice is the selection of the dominant quality and thus the method is related to Halpern's technique for taste quality tracking (Halpern, 1991). Panelists are instructed to attend to and choose only the "dominant" sensation at any one time after tasting the sample and clicking on a start button. "Dominant" has been described as the "most striking perception" (Pineau et al., 2009), "the most intense sensation" (Labbe et al., 2009), "the sensation catching the attention," or "new sensation popping up" at a given time (and not necessarily the most intense) (Pineau et al., 2009) and by no additional definition (Le Reverend et al., 2008). To the extent that one is scoring attributes in order of appearance, the method has some precedent in the original Flavor Profile method.

After sipping or swallowing the sample, the panelist is instructed to click on the start button and immediately choose which attribute on the screen is the dominant one and to rate its intensity, usually on a 10 point or 10 cm line scale. The computer continues to record that intensity until something changes, and a new dominant attribute is selected. In one version of this method, multiple changing attributes can be scored at various time intervals (Le Reverend et al., 2008) "until all sensations have been scored as chronologically perceived." Other publications seem to imply that only one dominant attribute is recorded at any given time (Pineau et al., 2009).

This technique produces a detailed time-by-panelist-by-attribute-by-intensity record. Curves for each attribute can be constructed then by summing across panelists and smoothing the curves. In two other procedures, the data are simplified. Labbe et al. (2009) describe the derivation of an overall TDS score by averaging the intensity multiplied by the duration of each choice, divided by the sum of the durations (i.e., weighted). This produces an integrated value similar to the area-under-the-curve TI scores or those recorded by the SMURF method (intensity multiplied by persistence) in Birch's group (Birch and Munton, 1981). Note that temporal information is used but lost in the scores, i.e., no curves can be constructed from these derived scores. These overall scores were found to correlate well with traditional profiling scores in a series of flavored gels (Labbe et al., 2009). A second derived statistical measure is the proportion of panelists reporting a given attribute as dominant at any given time (Pineau et al., 2009). This ignores the intensity information but produces a simple percentage measure that can be plotted over time to produce a (smoothed) curve for each attribute. "Significance" of an attribute's proportion versus chance is evaluated using a simple one-tailed binomial statistic against a baseline proportion of $1/k$ where k is the number of attributes. The required significance level can be plotted as a horizontal line on the plot of dominance curves to show where attributes are significantly "dominant." Any two products can be compared using the simple binomial test for the difference of two proportions (Pineau et al., 2009). The other information that is available from this procedure is the computation of a difference score for pairs of products, which when plotted over time provides potentially useful information on differences in the dominance of each attribute and how the pattern changes.

The purported advantages of this method are that (1) it is more time and cost efficient than the one-at-a-time TI tracking methods because multiple attributes are rated in each trial, (2) that it is simple and easy to do, requiring little or no training, and (3) that it provides a picture of enhanced differences relative to the TI records. Because panelists are forced to respond to only one attribute at a time, differences in the temporal profile may be accentuated. However, at the time of this publication, no standard procedure seemed to be agreed upon. The technique requires specialized software to collect the information, but at least one of the major sensory software systems has implemented a TDS option. Attributes are assumed to have a score of zero before they become dominant, and some attributes may never be rated. This would seem to necessarily lead to an incomplete record. Different panelists are contributing at different times to different attributes, so statistical methods for comparison of differences between products using the raw data set are difficult. However, products can be statistically compared using the simplified summary scores (summed intensity by duration measures, but losing time information) or by comparison of the proportion of responders (losing intensity information) as in Labbe et al. (2009). Qualitative comparisons can be made from inspecting the curves, such as "this product is initially sweet, then becoming more astringent, compared to product X which is initially sour, then fruity."

Is information provided by TDS and by traditional TI tracking methods different? One study found a strong resemblance of the constructed time curves for both methods, providing virtually redundant information for some attributes (Le Reverend et al., 2008). In another study, correlations with TI parameters were high for intensity maxima versus dominance proportion maxima, as might be expected since a higher number of persons finding an attribute dominant should be related to the mean intensity in TI. Correlations with other time-dependent parameters such as time to I_{max} (T_{max}) and duration measures were low, due to the different information collected in TDS and the limited attention to one attribute at a time (Pineau et al., 2009).

8.4 Recommended Procedures

8.4.1 Steps in Conducting a Time–intensity Study

The steps in conducting a time–intensity study are similar to those in setting up a descriptive analysis procedure. They are listed in Table 8.2. The first

Table 8.2 Steps in conducting a time–intensity study

1. Determine project objectives: Is TI the right method?
2. Determine critical attributes to be rated.
3. Establish products to be used with clients/researchers.
4. Choose system and/or TI method for data collection.
   a. What is the response task?
   b. What visual feedback is provided to the panelists?
5. Establish statistical analysis and experimental design.
   a. What parameters are to be compared?
   b. Are multivariate comparisons needed?
6. Recruit panelists.
7. Conduct training sessions.
8. Check panelist performance.
9. Conduct study.
10. Analyze data and report.

important question is to establish whether TI methods are appropriate to the experimental research objective. Is this a product with just one or a few critical attributes that are likely to vary in some important way in their time course? Is this difference likely to impact consumer acceptance of the product? What are these critical attributes? Next, the product test set should be determined with the research team, and this will influence the experimental design. The TI method itself must be chosen, e.g., discrete point, continuous tracking, or TDS. The sensory specialist should by this time have an idea of what the data set will look like and what parameters can be extracted from the TI records for statistical comparisons such as intensity maxima, time to maxima, areas under the curve, and total duration. Many of the TI curve parameters are often correlated, so there is little need to analyze more than about ten parameters. Practice is almost always essential. You cannot assume that a person sitting down in a test booth will know what to do with the TI system and feel comfortable with the mouse or other response device. A protocol for training TI panelists was outlined by Peyvieux and Dijksterhuis (2001) and this protocol or similar versions have been widely adopted. It is also wise that some kind of panel checking be done to make sure the panelists are giving reliable data (see Bloom et al., 1995; Peyvieux and Dijksterhuis, 2001) and to examine the reasonableness of their data records. At this time the researchers and statistical staff should also decide how to handle missing data or records that may have artifacts or be incomplete. As in any sensory study, extensive planning may save a lot of headaches and problems, and this is especially true for TI methods.

8.4.2 Procedures

If only a few attributes are going to be evaluated, the continuous tracking methods are appropriate and will provide a lot of information. This usually requires the use of computer-assisted data collection. Many, if not all, of the commercial software packages for sensory evaluation data collection have TI modules. The start and stop commands, sampling rate, and inter-trial intervals can usually be specified. The mouse movement will generally produce some visual feedback such as the motion of a cursor or line indicator along a simple line scale. The display often looks like a vertical or horizontal thermometer with the cursor position clearly indicated by bar or line that rises and falls. The computer record can be treated as raw data for averaging across panelists. However, statistical analysis by simple averaging raises a number of issues (discussed in Section 8.5). A simple approach is to pull characteristic curve parameters off each record for purposes of statistical comparisons such as intensity maximum (I_{max}), time to maximum (T_{max}), and area under the curve (AUC). These are sometimes referred to as "scaffolding parameters" as they represent the fundamental structure of the time records. Statistical comparison of these parameters can provide a clear understanding of how different products are perceived with regard to the onset of sensations, time course of rising and falling sensations, total duration, and total sensory impact of that flavor or textural aspect of the products. If a computer-assisted software package is not available or cannot be programmed, the research can always choose to use the cued/discontinuous method (e.g., with a stopwatch and verbal commands). This may be suitable for products in which multiple attributes must be rated in order to get the full picture. However, given the widespread availability of commercial sensory data-collection systems in major food and consumer product companies, it is likely that a sensory professional will have access to a continuous tracking option.

The starting position of the cursor on the visible scale or computer screen should be considered carefully. For most intensity ratings it makes sense to start at the lower end, but for hedonics (like/dislike) the cursor should begin at the neutral point. For meat tenderness or product melting, the track is usually unidirectional, so the cursor should start at "not tender" or "tough" for meat and "not melted" for a product that melts. If the cursor is started at the wrong end of a unidirectional tracking situation, a falsely bi-directional record may be obtained due to the initial movement.

An outline of panel training for TI studies was illustrated in a case study by Peyvieux and Dijksterhuis (2001) and the sensory specialist should consider using this or a similar method to insure good performance of the panelists. These authors had panelists evaluate flavor and texture components of a complex meat product. The prospective panelists were first introduced to the TI method, and then given practice with basic taste solutions over several sessions. The basic tastes were considered simpler than the complex product and more suitable for initial practice. Panelist consistency was checked and a panelist was considered reliable if they could produce two out of three TI records on the same taste stimulus that did not differ more than 40% of the time. A vertical line scale was used and an important specification was when to move the cursor back to zero (when there was no flavor or when the sample was swallowed for the texture attribute of juiciness). Problems were noted with (1) nontraditional curve shapes such as having no return to zero, (2) poor replication by some panelists, (3) unusable curves due to lack of a landmark such as no I_{max}. The authors also conducted a traditional profiling (i.e., descriptive analysis) study before the TI evaluations to make sure the attributes were correctly chosen and understood by the panelists. If TI panelists are already chosen from an existing descriptive panel, this step may not be needed. The authors conducted several statistical analyses to check for consistent use of the attributes, to look for oddities in curve shapes by some panelists, and to examine individual replicates. Improvements in consistency and evidence of learning and practice were noted.

8.4.3 Recommended Analysis

For purposes of comparing products, the simplest approach is to extract the curve parameters such as I_{max}, T_{max}, AUC, and total duration from each record. Some sensory software systems will generate these measures automatically. Then these curve parameters can be treated like any data points in any sensory evaluation and compared statistically. For three or more products, analysis would be by ANOVA and then planned comparisons of means (see Appendix C). Means and significance of differences can be reported in graphs or tables for each curve attribute and product. If a time by intensity curve is desired, the curves can be averaged in the time direction by choosing points at specific time intervals. This averaging method is not without its pitfalls; however, a number of alternative methods are given in the next section. An example of how to produce a simplified averaged curve is given below in the case study of the trapezoidal method (Lallemand et al., 1999).

8.5 Data Analysis Options

8.5.1 General Approaches

Two common statistical approaches have been taken to perform hypothesis testing on TI data. Perhaps the most obvious test is simply to treat the raw data at whatever time intervals were sampled as the input data to analyses of variance (ANOVA) (e.g., Lee and Lawless, 1991). This approach results in a very large ANOVA with at least three factors—time, panelists, and the treatments of interest. Time and panelists effects may not be of primary interest but will always show large F-ratios due to the fact that people differ and the sensations change over time. This is not news. Another common pattern is a time-by-treatment interaction since all curves will tend to converge near baseline at later time intervals. This is also to be

expected. Subtle patterns in the data may in fact be captured in other interaction effects or other causes of time-by-treatment interaction. However, it may be difficult to tell whether the interaction is due to the eventual convergence at baseline or to some more interesting effects such as a faster onset time or decay curves that cross over.

As noted above, researchers often select parameters of interest from the TI curve for analysis and comparisons. Landmarks on the curve included the perceived maximum intensity of the sensation, the time needed to reach maximum intensity and duration or time to return to baseline intensity. With computer-assisted data collection many more parameters are easily obtained and parameters such as the area under the curve, the area under the curve before and after perceived maximum intensity, as well as rate from onset to maximum and rate of decay from maximum to endpoint. Additional parameters include plateau time at perceived maximum intensity, lag time prior to start of responses, and the time needed to reach half of the perceived maximum intensity. A list of parameters is shown in Table 8.3.

Thus a second common approach is to extract the curve parameters on each individual record and then perform the ANOVA or other statistical comparison on each of the aspects of the TI curve, as recommended above. For an example, see Gwartney and Heymann (1995) in a study of the temporal perception of menthol. One advantage of this method is that it captures some (but probably not all) of the individual variation in the pattern of the time records. Individual judges' patterns are unique and reproducible within individuals (Dijksterhuis, 1993; McGowan and Lee,

Table 8.3 Parameters extracted from time–intensity curves

Parameter	Other names	Definition
Peak intensity	I_{max}, I_{peak}	Height of highest point on TI record
Total duration	DUR, D_{total}	Time from onset to return to baseline
Area under the curve	AUC, A_{total}	Self-explanatory
Plateau	D_{peak}	Time difference between reaching maximum and beginning descent
Area under plateau	A_{peak}	Self-explanatory
Area under descending phase	P_{total}	Area bounded by onset of decline and reaching baseline
Rising slope	R_i	Rate of increase (linear fit) or slope of line from onset to peak intensity
Declining slope	R_f	Rate of decrease (linear fit) or slope of line from initial declining point to baseline.
Extinction		Time at which curve terminates at baseline
Time to peak	T_{max}, T_{peak}	Time to reach peak intensity
Time to half peak	Half-life	Time to reach half maximum in decay portion

Modified from Lundahl (1992)
Other shape parameters are given in Lundahl (1992), based on a half circle of equivalent area under the curve and dividing the half circle into rising and falling phase segments

2006), an effect that is sometimes described as an individual "signature." Examples of individual signatures are shown in Fig. 8.3. The causes of these individual patterns are unknown but could be attributed to differences in anatomy, differences in physiology such as salivary factors (Fischer et al., 1994), different types of oral manipulation or chewing efficiency (Brown et al., 1994; Zimoch and Gullet, 1997), and individual habits of scaling. Some of this information may be lost when analyzing only the extracted parameters.

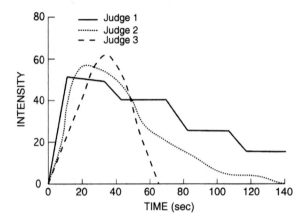

Fig. 8.3 Examples of time–intensity records showing characteristic signatures or shapes. Judge 1 shows a record with multiple plateaus, a common occurrence. Judge 2 shows a smooth and continuous curve. Judge 3 shows a steep rise and fall.

A third approach is to fit some mathematical model or set of equations to each individual record and then use the constants from the model as the data for comparisons of different products (Eilers and Dijksterhuis, 2004; Garrido et al., 2001; Ledauphin et al., 2005, 2006; Wendin et al., 2003). Given the increasing activity and ingenuity in the area of sensometrics, it is likely that such models will continue to be developed. The sensory scientist needs to ask how useful they are in the product testing arena and whether the model fitting is useful in differentiating products. Various approaches to modeling and mathematical description of TI curves are discussed in the next section.

8.5.2 Methods to Construct or Describe Average Curves

The analysis of TI records has produced a sustained interest and response from sensometricians, who have

proposed a number of schemes for curve fitting and summarization of individual- and group-averaged TI records. Curve-fitting techniques include fitting by spline methods (Ledauphin et al., 2005, 2006) and various exponential logistic or polynomial equations (Eilers and Dijksterhuis, 2004; Garrido et al., 2001; Wendin et al., 2003). An important question for anyone attempting to model TI behavior by a single equation or set of equations is how well it can take into account the individual "signatures" of panelists (McGowan and Lee, 2006). What appears to be a good approximation to a smooth TI record from one panelist may not be a very good model for the panelist who seems to show a step function with multiple plateaus. It is unknown at this time how many of these schemes have found use in industry or whether they remain a kind of academic exercise. Their penetration into the mainstream of sensory practice may depend on whether they are incorporated into one of the commercial sensory evaluation software data-collection systems.

In the simplest form of averaging, the height of each curve at specific time intervals is used as the raw data. Summary curves are calculated by averaging the intensity values at given times and connecting the mean values. This has the advantage of simplicity in analysis and keeps a fixed-time base as part of the information. However, with this method there may be no detection of an atypical response. As noted above, judges will have characteristic curve shapes that form a consistent style or signature in their response, but that differ in shape from other judges. Some rise and fall sharply. Others form smooth rounded curves while others may show a plateau. Simple averaging may lose some information about these individual trends, especially from outliers or a minority pattern. Furthermore, the average of two different curves may produce a curve shape that does not correspond to either of the input curves. An extreme (and hypothetical) example of this is shown in Fig. 8.4, where the two different peak intensity times lead to an average curve with two maxima (Lallemand et al., 1999). Such a double-peaked curve is not present in the contributing original data.

To avoid these problems, other averaging schemes have been proposed. These approaches may better account for the different curve shapes exhibited by different judges. To avoid irregular curve shapes, it may be necessary or desirable to group judges with similar responses before averaging (McGowan and Lee, 2006). Judges can be subgrouped by "response

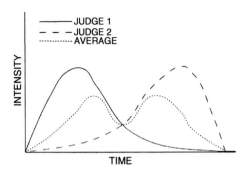

Fig. 8.4 Two curves with different peak times, if averaged, can lead to a double-peaked curve that resembles neither original data record.

style" either by simple visual inspection of the curves or by a clustering analysis or other statistical methods (van Buuren, 1992; Zimoch and Gullet, 1997). Then these subgroups can be analyzed separately. The analysis may proceed using the simple fixed-time averaging of curve heights or one of the other methods described next.

An alternative approach is to average in both the intensity and time directions, by setting each individual's maximum of the mean time to maximum across all curves, and then finding mean times to fixed percentages of maximum in the rising and falling phases of each curve. This procedure was originally published by Overbosch et al. (1986) and subsequent modifications were proposed by Liu and MacFie (1990). The steps in the procedure are shown in Fig. 8.5. Briefly, the method proceeds as follows: In the first step, the geometric mean value for the intensity maximum is found. Individual curves are multiplicatively scaled to have this I_{max} value. In the second step, the geometric mean time to I_{max} is calculated. In the next steps, geometric mean times are calculated for fixed percentage "slices" of each curve, i.e., at fixed percentages of I_{max}. For example, the rising and falling phases are "sliced" at 95% of I_{max} and 90% of I_{max} and the geometric mean time values to reach these heights are found.

This procedure avoids the kind of double-peaked curve that can arise from simple averaging of two distinctly different curve shapes as shown in Fig. 8.4. The method results in several desirable properties that do not necessarily occur with simple averaging at fixed times. First, the I_{max} value from the mean curve is the geometric mean of the I_{max} of the individual curves. Second, the T_{max} value from the mean curve is the geometric mean of the T_{max} of the individual

curves. Third, the endpoint is the geometric mean of all endpoint times. Fourth, all judges contribute to all segments of the curve. With simple averaging at fixed times, the tail of the curve may have many judges returned to zero and thus the mean is some small value that is a poor representation of the data at those points. In statistical terms, the distribution of responses at these later time intervals is positively skewed and left-censored (bound by zero). One approach to this problem is to use the simple median as the measure of central tendency (e.g., Lawless and Skinner, 1979). In this case the summary curve goes to zero when over half the judges go to zero. A second approach is to use statistical techniques designed for estimating measures of central tendency and standard deviations from left-censored positively skewed data (Owen and DeRouen, 1980).

Overbosch's method works well if all individual curves are smoothly rising and falling with no plateaus or multiple peaks and valleys, and all data begin and end at zero. In practice, the data are not so regular. Some judges may begin to fall after the first maximum and then rise again to a second peak. Due to various common errors, the data may not start and end at zero, for example, the record may be truncated within the allowable time of sampling. To accommodate these problems, Liu and MacFie (1990) developed a modification of the above procedure. In their procedure, I_{max} and four "time landmarks" were averaged, namely starting time, time to maximum, time at which the curve starts to descend from I_{max} and ending time. The ascending and descending phases of each curve were then divided into about 20 time interval slices. At each time interval, the mean I value is calculated. This method then allows for curves with multiple rising and falling phases and a plateau of maximum intensity that is commonly seen in some judges' records.

8.5.3 Case Study: Simple Geometric Description

A simple and elegant method for comparing curves and extracting parameters by a geometric approximation was described by Lallemand et al. (1999). The authors used the method with a trained texture panel to evaluate different ice cream formulations. The labor-intensive nature of TI studies was illustrated in the fact that

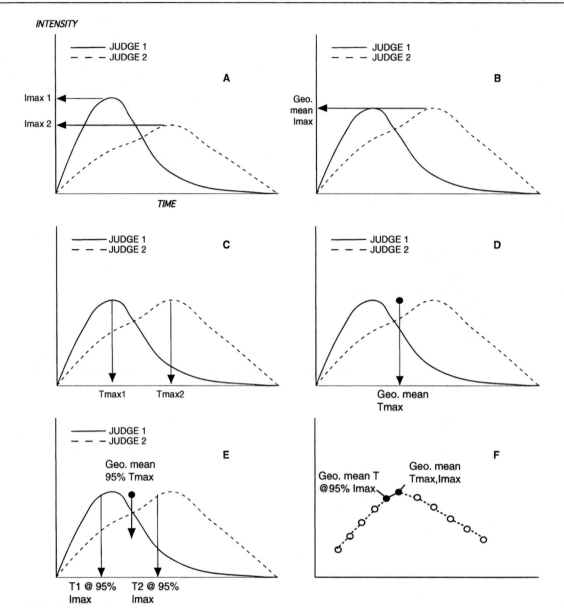

Fig. 8.5 Steps in the data analysis procedure recommended by Overbosch et al. (1986). (**a**) Two hypothetical time–intensity records from two panelists showing different intensity maxima at different times. (**b**) The geometric mean value for the intensity maximum is found. Individual curves may then by multiplicatively scaled to have this I_{max} value. (**c**) The two T_{max} values. (**d**) The geometric mean time to maximum (T_{max}) is calculated.

(**e**) Geometric mean times are calculated for fixed percentage "slices" of each curve, i.e., at fixed percentages of I_{max}. The rising phase is "sliced" at 95% of I_{max} and the time values determined. A similar value will be determined at 95% of maximum for the falling phase. (**f**) The geometric mean times at each percent of maximum are plotted to generate the composite curve.

12 products were evaluated on 8 different attributes in triplicate sessions, requiring about 300 TI curves from each panelist! Texture panelists were given over 20 sessions of training, although only a few of the final sessions were specifically devoted to practice

with the TI procedure. Obviously, this kind of extensive research program requires a significant time and resource commitment.

Data were collected using a computer-assisted rating program, where mouse movement was linked to the position of a cursor on a 10 cm 10-point scale. The authors noted a number of issues with the data records due to "mouse artifacts" or other problems. These included sudden unintended movement of the mouse leading to false peaks or ratings after the end of the sensation, mouse blockage leading to unusable records, and occasional inaccurate positioning by panelists causing data that did not reflect their actual perceptions. Such ergonomic difficulties are not uncommon in TI studies although they are rarely reported or discussed. Even with this highly trained panel, from 1 to 3% of the records needed to be discarded or manually corrected due to artifacts or inaccuracies. Sensory professionals should not assume that just because they have a computer-assisted TI system, the human factors in mouse and machine interactions will always work smoothly and as planned. Examples of response artifacts are shown in Fig. 8.6.

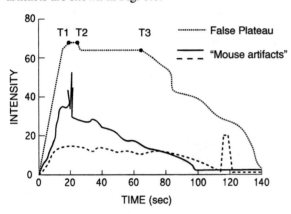

Fig. 8.6 Response artifacts in TI records. The *solid line* shows some perhaps unintended mouse movement (muscle spasm?) near the peak intensity. The *dashed line* shows a bump in the mouse after sensation ceased and returned to zero. The *dotted line* illustrates an issue in determining at what point the intensity plateau has ended. The short segment between *T1* and *T2* may have simply been an adjustment of the mouse after the sudden rise, when the panelist felt they overshot the mark. The actual end of the plateau might more reasonably be considered to occur at *T3*. (see Lallemand et al. 1999).

Lallemand and coworkers noted that TI curves often took a shape in which the sensation rose to a plateau near peak intensity for a period during which intensity ratings changed very little and then fell to the baseline. They reasoned that a simple geometric approximation by a trapezoid shape might suffice for extracting curve parameters and finding the area under the curve (not unlike the trapezoidal approximation method used for integration in calculus). In principle, four points could

be defined that would describe the curve: the onset time, the time at the intensity maximum or beginning of the plateau, the time at which the plateau ended and the decreasing phase began, and the time at which sensation stopped. These landmarks are those originally proposed by Lui and MacFie (1990). In practice, these points turned out to be more difficult to estimate than expected, so some compromises were made. For example, some records would show a gradually decreasing record during the "plateau" and before the segment with a more rapidly falling slope was evident. How much of a decrease would justify the falling phase or conversely, how little of a decrease would be considered still part of the plateau (see Fig. 8.6)? Also, what should one do if the panelist did not return to zero sensation or unintentionally bumped the mouse after reaching zero? In order to solve these issues, the four points were chosen at somewhat interior sections of the curve, namely the times at 5% of the intensity maximum for the onset and endpoint of the trapezoid, and the times at 90% of the intensity maximum for the beginning and end of the plateau.

This approximation worked reasonably well, and its application to a hypothetical record is shown in Fig. 8.7. Given the almost 3,000 TI curves in this one study, the trapezoid points were not mapped by hand or by eye, but a special program written to extract the points. However, for smaller experiments it should be quite feasible to do this kind of analysis "by hand" on any collection of graphed records. The establishment of the four trapezoid vertices now allows extraction of the six basic TI curve parameters for statistical analysis (5 and 90% of maximum intensity points, the four times at those points), as well as the intensity maximum from the original record, and derived (secondary) parameters such as rising and falling slopes and the total area under the curve. Note that the total area becomes simply the sum of the two triangles and the rectangle described by the plateau. These are shown in the lower section of Fig. 8.7. A composite trapezoid can be drawn from the averaged points.

The utility and validity of the method was illustrated in one sample composite record, showing the fruity flavor intensity from two ice creams differing in fat content. Consistent with what might be expected from the principles of flavor release, the higher fat sample had a slower and more delayed rise to the peak (plateau) but a longer duration. This would be predicted if the higher fat level was better able to sequester

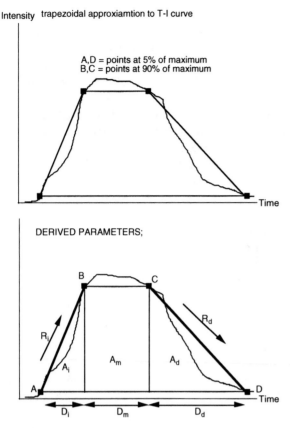

Fig. 8.7 The trapezoidal method of Lallemand et al. (1999) for assessing curve parameters on TI records. The *upper panel* shows the basic scheme in which four points are found when the initial 5% of the intensity maximum (I_{max}) occurs, when 90% of I_{max} is first reached on the ascending segment, when the plateau is finished at 90% of I_{max} on the descending phase and the end-point approximation at 5% of I_{max} on the descending phase. The *lower panel* shows the derived parameters, namely R_i, A_i, and D_i for the rate (slope), area, and duration of the initial rising phase; A_m and D_m for the area and duration of the middle plateau section; and R_d, A_d, and D_d for the rate (slope), area, and duration of the falling phase. A total duration can be found from the sum of D_i, D_m, and D_d. The total area is given by the sum of the A parameters or by the formula for the area of a trapezoid: Total area $= (I_{90}–I_5) (2D_m + D_i + D_d)/2$. (Height times the sum of the two parallel segments, then divided by 2).

a lipophilic or nonpolar flavor compound and thus delay the flavor release. They also examined the correlation with a traditional texture descriptive analysis and found very low correlations of individual TI parameters with texture profiling mean scores. This would be expected if the TI parameters were contributing unique information or if the texture profilers were integrating a number of time-dependent events in coming up

with their single-point intensity estimates. Consistent with the latter notion, the profiling scores could be better modeled by a combination of several of the TI parameters. The simplicity and validity of this analysis method suggests that it should find wider application in industrial settings.

8.5.4 Analysis by Principal Components

Another analysis uses principal components analysis (PCA, discussed in Chapter 18) (van Buuren, 1992). Briefly, PCA is a statistical method that "bundles" groups of correlated measurements and substitutes a new variable (a factor or principal component) in place of the original variables, thus simplifying the picture. In studying the time–intensity curves for bitterness or different brands of lager beer, van Buuren noticed that individuals once again produced their own characteristic "style" of curve shape. Most people showed a classic TI curve shape, but some subjects were classified as "slow starters," with a delayed peak and some showed a tendency to persist and not come back down to baseline within the test period. Submission of the data to PCA allowed the extraction of a "principal curve" which captured the majority trend. This showed a peaked TI curve and a gradual return to baseline. The second principal curve captured the shape of the minority trends, with slow onset, a broad peak and slow decline without reaching baseline. The principal curves were thus able to extract judge trends and provide a cleaned-up view of the primary shape of the combined data (Zimoch and Gullet, 1997). Although a PCA program may extract a number of principal components, not all may be practically meaningful (for an example, see Reinbach et al., 2009), and the user should examine each one for the story it tells. Reasonable questions are whether the component reflects something important relative to the simple TI curve parameters, and whether it shows any patterns related to individual differences among panelists.

Dijksterhuis explored the PCA approach in greater detail (Dijksterhuis, 1993; Dijksterhuis and van den Broek, 1995; Dijksterhuis et al., 1994). Dijksterhuis (1993) noted that the PCA method as applied by van Buuren was not discriminating of different bitter stimuli. An alternative approach was "non-centered PCA" in which curve height information was retained during data processing, rather than normalizing curves to a common scale. The non-centered approach works on the raw data matrix. Stimuli or treatments were better distinguished. The first principal curve tends to look like the simple average, while the second principal curve contains rate information such as accelerations or inflection points (Dijksterhuis et al., 1994). This could be potentially useful information in differentiating the subtle patterns of TI curves for different flavors. The PCA approach also involves the possibility of generating weights for different assessors that indicate the degree to which they contribute to the different principal curves. This could be an important tool for differentiating outliers in the data or panelists with highly unusual TI signatures (Peyvieux and Dijksterhuis, 2001).

8.6 Examples and Applications

A growing number of studies have used TI methods for the evaluations of flavor, texture, flavor release, hedonics, and basic studies of the chemical senses. A short review of these studies follows in this section, although the reader is cautioned that the list is not exhaustive. We have cited a few of the older studies to give credit to the pioneers of this field as well as some of the newer applications. The examples are meant to show the range of sensory studies for which TI methods are suitable.

8.6.1 Taste and Flavor Sensation Tracking

A common application of continuous time–intensity scaling is tracking the sensation rise and decay from important flavor ingredients, such as sweeteners (Swartz, 1980). An early study of Jellinek and Pangborn reported that addition of salt to sucrose extended the time–intensity curve and made the taste "more rounded" in their words (Jellinek, 1964). One of the salient characteristics of many intensive or non-carbohydrate sweeteners is their lingering taste that is different from that of sucrose. Time–intensity tracking of sweet tastes was an active area of study (Dubois and Lee, 1983; Larson-Powers and Pangborn, 1978; Lawless and Skinner, 1979; Yoshida, 1986) and

remains of great interest to the sweetener industry. Another basic taste that has often been scaled using time–intensity methods is bitterness (Dijksterhuis, 1993; Dijksterhuis and van den Broek, 1995; Leach and Noble, 1986; Pangborn et al., 1983). Beer flavor and bitterness were two of the earliest attributes studied by time–intensity methods (Jellinek, 1964; Pangborn et al., 1983; Sjostrom, 1954; van Buuren, 1992). Robichaud and Noble (1990) studied the bitterness and astringency of common phenolics present in wine and found similar results using traditional scaling and the maximum intensity observed in time–intensity scaling.

Taste properties have been studied in foods and model systems and how they change with other food ingredients and/or flavors present. Sweetness and other flavors may also change in their temporal properties with changes in food formulation, such as changes in viscosity caused by addition of thickening agents or changes due to addition of fat substitutes (Lawless et al., 1996; Pangborn and Koyasako, 1981). In breads, Barylko-Pikielna et al. (1990) measured saltiness, sourness, and overall flavor. TI parameters such as maximum intensity, total duration, and area under the TI curve increased monotonically with salt added to wheat breads. Lynch et al. (1993) found evidence for suppression of taste in gelatin samples when the mouth was precoated with various oils. For some tastes, especially bitter, suppression was evident both in decreased peak intensity and shorter overall duration. Recently, several sensory scientists have applied TI studies to examine flavor intensity in different media and via different routes of olfactory perception. Shamil et al. (1992) showed that lowering the fat content of cheeses and salad dressings caused increases in persistence time and alters the rate of flavor release. Kuo et al. (1993) examined differences in citral intensity comparing orthonasal and retronasal conditions in model systems of citral and vanillin with different tastants or xanthan gum added. In time–intensity study of flavors in different dispersion media, Rosin and Tuorila (1992) found pepper to be more clearly perceived in beef broth than potato, while garlic was equally potent in either medium. Another active area for time-related judgments has been in the study of flavor interactions. Noble and colleagues have used time–intensity measurements to study the interactions of sweetness, sourness, and fruitiness sensations in beverages and simple model systems (Bonnans and Noble, 1993;

Cliff and Noble, 1990; Matysiak and Noble, 1991). Using these techniques, enhancement of sweetness by fruity volatiles has been observed, and differences in the interactions were seen for different sweetening agents.

8.6.2 Trigeminal and Chemical/Tactile Sensations

Reactions to other chemical stimuli affecting irritation or tactile effects in the mouth have been a fertile area for time-related sensory measurements. Compounds such as menthol produce extended flavor sensations and the time course is concentration dependent (Dacanay, 1990; Gwartney and Heymann, 1995). A large number of studies of the burning sensation from hot pepper compounds have used time–intensity scaling, using repeated ratings at discrete time intervals as well as continuous tracking (Cliff and Heymann, 1993a; Green, 1989; Green and Lawless, 1991; Lawless and Stevens, 1988; Stevens and Lawless, 1986). Given the slow onset and extended time course of the sensations induced by even a single sample of a food containing hot pepper, this is a highly appropriate application for time-related judgments. The repeated ingestion paradigm has also been used to study the short- and long-term desensitization to irritants such as capsaicin and zingerone (Prescott and Stevenson, 1996). The temporal profile of different irritative spice compounds is an important point of qualitative differentiation (Cliff and Heymann, 1992). Reinbach and colleagues used TI tracking to study the oral heat from capsaicin in various meat products (Reinbach et al., 2007, 2009) as well as the interactions of oral burn with temperature (see also Baron and Penfield, 1996). When examining the decay curves following different pepper compounds, different time courses can be fit by different decay constants in a simple exponential curve of the form

$$R = R_0 e^{-kt} \qquad (8.1)$$

or

$$\ln R = \ln R_0 - kt \qquad (8.2)$$

where R_0 is the peak intensity and t is time, k is the value determining how rapid the sensation falls off

from peak during the decay portion of the time curve (Lawless, 1984).

Another chemically induced tactile or feeling factor in the mouth that has been studied by TI methods is astringency. Continuous tracking has been applied to the astringency sensation over repeated ingestions (Guinard et al., 1986). Continuous tracking can provide a clear record of how sensations change during multiple ingestions, and how flavors may build as subsequent sips or tastings add greater sensations on an existing background. An example of this is shown in Fig. 8.8 where the saw-tooth curve shows an increase and builds astringency over repeated ingestions (Guinard et al., 1986). Astringency has also been studied using repeated discrete-point scaling. For example, Lawless and coworkers were able to show differences in the time profiles of some sensory sub-qualities related to astringency, namely dryness, roughness in the mouth, and puckery tightening sensations (Lawless et al., 1994; Lee and Lawless, 1991) depending on the astringent materials being evaluated.

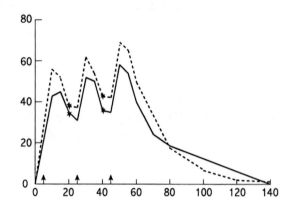

Fig. 8.8 Continuous tracking record with multiple ingestions producing a "sawtooth" curve record. The *abscissa* shows the time axis in seconds and the ordinate the mean astringent intensity from 24 judges. The *dashed curve* is from a 15 ml sample of a base wine with 500 gm/l added tannin and the *solid curve* from the base wine with no added tannin. Sample intake and expectoration are indicated by *stars* and *arrows*, respectively. From Guinard et al. (1986), reprinted with permission of the American Society of Enology and Viticulture.

8.6.3 Taste and Odor Adaptation

The measurement of a flavor sensation by tracking over time has a close parallel in studies of taste and odor adaptation (Cain, 1974; Lawless and Skinner,

1979; O'Mahony and Wong, 1989). Adaptation may be defined as the decrease in responsiveness of a sensory system to conditions of constant stimulation providing a changing "zero point" (O'Mahony, 1986). Adaptation studies have sometimes used discrete, single bursts of stimulation (e.g., Meiselman and Halpern, 1973). An early study of adaptation used a flowing system through the subject's entire mouth using pipes inserted through a dental impression material that was held in the teeth (Abrahams et al., 1937). Disappearance of salt taste was achieved in under 30 s for a 5% salt solution, although higher concentrations induced painful sensations over time that were confusing to the subjects and sometimes masked the taste sensation. By flowing a continuous stream over a section of the subject's tongue or stabilizing the stimulus with wet filter paper, taste often disappears in under a few minutes (Gent, 1979; Gent and McBurney, 1978; Kroeze, 1979; McBurney, 1966). Concentrations above the adapting level are perceived as having the characteristic taste of that substance, e.g., salty for NaCl. Concentrations below that level, to which pure water is the limiting case, take on other tastes so that water after NaCl, for example, is sour–bitter (McBurney, and Shick, 1971). Under other conditions adaptation may be incomplete (Dubose et al., 1977; Lawless and Skinner, 1979; Meiselman and Dubose, 1976). Pulsed flow or intermittent stimulation causes adaptation to be much less complete or absent entirely (Meiselman and Halpern, 1973).

8.6.4 Texture and Phase Change

Tactile features of foods and consumer products have been evaluated using time-related measurements. Phase change is an important feature of many products that undergo melting when eaten. These include both frozen products like ice cream and other dairy desserts and fatty products with melting points near body temperature such as chocolate. Using the chart-recording method for time–intensity tracking, Moore and Shoemaker (1981) evaluated the degree of coldness, iciness, and sensory viscosity of ice cream with different degrees of added carboxymethyl cellulose. The added carbohydrate shifted the time for peak intensity of iciness and extended the sensations of coldness on the tongue. Moore and Shoemaker also studied melting behavior. Melting is an example of a TI curve that does not reach a maximum and then decline, since items placed in the mouth do not re-solidify after melting. In other words, it is unidirectional, from not melted to completely melted. Melting rates have also been studied in table spreads of varying fat composition, with similar unidirectional time curves (Tuorila and Vainio, 1993; see also Lawless et al., 1996).

Other reports have been published applying time–intensity methods to texture evaluation. Larson-Powers and Pangborn (1978) used the strip-chart method to evaluate a number of taste properties of gelatins sweetened with sugar or intensive sweeteners and also evaluated hardness. The hardness curves showed the one-directional decay as expected. Rine (1987) used time–intensity techniques to study the textural properties of peanut butter. Pangborn and Koyasako (1981) were able to track differences in perceived viscosity of chocolate pudding products that differed in their thickening agents. Meat texture has been evaluated during chewing, and tenderness is usually an example of a unidirectional TI curve (Duizer et al., 1995, but see also Zimoch and Gullett (1997) which is bi-directional). Juiciness is another meat texture variable that is well suited to TI evaluation (Peyvieux and Dijksterhuis, 2001; Zimoch and Gullett, 1997). In one early application of TI methods to meat texture, Butler et al. (1996) noted the tendency toward individual "signatures" in the TI records, now a common finding (Zimoch and Gullett, 1997). Jellinek (1985, p. 152) gave an interesting example of how texture may be judged by auditory cures for crispness, and showed time–intensity records of two product differing in both initial crispness and duration. One product made louder sounds initially, but dropped off a steeper rate than another during chewing. Jellinek pointed out that the maintenance of crispness during chewing could be important. Time–intensity methods can thus provide potentially important sensory information about the rate of destruction of a product and texture change during deformation.

8.6.5 Flavor Release

A potentially fertile area for the application of time-related sensory measurements is in flavor release from foods during eating. Not only are texture changes obvious during chewing but also a number of factors operate to change the chemical nature of the matrix within which food flavors exist once the food enters the mouth. The degree to which the individual flavor compounds are held in the matrix as opposed to being released into the mouth space and into the breath and nose of the judge will depend on their relative solubility and binding to the matrix of the food bolus and saliva (McNulty, 1987; Overbosch, 1987). Saliva is a complex mixture of water, salts, proteins, and glycoproteins. It has pH buffering capacity and active enzymes. The flavor volatilization changes as a function of mixing with saliva, pH change, enzymatic processes such as starch breakdown by salivary amylase, warming, mechanical destruction of the food matrix, and changes in ionic strength (Ebeler et al., 1988; Haring, 1990; Roberts and Acree, 1996). Temperature may not only affect the partial vapor pressure of a flavor above the liquid phase but also the degree of association of flavor compounds to proteins and other components of the food matrix (O'Keefe et al., 1991). The flavor balance, interactions with other tastes, and the time properties of release may all be different as a function of sniffing (the orthonasal route to the olfactory receptors) as opposed to sipping (the retronasal route) (Kuo et al., 1993).

A number of devices have been developed to study how volatile flavors are released from food in simulated conditions of oral breakdown (Roberts and Acree, 1996). These usually involve some degree of mechanical agitation or stirring, warming, and dilution in some medium that is designed to reflect the chemical composition of saliva to some degree. Some research has focused on chemical sampling of this altered "headspace," i.e., the vapor phase above the simulated oral mixture (Lee, 1986; Roberts et al., 1996). de Roos (1990) examined flavor release in chewing gum, a matrix from which different flavor compounds are released at different rates, changing the flavor character and detracting from consumer appeal. Large individual differences were observed for common flavors such as vanillin. de Roos was further able to divide his groups into efficient, highly efficient, and inefficient chewers, who differed in their degree and rate of flavor release. This serves to remind us that not everyone masticates in the same way and that mechanical breakdown factors will be different among individuals. Mastication and salivation variables were related to inter-individual differences in flavor release in model cheese systems (Pionnier et al., 2004) using a discrete point TI method.

8.6.6 Temporal Aspects of Hedonics

Since the pleasantness or appeal of a sensory characteristic is largely dependent on its intensity level, it is not surprising that one's hedonic reaction to a product might shift over time as the strength of a flavor waxes and wanes. Time-related shifts in food likes and dislikes are well known. In the phenomenon known as alliesthesia, our liking for a food depends a lot on whether we are hungry or replete (Cabanac, 1971). The delightful lobster dinner we enjoyed last night may not look quite so appealing as leftovers at lunchtime the next day. Wine tasters may speak of a wine that is "closed in the glass, open on the palate, and having a long finish." Accompanying such a description is the implicit message that this particular wine got better over the course of the sensory experience. Given the shorter time span of flavor and texture sensations in the mouth, we can ask whether there are shifts in liking and disliking. This has been measured in several studies. Taylor and Pangborn (1990) examined liking for chocolate milk with varying degrees of milk fat. Different individual trends were observed in liking for different concentrations, and this affected the degree of liking expressed over time. Another example of hedonic TI scaling was in a study of the liking/disliking for the burning oral sensation from chili (hot) pepper (Rozin et al., 1982). They found different patterns of temporal shifting in liking as the burn rose and declined. Some subjects liked the burn at all time intervals, some disliked the burn at all time intervals, and some shifted across neutrality as strong burns became more tolerable. This method was revisited by Veldhuizen et al. (2006) who used a simple bipolar line scale for pleasantness and had subjects evaluated both intensity and hedonic reactions to a citrus beverage flowed over the tongue from a computer-controlled delivery system (see Fig. 8.2 for an early example of this kind of device). Note that with a bipolar hedonic scale, the mouse and cursor positions must begin at the center of the scale and not the lower end as with intensity scaling. The authors found a delayed pleasantness response compared to the intensity tracking, a similar time to maximum, but an unexpectedly quicker offset of response for pleasantness tracking. Some panelists produced a double-peaked pleasantness response, as the sensation could rise in pleasantness, but then become too intense, but become more pleasant again as adaptation set in and the perceived strength decreased.

8.7 Issues

Sensory scientists who wish to use time–intensity methods for any particular study need to weigh the potential for obtaining actionable information against the cost and time involved in collecting these data. Some orientation or training is required (Peyvieux and Dijksterhuis, 2001) and in some published studies, the training and practice is quite extensive. For example, Zimoch and Gullet (1997) trained their meat texture panel for 12 h. Panelists must be trained to use the response device and sufficiently practiced to feel comfortable with the requirements of the task in terms of maintaining focused attention to momentary sensation changes. With the use of online data collection the tabulation and processing of information is generally not very labor intensive; but without computer-assisted collection the time involved can be enormous. Even with computerized systems, the data collection is not foolproof. Responses may be truncated or fail to start at zero in some records (Liu and MacFie, 1990; McGowan and Lee, 2006) making automatic averaging of records infeasible. In one study of melting behavior (Lawless et al., 1996), some subjects mistakenly returned the indicating cursor to zero as soon as the product was completely melted, instead of leaving the cursor on maximum, producing truncated records. Such unexpected events remind us to never assume that panelists are doing what you think they should be doing.

A fundamental issue is information gain. In the case where changes in duration are observed at equal maximum intensities, it can be argued that the traditional scaling might have missed important sensory differences. For example, TI can capture information such as when the TI curves cross over, e.g., the interesting case when a product with a lower peak intensity has a longer duration (e.g., Lallemand et al., 1999; Lawless et al., 1996). However, this pattern is not often seen. Usually products with stronger peak height have longer durations. In general there is a lot of redundant information in TI parameters. Lundahl (1992) studied the correlation of 15 TI parameters associated with TI

curves' shapes, sizes, and rates of change. Curve size parameters were highly correlated and usually loaded on the first principal component of a PCA, capturing most of the variance (see also Cliff and Noble, 1990). Curve size parameters, including peak height, were correlated with simple category ratings of the same beverages. So an open question for the sensory scientist is whether there is any unique information in the TI parameters extracted from the records and whether there is information gain over what would be provided by more simple direct scaling using a single intensity rating.

A potential problem in time–intensity methods is that factors affecting response behavior are not well understood. In TI measurements, there are dynamic physical processes (chewing, salivary dilution) leading to changes in the stimulus and resulting sensations (Fischer et al., 1994). A second group of processes concerns how the participant translates the conscious experience into an overt response, including a decision mechanism and motoric activation (Dijksterhuis, 1996). The notion that TI methods provide a direct link from the tongue of the subject to the hand moving the mouse is a fantasy. Even in continuous tracking, there must be some decision process involved. There is no information as to how often a panelist in the continuous procedure reflects upon the sensation and decides to change the position of the response device. Decisions are probably not continuous even though some records from some subjects may look like smooth curves.

An indication that response tendencies are important is when the conditions of stimulation are held constant, but the response task changes. For example, using the graphic chart-recorder method, Lawless and Skinner (1979) found median durations for sucrose intensity that were 15–35% shorter than the same stimuli rated using repeated category ratings. Why would the different rating methods produce apparently different durations? Very different patterns may be observed when taste quality and intensity are tracked. Halpern (1991) found that tracked taste quality of 2 mM sodium saccharin had a delayed onset (by 400 ms) compared with tracked intensity. This might be understandable from the point of requiring a more complex decision process in the case of tracking intensity. However, it still alerts us to the fact that the behavior probably trails the actual experience by some unknown amount. What is more surprising in Halpern's data is that tracked quality also stopped well before tracked intensity (by

600 ms). Can it be possible that subjects are still experiencing a taste of some definable intensity and yet the quality has disappeared? Or is there another response generation process at work in this task?

A third area where potential response biases can operate is in contextual effects. Clark and Lawless (1994) showed that common contextual affects like successive contrast also operated with TI methods, as they do with other scaling tasks. Also, some ratings could be enhanced when a limited number of scales were used by subjects. As observed in single-point scaling, enhancement of sweetness by fruity flavors tends to occur when only sweetness is rated. When the fruity flavor is also rated, the sweetness enhancement often disappears (Frank et al., 1989), an effect sometimes referred to as halo dumping or simply "dumping." Using the discrete-point version of TI scaling, so that multiple attributes could be rated, Clark and Lawless showed a similar effect. This is potentially troublesome for the continuous tracking methods, since they often limit subjects to responding to only one attribute at a time. This may explain in part why sweetness enhancement by flavors can occur so readily in TI studies (e.g., Matysiak and Noble, 1991).

A final concern is the question of whether the bounded response scales often used in TI measurement produces any compression of the differences among products. In analog tracking tasks, there is a limit as to how far the joystick, mouse, lever, dial, or other response device can be moved. With some practice, judges learn not to bump into the top. Yet the very nature of the tracking response encourages judges to sweep a wide range of the response scale. If this were done on every trial, it would tend to attenuate the differences in maximum tracked intensity between products. As an example, Overbosch et al. (1986, see Fig. 2) showed curves for pentanone where doubling the concentration changed peak heights by only about 8%. A similar sort of compression is visible in Lawless and Skinner's (1979) data for sucrose, compared to the psychophysical data in the literature.

8.8 Conclusions

In most cases TI parameters show similar statistical differentiation as compared to traditional scales, but this is not universally the case (e.g., Moore and Shoemaker, 1981).

Many sensory evaluation researchers have supported increased application of time–intensity measurements for characterization of flavor and texture sensations. In particular, the method was championed by Lee and Pangborn, who argued that the methods provide detailed information not available from single estimates of sensation intensity (Lee, 1989; Lee and Pangborn, 1986). TI methods can provide rate-related, duration, and intensity information not available from traditional scaling. However, the utility of the methods must be weighed against the enhanced cost and complexity in data collection and analysis. In deciding whether to apply TI methods over conventional scaling, the sensory scientist should consider the following criteria:

(1) Is the attribute or system being studied known to change over time? Simply eating the food can often settle this issue; in many cases it is obvious.
(2) Will the products differ in sensory time course as a function of ingredients, processing, packaging, or other variables of interest?
(3) Will the time variation occur in such a way that it will probably not be captured by direct single ratings?
(4) Is some aspect of the temporal profile likely to be related to consumer acceptability?
(5) Does the added information provided by the technique outweigh any additional costs or time delays in panel training, data acquisition, and data analysis?

Obviously, when more answers are positive on these criteria, a stronger case can be made for choosing a TI method from the available set of sensory evaluation tools.

References

Abrahams, H., Krakauer, D. and Dallenbach, K. M. 1937. Gustatory adaptation to salt. American Journal of Psychology, 49, 462–469.

Ayya, N. and Lawless, H. T. 1992. Qualitative and quantitative evaluation of high-intensity sweeteners and sweetener mixtures. Chemical Senses, 17, 245–259.

Baron, R. F. and Penfield, M. P. 1996. Capsaicin heat intensity – concentration, carrier, fat level and serving temperature effects. Journal of Sensory Studies, 11, 295–316.

Barylko-Pikielna, N., Mateszewska, I. and Helleman, U. 1990. Effect of salt on time–intensity characteristics of bread. Lebensmittel Wissenschaft und Technologie, 23, 422–426.

Birch, G. G. and Munton, S. L. 1981. Use of the "SMURF" in taste analysis. Chemical Senses, 6, 45–52.

Bloom, K., Duizer, L. M. and Findlay, C. J. 1995. An objective numerical method of assessing the reliability of time–intensity panelists. Journal of Sensory Studies, 10, 285–294.

Bonnans, S. and Noble, A. C. 1993. Effect of sweetener type and of sweetener and acid levels on temporal perception of sweetness, sourness and fruitiness. Chemical Senses, 18, 273–283.

Brandt, M. A., Skinner, E. Z. and Coleman, J. A. 1963. Texture profile method. Journal of Food Science, 28, 404–409.

Brown, W. E., Landgley, K. R., Martin, A. and MacFie, H. J. 1994. Characterisation of patterns of chewing behavior in human subjects and their influence on texture perception. Journal of Texture Studies, 15, 33–48.

Butler, G., Poste, L. M., Mackie, D. A., and Jones, A. 1996. Time–intensity as a tool for the measurement of meat tenderness. Food Quality and Preference, 7, 193–204.

Cabanac, M. 1971. Physiological role of pleasure. Science, 173, 1103–1107.

Cain, W. S. 1974. Perception of odor intensity and time-course of olfactory adaptation. ASHRAE transactions, 80, 53–75.

Clark, C. C. and Lawless, H. T. 1994. Limiting response alternatives in time–intensity scaling: An examination of the Halo-Dumping effect. Chemical Senses, 19, 583–594.

Cliff, M. 1987. Temporal perception of sweetness and fruitiness and their interaction in a model system. MS Thesis, University of California, Davis, USA.

Cliff, M. and Heymann, H. 1992. Descriptive analysis of oral pungency. Journal of Sensory Studies, 7, 279–290.

Cliff, M. and Heymann, H. 1993a. Time–intensity evaluation of oral burn. Journal of Sensory Studies, 8, 201–211.

Cliff, M. and Heymann, 1993b. Development and use of time–intensity methodology for sensory evaluation: A review. Food Research International, 26, 375–385.

Cliff, M. and Noble, A. C. 1990. Time–intensity evaluation of sweetness and fruitiness in a model solution. Journal of Food Science, 55, 450–454.

Dacanay, L. 1990. Thermal and concentration effects on temporal sensory attributes of L – menthol. M.S. Thesis, University of California, Davis, USA.

de Roos, K. B. 1990. Flavor release from chewing gums. In: Y. Bessiere and A. F. Thomas (eds.), Flavour Science and Technology. Wiley, Chichester, pp. 355–362.

DeRovira, D. 1996. The dynamic flavor profile method. Food Technology, 50, 55–60.

Dijksterhuis, G. 1993. Principal component analysis of time–intensity bitterness curves. Journal of Sensory Studies, 8, 317–328.

Dijksterhuis, G. 1996. Time–intensity methodology: Review and preview. Proceedings, COST96 Meeting: Interaction of Food Matrix with Small Ligands Influencing Flavour and Texture, Dijon, France, November 20, 1995.

Dijksterhuis, G. and van den Broek, E. 1995. Matching the shape of time–intensity curves. Journal of Sensory Studies, 10, 149–161.

Dijksterhuis, G., Flipsen, M. and Punter, P. H. 1994. Principal component analysis of time–intensity data. Food Quality and Preference, 5, 121–127.

DuBois, G. E. and Lee, J. F. 1983. A simple technique for the evaluation of temporal taste properties. Chemical Senses, 7, 237–247.

Dubose, C. N., Meiselman, H. L., Hunt, D. A. and Waterman, D. 1977. Incomplete taste adaptation to different concentrations of salt and sugar solutions. Perception and Psychophysics, 21, 183–186.

Duizer, L. M., Findlay, C. J. and Bloom, K. 1995. Dual-attribute time–intensity sensory evaluation: A new method for temporal measurement of sensory perceptions. Food Quality and Preference, 6, 121–126.

Duizer, L. M., Bloom, K. and Findlay, C. J. 1996. Dual attribute time–intensity measurements of sweetness and peppermint perception of chewing gum. Journal of Food Science, 61, 636–638.

Ebeler, S. E., Pangborn, R. M. and Jennings, W. G. 1988. Influence of dispersion medium on aroma intensity and headspace concentration of menthone and isoamyl acetate. Journal of Agricultural and Food Chemistry, 36, 791–796.

Eilers, P. H. C. and Dijksterhuis, G. B. 2004. A parametric model for time–intensity curves. Food Quality and Preference, 15, 239–245.

Fischer, U., Boulton, R. B. and Noble, A. C. 1994. Physiological factors contributing to the variability of sensory assessments: Relationship between salivary flow rate and temporal perception of gustatory stimuli. Food Quality and Preference, 5, 55–64.

Frank, R. A., Ducheny, K. and Mize, S. J. S. 1989. Strawberry odor, but not red color, enhances the sweetness of sucrose solutions. Chemical Senses, 14, 371–377.

Garrido, D., Calviño, A. and Hough, G. 2001. A parametric model to average time–intensity taste data. Food Quality and Preference, 12, 1–8.

Gent, J. F. 1979. An exponential model for adaptation in taste. Sensory Processes, 3, 303–316.

Gent, J. F. and McBurney, D. H. 1978. Time course of gustatory adaptation. Perception and Psychophysics, 23, 171–175.

Green, B. G. 1989. Capsaicin sensitization and desensitization on the tongue produced by brief exposures to a low concentration. Neuroscience Letters, 107, 173–178.

Green, B. G. and Lawless, H. T. 1991. The psychophysics of somatosensory chemoreception in the nose and mouth. In: L. M. B. T.V. Getchell and J. B. Snow (eds.), Smell and Taste in Health and Disease. Raven, New York, pp. 235–253.

Guinard, J.-X., Pangborn, R. M. and Shoemaker, C. F. 1985. Computerized procedure for time–intensity sensory measurements. Journal of Food Science, 50, 543–544, 546.

Guinard, J.-X., Pangborn, R. M. and Lewis, M. J. 1986. The time course of astringency in wine upon repeated ingestions, American Journal of Enology and Viticulture, 37, 184–189.

Gwartney, E. and Heymann, H. 1995. The temporal perception of menthol. Journal of Sensory Studies, 10, 393–400.

Haring, P. G. M. 1990. Flavour release: From product to perception. In: Y. Bessiere and A. F. Thomas (eds.), Flavour Science and Technology. Wiley, Chichester, pp. 351–354.

Halpern, B. P. 1991. More than meets the tongue: Temporal characteristics of taste intensity and quality. In: H. T. Lawless and B. P. Klein (eds.), Sensory Science Theory and Applications in Foods. Marcel Dekker, New York, pp. 37–105.

Holway, A. H. and Hurvich, L. M. 1937. Differential gustatory sensitivity to salt. American Journal of Psychology, 49, 37–48.

Janusz, J. M., Young, P. A., Hiler, G. D., Moese, S. A. and Bunger, J. R. 1991. Time–intensity profiles of dipeptide sweeteners. In: D. E.Walters, F. T. Orthoefer and G. E.

DuBois (eds.), Sweeteners: Discovery, Molecular Design and Chemoreception. ACS Symposium Series #450. American Chemical Society, Washington, DC, pp. 277–289.

Jellinek, G. 1964. Introduction to and critical review of modern methods of sensory analysis (odor taste and flavor evaluation) with special emphasis on descriptive analysis. Journal of Nutrition and Dietetics, 1, 219–260.

Jellinek, G. 1985. Sensory Evaluation of Food, Theory and Practice. Ellis Horwood, Chichester, England.

Kroeze, J. H. A. 1979. Masking and adaptation of sugar sweetness intensity. Physiology and Behavior, 22, 347–351.

Kuo, Y.-L., Pangborn, R. M. and Noble, A. C. 1993. Temporal patterns of nasal, oral and retronasal perception of citral and vanillin and interactions of these odorants with selected tastants. International Journal of Food Science and Technology, 28, 127–137.

Labbe, D., Schlich, P., Pineau, N., Gilbert, F. and Martin, N. 2009. Temporal dominance of sensations and sensory profiling: A comparative study. Food Quality and Preference, 20, 216–221.

Lallemand, M., Giboreau, A., Rytz, A. and Colas, B. 1999. Extracting parameters from time–intensity curves using a trapezoid model: The example of some sensory attributes of ice cream. Journal of Sensory Studies, 14, 387–399.

Larson-Powers, N. and Pangborn, R. M. 1978. Paired comparison and time–intensity measurements of the sensory properties of beverages and gelatins containing sucrose or synthetic sweeteners. Journal of Food Science, 43, 41–46.

Lawless, H. T. 1980. A computerized system for assessing taste intensity over time. Paper presented at the Chemical Senses and Intake Society, Hartford, CT, April 9, 1980.

Lawless, H. T. 1984. Oral chemical irritation: Psychophysical properties. Chemical Senses, 9, 143–155.

Lawless, H. T. and Clark, C. C. 1992. Psychological biases in time–intensity scaling. Food Technology, 46(11), 81, 84–86, 90.

Lawless, H. T. and Skinner, E. Z. 1979. The duration and perceived intensity of sucrose taste. Perception and Psychophysics, 25, 249–258.

Lawless, H. T. and Stevens, D. A. 1988. Responses by humans to oral chemical irritants as a function of locus of stimulation. Perception and Psychophysics, 43, 72–78.

Lawless, H. T., Corrigan, C. L. and Lee, C. L. 1994. Interactions of astringent substances. Chemical Senses, 19, 141–154.

Lawless, H. T., Tuorila, H., Jouppila, K., Virtanen, P. and Horne, J. 1996. Effects of guar gum and microcrystalline cellulose on sensory and thermal properties of a high fat model food system. Journal of Texture Studies 27, 493–516.

Le Reverend, F. M., Hidrio, C., Fernandes, A. and Aubry, V. 2008. Comparison between temporal dominance of sensation and time intensity results. Food Quality and Preference, 19, 174–178.

Leach, E. J. and Noble, A. C. 1986. Comparison of bitterness of caffeine and quinine by a time–intensity procedure. Chemical Senses, 11, 339–345.

Ledauphin, S., Vigneau, E. and Causeur, D. 2005. Functional approach for the analysis of time intensity curves using B-splines. Journal of Sensory Studies, 20, 285–300.

Ledauphin, S., Vigneau, E. and Qannari, E. M. 2006. A procedure for analysis of time intensity curves. Food Quality and Preference, 17, 290–295.

Lee, C. B. and Lawless, H. T. 1991. Time-course of astringent materials. Chemical Senses, 16, 225–238.

Lee, W. E. 1985. Evaluation of time–intensity sensory responses using a personal computer. Journal of Food Science, 50, 1750–1751.

Lee, W. E. 1986. A suggested instrumental technique for studying dynamic flavor release from food products. Journal of Food Science, 51, 249–250.

Lee, W. E. 1989. Single-point vs. time–intensity sensory measurements: An informational entropy analysis. Journal of Sensory Studies, 4, 19–30.

Lee, W. E. and Pangborn, R. M. 1986. Time–intensity: The temporal aspects of sensory perception. Food Technology, 40, 71–78, 82.

Liu, Y. H. and MacFie, H. J. H. 1990. Methods for averaging time–intensity curves. Chemical Senses, 15, 471–484.

Lundahl, D. S. 1992. Comparing time–intensity to category scales in sensory evaluation. Food Technology, 46(11), 98–103.

Lynch, J., Liu, Y.-H., Mela, D. J. and MacFie, H. J. H. 1993. A time–intensity study of the effect of oil mouthcoatings on taste perception. Chemical Senses, 18, 121–129.

Matysiak, N. L. and Noble, A. C. 1991. Comparison of temporal perception of fruitiness in model systems sweetened with aspartame, aspartame + acesulfame K blend or sucrose. Journal of Food Science, 65, 823–826.

McBurney, D. H. 1966. Magnitude estimation of the taste of sodium chloride after adaptation to sodium chloride. Journal of Experimental Psychology, 72, 869–873.

McBurney, D. H. and Shick, T. R. 1971. Taste and water taste of 26 compounds for man. Perception and Psychophysics, 11, 228–232.

McGowan, B. A. and Lee, S.-Y. 2006. Comparison of methods to analyze time–intensity curves in a corn zein chewing gum study. Food Quality and Preference 17, 296–306.

McNulty, P. B. 1987. Flavour release—elusive and dynamic. In: J. M. V. Blanshard and P. Lillford (eds.), Food Structure and Behavior. Academic, London, pp. 245–258.

McNulty, P. B. and Moskowitz, H. R. 1974. Intensity -time curves for flavored oil-in-water emulsions. Journal of Food Science, 39, 55–57.

Meiselman, H. L. 1968. Magnitude estimation of the time course of gustatory adaptation. Perception and Psychophysics, 4, 193–196.

Meiselman, H. L. and Dubose, C. N. 1976. Failure of instructional set to affect completeness of taste adaptation. Perception and Psychophysics, 19, 226–230.

Meiselman, H. L. and Halpern, B. P. 1973. Enhancement of taste intensity through pulsatile stimulation. Physiology and Behavior, 11, 713–716.

Moore, L. J. and Shoemaker, C. F. 1981. Sensory textural properties of stabilized ice cream. Journal of Food Science, 46, 399–402, 409.

Neilson, A. J. 1957. Time–intensity studies. Drug and Cosmetic Industry, 80, 452–453, 534.

O'Keefe, S. F., Resurreccion, A. P., Wilson, L. A. and Murphy, P. A. 1991. Temperature effect on binding of volatile flavor compounds to soy protein in aqueous model systems. Journal of Food Science, 56, 802–806.

O'Mahony, M. 1986. Sensory adaptation. Journal of Sensory Studies, 1, 237–257.

O'Mahony, M. and Wong, S.-Y. 1989. Time–intensity scaling with judges trained to use a calibrated scale: Adaptation, salty and umami tastes. Journal of Sensory Studies, 3, 217–236.

Ott, D. B., Edwards, C. L. and Palmer, S. J. 1991. Perceived taste intensity and duration of nutritive and non-nutritive sweeteners in water using time–intensity (T–I) evaluations. Journal of Food Science, 56, 535–542.

Overbosch, P. 1987. Flavour release and perception. In: M. Martens, G. A. Dalen and H. Russwurm (eds.), Flavour Science and Technology. Wiley, New York, pp. 291–300.

Overbosch, P., Van den Enden, J. C., and Keur, B. M. 1986. An improved method for measuring perceived intensity/time relationships in human taste and smell. Chemical Senses, 11, 315–338.

Owen, W. J. and DeRouen, T. A. 1980. Estimation of the mean for lognormal data containing zeroes and left-censored values, with application to the measurement of worker exposure to air contaminants. Biometrics, 36, 707–719.

Pangborn, R. M. and Koyasako, A. 1981. Time-course of viscosity, sweetness and flavor in chocolate desserts. Journal of Texture Studies, 12, 141–150.

Pangborn, R. M., Lewis, M. J. and Yamashita, J. F. 1983. Comparison of time–intensity with category scaling of bitterness of iso-alpha-acids in model systems and in beer. Journal of the Institute of Brewing, 89, 349–355.

Peyvieux, C. and Dijksterhuis, G. 2001. Training a sensory panel for TI: A case study. Food Quality and Preference, 12, 19–28.

Pineau, N., Schlich, P., Cordelle, S., Mathonniere, C., Issanchou, S., Imbert, A., Rogeaux, M., Eteviant, P. and Köster, E. 2009. Temporal dominance of sensations: Construction of the TDS curves and comparison with time–intensity. Food Quality and Preference, 20, 450–455.

Pionnier, E., Nicklaus, S., Chabanet, C., Mioche, L., Taylor, A. J., LeQuere, J. L. and Salles, C. 2004. Flavor perception of a model cheese: relationships with oral and physico-chemical parameters. Food Quality and Preference, 15, 843–852.

Prescott, J. and Stevenson, R. J. 1996. Psychophysical responses to single and multiple presentations of the oral irritant zingerone: Relationship to frequency of chili consumption. Physiology and Behavior, 60–617–624.

Reinbach, H. C., Toft, M. and Møller, P. 2009. Relationship between oral burn and temperature in chili spiced pork patties evaluated by time–intensity. Food Quality and Preference, 20, 42–49.

Reinbach, H. C., Meinert, L., Ballabio, D., Aayslyng, M. D., Bredie, W. L. P., Olsen, K. and Møller, P. 2007. Interactions between oral burn, meat flavor and texture in chili spiced pork patties evaluated by time–intensity. Food Quality and Preference, 18, 909–919.

Rine, S. D. 1987. Computerized analysis of the sensory properties of peanut butter. M. S. Thesis, University of California, Davis, USA.

Roberts, D. D. and Acree, T. E. 1996. Simulation of retronasal aroma using a modified headspace technique: Investigating the effects of saliva, temperature, shearing, and oil on flavor release. Journal of Agricultural and Food Chemistry, 43, 2179–2186.

Roberts, D. D., Elmore, J. S., Langley, K. R. and Bakker, J. 1996. Effects of sucrose, guar gum and carboxymethylcellulose on the release of volatile flavor compounds under dynamic conditions. Journal of Agricultural and Food Chemistry, 44, 13221–1326.

Robichaud, J. L. and Noble, A. C. 1990. Astringency and bitterness of selected phenolics in wine. Journal of the Science of Food and Agriculture, 53, 343–353.

Rosin, S. and Tuorila, H. 1992. Flavor potency of garlic, pepper and their combination in different dispersion media. Lebensmittel Wissenschaft und Technologie, 25, 139–142.

Rozin, P., Ebert, L. and Schull, J. 1982. Some like it hot: A temporal analysis of hedonic responses to chili pepper. Appetite, 3, 13–22.

Shamil, S., Wyeth, L. J. and Kilcast, D. 1992. Flavour release and perception in reduced-fat foods. Food Quality and Preference, 3, 51–60.

Sjostrom, L. B. 1954. The descriptive analysis of flavor. In: Food Acceptance Testing Methodology, Quartermaster Food and Container Institute, Chicago, pp. 4–20.

Stevens, D. A. and Lawless, H. T. 1986. Putting out the fire: Effects of tastants on oral chemical irritation. Perception and Psychophysics, 39, 346–350.

Swartz, M. 1980. Sensory screening of synthetic sweeteners using time–intensity evaluations. Journal of Food Science, 45, 577–581.

Taylor, D. E. and Pangborn, R. M. 1990. Temporal aspects of hedonic response. Journal of Sensory Studies, 4, 241–247.

Tuorila, H. and Vainio, L. 1993. Perceived saltiness of table spreads of varying fat compositions. Journal of Sensory Studies, 8, 115–120.

time–intensity van Buuren, S. 1992. Analyzing time–intensity responses in sensory evaluation. Food Technology, 46(2), 101–104.

Veldhuizen, M. G., Wuister, M. J. P. and Kroeze, J. H. A. 2006. Temporal aspects of hedonic and intensity responses. Food Quality and Preference, 17, 489–496.

Wendin, K., Janestad, H. and Hall, G. 2003. Modeling and analysis of dynamic sensory data. Food Quality and Preference, 14, 663–671.

Yoshida, M. 1986. A microcomputer (PC9801/MS mouse) system to record and analyze time–intensity curves of sweetness. Chemical Senses, 11, 105–118.

Zimoch, J. and Gullett, E. A. 1997. Temporal aspects of perception of juiciness and tenderness of beef. Food Quality and Preference, 8, 203–211.

Chapter 9

Context Effects and Biases in Sensory Judgment

Abstract   Human judgments about a sensation or a product are strongly influenced by items that surround the item of interest, either in space or in time. This chapter shows how judgments can change as a function of the context within which a product is evaluated. Various contextual effects and biases are described and categorized. Some solutions and courses of action to minimize these biases are presented.

By such general principles of action as these everything looked at, felt, smelt or heard comes to be located in a more or less definite position relatively to other collateral things either actually presented or only imagined as possibly there.

— James (1913, p. 342)

Contents

9.1 Introduction: The Relative Nature of Human Judgment

This chapter will discuss context effects and common biases that can affect sensory judgments. Context effects are conditions in which the judgment about a product, usually a scaled rating, will shift depending upon factors such as the other products that are evaluated in the same tasting session. A mediocre product evaluated in the context of some poor-quality items may seem very good in comparison. Biases refer to tendencies in judgment in which the response is influenced in some way to be an inaccurate reflection of the actual sensory experience. In magnitude estimation ratings, for example, people have a tendency to use numbers that are multiples of 2, 5, and 10, even though they can use any number or fraction they wish. At the

H.T. Lawless, H. Heymann, *Sensory Evaluation of Food*, Food Science Text Series,
DOI 10.1007/978-1-4419-6488-5_9, © Springer Science+Business Media, LLC 2010

end of the chapter, some solutions to these problems are offered, although a sensory scientist should realize that we can never totally eliminate these factors. In fact, they are of interest and deserve study on their own for what they can tell us about human sensory and cognitive processes.

An axiom of perceptual psychology has it that humans are very poor absolute measuring instruments but are very good at comparing things. For example, we may have difficulty estimating the exact sweetness level of our coffee, but we have little trouble in telling whether more sugar has been added to make it sweeter. The question arises, if people are prone to making comparisons, how can they give ratings when no comparison is requested or specified? For example, when asked to rate the perceived firmness of a food sample, how do they judge what is firm versus what is soft? Obviously, they must either choose a frame of reference for the range of firmness to be judged or be trained with explicit reference standards to understand what is high and low on the response scale. In other words, they must relate this sensory judgment to other products they have tried. For many items encountered in everyday life, we have established frames of reference based on our experiences. We have no trouble forming an image of a "large mouse running up the trunk of a small elephant" because we have established frames of reference for what constitutes the average mouse and the average elephant. In this case the judgment of large and small is context dependent. Some people would argue that all judgments are relative.

This dependence upon a frame of reference in making sensory judgments demonstrates the influence of contextual factors in biasing or changing how products are evaluated. We are always prone to see things against a background or previous experience and evaluate them accordingly. A 40° (Fahrenheit) day in Ithaca, New York, in January seems quite mild against the background of the northeastern American winter. However, the same 40°C temperature will feel quite cool on an evening in August in the same location. This principle of frame of reference is the source of many visual illusions, where the same physical stimulus causes very different perceptual impressions, due to the context within which it is embedded. Examples are shown in Fig. 9.1.

A simple demonstration of context is the visual afterimage effect that gave rise to Emmert's law (Boring, 1942). In 1881, Emmert formalized a

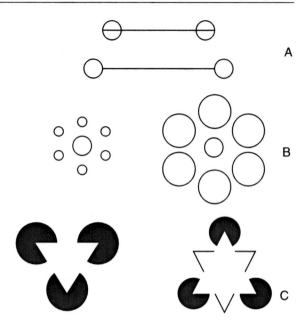

Fig. 9.1 Examples of contextual effects from simple visual illusions. (**a**) The dumbell version of the Muller–Lyer illusion. (**b**) The Ebbinghaus illusion. (**c**) Illusory contours. In this case the contexts induce the perceptions of shapes.

principle of size constancy based on the following effect: Stare for about 30 s at a brightly illuminated colored paper rectangle (it helps to have a small dot to aid in fixation in the center) about a meter away. Then shift your gaze to a white sheet on the table in front of you. You should see the rectangle afterimage in a complementary color and somewhat smaller in size as compared to the original colored rectangle. Next, shift your gaze to a white wall some distance off. The afterimage will now appear much larger, as the brain finds a fixed visual angle at greater distance to represent larger physical objects. Since the mind does not immediately recognize that the afterimage is just a creation of the visual sensory system, it projects it at the distance of the surface upon which it is "seen." The more distant frame of reference, then, demands a larger size perception.

The close link between sensory judgments and context presents problems for anyone who wants to view ratings as absolute or comparable across different times, sessions, or settings. Even when the actual sensory impression of two items is the same, we can shift the frame of reference and change the overt behavior of the person to produce a different response. This problem (or principle of sensory function) was glossed over by early psychophysical scientists. In psychological terms, they used a simple stimulus–response (S–R) model, in which response was considered a direct and unbiased representation of sensory experiences. Certain biases were observed, but it was felt that suitable experimental controls could minimize or eliminate them (Birnbaum, 1982; Poulton, 1989).

A more modern view is that there are two or three distinct processes contributing to ratings. The first is a psychophysical process by which stimulus energy is translated into physiological events that result in a subjective experience of some sensory intensity. The second, equally important process is the function by which the subjective experience is translated into the observed response, i.e., how the percept is translated onto the rating scale (see Fig. 9.2). Many psychophysical researchers now consider a "judgment function" to

be an important part of the sequence from stimulus to response (Anderson, 1974, Birnbaum, 1982; McBride and Anderson, 1990; Schifferstein and Frijters, 1992; Ward, 1987). This process is also sometimes referred to as a response output function. A third intermediate step is the conversion of the raw sensory experience into some kind of encoded percept, one that is available to memory for a short time, before the judgment is made (Fig. 9.2c).

Given this framework, there are several points at which stimulus context may influence the sensory process. First, of course, the actual sensation itself may change. Many sensory processes involve interaction effects of simultaneous or sequential influences of multiple items. An item may be perceived differently due to the direct influence of one stimulus upon another that is nearby in time or space. Simultaneous color contrast is an example in color vision and some types of inhibitory mixture interactions and masking in taste and smell are similarly hard wired. Quinine with added salt is less bitter than quinine tasted alone, due to the ways that sodium ions inhibit bitterness transduction. Sensory adaptation weakens the perception of a stimulus because of what has preceded it. So the psychophysical process itself is altered by the milieu in which the stimulus is observed, sometimes because of physical effects (e.g., simple buffering of an acid) or physiological effects (e.g., neural inhibition causing mixture suppression) in the peripheral sensory mechanisms. A second point of influence is when the context shifts the frame of reference for the response output function. That is, two sensations may have the same subjective intensity under two conditions, but because of the way the observer places them along the response continuum (due to different contexts), they are rated differently. A number of studies have shown that contextual factors such as the distribution of stimuli along the physical continuum affect primarily (although not exclusively) the response output function (Mellers and Birnbaum, 1982, 1983). A third process is sometimes added in which the sensation itself is translated into an implicit response or encoded image that may also be affected by context (Fig. 9.2c). This would provide another opportunity to influence the process if contextual factors affect this encoding step.

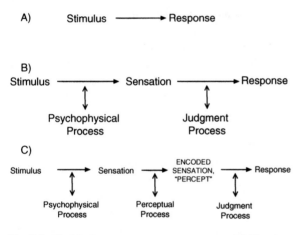

Fig. 9.2 Models for sensory-response processes. (**a**) The simple stimulus–response model of twentieth-century behavioral psychology. (**b**) Two processes are involved in sensation and response, a psychophysical process and then a response output or a judgment process in which the participant decides what response to give for that sensation. (**c**) A more complex model in which the sensation may be transformed before the response is generated. It may exist in short-term memory as an encoded percept, different from the sensation. Contextual effects of simultaneous or sequential stimuli can influence the stimulus–response sequence in several ways. Peripheral physiological effects such as adaptation or mixture inhibition may change the transduction process or other early stages of neural processing. Other stimuli may give rise to separate percepts that are integrated into the final response. Contextual factors may also influence the frame of reference that determines how the response output function will be applied. In some models, an additional step allows transformation of the percept into covert responses that are then translated as a separate step into the overt response R.

Contextual change can be viewed as a form of bias. Bias, in this sense, is a process that causes a shift or a change in response to a constant sensation. If one situation is viewed as producing a true and accurate response, then the contextual conditions that cause shifts away from this accurate response are "biased." However, bias need only have a negative connotation if there is a reason to presume that one condition of judgment is more accurate than all others. A broader view is to accept the idea that all judgments are a function of observing conditions and therefore all judgments are biased from one another in different ways. Fortunately, many of these biases and the conditions that cause them are predictable and well understood, and so they can be eliminated or minimized. At the very least, the sensory practitioner needs to understand how these influences operate so as to know when to expect changes in judgments and ratings. An important endpoint is the realization that few, if any, ratings have any absolute meaning. You cannot say that because a product received a hedonic rating of 7.0 today, it is better than the product that received a rating of 6.5 last week. The context may have changed.

9.2 Simple Contrast Effects

By far the most common effect of sensory context is simple contrast. Any stimulus will be judged as more intense in the presence of a weaker stimulus and as less intense in the presence of a stronger stimulus, all other conditions being equal. This effect is much easier to find and to demonstrate than its opposite, convergence or assimilation. For example, an early sensory worker at the Quartermaster Corp., Kamenetsky (1957), noticed that the acceptability ratings for foods seemed to depend upon what other foods were presented during an evaluation session. Poor foods seemed even worse when preceded by a good sample. Convergence is more difficult to demonstrate, although under some conditions a group of items may seem more similar to each other when they are in the presence of an item that is very different from that group (Zellner et al., 2006).

9.2.1 A Little Theory: Adaptation Level

As we noted above, a 40° day in January (in New York) seems a lot warmer than the same temperature in August. These kinds of effects are predicted Helson's theory of adaptation level. Helson (1964) proposed that we take as a frame of reference the average level of stimulation that has preceded the item to be evaluated. The mild temperature in the middle of a hot and humid summer seems a lot more cool and refreshing than is the mild temperature after a period of cold and icy weather. So we refer to our most recent experiences in evaluating the sensory properties of an item. Helson went on to elaborate the theory to include both immediate and distant predecessors. That is, he appreciated the fact that more recent items tend to have a stronger effect on the adaptation level. Of course, mere reference to the mean value of experience is not always sufficient to induce a contrast effect—it is more influential if the mean value comes to be centered near the middle of the response scale, an example of a centering bias, discussed below (Poulton, 1989).

The notion of adaptation, a decrease in responsiveness under conditions of constant stimulation, is a major theme in the literature on sensory processes. Physiological adaptation or an adjustment to the ambient level of stimulation is obvious in light/dark adaptation in vision. The thermal and tactile senses also show profound adaptation effects—we become easily adjusted to the ambient room temperature (as long as it is not too extreme) and we become unaware of tactile stimulation from our clothing. So this mean reference level often passes from consciousness or becomes a new baseline from which deviations in the environment become noticeable. Some workers have even suggested that this improves discrimination—that the difference threshold is smallest right around the adaptation level or physiological zero, in keeping with Weber's law (McBurney, 1966). Examples of adaptation effects are discussed in Chapter 2 for the senses of taste and smell. In the chemical, thermal, and tactile senses, adaptation is quite profound.

However, we need not invoke the concept of neural adaptation to a preceding item or a physiological effect to explain all contrast effects. It may be simply that more or less extreme stimuli change our frame of reference or the way in which the stimulus range and response scales are to be mapped onto one another. The general principle of context is that human observers act like measuring instruments that constantly recalibrate themselves to the experienced frame of reference. What we think of as a small horse may depend upon whether the frame of reference includes Clydesdales, Shetland ponies, or tiny prehistoric equine species. The following examples show simple effects of context on intensity, sensory quality, and hedonics or acceptability. Most of these examples are cases of perceptual contrast or a shift in judgment *away from other stimuli* presented in the same session.

9.2.2 Intensity Shifts

Figure 9.3 shows a simple contrast effect of soups with varying salt levels presented in different contexts (Lawless, 1983). The central stimulus in the series was presented either with two lower or two higher concentrations of salt added to a low sodium soup. Ratings of saltiness intensity were made on a simple nine-point category scale. In the lower context, the central soup received a higher rating, and in the higher context, it received a lower rating, analogous to our perception of a mild day in winter (seemingly warmer) versus a mild day in summer (seemingly cooler). Note that the shift is quite dramatic, about two points on the nine-point scale or close to 25% of scale range.

A simple classroom demonstration can show a similar shift for the tactile roughness of sandpapers varying in grit size. In the context of a rougher sample, a medium sample will be rated lower than it is in the context of a smoother sample. The effects of simple contrast are not limited to taste and smell.

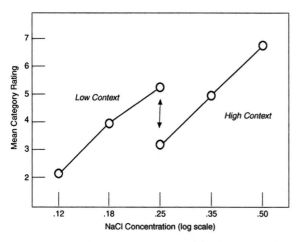

Fig. 9.3 Saltiness ratings of soups with added NaCl. The sample at 0.25 M was evaluated in two contexts, one with higher concentrations and one with lower concentrations. The shift is typical of a simple contrast effect of contrast. Replotted from Lawless (1983). Copyright ASTM, used with permission.

Contrast effects are not always observed. In some psychophysical work with long series of stimuli, some item-to-item correlations have been observed. The effects of immediately preceding versus remotely preceding stimuli have been measured and a positive correlation among adjacent responses in the series was found. This can be taken as evidence for a type of assimilation, or underestimation of differences (Ward, 1979, 1987; but see also Schifferstein and Frijters, 1992).

9.2.3 Quality Shifts

Visual examples such as color contrast were well known to early psychologists like William James: "Meanwhile it is an undoubted general fact that the psychical effect of incoming currents does depend on what other currents may be simultaneously pouring in. Not only the perceptibility of the object which the current brings before the mind, but the quality of it is changed by the other currents." (1913, p. 25). A gray line against a yellow background may appear somewhat bluish, and the same line against a blue background may seem more yellowish. Paintings of the renowned artist Josef Albers made excellent use of color contrast. Similar effects can be observed for the chemical senses. During a descriptive panel training period for fragrance evaluation, the terpene aroma compound dihydromyrcenol was presented among a set of woody or pine-like reference materials. The panelists complained that the aroma was too citrus-like to be included among the woody reference materials. However, when the same odor was placed in the context of citrus reference materials, the same panelists claimed that it was far too woody and pine-like to be included among the citrus examples. This contextual shift is shown in Fig. 9.4. In a citrus context, the item is rated as more woody in character than when placed in a woody context. Conversely, ratings for citrus intensity decrease in the citrus context and increase in the woody context. The effect is quite robust and is seen whether or not a rest period is included to undo the potential effects of sensory adaptation. It even occurs when the contextual odor follows the target item and judgments are made after both are experienced (Lawless et al., 1991)! This striking effect is discussed further in Section 9.2.5.

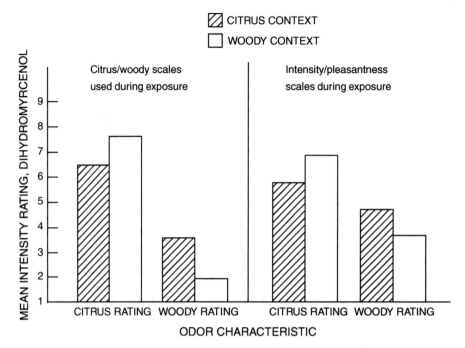

Fig. 9.4 Odor quality contrast noted for the ambiguous terpene aroma compound dihydromyrcenol. In a citrus context, woody ratings increase and citrus character decreases. In a woody context, the woody ratings decrease. The group using different scales did not rate citrus and woody character during the contextual exposure phase, only overall intensity and pleasantness were rated. From Lawless et al. (1991) by permission of Oxford University Press.

Context effects can also alter how items are identified and characterized. When people categorize speech sounds, repeated exposure to one type of simple phoneme changes the category boundary for other speech sounds. Repeated exposure to the sound of the phoneme "bah," which has an early voice onset time, can shift the phoneme boundary so that speech sounds near the boundary are more likely classified as "pah" sounds (a later voice onset) (Eimas and Corbit, 1973). Boundary-level examples are shifted across the boundary and into the next category. This shift resembles a kind of contrast effect.

9.2.4 Hedonic Shifts

Changes in the preference or acceptance of foods can be seen as a function of context. Hedonic contrast was a well-known effect to early workers in food acceptance testing (Hanson et al., 1955; Kamenetzky, 1959). An item seems more appealing if it followed an item of poor quality and less appealing if it followed something of better quality. The effect was known to Beebe-Center (1932), who also attributed it to Fechner in 1898. This kind of contrast has been observed for tastes (Riskey et al., 1979; Schifferstein, 1995), odors (Sandusky and Parducci, 1965), and art (Dolese et al., 2005). Another effect observed in these kinds of experiments is that a contrasting item causes other, generally lower rated stimuli to become more similar or less discriminable, an effect termed condensation (Parker et al., 2002; Zellner et al., 2006;). In the study by Zellner et al. (2006), pre-exposure to a good-tasting juice reduced the magnitude of preference ratings among less appealing juices. Mediocre items were both worse and more similar.

An example of hedonic shifting was found in a study on the optimization of the saltiness of tomato juice and also the sweetness of a fruit beverage using the method of adjustment (Mattes and Lawless, 1985). When trying to optimize the level of sweetness or saltiness in this study, subjects worked in two directions. In an ascending series, they would concentrate a dilute solution by mixing the beverage with a more concentrated version having the same color, aroma, and other flavor materials (i.e., only sweetness or saltiness was different). In a second descending series, they would be given a very intense sample as the starting point and then dilute down to their preferred level. This effect is shown in Fig. 9.5. The adjustment stops too soon and the discrepancy is remarkable, nearly a concentration range of 2:1. The effect was also robust—it could not be attributed to sensory adaptation or lack of discrimination and persisted even when subjects were financially motivated to try to achieve the same endpoints in both trials. This is a case of affective contrast. When compared to a very sweet or salty starting point, a somewhat lower item seems just about right, but when starting with a relatively sour fruit beverage or bland tomato juice, just a little bit of sugar or salt helps quite a bit. The stopping point contrasts with the starting material and seems to be better than it would be perceived in isolation. In an ascending or descending sequence of products, a change in responses that happens too soon is called an "error of anticipation."

9.2.5 Explanations for Contrast

At first glance, one is tempted to seek a physiological explanation for contrast effects, rather than a psychological or a judgmental one. Certainly sensory adaptation to a series of intense stimuli would cause any subsequent test item to be rated much lower. The importance of sensory adaptation in the chemical senses of taste and smell lends some credence to this explanation. However, a number of studies have shown that precautions against sensory adaptation may be taken, such as sufficient rinsing or time delays between stimuli and yet the context effects persist (Lawless et al., 1991; Mattes and Lawless, 1985; Riskey, 1982). Furthermore, it is difficult to see how sensory adaptation to low-intensity stimuli would cause an increase in the ratings for a stronger item, as adaptation necessarily causes a decrement in physiological responsiveness compared to a no-stimulation baseline.

Perhaps the best evidence against a simple adaptation explanation for contrast effects is from the *reversed-pair* experiments in which the contextual item follows the to-be-rated target item and therefore can have no physiologically adapting effect on it. This paradigm calls for a judgment of the target item from memory after the presentation of the contextual item, in what has been termed a reversed-pair procedure (Diehl et al., 1978). Due to the reversed order, the context effects cannot be blamed on physiological adaptation of receptors, since the contextual item follows rather than precedes the item to be rated. Reversed-pair effects are seen for shifts in odor quality of aroma compounds like dihydromyrcenol and are only slightly smaller in magnitude than the contextual shift caused when the contextual item comes first (Lawless et al., 1991). The reversed-pair situation is also quite capable of causing simple contrast effects in sensory intensity. A sweetness shift was observed when a higher or a lower sweetness item was interpolated between the tasting and rating (from memory)

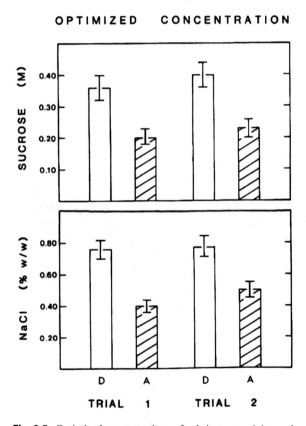

Fig. 9.5 Optimized concentrations of salt in tomato juice and sucrose in a fruit beverage. In the trials labeled D, the concentration was diluted from a concentrated version of the test sample. In the trials marked A, the concentration was increased from a dilute version of the test sample. Concentrations of other ingredients were held constant. The contextual shift is consistent with reaching the apparent optimum too soon as if the apparent optimum was shifted in contrast to the starting point. From Mattes and Lawless (1985) with permission.

of a normal-strength fruit beverage (Lawless, 1994). Looking back at Fig. 9.2, it seems more likely that the effect changes the response function. However, not all workers in the field agree. In particular, Marks (1994) has argued that the contextual shifts are much like an adaptation process and that for auditory stimuli this is a peripheral event. It is possible that what changes is not the sensation/experience, but to some encoded version of the sensation, or to some kind of implicit response, not yet verbalized. If a person, when rating, is evaluating some memory trace of the experience, it is possible that this memory, for example, could be altered.

9.3 Range and Frequency Effects

Two of the most common factors that can affect ratings are the sensory range of the products to be evaluated and the frequency with which people use the available response options. These factors were nicely integrated into a theory that helped to explain shifts in category ratings. They are also general tendencies that can affect just about any ratings or responses.

9.3.1 A Little More Theory: Parducci's Range and Frequency Principles

Parducci (1965, 1974) sought to go beyond Helson's (1964) simple idea that people respond to the mean or the average of their sensory experiences in determining the frame of reference for judgment. Instead, they asserted that the entire distribution of items in a psychophysical experiment would influence the judgments of a particular stimulus. If this distribution was denser (bunched up) at the low ends and a lot of weak items were presented, product ratings would shift up. Parducci (1965, 1974) proposed that behavior in a rating task was a compromise between two principles. The first was the range principle. Subjects use the categories to sub-divide the available scale range and will tend to divide the scale into equal perceptual segments. The second was the frequency principle. Over many judgments, people like to use the categories an equal number of times (Parducci, 1974). Thus it is not only

the average level that is important but also how stimuli may be grouped or spaced along the continuum that would determine how the response scale was used. Category scaling behavior could be predicted as a compromise between the effects of the range and frequency principles (Parducci and Perrett, 1971).

9.3.2 Range Effects

The range effect has been known for some time, both in category ratings and other judgments including ratio scaling (Engen and Levy, 1958; Teghtsoonian and Teghtsoonian, 1978). When expanding or shrinking the overall range of products, subjects will map their experiences onto the available categories (Poulton, 1989). Thus short ranges produce steep psychophysical functions and wide ranges produce flatter functions. An example of this can be seen in two published experiments on rating scales (Lawless and Malone, 1986a, b). In these studies, four types of response scales and a number of visual, tactile, and olfactory continua were used to compare the abilities of consumers to use the different scales to differentiate products. In the first study, the consumers had no trouble in differentiating the products and so in the second study, the stimuli were spaced more closely on the physical continua so that the task would be more challenging. However, when the experimenters closed the stimuli in, the range principle took over, and participants used more of the rating scale than expected. In Fig. 9.6, ratings for perceived thickness of (stirred) silicone samples are shown in the wide and narrow stimulus ranges. Note the steepening of the response function. For the same one log unit change in physical viscosity, the range of responses actually doubled from the wide range to the narrower range.

Another kind of stimulus range effect occurs with anchor stimuli. Sarris (1967) previously showed a strong effect of anchor stimuli on the use of rating scales, unless the anchors were very extreme, at which point their influence would tend to diminish, as if they had become irrelevant to the judgmental frame of reference. Sarris and Parducci (1978) found similar effects of both single and multiple end anchors that generally take the form of a contrast effect. For example, a low anchor stimulus, whether rated or unrated, will cause stronger stimuli to receive higher ratings than

Fig. 9.6 A simple range effect. When products are presented over a wide range, a shallow psychophysical function is found. Over a narrow range, a steeper psychophysical function will be observed. This is in part due to the tendency of subjects to map the products (once known) onto the available scale range. From Lawless and Malone (1986b), with permission.

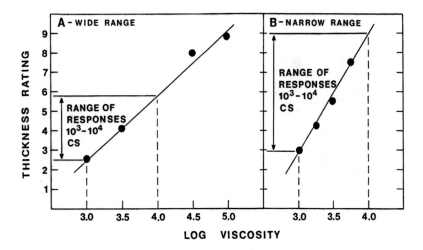

9.3.3 Frequency Effects

The frequency effect is the tendency of people to try to use the available response options about the same number of times across a series of products or stimuli to be rated. The frequency effect can cause shifts that look like simple contrast and also a local steepening of the psychophysical function around points where stimuli were closely spaced or very numerous (compared with less "dense" portions of the stimulus range). The frequency principle dictates that when judging many samples, products that are numerous or bunched at the low or high ends of the distributions tend to be spread out into neighboring categories. This is illustrated in the two panels of Fig. 9.7. The upper panel shows four hypothetical experiments and how products might be bunched in different parts of the range. In the upper left panel, we see how a normal replicated psychophysical experiment would be conducted with

equal presentations of each stimulus level. The common outcome of such a study using category ratings would be a simple linear function of the log of stimulus intensity. However, if the stimulus presentations were more frequent at the high end of the distribution, i.e., negative skew, the upper categories would be overused, and subjects would begin to distribute their judgments into lower categories. If the samples were bunched at the lower end, the lower response categories would be overused and subjects would begin to move into higher categories. If the stimuli were bunched in the midrange, the adjacent categories would be used to take on some of the middle stimuli, pushing extreme stimuli into the ends of the response range, as shown in the panel for a quasi-normal distribution.

Such behavior is relevant to applied testing situations. For example, in rating the presence of off-flavors or taints, there may be very few examples of items with high values on the scale and lots of weak (or zero) sensations. The frequency effect may explain why the low end of the scale is used less often than anticipated, and higher mean values are obtained than one would deem appropriate. Another example is screening a number of flavor or fragrance candidates for a new product. A large number of good candidates are sent for testing by suppliers or a flavor development group. Presumably these have been pre-tested or at least have received a favorable opinion from a flavorist or a perfumer. Why do they then get only mediocre ratings from the test panel? The high end of the distribution is over-represented (justifiably so and perhaps on purpose), so the tendency for the panel is to drop into lower categories. This may partly explain why in-house testing

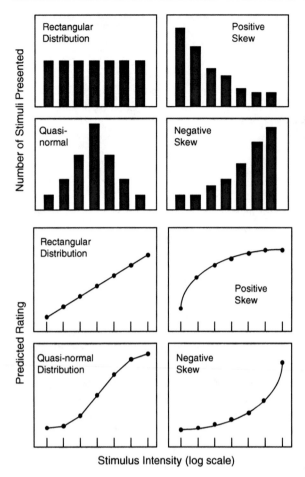

Fig. 9.7 Predictions from the Parducci range–frequency theory. Distributions of stimuli that are concentrated at one part of the perceptual range (*upper quartet*) will show local steepening of the psychophysical functions (*lower quartet*). This is due to subjects' tendencies to use categories with equal frequency, the resulting shifting into adjacent categories from those that are overused.

panels are sometimes more critical or negative than consumers when evaluating the same items.

Although the great majority of experiments on the range and frequency effects have been performed with simple visual stimuli, there are also examples from taste evaluation (Lee et al., 2001; Riskey et al., 1979; Riskey, 1982; Schifferstein and Frijters, 1992; Vollmecke, 1987). Schifferstein and Frijters found similar effects of skewed distributions with line-marking responses as seen in previous studies with category ratings. Perhaps line marking is not a response scale with infinite divisions, but panelists sub-divide the line into discrete sub-ranges as if they were using a limited

number of zones or categories. The effect of grouping or spacing products also intensifies as the exposure to the distributions increases. Lawless (1983) showed that the shift that occurred with a negative skew (bunching at the upper end) into lower response categories would intensify as the exposure to the skewed distribution went from none to a single exposure to three exposures. Thus the contextual effects do not suddenly appear but will take hold of the subjects' behavior as they gain experience with the sample set.

9.4 Biases

9.4.1 Idiosyncratic Scale Usage and Number Bias

People appear to have preferred ranges or numbers on the response scale that they feel comfortable using. Giovanni and Pangborn (1983) noted that people using magnitude estimation very often used numbers that were multiples of 2 and 5 (or obviously 10), an effect that is well known in the psychophysical literature (Baird and Noma, 1978). With magnitude estimation, the idiosyncratic usage of a favorite range of numbers causes a correlation of the power function exponents across different sensory continua for an individual (Jones and Marcus, 1961; Jones and Woskow, 1966). This correlation can be explained if people are more or less expansive (versus restrictive) in their response output functions, i.e., in how they apply numbers to their sensations in magnitude estimation studies. Another version of such personal idiosyncrasy is the common observation in time–intensity scaling that people show a kind of personal "signature" or a characteristic curve shape (Dijksterhuis, 1993; McGowan and Lee, 2006).

Another version of self-induced response restriction can be seen when people use only selected portions of the scale in a line-marking rating task. On a line scale with verbal labels, people may choose to make markings only near the verbal labels, rather than distributing them across the response scale. This was first observed by Eng (1948) with a simple hedonic line scale labeled Like Very Highly at one end, Dislike Very Highly at the other, and Neither Like nor Dislike at the center. In a group of 40 consumers, 24 used only the three labeled parts of the scale, and Eng deleted them from the data analysis! This kind of behavior was also noted with the labeled affective magnitude scale (LAM scale) by Cardello et al. (2008) with both Army laboratory and student groups. Lawless et al. (2010a) found a very high frequency (sometimes above 80%) of people making marks within ±2 mm of a phrase mark on the LAM scale in a multi-city consumer central location test. Lawless et al. (2010b) found that instructions did not seem to change this behavior much but that expanding the physical size of the scale on the ballot (from about 120 to 200 mm) decreased the "categorical" behavior somewhat. Categorical rating behavior can also be seen as a step function in time–intensity records (rather than a smooth continuous curve).

Finding product differences against the background of these individual response tendencies can be facilitated by within-subject experimental designs. Each participant is used as his or her own baseline in comparisons of products, as in dependent *t*-tests or repeated measure analysis of variance in complete block designs. Another approach is to compute a difference score for a comparison of products in each individual's data, rather than merely averaging across people and looking at differences between mean values.

9.4.2 Poulton's Classifications

Poulton (1989) published extensively on biases in ratings and classified them. Biases in Poulton's system go beyond Parducci's theory but are documented in the psychophysical literature. These include centering biases, contraction biases, logarithmic response bias with numerical ratings, and a general transfer bias that is seen when subjects carry the context from a previous session or study into a new experiment. The centering bias is especially relevant to just-right scales and is discussed in a later section. The response range bias is also a special case and follows this section.

The contraction biases are all forms of assimilation, the opposite of contrast. According to Poulton, people may rate a stimulus relative to a reference or a mean value that they hold in memory for similar types of sensory events. They tend to judge new items as being close (perhaps too close) to this reference value, causing underestimation of high values and overestimation of low values. There may also be overestimation of an item when it follows a stronger standard stimulus or underestimation when it follows a weaker standard stimulus, a sort of local contraction effect. Poulton also classifies the tendency to gravitate toward the middle of the response range as a type of contraction effect, called a response contraction bias. While all of these effects undoubtedly occur, the question arises as to whether contrast or assimilation is a more common and potent process in human sensory judgment. While some evidence for response assimilation has been found in psychophysical experiments through sequential analysis of response correlations (Ward, 1979), contrast seems much more to be the rule with taste stimuli (Schifferstein and Frijters, 1992) and foods (Kamenetzky, 1959). In our experience, assimilation effects are not as prevalent as contrast effects, although assimilation has certainly been observed in experiments on consumer expectation (e.g., Cardello and Sawyer, 1992). In that case, the assimilation is not toward other actual stimuli but toward expected levels.

The logarithmic response bias can be observed with open-ended response scales that use numbers, such as magnitude estimation. There are several ways to view this type of bias. Suppose that a series of stimuli have been arranged in increasing magnitude and they are spaced in subjectively equal steps. As the intensity increases, subjects change their strategy as they cross into ranges of numerical responses where there are more digits. For example, they might be rating the series using numbers like 2, 4, 6, and 8, but then when they get to 10, they will continue by larger steps, perhaps 20, 40, 60, 80. In Poulton's view they proceed through the larger numerical responses "too rapidly." A converse way of looking at this problem is that the perceived magnitude of the higher numbers is in smaller arithmetic steps as numbers get larger. For example, the difference between one and two seems much larger compared to the difference between 91 and 92. Poulton also points out that in addition to contraction of stimulus magnitude at very high levels, the converse is also operating and that people seem to illogically expand their subjective number range when using responses smaller than the number 3. One obvious way to avoid the problems in number bias is to avoid numbers altogether or to substitute line scaling or cross-modality matching to line length as a response instead of numerical rating techniques like magnitude estimation (Poulton, 1989).

Transfer bias refers to the general tendency to use previous experimental situations and remembered judgments to calibrate oneself for later tasks. It may involve any of the biases in Poulton's or Parducci's theories. The situation is common when subjects are used in multiple experiments or when sensory panelists are used repeatedly in evaluations (Ward, 1987). People have memories and a desire to be internally consistent. Thus the ratings given to a product on one occasion may be influenced by ratings given to similar products on previous occasions. There are two ways to view this tendency. One is that the judgments may not shift appropriately when the panelists' sensory experience, perception, or opinion of the product has in fact changed. On the other hand, one of the primary functions of panelist training and calibration in descriptive analysis is to build up exactly those sorts of memory references that may stabilize sensory judgments. So there is a positive light to this tendency as well. An open question for sensory evaluation is whether exposure to one continuum of sensory intensities or one type of product will transfer contextual effects to another sensory attribute or a related set of products (Murphy, 1982; Parducci et al., 1976; Rankin and Marks, 1991). And if so, how far does the transfer extend?

9.4.3 Response Range Effects

One of Poulton's biases was called the "response range equalizing bias" in which the stimulus range is held constant but the response range changes and so do the ratings. Ratings expand or contract so that the entire range is used (minus any end-category avoidance). This is consistent with the "mapping" idea mentioned for stimulus range effects (stimuli are mapped onto the available response range). Range stabilization is implicit in the way some scaling studies have been set up and in the instructions given to subjects. This is similar to the use of physical reference standards in some descriptive analysis training (Muñoz and Civille, 1998) and is related to Sarris's work on anchor stimuli (Sarris and Parducci, 1978). In Anderson's work with 20-point category scales and line marking, high and low examples or end anchors are given to subjects to show them the likely range of the stimuli to be encountered. The range of responses is known since

it is visible upon the page of the response sheet or has been pre-familiarized in a practice session (Anderson, 1974). Thus it is not surprising that subjects distribute their responses across the range in a nicely graded fashion, giving the appearance that there is a reasonably linear use of the scale. Anderson noted that there are end effects that work against the use of the entire range (i.e., people tend to avoid using the endpoints) but that these can be avoided by indenting the response marks for the stimulus end anchors, for example, at points 4 and 16 on the 20-point category scale. This will provide psychological insulation against the end effects by providing a comfort zone for unexpected or extreme stimuli at the ends of the scale while leaving sufficient gradations and room to move within the interior points. The "comfort zone" idea is one reason why early workers in descriptive analysis used line scales with indented vertical marks under the anchor phrases.

An exception to the response range mapping rule is seen when anchor phrases or words on a scale are noted and taken seriously by participants. An example is in Green's work on the labeled magnitude scale, which showed a smaller response range when it was anchored to "greatest imaginable sensation" that included all oral sensations including pain, as opposed to a wider range when the greatest imaginable referred only to taste (Green et al., 1996). This also looks like an example of contrast in which the high-end anchor can evoke a kind of stimulus context, at least in the participant's mind. If the image evoked by the high-end phrase is very extreme, it acts like a kind of stimulus that compresses ratings into a smaller range of the scale. A similar kind of response compression was seen with the LAM scale when it was anchored to greatest imaginable liking for "sensations of any kind" as opposed to a more delimited frame such as "foods and beverages" (Cardello et al., 2008). A sensory scientist should consider how the high anchor phrase is interpreted, especially if he or she wants to avoid any compression of ratings along the response range. As Muñoz and Civille (1998) pointed out, the use of a descriptive analysis scale also depends a lot on the conceptualization of the high extreme. Does "extremely strong" refer to the strongest possible taste among all sensations and products, the strongest sensation in this product type, or just how strong this particular attribute can become in this particular product? The strongest sweetness in this product might be more intense than the strongest saltiness. The definition needs to be a deliberate choice of the panel leader and an explicit instruction to the panelist to give them a uniform frame of reference.

9.4.4 The Centering Bias

The centering bias arises when subjects become aware of the general level of stimulus intensity they are likely to encounter in an experiment and tend to match the center or midpoint of the stimulus range with the midpoint of the response scale. Poulton (1989) distinguished a stimulus centering bias from a response centering bias, but this distinction is primarily a function of how experiments are set up. In both cases, people tend to map the middle of the stimulus range onto the middle of the response range and otherwise ignore the anchoring implications of the verbal labels on the response scale. Note that the centering bias works against the notion that respondents can use unbalanced scales with any effectiveness. For example, the "Excellent–very good–good–fair–poor" scale commonly used in marketing research with consumers is unbalanced. The problem with unbalanced scales is that over many trials, the respondents will come to center their responses on the middle category, regardless of its verbal label.

The centering bias is an important problem when there is a need to interpolate some value on a psychophysical function or to find an optimal product in just-right scaling. Poulton gives the example of McBride's method for considering bias in the just-about-right (JAR) scale (McBride, 1982; see also Johnson and Vickers, 1987). In any series of products to be tested, say for just-right level of sweetness, there is a tendency to center the series so that the middle product will come out closest to the just-right point. The function shifts depending upon the range that is tested. One way to find the true just-right point would be to actually have the experimental series centered on that value, but then of course you would not need to do the experiment. McBride gives a method for interpolation across several experiments with different ranges. The point at which the just-right function and the median of the stimulus series will cross shows the unbiased or true just-right level. This method of interpolation is shown in Fig. 9.8. In this method, you

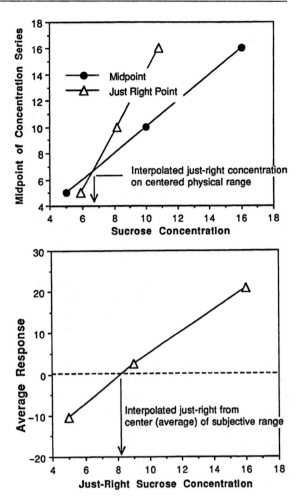

Fig. 9.8 Adjusting for the centering bias in just-right ratings. Three series of sucrose concentrations in lemonade were tested, a low series (2–8%), a middle series (6–14%), and a high range (10–22%). In the *upper panel*, the method of Poulton is used to interpolate the unbiased just-right point from the series where the midpoint concentration would correspond to the just-right item. In the *lower panel*, the method of McBride is used to interpolate the just-right point from a series in which the average response would correspond to the just-right point. When the average response would be just right (zero on this scale), the hypothetical stimulus range would have been centered on the just-right level. Replotted from Johnson and Vickers (1987), with permission.

present several ranges of the products in separate sessions and plot how the judgments of the JAR point shift up and down. You can then interpolate to find the range in which the just-right point would have been from the center product in the series. This obviously takes more work to do the test a couple of times, but it could avoid a mistaken estimate of the JAR level.

9.5 Response Correlation and Response Restriction

Early experimental psychologists like Thorndike (1920) noted that one very positive attribute of a person could influence judgments on other, seemingly unrelated characteristics of that individual. In personnel evaluations of military officers, Thorndike noted a moderate positive correlation among the individual rated factors. People evaluate others like this in real life. If achievement in sports is influential in our assessment of a person, we might suppose a gifted athlete to also be kind to children, generous to charities, etc., even though there is no logical relationship between these characteristics. People like to have cognitive structures that form consistent wholes and are without conflicts or contradictions (called cognitive dissonance) that can make us uncomfortable. The halo effect has also been described as a carry-over from one positive product to another (Amerine et al., 1965), but its common usage is in reference to a positive correlation of unrelated attributes (Clark and Lawless, 1994). Of course, there can also be negative or horns effects, in which one salient negative attribute causes other, unrelated attributes to be viewed or rated negatively. If a product makes a mess in the microwave, it might be rated negatively for flavor, appearance, and texture as well.

9.5.1 Response Correlation

A simple example of a halo effect is shown in Fig. 9.9. In this case, a small amount of vanilla extract was added to low-fat milk, near the threshold of perceptibility. Ratings were then collected from 19 milk consumers for sweetness, thickness, creaminess and liking for the spiked sample, and for a control milk. In spite of the lack of relationship between vanilla aroma and sweet taste and between vanilla and texture characteristics, the introduction of this one positive aspect was sufficient to cause apparent enhancement in sweetness, creaminess, and thickness ratings.

Apparent enhancement of sweetness is an effect long known for ethyl maltol, a caramelization product that has an odor similar to heated sugar (Bingham

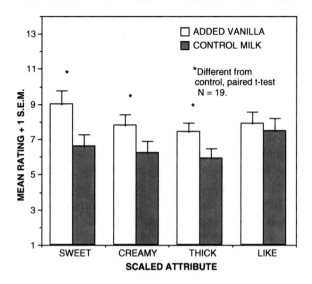

Fig. 9.9 Adding a just perceivable level of vanilla extract to low-fat milk causes increases in rated sweetness, thickness, creaminess, and liking, an example of the Halo effect. From Lawless and Clark (1994), with permission.

et al., 1990). When maltol is added to various products, sweetness ratings may rise compared to products lacking this flavor. However, the effect seems to be a case of the misattribution of olfactory stimulation to the taste sense. Murphy and Cain (1980) showed that citral (a lemon odor) could enhance taste ratings, but only when the nostrils were open, which allows diffusion of the odor into the nose and stimulation of the olfactory receptors (i.e., retronasal smell). When the nostrils are pinched shut, the diffusion is effectively eliminated and the enhancement disappears. Murphy and Cain interpreted this as convincing evidence that there was no true enhancement of taste intensity by citral, but only olfactory referral, a kind of confusion between taste and smell. Studies with other odors have also shown that the taste enhancement effect from volatile flavors can be eliminated by nose pinching (Frank and Byram, 1988) even for maltol (Bingham et al., 1990). The maltol effect is also minimized by training subjects who then learn to more effectively separate or localize their odor experiences from taste (Bingham et al., 1990). The sweetness enhancement may arise as a function of conditioning or experience with the pairing of sweet tastes with some odors in foods (Stevenson et al., 1995).

Several lessons can be learned from the vanilla halo effect shown in Fig. 9.9. First, untrained consumers cannot be trusted to provide accurate sensory specifications of product characteristics. While it may be common practice to collect some diagnostic attribute ratings from consumers in central location or home use tests, such information must be viewed with caution. There are well-known correlations among attributes (easily shown by principal components analysis, see Chapter 18), halo effects, and taste–smell confusions, all of which can bias the obtained ratings. Second, consumers see products as Gestalten, as whole patterns. They do not act analytically in short product tests. They do not learn to separate their sensations and attend effectively to individual product attributes. Third, if consumers do not have a chance to comment on a salient product characteristic, they may find some other place on the questionnaire to voice that feeling, perhaps in an inappropriate place. This last tendency was taken advantage of in our milk example—note that no scale for vanilla flavor was provided. The effect of misusing response scales in this way on a questionnaire is called response restriction or simply the "dumping effect."

9.5.2 "Dumping" Effects: Inflation Due to Response Restriction in Profiling

It is part of the folklore of consumer testing that if there is one very negative and salient attribute of a product, it will influence other attributes in a negative direction, an example of a horns effect. The effect is even worse when the salient negative attribute is omitted from the questionnaire. Omission could be due to some oversight or failure to anticipate the outcome in a consumer test or simply that it was not observed in the laboratory conditions of preliminary phases of testing. In this case, consumers will find a way to dump their frustration from not being able to report their dissatisfaction by giving negative ratings on other scales or reporting negative opinions of other even unrelated attributes. In other words, restricting responses or failure to ask a relevant question may change ratings on a number of other scales.

A common version of this restriction effect can be seen in sweetness enhancement. Frank et al. (1993) found that the enhancement of sweet ratings in the presence of a fruity odor was stronger when ratings were restricted to sweetness only. When

both sweetness and fruitiness ratings were allowed, no enhancement of sweetness was observed. Exactly the same effect was seen for sweetness and fruitiness ratings and for sweetness and vanilla ratings (Clark and Lawless, 1994). So allowing the appropriate number of attributes can address the problem of illusory enhancement. Schifferstein (1996) gave the example of hexenol, a fresh green aroma, which when added to a strawberry flavor mixture caused mean ratings in several other scales to increase. The enhancement of the other ratings occurred only when the "green" attribute was omitted from the ballot. When the "green" attribute was included in the ballot, the response was correctly assigned to that scale, and there was no apparent enhancement in the other attributes in the aroma profile.

There is good news and bad news in these observations. From a marketing perspective, ratings can be easily obtained from consumers that will show apparent sweetness enhancements if the questionnaires cleverly omit the opportunity to report on sensations other than sweetness. However, the nose pinch conditions and the use of complete sets of attributes show us that these volatile odorants such as maltol are not sweet taste enhancers but they are sweet *rating* enhancers. That is, they are not affecting the actual perception of sweet taste intensity but are changing the response output function or perhaps broadening the concept of sweetness to go beyond taste and include pleasant aromas as well. It would not be wise to try to use maltol to sweeten your coffee.

Are there other situations in sensory testing where the dumping effect can show up? One area in which responses are usually restricted to one attribute at a time is in time–intensity scaling (Chapter 8). In a common version of this technique, the subject moves a pointer, a mouse, or some other response device to provide a continuous record of sensory intensity for a specified attribute. Usually, just one attribute at a time is rated since it is very difficult to attend continuously or even by rapid shifting of attention to more than one attribute. This would seem to be a perfect opportunity for the dumping tendency to produce illusory enhancements (e.g., Bonnans and Noble, 1993). This idea was tested in experiments with repeated category ratings across time, a time–intensity procedure that allows for ratings of multiple attributes. These studies showed sweetness enhancement in sweet–fruity mixtures when sweetness alone was rated, but little or no enhancement when both sweetness and fruit intensity were rated over time (Clark and Lawless, 1994). This is exactly parallel to the sweetness enhancement results seen by Frank and colleagues. Workers using single-attribute time–intensity ratings should be wary of apparent enhancements due to response restriction.

9.5.3 Over-Partitioning

In the data of van der Klaauw and Frank (1996), one can also see cases in which having too many attributes causes a deflation in ratings. As in the dumping examples, their usual paradigm was to compare the sweetness ratings of a simple sucrose solution to the same concentration with a fruity odor added. When rating sweetness only, the rating is higher than when rating sweetness and fruitiness, the common dumping effect. But when total intensity and six additional attributes were rated, the sweetness rating was significantly lower than either of the other two conditions. In another example, including a bitterness rating (in addition to the sweetness and fruitiness ratings) lowered the sweetness rating compared to rating sweetness (the highest condition) and also compared to rating sweetness and fruitiness (an intermediate sweetness rating was obtained). This effect appears to be a deflation due to people over-partitioning their sensations into too many categories. The specific choices may be important in this effect. Adding only a bitter or a bitter and a floral rating had little or no effect and dumping inflation was still observed probably because there was no fruity rating.

In the study by Clark and Lawless (1994), the control condition (sweetener only) showed some evidence of a decrement when the attributes for volatiles were also available. Even more dramatic was the complete reversal of sweet enhancement to inhibition when a large number of response categories were provided for simple mixtures (Frank et al., 1993).

Although this effect has not been thoroughly studied, it serves to warn the sensory scientist that the number of choices given to untrained consumers may affect the outcome and that too many choices may be as dangerous as too few. Whether this effect might be seen with trained panels remains an open question. It is sometimes difficult to predetermine the correct number of attributes to rate in order to guard against

the dumping effect. Careful pre-testing and discussion phases in descriptive training may help. It is obviously important to be inclusive and exhaustive, but also not to waste the panelists' time with irrelevant attributes.

9.6 Classical Psychological Errors and Other Biases

A number of psychological errors in judgment have been described in the literature and are commonly listed in references in sensory evaluation (e.g., Amerine et al., 1965; Meilgaard et al., 2006). They are only briefly listed here, as they serve primarily as empirical descriptions of behavior, without much reference to cause or any theoretical bases. It is important to distinguish between description and explanation, and not to confuse naming something with trying to explain why it occurred in terms of mechanism or larger theory. The sensory evaluation practitioner needs to be aware of these errors and the conditions under which they may occur.

9.6.1 Errors in Structured Sequences: Anticipation and Habituation

Two errors may be seen when a non-random sequence of products is presented for evaluation, and the observer is aware that a sequence or a particular order of items is going to be presented. The error of anticipation is said to occur when the subject shifts responses in the sequence before the sensory information would indicate that it is appropriate to do so (Mattes and Lawless, 1985). An example is found in the method of limits for thresholds, where an ascending sequence is presented and the observer expects a sensation to occur at some point and "jumps the gun." The opposite effect is said to be the error of habituation, in which the panelist stays put too long with one previous response, when the sensory information would indicate that a change is overdue. Obviously, the presentation of samples in random order will help to undo the expectations involved in causing the error of anticipation. Perseveration is a little bit harder to understand but may have to do with lack of attention or motivation on the part of the observer or having an unusually strict criterion for changing responses. Attention and motivation can be addressed by sufficient incentives and keeping the test session from being too long.

9.6.2 The Stimulus Error

The stimulus error is another classical problem in sensory measurement. This occurs when the observer knows or presumes to know the identity of the stimulus and thus draws some inference about what it should taste, smell, or look like. The judgment is biased due to expectations about stimulus identity. In the old parlor game of trying to identify the origin and vintage of a wine, stimulus error is actually a big help. It is much easier to guess the wine if you know what the host is prone to drink or you have taken a peek at the bottles in the kitchen beforehand. In sensory evaluation, the principle of blind testing and the practice using random three-digit coding of samples mitigate against the stimulus error. However, panelists are not always completely in the dark about the origin or the identity of samples. Employee panels may have a fair amount of knowledge about what is being tested and they may make inferences, correctly or incorrectly. For example, in small-plant quality assurance, workers may be aware of what types of products are being manufactured that day and these same workers may serve as sensory panelists. In the worst possible scenario, the persons drawing the samples from production are actually doing the tasting. As much as possible, these situations should be avoided. In quality control panels, insertion of blind control samples (both positive controls and flawed samples) will tend to minimize the guesswork by panelists.

9.6.3 Positional or Order Bias

Time-order error is a general term applied to sequential effects in which one order of evaluating two or more products produces different judgments than does another order (Amerine et al., 1965). There are two philosophies for dealing with this problem. The first approach is to provide products in all possible orders, counterbalanced orders, or randomized orders so that

the sequential effects may be counterbalanced or averaged out in the group data. The second approach is to consider the order effects of interest. In this case, different orders are analyzed as a purpose of the experiment and, if order effects are observed, they are duly noted and discussed. Whether order effects are of interest will depend upon the circumstances of the product evaluation and its goals. If counterbalanced orders or randomization cannot be worked into the experimental design, the experimenter must consider whether product differences are true sensory differences or are artifacts of stimulus order. Purposeful experimentation and analysis of order effects in at least some of the routine evaluations may give one an appreciation for where and when these effects occur in the product category of interest.

Another well-known order effect in acceptance testing is the reception of a higher score for the first sample in a series (Kofes et al., 2009). Counterbalancing orders is of course appropriate, but one can also give a "dummy" product first to absorb the first product's score. With monadic (single product) tests, such inflation could be misleading (Kofes et al., 2009). Positional bias was also of concern in early paired tests and also the triangle procedure (Amerine et al., 1965). Another bias was seen when preference questions were asked following the triangle difference test (Schutz, 1954). Following a difference test with a preference test is not recommended in good sensory practice, in part because these early studies showed a bias against the sample that was considered the odd one in the triangle test. Recent research indicates that this effect may not be so robust—a replication of Schutz's original experiment but using the words "different" rather than "odd" did not find much evidence for this bias (El Gharby, 1995). Perhaps the meaning of the term "odd" in the 1950s was in itself sufficiently biasing.

9.7 Antidotes

9.7.1 Avoid or Minimize

At first glance, one way to avoid contextual effects would seem to be only to present products as single items, in other words to perform only monadic tests. This may be appropriate in some consumer testing situations where the test itself changes the situation so dramatically that future evaluation would be unduly influenced by the first product tested. Examples occur in testing consumer products such as insecticides or hair conditioners. However, monadic testing is rarely practical for analytical sensory tests such as descriptive analyses. It would be highly inefficient both financially and statistically to have trained panels evaluate only single samples in a session. More importantly, monadic testing does not take any advantage of the inherent comparative abilities of human observers. Furthermore, because of transfer bias (panelists, after all, have memories), this solution may be illusory for an ongoing testing program. Even without an immediate frame of reference in the experimental session, people will evaluate products based on their memory of similar items that have been recently experienced. So they will adopt a frame of reference if none is explicitly provided.

Poulton (1989) asserted that in most Western cultures, this baseline would be fairly constant (perhaps for consumers) and neutral based on the comparable experiences of experimental participants. He went on to suggest monadic testing as a way to avoid frequency and centering biases. However, experimental evidence for this assertion is lacking. Also, given the high degree of idiosyncratic food preferences and food habits, a constant baseline across individuals seems rather unlikely in sensory evaluation of foods. So monadic testing could actually add noise to the data. Furthermore, monadic test designs necessitate the use of between-subject comparisons and lose the statistical and scaling advantages inherent in using subjects as their own controls or baselines for comparison.

The problem can be rephrased as to how sensory evaluation specialists can control for contextual biases or minimize them. There are four approaches to dealing with context effects: randomization (including counterbalancing), stabilization, calibration, and interpretation. Stabilization refers to the attempt to keep context the same across all evaluation sessions so that the frame of reference is constant for all observers. Calibration refers to the training of a descriptive panel so that their frame of reference for the scale is internalized through training with reference standards. Interpretation is simply the careful consideration of whether ratings in a given setting may have been influenced by experimental context, e.g., by the specific items that were also presented in that session. Each of these approaches is considered below.

9.7.2 Randomization and Counterbalancing

The use of different random or counterbalanced orders has long been a principle of good practice in applied sensory testing. Simple order effects, sequential dependencies, and contrast between any two items can be counteracted by using sufficient orders so that the immediate frame of reference for any one product is different across a group of respondents. Using an order that mixes up products from different positions on the scale range will also help avoid the occurrence of a local frequency bias. That is, if too many samples are given from the high end of the sensory continuum to be rated, it may give the respondent the impression that there is bunching at that end, even though there may not be in the product set as a whole. Poulton (1989) noted that a "controlled balanced order" might help to avoid this. As in the discussion of the classical time-order effect above, there are two philosophies here. Whether one randomizes and ignores the local sequential effects, or systematically counterbalances orders and analyzes the order dependencies as effects of experimental interest will depend upon the experimental objectives, the resources of the experimenter and the information needed by the end users of the experimental results.

However, using random or counterbalanced orders will not in itself undo the broader effects of context that develop during an experiment with repeated ratings. The broader frame of reference still exists and is still used by judges to frame the range of products and to map the products onto the known scale range. Thus the use of randomization or counterbalancing of orders does not get around the problem of altered context when results from two different experimental sessions are compared.

Note that the examination of multiple stimulus ranges is similar philosophically to the approach of randomization or counterbalancing. One approach to avoiding the centering bias in just-right scaling is to use different ranges so that the stimulus set that would be centered on the true just-right point can be found by interpolation (Johnson and Vickers, 1987; McBride, 1982) (see Fig. 9.8). Multiple contexts become part of the experimental design. The goal is to purposefully examine the range effects, rather than averaging or neutralizing them through randomization. As a general principle, the greater the number of contexts within which a product is evaluated, the greater the understanding and accuracy of the final interpretation by the sensory scientist.

9.7.3 Stabilization and Calibration

The second approach to dealing with context effects is to try and hold the experimental context constant across all sessions containing products that will be compared. This can be difficult if the products are fatiguing or if adaptation or carryover effects are likely to restrict the number of products that can be given. In its simplest form, this strategy takes the form of a simple comparison of all test products to one control item (e.g., Stoer, 1992). Difference scores may then be constructed and serve as the primary form of the data. Alternatively, difference-from-reference ratings (Larson-Powers and Pangborn, 1978) or magnitude estimation with a constant reference item may be used. The presentation of a warm-up or practice sample probably has some stabilizing effect, another good reason to use such "throw-away" samples if the product is not too fatiguing. Another approach is to attempt to stabilize the endpoints by using reference standards for high and low stimuli that appear in every session. Also, high and low examples can be given as blind "catch trials" and judges suitably motivated to expand their scale usage if they are caught exhibiting contraction bias, end-of-scale avoidance, or simply gravitating toward the middle of the scale as sometimes occurs in repeated testing. In magnitude estimation and also in the use of the labeled magnitude scale, the reference standard used for comparison may have a stabilizing effect and reduce some of the contrast effects seen with all scaling methods (Diamond and Lawless, 2001; Lawless et al., 2000).

Calibration can refer to the use of bracketing reference standards in the experimental session, or reference standards given in training. Considering the context within the evaluation session is an important part of collecting good sensory judgments. Anderson (1974), for example, in discussing the use of category scales gives the following advice: "Several precautions have been standard with ratings scales in functional measurement. First, is the use of preliminary practice, which has several functions. The general range of

stimuli is not known to the subject initially, and the rating scale is arbitrary. Accordingly, the subject needs to *develop a frame of reference* for the stimuli and correlate it with the given response scale." (emphasis added; pp. 231–232). Anderson notes that the effect of such practice is a decrease in variability. His practice in the psychological laboratory for stabilizing scale usage is similar to the training of descriptive panelists with examples of products to be evaluated. Anderson goes on to describe end-anchor stimuli, which serve as low- and high-intensity standards on his 20-point scale: "Stimulus end-anchors are extremely important. These are additional stimuli that are more extreme than the experimental stimuli to be studied. One function of the end-anchors is to help define the frame of reference." (p. 232). In this view then, the proper use of rating scales includes a definition of the context in which the sample is to be judged. In practice, this is achieved by presentation of specific examples that bracket the sensory range. An explicit (if extreme) example of this is the relative scaling method of Gay and Mead (Gay and Mead 1992; Mead and Gay, 1995) in which the samples are inspected, then the highest and lowest placed at the endpoints, and all others distributed along the scale. This does insure usage of the whole scale but renders the data totally relative and totally specific to that context and set of samples.

Calibration of observers is a common practice in descriptive analysis, especially in techniques with intensity reference standards such as the texture profile and the Spectrum method (Meilgaard et al., 2006). Anderson (1974) warned that endpoint calibration was in fact a necessary practice with category scales in order to fix the linear usage of the scale by experimental subjects. What appears to be a problem in obtaining unbiased scale usage may be turned to an advantage if a stable frame of reference can be induced in subjects through training. Evidence shows that contextual effects do not appear all at once but are contingent upon and strengthened by experience (e.g., Lawless, 1983). So the question arises whether it is possible to "inoculate" a trained panelist against the day-to-day context of the experimental session by sufficient training. That is, is it possible to calibrate and stabilize judges' frame of reference to make them immune to contextual effects? Some of the examples of transfer bias cited by Poulton (1989) certainly make this seem reasonable. However, a recent study using extensive training on a 15-point sweetness scale failed to eliminate simple contrast effects (Olabi and Lawless, 2008). Reference standards were similarly ineffective. Perhaps it is asking too much of human nature to get panelists to act like absolute measuring instruments.

9.7.4 Interpretation

The last approach to context effects and biases is to be aware that they are operating and to draw conclusions about product differences with suitable caution. It is not appropriate to conclude that two products evaluated in different sessions are in fact different unless they were examined in similar contexts. The sensory professional must look at the whole experimental context and not just the summary statistics for the products in question in order to draw accurate conclusions. In drawing conclusions about product differences, it is necessary to question whether the observed difference could possibly have arisen from contextual effects or whether it is likely to be a true sensory-based difference. In situations of apparent enhancement or synergistic influence of one flavor upon another, one must always question how the data were gathered. Were there sufficient and appropriate response categories or was there the possibility that the apparent enhancement occurred due to dumping into the available but restricted set of response attributes?

9.8 Conclusions

Notes make each other sweeter in a chord, and so do colors when harmoniously applied. A certain amount of skin dipped in hot water gives the perception of a certain heat. More skin immersed makes the heat much more intense although of course the water's heat is the same. (James, 1913, p. 25).

William James reminds us that complex patterns of stimulation alter the perception of things around us. What something looks like, feels like, or tastes like depends upon the other patterns of stimulation that are present and that have come before the stimulus to be judged. Context effects are present in all scaling methods, and human behavior appears to be a sort of compromise between a completely relative system of

judgment and one that retains some absolute or calibrated properties (Ward, 1987). A main point of this chapter has been to show the importance of a frame of reference in influencing people's judgments about products. We have seen that other products in the same session can have a marked effect, usually one of contrast. The word anchors or phrases on the scale can influence the frame of reference (what you ask and how you ask it) and, from the dumping effect, we find that even what you *do not ask* can also influence responses. One goal of the sensory evaluation specialist should be to minimize unwanted effects that endanger the accuracy of results. Poulton (1989) discussed range biases as potential problems and stated that they were unavoidable in category scaling. Another perspective is to consider them interesting human phenomena, worthy of study for what they tell us about the judgment process. A third approach is to embrace the relative nature of human judgment and reduce all data to difference scores or similar explicit comparisons. This would be difficult for most practitioners of descriptive analysis methods.

From a practical perspective, a very real danger exists in sensory evaluation when people try to compare ratings given to products in different settings or from different experimental sessions. Unless the context and frame of reference is the same in both sessions, it is not possible to say whether differences between the products arose from true sensation differences or from differences in ratings due to contextual effects. A difference in the data set might occur merely because the two items were viewed among higher or lower valued items in their test sessions. Conversely, two items might appear similar in ratings across two sessions, but their ratings might be similar only due to range or centering effects. How can a sensory practitioner know whether the scale value of 5.3 for this week's product is actually superior to the value of 4.9 given to the prototype evaluated last week?

Unless the sensory professional is aware of context effects and guards against them, inappropriate conclusions may be drawn from evaluation sessions, especially if products evaluated in different contexts are to be compared. Sometimes the effects can be subtle and insidious. Consider the context effect discussed above in optimization (Mattes and Lawless, 1985). Ascending in concentration results in the estimation of a preferred sensory optimum that is too low relative to the peak that would be obtained from a randomized order of samples. Yet adding a flavor ingredient "to taste" is what product developers commonly do at the benchtop or chefs do in a research kitchen when they come up with a seemingly optimized ingredient level. However, even an informal tasting has its own context. We cannot assume that the results of such informal tastings are accurate—only properly randomized or counterbalanced sensory testing using acceptability ratings or a just-right scale with precautions against centering bias would give accurate direction as to the appropriate flavor level.

It is sometimes difficult to pin down exactly which of the many biases discussed in this chapter may be operating in an experiment. For example, the contrast effect and response contraction effects can work against each other so that there may be little evidence that these tendencies are present in a particular evaluation session (Schifferstein and Frijters, 1992). Communication with test panelists, careful examination of the data, and looking at historical records can all give hints about what may be going on with a panel that does repeated testing. Is there evidence of response contraction or gravitation toward the center of the scale? If these conservative tendencies are creeping in, they should show up through examination of data sets over time. The contraction effect is insidious since the trend may appear as reduced standard deviations that may give the false impression that the panel is achieving higher levels of calibration or consensus. However, the reduced error will not be accompanied by a higher rate of significant differences among products, since the product means will also gravitate toward the middle of the scale and differences between means will be smaller. This type of insight highlights the depth of analysis that is needed in good sensory practice and that a sensory professional must be more than a person who merely conducts tests and reports results. They must be in touch with the trends and finer grain of the data set along with the psychological tendencies of the respondents that may arise as a function of changes in frame of reference and other biases.

References

Amerine, M. A., Pangborn, R. M. and Roessler, E. B. 1965. Principles of Sensory Evaluation of Food. Academic, New York.

Anderson, N. 1974. Algebraic models in perception. In: E. C. Carterette and M. P. Friedman (eds.), Handbook of Perception. II. Psychophysical Judgment and Measurement. Academic, New York, pp. 215–298.

Baird, J. C. and Noma, E. 1978. Fundamentals of Scaling and Psychophysics. Wiley, New York.

Beebe-Center, J. G. 1932. The Psychology of Pleasantness and Unpleasantness. Russell & Russell, New York.

Bingham, A. F., Birch, G. G., de Graaf, C., Behan, J. M. and Perring, K. D. 1990. Sensory studies with sucrose maltol mixtures. Chemical Senses, 15, 447–456.

Birnbaum, M. H. 1982. Problems with so called "direct" scaling. In: J. T. Kuznicki, A. F. Rutkiewic and R. A. Johnson (eds.), Problems and Approaches to Measuring Hedonics (ASTM STP 773). American Society for Testing and Materials, Philadelphia, pp. 34–48.

Bonnans, S. and Noble, A. C. 1993. Effects of sweetener type and of sweetener and acid levels on temporal perception of sweetness, sourness and fruitiness. Chemical Senses, 18, 273–283.

Boring, E. G. 1942. Sensation and Perception in the History of Experimental Psychology. Appleton-Century-Crofts, New York.

Cardello, A. V. and Sawyer, F. M. 1992. Effects of disconfirmed consumer expectations on food acceptability. Journal of Sensory Studies, 7, 253–277.

Cardello, A. V., Lawless, H. T. and Schutz, H. G. 2008. Effects of extreme anchors and interior label spacing on labeled magnitude scales. Food Quality and Preference, 21, 323–334.

Clark, C. C. and Lawless, H. T. 1994. Limiting response alternatives in time–intensity scaling: An examination of the halo-dumping effect. Chemical Senses, 19, 583–594.

Diamond, J. and Lawless, H. T. 2001. Context effects and reference standards with magnitude estimation and the labeled magnitude scale. Journal of Sensory Studies, 16, 1–10.

Diehl, R. L., Elman, J. L. and McCusker, S. B. 1978. Contrast effects on stop consonant identification. Journal of Experimental Psychology: Human Perception and Performance, 4, 599–609.

Dijksterhuis, G. 1993. Principal component analysis of time–intensity bitterness curves. Journal of Sensory Studies, 8, 317–328.

Dolese, M., Zellner, D., Vasserman, M. and Parker, S. 2005. Categorization affects hedonic contrast in the visual arts. Bulletin of Psychology and the Arts, 5, 21–25.

Eimas, P. D. and Corbit, J. D. 1973. Selective adaptation of linguistic feature detectors. Cognitive Psychology, 4, 99–109.

El Gharby, A. 1995. Effect of Nonsensory Information on Sensory Judgments of No-Fat and Low-Fat Foods: Influences of Attitude, Belief, Eating Restraint and Label Information. M.Sc. Thesis, Cornell University.

Eng, E. W. 1948. An Experimental Study of the Reliabilities of Rating Scale for Food Preference Discrimination. M. S. Thesis, Northwestern University, and US Army Quartermaster Food and Container Institute, Report # 11–50.

Engen, T. and Levy, N. 1958. The influence of context on constant-sum loudness judgments. American Journal of Psychology, 71, 731–736.

Frank, R. A. and Byram, J. 1988. Taste–smell interactions are tastant and odorant dependent. Chemical Senses, 13, 445.

Frank, R. A., van der Klaauw, N. J. and Schifferstein, H. N. J. 1993. Both perceptual and conceptual factors influence taste–odor and taste–taste interactions. Perception & Psychophysics, 54, 343–354.

Gay, C. and Mead, R. 1992 A statistical appraisal of the problem of sensory measurement. Journal of Sensory Studies, 7, 205–228.

Giovanni, M. E. and Pangborn, R. M. 1983. Measurement of taste intensity and degree of liking of beverages by graphic scaling and magnitude estimation. Journal of Food Science, 48, 1175–1182.

Green, B. G., Dalton, P., Cowart, B., Shaffer, G., Rankin, K. and Higgins, J. 1996. Evaluating the 'labeled magnitude scale' for measuring sensations of taste and smell. Chemical Senses, 21, 323–334.

Hanson, H. L., Davis, J. G., Campbell, A. A., Anderson, J. H. and Lineweaver, H. 1955. Sensory test methods II. Effect of previous tests on consumer response to foods. Food Technology, 9, 56–59.

Helson, H. H. 1964. Adaptation-Level Theory. Harper & Rowe, New York.

James, W. 1913. Psychology. Henry Holt and Company, New York.

Johnson, J. and Vickers, Z. 1987. Avoiding the centering bias or range effect when determining an optimum level of sweetness in lemonade. Journal of Sensory Studies, 2, 283–291.

Jones, F. N. and Marcus, M. J. 1961. The subject effect in judgments of subjective magnitude. Journal of Experimental Psychology, 61, 40–44.

Jones, F. N. and Woskow, M. J. 1966. Some effects of context on the slope in magnitude estimation. Journal of Experimental Psychology, 71, 177–180.

Kamenetzky, J. 1959. Contrast and convergence effects in ratings of foods. Journal of Applied Psychology, 43(1), 47–52.

Kofes, J., Naqvi, S., Cece, A. and Yeh, M. 2009. Understanding Presentation Order Effects and Ways to Control Them in Consumer Testing. Paper presented at the 8th Pangborn Sensory Science Symposium, Florence, Italy.

Larson-Powers, N. and Pangborn, R. M. 1978. Descriptive analysis of the sensory properties of beverages and gelatins containing sucrose or synthetic sweeteners. Journal of Food Science, 43, 47–51.

Lawless, H. T. 1983. Contextual effect in category ratings. Journal of Testing and Evaluation, 11, 346–349.

Lawless, H. T. 1994. Contextual and Measurement Aspects of Acceptability. Final Report #TCN 94178, US Army Research Office.

Lawless, H. T. and Malone, G. J. 1986a. A comparison of scaling methods: Sensitivity, replicates and relative measurement. Journal of Sensory Studies, 1, 155–174.

Lawless, H. T. and Malone, J. G. 1986b. The discriminative efficiency of common scaling methods. Journal of Sensory Studies, 1, 85–96.

Lawless, H. T., Glatter, S. and Hohn, C. 1991. Context dependent changes in the perception of odor quality. Chemical Senses, 16, 349–360.

Lawless, H. T., Horne. J. and Speirs, W. 2000. Contrast and range effects for category, magnitude and labeled magnitude scales. Chemical Senses, 25, 85–92.

Lawless, H. T., Popper, R. and Kroll, B. J. 2010a. Comparison of the labeled affective magnitude (LAM) scale, an 11-point category scale and the traditional nine-point Hedonic scale. Food Quality and Preference, 21, 4–12.

Lawless, H. T., Sinopoli, D. and Chapman, K. W. 2010b. A comparison of the labeled affective magnitude scale and the nine point hedonic scale and examination of categorical behavior. Journal of Sensory Studies, 25, S1, 54–66.

Lee, H.-S., Kim, K.-O. and O'Mahony, M. 2001. How do the signal detection indices react to frequency context bias for intensity scaling? Journal of Sensory Studies, 16, 33–52.

Marks, L. E. 1994. Recalibrating the auditory system: The perception of loudness. Journal of Experimental Psychology: Human Perception & Performance, 20, 382–396.

Mattes, R. D. and Lawless, H. T. 1985. An adjustment error in optimization of taste intensity. Appetite, 6, 103–114.

McBride, R. L. 1982. Range bias in sensory evaluation. Journal of Food Technology, 17, 405–410.

McBride, R. L. and Anderson, N. H. 1990. Integration psychophysics. In R. L. McBride and H. J. H. MacFie (eds.), Psychological Basis of Sensory Evaluation. Elsevier Applied Science, London, pp. 93–115.

McBurney, D. H. 1966. Magnitude estimation of the taste of sodium chloride after adaptation to sodium chloride. Journal of Experimental Psychology, 72, 869–873.

McGowan, B. A. and Lee, S.-Y. 2006. Comparison of methods to analyze time–intensity curves in a corn zein chewing gum study. Food Quality and Preference, 17, 296–306.

Mead, R. and Gay, C. 1995. Sequential design of sensory trials. Food Quality and Preference, 6, 271–280.

Meilgaard, M., Civille, G. V. and Carr, B. T. 2006. Sensory Evaluation Techniques, Third Edition. CRC, Boca Raton, FL.

Mellers, B. A. and Birnbaum, M. H. 1982. Loci of contextual effects in judgment. Journal of Experimental Psychology: Human Perception and Performance, 4, 582–601.

Mellers, B. A. and Birnbaum, M. H. 1983. Contextual effects in social judgment. Journal of Experimental Social Psychology, 19, 157–171.

Muñoz, A. M. and Civille, G. V. 1998. Universal, product and attribute specific scaling and the development of common lexicons in descriptive analysis. Journal of Sensory Studies, 13, 57–75.

Murphy, C. 1982. Effects of exposure and context on hedonics of olfactory-taste mixtures. In: J. T. Kuznicki, R. A. Johnson and A. F. Rutkeiwic (eds.), Selected Sensory Methods: Problems and Applications to Measuring Hedonics. American Society for Testing and Materials, Philadelphia, pp. 60–70.

Murphy, C. and Cain, W. S. 1980. Taste and olfaction: Independence vs. interaction. Physiology and Behavior, 24, 601–605.

Olabi, A. and Lawless, H. T. 2008. Persistence of context effects with training and reference standards. Journal of Food Science, 73, S185–S189.

Parducci, A. 1965. Category judgment: A range-frequency model. Psychological Review, 72, 407–418.

Parducci, A. 1974. Contextual effects: A range-frequency analysis. In: E. C. Carterette and M. P. Friedman (eds.), Handbook of Perception. II. Psychophysical Judgment and Measurement. Academic, New York, pp. 127–141.

Parducci, A. and Perrett, L. F. 1971. Category rating scales: Effects of relative spacing and frequency of stimulus values. Journal of Experimental Psychology (Monograph), 89(2), 427–452.

Parducci, A., Knobel, S. and Thomas, C. 1976. Independent context for category ratings: A range-frequency analysis. Perception & Psychophysics, 20, 360–366.

Parker, S., Murphy, D. R. and Schneider, B. A. 2002. Top-down gain control in the auditory system: Evidence from identification and discrimination experiments. Perception & Psychophysics, 64, 598–615.

Poulton, E. C. 1989. Bias in Quantifying Judgments. Lawrence Erlbaum Associates, Hillsdale, NJ.

Rankin, K. M. and Marks, L. E. 1991. Differential context effects in taste perception. Chemical Senses, 16, 617–629.

Riskey, D. R. 1982. Effects of context and interstimulus procedures in judgments of saltiness and pleasantness. In: J. T. Kuznicki, R. A. Johnson and A. F. Rutkeiwic (eds.), Selected Sensory Methods: Problems and Applications to Measuring Hedonics. American Society for Testing and Materials, Philadelphia, pp. 71–83.

Riskey, D. R., Parducci, A. and Beauchamp, G. K. 1979. Effects of context in judgments of sweetness and pleasantness. Perception & Psychophysics, 26, 171–176.

Sandusky, A. and Parducci, A. 1965. Pleasantness of odors as a function of the immediate stimulus context. Psychonomic Science, 3, 321–322.

Sarris, V. 1967. Adaptation-level theory: Two critical experiments on Helson's weighted-average model. American Journal of Psychology, 80, 331–344.

Sarris, V. and Parducci, A. 1978. Multiple anchoring of category rating scales. Perception & Psychophysics, 24, 35–39.

Schifferstein, H. J. N. 1995. Contextual shifts in hedonic judgment. Journal of Sensory Studies, 10, 381–392.

Schifferstein, H. J. N. 1996. Cognitive factors affecting taste intensity judgments. Food Quality and Preference, 7, 167–175.

Schifferstein, H. N. J. and Frijters, J. E. R. 1992. Contextual and sequential effects on judgments of sweetness intensity. Perception & Psychophysics, 52, 243–255.

Schutz, H. G. 1954. Effect of bias on preference in the difference-preference test. In: D. R. Peryam, J. J. Pilgram and M. S. Peterson (eds.), Food Acceptance Testing Methodology. National Academy of Sciences, Washington, DC, pp. 85–91.

Stevenson, R. J., Prescott, J. and Boakes, R. A. 1995. The acquisition of taste properties by odors. Learning and Motivation, 26, 433–455.

Stoer, N. L. 1992. Comparison of Absolute Scaling and Relative-To-Reference Scaling in Sensory Evaluation of Dairy Products. Master's Thesis, Cornell University.

Teghtsoonian, R. and Teghtsoonian, M. 1978. Range and regression effects in magnitude scaling. Perception & Psychophysics, 24, 305–314.

Thorndike, E. L. 1920. A constant error in psychophysical ratings. Journal of Applied Psychology, 4, 25–29.

van der Klaauw, N. J. and Frank, R. A. 1996. Scaling component intensities of complex stimuli: The influence of response alternatives. Environment International, 22, 21–31.

Vollmecke, T. A. 1987. The Influence of Context on Sweetness and Pleasantness Evaluations of Beverages. Doctoral dissertation, University of Pennsylvania.

Ward, L. M. 1979. Stimulus information and sequential dependencies in magnitude estimation and cross-modality matching. Journal of Experimental Psychology: Human Perception and Performance, 5, 444–459.

Ward, L. M. 1987. Remembrance of sounds past: Memory and psychophysical scaling. Journal of Experimental Psychology: Human Perception and Performance, 13, 216–227.

Zellner, D. A., Allen, D., Henley, M. and Parker, S. 2006. Hedonic contrast and condensation: Good stimuli make mediocre stimuli less good and less different. Psychonomic Bulletin and Review, 13, 235–239.

Chapter 10

Descriptive Analysis

Abstract   This chapter describes the potential uses for descriptive analysis in sensory evaluation. We then discuss the use of language and concept formation as well as the requirements for appropriate sensory attribute terms. This is followed by a historical review of the first descriptive analysis technique, the Flavor Profile. We then describe the Texture Profile, as well as proprietary descriptive methods such as Quantitative Descriptive Analysis and the Spectrum method. We then lead the reader through a step-by-step application of consensus and ballot-trained generic descriptive analyses. We then highlight and discuss some of the studies comparing conventional descriptive analysis technique. This is followed by an in-depth discussion of the variations on the theme of descriptive analysis such as free choice profiling and flash profiling.

> *I want to reach that state of condensation of sensations which constitutes a picture.*
>
> —Henri Matisse

Contents

10.1 Introduction

Descriptive sensory analyses are the most sophisticated tools in the arsenal of the sensory scientist. These techniques allow the sensory scientist to obtain complete sensory descriptions of products, to identify underlying ingredient and process variables, and/or to determine which sensory attributes are important to acceptance. A generic descriptive analysis would usually have between 8 and 12 panelists that would have been trained, with the use of reference standards, to understand and agree on the meaning of the attributes used. They would usually use a quantitative scale for intensity which allows the data to be statistically analyzed. These panelists would not be asked for their hedonic responses to the products. However, as we will see in this chapter, there are several different descriptive analysis methods and, in general, these reflect very different sensory philosophies and approaches. Usually, descriptive techniques produce objective descriptions of products in terms of the perceived sensory attributes. Depending on

H.T. Lawless, H. Heymann, *Sensory Evaluation of Food*, Food Science Text Series,
DOI 10.1007/978-1-4419-6488-5_10, © Springer Science+Business Media, LLC 2010

the specific technique used, the description can be more or less objective, as well as qualitative or quantitative.

10.2 Uses of Descriptive Analyses

Descriptive analyses are generally useful in any situation where a detailed specification of the sensory attributes of a single product or a comparison of the sensory differences among several products is desired. These techniques are often used to monitor competitors' offerings. Descriptive analysis can indicate exactly how in the sensory dimension the competitor's product is different from yours. These techniques are ideal for shelf-life testing, especially if the judges were well trained and are consistent over time. Descriptive techniques are frequently used in product development to measure how close a new introduction is to the target or to assess suitability of prototype products. In quality assurance, descriptive techniques can be invaluable when the sensory aspects of a problem must be defined. Descriptive techniques tend to be too expensive for day-to-day quality control situations, but the methods are helpful when troubleshooting major consumer complaints. Most descriptive methods can be used to define sensory–instrumental relationships. Descriptive analysis techniques should never be used with consumers because in all descriptive methods, the panelists should be trained at the very least to be consistent and reproducible.

10.3 Language and Descriptive Analysis

There are three types of language, namely everyday language, lexical language, and scientific language. Everyday language is used in daily conversations and may vary across cultural subgroups and geographical regions. Lexical language is the language found in the dictionary and this language may be used in everyday conversations. However, few people use primarily lexical language in conversation. For most of us lexical language is best represented in our written documents. Scientific language is specifically created for scientific purposes and the terms used are usually very precisely defined. This is frequently the "jargon" associated with a specific scientific discipline.

The training phase of most descriptive analysis techniques includes an effort to teach the panel or to have the panel create their own scientific language for the product or product category of interest. Psychologists and anthropologists have argued for years about the interplay between language and perception. An extreme view is that of Benjamin Whorf (1952) who said that language both reflects and determines the way in which we perceive the world. On the other side of the coin are psychologists who say that perception is largely determined by the information and structure offered by stimulation from the environment. Words serve merely as instruments to convey our perceptions to other people. There is evidence that humans learn to organize patterns of correlated sensory characteristics to form categories and concepts. The concepts formed are labeled (given language descriptions) to facilitate communication.

Concept formation is dependent on prior experience. Thus different people or cultures may form different concepts from the same characteristics. Concepts are formed by a process involving both abstraction and generalization (Muñoz and Civille, 1998). A number of studies have shown that concept formation may require exposure to many similar products, certainly if the end result is a desire to align a concept among a group of people (Ishii and O'Mahony, 1991). A single example may define the prototype for the concept (usually called a descriptor in sensory studies) but does not necessarily allow the panelists to generalize, abstract, or learn where the concept boundaries are. To generalize and learn to distinguish weakly structured concepts (such as creaminess) the panelists should be exposed to multiple reference standards (Homa and Cultice, 1984).

In practice this means that when we train a descriptive panel, we must be careful to facilitate meaningful concept formation by exposing the panel to as many standards as feasible. However, if the concept boundaries are very clear and narrow (for example, sweetness) a single standard may be adequate. Concept formation is improved when it occurs within the product category under study. For example, Sulmont et al. (1999) working with orange juice found that panels receiving spiked orange juice samples as reference standards were more discriminant and homogeneous than panelists receiving either a single reference standard for each attribute or three reference standards per attribute. In their case it seemed that multiple reference standards actually had a negative effect on the panel performance. But, Murray et al. (2001) caution that reference standards external to the product category also have a role to play in concept formation. It is important to note that the use of reference standards does not necessarily eliminate contrast effects and the sensory scientist should keep that in mind.

If panelists are to use exact sensory descriptors' descriptions they must be trained. And untrained panelists frequently realize this when they are asked to evaluate products on attributes for which they have no clear concept. Armstrong et al. (1997) quote one of their untrained panelist: "I would rather we sat down and decided on what certain words and descriptions meant." The goal is for all panelists to use the same concepts and to be able to communicate precisely with one another; in other words the training process creates a "frame of reference" for the panel as a group (Murray et al., 2001). Thus, almost as an a priori assumption, descriptive analysis requires precise and specific concepts articulated in carefully chosen scientific language. The language used by consumers to describe sensory characteristics is almost always too imprecise and non-specific to allow the sensory specialist to measure and understand the underlying concepts in a way that will provide meaningful data.

Concept formation and definition can be illustrated as follows. In the United States and most Western countries our everyday concepts of colors are very similar, because we are taught as children to associate certain labels with certain stimuli. In other words, if a child says that the leaves of an oak tree are blue, the child will be told that the leaves are green. If the child persists in misnaming the color then the child would be tested for vision and/or other problems. Color is thus a well-structured concept for most individuals and possesses a widely understood scientific language for its description. However, with other sensory attributes such as flavor this is not true. In our culture we rarely describe the flavor of a food in precise terms. We usually say things like "the freshly baked bread smells good" or "the cough syrup tastes bad." There are charts with standard colors with coded labels (for example, the Munsell Book of Colors) but for taste, odor, and texture there is no "Munsell Book" and thus when we want to do research on these concepts we need to precisely define (preferably with reference standards) the scientific language used to describe the sensory sensations associated with the products studied.

When selecting terms (descriptors) to describe the sensory attributes of products we must keep the several desirable characteristics of descriptors in mind (Civille and Lawless, 1986). The desirable characteristics discussed by Civille and Lawless and others are listed in order of approximate importance in Table 10.1. We will consider each of these characteristics in turn. The selected descriptors should discriminate among the samples; therefore, they should indicate perceived differences among the samples. Thus, if we are evaluating cranberry juice samples and all the samples are the exact same shade of red then "red color intensity" would not be a useful descriptor. On the other hand, if the red color of the cranberry juice samples differs, due to processing conditions for example, then "red color intensity" would be an appropriate descriptor.

Table 10.1 Desirable characteristics that should be remembered when choosing terms for descriptive analysis studies (in order of importance)

Discriminate	More important
Non-redundant	
Relate to consumer acceptance/rejection	
Relate to instrumental or physical measurements	
Singular	
Precise and reliable	
Consensus on meaning	
Unambiguous	
Reference easy to obtain	
Communicate	
Relate to reality	Less important

The chosen term should be completely non-redundant with other terms; an example of redundancy is when the panelists are evaluating a steak and they are asked to rate both the perceived tenderness and the toughness of the meat (Raffensperger et al., 1956) since they both indicate the same concept in meat. It would be much better to decide that either the term "toughness" or the term "tenderness" should be used in the evaluation of the meat samples. Additionally the terms should be orthogonal. Orthogonal descriptors are not correlated with each other. Non-orthogonal descriptors overlap; for example, asking a panel to score the "red fruit intensity" of a Pinot noir wine and to score "cherry intensity" would be asking them to score non-orthogonal terms. It is very confusing, de-motivating, and mentally frustrating to the panelists when they are asked to score redundant and non-orthogonal terms. Sometimes, it is impossible to entirely eliminate term redundancy and to ensure that all terms are orthogonal. For example, in a study describing differences among vanilla essences, Heymann (1994a) trained a panel to evaluate both butterscotch odor and sweet milk flavor. The panel was convinced that these two terms described different sensations. Yet, during the data analysis it became clear from the principal component analysis that the two terms were redundant and that they exhibited a great deal of overlap. But, it is possible that while these terms were correlated in this product category, they may not be for another set of products!

Panelists often have preconceived notions about which terms are correlated and which are not. During training it is often necessary to help panelists "decorrelate" terms (Civille and Lawless, 1986; Lawless and Corrigan, 1993). In texture analysis panelists frequently cannot grasp the differences between denseness and hardness, since these terms are correlated in many foods but not all. Some foods are dense but not hard (cream cheese, refrigerated butter) and other foods are hard but not dense (American malted milk bars, refrigerated "aerated chocolate bars" in the United Kingdom). Exposing panelists to these products would help de-correlate these terms, allowing the panel to understand that the two terms do not always have to vary together.

The data from descriptive analyses are often used to interpret consumer hedonic responses to the same samples. Therefore, it is very helpful if the descriptors used in the descriptive analysis can be related to concepts that lead consumers to accept or reject the product. In a sensory profiling of aged natural cheeses the panel trained by Heisserer and Chambers (1993) chose to use the term "butyric acid" (a chemical name) instead of the panel's consensus term for the sensory odor impression, namely "baby vomit." In this case the term that they discarded might have been more helpful in relating consumer acceptance or rejection of the cheese than the more precise chemical term chosen. Also, the ideal descriptors can be related to the underlying natural structure (if it is known) of the product. For example, many terms associated with the texture profile are tied to rheological principles (Szczesniak et al., 1963). It is also possible to use terms that are related to the chemical nature of the flavor compounds found in the product. For example, Heymann and Noble (1987) used the term "bell pepper" to describe

the odor sensation in Cabernet sauvignon wines associated with the chemical 2-methoxy-3-isobutyl pyrazine. The pyrazine odorant is present in Cabernet sauvignon wines and it is also the impact compound for bell pepper aroma. The use of "butyric acid" by Heisserer and Chambers (1993) to describe a specific odor in aged cheese is tied to the compound probably responsible for the odor.

Descriptors should be singular rather than combinations of several terms. Combination or holistic terms such as creamy, soft, clean, fresh are very confusing to panelists. Integrated terms may be appropriate in advertising but not in sensory analysis. These terms should be broken down into their elemental, analytical, and primary parts. For example, a number of scientists have found that creaminess perception is a function of smoothness, viscosity, fatty mouth feel, and cream flavor (see Frøst and Janhøj, 2007, for an excellent overview). A study involving creaminess will likely be more easily interpreted and understood if most or all of these terms are examined. Also, the term acrid is a combination of aroma and tactile sensations (Hegenbart, 1994), and panelists should be trained to evaluate the components of acrid rather than the integrated term itself. The term soft, as used with fabrics, is a combination of compressibility, springiness, smoothness to touch, and a lack of crisp edges when folded. The problem with compound descriptors like creamy is that they are not actionable. Product developers do not know what to fix if the data indicate that there is a problem with this descriptor. Do they change the viscosity? The particle size? The aroma? It is possible that the term is not weighted similarly by all panelists; some may emphasize the thickness concept and others the cream aroma which often vary independently, thus "muddling up" the analysis. This is clearly not a good state of affairs for a descriptive analysis panel.

Suitable descriptors are ones that can be used with precision and reliability by the panelists. Panelists should fairly easily agree on the meaning of a specified term, the term should thus be unambiguous. They should be able to agree on the prototypical examples related to the descriptor and they should agree on the boundaries of the descriptor. Using reference standards to signify these boundaries is encouraged. It simplifies the life of the panel leader if the physical reference standards for the descriptor are easy to obtain. However, difficulties in obtaining physical reference standards should not prevent the panel leader or the panelists from using terms that are ideal in every other way.

The chosen descriptors should have communication value and should not be jargon. In other words, the terms should be understandable to the users of the information obtained in the study and not only to the descriptive panel and their leader. It is also helpful if the term had been used traditionally with the product or if it can be related to the existing literature. The reference standards associated with each descriptor have a two-fold purpose: to align the underlying concepts for the panelists and to act as "translation" devices for users of the information obtained from the study. Giboreau et al. (2007) stressed that circularity should be avoided in defining sensory descriptors, for example, "noisy" should not be defined as "that which makes noise" but rather as "that which produces sound when it is bitten." These authors also stress that reference standards would increase the utility of these definitions and that definitions should be exact substitutes for the defined terms. An example would be "This piece of meat is very tough" and substituting the definition for "tough" one would say "This piece of meat is very difficult to chew."

Krasner (1995) working with water taints showed that some reference standards, for example, a hypochlorite solution for chlorine odor in water or a piece of boiled rubber hose for a rubbery smell, were distinctive and a large percentage of panelists agreed on the odor. Other chemicals were not successful as reference standards, for example, hexanal evoked a grassy odor descriptor from about 24% of his panelists and a lettuce aroma descriptor from 41% of the panelists with the rest divided between celery, olives, tobacco smoke, and old produce. We are of the opinion that this occurs relatively frequently with single chemical compounds.

The use of multiple reference standards for a single concept enhances learning and use of the concept (Ishii and O'Mahony, 1991). Additionally, panel leaders with a broad sensory reference base facilitate learning. For example, panelist responses to the odor of oil of bitter almonds may include descriptors such as almond, cherry, cough drops, Amaretto, and Danish pastries. All of these descriptors refer to the underlying benzaldehyde character in all these products. In another study the panelists may state that the product reminds them of cardboard, paint, and linseed oil. The experienced panel leader will realize that all these

terms are descriptive of the sensation associated with the oxidation of lipids and fatty acids. It is also helpful if the panel leader has background knowledge of the product category.

10.4 Descriptive Analysis Techniques

In the following section we will review the major approaches and philosophies of descriptive analysis techniques. Reviews can be found in Amerine et al. (1965), Powers (1988), Einstein (1991), Heymann et al. (1993), Murray et al. (2001), Stone and Sidel (2004), and Meilgaard et al. (2006). Additionally, Muñoz and Civille (1998) clearly explained some of the philosophical differences with respect to panel training and scale usage among the different techniques.

10.4.1 Flavor Profile®

In its original incarnation the Flavor Profile (FP) is a qualitative descriptive test. The name and the technique were trademarked to Arthur D. Little and Co., Cambridge, MA. This technique was developed in the late 1940s and early 1950s at Arthur D. Little by Lören Sjostrom, Stanley Cairncross, and Jean Caul. FP was first used to describe complex flavor systems measuring the effect of monosodium glutamate on flavor perception. Over the years FP was continually refined. The latest version of FP is known as Profile Attribute Analysis (Cairncross and Sjöstrom, 1950; Caul, 1957, 1967; Hall, 1958; Meilgaard et al., 2006; Moskowitz, 1988; Murray et al., 2001; Powers, 1988; Sjöström, 1954).

Flavor profiling is a consensus technique. The vocabulary used to describe the product and the product evaluation itself is achieved by reaching agreement among the panel members. The FP considers the overall flavor and the individual detectable flavor components of a food system. The profile describes the overall flavor and the flavor notes and estimates the intensity of these descriptors and the amplitude (overall impression). The technique provides a tabulation of the perceived flavors, their intensities, their order of perception, their aftertastes, and their overall impression (amplitude). If the panelists are trained appropriately this tabulation is reproducible.

Using standardized techniques of preparation, presentation, and evaluation, the four to six judges are trained to precisely define the flavors of the product category during a 2- to 3-week program. The food samples are tasted and all perceived notes are recorded for aroma, flavor, mouth feel, and aftertaste. The panel is exposed to a wide range of products within the food category. After this exposure the panelists review and refine the descriptors used. Reference standards and definitions for each descriptor are also created during the training phase. Use of appropriate reference standards improves the precision of the consensus description. At the completion of the training phase the panelists have defined a frame of reference for expressing the intensities of the descriptors used.

The samples are served to the panelists in the same form that they would be served to the consumer. Thus, if the panel was studying cherry pie fillings the filling would be served to the panel in a pie.

Originally, the intensities of the perceived flavor notes were rated on the following scale (this scale has subsequently been expanded with up to 17 points including the use of arrows, $\frac{1}{2}$'s, or plus and minus symbols):

Rating	Explanation
0	Not present
)(	Threshold or just recognizable
1	Slight
2	Moderate
3	Strong

The order in which the flavor notes are perceived is also indicated on the tabulated profile. The aftertaste is defined as one or two flavor impressions that are left on the palate after swallowing. The panel rates the aftertaste intensities 1 min after the product is swallowed.

The amplitude is the degree of balance and blending of the flavor. It is not supposed to be indicative of the overall quality of the product nor is it supposed to include the panelists' hedonic responses to the product. Proponents of FP admit that it is very difficult for novice panelists to divorce their hedonic responses from the concept of amplitude. However, panelists do reach an understanding of the term with training and exposure to the FP method and the product category. The amplitude is defined as an overall impression of balance and blending of the product. In a sense, the amplitude is not to be understood, just

to be experienced. For example, heavy cream, when whipped, has a low amplitude; heavy cream whipped with the addition of some sugar has a higher amplitude; and heavy cream whipped with the addition of some sugar and vanilla essence has a much higher amplitude. Usually, FP panelists determine the amplitude before they concentrate on the individual flavor notes of the product. However, the amplitude may be placed last in the tabular profile. The following scale is used to rate amplitude:

Rating	Explanation
)(	Very low
1	Low
2	Medium
3	High

The panel leader derives a consensus profile from the responses of the panel. In a true FP this is not a process of averaging scores, but rather that the consensus is obtained by discussion and re-evaluation of the products by the panelists and panel leader. The final product description is indicated by a series of symbols. As described earlier, these are a combination of numerals and other symbols that are combined by the panelists into potentially meaningful patterns, whether as a descriptive table (Table 10.2) or as a graphic, the "sunburst."

The "sunburst," which is not used currently, was a graphical representation of FP results (Cairncross and Sjöstrom, 1950). A semi-circle indicates the threshold intensity and the radiating line lengths indicated the consensus intensity of each attribute evaluated. The order in which various characteristics "emerge" from the sample is noted by the order (from left to right) on the graph. While these symbols can be used to describe the product, it is impossible to analyze the data obtained in this way by conventional statistical procedures. Therefore, the FP is classified as a qualitative descriptive technique.

With the introduction of numerical scales, between 1 and 7 points, (Moskowitz, 1988), the Flavor Profile was renamed the Profile Attribute Analysis (PAA). Data derived from PAA may be statistically analyzed but it is also possible to derive a FP-type consensus description. The use of numerical scales allows researchers employing this method to use statistical techniques to facilitate data interpretation. PAA is more quantitative than FP (Hanson et al., 1983).

Table 10.2 Example of the consensus result of a flavor profile study. Composite flavor profile for turkey patties with 0.4% added phosphate

Flavor	Attributes
	Intensity[a]
Protein	2–
Meat identity	1
Serumy	1
[pause]	
Metallic (aromatic and feel)	1+
(Carries through)	1–
Poultry	1+
Brothy	1–
[lag]	
Turkey	1
Organ meat	1–
Metallic (aromatic and feel)	1
Bitter	)(
Aftertaste	Intensity[a]
Metallic feel	2–
Poultry	1–
Other[b]	
Turkey	)(+
Organ meat	)(+

Adapted from Chambers et al. (1992)

[a]Scale:)(= threshold, 1 = slight, 2 = moderate, 3 = strong

[b]"Other" characteristics in the aftertaste were not found by the entire panel

Syarief and coworkers (1985) compared flavor profile results derived through consensus with flavor profile results derived by calculating mean scores. The mean score results had a smaller coefficient of variation than the consensus results and the principal component analysis (PCA) of the mean score data accounted for a higher proportion of the variance than the PCA of the consensus scores. Based on these criteria the authors concluded that the use of mean scores gave superior results to that of consensus scores. Despite these results, some practitioners still use both the FP and PAA as a consensus technique.

Proponents of FP state that the data are accurate and reproducible if the panelists are well trained. The necessity for vocabulary standardization among panelists cannot be overestimated. Detractors of these procedures complain that the derived consensus may actually be the opinion of the most dominant personality in the group or the panel member perceived to have the greatest authority, often the panel leader. Advocates of the techniques counter that with proper training the panel leader will prevent this from occurring. Additionally, champions of the method

maintain that a trained FP panel produces results rapidly. Proper training is critical when using these techniques successfully.

True FP resists most attempts for mathematical characterization of the data. Usually a series of symbols must be interpreted using intuition and experience on the part of the researcher. PAA, on the other hand, can be analyzed using parametric techniques such as analysis of variance and suitable means separation procedures. Currently, the FP technique is used extensively in the evaluation of water, probably because water utilities usually only have three to four people to troubleshoot taste and odor complaints (AWWA, 1993; Bartels et al., 1986, 1987; Ömür-Özbek and Dietrich, 2008).

10.4.1.1 Flavor Profile Judge Selection

Flavor Profile judges should be screened for long-term availability. It takes time, effort, and money to train a panel and the panelists should make a commitment to be available for years, if possible. It is not unusual to find FP panelists who have served on the same panel for more than 10 years. Potential panelists should have a keen interest in the product category and it is helpful if they have some background knowledge on the product type. These panelists should be screened to have normal odor and taste perceptions. Panelists are screened for normal acuity using solutions and pure diluted odorants (see Chapter 2). They should be very articulate and sincere with an appropriate personality (not timid or overly aggressive).

The panel leader is an active participant in both the language development and evaluation phases of the study. The panel leader must moderate the interactions between panelists, leading the entire group toward some unanimity of opinion. It is clear that the key element in a FP panel is the panel leader. This person coordinates the sample production, directs the panel evaluations, and finally verbalizes the consensus conclusions of the entire panel. The panel leader will often resubmit samples until reproducible results are obtained. Therefore, the panel leader should be especially articulate and knowledgeable about the product type. This person will also be responsible for communication with the panel and preparation of samples and reference standards. The panel leader should also be infinitely patient, socially sensitive, and diplomatic since he/she will be responsible for moving the panel to a consensus description of the product.

10.4.2 Quantitative Descriptive Analysis®

Quantitative Descriptive Analysis (QDA) was developed during the 1970s to correct some of the perceived problems associated with the Flavor Profile analysis (Stone and Sidel, 2004; Stone et al., 1974). In contrast to FP and PAA, the data are not generated through consensus discussions, panel leaders are not active participants, and unstructured line scales are used to describe the intensity of rated attributes. Stone et al. (1974) chose the linear graphic scale, a line that extends beyond the fixed verbal end points, because they found that this scale may reduce the tendency of panelists to use only the central part of the scale avoiding very high or very low scores. Their decision was based in part on Anderson's studies (1970) of functional measurement in psychological judgments. As with FP, QDA has many advocates and the technique has been extensively reviewed (Einstein, 1991; Heymann et al., 1993; Meilgaard et al., 2006; Murray et al., 2001; Powers, 1988; Stone and Sidel, 2004; Stone et al., 1980; Zook and Wessman, 1977).

During QDA training sessions, 10–12 judges are exposed to many possible variations of the product to facilitate accurate concept formation. The choice of range of samples is dictated by the purpose of the study and, similar to FP, panelists generate a set of terms that describe differences among the products. Then through consensus, panelists develop a standardized vocabulary to describe the sensory differences among the samples. The panelists also decide on the reference standards and/or verbal definitions that should be used to anchor the descriptive terms. Actual reference standards are only used about 10% of the time; usually, only verbal definitions are used (Murray et al.,

2001). In addition, during the training period the panel decides the sequence for evaluating each attribute. Late in the training sequence, a series of trial evaluations are performed. This allows the panel leader to evaluate individual judges based on statistical analysis of their performance relative to that of the entire panel. Evaluations of panelist performance may also be performed during the evaluation phase of the study.

Panelists begin training by generating a consensus vocabulary. During these early meetings, the panel leader acts *only* as a facilitator by directing discussion and supplying materials such as reference standards and product samples as required by the panel. The panel leader does not participate in the final product evaluations.

Unlike FP, QDA samples may not be served exactly as seen by the consumer. For example, if a Flavor Profile panel is to evaluate pie crusts, they would receive samples of pie crust filled with a standard pie filling. The QDA philosophy states that the pie filling could affect the discrimination of the crust samples. However, a case could also be made that crust baked without filling may perform differently than crust baked with filling. Depending on the situation, the QDA panelists may receive two different pie crust samples, one baked without filling and the other baked with filling, which was removed before the panelists received the crust samples.

The actual product evaluations are performed by each judge individually, usually while seated in isolated booths. Standard sensory practices such as sample coding, booth lighting, expectorating, and rinsing between samples are used for the evaluation phase. A 6 in. graphic line scale anchored with words generated by the panel is used (Fig. 10.1).

The resulting data can be analyzed statistically using analysis of variance and multivariate statistical techniques. It is necessary for judges to replicate their judgments, up to six times in some cases, to allow the sensory scientist to check the consistency of the individual panelists and of the whole panel.

Fig. 10.1 An example of the QDA graphic line scale. The mark made by the panelist is converted to a numerical value by measuring from the *left end of the line*.

Word anchor   Word anchor

Fig 10.2 Different uses of the line scale by panelists that are calibrated *relative* to each other. All ratings were plotted on the same line scale for illustration purposes.

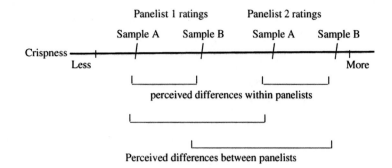

Replications also allow one-way analysis of variance of individual panelists across products. This allows the sensory specialist to determine whether the panelists can discriminate or need more training. The number of repeat judgments is somewhat product dependent and should be decided before the study is initiated. Studies where repeated evaluations are not performed should be viewed with extreme caution.

QDA may be used to completely describe the sensory sensations associated with a product from initial visual assessment to aftertaste, or panelists may be instructed to focus on a narrow range of attributes such as texture descriptors. However, limiting the range of attributes evaluated may lead to the "dumping" effect (see Chapter 9). This effect is especially important when a conspicuous sensory attribute that varies across the samples was omitted from the ballot. When this occurs panelists will, probably sub-consciously, express their frustration by modulating the scores on some of the scales used in the study. For this reason, the sensory scientist should be extremely careful about restricting the type and number of descriptors used in a descriptive analysis study. Sometimes, simply adding a scale labeled "other" can prevent this effect and if the panelists are allowed to describe the "other" characteristic valuable information may also be obtained.

During training, one of the challenges faced by the panel leader is how to help judges sort out the individual intensity characteristics of a product from overall impressions of quality or liking (Civille and Lawless, 1986). All descriptive evaluations should only be based on perceived intensities and should be free of hedonic responses.

Despite the extensive training employed in this method, most researchers assume that judges will use different parts of the scale to make their determinations. Thus, the absolute scale values are not important. It is the relative differences among products that provide valuable information. For example, Judge A scores the crispness of potato chip sample 1 as an 8, but Judge B scores the same sample as a 5; this does not mean that the two judges are not measuring the same attribute in the same way, but may mean that they are using different parts of the scale (Fig. 10.2). The relative responses of these two judges on a second different sample (say 6 and 3, respectively) would indicate that the two judges are calibrated with respect to the *relative* differences between the samples. Judicious choices of statistical procedures such as dependent *t*-tests and ANOVA allow the researcher to remove the effect of using different parts of the scale.

QDA training often takes less time than that required by FP. Consensus by personality domination, a potential problem with FP, is unlikely to occur since individual judgments are used in the data analysis. In addition, QDA data are readily analyzed by both univariate and multivariate statistical techniques. Statistical procedures such as multivariate analysis of variance, principal component analysis, factor analysis, cluster analysis have found application in the analysis of data generated by QDA-type procedures (Martens and Martens, 2001; Meullenet et al., 2007; Piggott, 1986). Graphical presentations of the data often involve the use of "cobweb" graphs (polar coordinate graphs a.k.a. radar plots, Fig. 10.3).

There is some argument about the assumption of normal distribution of the data set and hence the use of parametric statistics such as analysis of variance, *t*-tests. A few authors feel that non-parametric statistical treatment of the data is required (O'Mahony, 1986; Randall, 1989), but this appears to be a minority opinion.

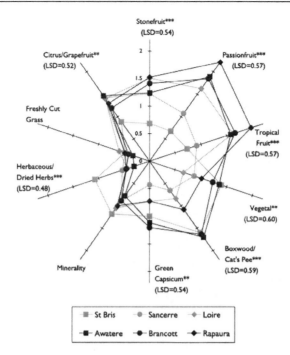

Fig. 10.3 An example of a cobweb or radar plot of Descriptive Analysis data. The data are from aroma profile descriptive analysis of Sauvignon blanc wines as a function of country of origin (France or New Zealand) and sub-region (France: Saint Bris, Sancerre, Loire; New Zealand: Awatere, Brancott, Rapaura). For each sensory attribute the perceived mean intensity increases outward from the center point. Sub-region means differing by more than the LSD value for that attribute differ in a Fisher's LSD multiple comparison test (Parr et al., 2009, used with permission).

The ease of data analysis using QDA may actually be considered one of the problems of the technique. The tendency to use the scales as absolute measures of an attribute rather than as a tool to see relative differences between samples is very common. Returning to the potato chip example, a decision may be made that samples scoring less than 5 on the crispness scale are not acceptable for sale. As we saw, Judge B's crispness intensity of 5 was very different from Judge A's 5. By extension, we can see that if the entire panel used the upper end of the scale, no samples would be considered unacceptable by this criterion. If another panel, analyzing the same samples, uses only the lower end of the scale, no sample is acceptable. The QDA data must therefore be viewed as *relative* values and not as absolutes. QDA studies should therefore be designed to include more than one sample and/or a benchmark or standard product as often as possible.

QDA has been extensively used, but often the experiments are not designed exactly as described by Stone and Sidel (2004). The relative simplicity of the technique allows it to be adapted in many different ways. However, any adaptation invalidates the use of the name QDA to describe the procedure.

Advantages cited by advocates of QDA include the ideas that the panelists perform independent judgments and that results are not consensus derived. Additionally, the data are easily analyzed statistically and graphically represented. Panel language development is free of panel leader influences and is, in general, based on consumer language descriptions. QDA suffers from the same disadvantage as FP, since in both cases the panels must be trained for the specific product category. Many US food companies maintain separate panels for their many product categories. This is very expensive and may limit the use of this technique by smaller firms. Unlike FP, the QDA results do not necessarily indicate the order of perception of sensations. However, the panel could be instructed to list the descriptors on the ballot in sequence of appearance, if that is needed to meet the objective of the study. Additionally, as indicated above, the results are relative and not absolute, since panelists may use different scale ranges.

10.4.2.1 Selection of Judges for Quantitative Descriptive Analysis

Similar to FP judges, Quantitative Descriptive Analysis (QDA) panelists should be screened for long-term availability. As with FP, it takes time, effort, and money to train a panel and the panelists should make a commitment to be available for years, if possible. This becomes a management support issue when the panelists are selected from within company since these employees may spend substantial time away from their main jobs. These panelists are screened for normal odor and taste perceptions using actual products from the category. The potential panelists should be very articulate and sincere.

Unlike FP, the panel leader is not an active participant in either the language development or the evaluation phases of the study. The panel leader acts as a facilitator only and does not lead or direct the panel. This person will be responsible for communication with the panel and preparation of samples and reference standards.

10.4.3 Texture Profile®

The Texture Profile was created by scientists working for General Foods during the 1960s and was subsequently modified by several sensory specialists (Brandt et al., 1963; Civille and Liska, 1975; Muñoz, 1986; Szczesniak, 1963, 1966, 1975; Szczesniak et al., 1963). The goal of the Texture Profile (TP) was to devise a sensory technique that would allow the assessment of all the texture characteristics of a product, from first bite through complete mastication, using engineering principles. The creators based the Texture Profile on the concepts pioneered by developers of the Flavor Profile. The texture profile was defined by Civille and Liska (1975) as

> the sensory analysis of the texture complex of a food in terms of its mechanical, geometrical, fat and moisture characteristics, the degree of each present and the order in which they appear from fist bite through complete mastication (p. 19).

The Texture Profile uses a standardized terminology to describe the textural characteristics of any product. Specific characteristics are described by both their physical and sensory aspects. Product-specific terms to be employed are chosen from the standardized terminology to describe the texture of a specific product. Definitions and order of appearance of the terms are decided through consensus by the TP panelists. Rating scales associated with the textural terms are standardized (Table 10.3).

Table 10.3 Example texture profile hardness[a] scale

Scale value	Product	Sample size	Temperature	Composition
1.0	Cream cheese	½" cube	40–45°C	Philadelphia cream cheese (Kraft)
2.5	Egg white	¼" cube	Room	Hard-cooked, 5 min
4.5	American cheese ½"' cube	40–45°C	Yellow pasteurized cheese (Land O Lakes)	
6.0	Olive	1 piece	Room	Stuffed, Spanish olives with pimentos removed (Goya Foods)
7.0	Frankfurter[b]	½" slice	Room	Beef Franks, cooked for 5 min in boiling water (Hebrew National Kosher Foods)
9.5	Peanut	1 piece	Room	Cocktail peanuts in vacuum tin (Planters, Nabisco Brands)
11.0	Almond	1 piece	Room	Shelled almonds (Planter, Nabisco Brands)
14.5	Hard candy	1 piece	Room	Life Savers (Nabisco Brands)

Adapted from Muñoz (1986)

[a]Hardness is defined as the force required to bite completely through sample placed between molar teeth

[b]Area compressed with molars is parallel to cut

Within each scale, the full range of a specific parameter is anchored by products having the specific characteristic as a major component. The reference product must be instrumentally evaluated to determine whether they conform to the intensity increments for the specified scale. The reference scales anchor both the range and the concept for each term (Szczesniak et al., 1963). For example, the hardness scale (Table 10.3) measures the compression force applied to the product between the molars. Note that the different foods (cream cheese, cooked egg white, cheese, olives, wieners, peanuts, raw carrots, almonds, and hard candies) used as reference points in the TP hardness scale increase in intensity from cream cheese to candy. However, these products alternately may shear, shatter, or compress when the compression force is applied. Thus when using the hardness reference scale panelists must understand that although all these products vary in a specific and definable dimension, hardness, they do not necessarily react in the same way to an applied compressive force.

It is crucial to the success of the TP that the frame of reference for all panelists be the same. All panelists must receive the same training in the principles of texture and TP procedures. Preparation, serving, and evaluation of samples should be rigidly controlled. Panelists should also be trained to bite, chew, and swallow in a standardized way. Usually, during panel training the panelists are first exposed to the Szczesniak (1963) classification of textural characteristics. They are subsequently exposed to a wide variety of food products and reference scales. During the third phase, the panelists refine their skills in recognizing, identifying, and quantifying degrees within each textural characteristic in a specific food category. This normally takes several weeks of daily training sessions but this may be worthwhile. Otremba et al. (2000), working with beef muscles, found that the extensive training led to greater consistency and accuracy.

The Texture Profile has been applied to many specific product categories including breakfast cereals, rice, whipped toppings, cookies, meat, snack foods. However, in many cases the experimenters will state that they had used TP in their studies but careful analysis of their methodology reveals that the exacting requirements of true TP were not adhered to during these studies. Often panelists are not trained using the standardized methodology to the degree recommended by the original proponents of this technique.

10.4.4 Sensory Spectrum®

Gail Civille, while working at General Foods in the 1970s, became an expert using the Texture Profile. She, subsequently, created the Sensory Spectrum technique using many of the ideas inherent to the Texture Profile. The Sensory Spectrum procedure is a further expansion of descriptive analysis techniques. The unique characteristic of the Spectrum approach is that panelists do not generate a panel-specific vocabulary to describe sensory attributes of products, but that they use a standardized lexicon of terms (Civille and Lyon, 1996). The language used to describe a particular product is chosen a priori and remains the same for all products within a category over time. Additionally, the scales are standardized and anchored with multiple reference points. The panelists are trained to use the scales identically; because of this, proponents of the Spectrum method state that the resultant data values are absolute. This means that it should be possible to design experiments that include only one sample and to compare the data from that sample with data derived in a different study. This philosophy suggests that since each panel is a unique group, allowing panels to generate their own consensus terms may lead to misleading results when attempting to apply the findings to a generalized population. The proponents of the method state that the descriptors used for the Spectrum method are more technical than the QDA descriptors. According to Sensory Spectrum users, QDA terms are generated by the panelists themselves and they are more likely to be related to consumer language. Reviews of the Spectrum method have been provided by Powers (1988), Murray et al. (2001), and Meilgaard and coworkers (2006).

Panelist training for the Spectrum method is much more extensive than QDA training and the panel leader has a more directive role than in QDA. As in QDA, the judges are exposed to a wide variety of the products in the specific product category. As in the Texture Profile, the panel leader provides extensive information on the product ingredients. The underlying chemical, rheological, and visual principles are explored by the panelists and the relationships between these principles and sensory perceptions of the products are considered. Similar to the Texture Profile the panelists are provided word lists (called lexicons by Sensory Spectrum) that may be used to describe perceived sensations associated with the samples. The ultimate goal is to develop an "…'expert panel' in a given field, … [to] demonstrate that it can use a concrete list of descriptors based on an understanding of the underlying technical differences among the attributes of the product" (Meilgaard et al., 2006). Additionally, panelists are supplied with reference standards. For attributes, specific singular references as well as standards in combination with a few other attributes are provided. An example would be vanilla and vanilla in milk and/or cream (Muñoz and Civille, 1998).

Panelists use intensity scales that are numerical, usually 15-point scales, and *absolute* also known as *universal* (Table 10.4; Muñoz and Civille, 1997). Civille (April 1996, personal communication) states that the scales are created to have equi-intensity across scales. In other words, a "5" on the sweetness scale is equal in intensity to a "5" on the salty scale and this is even equal in intensity to a "5" on the fruity scale (Table 10.5). Civille (April 1996, personal communication) says this goal has been achieved for fragrance, aroma, and flavor scales but not for texture scales. We are somewhat skeptical of this equi-intensity claim since there are no published data

Table 10.4 Example of aromatic reference samples used for spectrum scales

Descriptor	Scale value[a]	Product
Astringency	6.5	Tea bags soaked for 1 h
	6.5	Grape juice (Welch's)
Caramelized sugar	3.0	Brown Edge Cookies (Nabisco)
	4.0	Sugar Cookies (Kroger)
	4.0	Social Tea Cookies (Nabisco)
	7.0	Bordeaux Cookies (Pepperidge Farm)
Egg	5.0	Mayonnaise (Hellmann's)
Egg flavor	13.5	Hard boiled egg
Orange complex	3.0	Orange Drink (Hi-C)
	6.5	Reconstituted frozen orange concentrate (Minute Maid)
	7.5	Freshly squeezed orange juice
	9.5	Orange concentrate (Tang)
Roastedness	7.0	Coffee (Maxwell House)
	14.0	Espresso coffee (Medaglia D'Oro)
Vanilla	7.0	Sugar Cookies (Kroger)

Adapted from Meilgaard et al. (2006)
[a]All of the above scales run from 0 to 15

Table 10.5 Intensity values used for spectrum scales assigned to the four basic tastes in assorted products

Descriptor	Scale value[a]	Product
Sweet	2.0	2% sucrose-water solution
	4.0	Ritz cracker (Nabisco)
	7.0	Lemonade (Country Time)
	9.0	Coca Cola Classic
	12.5	Bordeaux Cookies (Pepperidge Farm)
	15.0	16% sucrose-water solution
Sour	2.0	0.05% citric acid-water solution
	4.0	Natural apple sauce (Motts)
	5.0	Reconstituted frozen orange juice (Minute Maid)
	8.0	Sweet pickle (Vlasic)
	10.0	Kosher dill pickle (Vlasic)
	15.0	0.20% citric acid-water solution
Salt	2.0	0.2% sodium chloride-water solution
	5.0	Salted soda cracker (Premium)
	7.0	American cheese (Kraft)
	8.0	Mayonnaise (Hellman's)
	9.5	Salted potato chips (Frito-Lay)
	15.0	1.5% sodium chloride-water solution
Bitter	2.0	Bottled grapefruit juice (Kraft)
	4.0	Chocolate bar (Hershey)
	5.0	0.08% caffeine-water solution
	7.0	Raw endive
	9.0	Celery seed
	10.0	0.15% caffeine-water solution
	15.0	0.20% caffeine-water solution

Adapted from Meilgaard et al. (2006)
[a]All the above scales run from 0 to 15

to support it. However, the concept of cross-modal matching may make the above claim reasonable for light and tones, tastants (sweetness and sourness), but it may not be reasonable for sweetness and hardness or fruitiness and chewiness (Stevens, 1969; Stevens and Marks, 1980; Ward, 1986).

Also, the stability of the absolute scale is not clear. Olabi and Lawless (2008) found contextual shifting in the 15-point scale even after extensive training.

As with the Texture Profile, scales are anchored by a series of reference points. In this schema at least two and preferably three to five references are recommended. The reference points are chosen to represent different intensities on the scale continuum. The reference points are used to precisely calibrate the panelists in the same way as pH buffers calibrate a pH meter. The panelists are "tuned" to act like true instruments.

After training, all panelists must use the scales in an identical fashion. Thus, they should all score a specific attribute of a specific sample at the same intensity. Testing is performed in isolated booths, using typical sensory practices.

The principal advantage claimed for the Spectrum method should be apparent after reading the discussion of the QDA procedure. In QDA, judges frequently use the scales provided in idiosyncratic but consistent ways. In contrast to the QDA, the Spectrum method trains all panelists to use the descriptor scales in the same way. Thus, scores should have absolute meaning. This means that mean scores could be used to determine if a sample with a specified attribute intensity fits the criterion for acceptability irrespective of panel location, history, or other variables. This has obvious advantages to organizations wishing to use a descriptive technique in routine quality assurance operations or in multiple locations and facilities.

Disadvantages of the procedure are associated with the difficulties of panel development and maintenance. Training of a Spectrum panel is usually very time consuming. Panelists have to be exposed to the samples and understand the vocabulary chosen to describe the product. They are asked to grasp the underlying technical details of the product and they are expected to have a basic understanding of the physiology and psychology of sensory perception. After all that, they must also be extensively "tuned" to one another to ensure that all panelists are using the scales in the same way. We are not sure that this level of calibration can be achieved in reality. In practice, individual differences among panelists related to physiological differences like specific anosmias, differential sensitivities to ingredients can lead to incomplete agreement among panelists. Theoretically, if panelists were in complete agreement one would expect the standard deviation (see Appendix) for any specific product–attribute combination to be close to zero. However, most Spectrum studies have attributes with non-zero standard deviations indicating that the panel is not absolutely calibrated. Civille (April 1996, personal communication) has stated that absolute calibration is feasible for most attributes but probably not for bitterness, pungency, and certain odor perceptions.

Data from the Spectrum technique are analyzed in a similar fashion to the QDA data. The deviation of mean values for particular attributes is of definite interest to the analyst, since these values can be directly related to the "tuning" or precision of the panel.

10.5 Generic Descriptive Analysis

QDA and Sensory Spectrum techniques have been adapted in many different ways. It is important to note, however, that any adaptations invalidate the use of the trademarked names "QDA" and "Sensory Spectrum." Unfortunately, it is often difficult to evaluate the effect that the myriad deviations from the standard methodologies have on the validity of the data. Academic researchers frequently employ the general guidelines of these methodologies to evaluate products. Table 10.6 shows the steps in conducting a generic descriptive analysis; these steps will be described in detail in the next sections. Additionally, some very interesting variations on the conventional generic descriptive analysis have been created and these will be discussed in Section 10.4.7.

10.5.1 How to Do Descriptive Analysis in Three Easy Steps

It is possible for any competent sensory scientist to perform a descriptive analysis study in three easy steps. These steps are train the judges, determine the judge reproducibility/consistency, and have the judges evaluate the samples. We will discuss each of these steps in more detail.

10.5.1.1 Training the Panelists

As we have seen with the QDA and Sensory Spectrum methods, there are two methods of judge training. The first is to provide the panelists with a wide range of products in the specific category. Panelists are asked to generate the descriptors and reference standards needed to describe differences among the products, usually by coming to some consensus. For simplicity we will call this "consensus training." The second method is to provide the panelists with a wide range of products within the category as well as a word

Table 10.6 Steps in conducting a generic descriptive analysis

1. Determine project objective: Is descriptive analysis the right method?
2. Establish products to be used with clients/researchers
3. Determine whether consensus or ballot training is most appropriate
4. Establish experimental design and statistical analyses
   a. Main effects and interactions for analyses of variance
   b. Multivariate techniques?
5. Select (and optionally, screen) panelists
   If choosing to do consensus training go to 6. If choosing to do ballot training go to 7
6. Consensus training
   a. During initial training sessions provide panelists with a wide range of products in the specific category
   b. Panelists generate descriptors (and ideas for reference standards)
   c. During subsequent training sessions panel leader provides potential reference standards as well as products
   d. Panelists reach consensus in terms of attributes, reference standards, and score sheet sequencing
7. Ballot training
   a. During initial training sessions provide panelists with a wide range of products in the specific category.
   b. Provide panelists with a word list (sample score sheet) and reference standards
   c. During subsequent training sessions panel leader provides reference standards as well as products
   d. Panelists indicate which attributes and reference standards from the word list should be used in the specific study. Panelists may also indicate sequence of attributes on score sheet
8. Once the training phase has been completed, panelists performance is checked
   a. A subset of samples are provided in duplicate (triplicate) under actual testing conditions
   b. Data are analyzed and any issues with reproducibility and/or attribute usage lead to additional training; testing may occur again after re-training.
9. Conduct study
10. Analyze and report data

list of possible descriptors and references that could be used to describe the products. We will refer to this method as "ballot training." In practice, both the consensus and the ballot methods have an application. However, keep in mind that Sulmont et al. (1999) found that panelists tended to perform better when trained by the "consensus" (trained by doing) rather than "ballot" (trained by being told) method. Frequently, however, a combination method is used. In the combination method panelists derive some descriptors on their own through consensus and others are

added through suggestions by the panel leader or from word lists. The panel leader may also reduce redundant terms. In our laboratories the consensus method is usually used in research studies with the exception of meat research studies. For meat we tend to use the ballot method, mostly because a multitude of studies in the area has convinced us only a limited number of descriptors are readily applicable to meat. In contract work for US food and consumer products companies, we tend to use the combination method, since the client companies often have certain terms that they deem important. These will then be suggested by the panel leader, if the panelists do not use them spontaneously.

A typical sequence of "consensus training" sessions would be the following:

Initially, the panelists are exposed to the entire range of the products. They are asked to evaluate the sensory differences among the samples and to write down the descriptors that describe these differences. This occurs in silence. When all panelists complete this portion of the assignment, the panel leader asks each panelist to list the words used to describe each sample. During this phase of the training it is extremely important that the panel leader must be cautious not to lead or to judge any descriptor from any panelist. However, the panel leader may ask for clarification, if needed. Usually, the panelists themselves will begin to move toward initial consensus when they see the total list of descriptors elicited.

Subsequently, the panel leader should attempt to provide potential reference standards based on the initial consensus. These reference standards are chemicals, spices, ingredients, or products that can be used to help the panelists identify and remember the sensory attribute found in the samples evaluated (Rainey, 1986). In general, the panel leader should strive to use actual physical objects as the reference standards but in some cases precise written description may be used instead (Table 10.7). At the next session, the panelists are exposed to the samples again and asked to decide on the possible reference standards. If reference standards are not feasible, the panelists can also be asked to verbally define the specific descriptor. This refinement of the consensus list of descriptors, reference standards, and definitions continues until the panelists are satisfied that they have the best possible list and that everyone understands each term completely. Murray and Delahunty (2000) had their panelists determine

Table 10.7 Composition of reference standards for aroma and flavor evaluations. These reference standards were used in a descriptive study of vanilla essences from different geographic locations (Woods, 1995)

Aroma attribute	Composition
Smoky	20"' of binding twine lit with lighter, allowed to burn and then blown out, smell smoke
Scotch[a]	15 ml of 5% solution of J&B Justerini & Brooks Ltd., rare blended scotch whiskies (London, England)
Bourbon	15 ml of 5% solution of Walker's deluxe straight Bourbon Whiskey (Hiram Walker & Sons Co., Bardstown, KY)
Rum	15 ml of 5% solution of Bacardi Superior Puerto Rican Rum (Bacardi Corp., San Juan, Puerto Rico)
Almond	15 ml of 1.25% solution of McCormick® Pure Almond extract (McCormick & Co., Inc., Hunt Valley, MD)
Kahlua	15 ml of 1.25% solution of original Mexican Kahlua (Kahlua S.A., Rio San Joaquin, Mexico)
Medicinal	15 ml of 20% solution of Cepacol® mouthwash (Merrell Dow Pharmaceuticals, Inc., Cincinnati, OH)
Buttery	One piece of Lifesavers® Butter Rum candy (©Nabisco Foods, Inc., Winston-Salem, NC)
Creme Soda	15 ml of 2% solution of Shasta® creme soda (Shasta Beverages Inc., Hayward, CA)
Fruity	15 ml of 30% (5:1) solution of Welch's Orchard® apple-grape-cherry fruit juice cocktail frozen concentrate and Welch's® 100% white grape juice from concentrate (no sugar added) (©Welch's, Concord, MA)
Prune	One Sunsweet® medium prune (Sunsweet Growers, Stockton, CA)
Tobacco	One pinch of large size Beech-nut Softer & Moister chewing tobacco (©National Tobacco, Louisville, KY)
Earthy	19 g of black dirt from Missouri
Musty	**Verbally defined as "a damp basement"**
Nutty	2–3 Kroner® salted pistachios (shelled and cut into pieces) (Kroner Co., Cincinnati, OH)
Flavor attribute[b]	**Composition**
Amaretto	15 ml of 5% solution of Disaronno-Amaretto Originale (Illva Saronno, Saronno, Italy)
Sweet	Panelists were not provided with a reference, but were given a 2 and 6% solution of sugar water during training to anchor the scale
Fruity	15 ml of 5% (5:1) solution of Welch's Orchard® apple-grape-cherry fruit juice frozen concentrate and Welch's® 100% white grape juice from concentrate (no sugar added) (©Welch's, Concord, MA)
Earthy	1 Campbell Soup Company fresh packaged mushrooms—diced (Camden, NJ)

Please note that for most of these attributes very precise reference standards were created—this is the ideal. But for the attribute in bold a definition and no reference standard is given—this is not an ideal situation

[a] All solutions made using Culligan sodium-free drinking water (Culligan Water Ltd., Columbia, MO)

[b] All other flavor standards were the same as those for aroma standards

the suitability of each potential reference standard for cheddar cheese by having them score the attributes on an appropriateness scale.

During the final training session the panelists create the score sheet. They may be allowed to decide on the scale to use, although in our laboratories we usually use either the unstructured line scale (similar to Fig. 10.1) or the 15-point unlabeled box scale (Fig. 10.4) for most studies.

Sweetness intensity

□ □ □ □ □ □ □ □ □ □ □ □ □ □ □

Weak   Strong

Fig. 10.4 Example of a 15-point unlabeled box scale.

The panelists are asked to decide on the words needed to anchor the scales such as none to extreme or slight to very strong. We also frequently allow the panelists to determine the sequence in which they would like to evaluate the attributes, for example, visual attributes first (unless these are performed separately in a color evaluation chamber such as the Gretag-MacBeth Judge II); then aroma; followed by taste, flavor-by-mouth, and mouth feel; and lastly, after expectoration or swallowing, after-taste. For some panels this order may be changed—for example they may choose to do the taste, flavor by mouth, and mouth feel terms prior to aroma. Once again, the panel leader makes sure that the panelists are comfortable with all the terms, references, and definitions used. At this point the panel leader will start to evaluate judge reproducibility.

A typical sequence of "ballot training" sessions would be the following: Initially, the panelists are exposed to the entire range of the products. They are asked to evaluate the sensory differences among the samples. This occurs in silence. When all panelists complete this portion of the assignment, the panel leader gives each panelist a word list (or sample score sheet) for the products. The word list contains words, definitions, and often the panel leader will also have reference standards available to anchor the descriptors. There are a number of published word lists (lexicons) available for a variety of food and personal care products. A non-exhaustive list is given at the end of this section. The panelists are then asked to indicate through consensus which of these words, reference standards, and definitions should be used in the specific study. The panelists are allowed to add or delete terms through consensus. They are also asked to sequence the descriptors on the ballot.

In subsequent sessions the panelists are exposed to the samples again and asked to look at the ballot that they previously created. They then have to decide if this is truly the score sheet they want to use with these products. Refinement of the score sheet, reference standards, and definitions continues until the panelists are satisfied that this is the best possible score sheet, best sequence, and that everyone understands each term completely. Now the panel leader is ready to determine judge reproducibility.

Some of the available sensory lexicons (vocabularies) are the ASTM publications that cover a large number of product categories (Civille and Lyon, 1996; Rutledge, 2004) as well as Drake and Civille (2003) which covers the creation of flavor lexicons and has numerous references to available word lists. A few recent word lists are Cliff et al. (2000) for apple juices, Dooley et al. (2009) for lip products, Drake et al. (2007) for soy and whey proteins in two countries, Retiveau et al. (2005) for French cheeses, Lee and Chambers (2007) for green tea, Krinsky et al. (2006) for edamame beans, and Riu-Aumatell et al. (2008) for dry gins. There are also published reports of generic descriptive analysis using terminology that are extremely localized. An example would be Nindjin et al. (2007) who trained a group of adult villagers in the Ivory Coast to use the local language to describe the sensory differences among samples of "foutou" (pounded yams).

10.5.1.2 Determining Panelist Reproducibility During Training

Immediately after the training phase the panelists are told that the evaluation phase of the study will begin. However, in reality, the first two or three sessions are used to determine judge consistency. A subset of samples to be used for the real study is served to the panelists in triplicate. The data from these sessions are analyzed; the sensory scientist will study the significance levels of the interaction effects associated with panelists. In a well-trained panel these effects would be not significantly different among judges. If there are significant panelist-associated interaction effects the sensory scientist will determine which judges should be further trained in the use of which descriptors. If all judges are not reproducible then they all need to return to the training phase. However, the results usually indicate that one or two subjects have problems with one or two descriptors. These problems can usually be resolved during a few one-on-one training sessions. Cliff et al. (2000) showed that as training progressed the standard deviations associated with 10 of their 16 attributes decreased. In some cases this decrease was large (0.90 on a 10 cm line scale for oxidized aroma and flavor) and in others smaller (<0.05 for green-grassy and sour). Their panelists anecdotally found that the biggest training effects occurred when the chosen reference standards were unambiguous. See below for a more in-depth discussion on panel performance monitoring.

Recently some work on the effect of feedback calibration on panel training has been published (Findlay et al., 2006, 2007). These authors found that immediate graphical computerized feedback on performance in the sensory booths during training led to reduced training time as well as excellent panel performance. McDonell et al. (2001) also found that feedback in the form of principal component analysis plots, analysis of variance shown to the panel after each descriptive analysis sped up the training process and made the panel more consistent. Nogueira-Terrones et al. (2008) trained a descriptive panel over the Internet to evaluate sausages. Their training process essentially involved feedback on performance at each session and increased training duration increased their Internet panelists' performance relative to the performance of panelists trained more conventionally. However, Marchisano et al. (2000) had found that feedback was positive for recognition tests, had no effect on discrimination tests (triangle tests), and may have been a negative for scaling tests. Clearly, additional studies are needed. There is an ongoing discussion in sensory circles as to whether panelists should be recruited from within or from outside companies, in other words, whether company employees should be expected to volunteer for panel duty as part of their other duties or whether panelists should be employed to only be on sensory panels. There is very little research to guide on in this discussion. One of the few studies was the one by Lund et al. (2009). They surveyed panelists in New Zealand, Australia, Spain, and the United States and found that the key drivers stimulating people to participate in sensory panels were a general interest in food and extra income. Additionally, panelists on external panels (those not otherwise employed by the company) were more intrinsically motivated than internal panelists (those otherwise employed by the company). Panelists' experience also improved their intrinsic motivation.

10.5.1.3 Evaluating Samples

Standard sensory practices, such as sample coding, randomized serving sequences, use of individual booths, should be employed during the evaluation phase of the study. The sample preparation and serving should also be standardized. The judges should evaluate all samples in at least duplicate, but preferably in triplicate. Samples are usually served monadically and all attributes for a specific sample are evaluated before the next sample is served. However, as shown by Mazzucchelli and Guinard (1999) and Hein (2005) there are no major differences between the results when samples are served monadically or simultaneously (all samples served together and attributes rated one at a time across samples). However, in both studies the actual time taken to do the evaluation increased for the simultaneous serving condition. Under ideal conditions, all samples will be served in a single session, with different sessions as the replicates. If it is not possible to do so then an appropriate experimental plan such as a Latin square, balanced incomplete block should be followed (Cochran and Cox, 1957; Petersen, 1985). The data are usually analyzed by analysis of variance. However, analysis by one

or more appropriate multivariate statistical techniques may yield additional information (see Chapter 18).

10.5.1.4 Panel Performance Monitoring

As stated in Section 10.5.1.2—Determining Panelist reproducibility during training—, the sensory scientist will usually have panelists evaluate a subset of products in replicate and then analyze that data to determine whether further training is warranted. However, one may also be interested in monitoring panelist performance over the life span of the panel. This is more usually done when a panel continues to be used over a number of projects or for a number of years, i.e., when one has a "permanent panel." For example, some of the panelists in the Kansas State University Sensory Analysis Center panel have been participating in the panel since 1982 (personal communication, Edgar Chambers, IV, October 2009). When one has a "temporary panel"—a panel that is trained for one specific project and then disbanded—it is more unusual to do ongoing panelist performance monitoring. One may also be interested in panelist performance monitoring when newly trained panelists are merged into an ongoing panel, something that occurs routinely in many commercial settings.

The techniques used to monitor panel performance are similar whether one is monitoring the panel toward the end of training or for the other reasons listed above. The key pieces of information the sensory scientist needs are (a) individual panelist discriminating ability; (b) individual panelist reproducibility; (c) individual panelist agreement with the panel as a whole; (d) panel discriminating ability; and (e) panel reproducibility. Numerous statistical analyses are available to find these pieces of information from the panel data. Please see Meullenet et al. (2007), Tomic et al. (2007), and Martin and Lengard (2005) for additional information on this topic. Derndorfer and coworkers (2005) published code in R to evaluate panel performance. Pineau et al. (2007) published a mixed-model and control chart approach using SAS (SI, Cary, NC). SensomineR (a freeware R-package) also contains panel performance techniques, as well as many sensory data analysis techniques (Lê and Husson, 2008). Additionally, Panel Check, another freeware R-based program, is available for download at http://www.panelcheck.com/ (Kollár-Hunek et al., 2007; Tomic et al., 2007). In this section we will briefly discuss four of these techniques. In order to simplify the discussion of panel performance monitoring we assume that each member of the sensory panel evaluated the entire set of products in triplicate.

Univariate Techniques

One-way analyses of variance with product as the main effect for each panelist and each attribute allow the sensory scientist to evaluate the individual panelists' discriminability as well as their repeatability. The assumption is that panelists with excellent discriminability for a specific attribute would have large F-values and small probability (p) values. Panelists with good repeatability would tend to have small mean square error (MSE) values. A plot of p-values by MSE values allows the sensory scientist to simultaneously evaluate both discriminability and repeatability (Fig. 10.5).

A three-way analysis of variance with main effects (product, panelist, and replication) and interaction effects (panelist by product, panelist by replication, and product by replication) will fairly quickly indicate

some trouble spots in panelist performance relative to the rest of the panel. The sensory scientist should be on the lookout for attributes with significant panelist by product interactions. These would indicate that at least one (and possibly more panelists) is not scoring these attributes similarly. One should always plot the data. If a panelist's results decrease (increase) while the panel means increase (decrease) it is called a cross-over interaction and it is a major problem. If a panelist's results decrease (increase) while the panel means decrease (increase) but at a different rate then the interaction is less of a problem.

Panelist performance relative to the panel as a whole for each attribute can also be visually shown with eggshell plots (Hirst and Næs, 1994). In this case the panelist's scores for each attribute are transformed into ranks. A consensus ranking for each attribute is then created by finding the mean rank over panelists for each product and then ranking these means. Each panelist's cumulative scores are then plotted relative to the consensus ranks. The resultant plot looks similar to an eggshell, and the intention is to have as few "cracks" as possible in the shell for each attribute (Fig. 10.6).

Multivariate Techniques

A principal component analysis (PCA) of each attribute for all the panelists will indicate the consonance (agreement) among the panelists for that attribute (Dijksterhuis, 1995). In this case the panelist scores for each product for the specified attribute are used as the variables (columns) in the analysis. If there is substantial agreement (consonance) among the panelists then the majority of the variance should be explained by the first dimension. In other words if the panelists use the specific attribute similarly then the PCA should tend to become unidimensional. Usually, for well-trained panels the amount of variance explained on the first dimension ranges from about 50 to 70% (Fig. 10.7).

Worch et al. (2009) found that for untrained consumers these values tend to be much lower, ranging from about 15 to about 24%. The sensory scientist can also calculate a consonance (C) score for each attribute from the PCA results. Dijksterhuis (1995) defined C as the ratio of the variance explained by

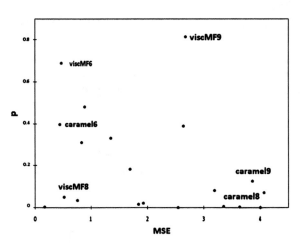

Fig. 10.5 An example of a p by MSE plot for all panelists (only some panelists are named) for caramel aroma (caramel) and viscous mouth feel (viscMF). Panelist 8 shows excellent discriminability (low p-value) as well as excellent repeatability (low MSE) for viscous mouth feel. For caramel this panelist also has excellent discriminability (low p-value) despite repeatability issues. Panelist 9 has repeatability issues, especially for both attributes but also has discriminability issues with viscous mouth feel and to a lesser extent with caramel aroma.

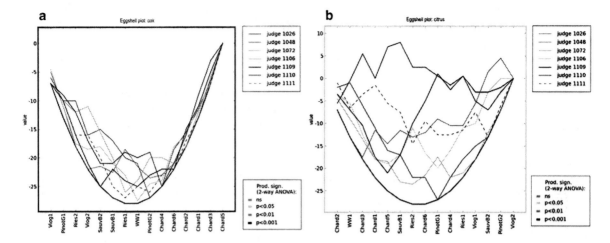

Fig. 10.6 Two examples of eggshell plots. The smooth line on the bottom of the plot is the consensus rank for the specific attribute. From the plots it is evident that the panelists were more in agreement with each other on the oak aroma attribute (**a**) than on the citrus aroma attribute (**b**).

Fig. 10.7 An example of two PCA panelist consonance plots. In the first plot (**a**) there is disagreement among the panelists in their usage of the specific term. In the second plot (**b**) there is more agreement among the panelists in terms of their use of the second term (reprinted with permission from Le Moigne et al., 2008).

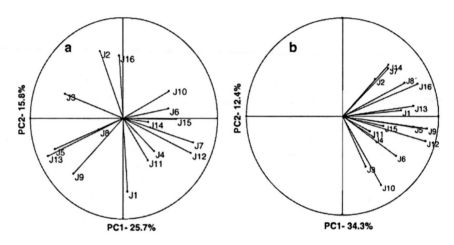

the first dimension to the sum of the remaining variances. Large values of C would indicate that there was agreement among the panelists in the usage of a specified term since the vectors for the terms would "point" in the same direction. The sensory scientist must be careful to not just blindly calculate C since large values of C are possible when there are large negative loadings on the first dimension as well as large positive ones. Thus prior to calculating C one should always plot the PCA for each attribute. Dellaglio et al. (1996) reported C values ranging from about 0.4 to 2.3 for a panel evaluating Italian dry-cured sausages. Carbonell et al. (2007) found C values ranging from 0.46 to 4.6 for a panel evaluating Spanish mandarin juices.

10.5.2 Studies Comparing Different Conventional Descriptive Analysis Techniques

Risvik et al. (1992) and Heymann (1994a) found that well-trained independent panels (in two different countries, Norway and Britain, and in the same university setting, respectively) gave very comparable results. A study by Lotong et al. (2002) on the evaluation of orange juices by two independently highly trained panels (one panel used individual judgments and the other created a consensus evaluation) showed that the results from the different panels were comparable. Drake et al. (2007) evaluated the descriptive sensory analyses of whey and soy proteins by two independently, and extensively, trained panels (one in the United States trained through the Spectrum method and one in New Zealand trained using a general descriptive analysis approach). They found that the two panels found distinct and consistent differences between the whey and soy proteins and that they performed similarly. As stated by the authors "…product differentiation was similar, but attribute usage was not. … A key result from this study … is that *well-trained* panels using independent sensory languages can produce comparable results." Italics are ours—the key to adequate consistency across languages and philosophies is in the training of the panels. A comparison of cheddar cheese descriptive analyses from Ireland, the United States, and New Zealand also indicated that highly trained panels using standardized, representative languages can provide comparable results (Drake et al., 2005). Bárcenas et al. (2007) had five highly trained panels (two in Spain, one each in Italy, France, and Portugal) evaluate European ewes milk cheeses and found that the panels all significantly discriminated among the cheeses, although there were some differences in the use of attributes. However, McEwan et al. (2002) found that the performances of 12 European panels evaluating red wines were not comparable and it seems that a lack of adequate training may be to blame for this result. Similarly studies comparing trained descriptive analysis panels and untrained consumer panels asked to rate attribute intensities tend to show non-comparable results (Gou et al., 1998). This study, once again, emphasizes the need to adequately train descriptive panels to ensure reliable, consistent valid results. There are numerous additional studies comparing descriptive panels from different countries, using different training techniques, different vocabulary generation, etc., and all essentially conclude that results are similar as long as the panels were highly trained.

10.6 Variations on the Theme

10.6.1 Using Attribute Citation Frequencies Instead of Attribute Intensities

Currently this technique has usually been used with wines, but it would be appropriate for other products as well. This technique involves a similar training schedule to normal generic descriptive analysis but in the case of the citation frequency descriptive analysis the aim is to have as many relevant terms with their appropriate consensus-derived reference standards as possible. The number of attributes retained by the trained panel has varied from an unusually low 10 attributes (McCloskey et al., 1996) to 73 terms (Campo et al., 2008) to 113 attributes (Campo et al., 2009) to 144 terms (Le Fur et al., 2003). The second difference from generic descriptive analysis is that the panels for citation frequency descriptive analysis are much larger than the usual 8–12 used in generic descriptive analysis. Panel sizes have ranged from an unusually low 14 (Le Fur et al., 2003) to 26 (McCloskey et al., 1996) to 28 (Campo et al., 2008) to 38 (Campo et al., 2009). Once they are trained the panelists are asked to evaluate the wines in duplicate or triplicate but unlike generic descriptive analysis they do not use a scale. Instead each panelist is asked to indicate a specified number of attributes that are most descriptively associated with each product. McCloskey et al. (1996) asked panelists to use between two and five terms per Chardonnay wine, Le Fur used five to six terms for each Chardonnay wine, Campo et al. (2009, 2008) used either a maximum of five terms for Burgundy Pinot noir wines or a maximum of six terms for Spanish monovarietal white wines, respectively.

It is possible to calculate an average reproducibility index (R_i) to assess individual panel performance across duplicate evaluations:

$$R_i = \sum [2 \times \text{des}_{\text{com}}/(\text{des}_{\text{rep1}} + \text{des}_{\text{rep2}})/n] \quad (10.1)$$

where

des_{com} = number of common terms used by the specific panelist in the two replicates

des_{rep1} = number of terms used by the specific panelist in Replicate 1

des_{rep2} = number of terms used by the specific panelist in Replicate 2

n = number of products

The R_i value can range from 0 (no reproducibility across replicates) to 1 (perfect agreement between replicates). It has been suggested that the data from panelists with R_i values less than 0.2 should not be used for further data analyses. Campo et al. (2008) found a mean R_i of 0.51 (on average 51% of the terms were used by panelists in both replicates), a median R_i of 0.32, and a low R_i of 0.17. In the Burgundy Pinot noir study Campo et al. (2009) found a mean R_i of 0.69 and a low R_i of 0.24.

For data analyses the terms are ranked by their citation frequency (C_f) to determine the most relevant attributes. Usually, only attributes with a C_f of at least 15% (in other words used in at least one wine by replication situation by at least 15% of the panelists) are used in the subsequent data analyses. A chi-square analysis is performed on the mean C_f (averaged across replications) of each attribute and wine to determine discriminating attributes. Correspondence analysis is then used to create two- or three-dimensional maps of the product–attribute spaces (Greenacre, 2007; Murtagh, 2005). Correspondence analysis requires contingency tables. In this case the data are organized into a contingency table of the mean C_f with rows as the products and attributes as the columns.

Currently (2009), only one study has been published that compared citation frequency descriptive analysis with conventional descriptive analysis. In this study Campo et al. (2009) found that there were some similarities between the two methods but that the citation frequency method, despite its longer training requirements and its requirement for more panelists and thus more products, may lead to more nuanced results. In other words the technique may detect more subtle differences than conventional descriptive analyses. This should be studied further.

10.6.2 Deviation from Reference Method

The Deviation from Reference method (Larson-Powers and Pangborn, 1978) uses a reference sample against which all other samples are evaluated. The scale is a degree of difference scale with the reference as the midpoint anchor. The example in Fig. 10.5 is an unstructured line scale but numerical scales are also used. The reference is often included as a sample (not identified as the reference) and used as an internal measurement of subject reliability. The results are evaluated in relation to the reference. Thus samples that score less than the reference for specified descriptors are indicated by negatives and those that score more are indicated with positives (Fig. 10.8).

Larson-Powers and Pangborn (1978) concluded that the deviation-from-reference scale improved the precision and accuracy of the responses in a descriptive analysis study. However, Stoer and Lawless (1993) found that the method did not necessarily increase precision. They felt that the technique would best be used when distinctions among sample are difficult or when the objective of the study involves comparisons to a meaningful reference. An example of a meaningful reference would be a control sample that had been stored in such a way as to undergo no changes, compared to samples that have undergone accelerated shelf-life testing (Labuza and Schmidl, 1985). This is exactly the protocol Boylston et al. (2002) used when evaluating the effects of radiation on papayas, rambutans, and kau oranges. The panelists used deviation from reference

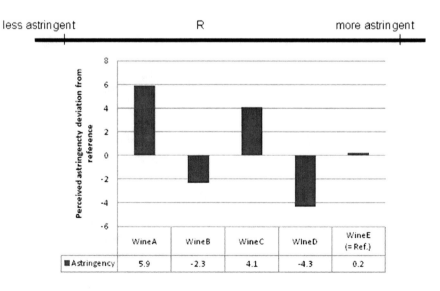

Fig. 10.8 (*Top*) Example of a Deviation from Reference scale and (*Bottom*) Graphical representation of deviation-from-reference results.

scales to evaluate the fruits with the reference being a control, un-irradiated sample.

10.6.3 Intensity Variation Descriptive Method

This method was developed by Gordin (1987) to provide the sensory scientist information about changes in descriptive attribute intensities as the sample is consumed. Specifically, the technique was created to quantify changes occurring in the sensory characteristics of a cigarette during consumption. Conventional time–intensity and conventional descriptive techniques were not suitable for this product since the variability in smoking rate would not allow the same portions of the cigarette to be evaluated by all panelists within the same time frame. The Intensity Variation Descriptive method concentrated panelist evaluations within specified locations of the product, rather than within specified time intervals. The cigarette was divided into sections by drawing lines on the tobacco rod with marking pen. Through consensus the panelists derived attributes that would be evaluated within each marked section of the cigarette. Panelist training, ballot development, and data analyses were standard descriptive methodology. As far as we can ascertain this method has only been used with cigarettes but it could be adapted for use with other products.

10.6.4 Combination of Descriptive Analysis and Time-Related Intensity Methods

10.6.4.1 Dynamic Flavor Profile Method

The Dynamic Flavor Profile Method (DeRovira, 1996) is a further extension of the combination of descriptive analysis and time–intensity methodology. As described by DeRovira the panelists are trained to evaluate the perceived intensities of 14 odor and taste components (acids, esters, green, terpenoid, floral, spicy, brown, woody, lactonic, sulfury, salt, sweet, bitter, and sour) over time. The data are graphically represented by isometric three-dimensional techniques, where a cross section of the graph at any specified time instant yields a spider-web profile for that time instant. The

technique seems to have some potential uses, yet we are concerned that the specification of 14 attributes may be too restrictive. It would be better to allow the panelists to determine the descriptors that they would like to use to describe the sensory characteristics associated with the product. It is conceivable that the panelists would decide that some attributes do not change over time and that others do. They would then do descriptive analysis of the attributes that do not change and a variation of the dynamic flavor profile for the attributes that do change.

10.6.4.2 Temporal Dominance of Sensations (TDS) Method

This method is described in the Time–Intensity chapter (Chapter 8). Briefly, the idea is to present a set of predetermined attributes together on the computer screen for the panelist's choice and scales for rating the intensity of each. Panelists are instructed to attend to and choose only the "dominant" sensation at any one time after tasting the sample and clicking on a start button.

10.6.4.3 Time-Scanning Descriptive Analysis

This method was devised by Seo and coworkers (2009) to evaluate hot beverages. They were concerned that panelists may take different amounts of time to do their evaluations and hence may be smelling and/or tasting the beverages at differing temperatures. In order to avoid this potential source of variability, they employed a time limit for the evaluation of each attribute. This ensured that all attributes across panelists were evaluated at the same time and at the same beverage temperature. One of us had done something similar many years ago in the evaluation of a series of body wash products (liquid soaps). Panelists were supplied with stop watches and were asked to evaluate the tackiness (stickiness) and smoothness of their skin at specified time intervals after stepping out of a shower and drying themselves.

10.6.5 Free Choice Profiling

During the 1980s British sensory scientists created and promoted a descriptive technique known as

Free Choice Profiling. The procedure was employed by several European researchers early on (Arnold and Williams, 1986; Langron, 1983; MacFie, 1990; Marshall and Kirby, 1988; Schlich, 1989; Williams and Langron, 19841984). The Free Choice Profiling technique shares much with the other techniques discussed previously. However, the method differs radically on at least two counts.

In the first place, vocabulary to describe the flavor notes is generated in a novel way. Instead of extensively training the panelists to create a consensus vocabulary for the product, FCP requires that each panelist creates his/her own idiosyncratic list of descriptive terms. Panelists are allowed to evaluate the product in different ways. They may touch, taste, or smell. They may evaluate shape, color, gloss, or other stimuli that interests them. Each sensation is rated on a scale using terms of the panelist's own devising. These individually generated terms need only be understood by the specific panelist. However, the individual must use the terms consistently when evaluating the products. Each panelist then uses his/her own unique list of terms to evaluate the products. As with QDA and the Spectrum methods, evaluations are performed in individual booths, under standard conditions.

The second unique feature of the FCP is in the statistical treatment of the scores from the panelists. These data are mathematically manipulated through the use of a procedure known as the Generalized Procrustes Analysis (Gower, 1975; Gower and Dijksterhuis, 2004; Oreskovich et al., 1991; Schlich, 1989). The Procrustes analysis usually provides a consensus picture of the data from each individual panelist in two- or three-dimensional space. It is possible to have a Procrustes solution with more than three dimensions but these are usually not interpretable.

The most distinct advantage of this technique is the avoidance of panel training. Experiments may be completed faster and at less expense. However, there is a significant time burden in creating a different ballot for each individual panelist. The ability of the panelist to use words that have unique meaning to that individual may allow for a more complete analysis of the sample flavor components. On the other hand, the idiosyncratic nature of the vocabularies may make interpretation of the sources of individual flavor notes difficult or impossible. For example, a panelist may use the word "camping" to describe the flavor note of a product. The researcher is forced to guess what

aspect of "camping" the panelist means: the smells of the woods, the musty leaves, the campfire smoke, etc. If this descriptor is in the same space as the musty, earthy, dirty descriptors of other panelists, the scientist has a clue to the flavor attribute being evaluated by that particular panelist. It is possible to imagine an analysis where all the words used by the individual panelists do not provide any clue as to their origin. For instance, in an evaluation of tea leaves performed in our laboratories, one panelist used the descriptor "coolstuff" to label one scale. Our curiosity was aroused, so we asked the panelist at the completion of the experiment what she meant by this descriptor. Cool Stuff is the name of a sundries store, in the town where one of us used to live, selling an eclectic collection of products, most notably to anyone walking into the store—incense. In this case it was possible to get to the underlying sensory characteristic being evaluated. However, in the same panel one individual used the descriptor "mom's cooking" which seems hopeless to characterize in more traditional sensory nomenclature.

As we have seen, in a FCP each individual panelist evaluates the products using their own idiosyncratic descriptors. Some panelists may use very few descriptors and others may use many descriptors. Additionally the terms rated by the panelists may not coincide. Therefore, standard univariate and multivariate statistical techniques like analysis of variance, principal component analysis, multi-linear regression cannot be used. The data from FCP studies are analyzed using generalized Procrustes analysis. The technique allows one to scale, reflect, and rotate multiple data matrices (one for each panelist for each replication) to obtain a consensus space (Gower, 1975). The iterative technique is called Procrustes analysis in reference to Hellenic mythology. Procrustes, a nickname for Damastes or Polypemon, meaning "the stretcher" was a robber who invited travelers to stay at his house (Kravitz, 1975). If the visitor did not fit his bed he would either stretch them or cut off their legs to make them fit the bed. His guests being thus incapacitated, Procrustes was able to help himself to his guests' possessions at his leisure.

The Procrustes analysis also in a sense force-fits the data matrices from the individual panelists into a single consensus space. The most important aspect of the Procrustes analysis is that it allows the analyst to determine the terms used by individual panelists that appear to be measuring the same sensory attributes as the other judges. With this technique each judge's data are transformed into individual spatial configurations. These configurations of individual judges are then matched by Procrustes analysis to a consensus configuration. The consensus configuration may be interpreted in terms of each individual assessor's vocabulary and the scientists may also determine how the different terms used by different assessors may be interrelated. For example, if the judges are evaluating vanilla ice creams, there might be a dimension where the judges' individual terms include "musty," "earthy," "dirty," "old refrigerator." In this case the sensory scientist may recognize these as attributes that are associated with the source of vanilla used in the flavoring. Thus the data analysis reduces the information to a few dimensions with a great loss of detail. This is a major problem with these studies. The results of FCP studies rarely allow the sensory scientist to give the product developer actionable information. In other words, the results show gross differences among samples but they do not indicate the subtle differences among products that are often very important to product developers. On the other hand, by allowing a "non-standard" evaluation technique creative or astute panelists may identify characteristics of a product that have not been considered using a more traditional approach. These novel dimensions may provide researchers new ways to differentiate products.

A few recent examples of FCP are Narain et al. (2003) for coffee, Kirkmeyer and Tepper (2003) to evaluate creaminess, and Aparicio et al. (2007) for orange juice.

Some authors have found that naïve panelists have difficulty in generating sensory terms (Heymann, 1994b; McEwan et al., 1989). A more structured FCP descriptor generation based on the repertory grid method may improve the term generation process (Kelly, 1955; Ryle and Lunghi, 1970; Stewart et al., 1981) and should be implemented. In the simplest terms, the repertory grid method is a way of eliciting multiple descriptive terms from panelists through a series of comparisons among groups of objects. In this method the panelists are presented with objects arranged in groups of three (triads) (Gains, 1994). The arrangement is such that each object appears in at least one triad and that an object from each triad is carried over to the next triad. Two objects in each triad are arbitrarily associated with each other and the panelist is asked to describe how these two objects (A and

B) are similar and, in the same way, different from the third (C). Once all possible similarities and differences, within the specified group of two with a third one as the odd object, have been exhausted, the researcher then presents the remaining two combinations (A and C with B as the odd object, as well as B and C, with A as the odd object) within the triad to the panelist who repeats the effort to describe similarities and differences. This is repeated for all triads. The descriptors used are placed on scales and the panelists then use their own sets of scales to describe the objects in the study. The data are then analyzed by Generalized Procrustes Analysis. McEwan et al. (1989) and Piggott and Watson (1992) compared the results from conventional FCP and FCP with attribute generation through the repertory grid method. They found that the repertory grid method of attribute generation did not have an advantage over conventional FCP.

Heymann (1994b) and Narain et al. (2003) also found that naïve panelists and panelists with mixed sensory experiences, respectively, did not use their individual vocabularies consistently. This resulted in non-significant GPA results. Narain et al. (2003) actually used the vocabulary derived by the panel individually for the FCP as a starting point for training the same panelists in a generic descriptive analysis. They felt that doing this improved the training of the panel for the descriptive analysis. However, we feel that this is a somewhat roundabout way to training a descriptive panel.

FCP is under scrutiny from sensory scientists who are somewhat skeptical that the results from this technique are not subject to the desired interpretation of the researcher. Williams and Arnold (1984) published the first comparison of FCP with other descriptive sensory procedures. Other authors have also compared FCP to more conventional descriptive techniques. For example, using canned cat food, Jones et al. (1989) compared FCP to generic descriptive analysis results. It seems that the best use for the FCP technique is in the area of perceptual mapping of product spaces. Perceptual maps are frequently created in marketing research and research in our and other laboratories has shown that FCP allows the creation of perceptual maps that are very similar to those created by traditional mapping techniques, such as multidimensional scaling of sorting techniques, principal component analysis of descriptive data, and principal component analysis of consumer attribute analysis (Elmore and Heymann,

1999; Steenkamp et al., 1988, 1994; Wright, 1994). These and subsequent studies have shown that under certain circumstances one would reach similar conclusions about differences among products as with other descriptive techniques (Gilbert and Heymann, 1995; Heymann, 1994b; Oreskovich et al., 1990; Skibba and Heymann, 1994a, b). There are indications that FCP is capable of revealing large differences among products but is less successful at more subtle differences. These subtle differences are more easily determined using a trained descriptive panel (Cristovam et al., 2000; Saint-Eve et al., 2004).

10.6.6 Flash Profiling

Flash profiling was invented in 2000 by Sieffermann. The technique combines individual panelist vocabulary generation through free choice profiling followed by a ranking of the simultaneously presented whole product set for each attribute. Proponents of this technique insist that the panelists chosen for the Flash profile have to be sensory evaluation experts and/or product experts (Dairou and Sieffermann, 2002; Rason et al., 2006). In the first session the panelists receive the entire product set and are asked to individually generate sensory descriptors that differentiate among the products. They are also instructed to avoid hedonic terms. At the session, the panelists are shown a pooled attribute list and are asked to update (add and/or subtract) their own individual lists if they wanted to. At the next session the panelists rank the whole product set for each of their individual attribute lists. Ties are usually allowed. At subsequent sessions the ranking process is repeated. It is preferable to do at least three replications (Delarue and Sieffermann, 2002) but some authors have done flash profiling with only duplicate rankings (Rason et al., 2006).

All products are served to the panelists simultaneously, this seems reasonable when the numbers of samples are relatively low. For example, Delarue and Sieffermann (2000) served 16 strawberry yogurts to their panel; Dairou and Sieffermann (2002) served 14 jams; Delarue and Sieffermann (2004) served either 6 strawberry yogurts or 5 apricot-flavored fresh cheeses; Rason et al. (2006) served 12 dry pork sausages; Blancher et al. (2007) served 20 jellies; Lassoued et al. (2008) served 15 wheat breads; and Jaros et al. (2009)

served 6 cloudy apple juices. However, Tarea et al. (2007) seriously pushed the limits of their panel by asking them to evaluate 49 pear and apple purees. The panel actually completed the task despite complaining about the tedium of the task and the results look valid. However, we would seriously question whether these numbers are sustainable with products that are more fatiguing.

Similar to free choice profiling the data are evaluated by generalized Procrustes analysis, where each panelist corresponds to a data matrix and a consensus configuration is produced. If the sensory scientist wants to evaluate the individual panelist's performance the data are analyzed by one-way analysis of variance (main effect: product) for each individual panelist. Keep in mind that the ranking data are nonparametric and should ideally be analyzed by a nonparametric test such as the Friedman test; however, the Friedman analysis does not handle replications and thus the individual panelists' data are usually analyzed by analysis of variance. It should be noted that doing this is contrary to the normality assumption of analysis of variance but Dairou and Sieffermann (2002) felt that the results were adequate to evaluate panelist reproducibility and attribute discriminability. Since Rason et al. (2006) did only two replications they could use the Spearman's correlation coefficient to determine each individual panelist's reproducibility.

According to Dairou and Sieffermann (2002) the advantages of flash profiling are (a) its speed because there is no training of the panelists and also all products are evaluated simultaneously, thus three replications only require three sessions and (b) that the technique allows for a diversity in panelist point of view since each panelist uses his/her own internalized vocabulary. However, this requires that the panelists either are experts in sensory description or are experts in terms of knowledge and experience with the product category. According to the above authors disadvantages of the technique are that (a) all products must be available simultaneously and thus it could probably not be used for shelf-life testing; (b) the panelists have to be experts; (c) interpretation of sensory characteristics is difficult since the terms are idiosyncratic to each panelist (this is similar to free choice profiling); and (d) the terminology cannot be used by a different panel.

Comparisons between flash profiling and generic descriptive analysis have been published. Dairou and Sieffermann (2002) found that the two procedures produced similar groupings of the jams. The flash profiling process was quicker but the standard generic descriptive analysis provided more explanatory descriptors. Delarue and Sieffermann (2002) found that for strawberry yogurts the two techniques provided very similar sensory spaces but that for apricot-flavored fresh cheeses the spaces were somewhat different. Blancher et al. (2007) also found that the conventional descriptive profile sensory spaces were similar to the flash profile spaces regardless of whether the flash profiling was done by French or Vietnamese panelists. We agree with Delarue and Sieffermann (2004) when they state "…, we think it would be misleading to consider flash profiling as a substitute for conventional profiling, which is certainly the most adapted and accurate profiling method to date. Furthermore, these two methods do not fulfill exactly the same objectives. We rather propose to consider flash profile as a convenient sensory mapping tool for conducting preliminary phases of thorough sensory studies…. One could also use it as a screening tool for selecting products or factors when designing a larger experiment."

References

Anderson, N. H. 1970. Functional measurement and psychological judgment. Psychological Review, 77, 153–170.

Aparicio, J. P., Medina, M. Á. T. and Rosales, V. L. 2007. Descriptive sensory analysis in different classes of orange juice by a robust free-choice profile method. Analytica Chimica Acta, 595, 238–247.

Armstrong, G., McIlveen, H., McDowell, D. and Blair, I. 1997. Sensory analysis and assessor motivation: Can computers make a difference? Food Quality and Preference, 8, 1–7.

Arnold, G. and Williams, A. A. 1986. The use of generalized procrustes technique in sensory analysis. In: J. R. Piggott (ed.), Statistical Procedures in Food Research. Elsevier Applied Science, London, UK, pp. 233–254.

AWWA (American Water Works Association) Manual, 1993. Flavor Profile Analysis: Screening and Training of Panelists. AWWA, Denver, CO.

Amerine, M. A., Pangborn, R. M. and Roessler, E. R. 1965. Principles of Sensory Evaluation of Foods. Academic, New York, NY.

Bárcenas, P., Pérez Elortondo, Albisu, M., Mège, K., Bivar Roseiro, L., Scintu, M. F., Torre, P., Loygorri, S. and Lavanchy, P. 2007. An international ring trial for the sensory evaluation of raw ewes' milk cheese texture. International Dairy Journal, 17, 1139–1147.

Bartels, J. H. M., Burlingame, G. A. and Suffett, I. H. 1986. Flavor profile analysis: Taste and odor control of the future. American Water Works Association Journal, 78, 50–55.

Bartels, J. H. M., Brady, B. M. and Suffet, I. H. 1987. Training panelists for the flavor profile analysis method. American Water Works Association Journal, 79, 26–32.

Blancher, G., Chollet, S., Kesteloot, R., Nguyen Hoang, D., Cuvelier, G. and Sieffermann, J.-M. 2007. French and Vietnamese: How do they describe texture characteristics of the same food? A case study with jellies. Food Quality and Preference, 18, 560–575.

Boylston, T. D., Reitmeier, C. A., Moy, J. H., Mosher, G. A. and Taladriz, L. 2002. Sensory quality and nutrient composition of three Hawaiian fruits treated by X-irradiation. Journal of Food Quality, 25, 419–433.

Brandt, M. A., Skinner E. Z. and Coleman J. A. 1963. The texture profile method. Journal of Food Science, 28, 404–409.

Cairncross, S. E. and Sjöstrom, L. B. 1950. Flavor profiles: A new approach to flavor problems. Food Technology, 4, 308–311.

Campo, E., Do. B. V., Ferreira, V. and Valentin, D. 2008. Aroma properties of young Spanish white wines: A study using sorting task, list of terms and frequency of citation. Australian Journal of Grape and Wine Research, 14, 104–115.

Campo, E., Ballester, J., Langlois, J., Dacremont, C. and Valentin, D. 2009. Comparison of conventional analysis and a citation frequency based descriptive method for odor profiling: An application to Burgundy Pinot noir wines. Food Quality and Preference, doi:10.1016/j.foodqual.2009.08.001.

Carbonell, L., Izquierdo, L. and Carbonell, I. 2007. Sensory analysis of Spanish mandarin juices. Selection of attributes and panel performance. Food Quality and Preference, 18, 329–341.

Caul, J. F. 1957. The profile method of flavor analysis. Advances in Food Research, 7, 1–40.

Caul, J. F. 1967. The profile method of flavor analysis. Advances in Food Research, 7, 1–40.

Chambers, E. IV, Bowers, J. R. and Smith, E. A. 1992. Flavor of cooked, ground turkey patties with added sodium tripolyphosphate as perceived by sensory panels with differing phosphate sensitivity. Journal of Food Science, 57, 521–523.

Civille, G. V. and Lyon, B. 1996. ASTM Lexicon Vocabulary for Descriptive Analysis. American Society for Testing and Materials, Philadelphia.

Civille, G. V. and Liska, I. H. 1975. Modifications and applications to foods of the general foods sensory texture profile technique. Journal of Texture Studies, 6, 19–31.

Civille, G. V. and Lawless, H. T. 1986. The importance of language in describing perceptions. Journal of Sensory Studies, 1, 217–236.

Cliff, M. A., Wall, K., Edwards, B. J. and King, M. C. 2000. Development of a vocabulary for profiling apple juices. Journal of Food Quality, 23, 73–86.

Cochran, W. G. and Cox, G. M. 1957. Experimental Designs. Wiley, New York.

Cristovam, E., Paterson, A. and Piggott, J. R. 2000. Differentiation of port wines by appearance using a sensory panel: Comparing free choice and conventional profiling. European Food Research and Technology, 211, 65–71

Dairou, V. and Sieffermann, J.-M. 2002. A comparison of 14 jams characterized by conventional profile and a quick original method, flash profile. Journal of Food Science, 67, 826–834.

Delarue, J. and Sieffermann, J-M. 2000. Use of Flash Profile for a quick sensory characterization of a set of sixteen strawberry yogurts. In: K. C. Persaud and S. van Toller (eds.), 13th International Symposium of Olfaction and Taste/14th European Chemoreception Research Organization Congress. ECRO, Brighton, UK, pp. 225–226.

Dellaglio, S., Casiraghi, E. and Pompei, C. 1996. Chemical, physical and sensory attributes for the characterization of an Italian dry-cured sausage. Meat Science, 42, 25–35.

DeRovira, D. 1996. The dynamic flavor profile method. Food Technology, 50, 55–60.

Derndorfer, E., Baierl, A., Nimmervoll, E. and Sinkovits, E. 2005. A panel performance procedure implemented in R. Journal of Sensory Studies, 20, 217–227.

Dijksterhuis, G. 1995. Assessing panel consensus. Food Quality and Preference, 6, 7–11.

Dooley, L. M., Adhikari, K. and Chambers, E. 2009. A general lexicon for sensory analysis of texture and appearance of lip products. Journal of Sensory Studies, 24, 581–600.

Drake, M. A. and Civille, G. V. 2003. Flavor lexicons. Comprehensive Reviews of Food Science and Food Safety, 1, 33–40.

Drake, M. A., Yates, M. D., Gerard, P. D., Delahunty, C. M., Sheehan, E. M., Turnbull, R. P. and Dodds, T. M. 2005. Comparison of differences between lexicons for descriptive analysis of Cheddar cheese flavor in Ireland, New Zealand, and the United States of America. International Dairy Journal, 15, 473–483.

Drake, M. A., Jones, V. S., Russell, T., Harding, R. and Gerard, P. D. 2007. Comparison of lexicons for descriptive analysis of whey and soy proteins in New Zealand and the U.S.A. Journal of Sensory Studies, 22, 433–452.

Einstein, M. A. 1991. Descriptive techniques and their hybridization. In: H. T. Lawless and B. P. Klein (eds.), Sensory Science Theory and Applications in Foods. Marcel Dekker, New York, NY, pp. 317–338.

Elmore, J. and Heymann, H. 1999. Perceptual maps of photographs of carbonated beverages created by traditional and free-choice profiling Food Quality and Preference, 10, 219–227

Findlay, C. J., Castura, J. C., Schlich, P. and Lesschaeve, I. 2006. Use of feedback calibration to reduce the training time for wine panels. Food Quality and Preference, 17, 266–276.

Findlay, C. J., Castura, J. C. and Lesschaeve, I. 2007. Feedback calibration: A training method for descriptive panels. Food Quality and Preference, 8, 321–328.

Frøst. M. B. and Janhøj, T. 2007. Understanding creaminess. International Dairy Journal, 17, 1298–1311.

Gains, N. 1994. The repertory grid approach. In: H. J. H. MacFie and D. M. H. Thomson (eds.), Measurement of Food Preferences. Blackie Academic and Professional, Glasgow, pp. 51–76.

Giboreau, A., Dacremont, C., Egoroff, C., Guerrand, S., Urdapilleta, I., Candol, D. and Dubois, D. 2007. Defining sensory descriptors: Towards writing guidelines based on terminology. Food Quality and Preference, 18, 265–274.

Gilbert J. M. and Heymann, H. 1995. Comparison of four sensory methodologies as alternatives to descriptive analysis for the evaluation of apple essence aroma. The Food Technologist (NZIFST), 24, 4, 28–32.

Gordin, H. H. 1987. Intensity variation descriptive methodology: Development and application of a new sensory evaluation technique. Journal of Sensory Studies 2, 187–198.

Gou, P., Guerrero, L. and Romero, A. 1998. The effect of panel selection and training on external preference mapping using a low number of samples. Food Science and Technology International, 4, 85–90.

Gower, J. C. 1975. Generalized procrustes analysis. Psychometrika, 40, 33–50.

Gower, J. C. and Dijksterhuis, G. B. 2004. Procrustes Problems. Oxford University Press, New York.

Greenacre, M. 2007. Correspondence Analysis in Practice, Second Edition. Chapman and Hall/CRC, New York.

Hall, R. L. 1958. Flavor study approaches at McCormick and Company, Inc., In: A. D. Little, Inc. (ed.), Flavor Research and Food Acceptance. Reinhold, New York, NY, pp. 224–240.

Hanson, J. E., Kendall, D. A. and Smith, N. F. 1983. The missing link: Correlation of consumer and professional sensory descriptions. Beverage World, 102, 108–116.

Hegenbart, S. 1994. Learning and speaking the language of flavor. Food Product Design, 8, 33, 34, 39, 40, 43, 44, 46–49.

Hein, K. A. 2005. Perception of Vegetative and Fruity Aromas in Red Wine and Evaluation of a Descriptive Analysis Panel Using Across Product Versus Across Attribute Serving. MS Thesis, University of California, Davis.

Heisserer, D. M. and Chambers, E., IV. 1993. Determination of sensory flavor attributes of aged natural cheese. Journal of Sensory Studies, 8, 121–132.

Heymann, H. 1994a. A comparison of descriptive analysis of vanilla by two independently trained panels. Journal of Sensory Studies, 9, 21–32.

Heymann, H. 1994b. A comparison of free choice profiling and multidimensional scaling of vanilla samples. Journal of Sensory Studies, 9, 445–453.

Heymann, H., Holt, D. L. and Cliff, M. A. 1993. Measurement of flavor by sensory descriptive techniques. In: C.-T. Ho and C. H. Manley (eds.), Flavor Measurement. Proceedings of the Institute of Food Technologists Basic Symposium, New Orleans, LA, Chapter 6, pp. 113–131, June 19–20, 1993.

Heymann, H. and Noble, A. C. 1987. Descriptive analysis of commercial Cabernet sauvignon wines in California. American Journal of Enology and Viticulture, 38, 41–44.

Hirst, D. and Næs, T. 1994. A graphical technique for assessing differences among a set of rankings. Journal of Chemometrics, 8, 81–93.

Homa, D. and Cultice, J. 1984. Role of feedback, category size, and stimulus distortion on the acquisition and utilization of ill-defined categories. Journal of Experimental Psychology, 10, 83–94.

Ishii, R. and O'Mahony, M. 1991. Use of multiple standards to define sensory characteristics for descriptive analysis: Aspects of concept formation. Journal of Food Science, 56, 838–842.

Jaros, D., Thamke, I., Raddatz, H. and Rohm, H. 2009. Single-cultivar cloudy juice from table apples: An attempt to identify the driving force for sensory preference. European Food Research and Technology, 229, 51–61.

Jellinek, G. 1964. Introduction to and critical review of modern methods of sensory analysis (odor, taste and flavor evaluation) with special emphasis on descriptive analysis (flavor profile method). Journal of Nutrition and Dietetics 1, 219–260.

Jones, P. N., MacFie, H. J. H. and Beilken, S. L. 1989. Use of preference mapping to relate consumer preference to the sensory properties of a processed meat product tinned cat food. Journal of the Science of Food and Agriculture, 47, 113–124.

Kelly, G. A. 1955. The Psychology of Personal Constructs. Norton, New York, NY.

Kirkmeyer, S. V. and Tepper, B. 2003. Understanding creaminess perception of dairy products using free-choice profiling and genetic responsivity to 6-n-propylthiouracil. Chemical Senses, 28, 527–536.

Kollár-Hunek, K., Heszberger, J., Kókai, Z., Láng-Lázi, M. and Papp, E. 2007. Testing panel consistency with GCAP method in food profile analysis. Journal of Chemometrics, 22, 218–226.

Krasner, S. W. 1995. The use of reference standards in sensory analysis. Water Science and Technology, 31, 265–272.

Krasner, S. W., McGuire, M. J. and Ferguson, V. B. 1985. Tastes and odors: The flavor profile method. American Water Works Association Journal, 77, 34–39.

Kravitz, D. 1975. Who's Who in Greek and Roman Methodology. Clarkson N. Potter, New York, p. 200.

Krinsky, B. F., Drake, M. A., Civille, G. V., Dean, L. L., Hendrix, K. W. and Sanders, T. H. 2006. The development of a lexicon for frozen vegetable soybeans (edamame). Journal of Sensory Studies, 21, 644–653.

Labuza, T. P. and Schmidl, M. K. 1985. Accelerated shelf-life testing of foods. Food Technology, 39, 57–64, 134.

Langron, S. P. 1983. The application of procrustes statistics to sensory profiling. In: A. A. Williams and R. K. Atkin (eds.), Sensory Quality in Foods and Beverages: Definition, Measurement and Control. Ellis Horwood, Chichester, UK, pp. 89–95.

Larson-Powers, N. M. and Pangborn, R. M. 1978. Descriptive analysis of the sensory properties of beverages and gelatin containing sucrose and synthetic sweeteners. Journal of Food Science, 43, 11, 47–51.

Lassoued, N., Delarue, J., Launay, B. and Michon, C. 2008. Baked product texture: Correlations between instrumental and sensory characterization using flash profile. Journal of Cereal Science, 48, 133–143.

Lawless, H. T. and Corrigan, C. J. 1993. Semantics of astringency. In: K. Kurihara, N. Suzuki, and H. Ogawa (eds.), Olfaction and Taste XI. Proceedings of the 11th International Symposium on Olfaction and Taste and of the 27th Japanese Symposium on Taste and Smell. Springer-Verlag, Tokyo, pp. 288–292.

Leach, E. J. and Noble, A. C. 1986. Comparison of bitterness of caffeine and quinine by a time-intensity procedure. Chemical Senses 11, 339–345.

Lee, J. and Chambers, D. 2007. A lexicon for flavor descriptive analysis of green tea. Journal of Sensory Studies, 22, 421–433.

Lê, S. and Husson, F. 2008. Sensominer: A package for sensory data analysis. Journal of Sensory Studies, 23, 14–25.

Le Fur, Y., Mercurio, V., Moio, L., Blanquet, J. and Meunier, J. M. 2003. A new approach to examine the relationships between sensory and gas chromatography-olfactometry data using generalized procrustes analysis applied to six French Chardonnay wines. Journal of Agriculture and Food Chemistry, 51, 443–452.

Le Moigne, M., Symoneaux, R. and Jourjon, F. 2008. How to follow grape maturity for wine professionals with a seasonal judge training? Food Quality and Preference, 19, 672–681.

Lotong, V., Chambers, D. H., Dus, C., Chambers, E. and Civille, G. V. 2002. Matching results of two independent highly trained sensory panels using different descriptive analysis methods. Journal of Sensory Studies, 17, 429–444.

Lund, C. M., Jones, V. S. and Spanitz, S. 2009. Effects and influences of motivation on trained panelists. Food Quality and Preference, 20, 295–303.

MacFie, H. J. H. 1990. Assessment of the sensory properties of food. Nutrition Reviews, 48, 87–93.

Marchisano, C., Vallis, L. and MacFie, H. J. H. 2000. Effect of feedback on sensory training: A preliminary study. Journal of Sensory Studies, 15, 119–135.

Marshall, R. J. and Kirby, S. P. 1988. Sensory measurement of food texture by free choice profiling. Journal of Sensory Studies, 3, 63–80.

Martens, H. and Martens, M. 2001. Multivariate Analysis of Quality: An Introduction. Wiley, Chichester, UK.

Martin, K. and Lengard, V. 2005. Assessing the performance of a sensory panel: Panelist monitoring and tracking. Journal of Chemometrics, 19, 154–161.

Matisse, H. 1908. In: J. Bartlett (ed.), Familiar Quotations, Fourteenth Edition. Little, Brown and Co., Boston, MA.

Mazzucchelli, R. and Guinard, J-X. 1999. Comparison of monadic and simultaneous sample presentation modes in a descriptive analysis of chocolate milk. Journal of Sensory Studies, 14, 235–248.

McCloskey, L. P., Sylvan, M. and Arrhenius, S. P. 1996. Descriptive analysis for wine quality experts determining appellations by Chardonnay aroma. Journal of Sensory Studies, 11, 49–67.

McDonell, E., Hulin-Bertaud, S., Sheehan, E. M. and Delahunty, C. M. 2001. Development and learning process of a sensory vocabulary for the odor evaluation of selected distilled beverages using descriptive analysis. Journal of Sensory Studies, 16, 425–445.

McEwan, J. A., Colwill, J. S. and Thomson, D. M. H. 1989. The application of two free-choice profiling methods to investigate the sensory characteristics of chocolate. Journal of Sensory Studies 3, 271–286.

McEwan, J. A., Hunter, E. A., van Gemert, L. J. and Lea, P. 2002. Proficiency testing for sensory panels: Measuring panel performance. Food Quality and Preference, 13, 181–190.

McTigue, M. C., Koehler, H. H. and Silbernagel, M. J. 1989. Comparison of four sensory evaluation methods for assessing cooked dry bean flavor. Journal of Food Science, 54, 1278–1283.

Meilgaard, M., Civille, C. V. and Carr, B. T. 2006. Sensory Evaluation Techniques, Fourth Edition. CRC, Boca Raton, FL.

Meng, A. K. and Suffet, I. H. 1992. Assessing the quality of flavor profile analysis data. American Water Works Association Journal, 84, 89–96.

Meullenet, J-F., Xiong, R. and Findlay, C. F. 2007. Multivariate and Probabilistic Analyses of Sensory Science Problems. Wiley-Blackwell, New York, NY.

Moore, L. J. and Shoemaker, C. R. 1981. Sensory textural properties of stabilized ice cream. Journal of Food Science, 46, 399–402, 409.

Moskowitz, H. R. 1988. Applied Sensory Analysis of Foods, Vols. I and II. CRC, Boca Raton, FL.

Muñoz, A. M. 1986. Development and application of texture reference scales. Journal of Sensory Studies 1, 55–83.

Muñoz, A. M. and Civille, G. V. 1998. Universal, product and attribute specific scaling and the development of common lexicons in descriptive analysis. Journal of Sensory Studies, 13(1), 57–75.

Murray, J. M. and Delahunty, C. M. 2000. Selection of standards to reference terms in a Cheddar cheese flavor language. Journal of Sensory Studies, 15, 179–199.

Murray, J. M., Delahunty, C. M. and Baxter, I. A. 2001. Descriptive analysis: Past, present and future. Food Research International, 34, 461–471.

Murtagh, F. (2005) Correspondence Analysis and Data Coding with Java and R. Chapman and Hall/CRC, New York, NY.

Narain, C., Paterson, A. and Reid, E. 2003. Free choice and conventional profiling of commercial black filter coffees to explore consumer perceptions of character. Food Quality and Preference, 15, 31–41.

Nindjin, C., Otokoré, D., Hauser, S., Tschannen, A., Farah, Z. and Girardin, O. 2007. Determination of relevant sensory properties of pounded yams (Dioscorea spp.) using a locally based descriptive analysis methodology. Food Quality and Preference, 18, 450–459.

Olabi, A. and Lawless, H. T. 2008. Persistence of context effects after training and with intensity references. Journal of Food Science, 73, S185–S189.

Nogueira-Terrones, H., Tinet, C., Curt, C., Trystam, G. and Hossenlop, J. 2008. Using the internet for descriptive sensory analysis: Formation, training and follow-up of a taste-test panel over the web. Journal of Sensory Studies, 21, 180–202.

O'Mahony, M. 1986. Sensory Evaluation of Food. Marcel Dekker, New York, NY.

Ömür-Özbek, P. and Dietrich, A. M. 2008. Developing hexanal as an odor reference standard for sensory analysis of drinking water. Water Research, 42, 2598–2604.

O.P.&P. 1991. Oliemans Punter and Partners. BV Postbus 14167 3508 SG Utrecht, The Netherlands.

Oreskovich, D. C., Klein, B. P. and Sutherland, J. W. 1991. Procrustes analysis and its applications to free-choice and other sensory profiling. In: H. T. Lawless and B. P. Klein (eds.), Sensory Science Theory and Applications in Foods. Marcel Dekker, New York, NY, pp. 317–338.

Oreskovich, D. C., Klein, B. P. and Sutherland, J. W. 1990. Variances associated with descriptive analysis and free-choice profiling of frankfurters. Presented at IFT Annual Meeting, Anaheim, CA, June 16–20, 1990.

Otremba, M. M., Dikeman, M. A., Milliken, G. A., Stroda, S. L., Chambers, E. and Chambers, D. 2000. Interrelationships between descriptive texture profile sensory panel and descriptive attribute sensory panel evaluations of beef Longissimus and Semitendinosus muscles. Meat Science, 54, 325–332

Parr, W., Valentin, D., Green, J. A. and Dacremont, C. 2009. Evaluation of French and New Zealand Sauvignon wines by experienced French Assessors. Food Quality and Preference, doi:10.1016/j.foodqual.2009.08.002.

Petersen, R. G. 1985. Design and Analysis of Experiments. Marcel Dekker, New York, NY.

Piggott, J. R. 1986. Statistical Procedures in Food Research. Elsevier Applied Science, London, UK.

Piggott, J. R. and Watson, M. P. 1992. A comparison of free-choice profiling and the repertory grid method in the flavor profiling of cider. Journal of Sensory Studies, 7, 133–146.

Pineau, N., Chabanet, C. and Schlich, P. 2007. Modeling the evolution of the performance of a sensory panel: A mixed-model and control chart approach. Journal of Sensory Studies, 22, 212–241.

Powers, J. J. 1988. Current practices and application of descriptive methods. In: J. R. Piggott (ed.), Sensory Analysis of Foods. Elsevier Applied Science, London, UK.

Raffensperger, E. L., Peryam, D. R. and Wood, K. R. 1956. Development of a scale for grading toughness-tenderness in beef. Food Technology, 10, 627–630.

Rainey, B. 1986. Importance of reference standards in training panelists. Journal of Sensory Studies. 1, 149–154.

Randall, J. H. 1989. The analysis of sensory data by generalized linear models. Biometrics Journal, 3, 781–793.

Rason, J., Léger, L., Dufour, E. and Lebecque, A. 2006. Relations between the know-how of small-scale facilities and the sensory diversity of traditional dry sausages from the Massif Central in France. European Food Research and Technology, 222, 580–589.

Retiveau, A., Chambers, D. H. and Esteve, E. 2005. Developing a lexicon for the flavor description of French cheeses. Food Quality and Preference, 16, 517–527.

Risvik, E., Colwill, J. S., McEwan, J. A. and Lyon, D. H. 1992. Multivariate analysis of conventional profiling data: A comparison of a British and a Norwegian trained panel. Journal of Sensory Studies, 7, 97–118.

Riu-Aumatell, M., Vichi, S., Mora-Pons, M., López-Tamames, E. and Buxaderas, S. 2008. Sensory characterization of dry gins with different volatile profiles. Journal of Food Science, 73, S286–S293.

Rutledge, K. P. 2004. Lexicon for Sensory Attributes Relating to Texture and Appearance. ASTM, CD-ROM, West Conshohocken, PA.

Ryle, A. and Lunghi, M. W. 1970. The dyad grid: A modification of repertory grid technique. British Journal of Psychology, 117, 323–327.

Saint-Eve, A., Paçi Kora, E. and Martin, N. 2004. Impact of the olfactory quality and chemical complexity of the flavouring agent on the texture of low fat stirred yogurts assessed by three different sensory methodologies. Food Quality and Preference, 15, 655–668.

Schlich, P. 1989. A SAS/IML program for generalized procrustes analysis. SEUGI '89. Proceedings of the SAS European Users Group International Conference, Cologne, Germany, May 9–12, 1989.

Seo, H.-S., Lee, M., Jung, Y.-J. and Hwang, I. 2009. A novel method of descriptive analysis on hot brewed coffee: Time scanning descriptive analysis. European Food Research and Technology, 228, 931–938.

Sieffermann, J.-M. 2000. Le profil Flash. Un outil rapide et innovant d'évaluation sensorielle descriptive, AGORAL 2000 – XIIèmes rencotres. In: Tec and Doc Paris (eds.), L'innovation: de l'idée au success,. Lavoisier, Paris, France, pp. 335–340.

Sjöström, L. B. 1954. The descriptive analysis of flavor. In: D. Peryam, F. Pilgrim, and M. Peterson (eds.), Food Acceptance Testing Methodology. Quartermaster Food and Container Institute, Chicago, pp. 25–61.

Skibba, E. A. and Heymann, H. 1994a. Creaminess Perception. Presented at ACHEMS Annual Meeting, Sarasota, FL, April 14, 1994.

Skibba, E. A. and Heymann, H. 1994b. The Perception of Creaminess. Presented at the IFT Annual Meeting, Atlanta, GA, June 23–26, 1994.

Steenkamp, J.-B. E. M. and van Trijp, H. C. M. 1988. Free choice profiling in cognitive food acceptance research. In: D. M. H. Thompson (ed.), Food Acceptability. Elsevier Applied Science, London, UK, pp. 363–376.

Steenkamp, J.-B. E. M., van Trijp, H. C. M. and ten Berge, M. F. 1994. Perceptual mapping based on idiosyncratic sets of attributes. Journal of Marketing Research, 31, 15–27.

Stevens, S. S. 1969. On predicting exponents for cross-modality matches. Perception and Psychophysics, 6, 251–256.

Stevens. S. S. and Marks, L. E. 1980. Cross-modality matching functions generated by magnitude estimation. Perception and Psychophysics, 27, 379–389.

Stewart, V., Stewart, A. and Fonda, N. 1981. Business applications of repertory grid. McGraw-Hill, London, UK.

Stoer, N. and Lawless, H. T. 1993. Comparison of single product scaling and relative-to-reference scaling in sensory evaluation of dairy products. Journal of Sensory Studies, 8, 257–270.

Stone, H., Sidel, J. L., Oliver, S., Woolsey, A. and Singleton, R. C. 1974. Sensory evaluation by quantitative descriptive analysis. Food Technology, 28, 24, 26, 28, 29, 32, 34.

Stone, H. and Sidel, J. L. 2004. Sensory Evaluation Practices, Third Edition. Academic, Orlando, FL.

Stone, H., Sidel, J. L. and Bloomquist, J. 1980. Quantitative descriptive analysis. Cereal Foods World, 25, 624–634.

Sulmont, C., Lesschaeve, I., Sauvageot, F. and Issanchou, S. 1999. Comparative training procedures to learn odor descriptors: Effects on profiling performance. Journal of Sensory Studies, 14, 467–490.

Syarief, H., Hamann, D. D., Giesbrecht, F. G., Young, C. T. and Monroe, R. J. 1985. Comparison of mean and consensus scores from flavor and texture profile analyses of selected products. Journal of Food Science, 50, 647–650, 660.

Szczesniak, A. S. 1966. Texture measurements. Food Technology, 20, 1292 1295–1296 1298.

Szczesniak, A. S. 1975. General foods texture profile revisited – ten years perspective. Journal of Texture Studies, 6, 5–17.

Szczesniak, A. S. 1963. Classification of textural characteristics. Journal of Food Science, 28, 385–389.

Szczesniak, A. S., Brandt, M. A. and Friedman, H. H. 1963. Development of standard rating scales for mechanical parameters of texture and correlation between the objective and the sensory methods of texture evaluation. Journal of Food Science, 28, 397–403.

Tarea, S., Cuvelier, G. and Sieffermann, J.-M. 2007. Sensory evaluation of the texture of 49 commercial apple and pear purees. Journal of Food Quality, 30, 1121–1131.

Tomic, O., Nilsen, A., Martens, M. and Næs, T. 2007. Visualization of sensory profiling data for performance monitoring. LWT- Food Science and Technology, 40, 262–269.

Ward, L. M. 1986. Mixed-modality psychophysical scaling: Double cross-modality matching for "difficult" continua. Perception and Psychophysics, 39, 407–417.

Whorf, B. L. 1952. Collected Papers on Metalinguistics. Department of State, Foreign Service Institute, Washington, DC, pp. 27–45.

Williams, A. A. and Arnold, G. M. 1984. A new approach to sensory analysis of foods and beverages. In: J. Adda (ed.), Progress in Flavour Research. Proceedings of the 4th Weurman Flavour Research Symposium, Elsevier, Amsterdam, The Netherlands, pp. 35–50.

Williams, A. A. and Langron, S. P. 1984. The use of free choice profiling for the examination of commercial ports. Journal of the Science of Food and Agriculture, 35, 558–568.

Woods, V. 1995. Effect of Geographical Origin and Extraction Method on the Sensory Characteristics of Vanilla Essences. MS Thesis, University of Missouri, Columbia, MO.

Worch, T., Lê, S. and Punter, P. 2009. How reliable are consumers? Comparison of sensory profiles from consumers and experts. Food quality and Preference, doi:10.1016/j.foodqual.2009.06.001.

Wright, K. 1994. Attribute Discovery and Perceptual Mapping. M.S. Thesis, Cornell University, Ithaca, New York.

Zook, K. and Wessman, C. 1977. The selection and use of judges for descriptive panels. Food Technology, 31, 56–61.

Chapter 11

Texture Evaluation

Abstract   In this chapter the sensory evaluation of texture is discussed. The concept of texture is defined and then the visual, auditory, and tactile textures related to food (and to some extent textiles) are described in detail. Sensory texture measurements, specifically the Texture Profile Method, are described followed by a relatively brief discussion of correlations between instrumental and sensory texture measurements.

Whenever I
Eat ravioli
I fork it quick
But chew it sloli.

—(Italian Noodles, 1992)

Contents

11.1 Texture Defined

Alina Szczesniak (2002) states that a generally accepted definition of texture is the following "texture is the sensory and functional manifestation of the structural, mechanical and surface properties of foods detected through the senses of vision, hearing, touch and kinesthetics." She then goes on to emphasize

a. "texture is a sensory property" which can only be perceived and described by humans (and animals) and any instrumental measurements must be related to sensory responses.
b. "texture is a multi-parameter attribute."
c. "texture derives from the structure of the food."
d. "texture is detected by several senses."

A number of texture review articles and textbooks have been published (Bourne, 2002; Chen, 2007, 2009; Christensen, 1984; Guinard and Mazzuccheli, 1996; Kilcast, 2004; McKenna, 2003; Moskowitz, 1987; Rosenthal, 1999; Szczesniak, 2002; Wilkinson et al., 2000; van Vliet et al., 2009).

H.T. Lawless, H. Heymann, *Sensory Evaluation of Food*, Food Science Text Series,
DOI 10.1007/978-1-4419-6488-5_11, © Springer Science+Business Media, LLC 2010

The texture of an object is perceived by the senses of sight (visual texture), touch (tactile texture), and sound (auditory texture), in some products only one of these senses is used to perceive the product texture and in other cases the texture is perceived by a combination of these senses. For example, the skin of an orange has a visual and tactile roughness that is absent in the skin of an apple. The crispness of a potato chip in the mouth is both a tactile and an auditory textural perception (Vickers, 1987b). The thickness (viscosity) of a malted milkshake can be assessed visually, in the glass, and then by proprioceptive sensations when stirring the milkshake with a straw as well as by tactile sensations in the mouth.

Ball and coworkers (1957) were among the first to distinguish between "sight" (visual) and "feel" (tactile) definitions of texture. Visual texture is often used by consumers as an indication of product freshness, for example, wilted spinach and shriveled grapes are deemed to be unacceptable in quality (Szczesniak and Kahn, 1971). Additionally, visual texture clues create expectations as to the mouth feel characteristics of the product. When the visual and tactile texture characteristics of a product are at variance the discrepancy causes a decrease in product acceptance.

Food texture can be extremely important to the consumer. Yet, unlike color and flavor, texture is frequently used by the consumer not as an indicator of food safety, but as an indicator of food quality. Szczesniak and Kahn (1971) found that socioeconomic class affected consumers' awareness of texture. Those individuals in higher socioeconomic classes were more aware of texture as a food attribute than those in lower socioeconomic classes. Also, consumers employed by a major food company placed relatively more emphasis on texture than the general population (Szczesniak and Kleyn, 1963). Szczesniak (2002) states that one of the main drivers of consumers' responses to food texture is that "people like to be in full control of the food placed in their mouth. Stringy, gummy or slimy foods or those with unexpected lumps or hard particles are rejected for fear of gagging or choking." Table 11.1 indicates the relative importance of consumers placed on texture versus flavor in a wide variety of foods.

In some foods, the perceived texture is the most important sensory attribute of the product. For these products a defect in the perceived texture would have

Table 11.1 Relative importance of texture to flavor for a wide variety of food products (texture/flavor index[a])

Item	American consumers[b]	Consumers employed by general foods[c]
Total group	0.89	1.20
Sex		
Male	0.76	1.10
Female	1.02	1.30
Socioeconomic class		
Upper lower	0.60	
Lower middle	0.95	
Upper middle	1.20	
Geographic location		
Chicago, IL	0.96	
Denver, CO	0.95	
Charlotte, NC	0.63	

Adapted from Szczesniak and Kahn (1971)
[a]Index values less than unity mean consumers placed relatively more emphasis on flavor, values larger than unity mean more emphasis was placed on texture
[b]One hundred and forty-nine consumers (three geographic areas) did a word-association test using the names of 29 foods (Szczesniak, 1971)
[c]One hundred consumers did a word-association test using the names of 74 foods (Szczesniak and Kleyn, 1963)

an extremely negative impact on consumers' hedonic responses to the product. Examples are soggy (not crisp) potato chips, tough (not tender) steak, and wilted (not crunchy) celery sticks. In other foods, the texture of the product is important but it is not the principal sensory characteristic of the product. Examples are candy, breads, and many vegetables. Lassoued et al. (2008) stated that about 20% of bread acceptability was related to crumb texture. In still other foods, the perceived texture has a minor role in the acceptance of the product and examples are liquids with relatively low viscosities such as wine and sodas.

The texture contrast within a food, on the plate, or across food products in a meal is important. A meal consisting of mashed potato, pureed winter squash, and ground beef sounds much less appetizing than one consisting of Salisbury steak, French fries, and chunks of winter squash, yet the difference between the two meals are all related to texture. Szczesniak and Kahn (1984) formulated general principles that should be kept in mind when creating textural contrasts in individual foods or across foods within a meal. Hyde and Witherly (1993) formulated the principle that dynamic contrast (the moment-to-moment change in sensory textural contrast in the mouth during chewing) is responsible for the high palatability of potato and corn chips and of ice cream. Additional examples of foods with dynamic contrast would be ice cream with candy inclusions and chocolate covered peanut M&M candies.

The importance of texture in the identification of foods was shown by Schiffman (1977) who blended and pureed 29 food products to eliminate their textural characteristics. She then asked her panelists to eat the food and to identify the food products. Overall about 40% of food products were identified correctly by normal weight college students. Only 4% of the panelists could correctly identify blended cabbage; 7% correctly identified pureed cucumber; 41% correctly identified blended beef; 63% correctly identified pureed carrots; and 81% correctly identified pureed apple. These data indicate that American consumers use texture information when they identify and classify food products.

In a word-association test Szczesniak and Kleyn (1963) found that foods elicited texture responses differentially. The percentage of texture-related responses was relatively high (over 20%) for peanut butter, celery, angel-food cake, and pie crust. Their panelists used a total of 79 texture words, with 21 words used 25 or more times by the 100 panelists to describe the 74 foods. The most frequently used words described hardness (soft, hard, chewy, and tender), crispness or crunchiness, and moisture content (dry, wet, moist, juicy). Yoshikawa et al. (1970) used the Szczesniak and Kleyn (1963) study as a basis to study the texture descriptions of female Japanese panelists. They found that the Japanese used many more words to describe texture (406) than the American panelists (79). This was probably not due to genetic differences between the two groups but more likely due to cultural differences since Japanese foods tend to have more textural variety than American foods. Additionally, the Japanese language is also very rich in subtle nuances and older respondents would likely have used even more terms since they "would have a greater knowledge of Japanese than younger people." Later, Szczesniak (1979a, b) commented on the onomatopoetic nature of Japanese texture terms. That is, the word tends to sound like the type of texture that is experienced.

Rohm (1990) also used the Szczesniak and Kleyn (1963) study as a basis to study Austrian texture descriptors. They found that Viennese students (100 males and 108 females) used 105 texture terms in a word association with 50 foods. Eighteen of these terms were used more than 25 times each while 47 terms were used less than 5 times each. When Rohm (1990) compared his data with Szczesniak and Kleyn (1963), Szczesniak (1971) and Yoshikawa et al. (1970), he found that five of the ten most frequently used terms were similar across studies (Table 11.2). Based on these studies we can thus state that certain textural terms and sensations are universal across cultures. However, there are some major exceptions. As pointed out by Roudaut et al. (2002) in France vegetables and fruits are not considered "croustillant" (crisp) yet in the United States these products, when fresh, are frequently described as crisp. The sensory specialist in any country, culture, or region should therefore pay attention not only to the perceived flavor, taste, and color dimensions of food products but also to the perceived textural characteristics. Drake (1989) published a list of textural terms in 23 languages. This list is invaluable when training panelists who are non-native English speakers or panels in different countries.

Table 11.2 The ten most frequently used texture terms in Austria[a], Japan[b], and the United States[c,d]

| Austria[a] | Japan[b] | United States | |
		1963[c]	1971[d]
Crisp	Hard	**Crisp**	**Crisp**
Hard	**Soft**	Dry	**Crunchy**
Soft	**Juicy**	**Juicy**	**Juicy**
Crunchy	Chewy	**Soft**	Smooth
Juicy	Greasy	**Creamy**	**Creamy**
Sticky	Viscous	Crunchy	Soft
Creamy	Slippery	Chewy	Sticky
Fatty	**Creamy**	Smooth	Stringy
Watery	**Crisp**	Stringy	Fluffy
Tough	**Crunchy**	Hard	Tender

Words in **bold** occurred in the top ten in all four studies
[a]Two hundred and eight Viennese students did a word-association test using the names of 50 foods (Rohm, 1990)
[b]One hundred and forty Japanese students did a word-association test using the names of 97 foods (Yoshikawa et al., 1970).
[c]One hundred and forty-nine consumers (three geographic areas) did a word-association test using the names of 29 foods (Szczesniak, 1971)
[d]One hundred consumers did a word-association test using the names of 74 foods (Szczesniak and Kleyn, 1963)

11.2 Visual, Auditory, and Tactile Texture

In this section we will discuss visual, auditory, and tactile perceptions of texture in more detail and then we will discuss how the sensory specialist can measure these perceived textures in food products. The usual sequence of texture perception when consuming a food product is visual evaluation of texture followed by direct (with the fingers) and/or indirect (with eating utensils such as knife, fork, or spoon) tactile evaluations followed by oral–tactile (with the lips, tongue, palate, saliva) evaluations. Concurrent with the oral–tactile evaluation (and sometimes also when cutting/stabbing the food with a utensil) are also the aural (sound) evaluations of crunchy, crispy, crackly, etc. (Kilcast, 1999).

11.2.1 Visual Texture

Many surface characteristics of a food product do not only affect the perceived appearance of the product but also affect the perception of the texture. Consumers know from prior experience that the lumps seen in tapioca pudding are also perceived as lumps in the mouth. Visual texture assessment has some overlap with appearance characteristics such as shine, gloss, and reflectance (discussed in Chapter 12). In this section we will discuss visual texture not related to these appearance terms. These visual texture terms would include roughness, uniformity, surface powderiness or bloom, oiliness, greasiness, flakiness, stringiness, smoothness, wilting, and surface wetness (Chen, 2007).

The surface roughness of an oatmeal or the cookie can be assessed both visually and through oral and hand tactile evaluations. The blister level of tortilla chips was assessed by Bruwer et al. (2007) who found that the blister level was negatively related to orally perceived denseness of the tortilla chip. In a bread crumb appearance study trained panelists have evaluated fineness ("... visual estimation of the amount of gas cells"), degree of homogeneity ("...refers to the degree of uniformity of the pore sizes"), and orientation ("...degree of orientation of the crumb grain") (Gonzalez-Barron and Butler, 2008b). Lassoued et al.

(2008) used flash profiling (see Chapter 10) to evaluate the visual crumb texture of wheat breads.

Using custards and a two level cup where the visible custards could be manipulated independently of the invisible ingested custards, de Wijk et al. (2004) found that the visual texture of the visible custards changed the oral texture ratings of the ingested custards. Carson et al. (2002) trained a descriptive panel to assess strawberry yogurts using visual texture terms including spoon impression ("the degree to which the product is jellified evaluated by looking at the impression left at the surface after lifting a spoonful from the unstirred product") and spoon covering ("the degree to which the product covers the back of the spoon evaluated by lifting a spoonful from the sample cup"). They found that both spoon impression and spoon covering were highly correlated with perceived oral thickness. The viscosity of a fluid can be assessed visually by pouring the fluid from a container, by tilting a container, or by evaluating the spreading of the fluid on a horizontal surface (Elejalde and Kokini, 1992; Kiasseoglou and Sherman, 1983; Sherman, 1977). Janhøj et al. (2006) trained a descriptive panel to evaluate low-fat yogurts using visual texture attributes such as grainy on lid and continuous flow from spoon. Lee and Sato (2001) used a paired comparison scaling technique to visually evaluate the perceived texture of real textile samples as well as photographic images of the samples. They found that the principal component spaces derived by the two methods were quite similar.

11.2.2 Auditory Texture

In some cases, consumers may find that the sounds (auditory texture) associated with eating a food product negatively impact the hedonic responses associated with the product. An example is the gritty sound of sand against the teeth when eating creamed spinach made with inadequately rinsed spinach leaves. On the other hand, auditory texture can add positively to consumers eating enjoyment as well, examples are the crisp sounds associated with many breakfast cereals or the crunchy sounds associated with eating a juicy apple. Consumers often use sound as an indication of food quality. Many of us have all thumped a watermelon to determine its ripeness (a hollow sound is indicative of a ripe watermelon) or broken a carrot to determine its crunchiness.

Auditory texture is to a large extent synonymous with crispness, crunchiness, and crackliness in foods. The early work in this area was done by Vickers and Bourne (1976). Lately there has been a resurgence of interest in the area with a review by Duizer (2001), and work by Luyten and van Vliet (2006), Salvador et al. (2009), and Varela et al. (2009). Sounds are produced by mechanical disturbances which generate sound waves which are propagated through the air or other media, such as bone conduction from the jaw bone to the bones of the middle ear (Dacremont, 1995).

Crisp and/or crunchy foods fall in two categories, namely wet foods and dry foods. Sound generation differs in these two types of foods (Vickers, 1979). Wet crisp foods, like fresh fruits and vegetables, are composed of living cells that are turgid if enough water is available. In other words, the cell contents exert an outward pressure against the cell walls. The tissue structure is thus similar to a collection of tiny water-filled balloons cemented together. When the structure is destroyed, by breaking or chewing, the cells pop and this produces a noise. In an air-filled balloon the popping sound is due to the explosive expansion of the air compressed inside the balloon. With turgid cells the noise is due to the sudden release of the turgor pressure. The amount of noise produced is less when the surface tension of the liquid is high. Exposing plant cells to sufficient moisture increases the turgor pressure of the cells and increases the perceived crispness of the product.

On the other hand, exposing dry crisp foods, like cookies, crackers, chips, and toast to moisture (humid air) decreases the perceived crispness of the food. These products have air cells or cavities surrounded by brittle cell or cavity walls. When these walls are broken any remaining walls and fragments snap back to their original shape. When the walls snap back vibrations are caused that generate sound waves (similar to a tuning fork). When the moisture content of dry crisp foods increases, the walls are less likely to snap back and the amount of sound generated is less.

Vickers (1981) and Christensen and Vickers (1981) showed that crispness and crunchiness of specified foods can be rated on the basis of sound alone, on the basis of oral–tactile clues alone, or on the basis of a combination of auditory and oral–tactile information. Crispness seemed to be acoustically related to

the vibrations produced by the food as it is deformed (Christensen and Vickers, 1981). However, later work by Edmister and Vickers (1985) indicated that auditory crispness is not redundant with oral–tactile crispness evaluations and Vickers (1987a) also indicated that the oral–tactile sensations are very important to evaluating crispness.

Vickers and Wasserman (1979) studied the sensory characteristics associated with food sounds. They had panelists evaluate the similarity between pairs of sounds produced by crushing the food with pliers (Table 11.3). The results of their study indicated that there may be two sensory characteristics separating food sounds, the evenness of the sound and the loudness of the sound. As the loudness of the sounds increased the panelists' perceptions of the intensities of crunchiness, crispness, crackliness, sharpness, brittleness, hardness, and snappiness also increased. When the sound is continuous (even) the panelists perceived

Table 11.3 Foods crushed with rubber-coated pliers to produce recorded sounds

Food	Description
Hard candy	1 whole Reeds Rootbeer candy
Fresh celery	1 cm piece cut perpendicular to stalk
Blanched celery	1 cm piece cut perpendicular to stalk and immersed in rapidly boiling water for 30 s
Cracker	1 whole Sunshine saltine cracker
Unripe pear	1 cm wedge
Peanut	1 whole Fisher's Virginia style peanut
Ginger snap	1 whole Nabisco Brands ginger snap
Fresh carrot	Crosswise section, 1 cm long and 1.5 cm wide
Blanched carrot	Crosswise section, 1 cm long and 1.5 cm wide, immersed in rapidly boiling water for 1 min
Potato chip	1 whole Pringles potato chip
Ruffled potato chip	1 whole Pringles ruffled potato chip
Unripe golden delicious apple	1 cm wedge
Ripe golden delicious apple	1 cm wedge
Graham cracker	1 whole (manufacturer unknown)
Milk chocolate	1 square of Hershey's milk chocolate, cold
Water chestnut	1 whole Geisha canned water chestnut
Shortbread cookie	1 whole Lorna Doone (Nabisco Brands) cookie
Shredded wheat	1 whole shredded wheat cake (Nabisco Brands)

Adapted from Vickers and Wasserman (1979)

the texture as popping or snappy and when the sound is not continuous the perception is of tearing or grinding. Zampini and Spence (2004) showed that potato chips were perceived as being crisper by panelists when the authors increased the overall sound level associated with biting the chip between the front teeth or when they increased was increased, or when they selectively amplified the high-frequency sounds (in the range of 2–20 kHz).

Dacremont (1995) found that crispy foods were characterized by high levels of air-conducted high-frequency sounds (5 kHz), crunchy foods were characterized by low pitched sounds with a peak in air-conduction at 1.25–2 kHz, and crackly foods were characterized by low-pitched sounds with a high level of bone conduction. Crunchiness is acoustically most related to a larger proportion of low-pitched sounds with frequencies less than 1.9 kHz, while a relatively larger proportion of high-pitched sounds, frequencies higher than 1.9 kHz, is related to crispness (Seymour and Hamann, 1988; Vickers, 1984a,b, 1985). It is more difficult to determine the crunchiness of a food through listening to someone else since many of the lower pitched sounds one hears while eating a crunchy food is conducted through the bones of the skull and jaw to the ear (Dacremont, 1995). The human jawbone and skull resonate at about 160 HZ and sounds in this frequency range are amplified by the bones, thus the panelists' own crunch sounds are perceived to be lower and louder than those of a person next to the panelist (Kapur, 1971). When training panelists to evaluate the perceived intensity of crunchiness one should train them to chew the food with the molars while the mouth is kept closed. Most of the high frequency sounds will be damped by the soft tissue and the crunchy sounds will be transmitted through the skull and jaw bones to the ear. Similarly, when training panelists to evaluate the perceived intensity of crispness one should train them to chew the food with the molars while the mouth is kept open (Lee et al., 1990). This method of chewing is seen as a violation of courtesy in some cultures but during training most panelists will succeed in chewing in this fashion. Most of the higher frequency sounds will travel undistorted through air to the ears (Dacremont et al., 1991).

Another view of crisp and crunchy foods looks at the time-sequence of breakage, the deformation and rupture of the food upon application of force (Szczesniak, 1991). Crisp foods break in a single stage whereas crunchy foods break in several successive stages. Thus, a crisp food will always be perceived as crisp regardless of the way the breaking force is applied, but a crunchy food may be perceived as crunchy or crisp depending on the applied force. A celery stick when chewed by the molars will be perceptibly crunchy since it will break in successive steps, but a celery stick snapped between the hands will be perceptibly crisp since the stalk will break in a single step.

Vickers (1981) found that it was possible to evaluate the perceived hardness of crisp foods based on sound alone. Castro-Prada et al. (2007) indicates that the best method of acquiring acoustical profiles of crispy foods to correlate with human sensory methods may be different from the best profiles to be used for fracture mechanical analyses. This may be because hardness is a component of crispness in these foods. Vickers (1984a, b) also evaluated the auditory component of the crackliness of foods. She found that like crispness and crunchiness, crackliness could be assessed by either sound or tactile evaluation. The number and amplitude of sharp repeated noises correlated with the perception of crackliness. However, oral–tactile sensations were more useful than auditory sensations for the assessment of hardness for most foods. As pointed out by Chen (2009) the vibrotactile perception of the teeth allows those hard of hearing to still enjoy crisp and crunchy foods.

11.2.3 Tactile Texture

Tactile texture can be divided into oral–tactile texture, mouth feel characteristics, phase changes in the oral cavity, and the tactile texture perceived when manipulating an object by hand (often used for fabric or paper and called "hand") or with utensils.

11.2.3.1 Oral–Tactile Texture

Oral–tactile texture encompasses all the textural sensations elicited in the mouth. The lips, teeth, oral mucosa, saliva, tongue, and the throat are involved in the perception of oral texture. Chen (2009), Lenfant et al. (2009), Xu et al. (2008), van der Bilt et al. (2006),

Bourne (2004), and Lucas et al. (2002) provide reviews of food oral processing, mastication, and the effects of oral physiology on the perception of food texture. According to van Vliet et al. (2009) and others the sequence of oral texture perception involves ingestion by the lips, biting by the front (incisor) teeth, chewing of hard foods by the molars, wetting with saliva and enzymatic breakdown, deformation of semi-solid foods between the tongue and hard palate, manipulation of the food into a bolus by the tongue and swallowing.

During ingestion the lips may signal that the food is sticky, slimy, hard, grainy, etc. For example, Engelen et al. (2007) had their panelists rate perceived roughness and slipperiness of custards and mayonnaises by rubbing the tongue against the inside of the lip.

The first bite allows the perceptions of hard, springy, cohesive, crumbly, etc., to occur. The force applied during the first bite is related to the food itself. Mioche and Peyron (1995) using pellet-shaped models found that for elastic food models (silicone elastomers) the bite force was symmetric, the food did not fracture and the perceived hardness was related to the perceived deformation under constant bite force. A food example of such a food probably does not exist but some foods such as gelatin gels come close. For food models that were more plastic (dental waxes) the biting force increases until a yield point is reached where the food begins to flow and then fracture. They found that the maximal bite force was highly correlated to perceived hardness ($r=0.96$). A real food example of a plastic food is butter. For brittle food models (pharmaceutical tablets) they found that the first bite biting cycle was the shortest with abrupt increases and decreases in force and again perceived hardness was highly correlated to maximal bite force ($r=0.99$). Cookies are a real world food example of a brittle product. Perceived hardness based on first bite increases with food thickness (Agrawal and Lucas, 2003). De Wijk et al. (2008) found that the bite size through a straw for a chocolate-flavored dairy semi-solid was significantly smaller (5.8 ± 0.3 g) than for a chocolate-flavored liquid dairy drink (8.7 ± 0.45 g). However, when these authors removed the bite effort (by using a pump) they found that these differences disappeared.

Chewing fragments solid and semi-solid foods into small enough particles to swallow and to mix these particles with saliva to form a lubricated bolus for swallowing. There is large variation in chewing cycles

and the length of chewing across individuals and across foods (Brown et al., 1994, 1995; Wintergerst et al., 2004, 2005, 2007). Engelen et al. (2005a) found that for 87 subjects with normal dentition the chewing cycles to ready 9.1 cm^3 peanuts for swallowing ranged from 17 to 110. In general, individuals producing more saliva tended to need fewer chewing cycles to ready a piece of dry toast for swallowing (Engelen et al., 2005a). These authors also found that buttering toast decreased the number of chew cycles prior to swallowing. Food hardness is also positively correlated to chewing length, chewing cycle, and muscle activity associated with chewing (Foster et al., 2006, Hutchings et al., 2009; Wintergerst et al., 2007). Blissett et al. (2007) showed that increased sample size (in their case 1, 2, or 4 orange-flavored Tooty-Frooties from Nestle, York, the United Kingdom) led to multiple changes in chewing behavior and that some of these changes were idiosyncratic.

A number of studies have shown wide ranges in salivary flow rates among individuals. Engelen et al. (2005a) found a mean flow rate of 0.45 $\pm$ 0.25 ml/min for unstimulated flow and a mean of 1.25 $\pm$ 0.67 ml/min for stimulated flows. Saliva has many functions but from an oral texture perspective it acts as a lubricant. The mucins (glycoproteins) are responsible for the lubrication effects of saliva. As shown by Prinz et al. (2007) salivary lubrication is increasingly efficient with high surface speeds and increased surface load. A few studies have shown that tougher meat samples lead to higher incorporation of saliva into the bolus prior to swallowing (Claude et al., 2005; Mioche et al., 2003). The salivary pH and α-amylase content also affects perceived texture. Engelen et al. (2007) found that α-amylase activity was negatively correlated to perceived thick mouth feel of custards and to perceived prickly mouth feel for mayonnaise.

11.2.3.2 Size and Shape

Tyle (1993) evaluated the effect of size, shape, and hardness of suspended particles on the oral perception of grittiness of syrups. He found that soft, rounded, or relatively hard, flat particles were not perceptually gritty up to about 80 μm. However, hard angular particles contributed to grittiness perception when they were above a size range of 11–22 μm. Richardson and Booth (1993) found that some of their panelists could distinguish between average fat-globule size and distance distributions of less than 1 μm (range: 0.5–3 μm, depending on the individual). Engelen et al. (2005b) found that polystyrene spheres between 2 and 80 μm decreased the perceived smoothness and slipperiness and increased perceived roughness of custards. Above 80 μm the perception of roughness decreased. In other studies the minimum individual particle size detectable in the mouth was less than 3 μm (Monsanto, 1994). Richardson and Booth (1993) working with milks and creams found that their panelists were sensitive to viscosity changes of about 1 mPa. Runnebaum (2007) working with wine found that his panelists could distinguish viscosity changes of about 0.057 mPa.

By definition (Peleg, 1983) a property is a characteristic of a material which is practically independent of the method of assessment. A property can only be called objective if its magnitude is independent of the particular instrument used and of the specimen mass and size. For example, the percentage of fat in an ice cream is the same regardless of the amount of the ice cream analyzed. However; sensory textural properties are affected by sample size. Large and small sample sizes may or may not be perceptually the same in the mouth. A debated question is whether humans compensate automatically for the difference in sample size or whether humans are only sensitive to very large changes in sample size. It is not known which of these happen, if either. One of the few studies to explicitly study the effect of sample size on texture perception was done by Cardello and Segars in 1989. They evaluated the effect of sample size on the perceived hardness of cream cheese, American cheese, and raw carrots and on the perceived chewiness of center cut rye bread, skinless all beef franks, and Tootsie roll candies. The sample sizes (volumes) evaluated were 0.125, 1.00, and 8.00 cm^3 and their experimental conditions were sequential versus simultaneous presentation of samples, sample presentation in random order or by ascending size; evaluation of samples by blindfolded and not blindfolded panelists; panelists allowed to handle the sample or not. These authors found both hardness and chewiness increased as a function of sample size independent of subject awareness of sample size. Therefore, texture perception does not appear to be independent of sample size. Additionally, as shown by Dan et al. (2008) the sensory perception of hardness varies with the specific definition associated with

the bite procedure. Initially panelists were instructed to evaluate the hardness of a cheese sample by biting the cheese normally with the molars on their habitual chewing side (Control condition). Subsequently, they were asked to evaluate hardness by either biting into the sample with the molar teeth (H1 condition) or to bite completely through the sample with the molar teeth (H2 condition). They found that the H2 condition led to high inter-panelist differences while the panelists were relatively homogeneous across the H1 condition. However, both conditions led to the same rank ordering of the cheese samples. For the sensory specialist the important "take-home" message is that all conditions such as sample dimensions, samples size, or volume must be specified since these could materially affect the results.

11.2.3.3 Mouth Feel

Mouth feel characteristics are tactile but often tend to change less dynamically than most other oral–tactile texture characteristics. For example, the mouth feel property astringency associated with a wine usually does not change perceptibly while the wine is manipulated in the mouth but the chewiness of a piece of steak or the consistency of ice cream will change during in-mouth manipulation. Often cited mouth feel characteristics are astringency, puckering (sensations associated with astringent compounds), tingling, tickling (associated with carbonation in beverages), hot, stinging, burning (associated with compounds that produce pain in the mouth such as capsaicin), cooling, numbing (associated with compounds that produce cooling sensations in the mouth such as menthol), and mouth coating by the food product. From this list it is clear that mouth feel characteristics are not necessarily related to the force of breakdown or to the rheological properties of the product. However, some mouth feel attributes are related to the rheology of the product and/or the force of breakdown, examples are viscosity, pulpy, sticky. Other mouth feel attributes are chemically induced tactile sensations such as astringency and cooling and these were discussed in Chapter 2.

As will be seen later (Brandt et al., 1963), the original Texture Profile method had only a single mouth feel-related attribute "viscosity." Szczesniak (1966) classified mouth feel attributes into nine groups: Viscosity-related (thin, thick); feel of soft tissue surfaces related (smooth, pulpy); carbonation related (tingly, foamy, bubbly); body related (watery, heavy, light); chemical related (astringent, numbing, cooling); coating of the oral cavity related (clinging, fatty, oily); related to resistance to tongue movement (slimy, sticky, pasty, syrupy); mouth after feel related (clean, lingering); physiological after feel related (filling, refreshing, thirst quenching); temperature related (hot, cold); and wetness related (wet, dry). Jowitt (1974) defined many of these mouth feel terms. Bertino and Lawless (1993) used multidimensional sorting and scaling to determine the underlying dimensions associated with mouth feel attributes in oral health-care products. They found that these clustered in three groups: astringency, numbing, and pain.

11.2.3.4 Phase Change (Melting) in the Oral Cavity

The melting behaviors of foods in the mouth and the associated textural changes have not been studied extensively. Many foods undergo a phase change in the mouth due to the increased temperature in the oral cavity. Examples are chocolates and ice cream. As mentioned earlier Hyde and Witherly (1993) proposed an "ice cream effect." They stated that dynamic contrast (the moment-to-moment change in sensory texture contrasts in the mouth) is responsible for the high palatability of ice cream and other products. The work by Hutchings and Lillford (1988) on emphasizing the dynamic breakdown of the sample in the mouth during mastication was a breakthrough that should (but has not yet) lead to the testing of a general physical and psychophysical hypothesis of perceived texture.

For some time the trend in food marketing and product development has been to eliminate as much fat as possible from food products. However, the fat is primarily responsible for the melting of ice cream, chocolates, yogurt, etc., in the oral cavity (Lucca and Tepper, 1994). Thus the characteristics associated with phase change should receive additional scrutiny as product developers attempt to replace the mouth feel characteristics of fats with fat replacer compounds.

In an early study Kokini and Cussler (1983, 1987) found that the perceived thickness of melting ice cream in the oral cavity was related to the following equation:

$$\text{Thickness} \propto \mu^{\frac{3}{4}} f^{\frac{1}{4}} V \left[\frac{2(1-\phi)\Delta H_i \rho}{3\, K \Delta T \pi R^4} \right]^{1/4}$$

where

μ = liquid phase viscosity
T = temperature difference between the solid phase (frozen ice cream) and the tongue
ϕ = volume fraction of air in the product (overrun)
H_i= heat of fusion of ice
ρ= density of ice
V = velocity of tongue movements
F = force applied by tongue
R = tongue radius (assuming a circle) in contact with the food
K = thermal conductivity of melted ice cream

As pointed out by Lawless et al. (1996) "while this equation may be useful to point out the various factors influencing melting systems, it is doubtful that all the parameters could be known in practice or standardized among sensory panelists." Thus, at this time, the study of melting is still being done empirically using panelists and descriptive sensory evaluation or time–intensity methodology. There has been a plethora of low-fat ice cream-related perceived texture and melt rate studies using generic descriptive analysis (Hyvönen et al., 2003; Liou and Grün, 2007; Roland et al., 1999). Lawless et al. (1996) studied the melting of a simple cocoa butter model food system and found that this system could be used to study the textural and melt properties of fat replacers. Changes in melting behavior, as assessed by descriptive analysis and by time–intensity measurements, were related to the degree of fat substitution by carbohydrate polymers. Mela et al. (1994) had found that panelists could not use the degree of melting in the oral cavity to accurately predict the fat content in oil-in-water emulsions (products similar to butter) with a melting range of 17–41°C.

11.2.3.5 Oral Crispness, Crunchiness, and Crackliness

As discussed in the section on auditory texture crispness, crunchiness, and crackliness clearly have an auditory component but these sensations also have an oral textural component. See the review by Roudaut et al. (2002) for a critical appraisal of the evaluation of crispness.

Vincent (1998) stated that these sensations are related to the sudden drop in force experienced by the teeth and the jaw muscles when a food item breaks between the teeth. Initially he thought that crumbliness, crispness, crunchiness, and hardness are descriptors falling on a continuous load-drop-size continuum. Subsequently (Vincent, 2004), he suggested that crack initiation and propagation in hard and crunchy foods are related to the force needed to fracture the sample and that crispness is a distinct and separate sensation related to fracturability of glassy cellular materials. Crispness decreases as product water activity (a_w) increases and at a water activity of 0.40–0.55 (depending on the product) the perceived crispness of the product decreases dramatically (Heidenreich et al., 2004). Primo-Martin et al. (2008) found that toasted rusk rolls lost 50% of their perceived crispness at critical water activities between 0.57 and 0.59.

11.2.4 Tactile Hand Feel

Tactile hand feel of foods are usually evaluated through the use of utensils (the amount of effort to cut a piece of steak, the ease of butter spreadability with a knife, the ease with which a fork penetrates a boiled potato, etc.) or by manipulation by hand (the ease of snapping a celery stalk, the difficulty in compressing a piece of cheese between the thumb and forefinger, etc.). Table 11.4 summarizes a few tactile hand feel attributes. Pereira et al. (2002) used a trained descriptive panel to evaluate cheese analogs and all of their texture attributes were through tactile hand feel. Ares et al. (2006) used non-oral texture evaluation to characterize dulce de leche. Dooley et al. (2009) used some tactile hand attributes in their evaluation of lip products. Darden and Schwartz (2009) found that their trained descriptive analysis panel could reproducibly score fabric abrasiveness, sensible texture, slipperiness, and fuzziness using their finger tips. Lassoued et al. (2008) used flash profiling to evaluate the tactile crumb texture of wheat breads.

The texture evaluation of fabric or paper frequently includes touching or manipulating the material with the fingers. Much of the work in this area comes from the textile literature; however, we feel that this area

Table 11.4 Examples of sensory hand tactile attributes

Texture attribute	Manipulation by hand
Fracturability	Extent to which a cheese slice (1 cm thick, 9 cm long) can be bent between the thumb and the index and middle fingers, until the ends touch, without breaking
Firmness (compression)	Amount of resistance to compression offered by a 1 cm thick slice of cheese, when pushed between the thumb and the index finger, until fingers touch each other (force required to deform the cheese structure)
Firmness (cutting)	Force required to cut through a 1 cm thick slice of cheese with a knife (pushed down on an angular, guillotine-like movement, from tip to full length of the knife)
Curdiness	Extent to which the original sample produces curdy lumps after being kneaded seven times between thumb and index and middle finger
Hardness	Force required scooping up a teaspoonful of the sample
Ropiness	The amount of threads or drops that fall down when introducing the spoon vertically into the sample and raising it vertically from the sample once
Spreadability	The ease with which the product can be manipulated on the surface of the forearm (Vaseline=5, Classic Chapstick=9; Johnson & Johnson 24-h Moisturizer = 13)
Tackiness	The degree to which fingers adhere to the product; amount of adhesiveness (Johnson & Johnson Baby Oil=0, Post-it note=7.5)

Adapted from Pereira et al. (2002), Ares et al. (2006), and Dooley et al. (2009)

of sensory evaluation has potential application in the food arena as well. We will thus describe some of the vocabulary associated with fabric or paper hand with the intention of stimulating food sensory specialists to allow their panelists "to play with their food" on occasion when it could lead to appropriate results. Most of the information in this section was drawn from Civille and Dus (1990), Meilgaard et al. (2006), and Civille (1996).

Civille and Dus (1990) describe the tactile properties associated with fabric and paper as mechanical characteristics (force to compress, resilience, and stiffness), geometrical characteristics (fuzzy, gritty), moisture (oily, wet) and thermal characteristics (warmth), and non-tactile properties (sound).

The fabric/paper methodology developed by Civille is based on the General Foods Texture Profile (described in the next section) and includes a series of standard scales with reference anchors and precise definitions for each attribute evaluated. Some of these are listed in Table 11.5.

In a series of studies Japanese textile scientists (Kawabata and Niwa, 1989; Kawabata et al., 1992a, b; Matsudaira and Kawabata, 1988) quantified and correlated sensory evaluation results of textiles with instrumental measurements. Their techniques have been extensively used, studied, and adapted within the textile industry (Bertaux et al., 2007; Cardello et al., 2003; Chen et al., 1992; Kim et al., 2005; Koehl et al., 2006; Sztandera, 2009; Weedall et al., 1995).

Other sensory textile measurements have also been developed. Paired comparison discrimination tests have been used to assess the stiffness, smoothness, and softness of cotton fabrics (Ukponmwan, 1988). Burns et al. (1995) found that subjects who viewed and felt fabrics described the sensory properties of fabrics differently than did subjects who only felt the fabrics for their hand. They cautioned that laboratory techniques that only concentrated on hand may not correlate with consumer perceptions of fabric textures. Bertaux et al. (2007) used a paired comparison method to evaluate roughness and prickle of woven and knitted fabrics. Hu et al. (1993) used Steven's law as a psychophysical description of fabric hand evaluations. In another study, the tactile qualities of fabrics were evaluated using bipolar descriptive scales (Jacobsen et al., 1992). The authors found good correlations between the values obtained by the panel and with instrumental bending and compression evaluations. Philippe et al. (2004) and Cardello et al. (2003) described the use of generic descriptive analysis in the evaluation of the textile hand of cotton fabric treated with different industrial finishes and in military clothing fabrics, respectively.

Mahar et al. (1990) found that there were cultural differences in the handle preferences for men's winter suit fabrics. The panelists from Australia, India, New Zealand, the United States, and Hongkong/Taiwan had consistent preferences based on their evaluation of the fabric hand using the descriptors sleekness, fullness firmness, and drape. The panelists from Japan and the People's Republic of China had internally consistent and somewhat opposite preferences to that of the first

Table 11.5 Selected fabric hand profile attribute definitions and reference anchors

Attribute	Definition	Scale value	Fabric type
Force to compress	Amount of force required to compress gathered sample in palm (low force to high force)	1.5	Polyester/cotton 50/50 knit tubular
		3.4	Cotton cloth greige
		9.3	Cotton terry cloth
		14.5	#10 Cotton duck greige
Resilience	Force with which sample presses against cupped hands (creased to folded original shape)	1.0	Polyester/cotton 50/50 knit tubular
		7.0	Filament nylon 6.6 semi-dull taffeta
		14.0	Dacron
Stiffness	Degree to which sample feels pointed, ridged, and cracked, not round, pliable, curved (pliable to stiff)	1.3	Polyester/cotton 50/50 knit tubular
		4.7	Mercerized cotton print cloth
		8.5	Mercerized combed cotton poplin
		14.0	Cotton organdy
Geometrical properties			
Fuzziness	Amount of pile, fiber, fuzz on surface of sample (bald to fuzzy or nappy)	0.7	Dacron
		3.6	Cotton crinkle gauze
		7.0	Cotton T-shirt, tubular
		15.0	Cotton fleece
Grittiness	Amount of small picky particles in surface of sample (smooth to gritty)	1.5	Filament arnel tricot
		6.0	Cotton cloth greige
		10.0	Cotton print cloth
		11.5	Cotton organdy

Adapted from Civille (1996)

group. Raheel and Liu (1991) used a mathematical technique called fuzzy sets logic to integrate sensory fabric hand data with instrumental assessments. This is one of the earliest uses of the fuzzy logic technique with sensory data; however, it is still in use (Koehl et al., 2006).

11.3 Sensory Texture Measurements

Many texture attributes can be measured using standard sensory techniques such as discrimination testing, ranking, and descriptive techniques. Textural differences between two samples can be determined using the two-alternative forced choice test. The panelists should be trained to discriminate between the samples based on the specified textural attribute. For example, panelists can be trained to evaluate viscosity as "the amount of force required to draw a liquid from a spoon over the tongue" (Szczesniak et al., 1963) and could then be asked to determine if the perceived viscosity of two maple syrup samples differed.

It is also possible to quantify texture attributes using ordinal or interval scales. Examples would be "rank the ...." or "score the ...." Visual texture, especially, lends itself well to simple intensity or ordinal scales, such as apparent roughness of the surface, size or number of surface indentations, and density or amount of sediment in a container of a liquid product. Most of these simple and concrete attributes require little training and can be easily worked into a descriptive profile of the product. Of course, as in any other descriptive or scaling technique, the scale becomes more calibrated and there is better agreement among panelists if the low and high ranges are shown to provide the frame of reference that anchors the scale.

Szczesniak et al. (1975) used consumers to evaluate foods using terminology developed for the General Foods Texture Profile method (see below) and they found that consumers could use the scales and were sufficiently aware of the texture of food products to do a rudimentary and "fuzzy" texture profile.

11.3.1 Texture Profile Method

The Texture Profile method was developed at General Foods Corporation in the early 1960s. The scientists

at General Foods based their texture evaluation approach on the Flavor Profile developed by A.D. Little. They were interested in developing a method that would allow the evaluation of texture and which would be built on a well-defined and rational foundation.

Szczesniak (1963) developed a texture classification system to bridge the gap between consumer texture terminology and the rheological properties of the product (Table 11.6). She categorized the perceived textural characteristics of products as three groups: mechanical characteristics, geometrical characteristics, and other characteristics (alluding mostly to the fat and moisture content of foods). This classification formed the basis of the Texture Profile method (Brandt et al., 1963). These authors defined their method as a technique that would allow the description of the mechanical, geometric, and other textural sensations associated with a product from the first bite through complete mastication. The technique therefore borrows the "order of appearance" principle from the Flavor Profile and is thus a time-dependent method. The time sequence is the "first bite" or initial phase, the "chewing" or masticatory second phase followed by the residual or third phase. The textural sensations were evaluated by extensively trained panelists using standard rating scales. The original standard rating scales were developed by Szczesniak et al. (1963) to cover the range of intensity sensations found in foods. They used

Table 11.6 Texture classification and the bridge to some consumer texture descriptions

Primary terms	Secondary terms	Consumer terms
Adhesiveness		Sticky, tacky, gooey
Cohesiveness	Brittleness	Crumbly, crunchy, brittle
	Chewiness	Tender, chewy, tough
	Gumminess	Short, mealy, pasty, gummy
Elasticity		Plastic, elastic
Hardness		Soft, firm, hard
Viscosity		Thin, thick
Particle shape and orientation		Cellular, crystalline, fibrous, etc.
Particle size and shape		Coarse, grainy, gritty, etc.
Fat content	Greasiness	Greasy
	Oiliness	Oily
Moisture content		Dry, moist, wet, watery

Adapted from Szczesniak (1963)

specified food products to anchor each scale point. The earliest standardized texture scales were developed for adhesiveness, brittleness, chewiness, gumminess, hardness, and viscosity. These authors validated their scales by correlating the results obtained by the sensory panelists to the results obtained instrumentally by viscometer and texturometer. A later section will discuss sensory and instrumental texture correlations.

The Texture Profile method was used extensively at General Foods and the number of standardized rating scales was expanded over time, for example, Brandt et al. (1963) added elasticity which was later changed to springiness (Szczesniak, 1975), Szczesniak and Bourne (1969) added firmness and later brittleness was renamed fracturability (Civille and Szczesniak, 1973). The original Texture Profile had scales of varying length, for example, the scale for chewiness had seven points, gumminess had five points, and hardness had nine points (Bourne, 1982). The article by Civille and Szczesniak (1973) uses a 14-point intensity scale and the paper by Muñoz (1986) describes a 15 cm line scale with the intensity anchors positioned on the scale.

Civille and Szczesniak (1973) succinctly described how a Texture Profile panel should be selected and trained. They suggested training about ten panelists with the goal of having at least six available at all times. The panelists should undergo a physiological screening to eliminate potential panelists with dentures and those without the ability to discriminate among textural differences. Panelists are also interviewed to assess interest, availability, attitude, and communication skills. During panel training, the panelists are exposed to the basic concepts associated with flavor and texture perception and the underlying principles of the Texture Profile. They are also trained to use the standard rating scales in a uniform and consistent fashion. The panel will practice using the rating scales on a series of food products. This practice may be quite extensive, lasting several months. Any inconsistencies among panelists are discussed and resolved.

Once the panel has been trained, which in some cases could mean a time commitment of 2–3 h daily sessions for 2 weeks followed by 6 months of 1 h session four to five times a week; the panel can begin evaluating test products. On the other end of the time scale one of us was trained as a fish texture panelist where the training lasted only about 2 weeks. A well-trained panel should be maintained by testing their reproducibility with blind samples and by reviewing their results regularly. During these review sessions any inconsistencies among panelists should be ironed out. Additionally, the panel leader should continually strive to keep the panel motivated.

The Texture Profile has been modified and refined since its original creation. Civille and Liska (1975) described the modifications to that date. These included modifying some of the food products used to anchor the standard intensity scales, adding the evaluation of the products surface properties to the initial stage of the evaluation, and adding standard scales to evaluate liquids and semi-solids. Additionally, the cohesiveness of mass standard scale was developed as was a scale for bounce or elasticity.

Muñoz (1986) published a paper describing the selection of new products to anchor the intensity points on the standard scales. Between 1963 and 1986 many products had changed in formulation and were no longer representative of a specific intensity on a specified Texture Profile scale and others were not available anymore. She also modified and fleshed out a number of the scale definitions. Tables 11.7 and 11.8 are principally based on the improvements to the Texture Profile made by Muñoz (1986).

Others have modified the standard scales to better suit their needs, see, for example, Chauvin et al. (2008) who created new scales for the wet and dry food attributes: crispness, crunchiness, and crackliness. In this case the authors used acoustical parameters and sensory panelists to determine the appropriate products to use on the standard scales. In a few cases the modifications of the Texture Profile standard scales were made because the American food products used as anchors were not available in other countries, for example, Bourne et al. (1975), or Otegbayo et al. (2005); for non-food products, see Schwartz (1975). The Schwartz paper is a useful starting place for skin care products and related personal care or cosmetic items that have important skin feel properties. The review by Skinner (1988) is a very complete treatise on the state of the texture profile to that date. The sensory texture profile is still in use, see, for example, Lee and Resurreccion (2001) who used the technique for peanut butter and Breuil and Meullenet (2001) who used it for cheeses. Chauvin et al. (2008) developed new standard scales for crispness,

Table 11.7 Texture profile attribute definitions

Texture attribute	Definition
Non-oral	
Manual adhesiveness	Force required to separate individual pieces adhering to each other using the back of a spoon, after placing entire contents of the standard cup on a plate
Viscosity	Degree of resistance when stirred by a spoon
	Rate at which sample flows down the side of a tilted container
Oral	
Initial lip contact	
Adhesiveness to lips	Degree to which the product stick/adheres to the lips. The sample is placed between the lips and compressed once slightly and released to assess lip adhesiveness
Wetness	Amount of moisture perceived on the surface of the product, when in contact with the upper lip
Initial insertion in mouth	
Roughness	Degree of abrasiveness of the product's surface, as perceived by the tongue
Self-adhesiveness	Force required to separate individual pieces with the tongue, when the sample is placed in the mouth
Springiness	Force with which the sample returns to its original size/shape, after partial compression (without failure) between the tongue and the palate
Initial bite	
Cohesiveness	Amount of deformation undergone by the material before rupture when biting completely through sample with molars
Adhesiveness to palate	Force required to remove product completely from palate, using tongue, after compression of the sample between tongue and palate
Denseness	Compactness of the cross section of the sample after biting completely through with molars
Fracturability	Force with which the sample ruptures when placed between molars and bitten completely down at a fast rate
Hardness	Force required to bite completely through sample placed between molars
After chewing	
Adhesiveness to teeth	Amount of product adhering on/in the teeth after mastication of the product
Cohesiveness of mass	Degree to which the mass holds together after mastication of product
Moisture absorption	Amount of saliva absorbed by the sample after mastication of product

Adapted from Muñoz (1986) and Sherman (1977)

crackliness, and crunchiness in dry and wet foods (Table 11.9).

Cardello et al. (1982) used free-modulus magnitude estimation to rescale the standard texture profile scales of adhesiveness, chewiness, fracturability, hardness, gumminess, and viscosity. They found that the category scales of the traditional Texture profile were concave downward when plotted against the magnitude estimation scales. This indicates that for these attributes the panelists exhibit a greater discrimination at the lower levels of intensity. This is a pattern consistent with Weber's law (see Chapter 2). Weber's law predicts smaller difference thresholds at low levels of intensity. The data also suggest that the results from category scales and magnitude estimation scales are different but similar.

11.3.2 Other Sensory Texture Evaluation Techniques

The sensory scientist does not have to train a panel use the sensory texture profile analysis technique. It is entirely possible to use generic sensory descriptive analysis to describe differences in the textures of products. For example, Weenen et al. (2003) used consensus training to train a panel to evaluate mayonnaises, salad dressings, custards, and warm sauces. They found that the panel grouped the sensory texture of these semi-solid foods into six clusters (visco-elastic-related attributes; surface feel-related attributes; bulk homogeneity-related attributes; adhesion/cohesion-related attributes; wetness/dryness-related attributes;

Table 11.8 Examples of texture attribute intensity anchors

Texture attribute	Scale	Product
Adhesiveness	Low	Hydrogenated vegetable oil
	Medium	Marshmallow topping
	High	Peanut butter
Adhesiveness to lips	Low	Tomato
	Medium	Bread stick
	High	Rice cereal
Adhesiveness to teeth	Low	Clam
	Medium	Graham cracker
	High	Jujubes
Cohesiveness	Low	Corn muffin
	Medium	Dried fruit
	High	Chewing gum
Cohesiveness of mass	Low	Licorice
	Medium	Frankfurter
	High	Dough
Denseness	Low	Whipped topping
	Medium	Malted milk balls
	High	Fruit jellies
Fracturability	Low	Corn muffin
	Medium	ginger snap (inside part)
	High	Hard candy
Hardness	Low	Cream cheese
	Medium	Frankfurter
	High	Hard candy
Manual adhesiveness	Low	Marshmallow
	Medium	Dough
	High	Nougat
Moisture absorption	Low	Licorice
	Medium	Potato chip
	High	Cracker
Roughness	Low	Gelatin dessert
	Medium	Potato chip
	High	Thin bread wafer
Self-adhesiveness	Low	Gumi-bear
	Medium	American cheese
	High	Caramel
Springiness	Low	Cream cheese
	Medium	Marshmallow
	High	Gelatin dessert
Wetness	Low	Cracker
	Medium	Ham
	High	Wafer

Adapted from Muñoz (1986) and Meilgaard et al. (2006)

and fat-related attributes). These authors subsequently used generic descriptive analysis panels to evaluate a wide range of semi-solid foods under different conditions (Engelen et al., 2003; Weenen et al., 2005). Others have also used generic descriptive analysis to describe the texture of cooked potatoes (Thybo et al., 2000), ketchup (Varela et al., 2003), oat breads (Salmenkallio-Marttila et al., 2004), creamy foods (Tournier et al., 2007), crisp and crunchy dry foods (Dijksterhuis et al., 2007), mango puree with added barium sulfate (Ekberg et al., 2009), and mayonnaises (Terpstra et al., 2009).

Table 11.9 Crispness, crackliness, and crunchiness standard scales for dry foods

Attribute	Reference	Manufacturer	Sample size and scale value
Crispness (dry food)			
2	Rice Krispies treats	Kellogg's, Battle Creek, MI	1/6 bar
5	Fiber rye bread	Wasa, Bannockburn, IL	1/3 slice
8	Multigrain mini rice cakes	Honey Graham, Quaker, Chicago, IL	1 cake
10	Bite size Tostitos tortilla chips	Frito Lay, Dallas, TX	1 chip
15	Kettle Chips	Frito Lay, Dallas, TX	1 chip
Crackliness (dry food)			
2	Club cracker	Keebler, Battle Creek, MI	$\frac{1}{2}$ cracker
7	Multigrain mini rice cakes	Honey Graham, Quaker, Chicago, IL	$\frac{1}{2}$ cake
9	Le Petit Beurre tea cookie	Lu, Barcelona Spain	1/8 square
12	Triscuit	Nabisco/Kraft Foods, Chicago, IL	$\frac{1}{4}$ broken with grain
15	Ginger snap	Archway, Battle Creek, MI	$\frac{1}{2}$ cookie

Adapted from Chauvin et al. (2008)

11.3.3 Instrumental Texture Measurements and Sensory Correlations

"Texture is a sensory property" (Szczesniak (2002) and thus the goal of instrumental "texture" measurements is to produce a mechanical test that can replace sensory panels as texture evaluation tools. The need to replace the sensory panel is usually due to cost or efficiency. Basic questions that should be asked are what are meant by objective mechanical "texture" properties and does a sensory textural property have universal meaning across food products? For example, is the sensory hardness of cheese the same as the hardness of a cookie, or is the perceived juiciness associated with a grape the same as that perceived in cooked steak?

A glance at the literature would indicate many examples where the authors use the same word (e.g., hardness) for their measurements of both sensory and instrumental texture parameters in the food product. The problem is that these measurements are often not highly correlated with one another. When this occurs the author of a protocol or paper should be extremely careful to distinguish between the sensory and the instrumental measurement. Figure 11.1 indicates a

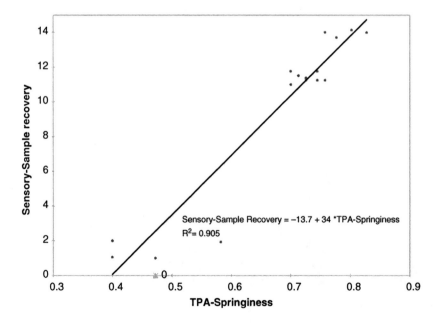

Fig. 11.1 Linear regression and correlation between the sensory texture attribute sample recovery and the instrumental texture parameter springiness (TPA= texture profile analysis). Redrawn in a different orientation from Kim et al. (2009).

Sensory-Sample Recovery = −13.7 + 34 *TPA-Springiness
R^2= 0.905

linear regression and correlation between a modified Texture Profile Analysis (TPA) and the sensory texture attribute (sample recovery) for cereal snack bars (Kim et al., 2009).

In this case the authors were very careful to use different terms for their sensory and instrumental measurements. Originally, many instrumental texture measurements were attempts to find a single parameter (or an overall value) to correlate with sensory texture evaluations. But "...it is often extremely difficult to predict sensory attributes from a unique instrumental parameter ..." (Breuil and Meullenet, 2001) and thus recently, many scientists have been exploring methods that would be more multivariate in nature (Varela et al., 2006).

One of the earliest papers to correlate instrumental texture parameters with sensory texture attributes was Friedman et al. (1963). These authors were part of the group developing the General Foods Texture Profile. They designed a new piece of equipment to translate the texture measurements defined by Szczesniak (1963) in physical measurements. The General Foods Texturometer had plungers which penetrated the food in two cycles, the penetration force was recorded and attributes of the instrumental texture profile were selected to correlate well with the sensory texture parameters rated by the trained Texture Profile panelists. Due to this careful selection of the instrumental texture parameters the authors had high correlation between their sensory and instrumental measurements. They continued to refine the Texturometer and published a number of papers correlating instrumental and sensory texture attributes (Szczesniak et al., 1963). The measurement technique based on the Texturometer became known as the Texture Profile Analysis (TPA) which is different from the sensory Texture Profile method. Later, the TPA techniques, developed with the Texturometer, were used with the Instron Universal Testing Machine and other related equipments (Breene, 1975; Finney, 1969; Szczesniak, 1966, 1969; Varela et al., 2006).

Szczesniak (1968, 1987) cautioned sensory specialists and food engineers against blindly correlating sensory and instrumental attributes. She cited a series of studies correlating sensory tenderness and shear force values obtained by Warner-Bratzler shear, the correlation coefficients ranged from −0.94 to −0.16. She stated that if one assumes that both the sensory and instrumental measurements were performed using standard good practices (not always an appropriate assumption) then these inconsistent correlations are due to other contributing conditions such as

(a) The correlation coefficient is dependent on the range and number of samples used. Additionally, the Pearson's correlation coefficient is based on a linear relationship, thus if the relationship is curvilinear the values may need to be logarithmically transformed.

(b) The instrumental measurement should mimic as far as possible the conditions used to evaluate the sensory attribute. Thus if tenderness is evaluated by a single bite through the sample with the incisors then a shear force measurement is more likely to be highly correlated. On the other hand, if the tenderness of the sample is evaluated by chewing with the molars then a shear force measurement may not be correlated with the perceived tenderness. If the sample is evaluated at above ambient temperature then the instrumental measurement should be made at the same temperature. Despite the evident obviousness of this statement this is not always done. Hyldig and Nielsen (2001) pointed out that in salmon-related studies the instrumental texture measurement is frequently performed on the raw fish and the sensory texture is measured on the cooked fish. It should not be surprising that resultant correlations between the two measurements are low.

(c) Since the sample is often destroyed during either measurement the same sample cannot be evaluated by both methods. Therefore the sample itself may be part of the problem, especially, if there is considerable variation in the texture attributes of samples from the same source. This is frequently a problem with meat samples where the tenderness within a single muscle can vary longitudinally (Cavitt et al., 2005). Newer non-destructive methods such as near-infrared spectroscopy (Blazquez et al., 2006) allow the sensory scientist to use the same sample for both the instrumental and texture measurements but many of these methods are in their infancy.

(d) Natural variability among panelists in terms of chewing cycles, dentition, salivary flow rates, etc., is a factor that will affect the quality of instrumental texture relationships.

Brennan and Jowitt (1977) categorized the instrumental texture measurement techniques as fundamental, imitative, and empirical. The fundamental techniques measure well-defined physical properties and at that time the authors felt that no measurement technique actually did a fundamental measurement. Recently, Ross (2009) stated that steady shear and dynamic viscometer measurements on fluids as well as measurements of deformation on solids are probably fundamental measurements of texture. Kim et al. (2009) stated that the 3-point bending test used to measure the fracturability of sheet-shaped foods was also a fundamental measurement. This technique was used with success by Rojo and Vincent (2008) to study perceived crispness in potato chips. With imitative techniques the measurement mimics the actions of the teeth and the jaws during the sensory measurement as closely as possible. Hyldig and Nielsen (2001) stated that the instrumental firmness evaluation of salmon by compression was an imitative method related to the sensory firmness evaluation of pressing the salmon with the index finger. Examples of imitative techniques are the puncture test which measures the force required to punch a hole in the food (a combination of shear and compression forces), the use of sound measurements (gnathosonics, Duizer, 2001; Ross, 2009; Kim et al., 2009), and electromyography (EMG). The early work by Vickers and coworkers (see Vickers, 1987b) on using the sounds associated with biting/chewing dry and wet crisp/crunchy food to determine perceived crispness and crunchiness has been expanded through the use of fast Fourier transform algorithms (Al-Chakra et al., 1996) and fractal analyses (Barrett et al., 1994, Gonzalez-Barron and Butler, 2008a) to analyze sound frequencies (de Belie et al., 2002). See González et al. (2001) for reviews of EMG in food texture evaluations. Additional information on EMG can be found in Foster et al., 2006; González et al., 2004; Ioannides et al., 2007, 2009.

Most instrumental texture measurements are empirical and do not necessarily "translate" across food products. This is not necessarily a problem since Drake et al. (1999) stated that "While fundamental rheological test reveal important information on network structure and molecular arrangement [in cheese], ... empirical texture evaluations work equally well or better at predicting sensory texture properties."

Image analyses and/or microscopy are also used in relationship to visual and, sometimes, oral and nonoral–tactile texture (Di Monaco et al., 2008; Gonzalez-Barron and Butler, 2008b; Lassoued et al., 2008; Martens and Thybo, 2000; Zheng et al., 2006). Chen (2007) reviews these instrumental techniques and their uses in the characterization of perceived surface texture.

11.4 Conclusions

The sensory evaluation of texture has advanced a great deal since the middle of this century, yet in 1991 Alina Szczesniak, surely the doyenne of food texture in the United States, could still state that "there are still many important gaps in the consumer/texture interface where progress has not kept up with that in the area of instrumental texture measurements." She continues "Quantitative measures of the relative importance of texture in specific food categories should be developed and related to the level of textural quality." This state of affairs is emphasized by Chen (2009) who stated that "... a thorough understanding of the principles and mechanisms involved in food oral processing will be essential. Without such knowledge, our studies of food texture probably would not go far." Given the importance of food texture in food quality and acceptance, there is still a great deal of work that must be done in this area.

References

Agrawal, K. R. and Lucas, P. W. 2003. The mechanics of the first bite. Proceedings of the Royal Society, London, B, Biological Science, 270, 1277–1282.

Al-Chakra, W., Allaf, K. and Jemai, A. B. 1996. Characterization of brittle food products: Application of acoustical emission method. Journal of Texture Studies, 27, 327–348.

Ares, G., Giménez, A. and Gámbaro, A. 2006. Instrumental methods to characterize nonoral texture of dulce de leche. Journal of Texture Studies, 37, 553–567.

Ball, C. O., Clauss, H. E. and Stier, E. F. 1957. Factors affecting quality of prepackaged meat. IB. Loss of weight and study of texture. Food Technology, 11, 281–284.

Barrett, A. H., Cardello, A. V., Lesher, L. L. and Taub, I. A. 1994. Cellularity, mechanical failure and textural perception of corn meal extrudates. Journal of Texture Studies, 25, 77–95.

Bertino, M. and Lawless, H. T. 1993. Understanding mouthfeel attributes: A multidimensional scaling approach. Journal of Sensory Studies, 8, 101–114.

Bertaux, E., Lewandowski, M. and Derler, S. 2007. Relationship between friction and tactile properties for woven and knitted fabrics. Textile Research Journal, 77, 387–396.

Blazquez, C., Downey, G., O'Callaghan, D., Howard, V., Delahunty, C., Sheehan, E., Everard, C. and O'Donnell, C. P. 2006. Modelling of sensory and instrumental texture parameters in processed cheese by near infrared reflectance spectroscopy. Journal of Dairy Research, 73, 58–69.

Blissett, A., Prinz, J. F., Wulfert, F., Taylor, A. J. and Hort, J. 2007. Effect of bolus size on chewing, swallowing, oral soft tissue and tongue movement. Journal of Oral Biology, 34, 572–582.

Bourne, M. 2004. Relation between texture and mastication. Journal of Texture Studies, 35, 125–143.

Bourne, M. C. 2002. Food Texture and Viscosity: Concept and Measurement, Second Edition. Academic, New York.

Bourne, M. C. 1982. Food Texture and Viscosity: Concept and Measurement. Academic, New York, NY.

Bourne, M. C., Sandoval, A. M. R., Villalobos, M. C. and Buckle, T. S. 1975. Training a sensory texture profile panel and development of standard rating scales in Colombia. Journal of Texture Studies, 1, 43–52.

Brandt, M. A., Skinner, E. Z. and Coleman, J. A. 1963. Texture profile method. Journal of Food Science, 28, 404–409.

Breene, W. M. 1975. Application of texture profile analysis to instrumental food texture evaluation. Journal of Texture Studies, 6, 53–82.

Brennan, J. G. and Jowitt, R. 1977. Some factors affecting the objective study of food texture. In: G. G. Birch, J. G. Brennan, and K. J. Parker (eds.), Sensory Properties of Foods. Applied Science, London, pp. 227–248.

Breuil, P. and Meullenet, J.-F. 2001. A comparison of three instrumental tests for predicting sensory texture profiles of cheese. Journal of Texture Studies, 32, 41–55.

Brown, W. E., Dauchel, C. and Wakeling, I. 1996. Influence of chewing efficiency on food texture and flavour perceptions in food. Journal of Texture Studies, 27, 433–450.

Brown, W. E., Langley, K. R., Martin, A. and MacFie, H. J. H. 1994. Characterisation of patterns of chewing behaviour in human subjects and their influence on texture perception. Journal of Food Texture, 25, 455–468.

Bruwer, M.-J., MacGregor, J. F., Bourg, W. M. Jr. 2007. Fusion of sensory and mechanical testing data to define measures of snack food texture. Food Quality and Preference, 18, 890–900.

Burns, L. D., Chandler, J., Brown, D. M., Cameron, B., Dallas, M. J. and Kaiser, S. B. 1995. Sensory interaction and descriptions of fabric hand. Perceptual and Motor Skills, 81, 120–122.

Cardello, A. V., Matas, A. and Sweeney, J. 1982. The standard scales of texture: Rescaling by magnitude estimation. Journal of Food Science, 47, 1738–1740. 1742.

Cardello, A. V. and Segars, R. A. 1989. Effects of sample size and prior mastication on texture judgments. Journal of Sensory Studies, 4, 1–18.

Cardello, A. V., Winterhalter, C. and Schutz, H. G. 2003. Predicting the handle and comfort of military clothing fabrics from sensory and instrumental data: Development and application of new psychophysical methods. Textile Research Journal, 73, 221–237.

Carson, K., Meullenet, J.-F. and Reische, D. W. 2002. Spectral stress strain analysis and partial least squares regression to predict sensory texture of yogurt using a compression/penetration instrumental method. Journal of Food Science, 67, 1224–1228.

Castro-Prada, E. M., Luyten, H., Lichtendonk, W., Hamer, R. J. and van Vliet, T. 2007. An improved instrumental characterization of mechanical and acoustic properties of crispy cellular solid food. Journal of Texture Studies, 38, 698–724.

Cavitt, L. C., Meullenet, J.-F. C., Xiong, R. and Owens, C. M. 2005. The relationship of razor blade shear, Allo-Kramer shear, Warner-Bratzler shear and sensory tests to changes in tenderness of broiler breast fillets. Journal of Muscle Foods, 16, 223–242.

Chauvin, M. A., Younce, F., Ross, C. and Swanson, B. 2008. Standard scales for crispness, crackliness and crunchiness in dry and wet foods: Relationship with acoustical determinations. Journal of Texture Studies, 39, 345–368.

Chen, J. 2007. Surface texture of foods: Perception and characterization. Critical reviews in Food Science and Nutrition, 47, 583–598.

Chen, J. 2009. Food oral processing – a review. Food Hydrocolloids, 23, 1–25.

Chen, P.-L., Barker, R. L., Smith, G. W. and Scruggs, B. 1992. Handle of weft knit fabrics. Textile Research Journal, 62, 200–211.

Christensen, C. M. 1984. Food texture perception. In: C. O. Chichester, E. M. Mrak and B. S. Schweigert (eds.), Advances in Food Research. Academic, Orlando, FL, pp. 159–199.

Christensen, C. M. and Vickers, Z. M. 1981 Relationships of chewing sounds to judgements of food crispness. Journal of Food Science, 46, 574–577.

Civille, G. V. and Liska, I. H. 1975. Modifications and applications to foods of the general foods sensory texture profile technique. Journal of Texture Studies, 6, 19–31.

Civille, G. V. Tactile-Fabric Feel Orientation. 1996. Workshop presented by Sensory Spectrum at Natick, November 18–21, 1996.

Civille, G. V. and Szczesniak, A. S. 1973. Guidelines to training a texture profile panel. Journal of Texture Studies, 4, 204–223.

Civille, G. V. and Dus, C. A. 1990. Development of terminology to describe the handfeel properties of paper and fabrics. Journal of Sensory Studies, 5, 19–32.

Claude, Y., Joseph, C. and Mioche, L. 2005. Meat bolus properties in relation with meat texture and chewing context. Meat Science, 70, 365–371.

Dacremont, C. 1995. Structural composition of eating sounds generated by crispy, crunchy and crackly foods. Journal of Texture Studies, 26, 27–43.

Dacremont, C., Colas, B. and Sauvageot, F. 1991. Contribution of air- and bone-conduction to the creation of sounds perceived during sensory evaluation of foods. Journal of Texture Studies, 22, 443–456.

Dan, H., Hayakawa, F. and Kohyama, K. 2008. Modulation of biting procedures induced by the sensory evaluation of cheese hardness with different definitions. Appetite, 50, 158–166.

Darden, M. A. and Schwartz, C. J. 2009. Investigation of skin tribology and its effects on the tactile attributes of polymer fabrics. Wear, 267, 1289–1294.

De Belie, N., Harker, F. R. and de Baerdemaeker, J. 2002. Crispness judgments of Royal gala apples based on chewing sounds. Biosystems Engineering, 81, 297–303.

De Wijk, R. A., Polet, I. A., Engelen, L., van Doorn, R. M. and Prinz, J. F. 2004. Amount ingested custard as affected by its color, odor and texture. Physiology and Behavior, 82, 397–403.

De Wijk, R. A., Zijlstra, N., Mars, M., de Graaf, C. and Prinz, J. F. 2008. The effects of food viscosity on bite size, bite effort and food intake. Physiology and Behavior, 95, 527–532.

Dijksterhuis, G., Luyten, H., de Wijk, R. and Mojet, J. 2007. A new sensory vocabulary for crisp and crunchy dry model foods. Food Quality and Preference, 18, 37–50.

Di Monaco, R., Giancone, T., Cavella, S. and Masi, P. 2008. Predicting texture attributes from microstructural, rheological and thermal properties of hazelnut spreads. Journal of Texture Studies, 39, 460–479.

Dooley, L. M., Adhikari, K. and Chambers, E. 2009. A general lexicon for sensory analysis of texture and appearance of lip products. Journal of Sensory Studies, 24, 581–600.

Drake, B. 1989. Sensory texture/rheological properties: A polyglot list. Journal of Texture Studies, 20, 1–27.

Drake, M. A., Gerard, P. D., Truong, V. D. and Daubert, C. R. 1999. Relationship between instrumental and sensory measurements of cheese texture. Journal of Texture Studies, 30, 451–476.

Duizer, L. M. 2001. A review of acoustical research for studying the sensory perception of crisp, crunchy and crackly texture. Trends in Food Science and Technology, 12, 17–24.

Edmister, J. A. and Vickers, Z. M. 1985. Instrumental acoustical measures of crispness in foods. Journal of Texture Studies, 16, 153–167.

Ekberg, O., Bülow, M., Ekman, S., Hall, G., Stading, M. and Wenden, K. 2009. Effect of barium sulfate contrast medium on rheology and sensory texture attributes in a model food. Acta Radiologica, 50, 131–138.

Elejalde, C. C. and Kokini, J. L. 1992. The psychophysics of pouring, spreading and in-mouth viscosity. Journal of Texture Studies, 23, 315–336.

Engelen, L., de Wijk, R., Prinz, J. F., Janssen, A. M., Weened, H. and Bosman, F. 2003. The effect of oral and product temperature on the perception of flavor and texture attributes of semi-solids. Appetite, 41, 273–281.

Engelen, L., Fontijn-Tekamp, F. A. and van der Bilt, A. 2005a. The influence of product and oral characteristics on swallowing. Archives of Oral Biology, 50, 739–746.

Engelen, L., de Wijk, R. A., van der Bilt, A., Prinz, J. F., Janssen, A. M. and Bosman, F. 2005b. Relating particles and texture perception. Physiology and Behavior, 86, 111–117.

Engelen, L., van den Keybus, P. A. M., de Wijk, R. A., Veerman, E. C. I., Nieuw Amerongen, A. V., Bosman, F., Prinz, J. F. and van der Bilt, A. 2007. The effect of saliva composition on texture perception of semi-solids. Archives of Oral Biology, 52, 518–525.

Finney, E. E. 1969. Objective measurements of texture in foods. Journal of Texture Studies, 1, 19–37.

Foster, K. D., Woda, A. and Peyron, M. A. 2006. Effect of texture of plastic and elastic model foods on the parameters of mastication. Journal of Neurophysiology, 95, 3469–3470.

Friedman, H. H., Whitney, J. E. and Szczesniak, A. S. 1963. The texturometer—a new instrument for objective texture measurement. Journal of Food Science, 28, 390–396.

González, R., Montoya, I. and Cárcel, J. 2001. Review: The use of electromyography on food texture assessment. Food Science and Technology International, 7, 461–471.

González, R., Montoya, I., Benedito, J. and Rey, A. 2004. Variables influencing chewing electromyography response in food texture evaluation. Food Reviews International, 20, 17–32.

Gonzalez-Barron, U. and Butler, F. 2008a. Prediction of panellists' perception of bread crumb appearance using fractal and visual texture features. European Food Research and Technology, 226, 779–785.

Gonzalez-Barron, U. and Butler, F. 2008b. Discrimination of crumb grain visual appearance of organic and non-organic bread loaves by image texture analysis. Journal of Food Engineering, 84, 480–488.

Guinard, J.-X. and Mazzuccheli, R. 1996. The sensory perception of texture and mouthfeel. Trends in Food Science and Technology, 7, 213–219.

Heidenreich, S., Jaros, D., Rohm, H. and Ziems, A. 2004. Relationship between water activity and crispness of extruded rice crisps. Journal of Texture Studies, 35, 621–633.

Hollingsworth, P. 1995. Lean times for U.S. food companies. Food Technology, 49, 1995.

Hu, J., Chen, W. and Newton, A. 1993. Psychophysical model for objective fabric hand evaluation: An application of Steven's law. Journal of the Textile Institute, 84, 354–363.

Hutchings, J. B. and Lillford, P. J. 1988. The perception of food texture—the philosophy of the breakdown path. Journal of Texture Studies, 19, 103–115.

Hutchings, S. C., Bronlund, J. E., Lentle, R. G., Foster, K. D., Jones, J. R. and Morgenstern, M. P. 2009. Variation of bite size with different types of food bars and implications for serving methods in mastication studies. Food Quality and Preference, 20, 456–460.

Hyde, R. J. and Witherly, S. A. 1993. Dynamic contrast: A sensory contribution to palatability. Appetite, 21, 1–16.

Hyldig, G. and Nielsen, D. 2001. A review of sensory and instrumental methods used to evaluate the texture of fish muscle. Journal of Texture Studies, 32, 219–242.

Hyvönen, L., Linna, M., Tuorila, H. and Dijksterhuis, G. 2003. Perception of melting and flavor release of ice cream containing different types and contents of fat. Journal of Dairy Science, 86, 1130–1138.

Ioannides, Y., Howarth, M. S., Raithatha. C., Deferenez, M., Kemsley, E. K. and Smith, A. C. 2007. Texture analysis of red delicious fruit: Towards multiple measurements on individual fruit. Food Quality and Preference, 18, 825–833.

Ioannides, Y., Seers, J., Raithatha, C., Howarth, M. S., Smith, A. and Kemsley, E. K. 2009. Electromyography of the masticatory muscles can detect variation in the mechanical and sensory properties of apples. Food Quality and Preference, 20, 203–215.

Italian Noodles. 1992. In: L. B. Hopkins (ed.), Pterodactyls and Pizza. The Trumpet Club, New York.

Jacobsen, M., Fritz, A., Dhingra, R. and Postle, R. 1992. Psychophysical evaluation of the tactile qualities of hand knitting yarns. Textile Research Journal, 62, 557–566.

Janhøj, T., Petersen, C. B., Frøst, M. B. and Ipsen, R. 2006. Sensory and rheological characterization of low-fat stirred yogurt. Journal of Texture Studies, 37, 276–299.

Jowitt, R. 1974. The terminology of food texture. Journal of Texture Studies, 351–358.

Kapur, K. 1971. Frequency spectrographic analysis of bone conducted chewing sounds in persons with natural and artificial dentitions. Journal of Texture Studies, 2, 50–61.

Kawabata, S. and Niwa, M. 1989. Fabric performance in clothing and clothing manufacture. Journal of the Textile Institute, 80, 19–51.

Kawabata, S. Inoue, M. and Niwa, M. 1992a. Non-linear theory of biaxial deformation of a triaxial- weave fabric. Journal of the Textile Institute, 83, 104–119.

Kawabata, S., Ito, K. and Niwa, M. 1992b. Tailoring process control. Journal of the Textile Institute, 83, 361–373.

Kiasseoglou, V. D. and Sherman, P. 1983. The rheological conditions associated with judgment of pourability and spreadability of salad dressings. Journal of Texture Studies, 14, 277–282.

Kilcast, D. 1999. Sensory techniques to study food texture. In: A. J. Rosenthal (ed.), Food Texture: Measurement and Perception. Springer, New York.

Kilcast, D. 2004. Texture in Food: Solid Foods, Vol. 2. CRC, New York.

Kim, J.-J., Yoo, S. and Kim, E. 2005. Sensorial property evaluation of scoured silk fabrics using quad analysis. Textile Research Journal, 75, 418–424.

Kim, E. H.-J., Corrigan, V. K., Hedderley, D. I., Motoi, L., Wilson, A. J. and Morgenstern, M. P. 2009. Predicting the sensory texture of cereal snack bars using instrumental measurements. Journal of Texture Studies, 40, 457–481.

Koehl, L., Zeng, X., Zhou, B. and Ding, Y. 2006. Subjective and objective evaluations on fabric hand: From manufacturers to consumers. International Nonwovens Technical Conference, INTC 2006, pp. 212–227.

Kokini, J. L. and Cussler, E. L. 1983. Predicting the texture of liquid and melting semi-solid food. Journal of Food Science, 48, 1221–1224.

Kokini, J. L. and Cussler, E. L. 1987. Psychophysics of fluid food texture. In: H. Moskowitz (ed.), Food Texture: Instrumental and Sensory Measurement. Marcel Dekker, New York, pp. 97–127.

Lassoued, N., Delarue, J., Launay, B. and Michon, C. 2008. Baked product texture: Correlations between instrumental and sensory characterization using flash profile. Journal of Cereal Science, 48, 133–143.

Lawless, H. T., Tuorila, H., Jouppila, K., Viratanen, P. and Horne, J. 1996. Effects of guar gum and microcrystalline cellulose on sensory and thermal properties of a high fat model food system. Journal of Texture Studies, 27, 493–516.

Lee, W. E., III, Schweitzer, M. A., Morgan, G. M. and Shepherd, D. C. 1990. Analysis of food crushing sounds during mastication: Total sound level studies. Journal of Texture Studies, 21, 165–178.

Lee, C. M. and Resurreccion, A. V. A. 2001. Improved correlation between sensory and instrumental measurement of peanut butter texture. Journal of Food Science, 67, 1939–1949.

Lee, W. and Sato, M. 2001. Visual perception of texture of textiles. Color Research and Application, 26, 469–477.

Lenfant, F., Loret, C., Pineau, N., Hartmann, C. and Martin, N. 2009. Perception of oral food breakdown. The concept of food trajectory. Appetite, 52, 659–667.

Liou, B. K. and Grün, I. U. 2007. Effect of fat level on the perception of five flavor chemicals in ice cream with or without fat mimetics using a descriptive test. Journal of Food Science, 72, S595–S604.

Lucas, P. W., Prinz, J. F., Agrawal, K. R. and Bruce, I. C. 2002. Food physics and oral physiology. Food Quality and Preference, 13, 203–213.

Lucca, P. A. and Tepper, B. J. 1994. Fat replacers and the functionality of fat in foods. Trends in Food Science and Technology, 5, 12–19.

Luyten, H. and van Vliet, T. 2006. Acoustic emission, fracture behavior and morphology of dry crispy foods: A discussion article. Journal of Texture Studies, 37, 221–240.

Mahar, T. J., Wheelwright, P., Dhingra, P. and Postle, R. 1990. Measuring and interpreting fabric low stress mechanical and surface properties. Part V. Fabric handle attributes and quality descriptors. Textile Research Journal, 60, 7–17.

Martens, H. J. and Thybo, A. K. 2000. An integrated microstructural, sensory and instrumental approach to describe potato texture. Lebensmittelwissenschaft und Technologie, 33, 471–482.

Matsudaira, M. and Kawabata, S. 1988. Study of the mechanical properties of woven silk fabrics. Part I. Fabric mechanical properties and handle characterizing woven silk fabrics. Journal of the Textile Institute, 79, 458–475.

McKenna, B. M. 2003. Texture in Food: Semi-Solid Foods, Vol. 1. CRC, New York.

Meilgaard, M., Civille, G. V. and Carr, B. T. 2006. Sensory Evaluation Techniques, Fourth Edition. CRC, Boca Raton, FL.

Mela, D. J., Langley, K. R. and Martin, A. 1994. No effect of oral or sample temperature on sensory assessment of fat content. Physiology and Behavior, 56, 655–658.

Mioche, L., Bourdiol, P. and Monier, S. 2003.Chewing behavior and bolus formation during mastication of meat with different textures. Archives of Oral Biology, 48, 193–200.

Mioche, L. and Peyron, M. A. 1995. Bite force displayed during assessment of hardness in various texture contexts. Archives of Oral Biology, 40, 415–423.

Monsanto. 1994. Simplesse Ingredient Overview, SB-5208. The NutraSweet Kelco Company.

Moskowitz, H. 1987. Food Texture: Instrumental and Sensory Measurement. Marcel Dekker, New York.

Muñoz, A. M. 1986. Development and application of texture reference scales. Journal of Sensory Studies, 1, 55–83.

Otegbayo, B., Aina, J., Sakyi-Dawson, E., Bokanga, M. and Asiedu, R. 2005. Sensory texture profiling and development of standard rating scales for pounded yam. Journal of Texture Studies, 36, 478–488.

Philippe, F., Schacher, L., Adolphe, D. C. and Dacremont, C. 2004. Tactile feeling: Sensory analysis applied to textile goods. Textile Research Journal, 74, 1066–1072.

Peleg, M. 1983. The semantics of rheology and texture. Food Technology, November 1983, 54–61.

Pereira, R. B., Bennett, R. J., McMath, K. L. and Luckman, M. S. 2002. In-hand sensory evaluation of textural characteristics in model processed cheese analogues. Journal of Texture Studies, 33, 255–268.

Primo-Martin, C., Castro-Prada, E. M., Meinders, M. B. J., Vereijken, P. F. G. and van Vliet, T. 2008. Effect of structure in the sensory characterization of the crispness of toasted rusk roll. Food Research International, 41, 480–486.

Prinz, J. F., de Wijk, R. A. and Huntjens, L. 2007. Load dependency of the coefficient of friction of oral mucosa. Food Hydrocolloids, 21, 402–408.

Raheel, M. and Liu, J. 1991. Empirical model for fabric hand. Part II. Subjective assessment. Textile Research Journal, 61, 79–82.

Richardson, N. J. and Booth, D. A. 1993. Multiple physical patterns in judgments of the creamy texture of milks and creams. Acta Psychologica, 84, 92–101.

Rohm, H. 1990. Consumer awareness of food texture in Austria. Journal of Texture Studies, 21, 363–373.

Rojo, F. J. and Vincent, J. F. V. 2008. Fracture properties of potato crisps. International Journal of Food Science and Technology, 43, 752–760.

Roland, A. M., Phillips, L. G. and Boor, K. J. 1999. Effects of fat replacers on the sensory properties, color, melting, and hardness of ice cream. Journal of Dairy Science, 82, 2094–2100.

Rosenthal, A. J. 1999. Food Texture: Measurement and Perception. Springer, New York.

Ross, C. F. 2009. Sensory science at the human-machine interface. Trends in Food science and Technology, 20, 63–72.

Roudaut, G., Dacremont, C., Vallès Pàmies, B., Colas B. and Le Meste, M. 2002. Crispness: A critical review on sensory and material science approaches. Trends in Food Science and Technology, 6/7, 217–227

Runnebaum, R. C. 2007. Key Constituents Affecting Wine Body: An Exploratory Study in White Table Wines. M. S. Thesis, University of California, Davis, USA.

Salmenkallio-Marttila, M., Roininen, K., Lindgren, J. T., Rousu, J., Autio, A. and Lähteenmäki, L. 2004. Applying machine learning methods in studying relationships between mouthfeel and microstructure of oat bread. Journal of Texture Studies, 35, 225–250.

Salvador, A., Varela, P., Sanz, T. and Fiszman, S. M. 2009. Understanding potato chips crispy texture by simultaneous fracture and acoustic measurements. LWT- Food Science and Technology, 42, 763–767

Schiffman, S. 1977. Food recognition by the elderly. Journal of Gerontology, 32, 586–592.

Schwartz, N. O. 1975. Adaptation of the sensory texture profile method to skin care products. Journal of Texture Studies, 1, 33–41.

Seymour, S. K. and Hamann, D. D. 1988. Crispness and crunchiness of selected low moisture foods. Journal of Texture Studies, 19, 79–95.

Sherman, P. 1977. Sensory properties of foods which flow. In: G. G. Birch, J. G. Brennan and K. J. Parker (eds.), Sensory Properties of Food. Applied Science, London, pp. 303–315.

Szczesniak, A. S. 1987. Correlating sensory with instrumental texture measurements—an overview of recent developments. Journal of Texture Studies, 18, 1–15.

Skinner, E. Z. 1988. The texture profile method. In: H. Moskowitz (ed.), Applied Sensory Analysis of Foods, Vol. I. CRC, Boca Raton, FL, pp. 89–107.

Szczesniak, A. S. 1963. Classification of textural characteristics. Journal of Food Science, 28, 385–389.

Szczesniak, A. S. 1966. Texture measurements. Food Technology, ctober 1966, 52, 55–56, 58.

Szczesniak, A. S. 1968. Correlations between objective and sensory texture measurements. Food Technology, August 1968, 49–51, 53–54.

Szczesniak, A. S. 1969. The whys and whats of objective texture measurements. Canadian Institute of Food Technology Journal, 2, 150–156.

Szczesniak, A. S. 1971. Consumer awareness of texture and of other food attributes, II. Journal of Texture Studies, 2, 196–206.

Szczesniak, A. S. 1975. General Foods texture profile revisited—ten years perspective. Journal of Texture Studies, 1, 5–17.

Szczesniak, A. S. 1979a. Recent developments in solving consumer-oriented texture problems. Food Technology, October 1979, 61–66.

Szczesniak, A. S. 1979b. Classification of mouthfeel characteristics of beverages. In: P. Sherman (ed.), Food Texture and Rheology. Academic, New York, pp. 1–20.

Szczesniak, A. S. 1991. Textural perceptions and food quality. Journal of Food Quality, 14, 75–85.

Szczesniak, A. S. 2002. Texture is a sensory property. Food Quality and Preference, 13, 215–225.

Szczesniak, A. S. and Bourne, M. C. 1969. Sensory evaluation of food firmness. Journal of Texture Studies, 1, 52–69.

Szczesniak, A. S., Brandt, M. A. and Friedman, H. H. 1963. Development of standard rating scales for mechanical parameters of texture and correlation between the objective and the sensory methods of texture evaluation. Journal of Food Science, 28, 397–403.

Szczesniak, A. S. and Kahn, E. L. 1971. Consumer awareness of and attitudes to food texture: Adults. Journal of Texture Studies, 2, 280–295.

Szczesniak, A. S. and Kahn, E. L. 1984. Texture contrasts and combinations: A valued consumer attribute. Journal of Texture Studies, 15, 285–301.

Szczesniak, A. S. and Kleyn, D. H. 1963. Consumer awareness of texture and other food attributes. Food Technology, 17, 74–77.

Szczesniak, A. S., Loew, B. J. and Skinner, E. Z. 1975. Consumer texture profile technique. Journal of Food Science, 40, 1253–1256.

Sztandera, L. M. 2009. Tactile fabric comfort prediction using regression analysis. WSEAS Transactions on Computers, 8, 292–301.

Terpstra, M. E. J., Jellema, R. H., Janssen, A. M., de Wijk, R. A., Prinz, J. F. and van der Linden, E. 2009. Prediction of texture perception of mayonnaises from rheological and novel instrumental measurements. Journal of Texture Studies, 40, 82–108.

Thybo, A., Bechmann, I. E., Martens, M. and Engelsen, S. B. 2000. Prediction of sensory texture of cooked potatoes using uniaxial compression, near infrared spectroscopy and low field ^{1}H NMR spectroscopy. Lebensmittelwissenschaft und Technologie, 33, 103–111.

Tournier, C., Martin, C., Guichard, E., Issanchou, S. and Sulmont-Rossé, C. 2007. Contribution to the understanding of consumers' creaminess concept: A sensory and a verbal approach. International Dairy Journal, 17, 555–564.

Tyle, P. 1993. Effect of size, shape and hardness of particles in suspension on oral texture and palatability. Acta Psychologica, 84, 111–118.

Ukponmwan, J. O. 1988. Assessment of fabric wear and handle caused by increments of accelerator abrasion in dry conditions. Part II. Correlation between objective and subjective methods of assessing fabric handle. Journal of the Textile Institute, 79, 580–587.

Van der Bilt, A., Engelen, L., Pereira, L. J., van der Glas, H. W. and Abbink, J. H. 2006. Oral physiology and mastication. Physiology and Behavior, 89, 22–27.

Van Vliet, T., van Aken, G. A., de Jongh, H. H. J. and Hamer, R. J. 2009. Colloidal aspects of texture perception. Advances in Colloid and Interface Science, 150, 27–40.

Varela, P., Chen, J., Fiszman, S. and Povey, M. J. 2006. Crispness assessment of roasted almonds by an integrated approach to texture description: Texture, acoustics, sensory and structure. Journal of Chemometrics, 20, 311–320.

Varela, P., Gámbaro, A., Giménez, A. M., Durán, I. and Lema. 2003. Sensory and instrumental texture measures on ketchup made with different thickeners. Journal of Texture Studies, 34, 317–330.

Varela, P., Salvador, A. and Fiszman, S. 2009. On the assessment of fracture in brittle foods II. Biting or chewing? Food Research International, doi:10.1016/j.foodres.2009. 08.004.

Vickers, Z. 1987a. Sensory, acoustical and force-deformation measurements of potato chip crispness. Journal of Food Science, 52, 138–140

Vickers, Z. 1987b. Crispness and crunchiness – textural attributes with auditory components. In: H. R. Moskowitz (ed.), Food Texture: Instrumental and Sensory Measurement. Dekker, New York, pp. 145–166.

Vickers, Z. 1979. Crispness and crunchiness of foods. In: P. Sherman (ed.), Food Texture and Rheology. Academic, London.

Vickers, Z. M. 1981. Relationships of chewing sounds to judgements of crispness, crunchiness and hardness. Journal of Food Science, 47, 121–124.

Vickers, Z. M. 1984a. Crackliness: Relationships of auditory judgments to tactile judgments and instrumental acoustical measurements. Journal of Texture Studies, 15, 59–58.

Vickers, Z. M. 1984b. Crispness and crunchiness—a difference in pitch? Journal of Texture Studies, 15, 157–163.

Vickers, Z. M. 1985. The relationship of pitch, loudness and eating technique to judgments of the crispness and crunchiness of food sounds. Journal of Texture Studies, 15, 85–95.

Vickers, Z. and Bourne, M. C. 1976. Crispness in foods—a review. Journal of Food Science, 41, 1153–1157.

Vickers, Z. M. and Wasserman, S. S. 1979. Sensory qualities of food sounds based on individual perceptions. Journal of Texture Studies, 10, 319–332.

Vincent, J. F. V. 1998. The quantification of crispness. Journal of the Science of Food and Agriculture, 78, 162–168.

Vincent, J. F. V. 2004. Application of fracture mechanics to the texture of food. Engineering Failure Analysis, 11, 695–704

Weedall, P. J., Harwood, R. J. and Shaw, N. 1995. An assessment of the Kawabata transformation equations for primary-hand values. Journal of the Textile Institute, 86, 47–475.

Weenen, H., van Gemert, L. J., van Doorn, J. M., Dijksterhuis, G. B. and de Wijk. 2003. Texture and mouthfeel of semisolid foods: Commercial mayonnaises, dressings, custard dessert and warm sauces. Journal of Texture Studies, 34, 159–179.

Weenen, H., Jellema, R. H. and de Wijk, R. A. 2005. Sensory sub-qualities of creamy mouthful in commercial mayonnaises, custard desserts and sauces. Food Quality and Preference, 16, 163–170.

Wilkinson, C., Dijksterhuis, G. B. and Minekus, M. 2000. From food structure to texture. Trends in Food Science and Technology, 11, 442–450.

Wintergerst, A. M., Buschang, P. H., Hutchins, B. and Throckmorton, G. S. 2005. Effect of auditory cue on chewing cycle kinematics. Archives of Oral Biology, 51, 50–57.

Wintergerst, A. M., Buschang, P. H. and Throckmorton, G. S. 2004. Reducing within-subject variation in chewing cycle kinematics – a statistical approach. Archives of Oral Biology, 49, 991–1000.

Wintergerst, A. M., Throckmorton, G. S. and Buschang, P. H. 2007. Effects of bolus size and hardness on the within-subject variability of chewing cycle kinematics. Archives of Oral Biology, 53, 369–375.

Xu, W. L., Bronlund, J. E., Potgieter, J., Foster, K. D., Röhrle, O., Pullan, A. J. and Kieser, J. A. 2008. Review of the human masticatory system and masticatory robotics. Mechanism and Machine Theory, 43, 1353–1375.

Yoshikawa, S., Nishimaru, S., Tashiro, T. and Yoshida, M. 1970. Collection and classification of words for description of food texture. I: Collection of words. Journal of Texture Studies, 1, 437–442

Zampini, M. and Spence, C. 2004. The role of auditory cues in modulating the perceived crispness and staleness of potato chips. Journal of Sensory Studies, 19, 347–363.

Zheng, C., Sun, D.-W. and Zheng, L. 2006. Recent applications of image texture for the evaluation of food qualities – a review. Trends in Food Science and Technology, 17, 113–128.

Chapter 12

Color and Appearance

Abstract   In this chapter we discuss what color is and then go on to describe color vision. We pay attention to variations in normal color vision due to genetic variations in the color receptor genes as well as to color blindness. We then discuss the measurement of appearance with attention to turbidity and glossiness. Instrumental color measurements are briefly described with special attention to the Munsell, RGB, and various CIE color systems.

Some days are yellow.
Some days are blue.
On different days I'm different too.
You'd be surprised how many ways
I change on different colored days.
On bright RED days how good it feels
to be a horse and kick my heels!

—(My Many Colored Days by Dr. Seuss)

Contents

12.1 Color and Appearance

In food products, especially meats, fruits, and vegetables, the consumer often assesses the initial quality of the product by its color and appearance. The appearance and color of these products are thus the primary indicators of perceived quality. The importance of color and appearance can be demonstrated when we think of drinking milk from a Coca-Cola bottle, when we choose bananas in the grocery store (a green–yellow–black continuum that indicates ripeness), when a friend serves green-colored bread and beer on St. Patrick's day, and when someone serves us a watermelon with yellow flesh instead of the more usual red. In food processing and cooking, color serves as a cue for the doneness of foods and is correlated with changes in aroma and flavor. Simple examples include the browning of baked and fried foods. For other foods, color or lightness is important to identity and to grading as in the lightness of canned tuna fish.

H.T. Lawless, H. Heymann, *Sensory Evaluation of Food*, Food Science Text Series,
DOI 10.1007/978-1-4419-6488-5_12, © Springer Science+Business Media, LLC 2010

Scientific studies have also shown that the color of the product affects our perception of other attributes, such as aroma, taste, and flavor. For example, DuBose et al. (1980) found that the number of correct identification of fruit-flavored beverage flavors decreased significantly when the beverage was atypically colored and that the number of correct identifications increased when the beverage was colored correctly. Shankar et al. (2009) studied the effect of color and label on perceived chocolate intensity and likability of brown and green milk and dark chocolate M&Ms (candy-coated chocolate buttons) and found that the color and the label affected the perceived chocolate intensity but not the likability. Additionally, they found no interaction effect of label and color. Christensen (1983) found that when sighted panelists scored the aroma intensity of appropriately and inappropriately colored cheese, soy analog bacon, margarine, raspberry-flavored gelatin and orange drink, the perceived intensity of the appropriately colored product was higher than for the inappropriately colored product. Interestingly, the bacon analog was a notable exception. The effect on perceived flavor intensity was less pronounced and there was no effect on perceived texture of the products.

Osterbauer et al. (2005) showed through functional magnetic resonance imaging (fMRI) of the brains of their subjects that as these subjects increased their rating of color–odor matches their brain activity in the caudal regions of the orbitofrontal cortex and in the insular cortex increased progressively with their perceptions of color-odor congruency. Therefore, these color–flavor interactions are likely "real."

Based on these studies and others (Demattè et al., 2009; Stevenson and Oaten, 2008) we can conclude that not only is the color and appearance of foods and products important to the consumer in and of themselves, but that color and appearance affect the consumers' perceptions of other sensory modalities in that food or product as well. Therefore it is very important that the sensory specialist knows how to ask panelists to evaluate product appearance and color and how to perform sensory tests to minimize the subjects' color and appearance biases from affecting the sensory results of other modalities.

12.2 What Is Color?

Color is the perception in the brain that results from the detection of light after it has interacted with an object. The perceived color of an object is affected by three entities: the physical and chemical composition of the object, the spectral composition of the light source illuminating the object, and the spectral sensitivity of the viewer's eye(s). As we will see in the following discussion changing any one of these entities can change the perceived color of the object.

The light striking an object may be refracted, reflected, transmitted, or absorbed by that object. If nearly all the radiant energy in the visible range of the electromagnetic spectrum is reflected from an opaque surface then the object appears white. If light through entire visible range of the electromagnetic spectrum is absorbed in part then the object appears gray. If light from the visible spectrum is absorbed almost completely then the object appears black. This also depends upon the surrounding conditions. The black type from this book in direct sunlight reflects more light than the white page under a reading lamp, yet they appear black and white under both conditions due to their relative reflectance of light.

The color of an object can vary in three dimensions, namely hue, this is typically what the consumer refers to as the "color" of the object (for example, green); lightness, also called the brightness of the object (light versus dark green); and saturation, also called the purity or chroma of the color (pure green versus grayish green). The perceived hue of an object is the perception of the color of the object and results from differences in the absorption of radiant energy at various wavelengths by the object. Thus if the object absorbs more of the longer wavelengths and reflects more of the shorter wavelengths (400–500 nm) then the object will be described as blue. An object with maximum light reflection in the medium wavelengths results in an object described as yellow-green in color and an object with maximal light reflection in the longer wavelengths (600–700 nm) will be described as red in color. The lightness (value) of the perceived color of an object indicates the relationship between reflected and absorbed light with no regard to specific wavelength(s) involved. The chroma (saturation or purity) of the color indicates how much a specified color differs from gray.

The visual perception of color arises from stimulation of photoreceptors in the retina by light in greater intensities at some wavelengths than others in the visible region (380–770 nm; Table 12.1) of the electromagnetic spectrum. The entire electromagnetic spectrum encompasses gamma rays (wavelengths of 10^{-5} nm) to radio waves (wavelengths at 10^{13} nm). However, the photoreceptors in the human eye only respond to a small range of this energy. Thus, color is an appearance property attributable to the spectral distribution of light interacting with the photoreceptors in the eye and visual color perception is the brain's response to this stimulus of the photoreceptors that results from the detection of light after it has interacted with an object. Or stated differently, wavelengths in the visual portion of the electromagnetic spectrum not absorbed by the viewed object are seen by the eye and interpreted by the brain as color.

Table 12.1 Visible portion of the electromagnetic spectrum

Color	Wavelength range (nm)
Violet	380–400
Blue	400–475
Green	500–570
Yellow	570–590
Orange	590–700
Red	700–770

Certainly color is an appearance property of an object attributable to the spectral distribution of light emanating from that object. However, gloss, transparency, haziness, and turbidity are appearance properties of materials attributable to the geometric manner in which light is reflected and transmitted. Something as simple as uneven reflection of light from a surface can make the object appear dull or matte. If the reflection is stronger at a specific angle or in a beam, then the resultant perception of gloss or sheen is a result of specular and/or directional reflectance. The reflectance is caused by the surface of the object. Smooth objects reflect in a directional manner and irregular, patterned, or particulate objects reflect light diffusely. The appearance of an object is affected by the optical properties associated with the object, namely the geometric light distribution, over the surface of the object and within the object if it is not opaque, the translucence of the object, the gloss, the size, shape, viscosity (Hutchings, 1999).

12.3 Vision

The light reflected from an object, or the light passing through an object, falls on the cornea of the viewer's eye(s), travels through the aqueous humor to the lens, and from there travels through the vitreous humor to the retina, where most of the light falls on or near a small hollow in the retina, the foveal pit. The visual receptors, the rods and cones, are located in the retina of the eye. These receptors contain light-sensitive pigments which change shape when stimulated by light energy, leading to the generation of electrical nerve impulses which travel along the optic nerves to the brain. There are approximately 120 million rods in the retina and the rods are capable of operating at extremely low light intensities (less than 1 lux). The rods yield only achromatic (black/white) information and under low-light conditions humans have scotopic vision with no color perception. This is why we cannot see colors by moonlight ("all cats are gray in the dark") although we can usually see well enough to move around. The maximum rod concentration is approximately 20° from the foveal area, this area is called the parafovea. Thus under low levels of illumination an object is more likely to be perceived when viewed slightly from the side than directly, called averted vision (Hutchings, 2002).

The 6 million cones operate at higher light intensities (levels of illumination) and provide chromatic information (color), allowing photopic vision. The cones are concentrated on the fovea, a small (2 mm diameter) depression located in a yellow colored spot (*macula lutea*) on the retina, where the highest color resolution occurs. When viewing an object, the unconscious movement of our eyes serves to bring the image of the object onto the foveal areas. The cones contain three color-sensitive pigments each responding most sensitively to red (two polymorphic variants at ~560 nm), the L-pigment also known as the ρ-receptors; to green (at ~530 nm), the M-pigment also known as the γ-receptors; or to blue (at ~420 nm), the S-pigment also known as the β-receptors (Deeb, 2006; Hutchings, 2002). A phenomenon called the Purkinje shift occurs under decreasing light conditions when humans become more sensitive to blue-green, with blues seemingly becoming brighter and reds relatively darker. Due to the Purkinje shift at very low light intensities the reds will appear almost black and the blues will appear gray.

12.3.1 Normal Human Color Vision Variations

It has been shown that variations in normal color vision are due to polymorphisms in the L- and M-pigments with amino acid substitutions at position 180 (alanine versus serine) accounting for most of the variations (Merbs and Nathans, 1992, 1993). There are additional amino acid substitutions at positions 277 and 285 but these are not as well studied, yet. In humans with normal color vision, Deeb (2005) found that among Caucasian males, 62% have serine at position 180 in the L-pigments (L_{serine}) and 38% have alanine ($L_{alanine}$). Using a color-matching test (the Rayleigh test) they asked their subjects to match a standard yellow (590 nm) light with a mixture of red (644 nm) and green (541 nm) lights. They found that males needing less red light to make the match (hence ones that were more sensitive to red light) were much more likely to have serine at position 180 of the L-pigment. The L-pigments are linked to the X-chromosome, thus men have two variants (about 60% express L_{serine} and about 40% $L_{alanine}$) and women have three variants (about 50% of women are heterozygous and express both L_{serine} and $L_{alanine}$; and the other 50% of women homozygously express either $L_{alanine}$ or L_{serine}). Pardo et al. (2007) showed that due to the above gender-related L-pigment expressions, on average women perceive some colors significantly differently from men. Jameson et al. (2001) specifically showed those women who were homozygous for the L- or M-pigments did not perform differently from men but those women who were heterozygous to L- and/or M-pigments had a relatively richer color experience. Additionally, ageing, glaucoma, and cataracts affect color vision. Older subjects (60–70 years old) perceive colored surfaces to be less chromatic ("colored") than subjects under 30 years of age (Hutchings, 2002).

12.3.2 Human Color Blindness

Humans either lacking one or more of the L-, M-, and S-pigments or having specific mutations in these pigments fall in various color-blind categories and comprise about 8% of males and 0.44% of females. Color-blind individuals are classified into different groups. The first group is the protanopes or protoanomalous trichromats who have no or a reduced ability, respectively, to see red due to absence or anomaly with ρ-receptors (L-pigments) and comprise about 1/4 of the color-blind population. The second group is the deuteranopes or deuteranomalous trichromats who have no or a reduced ability, respectively, to see green due to absence or anomaly with γ-receptors (M-pigments) and comprise about 3/4 of the color-blind population. The last and by far the smallest group is the tritanopes who have no or a reduced ability to see blue due to absence or anomaly with β-receptors (S-pigments). The genes for the more common forms of color blindness are recessive and carried on the X-chromosome. Thus the trait is seen much more frequently with men than with women.

It is possible to test panelists for color blindness and all panelists should be screened if they will be evaluating the color of samples. Techniques include pseudo-isochromatic plates such as the Ishihara plates, created in 1917, the Farnsworth Dichotomous Test for Color Blindness, or the Farnsworth–Munsell 110 Hue test (Farnsworth, 1943). Ishihara pseudo-isochromatic plates and the various Farnsworth tests can be obtained from any reputable optometric supply company.

12.4 Measurement of Appearance and Color Attributes

12.4.1 Appearance

Some scientists (Hutchings, 1999) maintain that product appearance is inclusive of product color and other appearance properties such as physical form (shape, size, and surface texture), temporal aspects (movement, etc.), and optical properties (reflectance, transmission, glossiness, etc.) For our purpose we will discuss color and appearance as separate entities, while keeping in mind that appearance attributes clearly affect perceived color.

Usually, physical appearance characteristics can easily be measured through sensory techniques. Standard descriptive techniques can quantify size, shape, and visual surface textures using simple intensity scales. An example would be "amount of chocolate chips visible on the surface of the cookie." In this case, "amount" might be rated from none to many, with examples being given in training to anchor the high and low ends of the scale. Visual texture is another example that lends itself well to simple intensity scales, such as apparent roughness of the surface, size or number of surface indentations, and density or amount of sediment in a container of a liquid product. Most of these simple and concrete attributes require little training and can be easily worked into a descriptive profile of the product. Of course, as in any other descriptive technique, the scale becomes more calibrated and there is better agreement among panelists if the low and high ranges are shown to provide the frame of reference that anchors the scale.

In food, temporal appearance characteristics are more rarely measured, even though they exist. Examples would be the viscosity of molasses as it drips from a spoon, the jiggle of JellO®, or the stringiness of pizza cheese. Optical properties (reflectance, transmission, glossiness, etc.) have been called "cesia" (Caivano et al., 2004); however, this term has not yet been widely used in the appearance research world. In the following section we will discuss few food-relevant appearance optical properties such as turbidity, translucency, and glossiness.

12.4.1.1 Turbidity (Cloudiness)

An important characteristic of many beverages is how clear versus how cloudy they appear. Turbidity (cloudiness or haze) occurs when small suspended particles divert light from a straight path through the material and scatter it in different directions. In physical terms, turbidity is the total light scattered from an incident beam as it transverses a suspension (Carrasco and Siebert, 1999). Consumers often expect beverages such as beer, fruit juices, and wines to be clear. In other beverages, for example, cider, cloudiness is expected and here again particulate matter is responsible for the light scattering. Various steps in beverage processing

may be aimed at reduction in turbidity and increasing the clarity, such as the use of fining agents in wine making. In some products such as beer, cider, and fruit juices, haze development is a function of polyphenol–protein interactions; others are due to carbohydrates and yet others are due to the growth of microorganisms (Siebert, 2009). Haze can also result from colloidal or larger particles that may precipitate in a container.

Instrumental methods for turbidity, such as nephelometers, use a focused light beam to measure light scattering at several angles. It is always prudent to cross-reference instrumental values to human perception. It is fairly simple to train a panel to evaluate turbidity. If the relationship between perceived turbidity and instrumental turbidity is not well known for a product, it is recommended that one performs the human testing to understand their sensory reactions to the product (Carrasco and Siebert, 1999). In other words, light scattering as a physically measured phenomenon may not tell you what you need to know about perceived turbidity. Relationships between instrumental measures of light scattering and human sensory ratings have been determined. Malcolmson et al. (1989) found a linear relationship between instrumentally measured turbidity and perceived clarity for commercial apple juices. Other studies have found relationships between physical measurements of cloudiness and sensory evaluations in different media including coffee (Pangborn, 1982) and beer (Hough et al., 1982; Leedham and Carpenter, 1977; Venkatasubramanian et al., 1975). Pieczonka and Cwiekala (1974, cited in Carrasco and Siebert, 1999) obtained an instrumental–sensory correlation between nephelometer values and a 5-point sensory scale of –0.81 in juices. Since light scattering is dependent upon particle size, it should be possible to measure a direct relationship between sensory clarity and the size and distribution of suspended matter in a product.

Clarity arises from the transmission of light, and fluids that transmit more light will appear more translucent. However, the relationship may be complicated by other factors such as the color of the medium (Siebert, 2009). Carrasco and Siebert (1999) addressed these issues in model systems and beverages, comparing turbidimeter results to those of human sensory panels. Haze perception thresholds were measured and while they varied with particle size and concentration depending upon the medium, the human sensory threshold was in a small range of instrumental haze values of about 0.5 Nephelos Turbidity Units. This suggested a good sensory–instrumental relationship at low levels. At ranges above threshold, perceived intensity followed the instrumental response until a saturation level was reached. After this point the instrument determined values continued to increase, but the sensory response was flat, even if panelists were allowed to use an open-ended magnitude estimation method of scaling (see Fig. 12.1). Sensory response (scaled intensity) was predicted on the basis of particle size, particle concentration, and suspension color.

Two situations arise when sensory–instrumental correlations break down. The most common example is when the human responds but the instrument does not, as in the case of olfactory sensitivities to some compounds which exceed the sensitivity of common analytical methods in chemistry. Another situation arises when the instrument responds, but the human does not. The scaling results in Carrasco and Siebert's study provide an interesting example of where the machine response has a broader dynamic range than the sensory judge. However, the upper range of turbidity becomes irrelevant when the sensory response does not change. This obviously imposes an upper limit on the utility of turbidimeter responses when a high level of cloudiness has been reached and the human eye no longer sees any further increase.

12.4.1.2 Glossiness (Shine)

Another important visual attribute is gloss or shine. Once again there are a variety of physical instruments to measure light reflectance, but the sensory data are still important to determine what humans will perceive in a specific situation. This becomes more important if the surface is non-uniform, since most instrumental reflectance measures are designed to work with uniform surfaces such as paints, waxes, and finishes. Many foods and consumer products will not conform well to these conditions. For example, the glaze on a cake or other baked product may not be a smooth surface or the shine on an apple may vary across the surface of the fruit. Just asking panelists about overall shine without appropriate training with reference standards may lead to different interpretations by different panelists, since there are two primary types of light reflectance. Specular reflectance refers to the mirror-like shine perceived when the actual image of a light source appears on the surface of the product (Beck and Prazdny, 1981). Obviously, standard angles and viewing conditions are necessary in order to test this in a reliable manner. Another important type of shine arises from diffuse reflectance. In this case the light is reflected, but it is scattered by the surface over such different angles that the reflected image of the light source is not seen. Buffing a metal surface with an abrasive cloth to produce many fine scratches will result in a good example of a surface with diffuse reflectance. The surface may seem quite shiny, but there is no mirror-like image, only the brightness of the light source. This type of shininess is also quite common with foods such as glazed doughnuts and egg-washed bread. A few example of studies on glossiness are Obein et al. (2004) and Xiao and Brainard (2008) where objects and pictures were used to determine

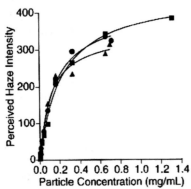
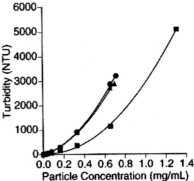

Fig. 12.1 Haze intensities (geometric means) perceived by sensory panelists using non-modulus magnitude estimation (*left*) and instrumentally measured turbidity (*right*) versus particle concentration for medium (2.600 mm diameter) particles in clear (■), yellow (●), and red (▲) liquids (Reprinted with permission from Carrasco and Siebert, 1999).

perceived glossiness. Chong et al. (2008) created a machine vision system to evaluate the surface gloss of eggplant fruit.

Translucency

Translucency is defined as the property of a specimen by which it transmits light diffusely without permitting a clear view of objects beyond the specimen (ASTM, 1987). Joshi and Brimelow (2002) gave a simple test to determine whether a sample is translucent or not. They suggest measuring the sample with a reflectance spectrophotometer at maximum area of illumination and with the maximum viewing aperture. Then repeat the measurement with the same viewing aperture but with a smaller area of illumination. If there is a large increase in the lightness reading (L* in CIELAB, see below) then the sample is translucent.

This property is important in orange juice (MacDougall, 2002), tomato skins (Hetherington et al., 1990), fresh-cut tomatoes (Lana et al., 2006), and pineapples (Chen and Paull, 2001) where flesh translucency is a defect associated with off-flavors and fruit fragility during harvest. Hetherington et al. (1990) found that increased sensory translucency scores of tomatoes were associated with increased opacity and that the translucency scores were inversely related to the L* values ($r=0.774$). Standard sensory techniques are used in the sensory assessment of translucency and instrumentally a reflectance spectrophotometer followed by the Kubelka-Munk data analysis is used (Talens et al., 2002).

The Kubelka-Munk theory is a relatively crude model to describe light scattering and its effect on translucency (see Nobbs (1985) as well as Vargas and Niklasson (1997) for excellent overviews of the theory and its applicability). Simply put a "scattering" coefficient (S) and an "adsorption" coefficient (K) are calculated and the ratio (K/S) is related to translucency of the object. For example, Lana et al. (2006) found that during storage the pericarp of tomato slices but not that of intact tomatoes became more translucent. The sensory translucency scores were related to changes in the K/S ratio of the Kubelka-Munk analysis of the reflection spectra of the sliced tomatoes. Additionally they found that removing the locular gel inhibited the development of translucency in the pericarp.

MacDougall (2002) gives an example that makes it very clear that only using instrumental values in measurement of translucent samples can give results that are totally inconsistent with visually observed values. In his example, 4-fold orange juice concentrate is diluted to a concentration of 0.2 and 4. When glasses of these oranges juices are viewed with overhead illumination they range from pale yellow (concentration less than 1) to deep orange (concentration of 4). Instrumentally, the most dilute juice had the lowest L* and it was the darkest according to the instrument. On the other hand, the most concentrated juice had the highest L* and was the lightest according to the instrument. This occurred due to the loss of light scatter in the more diluted samples. He cautions that one should remember that the instrument only sees light reflected from a limited solid angle while the human "is influenced by the multidirectionality of illumination, which makes coloured translucent materials glow."

One can do a simple experiment to visually demonstrate the above effect. Pour an equal amount of orange juice into two identical transparent glasses. Cover both glasses completely with white paper. The paper covering the side of one glass should have a circular hole cut into it the size of a dime (approximately 1.5 cm in diameter). The paper covering the side of the other glass should have a circular hole cut into it the size of a quarter (approximately 2.5 cm in diameter). Then evaluate the color of the juices by viewing the visible juice through the holes at a 90° angle. The juice in the glass covered with the paper with the small hole seems darker because much of the scattered light is "trapped" within the glass and not seen by the viewer.

12.4.2 Visual Color Measurement

Sensory evaluation of color is frequently performed. Sensory scientists have used the whole range of sensory testing tools to do visual color measurements. For example, Whiting et al. (2004) used triangle and two-out-of-five difference tests to investigate perceived color differences in liquid foundation cosmetics; and Eterradossi et al. (2009) used descriptive analysis and consumer satisfaction scales to evaluate red and blue automotive paints with different levels of quality.

When doing sensory color evaluation it is even more important than usual to standardize all factors that can affect the perceived color. In general the sensory scientist performing color assessment should carefully standardize, control, and report the following:

(a) the background color in the viewing area. Ideally the background color should be non-reflective and neutral, usually a matte gray, cream, or off-white is used (ASTM 1982).

(b) the light source (Table 12.2) in Kelvin and its intensity (in lux or foot candles) at the product surface. Eggert and Zook (2008) recommend a light intensity between 750 and 1,200 lux. Also, the light source (if it is not a standard illuminant) should be chosen to have a high color rendering index (R_a, see below) (Hutchings, 1999).

(c) the panelists' viewing angle and the angle of light incidence on the sample. These should not be the same since that leads to specular reflection of the incident light and a potential glossiness that may be an artifact of the method. Usually the booth area is set up with the light source vertically above the samples and the panelists viewing angle when they are seated is about 45° to the sample, this minimizes specular reflection effects.

Table 12.2 Light sources, color temperatures, and color rendering indices[a]

Light source	Color temperature (K)	Ambiance description	Color rendering Index (R_a)	Quality
Candle	1,800	Very warm		
High-pressure sodium lamp	2,100	Very warm	22	Poor
40 watt incandescent light bulb	2,770	Warm	Close to 100	Excellent
100 watt incandescent light bulb	2,870	Warm	Close to 100	Excellent
CIE source A	2,856	Warm	Close to 100	Excellent
Warm white fluorescent light Sylvania T5-warm	3,000	Warm	82	Very good
Metal halide lamp Sylvania MetalArc ProTech	3,000	Warm	85+	Very good
GroLux Wide Spectrum lamp	3400	Neutral	89	Excellent
Neutral fluorescent light PureLite	3,500	Neutral	85	Very good
Cool white fluorescent light Sylvania T5-cool	4,100	Cool	82	Very good
Tungsten/halogen light SoLux	4,700	Cool	99	Excellent
CIE source B (direct sunlight)	4,870	Cool		
Full spectrum fluorescent light DuroTest Vitalite	5,500	Cool	90	Excellent
Daylight fluorescent light Sylvania F40D	6,300	Cool-blue	76	Good
CIE source D65	6,500	Cool-blue	100	Excellent
Daylight fluorescent light DuroTest DayLite 65	6,500	Cool-blue	92	Excellent
CIE source C (overcast daylight)	6,774	Cool-blue		
CIE source D (daylight)	7,500	Cool-blue		

[a]Values collated from commercial literature and Hutchings (1999)

(d) the distance from the light source and the product. This will affect the amount of light incident on the sample. The light intensity should be measured at the product surface.

(e) whether the sample is lit with reflected or transmitted light.

Frequently, very little or none of the above information appears in the literature associated with food or personal care product color evaluations. Whiting et al. (2004) were exceptional in explicitly indicating the color of the sensory booth wall and table (gray with specified color system values); the color of the sample tray bottoms (a gray-woven fabric with specified color system values); the light source (D_{65} at 1,000 lux); the viewing distance (60 cm); and the viewing angle (each sample was subtended at 6°).

In color and appearance evaluations the light source is usually specified by its color temperature. The color temperature is determined from the temperature in Kelvin to which a black body that absorbs all energy that falls onto it needs to be heated to emit light of a spectral distribution characteristic of the specific light source (Table 12.2). The light emitted by the black body changes as the color temperature changes. At lower temperatures (2,000 K) the light emitted is redder, at higher temperatures (about 4,000–5,000 K) the light is whiter, and at high temperatures (8,000–10,000 K) the light becomes bluer (1999). Standard lights used in food color evaluation tend to be illuminants A (with a color temperature of 2,856 K), C (6,774 K), D_{65} (6,500 K), and D (7,500 K). These illuminants are all based on tungsten filaments. The spectral distribution of illuminant A is very different from the spectral distribution of illuminants B and C (Fig. 12.2). The spectral distribution of illuminant A is high in red-yellow wavelengths while it is low in blue-violet wavelengths. Illuminants C and D_{50}, through D_{65}, are high in blue wavelengths. Illuminants C, D_{65}, and the other D variants are designed to mimic variations of daylight. Standard fluorescent lights have very different spectral distributions (they tend to be more spiky and less smooth, see F11 in Fig. 12.2) than those from tungsten and incandescent lamps. The result is that objects viewed under fluorescent and tungsten lights often have differences in perceived color than when the same objects are viewed under say illuminant C. These differences in perceived color occur because the color depends on the absorption of light by the product and the incident spectrum's wavelengths. For example, under a standard illuminant if the product absorbs red wavelengths and not those in the green area of the spectrum the object would look green. However, if the incident light only has red wavelengths then the object would not appear green since there were no green wavelengths to reflect to the eye. Depending on the light source this object may appear black.

The color rendering index (R_a) is a measure of the effect of an illuminant on the perceived color of an object (CIE, 1995a). The R_a is measured by assessing the size of the color change of eight Munsell color samples under the light of interest versus a reference light, usually an incandescent light (a 60 W tungsten lamp, 2,900 K). Lights with a 100 R_a index exactly reproduce the perceived color of the reference light (Table 12.2).

Panelists should be tested for color blindness (see above). If reference color standards are desired they can be paint chips, Munsell spinning disks, model products, or digital images (Hernández et al., 2004; Kane et al., 2003). However, when using these standards the sensory specialist should keep in mind that the color of the standard and the sample may only be a metameric match. A metameric match is an apparent match in the colors of two objects when viewed under one light source but the colors of the objects are not matched when viewed under most other light sources (MacKinney and Little, 1962). Metameric matches also occur when two objects match under a specified light source when viewed by one observer but not when viewed by a second observer (Kuo and Luo, 1996).

Recently, a number of studies have been published on the use of digital images as reference standards or the use of virtual product images to evaluate color differences in foods. It could be very useful if it were possible to obtain accurate reproductions of color and appearance of products as images. These can then be displayed to the panelists (anywhere in the world) as long as they are viewed using the identical reference display and viewing conditions. Kane et al. (2003) studied the possibility of using digital references for the brownness of cookies and found that the panelists' scores when using either digital references or physical references led to the same trends in differences among cookie formulations but in some cases the panelists' scores were lower when using the digital references than when using the physical references. Hernández et al. (2004) created digitally processed color charts of

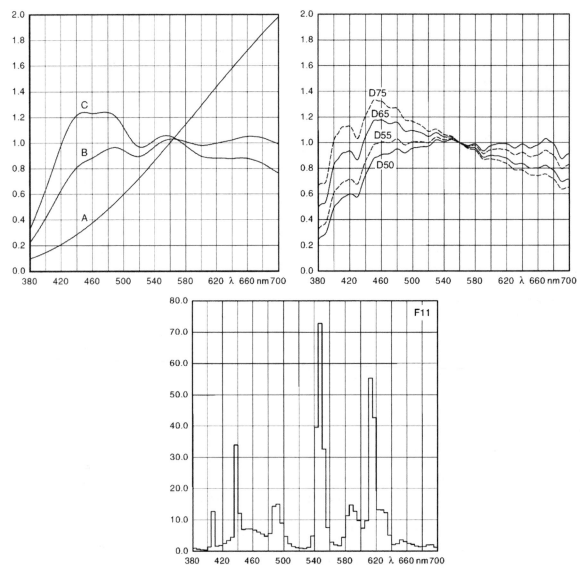

Fig. 12.2 The relative wavelength distributions for CIE standard illuminants A, B, C, D$_{65}$, and D variants and F11. Illuminant A has more yellow-red wavelengths, illuminant D$_{65}$ and the D variants have more blue wavelengths, and illuminant F11 (a fluorescent light) has a more spiky distribution in terms of wavelengths (Reprinted with permission from Gernot Hoffman, University of Applied Sciences, Emden, Germany).

Piquillo peppers to use as a color reference standard and they found that the repeatability of the visual color chart scores was satisfactory. Examples of digital images used as reference standards: Pointer et al. (2002) successfully used digital images of bananas, tomatoes, oranges, peas, and biscuits (cookies) that had been perturbed in terms of lightness or color in triangle tests; Valous et al. (2009) similarly used digital images of ham slices and successfully determined the CIE color characterization of these slices from the digital images using a computer vision system; Kang et al. (2008) successfully did something similar with a more complex product—bicolor mango fruit.

When asking panelists to evaluate color the sensory scientist has to keep in mind that humans are very good at evaluating color differences when samples are side by side or when they have access to color standards but humans are not good at evaluating color differences from memory. Additionally, research has shown that humans are quite good at evaluating hue (see Munsell color solid) and lightness (value) changes in objects but not good at discriminating chroma (saturation of color) changes (Melgosa et al., 2000). Additionally, Zhang and Montag (2006) confirmed Melgosa and coworkers' results and conclude with the following statement: "...people do not have ready access to the lower level color descriptors such as the common attributes used to define color spaces, and that higher level psychological processing involving cognition and language may be necessary for even apparently simple tasks involving color matching and describing color differences."

12.5 Instrumental Color Measurement

"There are a bewildering variety of methods and instruments available to the food technologist in the field of colour measurement. When one is approaching the subject for the first time or when attempting to devise a method for a material outside the normal experience the wealth of possibilities available sometimes makes the choice difficult" (Joshi and Brimelow, 2002). In the next section we will endeavor to shed some light on color measurement. For additional information the following are suggested: Hutchings (1999, 2003), MacDougall (2002), and Lee (2005).

12.5.1 Munsell Color Solid

Prior to the advent of instrumental techniques, several visual color solids were developed to describe color; one of the more famous was the Munsell color solid. The Munsell color solid was developed by A.H. Munsell around 1900 (Clydesdale, 1978). The Munsell system had three attributes: hue (H), value (V), and chroma (C). A specific color was described as a point in the three-dimensional hue–value–chroma space. In the Munsell color solid (or color space) the hue–value and chroma values for each color were arranged in a sphere composed of individual color "plates" separated by equal visual steps (Fig. 12.3). Hues are spaced around the circumference with ten

major hues (grouped into major divisions of red, yellow-red, yellow, green-yellow, green, blue-green, blue, purple-blue, purple, and red-purple), each being ten hue steps apart. These hue steps were supposed to be equal but research has shown that the hue spacing in the yellow-red, yellow-green, and blue regions is actually not equally spaced (Oleari, 2001). The value is a darkness or lightness scale with absolute black (at the bottom of the sphere) to absolute white (at the top of the sphere). The chromatic colors are positioned at the value that is equally spaced between absolute black and absolute white. The chroma is the amount by which a given hue deviates from a neutral gray of the same value. The chroma of a hue is imagined as a line of constant hue drawn from the center of the sphere to the edge of the sphere at a constant value.

Visual color solid systems are useful when one wants to specify a color but one always needs a human to do the matching of the sample color to the color solid (usually a color chip). However, due to the idiosyncratic nature of color vision, it was not possible to have an instrument measure color as specified in Munsell

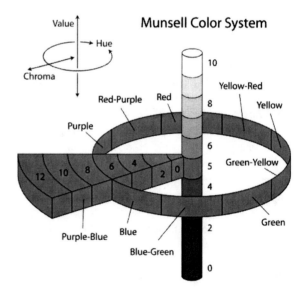

Fig. 12.3 A schematic of the Munsell color solid indicating the three dimensions of hue, chroma, and lightness (From Jacobolus, Wikimedia Commons, http://en.wikipedia.org/wiki/File:Munsell-system.svg. This file is licensed under the Creative Commons Attribution ShareAlike 2.5 License. In short: you are free to share and make derivative works of the file under the conditions that you appropriately attribute it and that you distribute it only under a license identical to this one).

notation. In order to develop instrumentation that could measure color, it was necessary to devise mathematical relationships to describe color (the so-called mathematical color solids).

12.5.2 Mathematical Color Systems

In order to develop meaningful mathematical color systems the approach used by Munsell had to be changed. Mathematical color systems are based on the physical laws related to the addition of lights and these are based on the existence of L-, M-, and S-receptor cones and rods in the human eye. The most used mathematical color systems are the CIE versions. The CIE acronym is based on the French name for the International Commission on Illumination or "La Commission Internationale de l'Eclairage" (CIE, 1978, 1986). In order to explain the CIE system it is easier to start with a less complex version, the so-called three lights system. The three lights system simply specifies color in terms of how colors are perceived by the human eye.

12.5.2.1 The R, G, B Mathematical Color System

Three projectors, one with a red filter (R), one with a green filter (G), and one with a blue filter (B), are set up to shine on a screen in such a way that they completely overlap. The sum of the wavelengths hitting the screen, the so-called spectral radiant flux, is perceived by an observer as a single color. Then, another projector with an unknown color filter is projected onto a separate portion of the same screen. It is now possible to adjust the energy (radiant flux) projected through the R, G, and B filters on the first three projectors until the combined radiant flux from these projectors matches the unknown color. One can then specify the unknown color as the energy combination from R, G, and B. The amounts of energy required to match the unknown from each of the three lights are the so-called tristimulus values. These values may be expressed as radiant flux (watts), luminous flux (lumens), or, more usually, in arbitrary psychophysical scales of red, blue, and green.

In practice this approach is overly simple leading to a number of problems. Some colors are too bright to match because no light source can project the required radiant flux. Other colors are too saturated. For example, some yellows cannot be matched using just red and green filters even if the blue filter is eliminated. "Matchable colors" are within the color gamut (or the acceptable color range) of a specific mathematical color system while "non-matchable colors" are outside the color range. Even if different filters had been chosen for the three projectors in this simple system it is still not possible to match all colors. In theory, the three lights system is based on the physiological response of the three cone types of the eye. In practice, it is further simplified by isolating the responses that are analogous to actual physiological responses. This simplification results in the unfortunate effect that there are always some colors outside the color gamut because nearly all parts of the color magnetic spectrum excite more than one of the cones to some extent. If it were possible to find a part of the spectrum that excited only one cone type while having no effect on the other two cone types, then a color gamut based on the three lights system would include all perceived colors. Despite its limitations, the three color system has been used extensively as the basis for other tristimulus color systems.

It is possible to express the color matching produced by the three lights algebraically (Clydesdale, 1978). If we assume that C is a color in the three-dimensional color space and its color is matched by the three lights red, green, and blue with tristimulus values R, G and B, then the following equation describes the color match:

$$C_{(R,G,B)} = R + G + B \qquad (12.1)$$

Based on the physical law of additivity of luminances, the intensity of color C (also known as the luminance L) in the three-dimensional space can be described by the next equation:

$$L = l_R + l_G + l_B \qquad (12.2)$$

where l_R, l_B, and l_G are the luminances (intensities) of the corresponding light primaries in their unit amount with $R = B = G = 1$. If the tristimulus values R, G, and B of color C are changed by a constant factor "a" then the luminance of C changes to "aL." If color D with tristimulus values R_D, G_D, and B_D is added to color C with tristimulus values G_C, B_C, and $\underline{R_C}$ then the new color E has tristimulus values of R_E, G_E, and B_E. This can be expressed algebraically:

$$E_{(R_E, G_E, B_E)} = (R_C + R_D) + (G_C + G_D) + (B_C + B_D) \tag{12.3}$$

So, the tristimulus values of a mixture of colors are equal to the sum of the tristimulus values of the component colors. Based on the above explanation it is possible to describe both the luminance (l) and tristimulus values r, g, b of a color in terms of three colored lights, if the color falls within the color gamut of the mathematical color solid.

It is also possible to define a unit plane within the three-dimensional mathematical color solid which has within it all colors with the same luminance. This unit plane is a plane of constant luminance in the three-dimensional mathematical color space and is similar to the plane of constant value in the Munsell color solid. Differences in colors within this plane are a function of hue and chroma of the specified colors. This unit plane is called a chromaticity diagram and a color point within the chromaticity diagram is not specified by the arbitrary tristimulus values R, G, and B but by fractions of their total:

$$r = \frac{R}{R + G + B} \tag{12.4}$$

$$g = \frac{G}{R + G + B} \tag{12.5}$$

$$b = \frac{B}{R + G + B} \tag{12.6}$$

A color may be therefore specified in the three-dimensional color by description of the luminance (l) and two of the color's three chromaticity coordinates. This will be illustrated in the next section (Fig. 12.4) for the CIExyz tristimulus system. This simple three-light system is the basis for all mathematical color solids like the CIE tristimulus system. However, this simple system does not work in reality because (1) some colors are outside the color gamut and a negative amount of radiant flux is needed to match these colors, (2) the color solid is not visually uniform, (3) a vector analysis is needed to calculate the luminance. The CIE system eliminates all of these problems.

12.5.2.2 CIE Mathematical Color Systems

In the CIE mathematical color system theoretical primaries were developed to remove the disadvantages of the actual lights (R, G, and B) while still retaining the advantages of the simple three-light system. The primaries are X, Y, and Z and their chromaticity coordinates are x, y, and z. The developers mathematically included luminance into one of the primaries (Y) and thus avoided the problem of needing vector analysis to calculate luminance. This was possible because the cones of the eyes are most sensitive to luminance in the green region of the spectrum. Careful choice allowed the theoretical primaries X, Y, and Z to cover the entire color gamut with positive values, thus the horseshoe-shaped CIE spectrum locus has a color gamut that includes all colors (Fig. 12.4).

In the CIE system it is possible to locate a color in the three-dimensional color space by specifying Y and two of the three possible chromaticity coordinates (x, y, and z). The chromaticity coordinates are related to each other by the following equation: $x+y+z = 1$. Thus, knowledge of two of the three possible values will define a specific color.

The CIE data are usually expressed as tristimulus values (X, Y, and Z) or as chromaticity coordinates (x, y, and z). The x, y chromaticity coordinates are often plotted on the horseshoe-shaped CIE spectrum locus with %Y superimposed (Fig. 12.5, please note that not all colors are present at all levels of %Y). The color can then be specified as x, y, and %Y. Since CIE spectrum locus is not based on Cartesian coordinates, it is difficult to express mathematically and even more difficult to explain to most people. One attempt to simplify the CIE system plots the CIE spectrum locus at constant %Y. Then x and y chromaticity coordinates, at a given Y value, appear on a unit plane.

The problem with the x, y, z chromaticity system is that the space looks like a horseshoe which makes any linear relationship calculations between these values and say sensory scales very difficult. Other color systems have been developed with more uniform diagrammatic representations of color spaces than the horseshoe-shaped CIE space. Early versions of these color spaces were the Gardner and the Hunter L,A,B spaces (which were associated with specific instruments) where the value (also known as the degree of whiteness or blackness) is represented by L. The

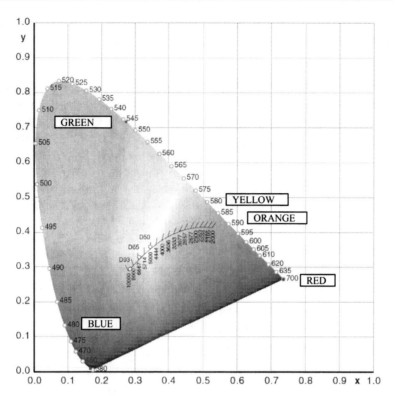

Fig. 12.4 Horseshoe-shaped chromaticity diagram (Reprinted with permission from Gernot Hoffman, University of Applied Sciences, Emden, Germany).

chromatic portion of the color space is based on rectangular Cartesian coordinates (*a*, *b*) with red represented by +*a*, green represented by –*a*, yellow represented by +*b*, and blue represented by –*b*. These systems made it easier to meaningfully communicate color data. Subsequently other spaces that were instrument invariant, like the CIELAB and CIELUV, also known as the L*a*b* and L*u*v*, respectively, were developed by CIE to improve the linearity of the CIE system (CIE, 1986). The L*u*v* system has been applied to food but was primarily devised for color additive mixing such as television and lighting. The L*a*b* space approximates the Munsell space. For both the L*u*v* and L*a*b* systems the three axes are mutually perpendicular. An increase in the value of +*a* indicates an increase in red; a larger –*a* value indicates an increase in green. An increase in +*b* indicates an increase in yellow and an increase in –*b* indicates an increase in blue. Increasing L* values indicate increasing lightness (or whiteness). One has to be careful not to oversimplify the space—this occurs when authors incorrectly describe *a* as redness and *b* as yellowness. In actuality (*a b*) are Cartesian coordinates that

together describe a point in space (Hutchings, 1999; Wrolstad et al., 2005).

In an effort to make the color coordinate values more intuitive the L*C*h* color space was devised (Sharma, 2003). This space uses the same diagram as the L*a*b* color space but uses angles rather than Cartesian coordinates for *a* and *b*. The L* in L*C*h* is identical to the L* in the L*a*b*. The C* indicates chroma (an indication of color saturation) and is equal to zero at the center of the color space and increases based on the distance from the center. The h* is the hue angle and it is expressed in degrees. Starting from the +*a* axis, 0° is +*a* (red), 90° is +*b* (yellow), 180° is –*a* (green), and 270° is –*b* (blue).

The above color systems are helpful in specifying a color but are not very useful when one wants to specify the differences between colors. Color difference can be calculated in the L*a*b*, the L*u*v*, and the L*C*h* systems. For the L*a*b* the equation for color difference between two samples is as follows:

$$\Delta E^* = \sqrt{(L_1 - L_2)^2 + (a_1 - a_2)^2 + (b_1 - b_2)^2}$$

$$(12.7)$$

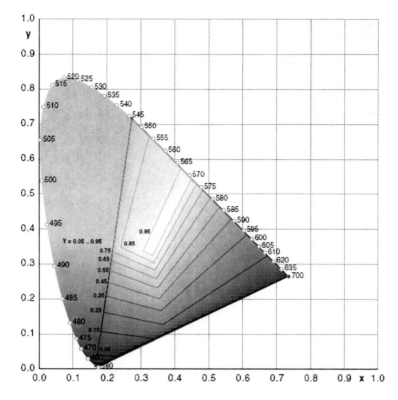

Fig. 12.5 Horseshoe-shaped chromaticity diagram with third dimension (*Y*). The third dimension is indicated by the tristimulus value *Y*. As previously mentioned, this value indicates the lightness or luminance of the color. The scale for *Y* extends from the white spot in a line perpendicular to the plane formed by *x* and *y* using a scale that runs from 0 to 1. The fullest range of color exists at 0 where the white point is equal to CIE illuminant *C*. As the *Y* value increases and the color becomes lighter, the range of color, or gamut, decreases so that the color space at 1 is just a sliver of the original area (Reprinted with permission from Gernot Hoffman, University of Applied Sciences, Emden, Germany).

It is important to note that once the ΔE is calculated the size of the difference is known but not whether it is due to *L*, *a*, *b* singly or in some combinations (Sharma, 2003). Because the *L*, *a*, *b* space is not uniform the ΔE is more accurate in some parts of the color space than others. In an attempt to improve the situation a number of other color difference equations have been proposed. The most popular are the CIE94 (also known as ΔE_{94}, CIE 1995b) and the CIEDE2000 (Luo et al., 2001; Sharma et al., 2005). The CIEDE2000 has been extensively studied and seems to be an improvement over the standard ΔE and CIE94 (Melgosa et al., 2008; Xu et al., 2002).

There are also mathematical color systems that may be less familiar to North American readers but very familiar to others, for example, the Swedish Natural Color System (NCS, Hard and Sivik, 1981), the DIN99 (Cui et al., 2002), and the CMC

(AATCC, 2005). Fortunately, values derived from any of these systems can be interconverted, provided conditions are appropriately specified. A few examples of color conversion tables and equations are listed in Table 12.3.

Interconversion between color systems can have problems. In food matrices there are frequently discrepancies when converting from the other systems to the CIE *XYZ* system, because the conversion calculations are based on the responses of opaque standards. Food systems, on the other hand, are often somewhat translucent and do not behave exactly as would be predicted by the standards.

Angela Little (MacKinney and Little, 1962) stated "Once we accept that color belongs to the realm of sensory perception, we must also accept that is can only be measured directly in psychological terms. From physical measurements, nevertheless, we can obtain

Table 12.3 Conversion equations and tables for some common color systems

Convert CIE XYZ to CIELUV L*u*v*[a]

$L^* = 116(Y/Y_n)^{1/3} - 16$ for $Y/Y_n > 0.008856$ where Y_n is the value for reference white

$L^* = 903.3(Y/Y_n)^{1/3}$ for $Y/Y_n \leq 0.008856$ where Y_n is the value for reference white

$u^* = 13L^*(u\prime - u\prime_n)$ where $u\prime$ is calculated as described below and $u\prime_n$ is for reference white

$v^* = 13L^*(v\prime - v\prime_n)$ where $v\prime$ is calculated as described below and $v\prime_n$ is for reference white

Calculation of $u\prime$ and $v\prime$:

$$u\prime = (4X)/(X+15Y+3Z) = (4x)/(-2x+12y+3)$$
$$v\prime = (9Y)/(X+15Y+3Z) = (9y)/(-2x+12y+3)$$

Convert CIE XYZ to CIELAB L*a*b*[b]

$L^* = 116(Y/Y_n)^{1/3}$ for $Y/Y_n > 0.008856$ where Y_n is the value for reference white

$a^* = 500\{(X/X_n)^{1/3} - (Y/Y_n)^{1/3}\}$ where X_n is the value for reference white

$b^* = 200\{(Y/Y_n)^{1/3} - (Z/Z_n)^{1/3}\}$ where Z_n is the value for reference white

Convert CIELAB L*a*b* to CIE XYZ[c]

$Y^{1/3} = (L^* + 16)/24.99$ if illuminant C was used

$X\%^{1/3} = (a^*/107.72) + Y^{1/2}$ if illuminant C was used

$Z\%^{1/3} = Y^{1/3} - (b^*/43.09)$ if illuminant C was used

Convert CIELAB L*a*b* to HunterLAB[c]

$L = 10Y^{\frac{1}{2}}$ if illuminant C was used

$a = 17(X\%-Y)/Y^{\frac{1}{2}}$ if illuminant C was used

$b = 7.0(Y-Z\%)/Y^{\frac{1}{2}}$ if illuminant C was used

Convert CIE XYZ to HunterLAB[d]

$L = 10Y^{\frac{1}{2}}$

$a = 175(1.02X-Y)/(Y^{\frac{1}{2}})$

$b = 70(Y-0.847Z)/(Y^{\frac{1}{2}})$

Convert Munsell values to CIE XYZ

Use tables by Glenn and Killian (1940)

Convert Munsell values to CIE xy

Use tables by Glenn and Killian (1940)

[a]Hutchings (1999) (CIELUV was intended for color additive mixing in the television and lighting industries, but it has been used in food color measurements)

[b]ASTM (1991)

[c]Pattee et al. (1991)

[d]Clydesdale (1978)

data which provide the basis for establishing psychophysical scales, from which we can predict visual color appearance." She suggests that usually the primary concern in color measurement is to measure what the eyes see. Thus it is necessary to produce data that correlate with human visual perception. Often the instrumental data (tristimulus values) do not correlate well with the data derived from panelists and further manipulation of the instrumental data may be needed to improve the correlation.

When the color of foods is measured instrumentally the scientist should keep in mind that the instrument was designed to measure the reflectance color of ideal samples, namely samples that are homogeneously pigmented, opaque, flat, and evenly light scattering (Clydesdale, 1975, 1978). Foods are far from the ideal sample. Nearly all foods have shape and texture irregularities and surface characteristics that scatter and transmit light. Additionally, the pigment distribution in most foods is also irregular. Instruments are also designed to measure the transmittance color of ideal samples, and in this case the ideal sample is clear and moderately light absorbing. Real liquids (where one usually measures transmittance color) tend to have hazes and may be very light absorbing (Clydesdale, 1978).

It is possible to obtain an approximate ideal reflectance color measuring surface for dry powders, such as flour and cocoa, by compressing the dry powdered sample into a pellet. Other dry foods such as instant coffee, potato flakes, dry gelatin crystals (dry Jell-O®) can be pressed into very thin wafers between

Teflon disks. When measuring the color of translucent liquids the area exposed should be much larger than the area illuminated. This allows any light entering the sample and traveling laterally within the sample to emerge in the direction where it will be measured. This minimizes the selective absorption effect that can change the hue of the liquid (see above under translucency).

12.6 Conclusions

Sensory color measurement is frequently neglected by sensory specialists or they add this measurement as an afterthought. We hope that this chapter has made the reader realize that the measurement of color, whether visually or by instrument, is no simple task. The sensory specialist should be very careful to standardize all possible conditions associated with these measurements and to carefully report the specific conditions used in a test. Additionally, it is important to realize that most (if not all) visual and appearance characteristics can be evaluated using standard descriptive analysis techniques.

References

AATCC. 2005. CMC: Calculation of small color differences for acceptability. AATCC Technical Manual, Test Method 173–1998, pp. 311–315.

ASTM. 1982. Committee D1. Standard practice for visual evaluation of color differences of opaque materials. American Society for Testing Materials Standards, Philadelphia, PA, USA.

ASTM. 1987. Color and Appearance Measurement, Second Edition. American Society for Testing Materials Standards, Philadelphia, PA, USA.

ASTM. 1991. Standard test method for computing the colors of objects using the CIE system. E 308–90. Annual Book of ASTM Standards, 14, 244–270.

Beck, J. and Prazdny, S. 1981. Highlights and the perception of glossiness. Perception and Psychophysics, 30, 407–410.

Caivano, J. L., Menghi, I. and Iadisernia, N. 2004. Casie and paints: An atlas of cesia with painted samples. Proceedings of the Interim Meeting of the International Color Association AIC 2004, pp. 113–116.

Carrasco, A. and Siebert, K. J. 1999. Human visual perception of haze and relationships with instrumental measurements of turbidity. Thresholds, magnitude estimation and sensory descriptive analysis of haze in model systems. Food Quality and Preference, 10, 421–436.

Chen, C.-C. and Paull, R. E. 2001. Fruit temperature and crown removal on the occurrence of pineapple fruit translucency. Scientia Horticulturae, 88, 85–95.

Christensen, C. M. 1983. Effect of color on aroma, flavor and texture judgements of foods. Journal of Food Science, 48, 787–790.

Chong, V. K., Nishi, T., Kondo, N., Ninomiya, K., Monta, M., Namba, K., Zhang, Q. and Shimizu, H. 2008. Surface gloss measurement on eggplant fruit. Applied Engineering in Agriculture, 24, 877–883.

CIE. 1978. Recommendations on uniform color spaces, color difference equations, psychometric color terms. Supplement #2 to CIE Publication 15 (E-1.3.1) 1971/(TC-1.3). Bureau Central de la CIE, 27 Kegel Strasse, A-1030 Vienna, Austria.

CIE. 1986. Colorimetry, Second Edition, CIE Publication 15.2. CIE Central Bureau, 27 Kegel Strasse, A-1030 Vienna, Austria.

CIE. 1995a. Method of measuring and specifying colour rendering properties of light sources. CIE Publication #13.3. CIE Central Bureau, 27 Kegel Strasse, A-1030 Vienna, Austria.

CIE. 1995b. Industrial colour-difference evaluation CIE Publication #116.CIE Central Bureau, 27 Kegel Strasse, A-1030 Vienna, Austria.

Cui, G., Luo, M. R., Rigg, B., Roesler, G. and Witt, K. 2002. Uniform color spaces based on the DIN99 colour-difference formula. Color Research and Application, 27, 282–290.

Clydesdale, F. J. 1975. Food Colorimetry: Theory and Applications. AVI, Westport, CT.

Clydesdale, F. M. 1978. Colorimetry—methodology and applications. CRC Critical Reviews in Food Science and Nutrition, 243–301.

Deeb, S. S. 2005. The molecular basis of variation in human color vision. Clinical Genetics, 67, 369–377.

Deeb, S. S. 2006. Genetics of variation in human color vision and the retinal cone mosaic. Current Opinion in Genetics and Development, 16, 301–307.

Demattè, M. L., Sanabria, D. and Spence, C. 2009. Olfactory discrimination: When vision matters. Chemical Senses, 34, 103–109.

DuBose, C. N., Cardello, A. V. and Maller, O. (1980). Effects of colorants and flavorants on identification, perceived flavor intensity and hedonic quality of fruit-flavored beverages and cake. Journal of Food Science, 45, 1393–1399, 1415.

Eggert, J. and Zook, K. 2008. Physical Requirement Guidelines for Sensory Evaluation Laboratories, Second Edition. ASTM Special Technical Publication 913. American Society for Testing and Materials, West Conshohocken, PA.

Eterradossi, O., Perquis, S. and Mikec, V. 2009. Using appearance maps drawn from goniocolorimetric profiles to predict sensory appreciation of red and blue paints. Color Research and Appreciation, 34, 68–74.

Farnsworth, D. (1943). The Farnsworth-Munsell 100-hue and dichotomous tests for color vision. Journal of the Optical Society of America, 33, 568–578.

Glenn, J. J. and Killian, J. T. 1940. Trichromatic analysis of the Munsell book of color. Journal of the Optical Society of America, 30, 609–616.

Hard, A. and Sivik, L. 1981. NCS—natural color system: A Swedish standard for color notation. Color Research and Application, 6, 129–138.

Hernández, B., Sáenz, C., Alberdi, C., Alfonso, S., Berrogui, M. and Diñeiro, J. M. 2004. Design and performance of a color chart based in digitally processed images for sensory evaluation of Piquillo peppers (*Capsicum annuum*). Color Research and Application, 29, 305–311.

Hetherington, M. J., Martin, A., MacDougall, D. B., Langley, K. R. and Bratchell, N. 1990. Comparison of optical and physical measurements with sensory assessment of the ripeness of tomato fruit *Lycopersicon esculenfum*. Food Quality and Preference, 2, 243–253.

Hough, J. S., Briggs, E. D., Stevens, R. and Young, T. W. 1982. Malting and Brewing Science. Chapman and Hall, London.

Hutchings, J. B. 1999. Food Colour and Appearance, Second Edition. Aspen, Gaithersburg, MD.

Hutchings, J. B. 2002. The perception and sensory assessment of colour. In: D. B. MacDougall (ed.), Colour in Foods: Improving Quality. Woodhead, Cambridge, England, pp. 9–32.

Hutching, J. B. 2003. Expectations and the Food Industry: The Impact of Color and Appearance. Springer, New York.

Jameson, K. A., Highnote, S. M. and Wasserman, L. M. 2001. Richer color experience in observers with multiple photopigment opsin genes. Psychonomic Bulletin and Review, 8, 244–261.

Joshi, P. and Brimelow, C. J. B. 2002. Colour measurements of foods by colour reflectance. In: D. B. MacDougall (ed.), Colour in Foods: Improving Quality. Woodhead, Cambridge, England, pp. 80–114.

Kane, A. M., Lyon, B. G., Swanson, R. B. and Savage, E. M. 2003. Comparison of two sensory and two instrumental methods to evaluate cookie color. Journal of Food Science, 68, 1831–1837.

Kang, S. P., East, A. R. and Tujillo, F. J. 2008. Colour vision system evaluation of bicolour fruit: A case study with "B74" mango. Postharvest Biology and Technology, 49, 77–85.

Kuo, W. G. and Luo, M. R. 1996. Methods for quantifying metamerism: Part I—Visual assessment. Journal of the Society of Dyers and Colourists, 112, 312–320.

Lana, M. M., Hogenkamp, M and Koehorst, R. B.M. 2006. Application of Kubelka-Munk analysis to the study of translucency in fresh-cut tomato. Innovative Food Science and Emerging Technologies, 7, 302–308.

Lee, H.-C. 2005. Introduction to Color Imaging Science. Cambridge University Press, England.

Leedham, P. A. and Carpenter, P. M. 1977. Particle size measurement and the control of beer clarity. Proceedings, European Brewery Convention 16th Congress. European Brewery Convention, London, pp. 729-744.

Little, A. 1962. Introduction. In: G. MacKinney and A. C. Little (eds.), Color of Foods. AVI, Westport, CT.

Luo, M. R., Cui, G. and Rigg, B. 2001. The development of the CIE 2000 colour difference formula: CIEDE2000. Color Research and Application, 26, 340–350.

MacDougall, D. B. 2002. Colour measurement of food. In: D. B. MacDougall (ed.), Colour in Foods: Improving Quality. Woodhead, Cambridge, England, pp. 33–63.

MacKinney, G. and Little, A. C. 1962. Color of Foods. AVI, Westport, CT.

Malcolmson, L., Jeffrey, L., Sharma, D. D. and Ng, P. K. W. 1989. Relationship between sensory clarity and turbidity values of apple juice. Canadian Institute of Science and Technology Journal, 22, 129-132.

Melgosa, M., Huertas, R. and Berns, R. S. 2008. Performance of recent advanced color-difference formulas using the standardized residual sums of squares index. Journal of the Optical Society of America A, 25, 1828–1834.

Melgosa, M., Rivas, M. J., Hita, E. and Viénot, F. 2000. Are we able to distinguish color attributes? Color Research and Application, 25, 356–367.

Merbs, S. L. and Nathans, J. 1992. Absorption spectra of human cone pigments. Nature (London), 356, 433–435.

Merbs, S. L. and Nathans, J. 1993. Role of hydroxyl-bearing amino acids in differently tuning the absorption spectra of the human red and green cone pigments. Photochemistry and Photobiology, 58, 706–710.

Nobbs, J. H. 1985. Kubelka-Munk Theory and the prediction of reflectance. Reviews in the Progress of Coloration, 15, 66–75.

Obein, G., Knoblauch, K. and Viénot, F. 2004. Difference scaling of gloss: Nonlinearity, binocularity, and constancy. Journal of Vision, 4(9), 4, 711–720.

Oleari, C. 2001. Comparisons between color-space scales, uniform-color-scale atlases, and color-difference formulae. Color Research and Application, 26, 351–361.

Osterbauer, R. A., Matthews, P. M., Jenkinson, M., Beckmann, C. F., Hansen, P. C. and Calvert, G. A. 2005. Color of scents: Chromatic stimuli modulate odor responses in the human brain. Journal of Neurophysiology, 93, 3434–3441.

Pangborn, R. M. 1982. Influence of water composition, extraction procedures and holding time and temperature on quality of coffee beverage. Lebensmittel Wissenschaft und Technologie, 15, 161-168.

Pardo, P. J., Pérez, A. L. and Suero, M. I. 2007. An example of sex-linked color vision differences. Color Research and Application, 32, 433–439.

Pattee, H. E., Giesbrecht, F. G. and Young, C. T. 1991. Comparison of peanut butter color determination by CIELAB L*a*b* and Hunter color-difference methods and the relationship of roasted peanut color to roasted peanut flavor response. Journal of Agriculture and Food Chemistry, 39, 519–523.

Pieczonka, W. C. E. 1974. [Nephelometric method for clarity assessment of beverage apple juice.] Przemysl Spozywczy, 28, 121-124.

Pointer, M. R., Attridge, G. G. and Jacobson, R. E. 2002. Food colour appearance judged using images on a computer display. The Imaging Science Journal, 50, 23–36.

Setser, C. S. 1984. Color: Reflections and transmissions. Journal of Food Quality, 42, 128–135.

Shankar, M. U., Levitan, C. A., Prescott, J. and Spence, C. 2009. The influence of color and label information on flavor perception. Chemosensory Perception, 2, 53–58.

Sharma, A. 2003. Understanding Color Management. Delmar Cengage Learning, Florence, KY.

Sharma, G., Wu, W. and Dalal, E. N. 2005. The CIEDE2000 color-difference formula: Implementation notes, supplementary test data, and mathematical observations. Color Research and Application, 30, 21–30.

Siebert, K. 2009. Haze in beverages. Advances in Food and Nutrition Research, 57, 53–86.

Siebert, K., Carrasco, A. and Lynn, P. Y. 1996. The mechanism of protein-polyphenol haze formation in beverages. Journal of Agricultural and Food Chemistry, 44, 1997-2005.

Stevens, M. A. and Scott, K. E. 1988. Analytical methods of tomato products. In: H. F. Linskens and J. F. Jackson (eds.), Analysis of Nonalcoholic Beverages. Springer, Berlin, pp. 134–165.

Stevenson, R. J. and Oaten, M. 2008. The effect of appropriate and inappropriate stimulus color on odor discrimination. Attention, Perception & Psychophysics, 70, 640–646.

Talens, P., Martínez-Navarrette, N., Fito, P. and Chiralt, A. 2002. Changes in optical and mechanical properties during osmodehydrofreezing of kiwi fruit. Innovative Food Science and Emerging Technologies, 3, 191–199.

Valous, N. A., Mendoza, F. Sun, D-W. and Allen, P. 2009. Colour calibration of a laboratory computer vision system for quality evaluation of pre-sliced hams. Meat Science, 81, 132–141.

Vargas, W. E. and Niklasson, G. 1997. Applicability conditions of the Kubelka-Munk theory. Applied Optics, 36, 5580–5586.

Venkatasubramanian, K., Saini, R. and Vieth, W. R. 1975. Immobilization of papain on collagen and the use of collagen-papain membranes in beer chill-proofing. Journal of Food Science, 40, 109-113.

Whiting, R., Murray, S., Ciantic, Z. and Ellison, K. 2004. The use of sensory difference tests to investigate perceptible colour-difference in a cosmetic product. Color Research and Application, 29, 299–304.

Wrolstad, R. E., Durst, R. W. and Lee, J. 2005. Tracking color and pigment changes in anthocyanin products. Trends in Food Science and Technology, 16, 423–428.

Xiao, B. and Brainard, D. H. 2008. Surface gloss and color perception of 3D objects. Neuroscience, 25, 371–385

Xu, H., Yaguchi, H. and Shiori, S. 2002. Correlation between visual and colorimetric scales ranging from threshold to large color difference. Color Research and Application, 27, 349–359.

Zhang, H. and Montag, E. D. 2006. How well can people use different color attributes? Color Research and Application, 31, 445–457.

Chapter 13

Preference Testing

Abstract Preference testing refers to consumer tests in which the consumer is given a choice and asked to indicate their most liked product, usually from a pair. Although these tests appear straightforward and simple, several complications are encountered in the methods, notably how to treat replicated data and how to analyze data that include a "no-preference" option as a response. Additional methods are discussed including ranking more than two products, choosing both the best and worst from a group, and rating the degree of preference.

The number of judges that are involved in a study may be such that rather unimportant differences may receive undue attention. It is quite possible to produce statistically significant differences in preference for product which have little practical value by simple increasing the number o judges that are utilized.

—H. G. Schutz (1971).

Contents

13.1 Introduction—Consumer Sensory Evaluation

Consumer sensory evaluation is usually performed toward the end of the product development or reformulation cycle. At this time the alternative product prototypes have usually been narrowed down to a manageable subset through the use of analytical sensory tests. Frequently, the sensory testing is followed by additional testing done through market research. The big difference between consumer sensory and marketing research testing is that the sensory test is generally conducted with coded, not branded, products while market research is most frequently done with branded products (van Trijp and Schifferstein, 1995). Also, in consumer sensory analysis the investigator is interested in whether the consumers like the product, prefer it over another product, or find the product acceptable based on its sensory characteristics. The consumer sensory specialist often has no interest in purchase intent, effect of branding, and/or cost factors. Thus, a product will not necessarily be financially successful just because it had high hedonic scores (was well liked) or because it was preferred over another product. Success in the marketplace is also affected by price, market image, packaging, niche, etc. However, a product that does not score well in a consumer acceptance test will probably fail despite great marketing.

Sensory techniques are widely used to assess the reactions of the public to a variety of stimuli, even environmental annoyances (Berglund et al., 1975). Historically, acceptance testing of foods with consumers represented an important departure from earlier methods based on the opinions of expert tasters or the assignment of quality scores by panels looking for product defects (Caul, 1957; Jellinek, 1964; Sidel et al., 1981). The growth of acceptance testing helped to foster the logical separation of the analytical sensory techniques from affective tests, a distinction that was lacking in the earlier traditions of expert grading and quality testing. Acceptability information is extremely useful. For example it can be combined with other sensory analyses, knowledge of consumer expectations, and product formulation constraints in determining the optimal design of food products (Bech et al., 1994; Moskowitz, 1983).

In foods and consumer products, there are two main approaches to consumer sensory testing, the measurement of preference and the measurement of acceptance (Jellinek, 1964). In preference measurement the consumer panelist has a choice. One product is to be chosen over one or more other products. In the measurement of acceptance or liking the consumer panelists rate their liking for the product on a scale. Acceptance measurements can be done on single products and do not require a comparison to another product. An efficient procedure is to determine consumers' acceptance scores in a multi-product test and then to determine their preferences indirectly from the scores. Both of these types of testing are called hedonic or affective tests. The term hedonic refers to pleasure. The goal of both types of tests is to assess the appeal of a product to a consumer on a sensory basis, i.e., to get the consumer's reaction on the basis of appearance, aroma, taste, flavor, mouthfeel, and texture. For

non-food products, other sensory factors may come into play as well as the perceived efficacy or performance of a product. Other factors related to the product appeal are discussed in the next chapters such as the appropriateness of a product for its intended use and the consumer's satisfaction with a product (performance relative to expectations). This chapter will deal with simple preference choice. Historically, the term preference test has also been used to refer to surveys of people's likes and dislikes based on presenting lists of food names (Cardello and Schutz, 2006). Such data are generally scaled and thus could be called an acceptance test as the term is used in the next chapter. In this chapter we will use the term preference only to refer to experiments in which a choice is made between two or more alternatives.

The key to a successful consumer test is finding the right participants. Persons in the test must be representative of the population to which the results will be generalized. They should be users of the product and probably frequent users. The sensory specialist will often negotiate with the client (the person requesting the test and who will use the results and conclusions) as to just how frequently the person must use the product to qualify for the test. Obviously, no trained panelists are used for such a test as they approach the product in a different frame of mind from the average consumer. Sometimes employees may be used in the early stages of testing, but they must be users of the product category being tested. Further information on qualifying and screening consumer test participants is found in Chapter 15, but this principle applies equally to the next three chapters on preference, acceptability scaling, and consumer field tests.

The sensory test is to some degree an artificial situation and may not always be a good predictor of consumer behavior in the real world regarding purchase and/or consumption (Lucas and Bellisle, 1987; Sidel et al., 1972; Tuorila et al., 1994; Vickers and Mullan, 1997). However, purchase decisions and even consumption involve many other influences than the simple sensory appeal of a product. In spite of this shortcoming, preference and acceptance tests can provide important information about the relative sensory appeal of a group of products that need blind-labeled consumer testing at some phase in the product development scenario. That is, the objective is to find the product among a group that has to be best potential for success, on a sensory basis. So in spite of the limitations of the methods imposed by the artificial context of the testing situation, they are still quite valuable to the sensory specialist and the clients who request the product test.

The following three chapters will discuss preference testing, acceptability testing, and then consumer field tests and questionnaire design, respectively. This chapter will focus on the simple paired preference test. In spite of its appeal and its apparent simplicity, sensory and market researchers have added additional variations that complicate the test and analysis. The two main variations involve replication and the offering of a "no-preference" option (or both). We will recommend as a good practice the simplest version of a paired preference test, but the other more complicated versions and ranking are also discussed in this chapter. Some worked examples are provided in the chapter itself and in the appendices that follow.

13.2 Preference Tests: Overview

13.2.1 The Basic Comparison

Preference tests are choices involving comparisons between two products or among several products. If there are two products, this is known as a paired preference test. It is the simplest (and most popular) type of test that looks at the appeal of products to consumers. Paired preference tests are some of the oldest sensory tests. A publication in 1952 described a mail panel maintained (for the preceding 20 years!) by the Kroger Company food retailers that would receive pairs of products for comparison in the mail along with a questionnaire (Garnatz, 1952). Paired tests are popular in part because of their simplicity, because they mimic what consumers do when purchasing (choosing among alternatives), and because some people believe they are more sensitive than scaled acceptance. We have seen no hard data to substantiate this latter belief although intuitively it is possible that two products might receive the same score on a scale, but one might be slightly preferred to the other. However, it is also possible that a product could win in a choice test, but still be unappealing on its own (like an election where you do not like either candidate but vote based on the lesser of two evils). This is one shortcoming of a preference test

that it gives you no absolute information on the overall appeal of a product. Acceptance testing with a scale is designed to do just that.

13.2.2 Variations

Variations on preference testing involve choosing a preferred product from multiple alternatives. One version of this is ranking, in which products are ordered from best liked to least liked. Another version gives products in small groups (usually three) and the consumer is asked which one is liked best and which one is worst. This is known as best–worst scaling, because the resulting data can be placed on a scale, even though the task itself involves a choice and not a response on a scale. Best–worst and paired tests are both special cases of ranking, i.e., you can think of a paired test as a ranking of only two products. Both ranking and best–worst scaling are discussed in this chapter in later sections.

Other important variations on the preference test involve the use of a "no-preference option" and the replication of the test on the same persons. The no-preference option provides more information, but complicates the statistical analysis. It is generally avoided by product researchers although in advertising claim substantiation, it may be required for legal reasons (ASTM, 2008). Replication is not common in preference tests. However, recent research has shown that replication will enhance the consumer's discrimination among products in an acceptance test. Also, replication can provide evidence as to whether there are stable segments of consumers who prefer different versions of a product. The primary goal of a preference test is to find a "winner," i.e., that product which has significantly higher appeal to consumers than other versions in the test.

There have been studies on the efficacy of the paired preference test with illiterate and semi-literate consumers (Coetzee, 1996; Coetzee and Taylor, 1996). These consumers, many of whom could not read or write, could reliably perform paired preference tests given verbal instructions and using paired symbols (the same symbol but one outlined and the other solid). When one of the authors tested the methods in a different country the method worked well with illiterate consumers (Coetzee, 1996). Paired preference tests using color codes have also been successfully used

(Beckman et al., 1984) and could be used with illiterate or semi-literate consumers. Preference tests are also suitable for young children as the task is straightforward and easily understood (Engen, 1974; Kimmel et al., 1994; Schmidt and Beauchamp, 1988; Schraidt, 1991).

Multiple paired preference tests form the basis for a new kind of threshold test, the consumer rejection threshold. Prescott et al. (2005) gave groups of wine consumers increasing levels of trichloroanisole in wine to try and find the level at which there was a consistent preference for wine without this "cork taint." This technique, discussed in Chapter 6, should find wide application in commodities in which the chemistry and origins of various taints and off-flavors are well understood (see Saliba et al., 2009, for another example).

13.2.3 Some Cautions

A common methodological problem comes from the temptation to add a preference test at the end of some other kind of sensory test. This should be avoided. It is very unwise to ask for preference choices after a difference test, for example. This is not recommended for a number of reasons. First, the participants in the two tests are not chosen on the same basis. In preference tests, the participants are users of the product, while in discrimination tests, panelists are screened for acuity, oriented to test procedures, and may even undergo rudimentary training. The discrimination panel is not representative of a consumer sample and it is usually not intended to be so. The emphasis instead is on providing a sensitive tool that will function as a safety net in detecting differences. Second, participants are in an analytic frame of mind for discrimination while they are in an integrative frame of mind (looking at the product as a whole, usually) and are reacting hedonically in preference tasks. Third, there is no good solution to the question of what to do with preference judgments from correct versus incorrect panelists in the discrimination test. Even if data are used only from those who got the test correct, some of them are most likely guessing (correct by chance). Exclusion of panelists from a consumer test on any other basis than their product usage is a poor practice. They have been selected based on their being a representative sample of the target group.

When doing a paired preference test keep in mind that the technique is designed to answer one and only one important question. Consumers are in an integrative frame of mind and are responding to the product as a whole. They are generally not analyzing it as to its individual attributes, although one or two salient characteristics may sometimes drive their decisions. However, these one or two salient characteristics may also cause other attributes to be viewed in a positive light, an example of what is called the "halo effect" (see Chapter 9). For these reasons, it is difficult to get consumers to accurately explain the basis for their choice. Although it is fairly common to ask for diagnostic information in a large and expensive multi-city consumer field test, one should recognize the "fuzziness" of this information that it can be difficult to interpret and that it may or may not be useful in any kind of important decision making.

Choice tests and rankings indicate the direction of preferences for the product but are not designed to find the relative differences in preference among the products. In other words, the results give no indication of the size of the preference. However, it is possible to derive Thurstonian scale values from proportions, giving some indication of the magnitude of the difference on a group basis. For example, Engen (1974) compared the hedonic range of odor preference among adults to the range of likes and dislikes for children using indirect scale values based on multiple paired comparisons. Adults showed a wider range of preferences on this basis.

13.3 Simple Paired Preference Testing

13.3.1 Recommended Procedure

In paired preference tests the participant receives two coded samples. The two samples are presented to the panelist simultaneously and the panelist is asked to identify the sample that is preferred. Often, to simplify the data analysis and interpretation, the subject must make a choice (forced choice) although it is possible to include a no-preference option (discussed later in this chapter). Figure 13.1 shows a sample score sheet without the no-preference option. The sensory specialist should make sure that the consumer panelist understands the task described by the score sheet.

Paired preference test

Orange Beverage

Name____________________________ Date________________

Tester Number____ ____ ____ ____ Session Code ___ ___ ___

Please rinse your mouth with water before starting

Please taste the two samples in the order presented, from left to right. You may drink as much as you would like, but you must consume at least half the sample provided.

If you have any questions, please ask the server now

Circle the number of the sample you prefer
(you must make a choice)

387 456

Thank you for your participation.
Please return your ballot through the window to the server

Fig. 13.1 Ballot example for a paired preference test when a choice is forced.

The paired preference test has two possible serving sequences (AB, BA). These sequences should be randomized across panelists with an equal number of panelists receiving either sample A or sample B first.

The steps in setting up and conducting a paired preference test are shown in Table 13.1. It is always appropriate to confirm the test objectives with the end user of the test results. Testing conditions should also be made clear in terms of the amount being served, temperature and other aspects of the physical setup. It is wise to write down these conditions and the procedures for serving as a standard operating procedure (SOP), so that the staff conducting the test has a clear understanding of what to do. Of course, the ballot has to be prepared, random codes assigned to products, and a counterbalancing scheme set up for the alternating positions of the two products. Consumers must be recruited and screened so that they are suitable for the test; usually frequent users of the product are appropriate. In this test, consumers are forced to make a choice. Responding with "no preference" or equally preferred is not an option.

13.3.2 Statistical Basis

For paired preference methods the probability of selection of a specific product is one chance in two. The null hypothesis states that in the long run (across all possible replications and samples of people) when the

Table 13.1 Steps in conducting a paired preference test

1. Obtain samples and confirm test purpose, details, timetable, and consumer qualifications (e.g., frequency of product usage) with client.
2. Decide testing conditions (sample size, volume, temperature, etc.).
3. Write instructions to the panelists and construct ballot.
4. Recruit potential consumers.
5. Screen for product usage to qualify.
6. Set up counterbalanced orders (AB, BA).
7. Assign random three digit codes and label sample cups/plates.
8. Conduct test.
9. Analyze results.
10. Communicate results to client or end-user.

underlying population does not have a preference for one product over the other, consumers will pick each product an equal number of times. Thus the probability of the null hypothesis is $P_{pop} = 0.5$. Remember that P_{pop}, the proportion that we are making an inference about, refers to the proportion we would see prefer one product over another in the underlying population. It is not the proportion preferring that sample in our data. Mathematically, this may be written as Ho: $p(A) = p(B) = 1/2$. The test is two tailed since prior to the study the experimenter does not know which sample will be preferred by the consumer population. There is no right answer; it is a matter of opinion. The alternative hypothesis for the paired preference test is written as Ha: $p(A) \neq p(B)$. Three data analyses can be used based on the binomial, chi-square, or normal distributions, respectively.

The binomial distribution allows the sensory specialist to determine whether the result of the study was due to chance alone or whether the panelists actually had a preference for one sample over the other. The following formula allows the exact probability of one specific outcome (but not the overall probability for the hypothesis test). This equation gives the probability of finding y judgments out of N possible, with a chance probability of one-half:

$$p_y = (1/2)^N \frac{N!}{(N-y)!y!} \qquad (13.1)$$

where

N = total number of judgments

y = total number of preference judgments for the sample that was most preferred

p_y = probability of making the number of preference choices for the most preferred sample

In this formula, $N!$ describes the mathematical factorial function which is calculated as $N\cdot(N-1)\cdot(N-2) \ldots 2\cdot 1$. For example, the exact probability of five out of eight people preferring one sample over another is $(1/2)^8 \bullet (8!)/(3!)(5!)$ or 56/256 or 0.219. Bear in mind that this is the probability of just one outcome (one term in a binomial expansion, see Appendix B) and two other considerations need to be taken into account in testing for the significance of preferences. The first of these is that we consider the probability of detecting an outcome this extreme *or more extreme* in testing for significance, so the other terms farther out in the tail of the distribution must also be added to the probability value. So in our example, we would also have to calculate the probability of getting 6 out of 8, 7 out of 8, and 8 out of 8 and add them all together. You can see that this becomes very cumbersome for large consumer tests and so this approach is rarely done by hand, although there are good statistical tables that use exact binomial probabilities (Roessler et al., 1978). The second consideration is that the test is two tailed, so once you have added all the necessary terms, the total probability in the tail should be doubled. Remember we have no a priori prediction that the test will go one way or the other and so the test is two tailed. These considerations lead to the calculated values shown in Table 13.2. The table gives the minimal values for statistical significance as a function of the number of people tested. If the obtained number preferring one product (the larger of the two, i.e., the majority choice) is equal to or exceeds the tabled value, there is a significant preference.

13.3.3 Worked Example

In a paired preference test using 45 consumer panelists, 24 panelists preferred sample A. In Table 13.2 we find that the table value for 45 consumer panelists with an alpha criterion of 5% is 30. This value is larger than 24 and therefore the consumer panelists did not have a preference for one sample over the other. Let us assume that in a different study, 35 of 50 consumer panelists

Table 13.2 Minimum value (X) required for a significant preference

N	X	N	X	N	X
20	15	60	39	100	61
21	16	61	39	105	64
22	17	62	40	110	66
23	17	63	40	115	69
24	18	64	41	120	72
25	18	65	41	125	74
26	19	66	42	130	77
27	20	67	43	135	80
28	20	68	43	140	83
29	21	69	44	145	85
30	21	70	44	150	88
31	22	71	45	155	91
32	23	72	45	160	93
33	23	73	46	165	96
34	24	74	46	170	99
35	24	75	47	175	101
36	25	76	48	180	104
37	25	77	48	185	107
38	26	78	49	190	110
39	27	79	49	195	112
40	27	80	50	200	115
41	28	81	50	225	128
42	28	82	51	250	142
43	29	83	51	275	155
44	29	84	52	300	168
45	30	85	53	325	181
46	31	86	53	350	194
47	31	87	54	375	207
48	32	88	54	400	221
49	32	89	55	425	234
50	33	90	55	450	247
51	34	91	56	475	260
52	34	92	56	500	273
53	35	93	57	550	299
54	35	94	57	600	325
55	36	95	58	650	351
56	36	96	59	700	377
57	37	97	59	800	429
58	37	98	60	900	480
59	38	99	60	1000	532

Notes: N is the total number of consumers

X is the minimum required in the larger of the two segments

Choice is forced

Values of X were calculated in Excel from the z-score approximation to the binomial distribution

Values of N and X not shown can be calculated from $X = 0.98\sqrt{N} + N/2 + 0.5$

Calculated values of X must be rounded up to the nearest whole integer

Tests with $N < 20$ are not recommended but critical values can be found by reference to the exact binomial (cumulative) probabilities in Table I

Values are based on the two-tailed Z-score of 1.96 for $\alpha = 0.05$

Critical minimum values for $\alpha = 0.01$ can be found in Table M

preferred sample A over sample B. In Table 13.2, we find that the table value for 50 panelists at alpha of 5% is 33. The obtained value of 35 is greater than this minimum and therefore the consumers had a significant preference for sample A over sample B.

13.3.4 Useful Statistical Approximations

Most sensory specialists use these simple lookup tables for finding the significance of a test outcome. Remember that the test is two tailed and that you cannot use the tables for the paired difference test, which are one tailed. If you do not have the tables handy or wish to calculate the probability for some value not in the table, you can also use the Z-score formula for proportions shown in Chapter 4 and Appendix B for discrimination tests. The binomial distribution begins to give values very near the normal distribution for large sample sizes, and since most consumer tests are large ($N > 100$), this approach is nearly mathematically equivalent. The following formula can be used to calculate the z-value associated with the results of a specific paired preference test (Stone and Sidel, 1978). The formula is based on the test for a difference of two proportions (the observed proportion preferring versus the expected number or one-half the sample, as follows:

$$z = \frac{(p_{obs} - p) - 1/2N}{\sqrt{pq/N}} = \frac{(X - Np) - 0.5}{\sqrt{Npq}} \quad (13.2a)$$

Or

$$z = \left[(X - N/2) - 0.5\right]\Big/0.5\sqrt{N} \quad (13.2b)$$

where

$X =$ number of preference judgments for the most preferred sample

$p_{obs} = X/N$

$N =$ total number of judgments (usually the number of panelists)

$p =$ probability of choosing the most preferred sample by chance

$q = 1 - p$ and for paired preference tests: $p = q = 1/2$

The probability associated with the paired preference test is two-tailed so that a Z-value of 1.96 is appropriate for a two-tailed test with alpha equal to 0.05. The obtained Z-score must be larger than 1.96 for the result to be statistical significant. For other alpha levels, the Z-table in the Appendix, Table A, may be consulted.

Another approach is to use the chi-square distribution which allows the sensory scientist to compare a set of observed frequencies with a matching set of expected (hypothesized) frequencies. The chi-square statistic can be calculated from the following formula (Amerine and Roessler, 1983), which includes the number –0.5 as a continuity correction. The continuity correction is needed because the χ^2 distribution is continuous and the observed frequencies from preference tests are integers. It is not possible for one-half of a person to have a preference and so the statistical approximation can be off by as much as 1/2, maximally. As in any chi-square test against expected values we have five steps: 1) subtract the observed value from the expected value and take the absolute value, 2) subtract the continuity correction (0.5), 3) square this value, 4) divide by the expected value, 5) sum all the values from step 4.

$$\chi^2 = \frac{[|(O_1 - E_1)| - 0.5]^2}{E_1} + \frac{[|(O_2 - E_2)| - 0.5]^2}{E_2} \tag{13.3}$$

O_1 = observed number choices for sample 1
O_2 = observed number choices for sample 2
E_1 = expected number choices for sample 1 (in a paired test = $N/2$)
E_2 = expected number of choices for sample 2 ($N/2$ again)

The critical value for χ^2 with one degree of freedom is 3.84. The obtained value must exceed 3.84 for the test to be significant. Note that this is the square of 1.96, our critical z-score. The binomial z-score and chi-square test are actually mathematically equivalent as long as both use or both do not use the continuity correction (see Proof, Appendix B, Section B.6).

13.3.5 The Special Case of Equivalence Testing

A parity or equivalence demonstration may be important for situations such as advertising claims.

Establishing an equivalence or parity situation for a preference test usually requires a much larger sample size (N) than a simple paired preference test for superiority. The theoretical basis for a statistical test of equivalence is given in Ennis and Ennis (2009) and tables derived from this theory are given in Ennis (2008). The theory states that a probability value can be obtained from an exact binomial or from a normal approximation as follows:

$$p = \Phi\left(\frac{|x| - \theta}{\sigma}\right) - \Phi\left(\frac{-|x| + \theta}{\sigma}\right) \tag{13.4}$$

where phi (Φ) symbolizes the cumulative normal distribution area (converting a z-score back to a p-value), theta (θ) is the half-interval, in this case 0.05, and sigma (σ) is the estimated standard error of the proportion (square root of pq/N). x in this case is the difference between the observed proportion and the null (subtract 0.5). Here is an example, modified from Ennis and Ennis (2009).

In a paired preference test of a soft drink, 295 of 600 consumers prefer one product and 305 choose the other. Using a boundary of 0.50±0.05 as the "equivalence" region (i.e., a true population proportion that lies between 45 and 55%), we can perform a simple test from Eq. (13.4).

First, we continuity correct the proportion of 295 to 294.5, giving a proportion of 0.4908. Subtracting 0.5 gives our x value of –0.0092. Then our standard error is estimated by

$$\sigma = \sqrt{\frac{0.45(0.55)}{600}} = 0.02031$$

and our p-calculation proceeds as follows:

$$p = \Phi\left(\frac{|0.0092| - 0.05}{0.02031}\right) - \Phi\left(\frac{-|0.0092| + 0.05}{0.02031}\right) = .0204$$

and so the value of 0.0204 ($p<0.05$) is good evidence that the true population proportion lies within the interval of 0.5±0.05, based on the obtained proportion in our data. Further information on preference tests in claim substantiation is given in Chapter 19.

13.4 Non-forced Preference

A no-preference option is sometimes included in a paired preference test, in spite of the fact that it complicates the analysis considerably. Many practitioners question whether it is worth the effort. However, it may be required due to the legal regulations for claim substantiation (ASTM, 2008). The size of the no-preference response group could be useful information in its own right. Also, some have proposed that it could help decide if an equal preference split was due to indifferent response or whether there might in fact be stable groups of about equal size with strong preferences (Gridgeman, 1959). In other words, a 50/50 split in a preference test is no clear win for either product,

but might be the result of two segments of consumers that each likes different versions of the product. Unfortunately this situation is not resolved by offering the no-preference option. A very robust finding is that people will avoid making this response, even with physically identical samples (Chapman and Lawless, 2005; Chapman et al., 2006; Marchisano et al., 2003). Although Gridgeman was correct in stating that the no-preference option offers additional information, the response option does not in fact resolve the question of stable segments. Nonetheless, the no-preference response may be required by someone requesting the test or be included for other legal considerations (ASTM, 2008). A ballot for a preference test with the no-preference option is shown in Fig. 13.2.

Fig. 13.2 Ballot for a paired preference test when a choice is not forced and a no-preference response is allowed.

Paired preference test
Orange Beverage

Name_____   Date_____

Tester Number____ ___ ___ ___   Session Code ___ ___ ___

Please rinse your mouth with water before starting

Please taste the two samples in the order presented, from left to right.
You may drink as much as you would like, but you must consume at least
half the sample provided.

If you have any questions, please ask the server now

Please indicate your preference by
Circling one of the following three answers:

387   456

No Preference

Thank you for your participation.
Please return your ballot through the window to the server

There are four approaches to dealing with the no-preference responses in a paired preference test: (1) eliminate them, (2) apportion them, (3) use a confidence interval analysis, (4) use a signal detection model analysis. To some degree the choice of how to deal with the responses depends on what you assume the basis for the no-preference choices was. For example, if you assume that people responding "no preference" really do not care, then it would make sense to apportion them 50/50 to each of the other response options. Unfortunately there is rarely any good basis (i.e., some additional information or data) for making such assumptions.

In the first approach, they are simply discounted. The analysis proceeds using the simple two-tailed binomial statistics for a difference of proportions. This approach lowers the sample size and thus the power of the test to finding significant preferences. This approach seems reasonable if the actual number is fairly low, say less than 10% of the responses. If the proportion of no-preference responses is high, say above 20%, but there is still a significant result in the remaining sample, the test result may be qualified in the report as follows: "*Among those expressing a preference*, there was a significant preference for product X" (see ASTM, 2008). The researcher should also state the raw percentages including the size of the no-preference response group.

In the second approach, the no-preference responses are apportioned. One way to do this is simply split them 50/50 into the existing preference groups. This maintains the sample size but does dilute the test result somewhat since the 50/50 split is what one expects by chance. Another option is to divide the no-preference votes in proportion to those who did express a preference. That is, if there is a 60/40 split among those who did express a preference, the no-preference votes would be apportioned to those groups in the same 60/40 ratio. This approach is based on some findings from Odesky (1967), who found that the proportions of people expressing a preference when the no-preference option was available were similar to the obtained proportions when choice was forced. However, this finding has been questioned and it may not be a valid generalization (Angulo and O'Mahony, 2005). In some cases for advertising claims of product superiority, they must be apportioned to the competitor's product, providing a very strict hurdle for proving a significant preference (ASTM, 2008).

In the third approach, a confidence interval may be constructed around each proportion of those who did express a preference. If the confidence intervals do not overlap, one may conclude that there is a significant preference win for the product with the larger proportion. This approach is justified if the sample size is large ($N > 100$) and the number of no-preference choices is relatively low (less than 20%). The formula for the confidence interval is

$$\text{CI} = \frac{\chi^2 + 2X \pm \sqrt{\chi^2 \left[\frac{\chi^2 + 4X(N-X)}{N} \right]}}{2\left(N + \chi^2\right)} \quad (13.5)$$

where X is the number of panelists preferring one of the two products, N is the total number of panelists, and χ^2 is the chi-square value for 2 df or 5.99. A worked example is shown in Appendix 3 at the end of this chapter.

A fourth approach is based on a Thurstonian model, which states that the degree of difference in liking must exceed a person's criterion, called tau, below which they will choose a no-preference option. This approach is based on an extension of the paired comparison test for difference with an "equal" option (see Braun et al., 2004). The degree of preference/difference is called d-prime (or in some literature, the Greek letter delta). See Chapter 5 for more information on Thurstonian models. The size of the difference, d-prime (d'), and the tau criterion can be solved by the following equations:

$$z_1 = (-\text{tau} - d')/\sqrt{2} \quad (13.6a)$$

$$z_2 = (+\text{tau} - d')/\sqrt{2} \quad (13.6b)$$

where z_1 is the z-score for the proportion preferring product A and z_2 is the z-score for the sum of the proportion for A *and* the no-preference votes.

This provides two equations in two unknowns, which can be solved for tau and d'. Once d' is obtained, a z-test can be performed which will tell if the d' value is significantly different from zero. The standard deviation (S) is found from the value $\sqrt{B/N}$. Then $z = d'/S$, which must be greater than 1.96. This would be taken as evidence of a significant preference. *B*-values (see Bi, 2006) are found for different d' values in Table O.

13.5 Replicated Preference Tests

Although replication is not often done with preference tests, there are a few good reasons to consider replicating in the design of a sensory test. First, the effort and cost to recruit, screen, and get a consumer panelist on-site is far larger than the time and cost associated with the conduct of the actual test. So why not get some additional information while the consumers are present? Second, there is evidence that many people will change their minds from trial to trial. Koster et al. (2003) summarized some data with children and novel foods and found, on the average, less than 50% consistency from trial to trial in the most liked (chosen) food and somewhat higher levels with adults. This result is similar to findings of Chapman and Lawless (2005) who found about 45% switching in a test of milks that allowed the no-preference option, although the marginal proportions of the groups that preferred each milk remained stable. Finally, repeated testing may be the only way to answer the question of whether a 50/50 split in preference represents equal appeal of the two products (or lack of preference) or whether there are two stable segments of equal size preferring each version of the product. In other words, repeated testing could yield evidence for stable segments, in which case both versions of the product become candidates for further consideration, development, and marketing.

Analysis of these data can be simple or more complex. A simple approach is to consider the expected value one would obtain on a replicated test if people were behaving randomly (i.e., had no real stable preference). For example, on a test with two trials, one would expect 25% of the consumers to choose product A twice, 25% to choose product B twice, and 50% to choose one of each. Given these expected frequencies, a simple chi-square analysis can be performed to see if the results differ from chance expectations (Harker et al., 2008; Meyners, 2007). The approach can obviously be extended to tests with more than two trials. In the case of two trials the data can also be cast in a 2 × 2 contingency table (showing the preferences on trial 1 in the columns and trial 2 in the rows) which classifies each consumer into one of four cells. A chi-square analysis can be conducted on this table as well. Individual binomial tests on proportions can also be done to compare the group consistently preferring A to that consistently preferring product B (but note

that this lowers the N by eliminating the people who switched choices).

A variety of other approaches have been suggested, and several are reviewed in the extensive paper by Cochrane et al. (2005). A beta-binomial approach is discussed by Bi (2006), similar to the beta-binomial analysis used with replicated difference tests. This approach is informative as it not only gives some overall statistical significance level but also calculates the gamma statistic which shows how random versus consistently grouped the panel appears to be behaving. If there are stable segments with consistent preferences, a significant (non-zero) gamma statistic could be obtained. Further information on gamma and the beta-binomial is given in Chapter 4. Also, this analysis formally recognizes that the data on the two trials are related, and that individual consistency or variation is part of the picture (i.e., "overdispersion"). In the next section we will see an approach that combines both replication and allowance for the no-preference option.

13.6 Replicated Non-forced Preference

Next we will look at perhaps the most complicated situation, involving both replication and the use of the no-preference response option. A simple yet elegant approach to the issue of stable preference segments is found in a 1958 paper by George E. Ferris (also discussed in Bi, 2006). An alternative analysis uses the Dirichlet-multinomial (DM) model, an extension of the beta-binomial approach to the multinomial situation. This is discussed in Gacula et al. (2009) and illustrated with a worked example. The DM model is attractive because it considers the heterogeneity of the consumer group, i.e., whether there are different pockets of individuals with consistent patterns or response across replications, much like the gamma statistic of the beta-binomial.

Ferris called his approach the k-visit method of consumer testing and the examples he gives are in sequential home use tests for paired preference, with a no-preference option (for our purposes, $k = 2$). In other words, there were at least two separate tests conducted using the same consumers, with a preference choice between products "A" and "B." There are several reasonable assumptions made, which are as follows: (1) persons with a consistent preference would choose A on both trials or B on both trials; (2) sometimes consumers would choose to respond A or B even if they had no preference or could not discriminate the samples, in order to please the experimenter; (3) thus the double response for A (or B) could include some people who had no preference (or could not discriminate); (4) the amount of switching responses (i.e., inconsistency) that goes on was a clue to the proportion of false preference expressed by nonpreferring consumers. Ferris was then able to get some good estimates of (1) the true proportion of consumers who consistently liked A, (2) the true proportion of consumers who consistently liked B, (3) and the true proportion of everyone left over (people with consistent non-preference, non-discriminators, and switchers). The calculations for this analysis and some simple tests for statistical significance are shown below. Note that if you wanted to just test for a difference between the two trials, you could do a Stuart–Maxwell test, but this would not answer the questions about a product win nor give any information about stable segments.

There are several areas for minor concern in using this analysis. First, it is possible that a person could have a real preference on the first trial, but change his or her mind on the second trial (Koster et al., 2003). For example, I might like product A on trial one, then get tired of the product, fatigued by the test, lose interest, etc., so that on trial two I really do not care any more and so I respond "no preference." Second, the opinion could actually change from trial to trial for good reason, e.g., perhaps with some very sweet foods which are liked at first, then become too cloying. Third, it is possible that the preference is not all-or-nothing (as assumed in this model) but rather that people have some percentage of the time they like A and another percentage they like B and perhaps even some of the time when the same person simply does not care. If any of these is the case, then this model is a bit too simple, but it can still be applied for decision-making purposes.

Here are the two experimental questions: Is there a significant preference? Is there a consistent segment even if there is not a "win"? The test design uses N consumers, who participate in each of two paired tests and must respond either "Prefer A," "Prefer B," or no preference on each questionnaire. The samples have different blind codes in each trial so the consumers are unaware that the test is repeated. The data are tabulated in a 3×3 table as shown below, with trial one

answers in the columns and trial two answers in the rows. We have retained the original notation of Ferris in the calculations (see also Bi, 2006). N's in the table are the actual frequency counts (not proportions). The subscript a means the consumer chose product A, b for product B, and o for no preference. Thus N_{ao} refers to the number of consumers who choose product A on the first trial and said "no preference" on the second. N without a subscript is the total number of consumers in the test (Table 13.3).

Table 13.3 Notation for analysis of non-forced replicated preference

	Trial 1 response		
	Prefer A	No preference	Prefer B
Trial 2 response			
Prefer A	N_{aa}	N_{oa}	N_{ba}
No preference	N_{ao}	N_{oo}	N_{bo}
Prefer B	N_{ab}	N_{ob}	N_{bb}

Calculations:

(1) Tabulate some "inconsistent behavior" totals. These will help to shorten some of the further calculations below.

$$N_y = N_{ao} + N_{oa} + N_{bo} + N_{ob} \qquad (13.7a)$$

(some people switching to or from no pref)

$$N_x = N_{ab} + N_{ba} \qquad (13.7b)$$

(some people switching products, A to B or B to A)

$$M = N - N_{aa} - N_{bb} \qquad (13.7c)$$

(all those showing no consistent preference)

(2) Calculate a consistency parameter, p.

$$p = \frac{M - \sqrt{\left[M^2 - (N_{oo} + N_y/2)(2N_x + N_y)\right]}}{2N_{oo} + N_y} \qquad (13.8)$$

(3) Figure the maximum likelihood estimates, fitted proportions π_A and π_B and π_o. These are the true or population estimates we are trying to obtain from the analysis. Note that π_A does not equal N_{aa}/N. In other words we are going to adjust the p_A estimate based on the notion that some of the consistent N_{aa} responders may have included some non-preferring consumers who were just trying to

please the experimenter. This was a remarkable insight on the part of Ferris, given the rediscovery of false preferences in the tests with identical products, just recently (see Chapman et al., 2006; Marchisano et al., 2003). This is the reason for the consistency parameter, p, in Eq. (13.8). Put another way, we can get a good estimate of the amount of random responding or false consistency by looking at the amount of no-preference votes and the amount of switching that went on in the data set.

$$\pi_A = \frac{[N_{aa}(1-p^2)] - [(N-N_{bb})p^2]}{N(1-2p^2)} \quad (13.9a)$$

$$\pi_B = \frac{[N_{bb}(1-p^2)] - [(N-N_{aa})p^2]}{N(1-2p^2)} \quad (13.9b)$$

$$\pi_o = 1 - \pi_A - \pi_B \quad (13.9c)$$

Note that it is these adjusted proportions π_A and π_B that we really want to know giving us estimates for the sizes of the consistent segments, not simply N_{aa}/N or N_{bb}/N which are the proportions of supposedly "consistent" consumers in the original data. The original raw data are probably tainted by some random responders or no-preference responders trying to please the experimenter.

(4) These are point estimates, so in order to get confidence limits and do some statistical testing, we need some variability estimates, too. Next, calculate some variance and covariance estimates using the parameters calculated so far:

$$\mathrm{Var}(\pi_A) = \frac{\pi_A(1-\pi_A) + (3\pi_o p^2)/2}{N} \quad (13.10a)$$

$$\mathrm{Var}(\pi_B) = \frac{\pi_B(1-\pi_B) + (3\pi_o p^2)/2}{N} \quad (13.10b)$$

$$\mathrm{COV}(\pi_A, \pi_B) = \frac{(\pi_o p^2/2) - (\pi_A \pi_B)}{N} \quad (13.10c)$$

(5) Now we can see if there is a win for product A or for product B. We test for difference of π_A versus π_B using a Z-test:

$$Z = \frac{\pi_A - \pi_B}{\sqrt{\mathrm{Var}(\pi_A) + \mathrm{Var}(\pi_B) - 2\mathrm{Cov}\pi_A\pi_B}} \quad (13.11)$$

This assumes we have a good-sized consumer test with $N > 100$ and preferably $N > 200$.

(6) If needed, test for the size of a consistent segment versus some benchmark. For example, if we needed to see the proportion of true preference for product A greater than 45% to make some advertising claim or take some action in further product marketing we would also use a Z-test against that benchmark, e.g., $\pi_A = 0.45$ (a 45% segment size).

$$Z = \frac{\pi_A - 0.45}{\sqrt{\mathrm{Var}(\pi_A)}} \quad (13.12)$$

We can also test to see whether we have surpassed a certain size of the no-preference segment by a Z-test using π_o and its benchmark. For example, we might have an action standard which states that the no-preference segment must be at or below 20%. We might also have an action standard that included several of these tests. For example, if π_o was less than 20% and π_A and π_B both higher than 35%, we might explore marketing two versions of the product.

A worked example of this analysis for the Ferris k-visit method is shown in Appendix 1 at the end of this chapter.

13.7 Other Related Methods

13.7.1 Ranking

In these tests the consumers are asked to rank several products in either descending or ascending order of preference or liking. The participants are usually not allowed to have ties in the ranking, thus the method is usually a forced choice. The paired preference test is a special case of a preference ranking, when the participant is asked to rank only two samples. Note that rankings do not give a direct estimate of the size of any difference in preference, although it is possible to derive some Thurstonian scale values from the proportions. Preference ranking is intuitively simple for the consumer, it can be done quickly and with relatively little effort. The ranks are based on a frame of reference that is internal to the specific set of products and thus the consumer does not have to rely on memory. A disadvantage of preference ranking is that the data from different sets of overlapping products cannot be compared since the rankings are based on this internal frame of reference. Visual and tactile preference rankings are relatively simple but the multiple comparisons involved in ranking samples by flavor or taste can be very fatiguing. A sample score sheet is shown in Fig. 13.3. An example of the use of ranking in sensory consumer testing is found in the study by Tepper et al. (1994).

13.7.2 Analysis of Ranked Data

The data are ordinal and are treated as nonparametric. Preference ranking data may be analyzed either by using the so-called Basker tables (Basker 1988a, b), those by Newell and MacFarlane (1987) (see Table J) or the Friedman test (Gacula and Singh, 1984). The tables require that the panelists were forced to make a choice and that there are no tied ranks. The Friedman test is tolerant of a small number of tied opinions. Examples follow.

Preference test - Ranking

Fruit Yogurt

Name_____   Date_____

Tester Number____ ____ ____ ____   Session Code___ ___ ___

Please rinse your mouth with water before starting.
You may rinse again at any time during the test if you need to.

Please taste the five samples in the order presented, from left to right.
You may re-taste the samples once you have tried all of them.

**Rank the samples from most preferred to least preferred
using the following numbers:
1 = most preferred, 5 = least preferred**

(If you have any questions, please ask the server now)

Sample	Rank (1 to 5) (ties are NOT allowed)
387	_____
589	_____
233	_____
694	_____
521	_____

Thank you for your participation,
Please return your ballot through the window to the server.

Fig. 13.3 Sample ballot for a ranking test.

To use Table J, assign numerical values to each of the n products (1–n) starting with one for the most preferred sample. Then sum the values across the group of panelists to obtain a rank sum for each sample. Next, consult a table for rank totals (Table J). In this example, six consumers ranked seven products using a rank scale with 1 = preferred most and 7 = preferred least. Clearly we would never, in real life, do a consumer ranking study with only six panelists, but this is only an example to illustrate the calculations associated with the tables. In this example rank totals for the products A through G are as follows:

Product	A	B	C	D	E	F	G
Rank total	18	28	20	10	26	32	34
Significance group	ab	ab	ab	a	ab	ab	b

The table indicates that the critical difference value for six consumers and seven products is 22. Products with the same letter below their rank sum are not significantly different by this test. Product D is thus significantly more preferred to Product G with no other preferences observed in this comparison.

The Friedman test is the nonparametric equivalent to the two-way analysis of variance without interaction. The Friedman test equation is based on the χ^2 distribution:

$$\chi^2 = \left\{ \frac{12}{[K(J)(J+1)]} \left[\sum_{j=1}^{J} T_j^2 \right] \right\} - 3K(J+1)$$
(13.13)

where

J = number of samples
K = number of panelists
T_j = rank total and
degrees of freedom for $\chi^2 = (J-1)$

Once we determine that the χ^2 test is significant then a comparison of rank total separation is done to determine which samples differ in preference from one another. Informally we have called the value that determines the significant difference in preference ranking the "least significant ranked difference" or LSRD, in an analogy to the LSD test used to test differences of means after analysis of variance.

$$\text{LSRD} = t\sqrt{\frac{JK(J+1)}{6}}$$
(13.14)

where

J = number of samples
K = number of panelists and
t is the critical t-value at $a = 5\%$ and degrees of freedom = 1

To return to the example previously used to explain the use of the Newell and MacFarlane or Basker tables, we will now use those data in a Friedman test. First, according to the overall test for differences, $\chi^2 = 15.43$. The critical χ^2-value for α at 5% and six degrees of freedom is 12.59. Therefore, the preference ranks for this data set differ significantly at $p < 5\%$. We now need to determine which products were ranked significantly higher in preference from one another. The least significant rank difference (LSRD) for the Friedman test is calculated from Eq. (13.14). In the example above, the LSRD = 14.7, giving the following pattern of statistical differences in preference:

Product	A	B	C	D	E	F	G
Rank total	18	28	20	10	26	32	34
Significance group	ab	bc	abc	a	bc	bc	c

Products sharing the same significance group letter show no difference in ranked preference. This pattern can be summarized as follows: Product D is significantly more preferred than products B, E, F, and G. Product G is significantly less preferred than products A and D. Note that the results from the Basker table were more conservative than the results from the Friedman test. Slight differences will sometimes occur in these two approaches.

13.7.3 Best–Worst Scaling

Best–worst scaling (also called maximum difference or max-diff) is a technique in which more than two products are given and the person chooses the one he or she likes best and the one he or she likes the least.

Although the data can result in scaled values for the overall appeal of each product, it is really a choice method and therefore falls into the same class as paired preference choice and ranking. This method has been popular in other areas like marketing research but has recently received some attention for food testing (Hein et al., 2008; Jaeger and Cardello, 2009; Jaeger et al., 2008). The psychometric models for the data from this method suggest that the data can yield scores on an interval and sometimes ratio scale (Finn and Louviere 1992). The method would seem to have some psychological benefit in terms of ease of use, under the notion that it is easier for people to differentiate products at the end of a continuum, as opposed to what is in the middle. However, it may not work well with very fatiguing products like wines (see Mueller et al., 2009).

The method works as follows: the set of products is partitioned into a series of three or more products, with each product representing an equal number of times. In one example, there are four products presented in triads, with four possible combinations of three products at a time (Jaeger et al., 2008). The consumer sees all four triads in random order and with each triad indicates which product is liked best and which is liked the least. There are then two ways of handling the data. The simplest is to sum up the number of times a product was liked best, and then subtract from that the number of times it was liked the least. This differencing procedure generates a score from every panelist which can then be submitted to analysis of variance or any other parametric statistical test. An alternative analysis is to fit the data by a multinomial logistic analysis, which will also yield scores and variance estimates, as well as test for differences among products and any other variables in the test. It is claimed that the simple difference scores have interval scale properties and that the multinomial logistic analysis has true ratio properties (one of the only scales for which this has any reasonable substantiation). See Finn and Louviere (1992) for the psychometric model upon which these claims are based.

Given the "natural" user-friendly nature of the task and the potential to get detailed interval or ratio scale data from the method, it has some appeal for food preference testing. The only drawback is that it requires much more testing to do all the possible combinations than a more straightforward acceptance tests, although the number of trials is somewhat more efficient than giving all possible pairs in a set of multiple

preference tests. Furthermore the number of combinations will decrease as more products are included in each trial (there is no law against using more than three), although the task may become more complex for the consumer as they have more choices to consider. Recent data in real food tests show the task to be easy to do, and the resulting data are as good or better than acceptance scales (9-point or labeled affective magnitude (LAM) scales, see Chapter 14) in differentiating among products (Jaeger and Cardello, 2009; Jaeger et al., 2008). This was confirmed by a comparison to the 9-point, LAM, unstructured line scale and ranking test by Hein et al. (2008). However, it should be noted that when the acceptance scales were replicated in that study, in order to better equate the total number of product exposures and judgments, the scaling data improved in terms of product differentiation, with the 9-point scale improving markedly on the second trial. Thus it seems that the better differentiation with best–worst may simply be a function of having more product comparisons.

The future for this choice method presents some opportunities. First, it may be well suited to food preference surveys, in which no foods are tasted (only names of foods presented) and thus the fatigue factor is not present. Second, incomplete block designs should be applicable to the situation where many products are tested, reducing the burden of a large number of triads that would be needed in a complete design. In their original study, Jaeger et al. (2008) also found that the best–worst scaling provided more useful data for preference mapping than did the line scaling for acceptance. That is the data were better fit and more information was forthcoming from the preference maps (Jaeger et al., 2008). There were also better fits to vectors from descriptive analysis attributes that were projected into the preference maps (see Chapter 18, under external preference mapping).

13.7.4 Rated Degree of Preference and Other Options

A few additional options are available for preference and ranking tests which we will briefly describe here, mostly because they appear in the literature. One option is to provide a rating scale for the degree of preference (Filipello, 1957). The simple preference test

```
+----------------------------------------------------------------------+
|                                                                      |
|                          Preference Test                             |
|                                                                      |
|                            Funistrada                                |
|                                                                      |
|         Name_____Date_____                  |
|                                                                      |
|         Tester number: ___ ___ ___ ___  Session Code____             |
|                                                                      |
|              Please rinse your mouth with water before starting      |
|                                                                      |
|         Please taste the two samples of Funistrada in the order      |
|         presented, from left to right. You may taste as much as      |
|         you would like but you must consume at least one-third        |
|         of the sample.                                               |
|                                                                      |
|              If you have any questions, please ask the server now.    |
|                                                                      |
|         Check one answer that best describes your preference:        |
|                                                                      |
|         _____ Strongly prefer 387 over 589                           |
|                                                                      |
|         _____ Prefer 387 over 589                                    |
|                                                                      |
|         _____ Slightly prefer 387 over 589                           |
|                                                                      |
|         _____ No preference                                          |
|                                                                      |
|         _____ Slightly prefer 589 over 387                           |
|                                                                      |
|         _____ Prefer 589 over 387                                    |
|                                                                      |
|         _____ Strongly prefer 589 over 387                           |
|                                                                      |
|                  Thank you for your participation.                   |
|          Please return your ballot through the window to the server. |
+----------------------------------------------------------------------+
```

Fig. 13.4 Sample ballot for a preference test with rated degree-of-preference and a no-preference option is allowed.

does not indicate the strength of the consumer's opinion. Their preferences might be a small matter or they might have a strong favorite among the alternatives. To get this information a rating scale could be used as shown in Fig. 13.4. If the sample size is large, as it is usually in a consumer test, the choices can be transcribed as −3 to +3 and the resulting distribution tested against a population value of zero by a simple t-test. As in the case of the just-right scale discussed

in the next chapter, it is important to look at the frequency counts in each category and not just the mean score, in case there are unexpected patterns of response (perhaps two groups that strongly prefer different products). Scheffe' (1952) proposed an analysis of variance model for these kinds of data.

Another option that is sometimes used is "dislike both equally" and "like both equally" as a variation of the no-preference option. This obviously provides some additional information. The data should be examined and then both of these combined and treated as if they were the usual no-preference option. An additional category is "do not care" which would provide a response for those individuals who have no preference and no liking or disliking anyway.

13.8 Conclusions

Preference testing is primarily based on a simple choice procedure. A consumer must choose from a pair of products which one is liked best. The analysis is straightforward: A two-tailed test is conducted against a binomial expectation of a 50/50 split under the null hypothesis. If one product has a significantly higher percentage than the expected 50%, a win is declared and the product may move forward in the development scenario.

Two complications arise, however, in common practice. The first is the use of a "no-preference" option. This non-forced preference task is desired for some advertising claim substantiation scenarios due to regulatory agency requirements. However, it complicates the analysis and under most product development situations, the option is not needed. A choice is forced under the assumption that if there is no clear preference (or people just do not care) they will split their votes according to the null expectations. Recent research demonstrates that people avoid the no-preference option, even with physically identical products, so its utility is questionable.

The second complication arises from replication. Replication is less troublesome than the non-forced or no-preference option. It offers a chance to examine the stability of preference choices. In the case where there is nearly an even split among the two products, replication can offer some evidence as to whether there are stable segments with strong loyalty to one of the two versions of the product or whether a substantial amount of shifting might occur. Various analyses of replicated preference are available, including a beta-binomial analysis similar to the one used for discrimination tests. The sensory scientist should weigh the advantages and disadvantages of these procedures carefully before choosing one of them over the simpler and more straightforward paired preference test.

Appendix 1: Worked Example of the Ferris *k*-Visit Repeated Preference Test Including the No-Preference Option

A consumer test with 900 respondents is completed with the following results.

Questions: (1) Is there a significantly higher preference for product A or product $\underline{B}$? (2) Is the preference for the winning product higher than 45%? (Example from Ferris (1958) and Bi (2006), pp. 72–76).

	Prefer A	No preference	Prefer B
Prefer A	$N_{aa} = 457$	$N_{oa} = 12$	$N_{ba} = 14$
No preference	$N_{ao} = 14$	$N_{oo} = 24$	$N_{bo} = 17$
Prefer B	$N_{ab} = 8$	$N_{ob} = 11$	$N_{bb} = 343$

$(N = 900)$

Here are the basic equations we need:

$$N_y = N_{ao} + N_{oa} + N_{bo} + N_{ob}$$
$$N_x = N_{ab} + N_{ba}$$
$$M = N - N_{aa} - N_{bb}$$

$$p = \frac{M - \sqrt{[M^2 - (N_{oo} + N_y/2)(2N_x + N_y)]}}{2N_{oo} + N_y}$$

So for this data set:

$M = 100$ (all those not showing AA or BB behavior, i.e., a consistent choice for one of the two products)

$$N_x = 8 + 14 = 22$$
$$N_y = 14 + 12 + 17 + 11 = 54$$
$$p = 0.257$$

Now we need the equations for the best estimates of each segment/proportion:

$$\pi_A = \frac{[N_{aa}(1 - p^2)] - [(N - N_{bb})p^2]}{N(1 - 2p^2)}$$

$$\pi_B = \frac{[N_{bb}(1 - p^2)] - [(N - N_{aa})p^2]}{N(1 - 2p^2)}$$

and

$$\pi_o = 1 - \pi_A - \pi_B$$

Now we can get our segment size estimates:

$\pi_A = 0.497$ or 49.7% true preference for product A.
$\pi_B = 0.370$ or 37% true preference for product B.
$\pi_0 = 0.133$ or 13.3% no real preference.

Next, we need the variability and covariance estimates for the Z-tests:

$$\text{Var}(\pi_A) = \frac{\pi_A(1 - \pi_A) + (3\pi_0 p^2)/2}{N}$$

$$\text{Var}(\pi_B) = \frac{\pi_B(1 - \pi_B) + (3\pi_0 p^2)/2}{N}$$

$$\text{COV}(\pi_A, \pi_B) = \frac{(\pi_0 p^2/2) - (\pi_A \pi_B)}{N}$$

$\text{Var}(\pi_A) = 0.000296$ (so π_A is 47.9% ± 1.7%, $0.017 = \sqrt{0.000296}$)
$\text{Var}(\pi_B) = 0.000297$
$\text{COV}(\pi_A, \pi_B) = -0.000198$

Now for the hypothesis tests:

$$Z = \frac{\pi_A - \pi_B}{\sqrt{\text{Var}(\pi_A) + \text{Var}(\pi_B) - 2\text{Cov}\pi_A\pi_B}}$$

Note that this is a little different from the simple binomial test for paired preference. In the simple case we test the larger of the two proportions against a null value of 0.5. In this case we actually test for a difference of the two proportions, since we do not expect a 50/50 split any more with the no-preference option.

So the Z for test of A versus B gives $Z = 4.067$, an obvious win for product A.

Finally, a test against a minimum required proportion or benchmark:

$$Z = \frac{\pi_A - 0.45}{\sqrt{\text{Var}(\pi_A)}}$$

Z-test for A versus benchmark of 0.45 (45%).

Appendix 2: The "Placebo" Preference Test

In this method a pair of identical samples are given on one of two preference test trials (Alfaro-Rodriguez et al., 2007; Kim et al., 2008). These physically identical samples are not expected to differ, hence the parallel to a placebo, or a sham medical intervention with no expected therapeutic value. In theory, this could provide a baseline or control condition, against which performance in the preference test (with a no-preference option) could be measured. However, the amount of information gained from this design is relatively small and once again the analysis becomes more complicated. For these reasons, the sensory professional should consider the potential cost, additional analysis, and interpretations that are necessary. A recommended analysis is given at the end of this section.

Issues and complications. The use of a no-preference option was proposed to be a solution to the problem of a 50/50 preference test result, which could result from two stable segments of consumers who have a (perhaps strong) preference for each of the versions, respectively. Hence the idea was to offer a no-preference option with the reasoning that if there were no preferences (rather than stable segments) respondents should opt for the no-preference response. However, persons given identical samples will avoid the no-preference option 70–80% of the time, as discussed earlier in this chapter. So an answer concerning the question of stable segments cannot be obtained by this approach. Evidence for stable segments could be found by replicated testing or by converging evidence from different kinds of tests and/or questions.

Possible analyses. It might be tempting to just eliminate those persons expressing a preference for one of the identical pair members, on the grounds that such respondents are biased. However, this could eliminate 70–80% of the consumers. It is generally not advisable to pre-select consumers on any other grounds than their product consumption, and this approach eliminates individual who are in fact a portion of the representative population we are trying to generalize the results to. Such persons may not be "biased" in any dysfunctional way. Even identical samples may seem different from moment to moment. Furthermore, individuals are clearly responding to the demands of the task (you are expecting them to state a preference in a preference test!).

If historical data are available concerning the frequencies of response to identical pairs, a chi-square test can be performed as shown below. Note that this analysis cannot be done for a test in which the same subjects provide the "placebo" judgments because the chi-square test assumes independent samples.

Placebo analysis #1. Using historical data for expected frequencies.

Cells in the top row (A1, NP1, and B1) are expected frequencies (expected proportion × N judges). Cells in row 2 (A2, NP2, and B2) are the obtained data, frequencies of response in the preference comparison for the actual (different) test samples.

	Prefer A	No preference	Prefer B
Historical data for identical samples	A1	NP1	B1
Test samples	A2	NP2	B2

Placebo analysis #2. The same consumers participate in the placebo trial and the test pair.

If the same people are used for the "placebo" trial and the normal preference comparison trial, an option is to recast the data into another 2 × 3 table, with the rows now representing whether the individual expressed a preference (or not) on the placebo trial. The columns remain Prefer A, no preference, Prefer B. A chi-square test will now tell you whether the proportions of preference changed comparing those who elected the no-preference option for the placebo pair versus those who expressed a "false preference" for the identical pair. This does not provide evidence for the existence of any stable segments nor will it tell you if there is a significant preference from this analysis alone.

If there is no significant chi square, you can feel justified in combining the two rows. If not, you may then analyze each row separately. The correct analyses are as stated in the no-preference Section 13.4: eliminating no-preference judgments, apportioning them, doing a test of d-prime values, or the confidence interval approach if the assumptions are met.

Each judge is classified into one of the six cells, A1, A2, B1, B2, NP1, or NP2. The first row contains the data from people who reported no preference on the

placebo pair. The second row contains the data from people who expressed a preference with the placebo pair. A chi-square test will now show whether the two rows have different proportions. If not, the rows may be combined. If they are different, separate analyses may be performed on each row, using the methods of analysis of the no-preference option discussed earlier in the chapter.

		Response—Trial 2:	
	Prefer A	No preference	Prefer B
Response, Trial 1:			
No preference on Trial 1	A1	NP1	B1
Preference on Trial 1	A2	NP2	B2

Appendix 3: Worked Example of Multinomial Approach to Analyzing Data with the No-Preference Option

This approach yields multinomial distribution confidence intervals for "no-preference" option, Data should be from a large test, $N > 100$, and the no-preference option was used rarely (<20%) (Quesenberry and Hurst, 1964, p. 193, Eq. (2.9)).

Upper and lower confidence interval boundaries are given by

$$CI = \frac{\chi^2 + 2X \pm \sqrt{\chi^2 \left[\frac{\chi^2 + 4X(N - X)}{N} \right]}}{2\left(N + \chi^2\right)}$$

where

$\chi^2_{critical} = 5.99$ for α at 5% and 2 df,
X = number of observed preference votes for one sample,
N = sample size.

Example: For: $N = 162$, $X_1 = 83$, $X_2 = 65$, and no preference $= 14$.

First find the confidence interval for product X_1 (choices = 83/162)

$$CI = \frac{5.99 + 2(83) \pm \sqrt{5.99\left[\frac{5.99+4(83)(162-83)}{162}\right]}}{2(162+5.99)}$$

$$= \frac{171.99 \pm 31.15}{335.98}$$

which gives an interval from 0.42 *to* 0.60 *for product* X_1.

Then find the confidence interval for product X_2 (choices = 65/162)

$$CI = \frac{5.99 + 2(65) \pm \sqrt{5.99\left[\frac{5.99+4(65)(162-65)}{162}\right]}}{2(162+5.99)}$$

$$= \frac{135.99 \pm 30.39}{335.98}$$

which gives an interval from 0.31 *to* 0.50 *for product* X_2. The lower bound of the higher proportion (0.42) overlaps with the upper bound of the lower proportion (0.50). We therefore conclude that there is not enough evidence for any difference in the preference proportions.

References

Amerine, M. A. and Roessler, E. B. 1983. Wines: Their Sensory Evaluation. Freeman, San Francisco.

Angulo, O. and O'Mahony, M. 2005. The paired preference test and the no preference option: Was Odesky correct? Food Quality and Preference, 16, 425–434.

ASTM International. 2008. Standard Guide for Sensory Claim Substantiation. Designation E 1958–07. Vol. 15.08 Annual Book of ASTM Standards. ASTM International, Conshohocken, PA, pp. 186–212.

Basker, D. 1988a. Critical values of differences among rank sums for multiple comparisons. Food Technology, February 1988, 79–84.

Basker, D. 1988b. Critical values of differences among rank sums for multiple comparisons. Food Technology, July 1988, 88–89.

Bech, A. C., Engelund, E., Juhl, H. J., Kristensen, K. and Poulsen, C. S. 1994. Qfood: Optimal design of food products. MAPP Working Paper 19, Aarhus School of Business, Aarhus, Denmark.

Beckman, K. J., Chambers, E. IV and Gragi, M. M. 1984. Color codes for paired preference and hedonic testing. Journal of Food Science, 49, 115–116.

Berglund, B., Berglund, U. and Lindvall, T. 1975. Scaling of annoyance in epidemiological studies. Proceedings, Recent Advances in the Assessments of the Health Effects of Environmental Pollution. Commission of the European Communities, Luxembourg, Vol. 1, pp. 119–137.

Bi, J. 2006. Sensory Discrimination Tests and Measurements. Blackwell, Ames, IA.

Braun, V., Rogeaux, M., Schneid, N., O'Mahony, M. and Rousseau, B. 2004. Corroborating the 2-AFC and 2-AC Thurstonian models using both a model system and sparkling water. Food Quality and Preference, 15, 501–507.

Cardello, A. V. and Schutz, H. G. 2006. Sensory science: Measuring consumer acceptance. In: Handbook of Food Science, Technology and Engineering, CRC Press, Boca Raton, FL. Vol. 2, Ch. 56.

Caul, J. 1957. The profile method of flavor analysis. Advances in Food Research, 7, 1–40.

Chapman, K. W., Grace-Martin, K. and Lawless, H. T. 2006. Expectations and stability of preference choice. Journal of Sensory Studies 21, 441–455.

Chapman, K. W. and Lawless, H. T. 2005. Sources of error and the no-preference option in dairy product testing. Journal of Sensory Studies 20, 454–468.

Cochrane, C.-Y. C., Dubnicka, S. and Loughin, T. 2005. Comparison of methods for analyzing replicated preference tests. Journal of Sensory Studies, 20, 484–502.

Coetzee, H. 1996. The successful use of adapted paired preference, rating and hedonic methods for the evaluation of acceptability of maize meal produced in Malawi. Abstract, 3rd Sensometrics Meeting, June 19–21, 1996, Nantes, France, pp. 35.1–35.3

Coetzee, H. and Taylor, J. R. N. 1996. The use and adaptation of the paired-comparison method in the sensory evaluation of Hamburger-type patties by illiterate/semi-literate consumers. Food Quality and Preference, 7, 81–85.

Ennis, D. M. 2008. Tables for parity testing. Journal of Sensory Studies, 32, 80–91.

Ennis, D. M., and Ennis, J. M. 2009. Equivalence hypothesis testing. Food Quality and Preference, doi:10.1016/j.foodqual.2009.06.005.

Engen, T. 1974. Method and theory in the study of odor preferences. In: A. Turk, J. W. Johnson, Jr. and D. G. Moulton (Eds.), Human Responses to Environmental Odors. Academic, New York, pp. 121–141.

Ferris, G. E. 1958. The k-visit method of consumer testing. Biometrics, 14, 39–49.

Finn, A. and Louviere, J. J. 1992. Determining the appropriate response to evidence of public concern: The case of food safety. Journal of Public Policy and Marketing, 11, 12–25.

Filipello, F. 1957. Organoleptic wine-quality evaluation. 1. Standards of quality and scoring vs. rating scales. Food Technology, 11, 47–51.

Gacula, M. C. and Singh, J. 1984. Statistical Methods in Food and Consumer Research. Academic, Orlando, FL.

Gacula, M., Singh, J., Bi, J. and Altan, S. 2009. Statistical Methods in Food and Consumer Research. Elsevier/Academic, Amsterdam.

Garnatz, G. 1952. Consumer acceptance testing at the Kroger food foundation. In: Proceeding of the Research Conference of the American Meat Institute, Chicago, IL, pp. 67–72.

Gridgeman, N. T. 1959. Pair comparison, with and without ties. Biometrics, 15, 382–388.

Harker, F. R., Amos, R. L., White, A., Petley, M. B. and Wohlers, M. 2008. Flavor differences in heterogeneous foods can be detected using repeated measures of consumer preferences. Journal of Sensory Studies, 23, 52–64.

Hein, K. A., Jaeger, S. R., Carr, B. T. and Delahunty, C. M. 2008. Comparison of five common acceptance and preference methods. Food Quality and Preference, 19, 651–661.

Jaeger, S. R. and Cardello, A. V. 2009. Direct and indirect hedonic scaling methods: A comparison of the labeled affective magnitude (LAM) scale and best-worst scaling. Food Quality and Preference, 20, 249–258.

Jaeger, S. R., Jørgensen, A. S., AAslying, M. D. and Bredie, W. L. P. 2008. Best-worst scaling: An introduction and initial comparison with monadic rating for preference elicitation with food products. Food Quality and Preference, 19, 579–588.

Jellinek, G. 1964. Introduction to and critical review of modern methods of sensory analysis (odour, taste and flavour evaluation) with special emphasis on descriptive sensory analysis (flavour profile method). Journal of Nutrition and Dietetics, 1, 219–260.

Kim, H. S., Lee, H. S., O'Mahony, M. and Kim, K. O. 2008. Paired preference tests using placebo pairs and different response options for chips, orange juices and cookies. Journal of Sensory Studies, 23, 417–438.

Kimmel, S. A., Sigman-Grant, M. and Guinard, J.-X. 1994. Sensory testing with young children. Food Technology, 48(3), 92–94, 96–99.

Koster, E. P., Couronne, T. Leon, F., Levy, C. and Marcelino, A. S. (2003) Repeatability in hedonic sensory measurement: A conceptual exploration. Food Quality and Preference, 14, 165–176.

Lucas, F. and Bellisle, F. 1987. The measurement of food preferences in humans: Do taste and spit tests predict consumption? Physiology and Behavior, 39, 739–743.

Marchisano, C., Lim, J., Cho, H. S., Suh, D. S., Jeon, S. Y., Kim, K. O. and O'Mahony, M. 2003. Consumers report preference when they should not: A cross-cultural study. Journal of Sensory Studies, 18, 487–516.

Meyners, M. 2007. Easy and powerful analysis of replicated paired preference tests using the c2 test. Food Quality and Preference, 18, 938–948.

Moskowitz, H. R. 1983. Product Testing and Sensory Evaluation of Foods. Marketing and R&D Approaches. Food and Nutrition, Westport, CT.

Mueller, S., Francis, I. L. and Lockshin, L. 2009. Comparison of best-worst and hedonic scaling for the measurement of consumer wine preferences. Australian Journal of Grape and Wine Research, 15, 1–11.

Newell, G. J. and MacFarlane, J. D. 1987. Expanded tables for multiple comparison procedures in the analysis of ranked data. Journal of Food Science, 52, 1721–1725.

Odesky, S. H. 1967. Handling the neutral vote in paired comparison product testing. Journal of Marketing Research, 4, 199–201.

Prescott, J., Norris, L., Kunst, M. and Kim, S. 2005. Estimating a "consumer rejection threshold" for cork taint in white wine. Food Quality and Preference, 18, 345–349.

Roessler, E. B., Pangborn, R. M., Sidel, J. L. and Stone, H. 1978. Expanded statistical tables for estimating significance in paired-preference, paired difference, duo-trio and triangle tests. Journal of Food Science, 43, 940–941.

Quesenberry, C. P. and Hurst, D. C. 1964. Large sample simultaneous confidence intervals for multinomial proportions. Technometrics, 6, 191–195.

Saliba, A. J., Bullock, J. and Hardie, W. J. 2009. Consumer rejection threshold for 1,8 cineole (eucalyptol) in Australian red wine. Food Quality and Preference, 20, 500–504.

Scheffe' H. 1952. On analysis of variance for paired comparisons. Journal of the American Statistical Association, 47, 381–400.

Schmidt, H. J. and Beauchamp, G. K. 1988. Adult-like odor preference and aversions in three-year-old children. Child Development, 59, 1136–1143.

Schraidt, M. F. 1991. Testing with children: Getting reliable information from kids. ASTM Standardization News, March 1991, 42–45.

Schutz, H. G. 1971. Sources of invalidity in the sensory evaluation of foods. Food Technology, 25, 53–57.

Sidel, J. L., Stone, H. and Bloomquist, J. 1981. Use and misuse of sensory evaluation in research and quality control. Journal of Dairy Science, 64, 2296–2302.

Sidel, J., Stone, H., Woolsey, A. and Mecredy, J. M. 1972. Correlation between hedonic ratings and consumption of beer. Journal of Food Science, 37, 335.

Stone, H. and Sidel, J. L. 1978. Computing exact probabilities in discrimination tests. Journal of Food Science, 43, 1028–1029.

Tepper, B. J., Shaffer, S. E. and Shearer, C. M. 1994. Sensory perception of fat in common foods using two scaling methods. Food Quality and Preference, 5, 245–252.

Tuorila, H., Hyvonen, L. and Vainio, L. 1994. Pleasantness of cookies, juice and their combinations rated in brief taste tests and following ad libitum consumption. Journal of Sensory Studies, 9, 205–216.

Van Trijp, H. C. M. and Schifferstein, H. N. J. 1995. Sensory analysis in marketing practice: Comparison and integration, Journal of Sensory Studies 10, 127–147.

Vickers, Z. and Mullan, L. 1997. Liking and consumption of fat free and full fat cheese. Food Quality and Preference, 8, 91–95.

Chapter 14

Acceptance Testing

Abstract   An alternative to choice procedures for assessing the consumer appeal of foods is to use a rating scale for the degree of liking or disliking, otherwise known as acceptability scaling or acceptance testing. This chapter illustrates procedures for acceptability scaling, starting with the traditional 9-point hedonic scale in widespread use. Alternative types of acceptance scales are shown. The just-about-right (JAR) scale is illustrated and its statistical analyses are discussed.

About 1930, Dr. Beebe-Center, psychologist at Harvard, wrote a book in which he reported the results of investigations of the pleasantness/unpleasantness of dilute solutions of sucrose and sodium chloride. He called his measurements hedonics. I liked the word, which is both historically accurate and now well installed, and used it in the first official report on the new scale.
—David Peryam, "Reflections" (1989)

Contents

14.1 Introduction: Scaled Liking Versus Choice

The previous chapter dealt with consumer tests involving choices among alternatives and ranking of alternative products. In this chapter we will look at methods for scaling the degree of acceptability of foods. Note that these methods do not require a choice between alternatives. In theory, acceptance scaling can be done on a single product, although a one-product test is usually not very informative and lacks any baseline for making comparisons. A scale that measures the sensory appeal of a product has distinct advantages over a simple choice task. Most importantly, it provides some information on whether the product is liked or disliked in some absolute sense. In a preference test, I might dislike both products and choose the least offensive. In such a case it would obviously not be a good idea to produce or try to sell either version of the product, but the preference test does not tell you this fact. In

H.T. Lawless, H. Heymann, *Sensory Evaluation of Food*, Food Science Text Series,
DOI 10.1007/978-1-4419-6488-5_14, © Springer Science+Business Media, LLC 2010

addition to the liking or disliking information, preference can be inferred from a superior acceptance score of one product over another. For these reasons, many sensory professionals consider acceptance tests to be a better choice than a preference test. Of course, there is no rule against asking both kinds of questions in a test with multiple products and this is often done in consumer field tests as discussed in Chapter 15.

Acceptability data from scales are useful for a number of additional purposes. It is also possible to convert the hedonic scale results to paired preference or rank data (Rohm and Raaber, 1991). Since the scaled acceptance data are "richer" in information, it is possible to derive these other simpler measures from them. Hedonic data can be used in preference mapping techniques (for examples, see Greenhoff and MacFie, 1994; Helgensen et al., 1997; McEwan, 1996). This is a useful technique that allows visualization of the directions for product preferences in spatial models of a product set (see Chapter 19). In spatial models from multivariate analyses, products are represented by points in the space and products that are similar are positioned close together. Dimensions or attributes that differentiate the products can be inferred from product positions, from opposites positioned at different sides, and from interpretation of the axes of the space. Preferences of individual consumers can be projected as vectors through the space to show directions of increased liking. These vectors can then suggest directions for product optimization. Also, differences in the preferred directions for different consumers can help discover market segments or groups with different likes and dislikes.

14.2 Hedonic Scaling: Quantification of Acceptability

The most common hedonic scale is the 9-point hedonic scale shown in Fig. 14.1. This is also known as a degree of liking scale. This scale has achieved wide popularity since it was first invented in the 1940s at the Food Research Division of the Quartermaster Food and Container Institute in Chicago, Illinois (Peryam and Girardot, 1952). David Peryam coined the name hedonic scale for the 9-point scale used to determine degree of liking for food products (Peryam and Girardot, 1952). The hedonic scale assumes consumer

LIKE EXTREMELY

LIKE VERY MUCH

LIKE MODERATELY

LIKE SLIGHTLY

NEITHER LIKE NOR DISLIKE

DISLIKE SLIGHTLY

DISLIKE MODERATELY

DISLIKE VERY MUCH

DISLIKE EXTREMELY

Fig. 14.1 The phrases for the 9-point hedonic scale for food acceptance testing. Responses on this scale are usually assigned values from 1 to 9, 1 for dislike extremely and 9 for like extremely.

preferences exist on a continuum and that preference can be categorized by responses based on like and dislike. The scientists at the Quartermaster Institute evaluated the scale using soldiers in the field, in the laboratory, and in attitude surveys (Peryam and Pilgrim, 1957). Samples were served to panelists monadically (one at a time) and the panelists were asked to indicate their hedonic response to the sample on the scale. Research at the Quartermaster Institute had indicated that the specific way the scale appeared on the score sheet, whether the scale was printed horizontally or vertically, or whether the like or dislike side was encountered first, did not affect results. Jones et al. (1955) found that the ideal number of categories was 9 or 11 and the researchers at the University of Chicago and the Quartermaster Institute decided to use the 9-point version, because it fits better on the typing paper of that era (Peryam, 1989).

> Why does the hedonic scale have nine categories, rather than more or less? Economy perhaps? Preliminary investigation had shown that discrimination between foods and reliability tended to increase up to eleven categories, but we encountered, in addition to the dearth of appropriate adverbs, a mechanical problem due to equipment limitations. Official government paper was only 8″ wide and we found that typing eleven categories horizontally was not possible. So we sacrificed a theoretical modicum of precision for a real improvement in efficiency at the moment. p. 23

The words chosen for each scale option were based on approximately equal differences as determined by Thurstonian methods (see Chapter 7). Thus the scale, psychologically, has ruler-like properties that are not necessarily present in other less carefully constructed liking scales. This equal-interval property is important in the assignment of numerical values to the response choices and to the use of parametric statistics in analysis of the data. Thus the sensory scientist should be cautious and avoid "tinkering" with the scale alternatives. It is important to resist pressure from non-sensory specialists or managers familiar with other scales to modify the scale or adopt alternatives.

The 9-point scale is very simple to use and is easy to implement. It has been widely studied and has been shown to be useful in the hedonic assessment of foods, beverages, and non-food products for decades. The US military has studied its applicability, validity, and reliability and the positive aspects of this scale have been widely accepted. Peryam and Pilgrim (1957) note that the hedonic rating can be affected by changes in environmental conditions (for instance, under field conditions versus cafeteria conditions) but the relative order of sample preference was usually not affected. In other words, the absolute magnitude of the hedonic score may increase or decrease but all samples had similar relative changes. Tepper et al. (1994) showed that consumers rank ordered and hedonically scored products similarly. It has been reported that the scale is reliable and has a high stability of response that is independent of region and to some extent of panel size. However, the applicability of the scale in other languages and cultures have not been as widely studied and it should be used cautiously in these situations.

The 9-point scale has been criticized on several grounds. Moskowitz (1980) suggested that the 9-point hedonic scale has potential problems associated with category scales such as the categories are not quite equally spaced, the neutral ("neither like nor dislike") category makes the scale less efficient and consumers tend to avoid the extreme categories. However, the initial calibration work indicated that this particular category scale has nearly equal-interval spacing although direct scaling methods seem to indicate that the distance from neutral to the like/dislike slightly categories is smaller than the other intervals (Schutz and Cardello, 2001). The neutral response category is important as it is a valid reaction to the product for some participants. Although many scales show "end use avoidance," this serves as a warning to those who are tempted to truncate the scale to fewer than nine

points. Truncating a scale to seven or five points may effectively reduce it to five or three useful categories since end-category avoidance may still come into play. This is one of the forms of "tinkering" to be avoided. The other temptation is to reduce the number of negative response options, often under the misplaced philosophy that the company does not make or test any really bad products. Due to some of these concerns, sensory researchers have used other scales for assessing liking, including various line scales and magnitude estimation, discussed further below.

A recent approach that is growing in popularity is a modification of the hedonic scale based on re-scaling of the word phrases using magnitude estimation and placing them on a line scale with the added end anchors, "greatest imaginable like" and "greatest imaginable dislike." This is the labeled affective magnitude scale or LAM scale discussed in detail later in this chapter (Schutz and Cardello, 2001). The scale development is based on the procedures for the labeled magnitude scale of Green and colleagues (see Chapter 7). Since the development of this scale, others have been developed using similar techniques, notably for oral pleasantness and wetness/dryness (Guest et al., 2007), clothing comfort (Cardello et al., 2003), taste pleasantness (Keskitalo et al., 2007), hedonics in general (Bartoshuk et al., 2006), and perceived satiety (Cardello et al., 2005).

14.3 Recommended Procedure

14.3.1 Steps

The procedure for conducting a scaled acceptance test is very similar to that for the simple paired preference test, except of course that the responses are required after each individual product, rather than each pair. The steps in conducting an acceptance test are shown in Table 14.1. Samples may be served one at a time, a response required after each sample and then the sample returned to the kitchen or prep area. Alternatively, the samples can be placed all on one tray, but this requires the panelist to match the correct test sample to the correct three-digit code on the questionnaire. This is usually done correctly but there are no guarantees. Therefore it is safest with truly naïve consumers to

Table 14.1 Steps in
conducting an acceptance test

1. Obtain samples and confirm test purpose, details, timetable, and consumer qualifications (e.g., frequency of product usage) with client
2. Decide testing conditions (sample size, volume, temperature, etc.)
3. Write instructions to the panelists and construct ballot
4. Recruit potential consumers
5. Screen for product usage to qualify
6. Set up counterbalanced orders
7. Assign random three-digit codes and label sample cups/plates
8. Conduct test
9. Analyze results
10. Communicate results to client or end user

serve products one at a time and retrieve them after each response. A sample ballot for acceptance scaling is shown in Fig. 14.2.

14.3.2 Analysis

The data from the 9-point scales are assigned the values one through nine, nine usually being the "like extremely" category. They are then analyzed using parametric statistics, *t*-tests on means for two products, or analysis of variance followed by comparisons of means for more than two products. Even though the scale may not achieve a true interval level of measurement, the parametric approach is usually justified based on the larger sample size in a consumer test.

14.3.3 Replication

Acceptance tests do not commonly involve replicated tastings on the same products by the same consumers. However, there are several reasons to consider replication. The first is that it may provide some additional information. Byer and Saletan (1961) used repeated tests on beers (judges were blind to the replication) over several days to see if there were systematic increasing or decreasing liking for some beers as opposed to others. Second, a replication may greatly increase the discrimination of products once consumers have a better idea of the range of products to be evaluated. In a study by Hein et al. (2008), the increase in product discrimination was especially pronounced for the 9-point hedonic scale. Third, the first judgment by a consumer may not be well predictive of later behaviors (Koster et al., 2003). Finally, replication will allow one to reduce the effects of serving order,

especially any advantage that might occur for the item in the first position (Hottenstein et al., 2008; Wakeling and MacFie, 1995).

14.4 Other Acceptance Scales

A number of other methods have been used to quantify consumer acceptance and this should not perhaps be surprising given the amount of consumer testing done by sensory evaluation personnel as well as marketing researchers. The 9-point scale itself has been modified in various ways in attempts to improve product discrimination. For example, Yao et al. (2003) found that an unstructured version of the 9-point scale (lacking the verbal labels) produced a somewhat wider range of scale usage among American and Japanese (but not Korean) consumers. In the early development and testing of various hedonic scales, Peryam (1989) noted that there could be room for expansion, especially at the positive end of the scale, stating, "An 8-point unbalanced scale with more 'like' than 'dislike' categories was shown to be somewhat better than the standard 9-point one, but only when dealing with relatively well-liked foods" (1989, p. 24). In the section below, we will briefly consider line scales, magnitude estimation, labeled magnitude scales, which are a combination of line marking and ratio-type scaling, and some relative scales which allow adjustment of previous ratings.

14.4.1 Line Scales

There are a number of studies where the panelists were asked to indicate their hedonic responses on

---

Acceptability Test
Braised Trake

Name_____   Date_____

Tester Number____ ____ ____ ____   Session Code ___ ___ ___

In a previous survey you indicated you are a consumer of BRAISED TRAKE.
Please check an answer below that describes your recent consumption of this
product.

In the last __3__ months, about how often have you eaten BRAISED TRAKE?
(check one)
_____   less than once a month
_____   more than once a month but less than every week
_____   once a week or more

Please rinse your mouth with water before starting.
Your can rinse at any time during the test if you need to.

Please taste the samples according to the number on each page.
Do NOT go back and re-taste the samples once you have turned the page.
If you have any questions, please ask the server now.

Check one phrase to indicate your overall opinion of the product.

Sample #___387___

___   Like extremely
___   Like very much
___   Like moderately
___   Like slightly
___   Neither like nor dislike
___   Dislike slightly
___   Dislike moderately
___   Dislike very much
___   Dislike extremely

PLEASE GO TO THE NEXT PAGE.

Fig. 14.2 A sample ballot for an acceptability test. The samples are evaluated on the 9-point balanced hedonic scale. Each subsequent page will have a new scale for a product with a new three-digit code. The order of evaluation is thus controlled by what is printed on each page, with randomization, rotation, or counterbalancing of orders. There is also a check on product usage frequency at the top of the page. This could be used to confirm that the panelists selected are still users of the product and are thus qualified to be in the test. Such confirmation is recommended for standing panels, such as employees or consumers chosen from a data bank, where the product-usage questionnaire might have been filled out some time in the past. In a consumer field test, the usage frequency would confirm what had been determined in a telephone screening or other recruiting interview (e.g., mall intercept).

unstructured line scales, sometimes anchored by like and dislike on each end (Hough, et al., 1992; Lawless, 1977; Rohm and Raaber, 1991). Line scales are sometimes referred to as visual analog scales (or VAS). That line scales would find some application in hedonics is not surprising, as they became the standard scaling method for descriptive analysis in the 1970s, and their extension into acceptance scaling would seem logical. Recently, a version of the line scale with pips or markers equally spaced along the line has been studied by Villanueva and colleagues and found to compare favorably against the 9-point scale in terms of product differentiation and identification of consumer segments (Villanueva and Da Silva, 2009; Villanueva et al., 2005). However, on a statistical basis any advantages were slight (Lawless, 2009). A simplified version of the labeled affective magnitude scale (LAM scale) was used by Wright (2007) in which the end anchors "greatest imaginable like(/dislike)" were used instead of the usual "like (dislike) extremely." Some line scales are shown in Fig. 14.3.

14.4.2 Magnitude Estimation

As discussed in Chapter 7, magnitude estimation is a scaling procedure in which people can use any numbers they wish and are asked to consider the ratios or proportions between products. In the case of acceptance, they would be told to make a mark twice as far from the origin if the product is liked twice as much. In bipolar magnitude estimation, there are positive and negative numbers used to indicate likes and dislikes. This is not done in unipolar magnitude estimation in which products are only scaled as a distance

from the bottom, which presumably represents a product which would not be liked at all. Given that people have likes and dislikes, a bipolar scale makes much more sense. In a study comparing the results of the 9-point hedonic scale to those obtained from a unipolar magnitude estimation scale and a bipolar magnitude estimation scale, Pearce et al. (1986) found that the three scales gave data that were very similar in terms of reliability, precision, and discrimination. However, the product category was fabric and the fabrics were evaluated by touching. It is possible that these results could have been different if a more fatiguing product category such as a tasted food was evaluated.

Magnitude estimation went through a period of some interest and has been used for evaluation of a number of food products (Lavanaka and Kamen, 1994; McDaniel and Sawyer, 1981). In an unusual combination of line scaling with ratio instructions, Lawless (1977) used a bipolar line scale with a zero or neutral point in the middle. Participants were instructed to consider ratios, for example, "if the next sample is liked twice as much, make a mark on the line twice as far from zero." The problem of having a bounded scale was circumvented by telling the subjects they could tape additional scales to the end of the strip of paper to extend the scale beyond the strip if needed.

These scales have not found much favor in industrial practice, in part because of the popularity of the 9-point scale and in part because of the complicated task of having consumers consider ratios of liking/disliking.

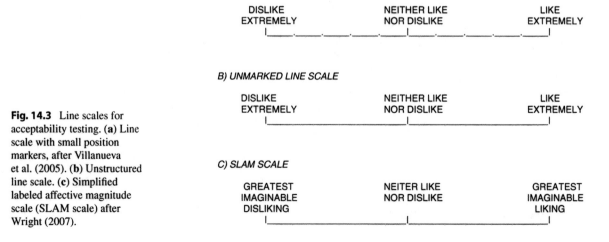

Fig. 14.3 Line scales for acceptability testing. (**a**) Line scale with small position markers, after Villanueva et al. (2005). (**b**) Unstructured line scale. (**c**) Simplified labeled affective magnitude scale (SLAM scale) after Wright (2007).

To make the process a little more user-friendly, other approaches have tried to simplify the task. In exploring the spacing of anchor terms for the LAM scale, a two-step process has been used, in which the magnitude of the hedonic reaction was considered, and then the sign, positive or negative, for the feeling evoked by the word (Cardello et al., 2008; Schutz and Cardello, 2001).

14.4.3 Labeled Magnitude Scales

The labeled affective magnitude scale (LAM, Fig. 14.4) was developed by Schutz and Cardello (2001) as an alternative to the commonly used 9-point category scale for measuring food acceptability (see also Cardello and Schutz, 2004). This scale was an extension of the procedure called the Labeled Magnitude Scale (LMS) that had been used for psychophysical intensity scaling. The LMS was developed by Green and colleagues and was based on earlier work by Borg for a hybrid "category–ratio scale" (Borg, 1982; Green et al., 1993). The LAM scale has been used for evaluation of consumer liking for teas (Chung and Vickers, 2007a, b) and in a comparison of young and older person's liking for different orange juices (Forde and Delahunty, 2004). The theoretical advantages to the LAM scale were proposed to be the following: First, because the word spacings were determined by magnitude estimation (ratio scaling instructions) one might presume that the data allow ratio-type conclusions ("Product A was liked twice as much as B."). In the published literature there are no examples in which this kind of conclusion has been drawn. Second, due to the high end anchors (greatest imaginable liking) people might have a similar idea of the intensity of reaction to this anchor (as proposed by Borg for his original intensity scale) and thus they would be working on the same or a similar psychological continuum.

Does the LAM scale provides any practical advantage over the traditional 9-point hedonic scale? The most important criterion for an advantage is whether one scale is better at finding differences among products (Lawless and Malone, 1986). In the original set of studies, performances of the LAM scale and the 9-point scale were similar (Schutz and Cardello, 2001). Two direct comparisons were conducted, one involving 51 food names and one involving 5 foods that were

The LAM Scale	Label Positions	
	−100 to +100	0 to 100
GREATEST IMAGINABLE LIKE	100.00	100.00
LIKE EXTREMELY	74.22	87.11
LIKE VERY MUCH	56.11	78.06
LIKE MODERATELY	36.23	68.12
LIKE SLIGHTLY	11.24	55.62
NEITHER LIKE NOR DISLIKE	0.00	50.00
DISLIKE SLIGHTLY	−10.63	44.69
DISLIKE MODERATELY	−31.88	34.06
DISLIKE VERY MUCH	−55.50	22.25
DISLIKE EXTREMELY	−75.51	12.25
GREATEST IMAGINABLE DISLIKE	−100.00	0.00

Fig. 14.4 The labeled affective magnitude (LAM) scale, from Schutz and Cardello (2001). For those who wish to construct the scale, the label positions are given on a 100-point and 200-point basis (−100 to +100), from Cardello and Schutz (2004).

actually tasted. Correlations between the mean values obtained on the two scales were +0.99 for the 51 food names and +0.98 for the tasted foods. Statistical differentiation was about the same in both cases. For the food names, there were about the same number of statistically different pairs of means (467 (LAM) versus 459 (9-point) out of 1,275 possible comparisons). A small advantage for the LAM scale was observed in comparing well-liked foods, i.e., those above the grand mean. The higher ends of the scale range were used more frequently with the LAM scale, consistent with the idea that it might be valuable for differentiating well-liked foods.

Several other studies have examined the performance of the two scales in direct comparisons. Greene et al. (2006) examined consumers' reactions to peanuts with and without fruity-fermented flavor defects. The

9-point hedonic scale only uncovered one significant pair of differences on one of the four scales, whereas the LAM scale showed four pairs of significant differences (out of 12 possible) and on three of the four scales. Rather than well-liked foods, these peanut samples scored very near the neutral point on both scales. El Dine and Olabi (2009) found that the LAM scale was as good and sometimes better than the 9-point scale in differentiating well-liked foods. However, in an extensive consumer study with several product categories, Lawless et al. (2009) found that in some cases the LAM was superior to the 9-point and in others the 9-point scale fared better. This was true for both product differentiation and correlation between the product that was best liked and the type of product the consumers said they most often purchased (a kind of validity check).

At this point it appears that the scales perform similarly, with a slight advantage to the LAM scale, which could be considered a viable alternative to the traditional 9-point scale, especially if well-liked foods are to be compared. There has been some discussion of whether the high end anchor should refer to the greatest imaginable like (dislike) for any kind of sensory experience (food or non-food) or whether the anchor should be more general or refer specifically to "foods like this." Using a more extreme end anchor (any imaginable sensory experience of any kind) will result in compression of the ratings toward the center of the scale, a context effect (Cardello et al., 2008). Because response compression is generally undesirable and it would be better to encourage fuller use of the scale, the choice of an extreme high end anchor (such as "greatest imaginable liking for any experience") is best avoided.

14.4.4 Pictorial Scales and Testing with Children

Hedonic scaling can also be achieved using face scales, frequently these are simple "smiley" faces (see Chapter 7) but they may also be more representational, involving animal cartoons (Moskowitz, 1986) or more realistic pictures of adults (Meilgaard et al., 1991). A variety of these pictorial scales can be found in Resurreccion (1998). These scales were invented for use by children

or illiterate persons (Coetzee, 1996). However, young children may not have the cognitive skills to infer that the picture is supposed to indicate their internal responses to the product. Additionally, they may be distracted by the pictures. Kroll (1990) showed that verbal descriptors, the so-called P&K scale worked better with children than either the 9-point hedonic scale or facial scales. The terms in this scale are shown in Fig. 14.5. Kroll urged further exploration of the

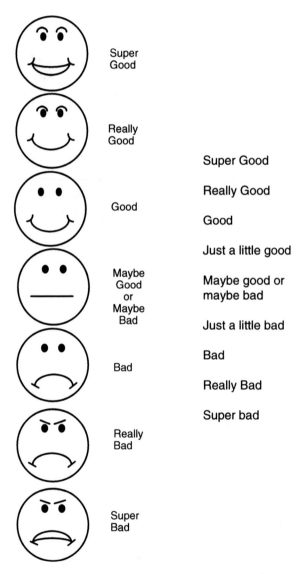

Fig. 14.5 Scales used with children. The *left side* shows an example of a facial scale constructed by Chen et al. (1996). The *right side* shows the super good–super bad verbal scale of Kroll (1990).

P&K scale with low-preference samples and with children under 5 years old (Schraidt, 1991). Chen et al. (1996) showed that 3-point facial hedonic scales with P&K verbal descriptors could be used with 36–47-month-old children, that a 5-point facial version could be used with 47–59-month-old children and that a 7-point version could be used with children 5 years and older. The facial scale used by Chen et al. is shown in Fig. 14.5. The facial scales have a long history of use in the study of food preferences and food habits among children (Birch 1979, Birch et al., 1980, 1982). Pagliarini et al. (2003) and Pagliarini et al. (2005) used an Italian version of the smile scale with verbal labels to study the acceptability of school lunch items and meal item combinations among Italian schoolchildren. Head et al. (1977) found that a 5-point scale (great, good, OK, bad, and terrible) was used reliably by elementary school children in grades 4–6 and secondary school children in grades 10–12.

An alternative to facial scales when testing children is to resort to simple paired preference. Kimmel et al. (1994) concluded that children as young as 2 years old could reliably perform a paired preference test if the appropriate environment and a one-on-one verbal test protocol were used. These authors also found that a 7-point facial hedonic scale anchored with words ranging from "super good" to "super bad" could be used consistently by children as young as 4 years old. Schmidt and Beauchamp (1988) also observed that 3-year-old children could reliably indicate their preference for odors using a paired test involving puppets. Bahn (1989) analyzed preference judgments for cereals made by 4- and 5-year-old children and by 8- and 9-year-old children using multidimensional scaling. Brand names had little effect on children's preferences and most preferences were based on sensory-affective responses to the cereals. Perceptual maps from the younger and older children were similar.

Preference or acceptance testing with children can be done with a few modifications from the adult methods. These often include the following: (1) one-on-one testing in most cases, to insure compliance, understanding, and to minimize social influences, (2) children can respond to either verbal scales or pictorial scales, (3) scales may need to be truncated for use with younger children, (4) paired preference testing is suitable for very young children in the ranges about 4–5 years. Below that age, likes and dislikes must be inferred from behaviors, such as counting oral contacts

in an ad lib situation (Engen, 1974, 1978; Lawless et al., 1982–1983) or from ingestion or sucking (Engen et al., 1974).

14.4.5 Adjustable Scales

Two kinds of adjustable scales have appeared in the literature although to our knowledge they have not found wide acceptance in industrial practice. Gay and Mead proposed a method of scaling in which consumers would look at all the products to be scaled, and place the highest at the top of the scale, the lowest at the bottom, and then partition all of the others at appropriate intermediate marks on the scale (Gay and Mead, 1992; Mead and Gay, 1995), much like a ranking. The advantage of this method is that it eliminates differences among respondents in their choice of what scale range to use, as everyone uses the whole range of the scale. The disadvantage is that the scale is truly relative, i.e., no absolute information about degree of liking is obtained, only the relative positions of products. Although this can be applied to evaluating the perceived intensity of an attribute (like sweetness), and perhaps is most sensible for that purpose, it can be used for hedonics as well (Villanueva and Da Silva, 2009).

Another adjustable scaling method is the "rank rating" method (Kim and O'Mahony, 1998). In this method, the category scale is represented pictorially on the table in front of the consumer. Each product is tasted and the cup or sample is placed on the table in its appropriate category. As the participant proceeds through the test, they are allowed to change the position of products already rated. There are thus two important aspects of this procedure that could potentially enhance product discrimination. First, the consumer can see where the previous products were rated and second, they can change their minds. If the first product was placed too high or too low on the scale, relative to the position of the second product, the situation can be remedied. Whether this procedure is advantageous remains to be seen as it does not have an extensive record at this point. The initial experiments using intensity ratings of salt solutions showed fewer reversals, defined as a high concentration of NaCl being rated lower on the scale (Kim and O'Mahony, 1998). Comparisons to the 9-point hedonic scale have shown only small advantages (if any) to the rank-rating procedure (Cordonnier and Delwiche, 2008; O'Mahony et al., 2004). Like the adjustable scale of Gay and Mead, rank rating may add some degree of relativity to the ratings (as opposed to product having absolute meanings regarding degree of liking). Both of these methods could be more susceptible to context effects and sequential dependencies (see Chapter 9).

14.5 Just-About-Right Scales

14.5.1 Description

A popular scale that combines intensity and hedonic judgments is the just-right or just-about-right (JAR) scales (Rothman and Parker, 2009). These scales, shown in Fig. 14.6 are bipolar, having opposite end-anchors and a center point. The end anchors are "Too little" and "Too much" of a specific attribute or a phrase such as "Too sweet" and "Not sweet enough." The center point can be labeled "just-right" but because it is felt that the choice of "just-right" may entail too strong a commitment on the part of the respondent, the center choice is usually rephrased as "just-about-right." The just-right scale is designed to measure the consumer's reaction to a specific attribute. For example, a just-right scale anchored with "Not salty enough" on the left, "Just-right" in the center and "Much too salty" on the right was used by Shepherd et al. (1989) to evaluate soups.

The just-right scales are popular for the direct information that they give on specific attributes to be optimized. Product developers like this information and so do managers. Furthermore, the concept of deviation from ideal taps into a basic decision when we react to products. For example, people commonly say that the coffee is too strong or too weak or that the wine is too sweet or too tannic. Whether we are aware of it or not, our opinions can be affected by what we expect and what we would like to obtain in terms of the sensory stimulation from a product. Many sensory continua have an optimum or "bliss point" (McBride, 1990). Booth has formulated a quantitative theory of food quality based upon the deviation of sensory attributes from their ideal levels (Booth, 1994, 1995). Using regular hedonic scaling, the "bliss

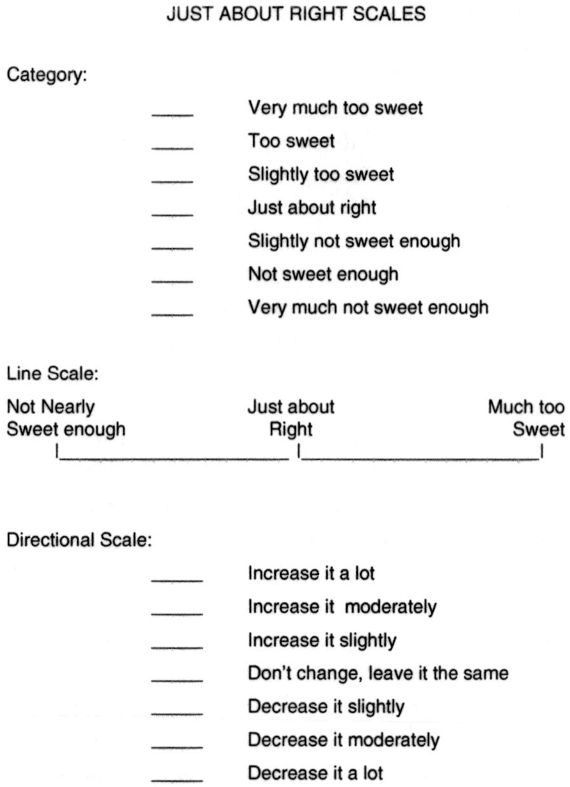

Fig. 14.6 Just-about-right scales. *Top*: a simple category scale. *Center*: an unstructured line scale. *Bottom*: a direction-of-change scale. Further examples can be found in Rothman and Parker (2009).

point" appears as a peak in a non-monotonic function. The just-right data, however, "unfold" this function as shown in Fig. 14.7. Sometimes the unfolded function is linear or at least monotonic, leading to simpler modeling or curve fitting.

The most obvious application of this information is in optimization of a product's key attributes. Intensity and hedonic judgments are combined to provide directional information for product re-formulation. In a consumer test, this can be part of the final field test to insure that no gross errors have been made in the product formulation. The JAR scale gives information that can be diagnostic or explanatory if the overall product appeal is lacking. Earlier in the product development process, the scales can be useful in comparing different versions of the product. Another useful piece of information can be the identification of different segments of consumers who prefer different levels of a sensory attribute. When combined

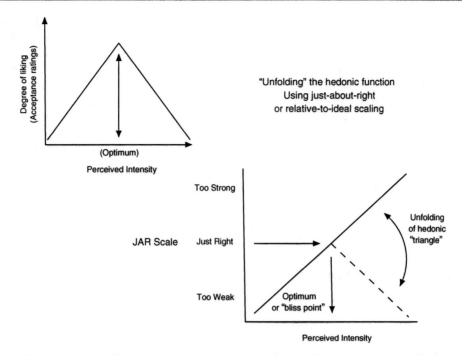

Fig. 14.7 "Unfolding" of the peaked hedonic function by the just-right scale. This can produce a linear relationship of just-right scores against sensory intensity or against ingredient con- centrations (usually on a log scale). The slope of this line, relative to error, is indicative of the tolerance of the consumer for deviations from the ideal level.

with hedonic judgments, the potential impact of being off from the just-right point can be estimated using "penalty analysis" (discussed below). Another advan- tage is that these scales give directional information for change and can do so by testing only a single product (no complicated designed experiments are required). Some direct comparisons have been made concern- ing the performance of JAR scales versus other more traditional methods and no consistent advantage is apparent at this time (Bower and Boyd, 2002; Epler et al., 1997; Popper and Kroll, 2005). A good review of the issues involved can be found in Van Trijp et al. (2007).

14.5.2 Limitations

There are several concerns and pitfalls to be aware of when using JAR scales. The use of JAR scales assumes that all the consumers understand what the attribute listed on the score sheet is referring to. In other words, the consumers must have a common idea or consen- sus understanding of the attribute in question. This limits the just-right scale to a few simple attributes that are widely understood such as sweetness and salti- ness. Other more technical descriptive attributes that require training would be unsuitable in a consumer test. Of course, JAR scales should not be used with trained panels as the judgment is about product likes and dislikes.

The endpoints must be true opposites. "Too thin" versus "Too thick" is a legitimate example. But "Too sour" versus "Too sweet" are not opposites even though they may show a negative correlation (as one goes up the other goes down) in a product. These should be separate scales. Avoid complex attributes like "creamy" which are made up of several com- bined qualities (smoothness, slipperyness, mouthcoat- ing, viscosity, and dairy aroma can all contribute to creaminess). Avoid inherent negatives like bitterness unless they are actually desired in a product such as beer. Bitterness in milk has no optimum. Avoid vague positive terms like "natural." As with all scales in con- sumer tests, avoid redundancy. It does not make sense to have separate JAR scales for thin and thick when they are true opposites.

One must also be careful with the actions that are taken after obtaining JAR scale information. Any attempt to reduce or increase the intensity of an attribute may decrease the product's acceptance among those people who felt it was just-right. Also, the JAR ratings do not indicate how much to change the product to get a better result. Finally, foods and beverages are complex systems and any change in one attribute is likely to affect others. It is difficult to change sweetness without altering the sourness of a product due to taste mixture interactions, for example. Other problems and issues are discussed below.

14.5.3 Variations on Relative-to-Ideal Scaling

There are several types of scales that have used this idea of a central optimal point for the intensity of an attribute. A simple line scale was used by Johnson and Vickers (1987) and Vickers (1988) to study the optimization of sweetness. It was labeled "not nearly sweet enough" at the left end, "just-right" at the center, and "much too sweet" at the right. Pokorny and Davidek (1986) gave examples of several just-right scales to optimize the most important attributes of a product. In one case, the scale points were labeled to show how the product should be changed to be improved. At one end the response label was (the attribute should be) "very much stronger" and the opposite end was "very much weaker." The center point was "without change; it is optimal" and intermediate points were labeled "very slightly stronger" (or weaker), "slightly stronger" (or weaker), and "much stronger" (or weaker). Data were depicted graphically in terms of the percent of respondents giving the "optimal" response for each of nine key characteristics and an improvement in an appetizer product after reformulation was shown. Note that this phrasing is reversed from the common just-right scale where the descriptors refer to difference from ideal whereas in Pokorny and Davidek's example, the descriptors refer to the direction of change that would bring the product back toward the ideal level.

Another variation is to present a normal intensity scale for judging the product, and then ask respondents to place a second mark on the scale for their "ideal" product (van Trijp et al., 2007). An early example of

this can be found in Szczesniak and Skinner (1975) in using a modified texture profile for consumers, where the ideal values for a whipped topping were also scored. This approach presents several advantages. First, the absolute intensity information is obtained as well as where the ideal product lies for that person on a scale. The second advantage is that the individual's scores can be expressed as deviation from ideal, so the just-right directional information can be obtained. Finally, and perhaps most importantly, an "ideal product profile" can be constructed if the data from the panelists are reasonably homogeneous, as well as mean deviations of each the test products from this ideal profile. The major limitation to this approach lies in the abilities of untrained consumers to act in such an analytical fashion, whether they can understand the terms that are being scaled and whether they can report their true feelings about an ideal product in the necessary detail.

A direct approach to product optimization is to have consumers adjust the level of some ingredient until they feel it is optimal (Pangborn and Pecore, 1982). This technique is commonly referred to as a "method of adjustment." However, this procedure must be performed in both directions, concentrating and diluting, to avoid context and/or adaptation effects which cause a consumer to stop too soon in the series (Mattes and Lawless, 1985). Hernandez and Lawless (1999) adapted the method of adjustment for liquid and solid food systems by sequentially weighing the amount added, then subtracting the amount consumed before the next addition in order to estimate the given concentration on any trial.

14.5.4 Analysis of JAR Data

When should product modifications be made based on JAR data? JAR data are actionable when there is an insufficient proportion of responses in the just-right category, when the data are asymmetric, or when there is evidence of a bimodal distribution. When should the current product be deemed acceptable at the current level? Obviously, a desirable set of responses on the just-right scale is centered on the optimum, is symmetric, and has low frequencies in the extremes of the scale. So the first step in analysis is to examine the frequencies across the scale, usually by plotting the data

Fig. 14.8 Graphing JAR data. The *upper panel* shows a simple histogram in which product 456 is centered on the JAR category and symmetric, suggesting no change needs to be made. Product 873 is not symmetric (skewed) and has a large segment of "somewhat not sweet enough" suggesting an increase in the sweetener level could be an improvement. The *lower panel* shows the same data graphed as a constant sum bar graph.

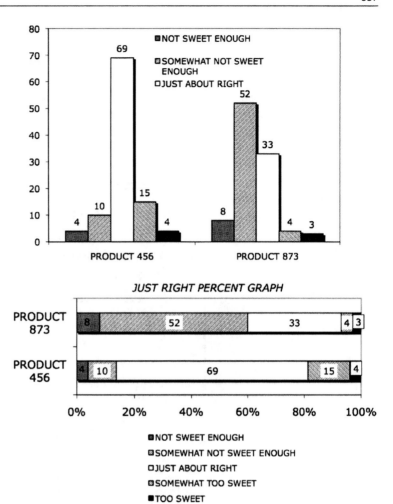

as a simple bar graph (histogram) or a fractionated bar as shown in Fig. 14.8. One could easily be misled by examining only the means. For example, there might be two segments of consumers, one group that prefers a stronger level of the attribute and another group that prefers less. Examining only the mean from such opinions could lead to a false impression that the product was at or near the optimum level. So plot your data. The second question is, "Do I have enough just-right votes to leave the product as-is?" A common benchmark is something like 80% JAR votes in the center category (Rothman and Parker, 2009). One quantitative test for being skewed away from the just-right point is a simple one-sample *t*-test of the product's mean versus the center point's value on the scale. A simple nonparametric test is to compare the frequencies of those

who are above JAR to those below JAR, while ignoring the JAR votes (Stone and Sidel, 2004). This can be a simple binomial probability test against an expected value of 0.5 (equal proportions). Both the *t*-test for the mean and the binomial test can provide evidence that the data are skewed higher or lower than the midpoint, for a single product.

The next consideration is how to compare multiple products to see if any are different or closer to the ideal. If each consumer has evaluated all the products (a within-subjects design or complete block), the chi-square statistic based on a cross-tabulation is not appropriate as it is based on the assumption that the data are from independent samples. Since the evaluations are related, several alternatives are available. Fritz (2009) provides a good discussion of these methods and worked examples. For more than two products, the Cochran-Mantel-Haenszel ("CMH") method can be used, but it is computationally intensive and requires statistical software. For comparing any two products, the data can be cross-tabulated by the frequencies in a 3×3 table, after collapsing the data into three categories (above just-right, just-right, below just-right)

for each product. This is shown in Fig. 14.9. Then the appropriate statistic is the Stuart–Maxwell test, as shown in the figure and Appendix B (Best and Raynor, 2001; Fleiss, 1981). If the data can be collapsed further, the simple McNemar test can be applied. For example, one might suspect that it is the "too sweet" category that is different for the two products. Then

Stuart Maxwell Calculations

1. Entries A thorough I are the cell totals.

2. Average the off-diagonal pairs, P1 - P3

3. Find differences of row and column totals:

$$D1=C1-R1 \quad D2=C2-R2 \quad D3=C3-R3$$

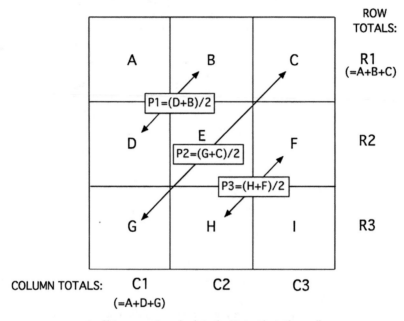

4. Chi-square is calculated. Note that the cell averages (P1, P2, P3) are multiplied by the squared differences (D1, D2, D3) of row and column totals in which they DO NOT participate.

$$\text{Chi-square} = \frac{[(P1)(D3)^2 + (P2)(D2)^2 + (P3)(D1)^2]}{2[(P1)(P2) + (P2)(P3) + (P1)(P3)]}$$

Fig. 14.9 The Stuart–Maxwell test for the differences between two products assessed on JAR scales. The data are first collapsed into three categories, those above just-right (i.e., too strong), those below just-right (i.e., too weak), and those falling in the just-right or just-about-right category. Entries A through I are the frequencies falling into each category considering their ratings for both products. The χ^2 test is done against

a critical value for 2 df, which is 5.99. This value must be exceeded for a significant difference to be obtained. After a finding of a significant difference, the rows and columns can be collapsed into a 2 × 2 table for further analysis by the McNemar test to see if there are particular cells driving the Stuart–Maxwell result. See Appendix B for a worked example.

the McNemar can be applied to a 2×2 table pitting "too sweet" against the combined frequencies of the other categories. If the scales are continuous like a line scale or have contain seven or nine categories, then one can treat them as any other scaled response, with parametric statistics like t-tests and analysis of variance to compare means. If there are only a few response categories the data should be treated as categorical or ordinal.

14.5.5 Penalty Analysis or "Mean Drop"

Another source of valuable information can be from overall acceptance ratings collected in the same questionnaire. The JAR data can be combined with this information to assess the potential impact of being non-JAR (off from just-right) on the overall acceptability of the product. The approach is simple and proceeds as follows (from Schraidt, 2009):

(1) Separate the data into groups that were above, below, and at the just-right category.
(2) Calculate the mean hedonic scores from the acceptability scales for each of the three groups.
(3) Subtract the mean of the above-JAR group from the JAR group and likewise subtract the mean of the below-JAR group from the JAR group. (Note: it is important to use the JAR group's mean and not the overall data mean for this purpose.)
(4) Plot the resulting difference scores, the "mean drop," or penalty in a scatter plot of the mean drop versus the percentage of the total consumer panel in each category.

In this plot, a point that shows a large mean drop and a large percentage is a cause for concern and suggests that the product be modified in the appropriate direction. An example is shown in Table 14.2 and plotted in Fig. 14.10. In this plot we see that there is a large group who felt the product was too sweet and had a large mean drop or reduction in the overall acceptance score. The product development team might want to increase the sweetness level in a new version. There were two large groups, one of which felt the product was too thick and one too thin, but their penalties were small so no action need be taken. Regarding the fruit

Table 14.2 Data for penalty analysis and mean drop in Fig. 14.10

	Mean	Drop	%
Attribute Sweetness			
Too sweet	6.2	–1.4	15
Just-about-right	7.6		50
Not sweet enough	5.2	–2.4	35
Thickness			
Too thick	6.5	–0.7	28
Just-about-right	7.2		37
Too thin	6.2	–1.0	35
Fruit flavor			
Too strong	5.2	–2.6	17
Just-about-right	7.8		63
Too weak	6.5	–1.3	20

flavor strength, there was a small but strongly dissatisfied group in the upper left corner. Further research might be warranted to see whether a different style of product with a milder flavor would appeal to this

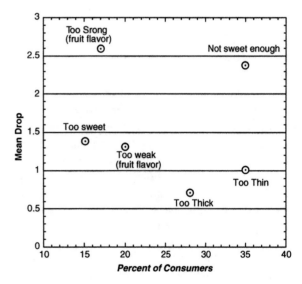

Fig. 14.10 Penalty analysis and mean drop for one hypothetical product. For each of the three scales shown in Table 14.2, consumers are categorized as rating each scale above, below or in the just-right category. The mean hedonic score (9-point scale value) is then calculated for each group. For groups above or below the JAR group, their scores are subtracted from the mean of the JAR group. This produces a "mean drop" score for each group. The percent of the consumer panel that fell into each category are plotted against their mean drops. Note that the category "not sweet enough" has both a high percentage and high mean drop, suggesting a potential improvement with increased sweetness.

group, perhaps leading to a lighter styled version for a consumer segment.

Various statistical tests and methods for modeling the penalty analysis are given in Rothman and Parker (2009). A simple approach is to make a 2×2 classification table of the above/below JAR versus likers/dislikers for the product and perform a chi-square test (Templeton, 2009). A significant result would suggest that one kind of JAR response was more detrimental to the product than the other. A parametric comparison could compare the group of above-JAR consumers to the below-JAR consumers to see whether their acceptance scores were in fact different (i.e., an independent groups t-test).

14.5.6 Other Problems and Issues with JAR Scales

In spite of their apparent simplicity, JAR scales are subject to a number of complications. These issues are discussed in Rothman and Parker (2009). Bear in mind that JAR scales only work with attributes where there is an optimum. If an attribute is a case of "more is always better" or "any of this is bad" then JAR scales are not appropriate. As noted before, one potential problem is that consumers may misinterpret the attribute. An example is the confusion that many people have concerning sour and bitter. A second concern is that consumers are generally integrative when they perceive a product as opposed to being analytical, and the JAR scale asks them to attend to a specific attribute in isolation. Consumers often show halo effects, in which one salient attribute can affect the ratings of other, logically unrelated attributes. Thus is it possible that a product could be rated less than ideal on sweetness, merely because the consumer is annoyed by some other taste or flavor. They can also bring in cognitive biases that have nothing to do with their own likes and dislikes. For example, if a person believes that salt is bad for you, the product could be rated as too salty, even if it is ideal. Some attributes are linked or may show trade-offs. Increasing sweetness may decrease sourness but the sweet/sour balance may be key as it is in some wines and fruit products. Some attributes may change with time or the hedonic reaction may change with time. Sweetness may seem acceptable at first, but may be cloying after consuming the entire

portion. Portion size effects may influence JAR ratings. What seems good in a small bite may not seem so appealing in a larger portion. Other problems can occur when JAR scales are used in isolation without asking any additional intensity-related questions. Thus two groups of respondents might both mark "just-right" but one might think the product is very strong (the level they prefer) while the other group thinks the product is fairly mild (the level they prefer). So the results might mislead product developers into thinking that there is a homogeneous population while there are really two consumer segments. The sensory professional should consider the usefulness of collecting both intensity and JAR information on the consumer questionnaire.

The centering tendency or bias is particularly problematic for JAR scales, especially in a multi-product test. The centering bias in this case is a tendency to put the product which is intermediate in intensity (i.e., the middle product) at the just-right point. This could lead to a false conclusion that the middle product was just-right when in fact the true optimum could still be higher or lower. Johnson and Vickers (1988) compared several methods for dealing with the centering problem, based on approaches of McBride (1982) and Poulton (1989). Both of these methods involve testing multiple ranges of products to interpolate the JAR point based on different ranges of the key ingredient. They are discussed further in Chapter 9. It may not always be possible to do the multiple sessions required to interpolate the true optimum, so the sensory scientist should be aware that a false reading for the middle product in a series can be a problem with this method.

14.6 Behavioral and Context-Related Approaches

It is difficult to measure a stable attitude toward a food when it is tasted in isolation. Our likes and dislikes may certainly change as a function of the context of a meal, the time of day, or the number of times we have consumed the food recently. Foods and beverages may be liked more or less depending upon the temperature at which they are served (Kahkonen et al., 1995) and whether that temperature is consistent with consumer's expectations (Cardello and Maller, 1982a). A person's recent history with similar foods can be an important influence. Some people are prone to seek a high degree of variety in their diet (van Trijp et al., 1992) and may tire of eating foods with consistent or similar sensory properties (Vickers, 1988). Eating steak or lobster may seem highly appealing in the abstract, but consuming steak or lobster for ten days in a row will certainly change a person's feelings about the item toward the end of that period. Preferences are also specific to foods that are combined. So although we may like ketchup in general, ketchup on mashed potatoes may or may not seem appealing to a given individual. A person's historical preferences across many foods may fail to predict their hedonic acceptance scores for items in an actual tasting (Cardello and Maller, 1982b). Context and expectations can affect simple hedonic judgments (Deliza and MacFie, 1996; Meiselman, 1992; Schifferstein, 1995), so we should expect them to have important effects in actual food choice and consumption situations. The eating situation can have a big effect (Edwards et al., 2003; King et al., 2007). Habit, experience, context, and attitudes are important contributors to the actual consumption of a food in a specific situation. To address some of these limitations of simple acceptance testing some more behaviorally oriented approaches have been utilized.

14.6.1 Food Action Rating Scale (FACT)

An example of a behaviorally oriented approach to scaling food acceptability was devised by Schutz (1965). He developed a scale based on attitudes and actions, combining some statements about frequency of consumption ("I would eat this food every opportunity I had") and some motivationally related statements ("I would eat this food only if I were forced to") to produce a more action-oriented index of food acceptance. This was called the Food Action Rating Scale or FACT. The complete list of descriptors is shown in Table 14.3. Although Schutz reasoned that the behaviors might not always match up with acceptance as scaled on a traditional 9-point hedonic scale, a study of the correspondence of the two measures showed a high degree of positive correlation ($r = +0.97$ in a questionnaire study of food likes). Data from the FACT scale gave lower mean values but less skew compared to the 9-point hedonic scale. In spite of this correlation, the scales are not necessarily interchangeable.

Table 14.3 Descriptors in the food action rating scale (FACT)

I would eat this food every opportunity I had
I would eat this very often
I would frequently eat this
I like this and would eat it now and then
I would eat this if available but would not go out of my way
I do not like it but would eat it on an occasion
I would hardly ever eat this
I would eat this only if there were no other food choices
I would eat this only if I were forced to

After Schutz (1965)

14.6.2 Appropriateness Scales

In related work, Shutz carried the contextual theme one step farther. Appropriateness ratings can be used to assess the effects of context associated with hedonic responses to food (Schutz, 1988). For example, one may really like pizza but when asked to rate one's liking at 8 A.M. may not find the pizza appealing since this is an inappropriate time to consume pizza, at least for most people. So pleasantness on a purely sensory basis and appropriateness in a context may not always be completely parallel. Inappropriate contexts for a given culture may override influences of sensory liking (Lahteenmaki and Tuorila, 1997). While it is certainly possible to poll consumers about the appropriateness of foods in different contexts, the sensory scientist should be sensitive to the testing burden that can arise in asking too many questions. If 15 foods are evaluated for appropriateness in 20 contexts each consumer panelist would have to rate 300 scales.

Appropriateness judgments are traditionally done for both a food and a usage statement like "eaten for breakfast." The scale commonly used is a 7-point scale ranging from "never" to "always." Note that this is not a scale that ranges from "not appropriate" to "very appropriate." Rather, it attempts to tap into a notion of the frequency with which an item would apply for that usage situation, in a graded fashion. The questionnaire often takes the form of a matrix or grid, with foods as rows and usage statements as columns. A good review of appropriateness questions and sample applications can be found in Schutz (1994). In some studies comparing appropriateness to the 9-point hedonic scale, Cardello and Schutz (1996) showed that foods with equal acceptability could differ dramatically in their appropriateness for various situations. Although this is perhaps not surprising, it should be noted that liking and appropriateness are often highly correlated. Foods that are highly liked find appropriate uses in a variety of situations. However, the work of Cardello and Schutz demonstrates that they are not equivalent. This study also noted that acceptability scores were unaffected by asking the appropriateness questions. So there seems to be no harm in asking for this additional information, keeping in mind the added burden on test subjects due to the length of the questionnaire.

Data from appropriateness judgments can be analyzed by principal component analysis to determine the underlying features common to the different foods and contexts. Jack et al. (1994) used a repertory grid triad method to derive all potential use occasions for cheeses. The appropriateness of the use of each of 16 cheeses was then evaluated on line scales anchored by suitable and unsuitable. They found that the melting characteristics and the texture of the cheeses were the major factors affecting the appropriateness of cheese for different occasions. In consumer products, it is often important to have fragrances that are consonant with the intended use of a product. The type of fragrance that is appropriate in a shampoo may not work well as a fragrance used to mask insecticide chemical smells and vice versa (Jellinek, 1975). So in screening candidate fragrances for specific product applications, it is necessary to go beyond simple hedonics and ask people about the fit of the smell to the intended product. A fruity-floral fragrance may be appealing as an air freshener, but may seem out of place in an institutional heavy-duty sanitizing cleaner. An appropriateness scale in this kind of situation would range from "not appropriate" to "very appropriate" in contrast to the frequency scale of Schutz.

14.6.3 Acceptor Set Size

Another variation in hedonic assessment brings the food acceptance measure back to a simple count of the proportion of people who find the product appealing. Norback and colleagues (Beausire et al., 1988; LaGrange and Norback, 1987) examined the proportion of acceptors as a variable in product optimization. The acceptor set size was defined as the proportion of

consumers who find the product acceptable. LaGrange and Norback (1987) reasoned quite logically that the causal chain in optimizing the sensory appeal of a product should consider an acceptability measure, such as the proportion of acceptors, to be driven by a set of sensory attributes. The sensory attributes in turn would be determined by the physical characteristics of the food, ingredient variables, and so on. This is simply a straightforward psychophysical model extended to include multiple attributes and a hedonic or behavioral consequence. They reasoned that those variables that have a strong impact on changing the acceptor set size (i.e., produce a steep slope or high rate of change) would be most influential in optimization. This is much in the tradition of modeling acceptance as a function of a set of contributing attributes (e.g., Moskowitz and Krieger, 1995). Beausire et al. (1988) made use of this approach in a linear programming model to explore the relationship between acceptor set size, toughness scores, and ingredients in turkey bratwurst. Various ingredient combinations were explored for their effects on the toughness attribute and the resulting acceptability function.

This approach makes use of some minimal information, basically whether people find the product acceptable or not. In the Beausire et al. (1988) study, participants were simply asked to decide "yes" or "no" as to whether they found the product acceptable. On the positive side, this measure taps into a fundamental concern for food marketers, the size of the pool of potential users. On the negative side, the measure is dichotomous and lacks the graded richness of information available on the 9-point acceptance scale. It is also possible to think of this measure as a kind of preference split, but based on only one product versus the consumer's minimal expectations for the category. As Stone and Sidel (2004) pointed out, preference may not always match up with acceptance. People who have no liking at all for the category do not belong in the test, but they may express a preference if mistakenly included in the study. It is also possible to have a preference win for a product based on the proportion of people preferring one product over another. However, a minority with strong opinions in the opposite direction may give the preferred product low hedonic ratings with the result that the mean acceptability scores are reversed from the preference split. For example, product A was higher than B in acceptance, but product B was preferred over A in simple choice (see Stone and Sidel, 2004, for numerical example). This same problem could potentially arise with the acceptor set. There might be fair proportion of people finding the product acceptable, but a group that dislikes it strongly since it is not in their favorite or accustomed style. Acceptor set size and simple preference/choice might miss this fact. Examination of acceptability scores, on the other hand, would uncover both the proportion and the degree of dislike in the minority segment.

14.6.4 Barter Scales

Another approach to food acceptance was developed for use in predicting meal combinations. Lawless (1994) examined meal combinations for military field rations. The dependent measure was based on the number of chocolate bars that a soldier would trade for individual items and for meal combinations. This measure was based on the observation that trading for food items within the prepackaged meals did occur in the field that the chocolate bar was a highly desirable item and that it could function as a kind of "common coinage" for swapping. The overall goal was to develop a measure that could predict the value or total appeal of meal combinations and then to use this measure along with nutritional information and other constraints in developing a linear optimization model for ration improvement. One concern of the product developers was that the 9-point scale was too constrained and would lack the additive properties necessary for an estimate of total hedonic value in meal combinations. For example, two items might be rated an eight and a five on the hedonic scale, but there is no scale point corresponding to 13 or the sum of the reactions to the two items since the scale is bounded by nine points. It was hypothesized that the barter value of the items and meal combinations would show more reasonable patterns of additivity that could be used in linear programming.

The values in chocolate bars of 33 individual items were estimated and then they were combined to form two-item combinations and five-item meals, similar or identical to actual field package combinations. The question of interest was whether the values of the meals and combinations could be predicted by the sum or some simple linear combination of the values of the items. The data showed almost perfect additivity with one notable exception. The data showed an "a la carte"

effect. That is, the value expected in trade for the meal was one chocolate bar less than the sum of the individual items. This is similar to the common pricing of meals versus a la carte or individual items where the sum of the individual prices would be higher than the price paid for the actual meal as a combination. The utility of this method may depend upon a consistent positive value of the bartered item among the participants.

14.7 Conclusions

The assessment of blind-labeled product acceptability is one of the cornerstones of sensory evaluation. Consumer acceptance is essential "bottom line" information for product developers and marketers alike. A variety of useful methods are available to the sensory scientist in order to assess the appeal of products and the relative preferences among a set of choices. Choice itself is fundamental to consumer behavior, as it is the decision process that we all make when faced with a number of different foods for purchase or for use in a meal. Acceptance can be related to other properties of foods such as the descriptive profile of a product or to physical ingredient, processing and packaging variables. We can study how the appeal of a product declines over periods of storage, in the distribution chain and over its shelf life.

In spite of the obvious importance of product acceptability, this group of sensory methodologies is prone to misuse, misinterpretation, and challenges from other fields. In particular, marketing researchers may not understand the importance of testing products on a blind-labeled basis, i.e., with a minimal conceptual framework. The rule of thumb for sensory research is to present only enough information so that the product is evaluated within the correct frame of reference, usually the product category. So participants in a sensory acceptance test are given a product with a random three-digit code, they know only that it is a test of scrambled eggs, and they are unaware that the eggs are frozen, reconstituted, microwaved, cholesterol reduced, or any number of other factors that might eventually form part of the product concept. The sole question is whether they are appealing to scrambled egg consumers on the basis of their sensory attributes (taste, texture, appearance, etc.).

A common complaint among marketers is that the blind test is unrealistic since the product will not appear on the store shelves in its unlabeled form without package and concept (see Garber et al., 2003, for a critique of sensory practice from a marketing perspective). However, they miss the point. The sensory test is the only way to assess true sensory appeal without the biasing effects of conceptual labeling (Gacula et al., 1986) and as such it gives essential feedback to product developers as to whether they have truly met their target profiles. Without the blind test, no one can tell. The product may succeed or fail in the concept-laden market research test for any number of reasons. Given the tendency of consumers to integrate their information, to show halo effects and other biases, you cannot always trust their stated reasons for liking in a questionnaire from a concept test. As stated above, high-sensory acceptability does not insure that the product will be a marketplace success. Purchase probability (and more importantly, repurchase) depends upon price, concept, positioning, promotions, advertising, package information, consumer awareness, nutritional characteristics, and many other factors (Garber et al., 2003). However, the sensory appeal is the essential "platform" without which the product is unlikely to succeed. This platform—of sensory-based acceptance—provides the foundation for successful marketers to then apply their artistry to sell the product to consumers in the real world.

References

Bahn, K. D. 1989. Cognitive and perceptually based judgments in children's brand discriminations and preferences. Journal of Business and Psychology, 4, 183–197.

Bartoshuk, L. M., Snyder, D. J. and Duffy, V. B. 2006. Hedonic gLMS: Valid comparisons for food liking/disliking across obesity, age, sex and PROP status. Paper presented at the 2006 Annual Meeting, Association for Chemoreception Sciences.

Beausire, R. L. W., Norback, J. P. and Maurer, A. J. 1988. Development of an acceptability constraint for a linear programming model in food formulation. Journal of Sensory Studies, 3, 137–149.

Best, D. J. and Rayner, J. C. W. 2001. Application of the Stuart test to sensory evaluation data. Food Quality and Preference, 12, 353–357.

Birch, L. L. 1979. Dimensions of preschool children's food preferences. Journal of Nutrition Education, 11, 77–80.

Birch, L. L., Birch, D., Marlin, D. W. and Kramer, L. 1982. Effects of instrumental consumption on children's food preferences. Appetite, 3, 125–143.

Birch, L. L., Zimmerman, S. I. and Hind, H. 1980. The influence of social-affective context on the formation of children's food preferences. Child Development, 51, 865–861.

Borg, G. 1982. A category scale with ratio properties for intermodal and interindividual comparisons. In: H.-G. Geissler and P. Petzold (eds.), Psychophysical Judgment and the Process of Perception. VEB Deutscherverlag der Wissenschaften, Berlin, pp. 25–34.

Booth, D. A. 1994. Flavour quality as cognitive psychology: The applied science of mental mechanisms relating flavour descriptions to chemical and physical stimulation patterns. Food Quality and Preference, 5, 41–54.

Booth, D. A. 1995. The cognitive basis of quality. Food Quality and Preference, 6, 201–207.

Bower, J. A. and Boyd, R. 2002. Effect of health concern and consumption patterns on measures of sweetness by hedonic and just right scales. Journal of Sensory Studies, 18, 235–248.

Byer, A. J. and Saletan, L. T. 1961. A new approach to flavor evaluation of beer. Wallerstein Laboratory Communications, 24, 289–300.

Cardello, A. V. and Maller, O. 1982a. Acceptability of water, selected beverages and foods as a function of serving temperature. Journal of Food Science, 47, 1549–1552.

Cardello, A. V. and Maller, O. 1982b. Relationships between food preferences and food acceptance ratings. Journal of Food Science, 47, 1553–1557, 1561.

Cardello, A. V. and Schutz, H. G. 1996. Food appropriateness measures as and adjunct to consumer preference/acceptability evaluation. Food Quality and Preference 7, 239–249.

Cardello, A. V. and Schutz, H. G. 2004. Research note. Numerical scale-point locations for constructing the LAM (Labeled affective magnitude) scale. Journal of Sensory Studies, 19, 341–346.

Cardello, A., Lawless, H. T. and Schutz, H. G. 2008. Effects of extreme anchors and interior label spacing on labeled magnitude scales. Food Quality and Preference, 21, 323–334.

Cardello, A. V., Schutz, H. G., Lesher, L. L. and Merrill, E. 2005. Development and testing of a labeled magnitude scale of perceived satiety. Appetite, 44, 1–13.

Cardello, A. V., Winterhaler, C. and Schutz, H. G. 2003. Predicting the handle and comfort of military clothing fabrics from sensory and instrumental data: Development and application of new psychophysical methods. Textile Research Journal, 73, 221–237.

Chen, A. W., Resurreccion, A. V. A. and Paguio, L. P. 1996. Age appropriate hedonic scales to measure food preferences of young children. Journal of Sensory Studies, 11, 141–163.

Chung, S.-J., and Vickers, A. 2007a. Long-term acceptability and choice of teas differing in sweetness. Food Quality and Preference, 18, 963–974.

Chung, S.-J., and Vickers, A. 2007b. Influence of sweetness on the sensory-specific satiety and long-term acceptability of tea. Food Quality and Preference, 18, 256–264.

Coetzee, H. 1996. The successful use of adapted paired preference, rating and hedonic methods for the evaluation of acceptability of maize meal produced in Malawi. Abstract, 3rd Sensometrics Meeting, June 19–21, 1996, Nantes, France, pp. 35.1–35.3.

Cordonnier, S. M. and Delwiche, J. F. 2008. An alternative method for assessing liking: Positional relative rating versus the 9-point hedonic scale. Journal of Sensory Studies, 23, 284–292.

Deliza, R. and MacFie, H. J. H. 1996. The generation of sensory expectation by external cues and its effect on sensory perception and hedonic ratings: A review. Journal of Sensory Studies, 11, 103–128.

Edwards, J. A., Meiselman, H. L., Edwards, A. and Lesher, L. 2003. The influence of eating location on the acceptability of identically prepared foods. Food Quality and Preference, 14, 647–652.

El Dine, A. N. and Olabi, A. 2009. Effect of reference foods in repeated acceptability tests: Testing familiar and novel foods using 2 acceptability scales. Journal of Food Science, 74, S97–S106.

Engen, T. 1974. Method and theory in the study of odor preferences. In: A. Turk, J. W. Johnson, Jr. and D. G. Moulton (eds.), Human Responses to Environmental Odors. Academic, New York, pp. 121–141.

Engen, T. 1978. The origin of preferences in taste and smell. In: J. H. A. Kroeze (ed.), Preference Behaviour and Chemoreception. Information Retrieval, London, pp. 263–273.

Engen, T., Lipsitt, L. P. and Peck, M. 1974. Ability of newborn infants to discriminate sapid substances. Developmental Psychology, 10, 741–744.

Epler, S., Chambers, E. and Kemp, K. E. 1997. Hedonic scales are better predictors than just right scales of optimal sweetness in lemonade. Journal of Sensory Studies, 13, 191–197.

Fleiss, J. L. 1981. Statistical Methods for Rates and Proportions, Second Edition. Wiley, New York.

Forde, C. G. and Delahunty, C. M. 2004. Understanding the role cross-modal sensory interactions play in food acceptability in younger and older consumers. Food Quality and Preference, 15, 715–727.

Fritz, C. 2009. Appendix G: Methods for determining whether JAR distributions are similar among products (Chi-square, Cochran-Mantel-Haenszel (VMH), Stuart-Maxwell, McNemar). In: L. Rothman and M. J. Parker (eds.), Just-About-Right Scales: Design, Usage, Benefits, and Risks. ASTM Manual MNL63, ASTM International, Conshohocken, PA, pp. 29–37.

Gacula, M. C., Rutenbeck, S. K., Campbell, J. F., Giovanni, M. E., Gardze, C. A. and Washam, R. W. 1986. Some sources of bias in consumer testing. Journal of Sensory Studies, 1, 175–182.

Garber, L. L., Hyatt, E. M. and Starr, R. G. 2003. Measuring consumer response to food products. Food Quality and Preference, 13, 3–16.

Gay, C., and Mead, R. 1992 A statistical appraisal of the problem of sensory measurement. Journal of Sensory Studies, 7, 205–228.

Green, B. G., Shaffer, G. S. and Gilmore, M. M. 1993. Derivation and evaluation of a semantic scale of oral sensation magnitude with apparent ratio properties. Chemical Senses, 18, 683–702.

Greene, J. L., Bratka, K. J., Drake, M. A. and Sanders, T. H. 2006. Effective of category and line scales to characterize consumer perception of fruity fermented flavors in peanuts. Journal of Sensory Studies, 21, 146–154.

Greenhoff, K. and MacFie, H. J. H. 1994. Preference mapping in practice. In: H. J. H. MacFie and D. M. H. Thomson (eds.), Measurement of Food Preferences. Blackie Academics, London, pp. 137–166.

Guest, S., Essick, G., Patel, A., Prajapati, R. and McGlone, F. 2007. Labeled magnitude scales for oral sensations of wetness, dryness, pleasantness and unpleasantness. Food Quality and Preference, 18, 342–352.

Head, M. K., Giesbrecht, F. G. and Johnson, G. N. 1977. Food acceptability research: Comparative utility of three types of data from school children. Journal of Food Science, 42, 246–251.

Hein, K. A., Jaeger, S. R., Carr, B. T. and Delahunty, C. M. 2008. Comparison of five common acceptance and preference methods. Food Quality and Preference, 19, 651–661.

Helgensen, H., Solheim, R. and Naes, T. 1997. Consumer preference mapping of dry fermented lamb sausages. Food Quality and Preference, 8, 97–109.

Hernandez, S. V. and Lawless, H. T. 1999. A method of adjustment for preferred levels of capsaicin in liquid and solid food systems among panelists of two ethnic groups and comparison to hedonic scaling. Food Quality and Preference, 10, 41–49.

Hottenstein, A. W., Taylor, R., and Carr, B. T. 2008. Preference segments: A deeper understanding of consumer acceptance or a serving order effect? Food Quality and Preference, 19, 711–718.

Hough, G., Bratchell, N. and Wakeling, I. 1992. Consumer preference of Dulce de Leche among students in the United Kingdom. Journal of Sensory Studies, 7, 119–132.

Jack, F. R., Piggott, J. R. and Paterson, A. 1994. Use and appropriateness in cheese choice, and an evaluation of attributes influencing appropriateness. Food Quality and Preference, 5, 281–290.

Jellinek, J. S. 1975. The Use of Fragrance in Consumer Products. Wiley, New York.

Johnson, J. and Vickers, Z. 1987. Avoiding the centering bias or range effect when determining an optimum level of sweetness in lemonade. Journal of Sensory Studies, 2, 283–292.

Johnson, J. R. and Vickers, Z. 1988. A hedonic price index for chocolate chip cookies. In: D. M. H. Thomson (ed.), Food Acceptability. Elsevier Applied Science, London, pp. 135–141.

Jones, L. V., Peryam, D. R. and Thurstone, L. L. 1955. Development of a scale for measuring soldiers' food preferences. Food Research, 20, 512–520.

Kahkonen, P., Tuorila, H. and Hyvonen, L. 1995. Dairy fact content and serving temperature as determinants of sensory and hedonic characteristics of cheese soup. Food Quality and Preference, 6, 127–133.

Keskitalo, K., Knaapila, A., Kallela, M., Palotie, A., Wessman, M., Sammalisto, S., Peltonen, L., Tuorila, H. and Perola, M. 2007. American Journal of Clinical Nutrition, 86, 55–63.

Kim, K.-O. and O'Mahony, M. 1998. A new approach to category scales of intensity I: Traditional versus rank-rating. Journal of Sensory Studies, 13, 241–249.

Kimmel, S. A., Sigman-Grant, M. and Guinard, J.-X. 1994. Sensory testing with young children. Food Technology, 48(3), 92–94, 96–99.

King, S. C., Meiselman, H. L., Hottenstein, A. W., Work, T. M. and Cronk, V. 2007. The effects of contextual variables on food acceptability: A confirmatory study. Food Quality and Preference, 18, 58–65.

Koster, E. P., Couronne, T. Leon, F. Levy, C. and Marcelino, A. S. 2003. Repeatability in hedonic sensory measurement: A conceptual exploration. Food Quality and Preference, 14, 165–176.

Kroll, B. J. 1990. Evaluating rating scales for sensory testing with children. Food Technology, 44(11), 78–80, 82, 84, 86.

Lagrange, V. and Norback, J. P. 1987. Product optimization and the acceptor set size. Journal of Sensory Studies, 2, 119–136.

Lahteenmaki, L. and Tuorila, H. 1997. Item by use appropriateness of drinks varying in sweetener and fat content. Food Quality and Preference, 8, 85–90.

Lawless, H. T. 1977. The pleasantness of mixtures in taste and olfaction. Sensory Processes, 1, 227–237.

Lawless, H. T. 1994. Contextual and measurement aspects of acceptability. Final Report #TCN 94178, U. S. Army Research Office.

Lawless, H. T. 2010. Commentary on "Comparative performance of the nine-point hedonic hybrid and self-adjusting scales in generation of internal preference maps." Food Quality and Preference, 21, 165–166.

Lawless, H. T. and Malone, G. J. 1986. A comparison of scaling methods: Sensitivity, replicates and relative measurement. Journal of Sensory Studies, 1, 155–174.

Lawless, H. T., Hammer, L. D. and Corina, M. D. 1982–1983. Aversions to bitterness and accidental poisonings among preschool children. Journal of Toxicology: Clinical Toxicology, 19, 951–964.

Lawless, H. T., Popper, R. and Kroll, B. 2010. Comparison of the labeled affective magnitude (LAM) scale, an 11-point category scale and the Traditional nine-point hedonic scale. Food Quality and Preference, 21, 4–12.

Lavanaka, N. and Kamen, J. 1994. Magnitude estimation of food acceptance. Journal of Food Science, 59, 1322–1324.

Mattes, R. D. and Lawless, H. T. 1985. An adjustment error in optimization of taste intensity. Appetite, 6, 103–114.

McBride, R. 1990. The Bliss Point Factor. Macmillan, South Melbourne, NSW (Australia).

McBride, R. L. 1982. Range bias in sensory evaluation. Journal of Food Technology, 17, 405–410.

McDaniel, M. R. and Sawyer, F. M. 1981. Preference testing and sensory evaluation: Magnitude estimation vs. the 9-point hedonic scale. Journal of Food Science, 46, 182–185.

McEwan, J. 1996. Preference mapping for product optimization. In: Multivariate Analysis of Data in Sensory Science. Elsevier Applied Science, London, pp. 71–102.

Mead, R. and Gay, C. 1995. Sequential design of sensory trials. Food Quality and Preference, 6, 271–280.

Meiselman, H. L. 1992. Methodology and theory in human eating research. Appetite, 19, 49–55.

Meilgaard, M., Civille, G. V. and Carr, B. T. 1991. Sensory Evaluation Techniques, Second Edition. CRC, Boca Raton.

Moskowitz, H. R. 1980. Psychometric evaluation of food preferences. Journal of Foodservice Systems, 1, 149–167.

Moskowitz, H. R. 1986. New Directions for Product Testing and Sensory Analysis of Foods. Food and Nutrition, Westport, CT.

Moskowitz, H. R. and Krieger, B. 1995. The contribution of sensory liking to overall liking: An analysis of six food categories. Food Quality and Preference, 6, 83–90.

O'Mahony, M., Park, H., Park, J. Y. and Kim, K.-O. 2004. Comparison of the statistical analysis of hedonic data using analysis of variance and multiple comparisons versus and R-index analysis of the ranked data. Journal of Sensory Studies, 19, 519–529.

Pagliarini, E., Gabbiadini, N. and Ratti, S. 2005. Consumer testing with children on food combinations for school lunch. Food Quality and Preference, 16, 131–138.

Pagliarini, E., Ratti, S., Balzaretti, C. and Dragoni, I. 2003. Evaluation of a hedonic method for measuring the acceptability of school lunches by children. Italian Journal of Food Science, 15, 215–224.

Pangborn, R. M. and Pecore, S. D. 1982. Taste perception of sodium chloride in relation to dietary intake of salt. American Journal of Clinical Nutrition, 35, 510–520.

Pearce, J. H., Korth, B. and Warren, C. B. 1986. Evaluation of three scaling methods for hedonics. Journal of Sensory Studies, 1, 27–46.

Peryam, D. R. 1989. Reflections. In: Sensory Evaluation. Celebration of our Beginnings. ASTM, Committee E-18 on Sensory Evaluation of Materials and Products, Philadelphia, pp. 21–30.

Peryam, D. R. and Girardot, N. F. 1952. Advanced taste test method. Food Engineering, 24, 58–61, 194.

Peryam, D. R. and Pilgrim, F. J. 1957. Hedonic scale method of measuring food preferences. Food Technology, September 1957, 9–14.

Pokorny, J. and Davidek, J. 1986. Application of hedonic sensory profiles for the characterization of food quality. Die Nahrung, 8, 757–763.

Popper, R. and Kroll, B. R. 2005. Just-about-right scales in consumer research. Chemo Sense, 7, 1–6.

Poulton, E. C. 1989. Bias in Quantifying Judgments. Lawrence Erlbaum Associates, Hillsdale, NJ.

Resurreccion, A. V. A. 1998. Consumer Sensory Testing for Product Development. Aspen, Gaithersburg, MD.

Rohm, H. and Raaber, S. 1991. Hedonic spreadability optima of selected edible fats. Journal of Sensory Studies, 6, 81–88.

Rothman, L. and Parker, M. J. 2009. Just-About-Right Scales: Design, Usage, Benefits, and Risks. ASTM Manual MNL63, ASTM International, Conshohocken, PA.

Schifferstein, H. J. N. 1995. Contextual shifts in hedonic judgment. Journal of Sensory Studies, 10, 381–392.

Schmidt, H. J. and Beauchamp, G. K. 1988. Adult-like odor preference and aversions in three-year-old children. Child Development, 59, 1136–1143.

Schraidt, M. F. 1991. Testing with children: Getting reliable information from kids. ASTM Standardization News, March 1991, 42–45.

Schraidt, M. 2009. Appendix L: Penalty analysis or mean drop analysis. In: L. Rothman and M. J. Parker (Eds.), Just-About-Right Scales: Design, Usage, Benefits, and Risks. ASTM Manual MNL63, ASTM International, Conshohocken, PA, pp. 50–47.

Schutz, H. G. 1965. A food action rating scale for measuring food acceptance. Journal of Food Science, 30, 365–374.

Schutz, H. G. 1988. Beyond preference: Appropriateness as a measure of contextual acceptance. In: D. M. H. Thomson (ed.), Food Acceptability. Elsevier, London, pp. 115–134.

Schutz, H. G. 1994. Appropriateness as a measure of the cognitive-contextual aspects of food acceptance. In: H. J. H. MacFie and D. M. H. Thomson (eds.), Measurement of Food Preferences. Chapman and Hall, pp. 25–50.

Schutz, H. G. and Cardello, A. V. (2001). A labeled affective magnitude (LAM) scale for assessing food liking/disliking. Journal of Sensory Studies, 16, 117–159.

Shepherd, R., Smith, K., and Farleigh, C. A. 1989. The relationship between intensity, hedonic and relative-to-ideal ratings. Food Quality and Preference 1, 75–80.

Stone, H. and Sidel, J. 2004. Sensory Evaluation Practices, Third Edition. Elsevier Academic, San Diego.

Szczesniak, A. S. and Skinner, E. Z. 1975. Consumer texture profile method. Journal of Food Science, 40, 1253–1256.

Templeton, L. 2009. Appendix R: Chi-square. In: L. Rothman and M. J. Parker (eds.), Just-About-Right Scales: Design, Usage, Benefits, and Risks. ASTM Manual MNL63, ASTM International, Conshohocken, PA, pp. 75–81.

Tepper, B. J., Shaffer, S. E. and Shearer, C. M. 1994. Sensory perception of fat in common foods using two scaling methods. Food Quality and Preference, 5, 245–252.

van Trijp, H. C. M., Lahtennmaki, L. and Tuorila, H. 1992. Variety seeking in the consumption of spread and cheese. Appetite, 18, 155–164.

van Trijp, H. C. M., Punter, P. H., Mickartz, F. and Kruithof, L. 2007. The quest for the ideal product: Comparing different methods and approaches. Food Quality and Preference, 18, 729–740.

Vickers, A. 1988. Sensory specific satiety in lemonade using a just right scale for sweetness. Journal of Sensory Studies, 3, 1–8.

Villanueva, N. D. M. and Da Silva, M. A. A. P. 2009. Performance of the nine-point hedonic, hybrid and self-adjusting scales in the generation of internal preference maps. Food Quality and Preference, 20, 1–12.

Villanueva, N. D. M., Petenate, A. J. and Da Silva, M. A. A. P. 2005. Comparative performance of the hybrid hedonic scale as compared to the traditional hedonic, self-adjusting and ranking scales. Food Quality and Preference, 16, 691–703.

Wakeling, I. N. and MacFie, H. J. H. 1995. Designing consumer trials balanced for first and higher orders of carry-over effect when only a subset of k samples from t may be tested. Food Quality and Preference, 6, 299–308.

Wright, A. O. 2007. Comparison of Hedonic, LAM, and other scaling methods to determine Warfighter visual liking of MRE packaging labels, includes web-based challenges, experiences and data. Presentation at the 7th Pangborn Sensory Science Symposium, Minneapolis, MN, 8/12/07. Supplement to Abstract Book/Delegate Manual.

Yao, E., Lim, J., Tamaki, K., Ishii, R., Kim, K.-O. and O'Mahony, M. 2003. Structured and unstructured 9-point hedonic scales: A cross cultural study with American, Japanese and Korean consumers. Journal of Sensory Studies, 18, 115–139.

Chapter 15

Consumer Field Tests and Questionnaire Design

Abstract   This chapter presents an introduction to consumer testing in various settings including central location and home use tests. The construction of a useful consumer questionnaire requires both skill and experience. General rules for questionnaire design and question construction are presented. Various question formats such as agree–disagree scales and open-ended questions are discussed.

Developing products is easy, developing products that appeal to consumers is less so, and developing products that appeal to a sufficient number of consumers and achieve commercial success based on specific business criteria is very difficult.

—Stone and Sidel (2007)

Contents

15.1 Sensory Testing Versus Concept Testing

Why do so many products fail in the competitive marketplace? To minimize new product failures, one strategy is to insure that the consumer perceives, through the senses and repeated experiences, the characteristics

Table 15.1 Sensory tests versus product concept tests

Test characteristic	Sensory test	Product concept test
Conducted by	Sensory Evaluation Department	Marketing research Department
Primary end user of information	Research and Development	Marketing
Product labeling	Blind-minimal concept	Full-conceptual presentation
Participant selection	Users of product category	Positive response to concept

that make the company's product superior to competitors' products and thus more desirable. Furthermore, this perception must be maintained to build brand loyalty and to insure repurchase. The sustainability of the perception of quality is important long after the initial rush of interest from advertising claims and promotions that surround a new product introduction. The purpose of this chapter is to discuss the techniques for consumer product testing in the field on a blind-labeled basis that insure the sustainable perception of positive product characteristics.

A consumer field test with new product prototypes or market candidates can provide several pieces of useful information to product developers. The blind-labeled sensory test can be an important step before a multi-city marketing research field test and product "launch." It provides an opportunity to determine consumer acceptability on a sensory basis, without the concepts and claims that will normally appear in advertising or on packaging. The sensory consumer test can facilitate diagnosis of problems before more expensive marketing tests. Costly mistakes can be avoided and problems uncovered that may not have been caught in laboratory tests or more tightly controlled central location tests. It can provide direction for re-formulation if needed. Multiple formulas or candidates can be compared on a blind-labeled, pure performance basis. Poorly performing products can be dropped from further consideration. Finally, since the tests are done with target consumers, the company can obtain data that may be used for claim substantiation. This can be valuable in defending challenges from competitors and responding to the requirements of advertising regulators.

At first glance, a consumer sensory evaluation field test looks a lot like the consumer tests done in marketing research. It is worth understanding some of the important differences. The research arm of a consumer product manufacturer may rely on the important technical support provided by the sensory consumer test. It provides validating data about reaching goals in terms of sensory factors and the perception of product performance by consumers. Important differences exist between sensory consumer tests and the typical marketing research consumer tests. Some of these are shown in Table 15.1. In both tests products will be placed with consumers and their opinions surveyed after a trial period. However, they differ in the amount of information given the consumer about the product and its conceptual features.

The marketing research "product concept" test usually proceeds as follows: participants are shown the product concept, via a storyboard or videotape mockup that often resembles a rudimentary advertisement for the idea. They are then questioned about their response and expectations of the product based on the presentation. Note that this is important strategic information for marketers. Next, *those that respond positively* are asked to take the product home and use it, and later to evaluate its sensory properties, appeal and performance relative to expectations. This selection may appear biased, but it is based on the idea that people who do not like your idea to begin with are probably not part of the target market. In the sensory-oriented consumer tests the conceptual information is kept to a minimum. The rule of thumb is to give only enough information to ensure proper use of the product and evaluation relative to the appropriate product category. For example, the product may be labeled simply "pizza" or "frozen pizza." In a market research test, it might be being evaluated after a conceptual presentation that communicates a number of features such as "new—improved—low-fat—high fiber—whole-wheat—stuffed-crust—convenient—microwavable—pizza." Ratings of attributes and product acceptance are contaminated by any additional information or by the expectations that are built up as a function of showing the detailed product concept. The sensory product test attempts to ascertain their perceptions about the sensory properties in isolation from other influences. These other influences can be quite profound. For example, the introduction of brand identity or other information can produce differences in the apparent acceptability of products where there is no differentiation on a blind-tested basis (Allison and Uhl, 1964; El Gharby, 1995; Gacula et al., 1986).

Sensory consumer tests and marketing product concept tests also differ in the participant selection. Only those who show an interest and react positively to the product concept are usually included in the actual product test in the marketing research scenario. As these participants have shown an initial positive bias, it is perhaps not surprising when the product receives good scores in the test. On the other hand, the sensory consumer test merely screens participants for being users of the product category. Given these differences, the two types of tests may give different evaluations of the consumer appeal of the product. The tests provide different types of information, viewed from different frames of reference by the consumers and they use different pools of respondents. In the concept test, product perception may be biased in the direction of assimilation toward their expectations (Cardello and Sawyer, 1992). Results from both types of tests may be equally "correct;" they are simply different techniques seeking different types of information. Both types of information should be weighed in management decisions to go forward or to seek further modifications to optimize the product.

Critics of the sensory approach often remark that the product will never be seen on the store shelf with a generic description and a three-digit code, so why bother evaluating it in the blind-labeled, concept-free form? The answer is simple. Suppose the product fails in the marketplace? How does one know what went wrong if only the product concept test was performed? Perhaps it did not have good sensory properties or perhaps the marketplace did not respond to the concept as predicted by the marketing tests. Without the sensory test, the situation is ambiguous and the direction for fixing the product is unclear. The research team may have designed a poor product that was only carried along by a catchy concept. However, after extended use, consumers may have figured out that the product does not deliver benefits in keeping with their expectations and they stopped buying it. On the other hand, the marketing team may have simply designed a poor concept that somehow moved forward in the rush of initial enthusiasm for a new product idea (Oliver, 1986). The research and development team has a need to know if their efforts at meeting sensory and performance targets were successful in an unambiguous blind test.

The following sections of this chapter are devoted to how consumer tests are conducted, emphasizing field testing and questionnaire construction. Although there is a substantial literature on survey techniques and questionnaire design for marketing research and for opinion polling, there is little published research on sensory field testing of consumer products. Product placement and interviewing in the field is a complicated, expensive, and time-consuming enterprise. Training is often obtained by "shadowing" an experienced researcher in industry. However, there are some general guides to consumer sensory tests (Schaefer, 1979; Sorensen, 1984). The book, *Consumer Sensory Tests for Product Development*, by Resurreccion (1998) contains guidelines and much practical advice. It provides detailed information on topics such as checklists for conducting various types of consumer tests, sample questionnaires, project management guidelines, and maintaining consumer databases of test participants for recruitment. Another resource dealing specifically with the issues and methods for claim substantiation is the extensive ASTM standard E-1958 (ASTM, 2008).

15.2 Testing Scenarios: Central Location, Home Use

15.2.1 Purpose of the Tests

The primary goal of a consumer field test is to assess the acceptability of a product or group of products or to determine whether a product is preferred over other products. Some typical situations that justify a consumer field test are (1) a new product entering the marketplace, (2) a reformulated product, that is, ingredient, process or packaging changes of a major nature, (3) entering a competitor's product category for the first time, or (4) competitive surveillance, as in a category appraisal (see Chapter 19). It is also an opportunity to collect some diagnostic information on the reasons behind consumer likes and dislikes. Reasons for liking are usually probed with a variety of techniques such as open-ended questions, agree–disagree scales, and just-right scales. Agreement with label claims of a perceptual nature (e.g., "crispier"), assessing consumer expectations and consumer satisfaction with the product can all be surveyed through questionnaires and interviews.

Four general categories of consumer sensory tests are often distinguished. The first is the use of internal panels checking acceptance using on-site testing, usually with employees. The second is the use of a local standing consumer panel. These are people, sometimes from social groups that are recruited for multiple tests for a period of time. We will refer to these two types of panels as "consumer models." The third is the central location test or CLT and the fourth is the home use test or HUT.

15.2.2 Consumer Models

A variety of consumer testing situations are used to assess the appeal and overall acceptability of products. Due to resource constraints in time, money, or security concerns, there are several types of acceptance testing that use what can be termed "consumer models." Such "consumer" groups may consist of employees or local residents, but often there is little or no attempt to insure that the group is representative of consumers at large. It is of course essential that the group be users of the product category. It would make no sense to ask people about the appeal of several extruded puffed breakfast cereals if they never consume such a product.

Internal consumer tests are tests conducted in the sensory test facility of a company or research department using employees. A major liability of employee panels is that they are not necessarily blind to the brand of the product and they may have potentially biasing information and assumptions about what is being tested (Resurreccion, 1998). Technical personnel may view the product quite differently from consumers, focusing on entirely different attributes. Such an internal consumer panel should be routinely compared to an outside sample of non-employee consumers by testing the same products with both groups. Stone and Sidel (2004) describe the use of a split-plot ANOVA with internal and external panelists evaluating the same products to assess the degree of agreement. Unfortunately this is rarely done during product development, because the diagnostic information from the early tests with the internal panel is used to make adjustments or optimizations before subsequent expensive field tests. So the product changes as it moves through the development process.

Another cost-efficient approach using consumer models is to use local standing consumer panels. The oldest record of a local standing panel that we could find was the use of 300 families in the Columbus, Ohio area by the Ohio State University and Ohio Agricultural Experiment Station in the 1950s (Gould et al., 1957). The families were stratified by socioeconomic class via census records of rental costs and they participated in simple preference tests. This panel was similar to the Kroger company's mail panel of the 1940s, described by Garnatz (1952), which had a more geographically diverse makeup. Test products were delivered to their homes and questionnaires returned by mail. Another way to recruit and set up standing panels is through community groups. These groups may be affiliated with schools, churches, fraternal, or hobby-oriented clubs (e.g., singing groups) or virtually any other organization that meets in a nearby location on a regular basis. Social groups can be used for central location tests, sometimes in their own facilities (Schaefer, 1979) or to facilitate distributions for home placement. They can be contacted through a panel leader or coordinator for product and questionnaire distribution, offering some time savings. Such panels are re-used for a period of time, so like the employee consumer models, they offer convenience and time savings in recruiting respondents and testing products on a routine basis. Incentives can be directed to the organization itself, so there can be social pressure, a powerful motivator, to participate.

However, there are several liabilities with a locally recruited and ongoing consumer panel. First, the sample is not necessarily representative of opinions beyond the delimited geographic area of the club or group. Second, the participants may know each other and talk to each other on a regular basis, so there is no guarantee that the opinions are completely independent. Using a variety of random codes for products may reduce this liability, but there is no airtight guarantee that their judgments are not influenced by others. Finally, unless an outside agency or a disguised testing lab is used for the contact and distribution, the participants may become aware of what company is conducting the test. Opinions or pre-existing attitudes about products from the originating company may bias results. If they view the company favorably, they may evaluate its products more favorably. As in any consumer acceptance test, the participants should be carefully screened for regular usage of the product category. That is, you must eliminate those members of the group who are not regular users even if it engenders some disappointment among people who were looking forward to being in the test. This possibility must be spelled out during orientation sessions after the group is recruited.

In spite of the apparent savings in recruitment and completion time for tests, the setup and maintenance of a local ongoing consumer panels can require considerable effort on the part of the sensory staff. If the program is big enough and has contact with ten or more consumer groups, a full-time staff member may be required to supervise such a program. Pickup and recall of products and questionnaires must be arranged, and returned questionnaires should be carefully examined for any evidence of cheating. People may fill out questionnaires and return them without actually trying the product. Containers returned full or nearly full, questionnaires with illogical answers or use of only one scale point on every question are some hints (Resurreccion, 1998). These respondents should be deleted from the data set and noted for future exclusion or monitoring. Maintaining good relations and close communication with a member of the group acting as the local contact person is key. These group coordinators or contact persons have many of the responsibilities for oversight that a field agency supervisor has, but bear in mind that this is not their profession. Cultivating this important contact person, orienting them to procedures and providing motivation may also take time and considerable social influence on the part of the sensory specialist. Such a program of local panels should also involve periodic furlough of each group and rotation of active participation cycles since some boredom and disinterest in the tests can set in over time. A 6-month active testing cycle within every 18 months to 2 years is reasonable.

In spite of their obvious problems in being non-representative of outside consumers, employee panels and local consumer panels can provide valuable information on a cost-efficient basis. For decades, the US Army has used employee panels at the Natick, Massachusetts laboratories to evaluate military rations and other foods. These panels have been reasonably well predictive of soldiers' opinions of the same foods (Meiselman and Schutz, 2003; Peryam and Haynes, 1957). However, more recent work showed that the correlations were higher for snack foods than main dishes and meal components and were more predictive when the laboratory test involved an element of choice

(de Graaf et al., 2005). In food companies, the risks from using an internal panel can be a bit higher. For a major roll-out of a new product with millions of dollars of advertising to be spent, it is much safer if not imperative to continue testing with a true consumer field test, in a home use scenario in multiple locations.

15.2.3 Central Location Tests

Probably the most popular type of consumer test with foods involves product trials at a central location. The central location test (CLT) is often conducted in the facilities of a field testing agency (a service provider), for example, in a shopping mall. However, there are just about as many variations on this theme as one could imagine including testing in retail outlets, recreational facilities, and schools (Resurreccion, 1998; Sorensen, 1984). A picture of a CLT setup classroom style is shown in Fig. 15.1. Consumers can come to a corporate sensory test lab, although that defeats the non-branded nature of the blind sensory test since the company identity is obvious. Having consumers on the company property may entail some security risks. If the testing program is extensive, it may be economically justifiable to set up a disguised test facility actually administered by the company's own sensory program, rather than subcontracting to outside testing services. A sensory group can use a mobile testing laboratory to change locations. The University of Georgia

Fig. 15.1 A consumer central location test being conducted in a classroom style arrangement. Photo courtesy of Peryam and Kroll Research.

maintained such a mobile testing laboratory moved by a truck (Resurreccion, 1998). This offers enormous flexibility in the opportunities for consumer contact. For example, foods targeted at summer picnics or outdoor cooking could be tested at or near campgrounds or parks. Such on-site testing can introduce a realistic element in the testers' frame of reference. Products aimed at children could be brought to school locations and the mobile test laboratory can provide a site for proper product preparation and controlled presentation. For some foods, special considerations for preparation are minimal and tests can be conducted at any site where people are gathered in large numbers and have flexible time. For example, sensory testing can be done at state fairs or other recreational events.

The central location test offers conditions of reasonably good control of product preparation as the staff can be trained in product preparation and handling. Compliance with instructions, manner of examining samples, and ways of responding may be monitored and controlled (Resurreccion, 1998; Schaefer, 1979). It is easier to isolate respondents in test booths or separate areas to minimize outside influence. Security can be maintained more easily than in a home placement. The tradeoff occurs against the necessarily limited product usage, i.e., participants' exposure to the product is much shorter than in the home placement and usually only limited amounts of the product are tasted or consumed (Schaefer, 1979). It is of course possible that the limited product interaction in a CLT can give erroneous results or lead to faulty conclusions (Oliver, 1986).

15.2.4 Home Use Tests (HUT)

The most expensive but most realistic situation is when consumers take the product home and try it under normal circumstances on several occasions. Home use tests are time consuming to set up and administer. They can be costly, especially if external field testing services are hired to do most of the work. However, HUTs offer tremendous advantages in terms of face validity of the data. This can be important in advertising claim support. Also, the opinions of other family members can enter the picture as they do in everyday use of purchased products. The primary advantage is that the consumer uses the product over a period of

time and can examine its performance on several occasions before forming an overall opinion. For foods, this becomes less of an issue where flavor, appearance, and texture are rapidly appreciated and the hedonic reaction of a person is virtually immediate upon tasting the item. For consumer products such as a shampoo or a floor wax, it may be critically important to have some extended use in the actual home and to see how the treated substrate (in this case the hair or the floor) holds up over time. Home placement provides a chance to look at the product in a variety of settings (Anon., 1974). Another important opportunity is the chance to test product and packaging interactions. Some products may be well or poorly suited to their package design (Gacula et al., 1986) and the home use test provides an excellent chance to probe this. Finally, the home use test can facilitate a more critical assessment of the product relative to the consumer's expectations.

In the case of product fragrance testing, the short exposure in a central location may overestimate the appeal of very sweet or perfume-like fragrances. When used in the home for extended periods, such fragrances may become cloying and a type of hedonic fatigue can set in, even though they score well in briefer laboratory sniffing tests. In general, it may be dangerous to screen fragrance candidates in lab-based sniffing tests, especially for functional products. A very appealing fragrance sniffed from a bottle may not communicate a message of efficacy with industrial cleaning products or insecticides. For functional products, the fragrance has to be chosen to support the perception of product efficacy. Mismatches can occur in flavors as well, for example, there is some resistance to candy-like flavors in toothpastes (Jellinek, 1975). Similarly, very sweet food products may score well in a central location, but do less well when used over an extended period.

In summary, four major categories of consumer tests are commonly used—employee consumer models, local standing consumer models, central location tests, and home placements. The remainder of the chapter will focus on field tests and on questionnaire design. Table 15.2 shows the characteristics of these levels of consumer testing and how they vary. The employee tests are the quickest, least expensive, and the most secure but have the greatest liabilities in terms of potential bias, lack of a representative sample, and lack of realism in the testing situation. The choice of test in any specific situation usually represents a compromise between time and expense on the one hand and

Table 15.2 Kinds of consumer tests

Type	Advantages	Disadvantages
Internal employee panel	Secure	Not representative
	Low cost	
	Rapid results	
Local standing panels	Reasonably secure	Not a random sample
	Lower (?) cost	Panelists may discuss products
	Easy distribution	
Central location test	Representative sample	Requires test agencies
	Control over product preparation	Cost, slower results
Home use test	Representative sample	Requires test agencies
	Realistic testing	Slowest to conduct, costliest
	Whole family input	Lack of product control
	Can test use directions	Security risk

the need for the most valid information on the other. The business risk of basing major decisions on less valid testing situations should be weighed against the cost of more extensive testing.

15.3 Practical Matters in Conducting Consumer Field Tests

15.3.1 Tasks and Test Design

A number of considerations will enter into the design of a home use test, many of which need to be negotiated with the client or end-users of the information and also with any field testing services that collect data. Some of the primary decisions of the sensory professional will include sample size, experimental design, qualification of participants, choice of locations and agencies, and structure of the interview or questionnaire. There are several dozen activities and decision points in setting up and conducting a field test, rendering this type of test as one of the most complicated projects that a sensory specialist may perform. The most important decisions that affect the experimental design include the number of consumer required, the number of products, and how the products will be compared. Statistical consultants, if needed, should be brought in at this stage. The specific tasks involved in conducting a consumer field test are discussed below. Resurreccion (1998) provides checklists for the various types of consumer tests, home use, central location, etc. These can be very useful for the sensory specialist who is new to consumer field tests and may not realize

the added levels of complexity compared to simple in-house acceptance/preference testing. Field tests are costly and require a high level of attention to detail. Simple mistakes can render the results of a test invalid (Schutz, 1971), at potentially great expense if the test must be repeated.

15.3.2 Sample Size and Stratification

In this instance, sample size refers to the numbers of consumer participants, not the size of the portion or amount of product served. How many people should be in the test? More powerful tests are less likely to miss a real difference or an important effect and having a sufficient sample size is the first concern in test design. A statistical consultant can help with estimates of test power, but there are ultimately some subjective decisions to be made about the size of a difference one is willing to miss, or conversely, that one must be sure to detect. This decision is akin to determining how big of a difference is practically meaningful or what small differences can safely be ignored. Once the effect size is specified, the probability of detection (one minus beta-risk) also must be chosen. This is called the power of the test (see Appendix E). These may be difficult concepts for management to understand, unless they have had extensive statistical training and a good deal of practical experience. The level of variability also affects the test power but it can be used as a kind of yardstick. The size of the difference you wish to detect can be stated in terms of standard deviations. A reasonable rule of the thumb for the level of error in consumer tests is that standard deviations will be in the range of 20–30% of the scale (or about two points on a 9-point scale, which has only eight intervals). The range of variability can be slightly lower for intensity scales than for hedonic scales. Variability will also be lower for "easy" attributes, like those having to do with appearance or some simple texture attributes, as opposed to taste characteristics or olfactory or aromatic attributes, which are the most difficult of all. Given this rule of thumb, it makes sense to have a sample size in the range of 75–150 persons (per product) for most tolerable levels of risk. A generally useful equation for evaluating the required sample size based on scaled data (like the 9-point hedonic scale) is shown here:

$$N = \frac{(Z_\alpha + Z_\beta)^2 S^2}{(\mu_1 - \mu_2)^2} \qquad (15.1)$$

where N is the number of consumers needed in the test, Z_α and Z_β are the Z-scores associated with your chosen levels of alpha- and beta-risk, S is the anticipated standard deviation of the scores (or a pooled estimate), and $\mu_1 - \mu_2$ is the difference between means or the size of the difference you want to be sure to detect.

Can a test be too big? Although some marketing groups are prone to do tests with hundreds or even thousands of respondents, this derives from a false sense of security in numbers (Stone and Sidel, 2004). There is a law of diminishing returns with sample size and statistical power, just as there is in interviewing in general. The largest amount of information is obtained from the first few interviews and additional testers yield less and less new information (Sorensen, 1984). It is also possible to have a test that is too sensitive, i.e., to show statistically significant results in an area that is of little practical consequence to consumers (Hays, 1973; Schutz, 1971). Stone and Sidel (2004) discuss "the curse of N" in testing in the sense that people put too much faith in large numbers. Statistical significance must be weighed against practical significance. It is unfortunate that the technical meaning of statistical significance refers to issues of confidence and likelihood, while the common everyday meaning is synonymous with "important" (Sorensen, 1984). Management must be reminded of the difference in these usages in order to keep from over-interpreting statistical significance, especially in large test populations. Finally, it is better to have a small test of high quality that is well designed, with careful attention to detail and close monitoring of field agencies than it is to have a sloppy test that uses large numbers of consumers to compensate for the added variability.

Of course, the sampling strategy may not be from a single group. It is sometimes desirable to look at different geographic locations, different demographic strata (e.g., age, gender, income) or groups with different product usage habits (Schaefer, 1979). There are two reasons for stratification of the sample group. The first is to insure a certain amount of diversity in the group so that it mirrors the target population. Thus there may be quotas for men, women, different age brackets, etc. This kind of quota sampling is very important in central location tests where the participants may be recruited through a mall intercept (recruitment on site). The second reason for stratification is that examining differences among these groups may be part of the research plan. If the test groups are stratified in this way, it becomes necessary to increase the total pool to maintain the minimum subgroup size in the range of 50–100 respondents. Obviously, such variables should be chosen very carefully and with solid justification, for they can dramatically increase the size and cost of the test.

15.3.3 Test Designs

There are three primary designs used in consumer testing. Side-by-side tests are sometimes done in which both products are placed simultaneously. These are more often performed in central location tests than in home placements. Under controlled circumstances, the side-by-side test will have great sensitivity, since the same people view both products. Difference scores (as in the dependent or paired t-test) or complete block (repeated measures) analysis of variance can be used to analyze the data. Comparisons are both statistically and perceptually direct. However, putting more than one product out simultaneously in a home placement test can lead to confusion for the participants. There are many chances for errors in following directions for product use, order of evaluation, and questionnaire usage if it is self-administered. The side-by-side evaluation is better suited to situations where the product–person interaction can be controlled and monitored.

More common designs in field tests are the monadic and monadic sequential placements. In the monadic test, only one product is placed with an individual. This is usually a faster test scenario, and can be completed more quickly with fewer drop-outs. However, it requires larger numbers of participants, one group for each product. It may not be practical if the incidence of product use is fairly low or participants are difficult to find and recruit. The statistical comparison between products is necessarily a "between-group" comparison. The opportunity to use consumers as their own baseline of comparison cannot be taken advantage of when different groups are compared. So there is a potential loss of sensitivity in this design due to high inter-individual variability. Conversely, the monadic sequential design permits the use of individuals as their own baseline. Scale usage habits or other individual peculiarities are the same for both products and can be statistically factored out of the analysis. This generally leads to a more sensitive test (discussed in the section on the paired t-test in Appendix A).

In the monadic sequential design, the products are placed one at a time in sequence. A questionnaire is normally administered at the end of each product's usage period, while the sensory characteristics and performance are fresh in the person's memory. Of course, careful counterbalancing of orders across the subgroups is necessary. Bear in mind that the first product used in a monadic sequential test will have the same frame of reference (or lack thereof) as a simple monadic design (Sorensen, 1984). So analysis of the first product used by each person can be informative if there is concern about sequential bias or any order effects on the second or third product. The monadic test leads to higher rates of attrition (non-completion). It does permit a preference question after completion of the second product placement.

Some situations arise in which monadic sequential tests are inapplicable. When the substrate or evaluation process is irreparably or severely altered by the initial product usage, the second placement becomes unworkable. For example, with a pharmaceutical, a personal care product like a hair conditioner or a home insecticide, use of the product may create such changes in the substrate that it is not practical to get a clear picture of product performance for the second product in a sequence. Of course, multiple products can be tested after a "wash-out" or recovery period, as is sometimes done in pharmaceutical tests. This may not be practical for market-driven new product tests where time is of the essence.

The number of products to include is also a consideration in test design. It is possible to test more than two products in sequence or to use incomplete sampling designs like balanced incomplete block designs (Cochran and Cox, 1957; Gacula and Singh, 1984; Gacula et al., 2009) to test a number of possible alternatives. Due to the expense and effort in conducting home placement or central location field tests, the number of alternative formulas should have been reduced to only a few highly promising candidates through earlier phases of testing. One design to avoid is the one-product monadic test, which is really not a test at all, but an exercise in confirmation of the intuition of the project supervisors. A one-product "test" puts far too much faith in the raw value of the scores received. As humans are very poor absolute measuring instruments (see Chapter 9) and prone to context effects, the absolute value of the scores is nearly meaningless, even in comparison to historical data with similar products due to a potential change of context. It is far safer and much more scientifically valid to include a baseline product for comparison. Examples of useful baselines for comparison are an alternative formula, the current product, or a repackaged sample of a competitor's most successful formula.

A final consideration in the test design is whether to include a question of paired preference. In a monadic sequential test, there may be considerable challenge to the participants' memory if a preference question is asked following the use of the last product in the sequence. Due to the possibility of sequence effects, it is also wise to look at preference ratios for each separate order of presentation and not just for the overall test population. Paired preference may still be confirmatory to comparison of (scaled) acceptability scores, so it can be used as an additional source of information in developing the "story" that is told by the test and the scorecard. It is also possible to get conflicting results from acceptance and preference questions from some individuals. This can occur if they change their basis for decision-making between the two questions. For example, the acceptability question may be answered with sensory properties like taste or texture in mind, while the preference question may consider preparation time or some convenience factor. It is also possible that a product may win in a preference comparison, but because there is a highly dissatisfied minority, the acceptance scores for the winning product are lower (an example is given in Stone and Sidel, 2004). There is a widespread belief that the paired preference question is somehow more sensitive than acceptability ratings, but this notion lacks empirical support. Nonetheless, considerable pressure may arise from clients to include a preference question. Practical considerations such as the length of the product usage period, results from pre-testing and needs for information for claim substantiation (claims such as superior to, unsurpassed, equally preferred) may determine whether or not to include a paired preference question.

15.4   Interacting with Field Services

15.4.1   Choosing Agencies, Communication, and Test Specifications

Field services are variously called agencies, vendors, suppliers, and field services. Some workers in this area prefer the term "suppliers" because they are supplying research information. Choosing a good field service or a test agency is largely a matter of experience. In a company with an ongoing product testing program, it makes sense to keep a record of those agencies that deliver timely and cooperative service, demonstrate quality in their interviewing, and show attention to detail in handling of products and questionnaires. The quality of service is not necessarily proportional to cost—high bidders may not always provide the best service (Schaefer, 1979). Costs of field services will depend upon their level of involvement. In some cases there may be two levels of contracting, a primary agency that administers the test, and subcontracting field test agencies in different cities that actually conduct the test under the direction of the primary contractor. It is important to distinguish between full-service suppliers and basic field test sites. Full-service suppliers can provide valuable input on the screening and product questionnaires, the design, execution, analysis, and reporting of results. They act as an extension of your professional team. In other cases, the subcontractors can merely provide a testing service, i.e., product placement and interviewing, and act according to your specific directions.

In each agency, it is important to identify a single person, sometimes called the field service supervisor or project manager, who is ultimately responsible for the conduct and quality of the test. Reporting to this person there are often many part-time employees who may have different degrees of training in interviewing techniques. The nature of interviewing is that it attracts a lot of freelance or part-time workers. They should have excellent interpersonal skills, the ability to follow directions, and a sense of caring and integrity about the quality of the job (Schaefer, 1979). The field service supervisor should visit any subcontracting sites and participate in or view the testing process if at all possible. There has to be good communication, a written test specification sheet, and a briefing of the field sites (well before the actual test) to answer any questions. Good agencies will provide training for the interviewers and a briefing for each test to review respondent qualifications, instructions for sampling, and placement and questionnaire structure. Supervision of the interviewers is important in quality control. Practice interviews have been found to predict field behavior and can be used as a screening device (Blair, 1980). Problem areas include cheating on screening and qualification of improper respondents and faking of part or all of the interview (Boyd et al., 1981). Checkups or validation of a given percentage of the completed questionnaires (usually by phone) should be requested and monitored by the field supervisor.

In using a field service for central location tests, facilities are very important. They must allow for proper experimental control, product serving and preparation, and provide an environment free from distractions and conducive to sensory testing (Schaefer, 1979). If there is a central location test involved, it may be necessary to hire an agency with facilities for product preparation. If agencies derive most of their business from servicing marketing research tests and/or focus groups, they may not be set up for food preparation and serving. So confirming the facilities, preferably by an in-person visit, can be an important detail. Considerations include the ability to isolate respondents to minimize interaction and the resulting loss of independence in their judgments. In a home placement, the agency test facilities are less important. Having a focus group room for follow-up discussions is one consideration if follow-up groups are a part of the project plan. As most testing companies maintain websites, some idea of the nature of the facilities can be assessed from the pictorial material on the website.

After the agency is hired, it is important to make sure all the test details and instructions are communicated in writing. Most of the general facts will have already been communicated in order for the agency to make a cost estimate. Further details are critical to a successful test. Test specification sheets should normally be sent to an agency, giving as many details as possible about the test design, qualification of respondents, quotas, deadlines and services to be delivered, including data tabulation and analysis, if any. Expectations about security, confidentiality, and the professional conduct of interviewers can also be spelled out. Arrangements for product retrieval and disposal can be specified, as well as shipping of completed questionnaires. A sample test specification sheet is shown in Appendix 1.

15.4.2 Incidence, Cost, and Recruitment

It is up to the sensory professional, in consultation with the clients and perhaps with input from the field supervisor, to determine the screening qualifications of the participants. Certainly participants should be users of the product category and usually people who also actually like the product—the two are often overlapping but not completely synonymous. In addition, one needs to determine what level of product usage is sufficient to qualify an individual. A screening questionnaire will normally include several usage frequency categories, in order to eliminate those consumers that only use the product so rarely that they are really not in the target market. A sample screening questionnaire is shown in Appendix 2. A major consideration in determining the cost of the test is the incidence of users of the product category (Sorensen, 1984). What percentage of consumers in the general population use this product or category of products? When hiring an agency or requesting a cost estimate, incidence figures will be required to estimate the time required and thus the cost of recruitment. Marketing data can be helpful in this regard.

Another consideration is whether to recruit participants by phone, by intercept or from existing subject pools using a database of product use information. Some recruiting may be done over the Internet. It

has become increasingly common to recruit from a national database and then ship the product to the consumer's home. E-mail or Internet recruiting may work well with younger consumers. Recruiting by telephone can be time consuming, but may provide the closest approximation to a random sample of the area. Unfortunately it may miss people with unlisted numbers or people with no land lines (only cell/mobile phones), who represent a demographically different population segment (Brunner and Brunner, 1971). Intercepting individuals at a site like a shopping mall has been popular when the field agency has a testing facility in a mall. However, the nature of the sample must be carefully scrutinized due to the biases inherent in sampling shoppers. If there are stratification quotas for age, gender, etc., these are very important. Some larger testing agencies may maintain databases on pools of local consumers who have been recruited for general service in multiple tests. They may have answered questions on product use and therefore can save a lot of time in locating regular users of the products. However, since habits and situations change (e.g., health, dietary restrictions, family members changing residence), it is necessary to confirm their current suitability through the normal screening questionnaire. Furthermore, it is important to guard against overuse of people from standing databases or retested subject pools. They can become jaded or take on characteristics of professional testers. The participants should be screened for testing frequency or having not participated within a given time frame, usually several months. Three months between tests is a common requirement although Resurreccion (1998) recommends 6 months between tests.

Some other requirements and choices will affect the activities of the field agencies. Consider the stability of the product relative to the holding time that may be encountered during recruitment, for example. This is especially pertinent if there is a low incidence of users. Recruiting time may be very long, but should not exceed the freshness limitations on the product. Distribution or shipping may be a factor as well. One of the authors supervised a test in which frozen pizzas were shipped to the United States from Europe, but due to unexpected delays in U.S. Customs clearance were ruined and useless by the time they reached the test site. The method of product distribution can also be an issue. The two primary choices are personal pickup or mailing. Delivery to the home is also possible but can be costly. If the individuals pick up the product at the agency facility it may be necessary to build in an extra incentive payment to cover their time and travel costs into the overall incentive for participation. Mailing a product risks mishandling, misdirection or delay, and the possibility of unknown temperature history, but it is a low-cost alternative if product stability is good.

15.4.3 Some Tips: Do's and Don'ts

A successful and useful relationship between a sensory researcher and a test agency requires good communication and a good working relationship. Here are some suggestions for situations to avoid when dealing with field services: First, resist the temptation to change the study design at the last minute. Do not expect to change your design, questionnaire, number of products, or recruitment criteria the day before the test. The testing service has scheduled the facilities and set up the test based on your specifications. It may not be possible to make the changes and keep the same schedule. It may be a simple matter to add or change a question or two, but it may not. Second, when requesting shelf life testing do not expect testing agencies to have a time machine. If you need shelf life data you must be willing to wait for it. Do not wait until 3 weeks before the product launch if you need 6 month's data. Even accelerated testing and the Arrhenius equation will not save you. Third, do not take any assumptions that the field service will fill in details the way you expect them to. Spell out all the test details in the test request or specification sheet (in writing) and re-visit them during the agency briefing (verbally) to field questions and resolve ambiguities. It is also a good idea to visit some of the test sites during actual testing. Finally, if the results do not turn out the way you would like, do not blame the field service.

15.4.4 Steps in Testing with Research Suppliers

Table 15.3 lists steps in conducting a test using subcontracting field services. Most of the items are self-explanatory and some are part of the normal testing process for any sensory test (such as problem identification). A few comments are made here as

guides to the sensory professional. Bear in mind that the exact nature of testing differs from product to product, among companies as a function of policies including security concerns. Electronic means of data collection will continue to supplant paper questionnaires.

The setup of the questionnaire takes considerable effort and involves negotiation with the clients and circulation of drafts. The contact person in the agency may be involved to some extent as appropriate to their level of expertise and the degree to which tasks are delegated to the field service. They may also be able to assist in some pre-testing of the questionnaire. Interviewers should be given explicit instructions. These usually include the following: (1) read the questions exactly as worded, (2) do not comment on meaning, (3) do not suggest any acceptable answers, (4) answer every question, even if recorded as "do not know," and (5) do not deviate from the sequence or skip pattern. In spite of such instructions, monitoring shows that many interviewers do not follow these rules (Boyd et al., 1981).

At this time, a number of details need to be arranged concerning the physical products to be tested. If the test is large enough, it may not be practical to have the products made in the laboratory, and arrangements for pilot plant or manufacturing time may be necessary. Storage conditions can also be arranged to mimic the conditions encountered in the normal distribution system (Schaefer, 1979). If competitive products are to be evaluated, they may need to be disguised or re-packaged for the test to insure the blind, unbranded nature of the test. It is important to obtain representative materials and to avoid samples that are abused or defective (Sorensen, 1984). Samples will have to be labeled with codes and the generic title of the product. The label usage directions should be carefully considered as part of the test design. Finally, shipping of products and questionnaires is arranged. For heat or cold-sensitive products, delivery, handling, packing, and unpacking are major considerations. Weekend delivery is sometimes difficult to coordinate. Delays in delivery may result in temperature abuse, as in our frozen pizza example.

Contact with the field agency is important before the testing begins. The questionnaires should be reviewed by their staff and any residual questions about the test, procedures, instructions, the qualifications of respondents, or the scorecard should be cleared up. A pre-test

Table 15.3 Steps in conducting a home use test using field services ("agencies")

Stage 1: Before the test
Identify problem, goals, and negotiate test design with client
Write proposal, including budget
Confer with statistical consultant if necessary (sample size, etc.)
Obtain approvals
Obtain bids and hire agencies
Send test specification sheets concerning participants, products, timing, questionnaires, etc.
Prepare questionnaire
Confer with client, marketing, etc. to make sure all issues are included
Pre-test, revise if necessary
Obtain samples, place request with pilot plant or other supplier
Obtain competitive product if necessary
Design labels
Choose coding system
Confirm usage instructions with developers
Get labels printed and affix to sample product
Prepare shipping orders and send product to agencies
Print questionnaires (if paper questionnaires are used)
Ship questionnaires to agencies with instructions
Visit agencies before the test or hold telephone briefings to review test details and to field any questions

Stage 2: During and after the data collection
Visit agencies to observe testing and/or participate in "callbacks"
Arrange for keypunching and data analysis confer with statistical consultant if necessary
Develop coding sheets for open-ended questions
Receive questionnaires (if paper), unpack and check for completeness, cull mistakes, incompletes
Arrange for data entry if paper
Conduct follow-up discussion groups if desired
Perform statistical analysis
Write report
Schedule presentation
Prepare visual aids for presentation
Present results
Revise, print, and distribute reports
Process bills from agencies
Archive questionnaires and data
Dispose of unused or returned product

visit to handle these matters in person is called a briefing. It may be appropriate to physically visit the agency and observe the test in progress and to participate in some of the interviews or "callbacks" if the product has already been placed and used. Personal visits ensure a quality check on the conduct of the data collection and can provide valuable insight into the degree of attention to detail paid by the agency's professional staff and interviewers.

Following the test, it is incumbent upon the sensory professional to guide and oversee the data entry and analysis, even if this has also been subcontracted. This does not mean that a physical presence is always necessary, but auditing of questionnaires, screening them

for potential fakes and eliminating botched interviews is part of the quality control of the test process. The person in charge of the test should also develop a coding scheme for answers to open-ended questions, to guide the data entry process. A second area of involvement comes in questionnaire validation. This is usually achieved by some telephone callbacks from the field supervisor to a proportion of the respondents, usually 10 or 15% to verify that their opinions were correctly recorded and the interview was not faked (Schaefer, 1979). The sensory project leader will set these quotas for validation and of course they enter into the cost analysis of the field agency. Common areas of difficulty occur from interviewers failing to follow the sampling plan or screening criteria and allowing people to slip in who are actually unqualified. Validate the respondent's qualifications as well as participation and responses. Close supervision of the interviewers by the field supervisor may help curtail many of these problems. A large percentage of the errors may be concentrated in just a few interviewers (Case, 1971).

A final opportunity to interact with the field service comes after analysis of some or all of the data. Answers to some questions, particularly open-ended probed questions, may suggest additional issues that require follow-up. Most agencies have or can provide focus group interview facilities. It may be advantageous to recruit a number of participants for group interviews to probe additional issues. For example, you can call back consumers who were very positive or very negative (perhaps in different discussion groups) to further probe their reasons for liking or disliking the product. Review of the questionnaire results may suggest potential alternative formulations or line extension opportunities or features that need to be added to meet consumers' expectations.

15.5 Questionnaire Design

15.5.1 Types of Interviews

The exact form and nature of the research instrument will depend upon the test objectives, constraints of funding or time and other resources, and the suitability of the interview format. Interviews may take place in person, be self-administered by paper, conducted via a website, or by phone. Each method has advantages and disadvantages (Schaefer, 1979). Self-administration is obviously the least expensive, but does not lend itself to probing of open-ended questions, is open to respondent confusion and mistakes in following directions, and is not suitable for complex issues that may require an explanation. There is no insurance that the person will not read ahead or scan the entire questionnaire before answering any questions. They may not follow the order of questions as printed on the survey. Cooperation and completion rates are poorer with self-administration (Schaefer, 1979). In addition, illiteracy is a problem so a self-administered questionnaire may simply be unusable. Many people will try to hide their inability to read.

Telephone interviews are a reasonable compromise, but may not lend themselves to complex multi-point scales—questions are necessarily short and direct. Respondents may also feel an urge to limit the time they spend on the phone and may produce shorter answers to open-ended questions (Groves, 1978). Telephone interviews are somewhat prone to early termination by the respondent. The in-person interview is the most flexible and the questionnaire can be complex and include a variety of scales since the interviewer and questionnaire are both present for clarification (Boyd et al., 1981). Visual aids can be brought along to illustrate scales and scale alternatives if the interviewer is reading the questionnaire to the respondent. Advantages of this method may be offset by the higher cost (Boyd et al., 1981).

Consider carefully the length of the interview. A good rule of thumb for the length of consumer surveys is about 15–20 min, the attention span of most adults. The issue is not the number of questions, but the time commitment. Boredom and extraneous factors will begin to take a toll on the quality of responses if the time required is too long (Schaefer, 1979). Questionnaires that are too long will annoy respondents and may generate negative responses due to declining interest and change of attitude toward the interview. Length problems can arise when many individuals, perhaps both in research and marketing, have input into the survey issues and thus the number of questions gets too large. A good test for the necessity of each question is whether it is an issue that *you need to know* or whether it would simply be *nice to know*. The test procedure should not incur any costs to the participant like return postage for a mailed questionnaire. Prepaid incentives can reduce nonresponse rates (Armstrong, 1975; Furse et al., 1981).

15.5.2 Questionnaire Flow: Order of Questions

When designing a questionnaire, it is useful to make a flow chart of the topics to be covered. The flow chart can be very detailed and include all skip patterns or it may simply list the general issues in order. A flow chart can be very helpful to clients and other personnel who review the instrument before the actual test. It allows them to see the overall plan of the interview.

The primary rule for questionnaire flow is to go from the general to the specific. With food and consumer product testing, this requires asking about the person's overall opinion of the product first. An overall opinion question is recommended using the 9-point balanced hedonic scale. It is immediately followed by open-ended probing of the reasons for liking or disliking with an appropriate "skip pattern." The skip pattern drops to reasons for liking if the respondent was positive and then probes any dislikes. Conversely, the skip pattern next probes any reasons for disliking if the person was negative and then follows up to see if there were any positive characteristics. Open-ended questions (discussed further below) provide an important opportunity to get at some reasons for likes and dislikes in the respondents own words before other issues are brought to mind. Next, more specific attributes are investigated through the use of intensity, just-right, or liking scales (e.g., appeal of taste, appearance, texture). Finally, overall satisfaction or some other correlated index of liking can be checked again at the end and a preference question asked if more than one product has been tested.

The important principle here is to ask overall acceptability first, before specific issues are raised. These issues may not have been on the person's mind and may take on false importance if asked before the overall acceptance question. Respondents will try to figure out what the issues are and give the right answer or try to please the interviewer (Orne, 1962; Orne and Whitehouse, 2000). As the interview is a social interaction, respondents will adjust their answers to what they feel is appropriate for the situation (Boyd et al., 1981). They will begin to use the terminology that you introduce in the specific attribute questions (Sorensen, 1984). Also, consumers are naturally in an integrative frame of mind when thinking about a product (Lawless, 1994). Questions about individual attributes may cause respondents to become unrealistically analytical.

In addition to the screening questions that qualify a person for participation, other personal information can be gathered at the time of the final interview. Demographic information about personal characteristics such as age and income, number of family members, residence, occupation, and so on can be collected. Some of this material is of a sensitive nature. Participants may feel some reluctance about disclosing their income level, for example. Before the test

proceeds, it is important to assure them of the confidential nature of the data and its end use. It is also best to ask sensitive demographic questions last. At that time the participant should feel comfortable and familiar with the interview process and may feel some rapport with the interviewer. Since they have already committed to answering questions during the product-oriented phase of the interview, it should seem natural to simply continue along and answer a few questions about their personal situation.

In summary, the questioning should follow this flow under most circumstances (Sorensen, 1984): (1) screening questions to qualify the respondent if they have not been previously qualified; (2) general acceptability; (3) open-ended reasons for liking or disliking; (4) specific attribute questions; (5) claims, opinions, and issues; (6) preference if a multi-sample test and/or rechecking acceptance via satisfaction or other scale; (7) sensitive personal demographics. An example of a well-designed questionnaire for a consumer product test is shown in Appendix 3.

15.5.3 Interviewing

Participating in a few interviews can be a valuable exercise to get an impression of how the questionnaire flows in practice, as well as providing an opportunity to interact with actual respondents and get an appreciation of their opinions and concerns first hand rather than exclusively in a data summary. Of course, this is a time-consuming process and the potential value must be weighed against other uses of professional time. If the sensory professional actually takes part in the interview process, there are several guidelines to keep in mind. Remember to introduce yourself. Establishing some rapport with the respondent is useful in getting them to volunteer more ideas. A small to moderate degree of social distance may provide the most unbiased results (Dohrenwend et al., 1968). Second, be sensitive to the time demands of the interview. Try not to take more time than expected. If asked, inform the respondent about the approximate length of the interview (Singer and Frankel, 1982). This may hurt the overall agreement rate (Sobal, 1982), but will result in fewer terminated interviews. Third, if conducting an in-person interview, be sensitive to body language. Probe issues if there are signs of discomfort.

Fourth, do not be a slave to the questionnaire. It is your response instrument. While the agency personnel should be cautioned about deviating from the flow, the sensory project leader can be more flexible.

Some of the response types may not be familiar to the participants. Be prepared to explain scales, using examples like "thermometers" or "ladders." Rank ordering can be described as analogous to the finishing order of a horse race or the order of Olympic medals. Interval scales can be likened to expressing the distance between the horses that finish a race (Schaefer, 1979). Ratio scales can be described as pails of water that are half full, twice as full, and so on. However, the concept of ratio scaling and magnitude estimation is sometimes difficult for consumers (Lawless and Malone 1986).

At the close of the interview, give the person a chance to add anything that has recently come to mind or that they might have omitted earlier in the structured interview. For example, "Is there anything else you'd like to tell me?" Successful interviewing, like focus group moderation, requires a fair amount of sensitivity and interpersonal skills. Being a good listener helps. Bear in mind that the interview is a social interchange. The interviewer should never "talk down" to the respondent or make them feel subordinate. Getting cooperation and honest responses can be achieved by rewarding the respondent by showing positive regard, verbal appreciation, and making the respondents feel their opinions are important (Dillman, 1978). Chances for embarrassment should be minimized in demographic questions.

15.6 Rules of Thumb for Constructing Questions

15.6.1 General Principles

There are a few general principles to keep in mind when constructing questions and setting up a questionnaire (shown in Table 15.4). One should never assume that people will know what you are talking about, that they will understand a question or that they will approach the issue from a given frame of reference (Schaefer, 1979). Pre-testing the instrument can expose faulty assumptions. Each of these rules is explained below. Resurreccion (1998) notes that it is wise to keep

Table 15.4 Ten guidelines for questionnaire construction

1. Be brief.
2. Use plain language.
3. Do not ask what they do not know.
4. Be specific.
5. Multiple choice questions should be mutually exclusive and exhaustive.
6. Do not lead the respondent.
7. Avoid ambiguity.
8. Beware effects of wording.
9. Beware of halos and horns.
10. Pre-testing is usually necessary.

the direction of all scales the same, to avoid confusion. A current trend in opinion polling is to vary the direction on scaled questions (e.g., good to bad, then bad to good) to keep the respondent from acting too mechanically without thoroughly reading and thinking about the question. In our opinion, the value of the reversals does not outweigh the liabilities and potential confusion.

15.6.2 Brevity

Keep the question as short as possible. Brevity affects both respondent motivation and comprehension. Brevity applies to the overall questionnaire as well. Good visual layout with lots of "white space" can help avoid problems. It has become fashionable in attitude surveys to have a list of issues or opinion statements that are followed by "importance" scales. The idea is that attitude should be weighted by importance in the subsequent analysis and modeling. As sensible as this sounds, the presentation of dozens of attitude statements for two sets of ratings (agreement and importance) can cause a good deal of apprehension on the part of the respondent. A monotonous format can also arise if too many products are rated on too many scales. This can look especially tedious if it takes for form of a matrix of boxes for the respondent to fill in. A matrix format may save space for repetitive questions but it may also result in non-response and incomplete questionnaires (Rucker and Arbaugh, 1979).

15.6.3 Use Plain Language

A common problem with consumer tests designed to support new product development or process optimization is that the research staff knows a technical language, complete with acronyms (e.g., UHT milk). Qualitative probing and pre-testing should indicate whether consumers understand the technical issues and terminology. If they do not, such terms are better avoided, even if it means that some issues are dropped.

15.6.4 Accessibility of the Information

Do not ask what they do not know. Only ask for information that a reasonable person has available and/or accessible in memory. Events that intervene can cause "juggling" of the facts. Respondents may have selective recall of different factors and not all memories decay at the same rates. The initial acts of perception themselves are selective—people usually note what they expected to see (Boyd et al., 1981). Since memory and perception present these challenges to the validity of the interview process, do not make things worse by placing unreasonable demands on the respondent. Common sense can be a useful guide here and also pre-testing to see if the question is reasonable. Asking for recall of exact amounts or about time frames that are too wide can create difficulties. Consider the question, how many times did you buy milk in the last year? Asking for product usage over a weekly or monthly period may bring answers to mind more easily. Resurreccion (1998) gives the example of "How much salt do you use when you prepare food?" This is not only hard to estimate, as it may differ for the specific food, but it is vague, an additional problem, as discussed in the next guideline.

15.6.5 Avoid Vague Questions

Be specific. A common mistake is the assumed frame of reference. The person that designs the interview may be addressing one issue, or the project leaders may assume some delimited frame of reference, but consumers may go in another direction or may range much more widely in their interpretations. Another common problem is the English pronoun, you, which is used both in singular and plural. For the singular meaning, it is better to say, "you, yourself," so the person does not interpret this as meaning "you and/or your

family." Questions about product use can be easily misconstrued. For example, "When did you last eat a pizza?" Does this include chilled pizzas, frozen pizzas, hot-delivered pizzas as well as those consumed in restaurants? Providing a checklist of alternatives can be informative, especially if a blank "other" category is left for listing what you may have left out. Questions about "usual" habits can also be vague. For example, "What brand of ice cream do you buy?" is vague. A more specific phrasing is, "What brand of ice cream have you, yourself most recently purchased?"

15.6.6 Check for Overlap and Completeness

Multiple choice questions should be mutually exclusive and exhaustive. Mistakes here are easy to make in demographic questions and screening items, such as marital status or educational levels. Questions about age and income level can have overlapping categories if one is not careful. This can also be avoided with careful pre-testing and review of the draft questionnaire by colleagues. Remember to allow for a "don't know" or "no answer" category, especially in demographics and attitude questions. If multiple combinations of the alternatives are possible, make sure these are listed or otherwise explicit or use a checklist with "check all that apply."

15.6.7 Do Not Lead the Respondent

Avoid questions that suggest a correct or desirable answer. Consider these questions: "Given the ease of preparation, what is your overall opinion of the product?" "Should we raise prices to maintain the quality of cheese or should we keep them about the same?" Both of these questions suggest an answer that the interviewer is looking for. There is always strong social pressure to give an appropriate answer just to please the experimenter (Orne, 1962; Orne and Whitehouse, 2000). Unbalanced value questions can have a leading influence. For example, "How much did you like the product?" may seem relatively harmless. However, the use of the word "like" suggests an acceptable response.

A balanced question would be "how much do you like or dislike the product?" "What is your overall opinion of the product?" is an even more neutral phrasing.

15.6.8 Avoid Ambiguity and Double Questions

Along with being specific, avoiding ambiguous questions is easier than it sounds. One common problem is that verbs in the English language have multiple meanings and we often determine which meaning is appropriate by the context. Consider the classic example "check your sex." Does "check" mean to mark, to restrain, or to verify? Another common problem is the construction of questions with multiple subjects or predicates. The question, "Do you think frozen yogurt and ice cream are nutritious?" is an example of the classic problem called a double-barreled question. Logic dictates that the use of the conjunction "and" requires that both parts be positive for there to be an overall positive response. However, consumers are not always logical and may respond positively to this question even though they think *only one* is nutritious. It is also easy for product developers to set up double-barreled questions for sensory characteristics that they see as always correlated. For example, if a cookie is usually soft and chewy but as aging occurs changes to hard and brittle, there is a strong temptation to combine these adjectives into a single question or a single scale. However, this ignores the possibility that some day a product may in fact be both hard and chewy.

15.6.9 Be Careful in Wording: Present Both Alternatives

As in the case of leading questions, wording a question with only positive or negative terminology can be influential on respondents. The literature on opinion polling shows that dichotomous questions will elicit different answers depending whether both alternatives are mentioned. If only one alternative is mentioned in the question, different frequencies of response are found depending upon which alternative is mentioned in the

question (Payne, 1951; Rugg, 1941). Less-educated respondents are more likely to go along with one-sided agree/disagree statements (Bishop et al., 1982). Try to give all explicit alternatives in asking opinion questions. For example, "Do you plan to buy another microwave in the future, or not?" If possible, balance the order of the alternatives so that one does not have precedence on all questionnaires.

15.6.10 Beware of Halos and Horns

Halo effects are biases involving the positive influence of one very important characteristic on other, logically unrelated characteristics. Someone might like the appearance of a product and consequently rate it more appealing in taste or texture as well. Asking questions only about good attributes can bias the overall rating in a positive direction. Conversely, asking questions about only defects can bias opinion in a negative way. As stated above, it is better to get overall opinion ratings before more specific issues are probed. Otherwise issues may be brought to mind and be given more weight than would otherwise occur.

15.6.11 Pre-test

Pre-testing questionnaires are necessary (Shaefer, 1979). At the very least, a few colleagues should review the draft for potential problems in interpretation. If the questionnaire will be administered by a group of interviewers, they should also look over the draft to see if there are potential problems in the flow, skip patterns, or interpretation. If at all possible, a small group of representative consumers should go through the items, even if it is a "mock" test with no actual product usage. A pre-test with consumers will also provide the opportunity to see whether items and issues are in fact applicable to all the potential respondents. For example, screening questions about food habits in the last month may not be appropriate for someone who was on vacation with small children and ate an unusually large amount of convenience foods.

15.7 Other Useful Questions: Satisfaction, Agreement, and Open-Ended Questions

15.7.1 Satisfaction

A consumer field test questionnaire can include some additional types of questions that may be useful to clients. A common issue is the degree of satisfaction with the sensory properties or performance of the product. This is often highly correlated with overall acceptance, but may be somewhat more related to performance relative to expectations than it is to acceptability. A typical phrasing would be "All things considered, how satisfied or dissatisfied were you with the product?" Typically, a short 5-point scale is used, as shown in Table 15.5. The appropriate analysis for a short scale like this is by frequency counts, sometimes collapsing the two highest alternatives into what is known as a "top two box score." Do not assign integer values to the response alternatives and assume the numbers have interval properties and then do parametric statistical analyses such as taking means and performing t-tests. Frequency counts and nonparametric tests like chi-square are appropriate.

Some variations on satisfaction scales include purchase intent and continue-to-use questions. Purchase intent is difficult to assess with a blind-labeled sensory test since the pricing and positioning of the product relative to the competition are not specified and label claims and advertising claims are not usually presented. So this is like trying to measure purchase intent in an information vacuum and it is not recommended. A variation that is a little more palatable is intent of continued use: "If this product were available to you at a reasonable price, how likely would you be to continue to use it?" A simple 3- or 5-point scale can be

constructed based on "very likely" to "very unlikely" and nonparametric frequency analysis done as in the case of the short satisfaction scale.

15.7.2 Likert (Agree–Disagree) Scales

Attitudes can also be probed during a consumer test. This is often done by assessing the degree of agreement or disagreement with statements about the product. The agree/disagree scales are sometimes referred to as "Likert" scales for the person who first studied them. An example follows: "Please check a box to indicate your feeling about the following statement: Product X ends dry skin." Scale points are shown in Table 15.5. Such information can be important in substantiating claims about consumer perception of the product in subsequent advertising and label information and for defense against any legal challenges of competitors. Broader issues can be studied such as product-usage situations, e.g., "I would allow my children to prepare this snack in the microwave."

15.7.3 Open-Ended Questions

An example of an open-ended question is "What did you like about the product?" No answers or any checklist of alternatives is given. The person can answer in his or her own words. The answers are generally probed if the interview is in-person or by phone. That is, when the person is done answering, the interviewer says something like "Is there anything else?" Such probes are generally recorded on the questionnaire with a code like w/e for "what else?"

There are different opinions about the usefulness of open-ended questions (Stone and Sidel, 2004).

Table 15.5 Examples of satisfaction and Likert scales

Overall, how satisfied or dissatisfied were you with the product? (Check one answer)	Do you agree or disagree with the following statement: This product ends dry skin. (Check one answer)
____ Very satisfied	____ Agree strongly
____ Somewhat satisfied	____ Agree
____ Neither satisfied nor dissatisfied	____ Agree slightly
____ Somewhat dissatisfied	____ Disagree slightly
____ Very Dissatisfied	____ Disagree
	____ Disagree strongly

Table 15.6 Advantages and disadvantages of open-ended questions

Advantages	Disadvantages
Easy to write	Difficult to code/combine answers
Participants use their own words	Favors verbal/intelligent consumers
Can uncover new issues; problems	Unintelligible/ambiguous answers
Can confirm other results	Statistical analysis difficult
Allows opinionated respondents to "vent"	
Amenable to probing, follow-up	

Open-ended questions have advantages and disadvantages (Resurreccion, 1998). A summary of these is listed in Table 15.6. A sense of their usefulness can be gained by experience and the sensory professional must decide whether they are worth the trouble. They can provide corroborating evidence for opinions voiced elsewhere in the questionnaire and can sometimes provide insights about issues that were not anticipated in setting up the more structured parts of the survey. They can also allow respondents a chance to vent dissatisfactions, which if left unprobed may lead to potential horns effects (negative halo effects) on other logically unrelated questions (see Chapter 9). One of our colleagues refers to this as "dumping" their frustrations. Aggregating responses or reporting only frequency counts can hide important information from a few insightful respondents. Sorensen stated that "sometimes one articulate respondent is worth a bucketful of illiterate rambling" (1984, p. 4). The sensory project leader should read over each of the verbatim remarks if time permits.

Their are several advantages to open-ended questions. They are easy to write. They are unbiased in the sense that they do not suggest specific responses, issues, or characteristics. They are courteous to opinionated respondents who may feel frustrated or limited by the issues in a more structured instrument. They allow for issues to come up that may have been omitted from the structured rating scales and fixed-answer questions. They are good for soliciting suggestions, for example, opportunities for product improvement, added features or variations on the theme of the product. They may confirm information gathered in the more structured part of the attribute questions (Sorensen, 1984).

The disadvantages of open-ended questions are similar to those that arise in qualitative research methods. First, they are difficult to code and tabulate. If one person says the product is creamy and another says smooth, they may or may not be responding to the same sensory characteristics. The project leader has to decide which answers to group and count together as the same response. Obviously, there may be experimenter bias in the coding and aggregation of different responses into categories. If the questionnaire is self-administered, it is sometimes difficult to read the handwriting of respondents. Responses may be ambiguous or misleading. Consider "not like real chocolate." Does the statement refer to taste or texture? Respondents may omit the obvious, that is feel that an issue does not require commentary since it is so clear. The open-ended question is more readily answered by more outspoken or better educated respondents. There will be a higher non-response rate to open-ended than to fixed-alternative questions. Finally, statistical analysis of the frequencies of response is not straightforward (Sorensen, 1984). Multiple responses can be given by the same person, so they cannot be treated as counts of independent observations.

Roughly the opposite situation holds for closed-option questions. There is tight control over the topic and possible responses. They are easily quantifiable and statistical analysis is straightforward. But they can give a false sense of security. Remember that people will answer any question, even if it is unimportant, misunderstood, nonsensical to them or even an imaginary issue. This is well known in opinion polling. The classic example in marketing research is the high level of responses obtained for the "Metallic Metals Act," a fictitious construction of a pollster's imagination (Boyd et al., 1981). Finally, the fixed-option question is not always amenable to further probing. The story ends.

15.8 Conclusions

Conducting a skillful consumer field test is a complex activity. For the sensory professional used to the analytical world of trained panels and laboratory control, the number of issues raised, the decisions in test design, the time and cost compromises, and the new skills needed can be quite daunting. Textbooks on marketing research (e.g. Boyd et al., 1981) can provide general guidance. Resurreccion (1998) contains many step-by-step guidelines for conducting consumer field tests. One approach is to subcontract the entire field test—from design to execution, analysis and reporting—to a full-service field agency. This is a costly solution, but one that may save the sensory project leader a lot of time and avoid common mistakes. Enlisting the guidance of experienced professionals from within the company or from the field agencies in the planning and design of the questionnaire can be very valuable. The sensory professional must be willing to accept criticism and advice and be willing to make changes in the test plan and questionnaire. Also bear in mind that the consumer interview is social process and that the quality of the data is determined in part by the social skills of the field interviewers. This underscores the need to have good field agencies with experienced supervisors and interviewers.

Reporting results from consumer field tests is much like any other sensory test, except that the amount and detail of the numerical results are often greater. The methodology and screening criteria (usually placed in an appendix) should be sufficiently detailed that another investigator, if given the same products and questionnaires, would obtain the same results (Sorensen, 1984). It is important to extract the important pattern of results rather than burying the reader in endless summaries of frequency counts and descriptive statistics. The sensory project leader should know the key issues and report conclusions and recommendations in order of importance. More detailed results that are broken down by location or demographic segments can be appended. Often multivariate analysis such as principal components analysis or regression of acceptance against key driving variables can focus the conclusions. Attributes are often correlated in consumer work and there is a strong tendency for "latent variables" or underlying causal factors to drive several items on the questionnaire.

As noted at the beginning of this chapter, sometime contentious issues related to company policy can arise in field testing, sometimes due to the apparent similarity of a consumer test to the activities of a marketing research group. It is important to communicate the value of blind "sensory" testing (with minimal concept) to upper management. Sensory field testing can provide vital information to product developers concerning the achievement of desired product characteristics and the potential for success. Sometimes management may decide to forego the sensory (blind-labeled) field test. Common justifications for skipping steps include the following: (1) other evidence of clear product superiority to competition, (2) high anticipated profitability, (3) unique attributes relative to competition, and (4) potential for competition launching a similar product first (Schaefer, 1979). The value of the research can be measured by its impact on business decisions. Timing may be critically important. If it takes too long to conduct the research, decisions may be made before results are reported (Sorensen, 1984). This can be an unfortunate and frustrating experience for sensory project leaders. Planning ahead and working with field agencies that can provide timely results can help avoid testing that is merely confirmatory, or in the worst case, not even considered at all.

Appendix 1: Sample Test Specification Sheet

Client_____   Field Service_____
Contact Person Info:

_____   _____

_____   _____

Number of products to be tested:_____3_____
Placement: _____monadic sequential, counterbalanced orders____
Coding: ____Random, 3-digits, supplied by client____
Target for placement: ____12/24/96____
Target for completion of test:___12/25/96___
Placement time: ___one week per product____
Interview procedure: telephone callback following completion of each use phase
Minimum number of respondents:____150____

RESPONDENT QUALIFICATION:
　　　Female, heads of household between 21 and 65 inclusive.
　　　Users of aerosol air fresheners, purchasing more than 2 cans per year
Disqualify: if any family members
　　　employees of consumer products companies
　　　marketing research, marketing or advertising companies
Disqualify if: participated in any product test or focus group in the last six months

Special handling requirements:___none_____
Unused/returned product:___ship to client_____

CLIENT PROVIDES:
　　　Questionnaires (screening and callback), labeled products, instructions for placement.
　　　Coding scheme for data tabulation including open-ended questions.

AGENCY PROVIDES:
　　　Placement, interviews (screening and callback), product retrieval.
　　　Tabulation and entry of data. NO STATISTICAL ANALYSIS REQUIRED.
　　　Assurance of confidentiality, adherence to MRA standards, timely completion and quota as
stated.

Appendix 2: Sample Screening Questionnaire

Thomas, Richard and Harry Research Services
Chicago, Illinois

SCREENER

Name:	Home Phone:
Address	Work Phone:
City/State/Zip	Gender: **Female Male Age: 1 2 3 4**
Day/Date of Study: Day: Date:	Group Time:
Date/Time Recruited:	Interviewer Initials:
Eth: AFRICAN AMERICAN CAUC, HIS, ASIAN, OTHER = ALL OTHER ETH. →	USER → USER 1 USER 2

- **USER 1 HEAVY USER – ONCE A WEEK OR MORE**
- **USER 2 LIGHT USER – AT LEAST ONCE A MONTH BUT NO MORE THAN ONCE A WEEK**

Hello, this is _____ from _____. We are conducting a survey and would like to ask you some questions.

1.   Are you the male/female head of your household?

YES	-- CONTINUE
NO	**-- TERMINATE**
DON'T KNOW	-- SCHEDULE CALL BACK

MALE	20 %
FEMALE	80 %

2a.   When it comes to grocery shopping, would you say…?

You do all or most of the grocery shopping for your household	(CONTINUE) – skip Q. 2b
You share the responsibility equally with someone else	(CONTINUE) – skip Q. 2b
Someone else in your household does most of the shopping	(CONTINUE) – ask Q. 2b
Someone else in your household does all of the shopping	(CONTINUE) – ask Q. 2b

2b.   Which of the following statements best describes you…?

Although someone else in my household does most of the grocery shopping, I request some specific products and brands that I want them to buy	(CONTINUE)
Although I request specific products to buy, the person in my household who does most of the grocery shopping generally decides which brands of these products to buy	(TERMINATE)
The person in my household responsible for the grocery shopping makes nearly all of the decisions on which products and brands to buy	(TERMINATE)

3.   Which of the following groups includes **your age**? AGE: ____   DOB: _____

Under 21	TERMINATE
21–29	
30–39	
40–49	
50–65	
66 and over	TERMINATE
refuse	TERMINATE

ALWAYS INCLUDE TO OBTAIN SOME DISTRIBUTION

4.   Have you participated in a food or beverage survey/study in the past 3 months? **(IF THREE MONTHS OR LESS, TERMINATE)**

_____

5.   Do you or any member of your immediate family or a close friend . . . ?

Work for an advertising agency	
Work for a marketing agency	
Work for a broadcast or print media	
Work for a public relations firm	
Work for a marketing research company or a marketing research department of a company	
Work for a food or beverage manufacturer, processor or wholesaler	
Have a management position in a company that sells food or beverages	
(IF YES TO ANY, TERMINATE)	

6.   Now, think about some specific companies, are you, any member of your household, relatives or close friends employed by . . . (READ LIST ONE AT A TIME.)

Con Agra	YES	NO
Clorox	YES	NO
Unilever	YES	NO
Nabisco	YES	NO
Altria	YES	NO
Borden	YES	NO
Sara Lee	YES	NO
Kraft Foods	YES	NO
Hellmann's	YES	NO
Ken's	YES	NO
Lipton	YES	NO
Procter & Gamble	YES	NO
Hunt's	YES	NO
Heinz	YES	NO
Reckitt Benckiser	YES	NO

(IF "YES" TO ANY -- **TERMINATE**)

7a.   Please tell me if any of the following applies to you. Do you have . . . ? (Read list – if any number circled, thank and terminate including don't know).

ANY MEDICAL RELATED DIETARY RESTRICTIONS	-- Thank and **TERMINATE**
DIABETES	-- Thank and **TERMINATE**
DON'T KNOW	-- Thank and **TERMINATE**

7b.   Are you allergic or sensitive to any food/food ingredients? (Do not read list).

YES	-- Thank and **TERMINATE**
NO	-- CONTINUE
DON'T KNOW	-- Thank and **TERMINATE**

7c.   Have you had gastric bypass surgery?

YES	-- Thank and **TERMINATE**
NO	-- CONTINUE

7d.   ASK FEMALES ONLY: Are you currently pregnant or nursing?

YES	-- Thank and **TERMINATE**
NO	-- CONTINUE

8. Which of the following products have you **purchased and eaten** in the past month? (READ LIST)

Mayonnaise		
Ketchup		
Meats/Chicken		
Bottled Salad Dressing		
Pasta		*(MUST BE CHECKED. IF NOT, TERMINATE)*
BBQ Sauce		
None (DO NOT READ)		*(TERMINATE)*

9. You mentioned that you purchase and eat pasta. Which of the following types of pasta have you personally purchased and eaten in the past month? (READ LIST AND RECORD BELOW)

	PAST MONTH
Finistrini	
Dopodomani	
Orecchini	
Stitichezza	
Fazzoletti	
Pantaloni	
Francobolli	

* TERMINATE IF NO TO ALL

10. How often you do personally **purchase and eat** pasta products?

    Three or more times a week   __   QUALIFIES AS USER 1
    At least once a week   __   QUALIFIES AS USER 1
    Once every two-three weeks   __   QUALIFIES AS USER 2
    At least once a month   __   QUALIFIES AS USER 2
    Less often than once a month   __   **TERMINATE**

11. For our study, we need to speak to people of different ethnic backgrounds. Which ethnic background do you consider yourself? **(READ LIST)**
                       White/Caucasian
                       African-American
                       Hispanic/Latino
                       Asian-American
                       Other _____

INCLUDE TO REPRESENT US CENSUS – NOT MANDATORY

12. Would you be interested in trying several samples of pasta products WITH SAUCE?

Yes	-- CONTINUE
No	-- Thank and **TERMINATE**

Thank you. We are conducting a test. The test will be conducted on _____ and will take _____. And you will be paid $____ for your time. Would you be interested?
                       (IF "NO," – **TERMINATE**)

(IF "YES," – **CONTINUE TO INVITATION**)

**IF RESPONDENT QUALIFIES, PLEASE READ THE
FOLLOWING VERBATIM:**
Because you would be participating in a research study on food products, it is important that you follow the following guidelines:
❖ Bring a picture ID that verifies your name and age.
❖ Do not eat a large meal before the study.
❖ Refrain from smoking or drinking coffee at least 30 minutes before the study.
❖ Do **not** wear any fragrances on this day. (Those that arrive with fragrances on will be asked to leave without compensation.)
❖ Arrive at **least** 10–15 minutes before your scheduled time to allow for check-in processes. Late arrivals **cannot** be guaranteed admittance or compensation.
❖ If you need glasses to read, please bring them as you will be reading and answering a questionnaire.
❖ To eliminate distractions and because we cannot provide supervision, no children under 12 will be allowed to wait alone at the facility.

Do you have any problems with these guidelines? (IF "YES", TERMINATE)

Appendix 3: Sample Product Questionnaire

Mary & Beverley's Research Services
Chicago, IL

Serial # _____

User # _____

Pasta Fazzoletti

NAME: _____	
ADDRESS: _____	
CITY: _____ STATE: _____ ZIP CODE: _____	
TELEPHONE #: (AREA CODE) _____	

City—Circle Below

Dallas—1	Urbana—2	Framingham -3
Seattle—4	Tampa—5	San Diego—6

Before you taste each product, take a sip of water and bite of cracker so that you remove any lingering tastes in your mouth. You will then be given a serving of the fazzoletti to eat. Please make sure to at least one-half of the product so that you can form an opinion. Once you are done with the entire questionnaire, there will be a 5-min break and then you will go on to the next product. The same procedures will be followed for the second product.

Circle the number on the item you are tasting.
Sample # 387   426   Serial # ___

PLEASE TASTE THE PRODUCT AND ANSWER THE FOLLOWING QUESTIONS. PLEASE EAT AT LEAST ONE-HALF OF THE PRODUCT SO THAT YOU WILL BE ABLE TO FORM AN OPINION.

---

1.   Everything considered, how much do you **LIKE** or **DISLIKE** this product **OVERALL**? ("X" ONE BOX BELOW)

Like it extremely............ ☐

Like it very much........... ☐

Like it moderately.......... ☐

Like it slightly................ ☐

Neither like nor dislike it ☐

Dislike it slightly............ ☐

Dislike it moderately ..... ☐

Dislike it very much ...... ☐

Dislike it extremely .... ☐

2a.   What, if anything, did you **LIKE** about the product? (WRITE IN BELOW) (INTERVIEWER PROBE FOR COMPLETENESS—THIS PROBE INSTRUCTION IS NOT SHOWN ON THE QUESTIONNAIRE)

_____

_____

2b. What, if anything, did you **DISLIKE** about the product? (WRITE IN BELOW)
(INTERVIEWER PROBE FOR COMPLETENESS—THIS PROBE
INSTRUCTION IS NOT SHOWN ON THE QUESTIONNAIRE)

_____

_____

**WE WOULD LIKE YOU TO EVALUATE THE PRODUCT ON A NUMBER OF
CHARACTERISTICS, USING THE SCALE INDICATED BELOW EACH
QUESTION.**

**THINKING SPECIFICALLY ABOUT THE WAY THE PASTA PRODUCT
LOOKS...**

3. How much do you **LIKE** or **DISLIKE** the **OVERALL APPEARANCE** of
thisP? ("X" ONE BOX BELOW)

Like it extremely .... ☐

Like it very much .... ☐

Like it moderately .... ☐

Like it slightly .... ☐

Neither like nor dislike it ☐

Dislike it slightly .... ☐

Dislike it moderately ..... ☐

Dislike it very much ...... ☐

Dislike it extremely .... ☐

4. And how would you **DESCRIBE** the **COLOR** of this fazzoletti? ("X" ONE
BOX BELOW)

Much too light .... ☐

Somewhat too light .... ☐

Just about right .... ☐

Somewhat too dark .... ☐

Much too dark .... ☐

**THINKING SPECIFICALLY ABOUT THE WAY THE TEXTURE OF THE
PASTA PRODUCT...**

5. How much do you **LIKE** or **DISLIKE** the **OVERALL TEXTURE** of this pasta
product? ("X" ONE BOX BELOW)

Like it extremely .... ☐

Like it very much .... ☐

Like it moderately .... ☐

Like it slightly .... ☐

Neither like nor dislike it ☐

Dislike it slightly .... ☐

Dislike it moderately ..... ☐

Dislike it very much ...... ☐

Dislike it extremely .... ☐

6.   Would you describe the **SOFTNESS/FIRMNESS** of this pasta product as . . .
     ("X" ONE BOX BELOW)

     Much too soft .... ☐
     Somewhat too soft................... ☐
     Just about right .... ☐
     Somewhat too firm................. ☐
     Much too firm .... ☐

7.   Thinking about the **STICKINESS OF THE PASTA,** would you say it is . . .
     ("X" ONE BOX BELOW)

     Not at all sticky .... ☐
     Slightly sticky .... ☐
     Moderately sticky................... ☐
     Very sticky .... ☐
     Extremely sticky .... ☐

8.   Would you describe the **MOISTNESS/DRYNESS** of this pasta product as . . .
     ("X" ONE BOX BELOW)

     Much too moist .... ☐
     Somewhat too moist............... ☐
     Just about right .... ☐
     Somewhat too dry .... ☐
     Much too dry .... ☐

**THINKING SPECIFICALLY ABOUT THE WAY THE PASTA PRODUCT
     TASTES . . .**

9.   How much do you **LIKE** or **DISLIKE** the **OVERALL FLAVOR/TASTE** of
     this product? ("X" ONE BOX BELOW)

     Like it extremely........... ☐
     Like it very much........... ☐
     Like it moderately......... ☐
     Like it slightly............... ☐
     Neither like nor dislike it ☐
     Dislike it slightly........... ☐
     Dislike it moderately ..... ☐
     Dislike it very much ...... ☐
     Dislike it extremely .... ☐

10. And how would you **DESCRIBE** the **OVERALL SALTINESS** of this product? ("X" ONE BOX BELOW)

Not at all salty enough .... ☐
Not quite salty enough .... ☐
Just about right...................... ☐
Somewhat too salty .... ☐
Much too salty....................... ☐

11. Did you experience an **AFTERTASTE**? ("X" ONE BOX BELOW)

Yes .... ☐   Continue
No.................................. ☐   STOP—Do **NOT** answer Q.12 - 13

12. Would you say that the **AFTERTASTE** was...? ("X" ONE BOX BELOW)

Pleasant .... ☐
Unpleasant.............................. ☐

13. How would you describe the **STRENGTH OF THE AFTERTASTE** of the pasta product? ("X" ONE BOX BELOW)

Very pleasant.......................... ☐
Somewhat pleasant................. ☐
Neither pleasant nor
unpleasant............................. ☐
Somewhat unpleasant............. ☐
Very unpleasant..................... ☐

STOP--PLEASE RAISE YOUR HAND TO LET US KNOW YOU ARE FINISHED.

Peryam & Kroll Research Corporation
Sample # _____   Serial # _____

Pasta
Preference

1. Now that you have tried both samples, which pasta did you prefer overall -- the sample you tried *first* or the sample you tried *second*? (CHECK ONE BOX ONLY)

Prefer the product you tried first
Prefer the product you tried second
No preference/both the same   (Skip 2.)

2. Why did you prefer that sample better overall? (WRITE IN)

References

Allison, R. L. and Uhl, K. P. 1964. Influence of beer brand identification on taste perception. Journal of Marketing Research, 1, 36–39.

Anon. 1974. Testing in the "real" world. The Johnson Magazine. S.C. Johnson Wax, Racine, WI.

Armstrong, J. S. 1975. Monetary incentives in mail surveys. Public Opinion Quarterly, 39, 111–116.

ASTM. 2008. Standard guide for sensory claim substantiation. Designation E-1958-07. Vol.15.08 2008. Annual Book of ASTM Standards. ASTM International, Conshohocken, PA, pp. 186–212.

Bishop, G. F., Oldendick, R. W. and Tuchfarber, A. J. 1982. Effects of presenting one vs. two sides of an issue in survey questions. Public Opinion Quarterly, 46, 69–85.

Blair, E. 1980. Using practice interviews to predict interviewer behaviors. Public Opinion Quarterly, 44, 257–260.

Boyd, H. W., Westfall, R. and Stasch, S. F. 1981. Marketing Research. Richard D. Irwin, Homewood, IL.

Brunner, J. A. and Brunner, G. A. 1971. Are voluntarily unlisted telephone subscribers really different? Journal of Marketing Research, 8, 121–124.

Cardello, A. V. and Sawyer, F. M. 1992. Effects of disconfirmed consumer expectations on food acceptability. Journal of Sensory Studies, 7(4), 253–277.

Case, P. B. 1971. How to catch interviewer errors. Journal of Marketing Research, 11, 39–43.

Cochran, W. G. and Cox, G. M. 1957. Experimental Designs. Wiley, New York.

De Graaf, C., Cardello, A. V., Kramer, F. M., Lesher, L. L., Meiselmand, H. L. and Schutz, H. G. 2005. A comparison between liking ratings obtained under laboratory and field conditions. Appetite, 44, 15–22.

Dillman, D. A. 1978. Mail and Telephone Surveys: The Total Design Method. Wiley, New York.

Dohrenwend, B. S., Colombos, J. and Dohrenwend, B. P. 1968. Social distance and interviewer effects. Public Opinion Quarterly, 3, 410–422.

El Gharby, A. E. 1995. Effects of nonsensory information on sensory judgments of no-fat and low-fat foods: Influences of attitude, belief, eating restraint and lavel information. M.Sc. Thesis, Cornell University.

Furse, D. H., Stewart, D. W. and Rados, D. L. 1981. Effects of foot-in-the-door, cash incentives and follow-ups on survey response, Journal of Marketing Research, 18, 473–478.

Gacula, M. C., Jr. and Singh, J. 1984. Statistical Methods in Food and Consumer Research. Academic, Orlando, FL.

Gacula, M., Singh, J., Bi, J. and Altan, S. 2009. Statistical Methods in Food and Consumer Research. Elsevier/Academic, Amsterdam.

Gacula, M. C. Rutenbeck, S. K. Campbell, J. F. Giovanni, M. E., Gardze, C. A. and Washam, R. W. 1986. Some sources of bias in consumer testing. Journal of Sensory Studies, 1, 175–182.

Garnatz, G. 1952. Consumer acceptance testing at the Kroger food foundation. In: Proceeding of the Research Conference of the American Meat Institute, Chicago, IL, pp. 67–72.

Gould, W. A., Stephens, J. A., DuVernay, G., Feil, J., Geisman, J. R., Prudent, I. and Sherman, R. 1957. Establishment and use of a consumer panel for the evaluation of quality of foods. Research Circular 40, January 1957, Ohio State Agricultural Experiment Station, Wooster, OH.

Groves, R. M. 1978. On the mode of administering a questionnaire and responses to open-end items. Social Science Research, 7, 257–271.

Hays, W. L. 1973. Statistics for the Social Sciences. Holt, Rinehart and Winston, New York.

Jellinek, J. S. 1975. The Use of Fragrance in Consumer Products. Wiley, New York.

Lawless, H. T. 1994. Getting results you can trust from sensory evaluation. Cereal Foods World, 39(11), 809–814.

Lawless, H. T. and Malone, G. J. 1986. A comparison of scaling methods: Sensitivity, replicates and relative measurement. Journal of Sensory Studies, 1, 155–174.

Meiselman, H. .L. and Schutz, H. G. 2003. History of food acceptance research in the US army. Appetite, 40, 199–216.

Oliver, T. 1986. The Real Coke, the Real Story. Random House, New York.

Orne, M. T. 1962. On the social psychology of the psychological experiment: With particular reference to demand characteristics and their implications. American Psychologist, 17, 776–783.

Orne, M. T. and Whitehouse, W. G. 2000. Demand characteristics. In: A. E. Kazdin (ed.), Encyclopedia of Psychology. American Psychological Association and Oxford, Washington, DC, pp. 469–470.

Payne, S. L. 1951. The Art of Asking Questions. Princeton University, Princeton, NJ.

Peryam, D. R. and Haynes, J. G. 1957. Prediction of soldiers' food preferences by laboratory methods. Journal of Applied Psychology, 41, 1–6.

Resurreccion, A. V. 1998. Consumer Sensory Testing for Product Development. Aspen, Gaithersburg, MD.

Rucker, M. H. and Arbaugh, J. E. 1979. A comparison of matrix questionnaires with standard questionnaires. Educational and Psychological Measurement, 39, 637–643.

Rugg, D. 1941. Experiments in wording questions: II. Public Opinion Quarterly, 5, 91–92.

Schaefer, E. E. 1979. ASTM Manual on Consumer Sensory Evaluation. STP 682. American Society for Testing and Materials, ASTM International, Conshohocken, PA.

Schutz, H. G. 1971. Sources of invalidity in the sensory evaluation of foods. Food Technology, 25, 53–57.

Singer, E. and Frankel, M. R. 1982. Informed consent procedures in telephone interviews. American Sociological Review, 47, 416–427.

Sobal, J. 1982. Disclosing information in interview introductions: Methodological consequences of informed consent. Sociological and Social Research, 66, 348–361.

Sorensen, H. 1984. Consumer Taste Test Surveys. A Manual. Sorensen Associates, Corbett, OR.

Stone, H. and Sidel, J. L. 2004. Sensory Evaluation Practices, Third Edition. Elsevier Academic, San Diego.

Stone, H. and Sidel, J. L. 2007. Sensory research and consumer-led food product development. In: H. Macfie (ed.).Consumer-Led Food Product Development. CRC, Boca Raton, FL.

Chapter 16

Qualitative Consumer Research Methods

Abstract Qualitative methods are used to probe issues in depth with small groups of consumers. They can provide valuable information about product concepts and prototypes. This chapter describes quantitative methods and especially the use of focus groups. The setup, conduct, analysis, and reporting of focus groups are discussed as well as moderator skills and techniques.

Discussing consumer perceptions of food quality is somewhat similar to exploring new and unknown land – it is not immediately clear where to begin or by what means to travel, and it is nearly impossible to foresee where one will end up.

—Schutz and Judge (1984)

Contents

H.T. Lawless, H. Heymann, *Sensory Evaluation of Food*, Food Science Text Series,
DOI 10.1007/978-1-4419-6488-5_16, © Springer Science+Business Media, LLC 2010

16.1 Introduction

16.1.1 Resources, Definitions, and Objectives

A number of techniques can be used to probe consumer responses to new products in addition to the traditional mode of inquiry using questionnaires and large statistical samples. Exploratory research methods often use small numbers of participants but allow for greater interaction and deeper probing of attitudes and opinions (Chambers and Smith, 1991). As a class of methods, they are referred to as qualitative techniques to distinguish them from quantitative survey work that stresses statistical treatment of numerical data and representative projectable sampling. This chapter reviews the principles and applications of qualitative research methods. We have drawn from the authors' experience and from the overviews of the area by Casey and Krueger (1994), Krueger (1994), Chambers and Smith (1991), Stewart and Shamdasani (1990), and Goldman and McDonald (1987) to which the reader is referred for further information. The updated and detailed guidebook on focus groups by Krueger and Casey (2009) is especially helpful for those planning to conduct qualitative research. The paper by Cooper (2007) reviews the history of qualitative research and its relationship to the forms of psychology that have been in vogue during recent decades.

Qualitative research methods are techniques that involve interviews or observations that are less structured than controlled laboratory experiments. They are also less structured than survey research based on fixed questionnaires. The methods are flexible in the sense that as new information arises, the flow and content of the investigation may change. This is one of the strengths of these methods. Qualitative consumer research is most applicable to the exploration and development of new concepts that go hand in hand with the development of successful products. Although this chapter will focus primarily on sensory research, the reader should bear in mind that discovery and/or optimization of sensory attributes may be only one part of the qualitative research done in any study. The process of concept development places qualitative research more traditionally in the bailiwick of marketing research than sensory research. However, the sensory scientist is often part of a team looking for a well-integrated sensory–conceptual product.

A variety of evolving techniques are used to interface with consumers in new product development. Many researchers now think of consumers as "co-designers" of products (Bogue et al., 2009; Moskowitz et al., 2006). Common methods include group interviews (focus groups), one-on-one interviews (also called "in-depth interviews"), observational methods (ethnography), focus panels that do repeated evaluations, and consumer immersion techniques where innovative and/or vocal consumers can work alongside product developers as the product prototypes are formulated and modified. These innovative consumers are sometimes referred to as "lead users," people with strong needs that will sooner or later be met by product innovations in the marketplace (von Hippel, 1986). The approach of making consumers part of the design team is the reverse of observational ethnographic research. Rather than making the researcher part of the consumer's situation, it makes the consumer part of the researchers' and designers' world. Sometimes combinations of methods are used. For example, enthographic observation coupled with in-depth interviews can yield insights about what consumers do as well as what they say and provide compelling real-life video clips to illustrate the main points and conclusions. Qualitative techniques are instrumental in starting projects off on the right foot and avoiding "type zero error," i.e., asking the wrong questions to begin with. Some researchers refer to this exploratory work as the "fuzzy front end." Companies are becoming increasingly consumer-centric although there is a danger that overreliance on the "average consumer's" input may miss some truly inspired, entrepreneurial, and/or creative opportunities (Cooper, 2007; von Hippel, 1986). New approaches and techniques continue to evolve. The Internet has opened a new area for mining product ideas, with blogs and websites that can be searched for innovative opinions, expectations, and/or points of dissatisfaction that could suggest new product opportunities.

16.1.2 Styles of Qualitative Research

The most common form of qualitative research is the group depth interview or focused group discussion, which has come to be known simply as a "focus group." This typically involves about ten consumers sitting around a table and discussing a product or idea with the seemingly loose direction of a professional moderator. The interview is focused in the sense that certain issues are on the agenda for discussion, so the flow is not entirely unstructured, but rather centered on a product, advertisement, concept, or perhaps promotional materials. The method has been widely used for over 50 years by social science researchers, government policy makers, and business decision makers. In 2007, Cooper estimated that there are about a half million focus groups conducted each year worldwide, with about half of those occurring in the United States. Of course, not all focus groups are concerned with new product development, and they are used for a variety of purposes such as research on advertising, assessing political opinions, and developing election strategies.

Historically, the focused group discussion grew from R.K. Merton's use of group interviews to assess audience reactions to radio programs in the 1940s, and later his use of the same techniques for analysis of Army training films (Stewart and Shamdasani, 1990). Currently, the methods are widely used in marketing research for probing of product concepts and advertising research concerning product presentation and promotion. Sensory evaluation departments have added these techniques to their repertoire. In 1987, Marlowe stated that many sensory evaluation groups in industry were already using these techniques to support product development, and that there was growing interest in professional organizations such as ASTM in these methods. This interest was generated by the realization that the methods could be used to develop insights and direction for sensory evaluation issues in early stages of new product development. This activity primarily serves product development clients, just as a marketing research department probes consumers' reactions to product concepts and potential advertising or promotions in order to provide information for their marketing clients. The main difference is that a sensory evaluation group is more likely to focus on product attributes, functional consumer needs, and perceptions of product performance, while a concept study done by marketing research addresses more of the ideas underlying a new product opportunity, i.e., its benefits, emotional connotations, and brand imagery. Obviously, there is often overlap. For example, both approaches usually involve probing of consumer attitudes toward the product category based on experience

and expectations. More and more often, a sensory specialist will be invited to "sit at the table" as the early qualitative work is done to initiate and then refine the product concept.

In general, qualitative methods are best suited for clarification of problems and consumer perspectives, identifying opportunities, and generating ideas and hypotheses (Stewart and Shamdasani, 1990). For example, a qualitative study of consumer attitudes toward irradiated poultry suggested directions for consumer education and label design (Hashim et al., 1996). The techniques are well suited to new product exploration and for follow-up to probe issues raised in other work, e.g., puzzling results from a consumer in-home test or survey. Groups can also function as a disaster check to make sure the conceptualization and realization of the product in the laboratory has not overlooked something important to consumers. Sometimes a high level of enthusiasm may follow a technical breakthrough in product research, but consumers may not share this enthusiasm. Conducting a few consumer groups to explore the new development may provide a sobering reality check (Marlowe, 1987). Qualitative research tends to be hypothesis generating but rarely stands alone to prove anything. It is good for exploration, rather than verification, and for creative stimulation and adding direction and deepening understanding. The techniques can be used to probe consumer opinion of a product category, to examine prototypes, to explore new product opportunities, to design questionnaires, and to examine motivations and attitudes about products (Marlowe, 1987).

The style of the interview, whether in a group or one-on-one, is characterized by careful probing of comments. The probing leads to deeper understanding of the reasons behind the comment. A classic question is "Why is that important to you?" This technique is commonly referred to as "laddering" because it takes steps down into the underlying reasons for an attitude, belief, or choice. Examples of laddering techniques can be found in Krystallis et al. (2008) and Ares et al. (2008) in studies of consumers' motivations in purchasing functional foods. Laddering may either be "hard" or "soft." Hard laddering refers to a fixed question sequence such as "Why did you choose that yogurt?" followed by "Why is that important to you?" and "Why is the latter important to you?" (Ares et al., 2008). Soft laddering refers to the same kind of question, but with more latitude given the interviewer to tailor the probing question to the specific comment or consumer. Bystedt et al. (2003) give an entire chapter to laddering techniques in consumer interviews. They stress that at the surface, there are a collection of desired attributes. Beneath these functional characteristics there are a set of objective benefits. Beneath the objective benefits is a set of emotional benefits. Beneath the emotional benefits are basic values (self-esteem, health, attraction to the opposite sex) that should be understood. Laddering works down this chain.

Because it is based on the reactions of small numbers of specifically recruited consumers who have limited interaction with the product, caution is justified in generalizing the findings to the population at large. Even if the respondents are selected on the basis of regular use of the product category, it is not possible to insure a representative sample of the public on all relevant demographic variables. This stands in contrast to a large-scale consumer home use test that may be conducted with hundreds of participants in several geographic areas. Other limitations are recognized in the method. Dominant members may have undue influence on expressed attitudes and the direction of the discussion. There is often only limited exposure to the product or it may not be used at all by the participants. Both the direction of the interview and the interpretation of results involve some subjectivity on the part of the moderator and analyst. Qualitative interview methods trade off a certain amount of objectivity and structure in favor of flexibility. Some differences of qualitative and quantitative research are shown in Table 16.1. Chambers and Smith (1991) point out that qualitative research may precede or follow quantitative research, and that both types of research gain in validity when they can be focused together on a research problem.

The trade-off between depth of understanding and the acknowledged limitations in sampling and projection was well stated in a study of food choice by Furst et al. (1996). This study used an interview method to uncover influences, valued aspects of each person's food choice system, and strategies used during purchase decisions. Important personal system values included sensory attributes, quality, convenience, health and nutrition concerns, cost, and interpersonal relationships. The study identified these consistent themes underlying food choice behavior and how they could interact with each other and with contextual factors. The rationale for the trade-off between extensive sampling and in-depth interaction was summarized as follows:

> In developing the conceptual model, depth of understanding was accorded a higher priority than breadth in sampling, and to this end a group of people in a particular food choice setting were invited to articulate their own thoughts and reflections on food choice. The sample was not designed to be representative, but was used to examine the range of factors involved in food choice among a group of diverse people. The component and processes represented by the model acknowledge and illuminate considerable variation in many dimensions, such as personal life course, extent of personal system, social setting and food context, even among a relatively small group of people operating within a specific context (Furst et al., 1996, p. 262).

16.1.3 Other Qualitative Techniques

In addition to the popular focus group method, other techniques are available. In some cases, one-on-one interviews are more appropriate for gathering the information of interest. This may be necessary when the issue is very personal, emotionally charged, or

Table 16.1 Some differences of qualitative and quantitative consumer research

Qualitative research	Quantitative research
Well suited to generate ideas and probe issues	Poorly suited to generate ideas, probe issue
Small numbers of respondents	Large projectable samples
($N < 12$ per group)	($N > 100$ per group)
Interactions among group members	Independent judgments
Flexible interview flow, modifiable content	Fixed and consistent questions
Analysis is subjective, non-statistical	Well suited to numerical analysis
Poorly suited to numerical analysis	Statistical analysis is appropriate
Difficult to assess reliability	Easy to assess reliability

Modified from Chambers and Smith (1991)

involves experts who are better probed individually. Experts include such individuals as culinary professionals, dieticians, physicians, lawyers, depending, of course, on the research question. Sometimes people with a high degree of ego involvement in the topic may give more complete information alone than in a group. Sometimes the topic lends itself more comfortably to individual interviews than to group discussion. Examples can be found in studies of food choice with consumers (Furst et al., 1996), older adults (Falk et al., 1996), and cardiac patients (Janas et al., 1996). Groups also run the risk of social competition, one-upsmanship, or unproductive arguing. One-on-one interviews are also better suited to groups that are extremely sensitive to social pressures. An example is teenagers, who are easily swayed by group influence (Marlowe, 1987). The limitation of one-on-one interviews is the loss of opportunity for synergistic discussion among participants. Of course, opinions from earlier interviews can be presented to later participants for their consideration. Thus the interview plan itself becomes dynamic and makes use of what is learned in further information gathering as the study progresses (Furst et al., 1996; Janas et al., 1996). This flexibility is a major point of separation from fixed quantitative questionnaire methods and is one advantage of the qualitative approach.

A third type of qualitative research is naturalistic observation, also known as ethnography (Bystedt et al., 2003; Eriksson and Kovalainen, 2008; Moskowitz et al., 2006). This is a process of observing and recording unguided behavior with the product, much in the ways that ethologists study animal behaviors by observing from concealed positions. This can be done by observing, videotaping, viewing from a one-way glass, or even going to live with a family to study their food habits, for example. Of course, people must be informed of the observation, but the goal is to be as unobtrusive as possible. Such methods are applicable to issues that involve behavior with the product, such as cooking and preparation; use of sauces, condiments, or other additions; spice usage, package opening and closure; time and temperature factors; whether directions are read; how food is actually served and consumed; plate waste; and the storage or use of leftovers. Data have high face validity since actual behavior is observed rather than relying on verbal report. However, data collection may be very slow and costly. Observational methods are well suited to

studying behaviors where consumers actively interact with the product (i.e., perhaps more than just eating it). Bystedt et al. (2003) give the example of observing women at a cosmetics counter in a department store. Suppose your company manufactured a nonstick spray product for barbecue grills. It would be appropriate to observe how grillers actually used the product, when they sprayed it, how much they used and how often, etc.

16.2 Characteristics of Focus Groups

16.2.1 Advantages

There are several advantages to qualitative research. The first is the depth of probing that is possible with an interactive moderator. Issues may be raised, attitudes probed, and underlying motivations and feelings uncovered. Beliefs may be voiced that would not easily be offered by consumers in a more structured and directed questionnaire study. Since the moderator is present (and often some of the clients, out of sight), issues that were not expected beforehand can be followed up on the spot, since the flow of the interview is usually quite flexible. The second advantage is the interaction that is possible among participants. One person's remark may bring an issue to mind in another person, who might never have thought about it in a questionnaire study. Often the group will take on a life of its own, with participants discussing, contrasting opinions, and even arguing about product issues, product characteristics, and product experiences. In a successful group interview, such interaction will occur with minimal direction from the moderator.

A perceived advantage of these methods is that they are quick and inexpensive to do. This perception is illusory (Krueger and Casey, 2009). In practice, multiple groups are conducted, often in several locations, so that moderators and observers may spend days in travel. Recruiting and screening participants also take time. The data analysis may be very time consuming if video or audiotapes must be reviewed. So time to completion of the report is no faster than other types of consumer research and professional hours involved may substantially add to costs. There are also some obvious efficiencies in the procedure for the users of the data who attend the groups. Consumer contact with 12 people can be directly observed, all within the space of an hour, and with the participants collected by appointment. So the rate of information transfer is very high once the groups are underway, as opposed to in-home individual interviews or waiting for a mail-in survey to be retrieved and tabulated.

16.2.2 Key Requirements

The environment is designed to be non-threatening and encourages spontaneity. One principle is the "strangers on a train" phenomenon. People may feel free to air their opinions because they will probably never meet these same people again. So there is nothing to lose in being candid, and there is no need to adopt socially expected postures as one might find in a group of neighbors. Of course, in every community there is some chance that people will be connected through a previous neighborhood or community group, but this is not a big problem. The commonly held belief that better data are given when the participants are total strangers has been opened to question (Stewart and Shamdasani, 1990). Commonly used warm-up procedures that are intended to facilitate acquaintance and interpersonal comfort would seem to contradict this notion of anonymity as a requirement for good data.

There are key requirements for a productive focus group study (Casey and Krueger, 1994; Chambers and Smith, 1991; Krueger, 2009). They include careful design, well thought-out questions, suitable recruiting, skillful moderating, prepared observers, and appropriate, insightful analysis. As in other sensory evaluation procedures, fitting the method to the questions of the client is key. For example, if the end users of the data want to say that over 55% of people prefer this product to the competition, then a quantitative test is needed and they must be dissuaded from using focus groups. The sensory professional must also consider the overall quality of the information produced and consider the reliability and validity of the method both in general and as practiced in their programs and research projects. The primary steps in conducting a focus group study were summarized by Stewart and Shamdasani (1990) as follows: define the problem, specify characteristics of participants and means of recruitment, choose the moderator, generate

and pre-test the discussion guide, recruit participants, conduct the study, analyze and interpret the data, and report the results (a more detailed list is given below). It should be fairly obvious from this list that the image of qualitative research as quick and easy is completely false. Conducting a good focus group study is as involved as any other behavioral research study or sensory test and it requires careful planning.

16.2.3 Reliability and Validity

Reliability and validity are issues in qualitative research, just like any other information gathering procedure or analysis tool. Concerns are often raised that the procedure would yield different results if conducted by a different moderator or if analyzed by a different person (Casey and Krueger, 1994). Having multiple moderators and more than one person's input on the analysis provides some protection. Reliability in a general sense is easy to judge although it is difficult to calculate in any mathematical way. When conducting several focus groups, common themes begin to emerge that are repeated in subsequent groups. After awhile, there is diminishing return in conducting additional groups since the same stories are repeated. This common observation tells us that the results from one group are incomplete, but that there is some retest reliability in the sense that additional groups yield similar information. Janas et al. (1996) framed this issue in terms of the "trustworthiness" of the data and cited three guiding processes during extended individual interviews that could be used to enhance trustworthiness: (1) peer debriefing where emerging concepts are questioned and discussed by co-investigators, (2) using return interviews that can assess consistency of emerging themes, and (3) checking conclusions and key findings with participants. The consistent themes also become part of the data coding and then provide a basis for categorization strategies and grouping of similar concepts. The principles for these guiding processes are found in "grounded theory" methods (discussed by Eriksson and Kovalainen, 2008).

The reliability of group interview information was examined by Galvez and Resurreccion (1992) in a study of attributes used to describe oriental noodles. Five consumer focus groups were run to generate important terms for the sensory aspects of the noodles and to sort them into positive and negative groups. Highly similar lists were generated by all five groups and they agreed on which terms were desirable versus undesirable. The lists of terms from the consumer groups included 12 of 14 terms used by a trained descriptive panel (apparently two attributes from the descriptive panel were not important to the consumers). In some cases, the words generated were not identical, but were synonyms, such as shiny and glossy. At least for this kind of sensory evaluation application the method seems to have good reliability. Galvez and Resurreccion were careful to screen their participants for product familiarity, which may have contributed to the consistency of results.

Validity is a little more difficult to judge. A study of consumer attitudes and self-reported behaviors found good directional agreement between the results of a series of 20 focus groups and the results of a quantitative mail survey (Reynolds and Johnson, 1978). Validity can also be sensed by the flow of the research process. If the qualitative attribute discovery process is complete, there will be few if any new issues rose on a subsequent consumer questionnaire in the open-ended questions. If the qualitative prototype exploration works well and changes are realized in product characteristics or even conceptual direction, consumer needs and expectations will be fulfilled in the later quantitative consumer test. Phased coordination of qualitative exploration and quantitative testing may enhance utility of results from both types of research (and note that this is a two-way street!) (Chambers and Smith, 1991; Moskowitz et al., 2006). Since conducting a number of groups provides similar information, there is validity in the sense that the information is projectable to additional consumer groups. Although we are careful to disclaim the ability to make any quantitative statistical inferences, the information must be representative of the larger consuming public or it would not be useful. Finally, one can examine the validity in terms of risk from making decisions based on the information. From a practical view, the question arises as to whether the end users of the data will make poor decisions or choose unwarranted courses of action. The sensory professional can aid in this regard in trying to keep product managers from overreaching in their deductions from the information.

The process of conducting focus groups or any kind of flexible interview can be thought of as a communication link (Krueger and Casey, 2009). There are at least five assumptions or key requirements for this process. First, the respondents must understand the question(s). Second the environment is conducive to an open honest answer. Third, the respondents know some answers, that is, they have information to provide. Fourth, the respondents are able to articulate their knowledge or beliefs. Finally, the researcher must understand the respondents' comments. A lack of accuracy or validity can creep in if any of these communication links is weak or poorly functioning. These concerns are key when developing good questions and a good discussion guide.

16.3 Using Focus Groups in Sensory Evaluation

How are qualitative methods employed for questions asked of sensory evaluation specialists? Here are some common applications.

Qualitative methods can be used for exploration of new product prototypes. While product concepts are usually explored by a marketing research group, product development groups that are most often the primary clients of sensory evaluation services may need early consumer input on the direction and success or shortcomings of newly developed variations. Rather than make mistakes that are not detected from the laboratory perspective, consumer opinions and concerns can be explored as part of the refinement and optimization process (Marlowe, 1987; Moskowitz et al., 2006). Prototypes can be evaluated in the group itself, or may be taken home to use, after which a group is convened. This may be very helpful in determining how a food product was prepared, served, and consumed and whether any abuse or unexpected uses and variations were tried (Chambers and Smith, 1991). Changes of direction and additional opportunities may be suggested in these interviews and this information should be shared with marketing managers. If they are partners in the development process they will use this information to the company's advantage. A key strategy is to explore consumer needs and expectations and whether the product in its early stages is moving toward meeting those needs and satisfying those expectations.

Consumer opinion may also help the sensory group focus on key attributes to evaluate in later descriptive analysis and quantitative consumer surveys. A common application of group interviews is in the identification and exploration of specific sensory characteristics. One issue is to try and define attributes that are strongly influential on consumer acceptance (Chambers and Smith, 1991). The early stages of a QDA procedure (Stone and Sidel, 1993) involving terminology discovery for descriptive scorecards resemble the information gathering nondirective approach in consumer focus groups. Attribute discovery can also be conducted with consumers or with technical personnel, e.g., technical sales support and quality assurance staff (Chambers and Smith, 1991). This can help insure that everyone is speaking the same language, or that the different languages can be related to one another, or at the very least that difficulties can be anticipated. In one such application, Ellmore et al. (1999) used qualitative interviews to explore dimensions related to product "creaminess" before further descriptive analysis and consumer testing. This phase was important to identify smoothness, thickness, melt rate, and adhesiveness as potential influences on consumer perception of the creaminess of puddings.

Such "ballot building" is very useful before a consumer questionnaire study. Research personnel may think that they have all the important attributes covered in a questionnaire, but it is likely that some additional consumer feedback will point out a few omissions. Consumers do not necessarily think like research staff. Chambers and Smith (1991) suggest that prescreening questionnaire items with qualitative interviews can address the following issues: Are questions understood? Are they likely to generate biased answers? Are questions ambiguous? Do they have more than one interpretation? Will they be viewed from the expected context? Were there unstated assumptions?

One can get an impression of the potential importance or weight that different product characteristics have in determining overall appeal of the product. This is a classic use of qualitative methods to explore variations in attributes that consumers might find appealing (or not). These variations can then be used in a designed study with a larger group of consumers to uncover the "hedonic algebra" of different attributes and combinations (Moskowitz et al., 2006). A case study linking focus groups with later quantitative consumer testing is given below. Of course, the exploratory groups can be integrated with the goal of getting consumer feedback on early prototypes, as

mentioned above. Insights may arise for new product opportunities here, too. For example, a discussion of flavor characteristics may easily lead to new directions in flavor variations that were not previously considered.

Another useful application is when groups are interviewed as a follow-up after a consumer test. After data have been analyzed or even partially analyzed, it is possible to convene groups of test participants, perhaps some of those who were generally positive toward the product and some who were negative. The interview can probe certain issues in depth, perhaps issues that were unclear from the quantitative questionnaire results or results that were puzzling, unexpected, and in need of further explanation. Chambers and Smith (1991) give the example of a barbecue sauce that received a low rating for spice intensity in a survey, where the actual problem was that the flavor was atypical. Interviews can help confirm or expand upon questionnaire results. For sensory professionals who need face-to-face consumer contact on a project, convening several focus groups is more cost efficient for company personnel than one-on-one interviews. Feedback, probing, and explanation from 20 or 30 consumers can be obtained in an afternoon. Groups are much more efficient than single interviews and have the interactive and synergistic idea-generation characteristics that are not present in other quantitative surveys.

If the company has standing panels of local consumers who regularly test products, it may be possible to bring them in at several points to discuss product characteristics. This is a special case of focus group research termed "focus panels" by Chambers and Smith (1991). This type of setup loses the anonymity factor that is seen as beneficial in most group interviews. However, it may work well in some cultures in which people do not feel comfortable speaking in front of strangers. Local standing consumer panels can be very cost efficient (see Chapter 15), as donations can be made to community groups rather than paying individuals for participation (Casey and Krueger, 1994).

16.4 Examples, Case Studies

The following case studies illustrate two appropriate uses of qualitative methods.

16.4.1 Case Study 1: Qualitative Research Before Conjoint Measurement in New Product Development

Raz et al. (2008) published a protocol for new product development using qualitative consumer information and quantitative data at several stages. Focus groups were used to identify the sensory factors that were used later in a large consumer study based on conjoint analysis principles. This is a classic application of qualitative research that is used to guide further quantitative research. Conjoint analysis is a technique in which consumers evaluate various combinations of attributes at different levels or options and rate their overall appeal. It seeks to find optimal combinations of key attributes at different levels and can estimate the individual contributions of each attribute (called "utilities") to the overall appeal of the product. Moskowitz et al. (2006) show several examples of how the mental algebra of product benefits can be uncovered using conjoint methods. Another example of qualitative interviews used before more structured concept development (also using conjoint measurement) can be found in the paper by Bogue et al. (2009) who looked at foods and beverages with a possible therapeutic or pharmacological function. In order to have meaningful variations of the product, the qualitative work must precede the construction of the product prototypes (or conceptual prototypes) to be evaluated.

The product in this case was a healthful juice drink targeted primarily at women. Two focus groups of nine consumers each were used in the initial stages. Consumers were selected based on socio-demographic characteristics of the target market and were users of the brand or people who switched among brands in the category. Groups were conducted by a psychologist and lasted from 2.5 to 3 h, an unusually lengthy session. The interviews consisted of three phases: an evocative phase involving free association, collage, product, and consumer profiling; a second phase involving presentation of the concept; and then exploration of sensory factors, use properties, and symbolic content of the product and package. The desired result was a set of key attributes with two to four levels or variants of each key attribute and an assessment of the potential products' fit to the concept (i.e., appropriateness).

The flow of the group interviews proceeded as follows: After the introductions, there was an exploration of the brand image and the imagined universe of the product category for that brand. Then the concept was explored and profiled without any additional stimuli. Next, actual samples of taste, odor, and mouthfeel "experiences" were presented to see how they might fit the concept. Visual images were presented that were evocative of the concept and finally packaging variables presented for tactile exploration. Next, the product identity was explored using a collage technique (patching together of visual images) to see if the product identity (as these consumers envisioned it) was in line with the brand image and with the target concept.

These results were used to set up product prototypes for evaluation in a conjoint design with a larger group of consumers generating liking scores for the various attribute combinations. "Importance" scores were also generated which reflected the degree of change across levels of an attribute. That is, attributes which showed a large change in liking scores as the attribute levels changed were high in "importance." A key part of the final analysis was an evaluation of potential segments of consumers, groups who might like different styles of the product, rather than assuming that there was just one overall optimal product from a single set of attribute combinations. The potential segments were explored using cluster analysis.

16.4.2 Case Study 2: Nutritional and Health Beliefs About Salt

Qualitative research is well suited to exploring and understanding consumers' attitudes, beliefs, and knowledge systems and understanding the vocabulary they use to talk about foods and nutritional issues. This study (Smith et al., 2006) examined beliefs and attitudes about salt and high salt-containing foods among at-risk groups of older Americans in the "stroke belt" of the rural southern United States. They conducted both in-depth (one-on-one) and focus group interviews with minority and white community-dwelling elders aged 60–90. The one-on-one interviews consisted of 60–90 min semi-structured interviews to uncover knowledge, beliefs, and folk phraseology. Themes discovered in these interviews were used to develop the interview guide for later focus groups. Seven groups with approximately eight to nine participants were conducted. Some groups were homogeneous in eth-nic makeup (African American, Native American, or white) and others were mixed.

A short demographic questionnaire was admin-istered before the group discussions. A moderator and note-taker participated and discussions were tape (audio) recorded. Tapes were transcribed verbatim. From the transcripts and notes, a codebook was devel-oped of key phrases consisting of core concepts and significant points. Multiple researchers checked the code system for accuracy. The combined transcripts and code system were submitted to ethnographic software analysis of the text. Segments of the text were extracted using the codes for further analysis. Researchers then reviewed the abstracted text samples for "themes." Themes were developed according to the level of consensus, strength, and depth of concepts and frequency. In the analysis and reporting, interpre-tation of the themes was supported by illustration with supporting quotes from individuals.

Results showed that participants believed that salt was an important element in their diet and regional cuisine (those foods loyally described as "southern") and that salt was important to counteract bland taste in fresh foods. Participants recognized a connection between discretionary (table salt, so-called raw salt) usage and high blood pressure, but less connection between salt used in cooking and blood pressure. There was also a connection to a folk condition termed "high blood" which included both dietary sugar and diabetes as linked concepts. The authors contrasted these folk systems and cultural beliefs with medical knowledge and common medical practice. The reader is referred to the full report for further details. A conceptual map depicting the results is shown in Figure 16.2, p. 400.

16.5 Conducting Focus Group Studies

16.5.1 A Quick Overview

A typical focus group procedure could be described as follows: At first glance, we have 8–12 people sitting around a table with an interviewer throwing out ques-tions for discussion. An example of the room setup for a focus group is shown in Fig. 16.1. Some smaller or "mini-groups" have become fashionable, although the probability of respondent synergy goes down with smaller groups, and the possibility of trouble from

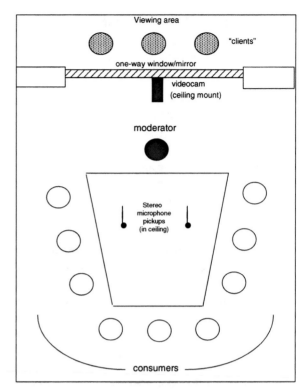

Fig. 16.1 A typical room setup for focus groups. Participants are seated in such a way that their facial expressions and body language are visible to clients behind the one-way glass and can be captured on videotape. Note the trapezoidal shaped table to facilitate this view. Stereo microphones are important to rein-troduce the spatial distribution of sounds on the audio track, as a single microphone may lose this and create difficulties in isolating comments if two people speak at the same time.

unresponsive participants goes up. A common style of questioning is the open-ended form that avoids the possibility of a simple yes/no response. For example, a moderator would probably not ask "Is this easy to prepare?" but rather "How do you feel about cooking this product?" Another useful question style is the "think back" type of question when probing previ-ous experience with the product category (Casey and Krueger, 1994). Probing can proceed by asking for examples, for clarification or simply admitting that you do not understand something. In summary, the visible activity in a focus group resembles the open-ended questions in a structured interview, but allows for more in-depth probing and, of course, considerable interaction of respondents.

A common rule of thumb is to conduct at least three groups (Casey and Krueger, 1994). In case two groups conflict one can get a sense of which group's opin-ions may be more unusual. However, since the method is not based on the need for quantitative projection, the usefulness of this rule is questionable. The discovery of differing opinions is in itself an important and reportable result. It may have later implications for market segments or the need to engineer different products for different consumer groups. Large marketing research projects often require multiple groups in multiple cities to insure some geographically diverse sampling in the United States. Exploratory projects for product development prototypes or language exploration as conducted by a sensory service will generally not be that extensive. However, if there is an important segmentation in the experimental design (e.g., ages, ethnic groups, gender, users versus non-users), then it may be necessary to conduct three groups from each segment (Krueger and Casey, 2009). As noted above, there is a marginal utility in increasing the number of groups, as repeated themes will emerge (Chambers and Smith, 1991).

The steps in conducting a focus group study are similar to those in other consumer research. They are outlined in Table 16.2 and another checklist can be found in Resurreccion (1998). A focus group study resembles a central location consumer test in many aspects of the setup and procedural details, with the notable exception that a moderator or trained interviewer is needed and that the activities are almost always recorded. The project team must be careful to insure that the facility is set up properly and that all recording equipment is pre-tested and functioning correctly. Do not leave this pre-testing up to the facility owners. The researchers may also have to arrange for transcriptions of the verbal record. It is advisable to send the tapes to the transcriptionist as they are completed because each 90 min group may require a day or more of transcription even from a professional.

16.5.2 A Key Requirement: Developing Good Questions

Questions and probes for focus groups are different from the structured questions one finds on a quantitative questionnaire. Examples of probing techniques and alternate methods to direct questions are given in Bystedt et al. (2003). Krueger and Casey (2009) list the following attributes of good questions in group interviews: A good question evokes conversation as well as a single response. It is phrased in common language (not technical jargon). It is short and easy to say/read, is open ended (not yes/no), and is specific, not double-barreled ("Do you think ice cream and frozen yogurt are healthful and nutritious?" is twice double-barreled). Often, focus group questions consider feelings and emotions. That is, they are not

Table 16.2 Steps in conducting focus group studies

1. Meet with clients, research team: Identify project goals and objectives
2. Determine best tools for meeting objectives
3. Identify, contact, and hire moderator
4. Develop screening criteria for participants
5. Develop questions, discussion guide, and sequence
6. Schedule room, facilities, taping equipment
7. Screen and recruit participants; send directions/map
8. Send reminders to participants, time/place/directions/parking
9. Identify and brief assistant moderator, if used
10. Arrange for incentive payments, refreshments
11. Pre-test recording equipment
12. Conduct groups
12a. Conduct de-briefings after each group
12b. Write summaries after each group
13. Arrange for transcriptions if used
14. Modify discussion guide as new information arises
15. Analyze information
15a. Review summaries
15b. Read transcripts or review audio or video tapes
15c. Select themes; find verbatim quotes to illustrate
15d. Confer with another team member to check themes and conclusions
16. Write report and present results

always about knowledge or factual issues. Even in laddering probes (searching for underlying benefits, emotions, values), focus group moderators will tend to avoid the simple question, "Why?" because it may be seen as a criticism or challenge. These can often be rephrased such as "What prompted you to buy X?" or "What aspect of the product motivated you to buy X?" Moderators should avoid giving any examples of answers, as this will tend to tell participants how to answer and get the group in a rut. If directed toward an action, the direction is detailed and specific, such as the following: "Take these magazines and clip out any images you associate with this concept. Put them in a pile in front of you on the table" (Bystedt et al., 2003).

Developing the questions and discussion guide (sequence) is not a solo activity. One should discuss important issues with the clients (people who are requesting the research), including any details, use of product prototypes, other sensory stimuli that might be used as "props," the concept, and review the general objectives. Questions should be brainstormed with five or six other researchers. One should seek appropriate phrasing (e.g., open ended, "think back"). Then the questions or topics can be sequenced. There are a few general rules, including the following: Proceed from general topics to more specific issues. Probing positive aspects generally should precede negatives. The researcher should estimate the time per topic or question area. Then the question guide or discussion flow guide can be drafted. Finally, it should be reviewed with the staff and clients. At this point, the client or research manager may think of all kinds of other issues to include. This can lead to length problems and the researcher has to remind people that this is a 90-min interview. The critical test is to separate what is simply nice to know from what you really need to know.

16.5.3 The Discussion Guide and Phases of the Group Interview

It is most common to have a scripted sequence of questions, but some highly skilled moderators may simply work from a list of issues. There are commonly about five distinct phases to the group and these will be organized on the discussion guide. The moderator may deal with the guide flexibly as new or unexpected potentially useful insights arise and call for probing.

The group begins with a warm-up phase. Turns may be taken or people will just go around the table in order and introduce themselves. One approach is the "introduce your neighbor" option. For example take 5 min to introduce yourself to your neighbor and get one or two interesting facts about this person next to you. Then participants go around and introduce their neighbor to the group. The purpose of the warm-up phase is to engage each participant and make him or her connect his or her thought process with the act of speaking. For many people, it is necessary to commit this simple act of speaking to be able to contribute later. There is otherwise a strong tendency to think about the issues being raised without actually engaging the vocal chords. The warm-up also helps the group members feel more comfortable with each other, as the other people become more of known entities rather than complete strangers, which can be inhibiting to some participants. The introduction phase should try to avoid status indicators. For example, it is better to get them to talk about hobbies than what they own or where they work.

Next comes the introduction to get the topic rolling. Sometimes they may be asked to say something about what products they use in the general category to be discussed that day. A common approach is to ask them to "think back," i.e., tell us about your latest or recent experience with the product. Some issues can be broached at this point, e.g., probe: "What comes to mind when you hear … X… about this kind of product?" The flow of the interview from the general to the specific is the normal trend, and occurs quite naturally in conversation. Stewart and Shamdasani (1990) refer to this as "the funnel" approach. The third phase is a transition phase that moves toward key issues. More specific questions are asked. A product concept may be introduced here or a sample prototype product explored.

The fourth phase gets to the meat of the key questions and issues. Now we get their overall reactions to product, concept, or issue(s) and any individual, personal reactions and thoughts, issues, concerns, and expectations. The bulk of the interview will occur in this phase and one must allow lots of time to probe issues and have them discussed. More specific issues are raised, finally focusing specifically on aspects of interest to the product developers. For example, what characteristics would you like to see in a microwaveable frozen pizza (browning? nutritional content? convenience? shelf life?). Often the critical issues will arise in the natural flow of the conversation. However, the moderator has a discussion guide, which will direct the flow in a general way and insure that all the issues get on the table.

Finally, there is an ending question period. The moderator can review issues and state tentative conclusions. He or she can ask again for overall opinions: "Given what you have heard…?" At this point, ask for comments or corrections of moderator summaries or conclusions that can be thrown out to the group. For example, are there any differing opinions that may not have been stated or included? Was there anything overlooked, something we should have explored but did not? Was there anything that should be included in the next group or should be done differently? Of course, after this phase the group is thanked, paid, and dismissed.

The discussion guide should be developed after a brainstorming session with key people involved in the project to get all potential issues included. The moderator can then draft the discussion guide and submit it for further revision. Examples of discussion guides are found in Table 16.3 and in Chambers and Smith (1991) and Resurreccion (1998). The keyword here is "guide," since flexibility is needed, especially when unexpected but potentially important issues come up. If the moderator recognizes such an opportunity, he or she can ignore the guide for the moment and change direction. Alternatively, the group could return to the issue later but the moderator must note this. If the discussion drifts in a totally unrelated direction (weather, politics, sports, TV shows are all common), the moderator can bring the group back to the main line of discussion.

16.5.4 Participant Requirements, Timing, Recording

As in most consumer tests, participants will be frequent users of the product category and have been carefully prescreened. One exception may be when the goal is to probe non-users, for example, when exploring what it would take for them to try the product or switch into the category or to another brand. In setting up the study, the project leaders should consider the demographic characteristics of the target consumers and set up screening mechanisms to recruit them based on such variables as gender, age, ethnic background; type of household; and location of residence. Generally, the participants will not know each other although in some areas it may be impossible not to get occasional acquaintances in the same group. The key for a group is not necessarily homogeneity, but compatibility (Stewart and Shamdasani, 1990). Some differences in background and opinion may facilitate

Table 16.3 Sample discussion guide: high fiber, microwave pizza	1. Introduce self, note ground rules, mention taping. 2. Warm up—go around table and state name and what type of pizzas you bought most recently (briefly). 3. Discuss pizza category. What is out there? What is most popular? What is changed in your pizza eating habits in the last 5 years? 4. When cooking pizza at home, what kinds do you make (frozen, chilled, baked, microwaved, etc.) Any related products? Probe issues: convenience, costs, variations, family likes and dislikes Probe issues: Any nutritional concerns? 5. Present concept. Better nutritional content from whole wheat and bran crust, high in dietary fiber. Strong convenience position due to microwavability. Competitive price. Several flavors available. Get reactions. Probe: Is fiber a concern? Target opportunity for some consumers? Probe: Is microwave preparation appealing? Concerns about browning, sogging/crispness? 6. Taste and discuss representative protypes. Discuss pros and cons. Probe important sensory attributes. Reasons for likes or dislikes. 7. Review concept and issues. Ask for clarification. Ask for new product suggestions or variations on the theme. Last chance for suggestions. False close (go behind mirror). 8. If further discussion or probes from clients, pick up thread and restart discussion. 9. Close, thanks, distribute incentives, dismissal.

the discussion. It is generally advisable to overbook the participants to anticipate no-shows (Resurreccion, 1998). No-shows can be minimized by sending maps, directions, and a follow-up reminder the day before the group.

The time necessary for most groups is about 90 min, and participants must be informed of this commitment. Motivating people to spend this time (plus travel to a facility) is not always easy. Participants are generally paid and may be provided with refreshments and child care. Incentives must be carefully considered so that the amount is neither too big nor too small, but just sufficient to motivate people to spend the time (Casey and Krueger, 1994). It is sometimes necessary to screen out people who enjoy the activity so much that they become professional participants in the recruiting pools of different testing services. On the other hand, it is sometimes desirable to screen for people who are more vocal. For example, a screening interview might ask, "Do you enjoy talking about _____?"

The discussion is almost always recorded on videotape and/or audiotape. In marketing research, some or all of the clients (those requesting the research) will view the proceedings from behind a one-way mirror. This unusual piece of equipment is not a necessity. Casey and Krueger (1994) point out that the environment will seem more natural without it, and there is a wider range of choices for facilities if you do not use one. Of course, it would be distracting to have clients sit in the same room with participants, so the alternative is to skip direct observation, which entails a good deal of faith in the skills and reporting abilities of the moderator. Participants must of course be told that they are being taped and viewed if that is the case. Usually they forget about this after the discussion starts. There is little reason to believe that the act of taping influences the discussion, since the opinions that are aired are being given publicly in any event (Stewart and Shamdasani, 1990). Debriefing the respondents after the interview about the general purpose of the study is considered polite as at least some participants always want to know. The amount of information disclosed may depend on the security concerns of the client.

Because there is remarkably little action involved, the question is sometimes raised whether videotape is necessary. The advantage is that facial expressions, gestures, and body language can be captured in this medium. This is information that is routinely lost in using only written transcripts for analysis (Stewart and Shamdasani, 1990). Whether non-verbal information is useful depends on the skill of people observing and interpreting the tapes. One or more people may be responsible for generating a report that summarizes the attitudes and opinions uncovered by the procedure. Often this responsibility falls to the moderator, but sometimes to another observer. It may be useful to have at least two independent interpreters view the tapes or proceedings, as a check on the subjective biases of the observers, as a kind of inter-judge reliability check. Tapes may be transcribed to facilitate the use of verbatim quotes to illustrate points and conclusions in the report (see section on reporting results, below). Backup systems for taping are a common recommendation to insure against equipment problems (Chambers and Smith, 1991).

16.6 Issues in Moderating

16.6.1 Moderating Skills

Like descriptive panel leadership, good moderating skills are developed with practice and training. First and foremost, a good moderator is a good listener (Chambers and Smith, 1991). People who like to talk a lot may not be able to suppress the temptation to give their own opinions. Social skills are also required, including the ability to put people at ease, and if necessary to be assertive but diplomatic (Casey and Krueger, 1994). Not everyone can develop a high level of facility in getting respondents to feel comfortable and express their candid personal opinions. Most moderators work well with consumers that are most like them. The same moderator may not fit both a group of female heads of household and a group of male sport fishermen. A weight loss product targeted at obese women should not have a professional racing cyclist as the moderator. Training and practice is important in the development of moderating skills. Videotaping and critique by experienced professionals can be beneficial.

Certain personality traits are helpful. A successful moderator is one who has a sense of humor, is interested in other people's opinions, is expressive and animated, is aware of his or her own biases, and is insightful about people (Stewart and Shamdasani,

1990). They will also show a good deal of flexibility, as the direction of a group can change rapidly. It is advisable to watch a trained moderator in several different groups and preferably in several different products in order to gain insights into the kinds of problems and opportunities that can occur in the flow of the discussion. A good next step is to moderate or co-moderate a session and have the tapes viewed and critiqued by an experienced moderator. Each group will differ so the key to doing this well is experience.

When moderating is done well the focus is on the participants and they discuss issues with one another, rather than answering probes of an interviewer in one direction only, i.e., back at the interviewer. Like a referee in a boxing match, a good moderator becomes invisible as the event progresses. Krueger and Casey (2009) list the following attributes of a good moderator: A good moderator understands the objectives of the project. The moderator has at least some basic familiarity with the product or product category. A good moderator communicates clearly, respects the participants (and shows it), and is open to new ideas. A moderator, by definition, is good at eliciting information. He or she gets people to talk and to elaborate on their comments. Three issues are keys to good moderating: nondirection, full participation, and coverage of issues.

16.6.2 Basic Principles: Nondirection, Full Participation, and Coverage of Issues

The primary goal of the moderator is to guide the discussion without suggesting answers or directing discussion toward a specific conclusion. In this respect, the moderator acts like a conceptual midwife, extracting ideas, perceptions, opinions, attitudes, and beliefs from the group without imparting his or her own opinions. The technique draws heavily from the client-centered interview techniques of psychologists such as Carl Rogers. Whenever participants look to the moderator for an answer or opinion, the question is thrown back to them, perhaps rephrased or in some general terms like "I hear what you are asking. Why is this issue important to you?" To avoid the temptation of subtle direction, many users of focus group information prefer to have a moderator who is

technically uninformed about the issues and has no preformed opinions. While this goes a certain distance in helping insure an unbiased discussion and report, it can sometimes miss an opportunity to probe important technical issues that arise, that only an informed moderator would recognize.

Much of the questioning from the moderator will take the form of probing for further thoughts. Sometimes silence is useful as a probe, as the recent participants may want to fill the gap with further elaboration. In general, an experienced moderator will use carefully placed silences to advantage (Stewart and Shamdasani, 1990). Silence is by its nature nondirective. Other useful probes are to ask for reasons behind feelings or to expand the discussion to other participants by asking whether anybody else "shares this view." However, it is important to avoid emotionally loaded phrases like "Does anybody agree (or disagree) with this?"

Moderator bias can easily creep in. This can arise from a need to please the client, reasons of personal bias on the issue at hand, or a need for consistency in the moderator's own thoughts and beliefs (Stewart and Shamdasani, 1990). It is a relatively simple matter to give undue support to the ideas of a participant you agree with by a number of different means: giving more eye contact, verbal affirmation, head nodding, being more patient, or calling on them first (Kennedy, 1976). Conversely, the unwanted opinions can be easily de-emphasized by failing to probe, summarize or reiterate contrasting, minority or unfavorable opinions. A good moderator will recognize these actions and avoid them as well as recognize when they may have occurred when viewing the taped records.

Moderators should also be sensitive to the answers that are generated due to social factors. Often respondents will choose to try and please the moderator by giving answers that they suppose are the desired ones. Chambers and Smith (1991) give the example of a discussion of brown bread, in which participants may claim to like or use it, but are actually fond of white bread.

A good moderator tries to encourage inclusion of all participants to insure that all sides of an issue are raised and aired. Depending on the culture, different techniques can be used by the moderator to this end. A good idea after some discussion of the idea is to probe the lack of consensus, encouraging new opinions by asking if anyone disagrees with what has been said.

Overly talkative or too quiet participants are a common problem. Dominant respondents may be experts, either real or self-appointed. True experts will tend to ruin a group since less knowledgeable participants will look to them for answers. These individuals can usually be screened out during recruiting. Self-appointed experts are a more difficult problem, and need to be controlled or they will have undue influence on the discussion. Dominating participants can be restrained somewhat by subtle negative reinforcement by the moderator. Non-verbal cues such as lack of eye contact (looking at the ceiling or floor), drumming fingers, reading the discussion guide, shuffling notes, getting up and doing something in the room, or even standing behind the person—each of these can provide negative feedback (Wells, 1974). Often a confrontational or aggressive individual will choose a seat directly opposite to the moderator. A change of seating, if this can be done comfortably (e.g., shifting name placards during a 5 min break) may help. Conversely, a shy person may choose to sit to the side, in a corner or facing the same direction as the moderator to avoid eye contact. Drawing such people out demands additional strategies. Nodding and smiling when they speak or leaning forward to show interest will reinforce their participation. Casey and Krueger (1994) suggest a combination of pausing with eye contact to elicit a response from a quiet participant. People feel a little uncomfortable with silences of even 5 s and will want to fill the gap.

Another goal of every moderator is to insure that all issues are covered. This entails careful development of the discussion guide with the people who request the groups. Interviewing them is no less important than interviewing the consumers in the actual group. All issues should be included, barring major time constraints. Time management is an important skill of the moderator as groups can become surly if held over the stated time (Stewart and Shamdasani, 1990). The moderator should have flexibility if some issues arise naturally and out of the order of the discussion guide. It is often best to keep the ball rolling. In some very good groups, the participants will anticipate the next issue and the stream of discussion will flow with little pushing and probing by the moderator. If a one-way viewing room is used, the moderator may wish to step out for a minute and visit with the observers to see if there are new issues or further probing that they desire. This may also provide the opportunity for a "false close," when the moderator

makes it appear that the discussion is ended and leaves the room, only to observe a sudden burst of discussion in his or her absence. People may voice opinions they did not feel comfortable saying in front of the moderator (Chambers and Smith, 1991). This should trigger some follow-up, as it is a clear indication that there are other issues on the people's minds that need to be probed. Data gathering on multiple occasions should be viewed as a learning process rather than a repeated experiment (e.g., Falk et al., 1996; Furst et al., 1996, Janas et al., 1996).

16.6.3 Assistant Moderators and Co-moderators

Some texts recommend the use of an assistant moderator (Krueger and Casey, 2009). This is not often seen in marketing research but makes a lot of practical sense. The assistant can check the equipment, arrange for food and incentive payments, check in the participants, and take care of any forms that need to be filled out. An assistant can take notes from behind the one-way mirror or off to the side of the room, paying careful attention to be unobtrusive and draw no attention from the participants. If there is only one professional moderator, the assistant can come from the research team. If so, it is important for other team members or clients to realize that they are *not allowed* to sit in on the groups in the same room. The assistant moderator's notes become an important source of data for the eventual report, along with the moderator summaries and any recorded and transcribed records. The person requesting the research act can also act as an assistant to the moderator. This insures that he or she sees a representative number of groups and does not rush to premature conclusions based on the first group alone.

Another variation of the group interview uses multiple moderators. It may be helpful in a group to have co-moderators to probe discussions from different points of view. The second moderator may also make some people feel more at ease, if they can identify better with that person in terms of gender and other social variables. Like an assistant moderator, an important function of a co-moderator is to assist in debriefing and constructing the immediate summary of the results. Co-moderators may also be technically experienced people (Marlowe, 1987). Such persons can clarify some issues or technical questions that arise and may recognize the potential importance of some unexpected lines of discussion to future projects. Not all focus group specialists recommend this approach. Casey and Krueger (1994), for example, feel that it may be confusing to respondents to have two moderators on an equal footing. They prefer the use of an assistant moderator, who can take care of latecomers, taping, refreshments, props, and other details.

16.6.4 Debriefing: Avoiding Selective Listening and Premature Conclusions

One advantage of viewing focus groups is to hear actual consumer comments in their own words, with tone of voice, gestures, and body language. The observation can be compelling. However, the act of viewing groups entails a major liability as well. Selective listening may occur and people will often remember the comments that tend to confirm their preconceived notions about the issues. The immediate and personal nature of viewing a group discussion can often be quite compelling and much more so than any written report or numerical summary of a quantitative survey. Some observers will tend to form their opinions long before they see the report, and sometimes without the information from subsequent groups, which may be contradictory. It is also quite possible to skew the reporting of comments by extracting them out of context, in order to confirm a client's favorite hypothesis. As noted by Stewart and Shamdasani, "Almost any contention can be supported by taking a set of unrepresentative statements out of the context in which they were spoken" (p. 110). One job of the sensory project leader should be to discourage selective listening and out-of-context reporting, as well as to caution against early conclusions and reports to management before the analyst can convert the data into information.

There are several ways to avoid this. If one observer is the biggest proponent of a project, concept, or prototype, that individual can be given the job of writing down every negative comment or consumer concern. Give the job of writing down positive information to the biggest skeptic. There is no rule against assigning tasks to observers, although whether you use the information later is optional. Of course, people who

are used to passively listening to focus groups (or worse yet, making comments or even jokes about the participants) may not be receptive to the idea that this experience requires their complete attention (cell phones off!). Marlowe (1987) suggests that listening behind the glass takes discipline, concentration, self-control, and objectivity, especially toward negative attitudes that are difficult to swallow. A debriefing session is held just after the group is concluded and can promote a balanced view of the proceedings (Bystedt et al., 2003; Chambers and Smith, 1991; Marlowe, 1987). Asking, "Did you hear what I heard?" about key points can also remind people of comments they may have missed, since human attention will eventually wander during the observation. Peer debriefing can be an important tool in enhancing the trustworthiness of conclusions (Janas et al., 1996).

16.7 Analysis and Reporting

16.7.1 General Principles

The type and style of analysis should be driven by the purpose of the study. Once again the objectives of project are the key concerns. The analysis must be systematic, verifiable, sequential, and ongoing (Krueger and Casey, 2009). By systematic, we mean the analysis follows a specified plan that is documented and understood. By verifiable we mean that there is a sufficient trail of evidence and documentation of how conclusions were arrived at. Another researcher would arrive at the same or very similar conclusions.

Analysis follows a sequence; it has phases including note taking, debriefing, and writing summary notes and transcript evaluation. Analysis may be ongoing, as the design and questioning may be modified as the group's progress. This flexibility is an asset. Eriksson and Kovalainen (2008) discuss various approaches to content analysis, including software programs used for text analysis in a variety of business applications. A detailed example of systematic and quantitative text analysis is given in Dransfield et al. (2004).

The data can take several forms including a full verbatim transcript, an abridged transcript, note-based analysis, and memory-based analysis. A full verbatim transcript is the most expensive and slowest but simplest for researchers. Transcripts can be useful for lifting verbatim quotes to illustrate points. Stewart and Shamdasani (1990) suggest transcribing as a first step in analysis. This allows the analyst to cut and paste (either physically or on a word processor) to group comments of a similar nature. An abridged transcript must be done by someone familiar with the project and its goals (they may not be a skilled transcriber). It cuts out introductory material and comments that are deemed irrelevant or off-topic. Note-based analysis depends on ability of the note-taker. If notes are the main source it is important to also have the audio record for review. Data will also consist of assistant moderator notes, moderator summary notes, and any debriefing notes. Memory-based analysis requires the most skill. It may be done with highly experienced moderators who offer an on-the-spot summary to those watching behind the one-way mirror.

16.7.2 Suggested Method ("Sorting/Clustering Approach"), also Called Classical Transcript Analysis

The systematic analysis of a verbatim transcript can be a detailed and objective approach to dealing with a collection of consumer discussions. However, it is time consuming and may be a bit slow for some marketing research requirements. In this section we describe a simple straightforward method that requires no specialized software, but will require a serious time commitment. It is based on groupings of similar ideas, sometimes called "affinity analysis" in new product design. The transcript analysis proceeds as follows (abridged from Krueger and Casey, 2009):

1. Setup: Obtain two verbatim transcripts, a large room with a large table (or similar functionality), flip chart paper (or similar, about 18×24 in. minimum) for each of the 8–10 key questions or thematic areas, scissors, and tape. Label each large paper sheet with the key question or theme from the discussion guide. Number the lines on the transcripts sequentially by word processor, to be able to refer to where they came from. If you have multiple groups with multiple transcripts: use different

colored paper for each transcript to be able to recover which group it came from.
2. Extracting quotes: Take one transcript and cut off the introductory material. Start with the first substantive comment. Does it offer any information? If so, slice it out and tape (or make a pile) under the appropriate key question or theme. If irrelevant or content free, discard. If uncertain, place aside in a pile to review later.
3. Organize: Continue to extract and categorize useful, information-rich quotes. As sub-themes emerge, organize quotes within sub-categories by dividing up each large sheet of paper into a matrix of sub-areas with a word or two describing each. If a quote seems to fall into two areas, photocopy it and place one in each area. Note: Do not worry if a sub-category has only one quote. One insightful idea may be quite valuable even if only one person thought of it.
4. After you have completed all transcripts, review the piles and sub-categories to make sure similar ideas are put together. Reorganize and make further sub-categories as needed. Be sure to review and contrast statements extracted earlier to insure that they are related, a method of "constant comparison" (Eriksson and Kovalainen, 2008).
5. If possible, have a second researcher review your work to see if there are ambiguities, points of disagreement, or outright differences in interpretation or groupings.

This simple cutting, pasting, and organizing task forms the basic information matrix from which you can begin to construct themes. Next, the analysis proceeds to written summaries and analysis of specific questions and themes as follows:

(1) Write a descriptive summary of responses to each question and organize by sub-theme or category.
(2) Contrast across groups if there are group differences (demographic, users versus non-users, gender, etc.).
(3) Weight the importance of each theme using the following criteria:

Frequency and extent: Important themes tend to emerge repeatedly. Frequency is the number of times something is mentioned, extent is the number of people who voice that opinion or comment. One person who continues to ramble may not be as significant as when similar comments emerge in different groups.

Specificity: Comments that are detailed, specific, and actionable rather than merely generalizations tend to be more useful.

Emotion: Opinions strongly voiced may be of greater importance.

Potential insight: Unanticipated, breakthrough, paradigm shifting, innovative, actionable.

(4) Develop the transcript summary using the weighted themes, organized by question or topic, then subtopic, and choose about three quotes to best illustrate each summary point. Use this transcript summary as the centerpiece of your written report.

There are a number of alternative analyses that can be done on a verbatim transcript. A similar approach can be done with a word processor instead of the physical cut-and-paste method. The researcher must be careful to "tag" the extracted quotes so that there is a record of what group they came from and what question or topic elicited that comment. An increasingly common technique is to use specialized software for text and content analysis (Dransfield et al., 2004; Eriksson and Kovalainen, 2008). A number of commercial packages are available for this purpose. They involve coding of various response types, categories, or sub-categories so that the text can be searched. This may require skill with the software program as well as skill in developing the codes. Programs also exist to analyze sound files. In a simple version of this, the researcher may be able to mark comments on the sound recording, to tag them for later analysis and sorting.

Once the transcripts are analyzed, one must return to the summaries that were written by the moderator or assistant back when the groups were initially conducted.

A good moderator will write a summary of key points immediately after each group (Casey and Krueger, 1994). These original summaries should be compared to the transcript summaries and combined and modified to begin the construction of the report. Any debriefing notes should also be considered. They should be reviewed one point at a time, usually in the order that issues arise in the discussion guide. In other words, read all of the summaries looking at a single issue and then write the overarching conclusion based on this impression.

16.7.3 Report Format

In general, for industrial reports, a bulleted style is recommended. Include a cover page, objectives, summary, key findings, interpretations (if needed), and recommendations. Append details of method, groups, locations, dates, etc. The discussion guide may be appended. A sample industrial report is shown in Appendix of this chapter. Some guidelines follow.

First, limit your points. Go from the most potentially important findings to the least. Within each category go from general ideas to more specific items (Casey and Krueger, 1994). The big ideas can form the basis for an executive summary if one is needed. Sometimes people will raise an issue in different contexts or different words, and these may fall under some general theme like packaging, convenience of preparation, concerns about nutritional content, and flavor. A good analyst will recognize overarching issues and organize the report around them. Such organization will make the report more digestible, actionable, and memorable to readers, as opposed to a disorganized list of opinions or quotes. Use verbatim quotes (limit of three) for each bullet point. This has high face validity and illustrates to readers of the report how consumers actually talked about the product or concept. The written report will normally summarize key points following the discussion guide and then raise new issues. If the report is an oral presentation or in electronic format, you may be able to illustrate with video clips which can be compelling.

Avoid the temptation to "count heads." It is a very natural tendency to report results using phrases like "the majority of participants," "most consumers agreed," or "a minority opinion was that …." Guard against such unintended quantitative comments, as these may sound like projections to the reader about the larger consumer population.

Concept maps or pictorial maps of the associational structure can be valuable. They can illustrate different results from different groups of individuals. For example, it may be desired to compare experts to novices for some products, or culinary professionals to consumers, or groups of regular and infrequent purchasers of a product. Concept mapping is a method by which the ideas (nouns, mostly) are presented as nodes (boxes or circles in the display) and their relationships (verb phrases) are pictured by labeled lines connecting the nodes (Novak and Gowin, 1984). A good example of this approach was a comparison of consumers to fishing industry experts in their approach to seafood quality (Bisogni et al., 1986). Experts were concerned with a wider array of technical issues and processing factors, while consumers were more focused on sensory attributes. This difference was plainly obvious in the pictorial concept maps. Another example can be found in Grebitus and Bruhn (2008) who examined consumers' concepts of pork quality. The individual maps were subjected to quantitative analysis to provide degrees of relationship between 15 key concepts. The complexity of the discussion, the information elicited, and its underlying conceptual structure can all be easily appreciated in this kind of pictorial display. A sample concept map from the previous case study on salt and health is shown in Fig. 16.2.

16.8 Alternative Procedures and Variations of the Group Interview

16.8.1 Groups of Children, Telephone Interviews, Internet-Based Groups

Krueger and Casey (2009) discuss several other variations on focus groups. Three areas are potentially useful in sensory evaluation and new product development: focus groups with children, telephone focus groups, and "discussion" groups held via the Internet. Each of these requires modification of the usual procedures for group discussions and a good deal of flexibility. As always, making sure the tool is appropriate for the project objectives is key. The following descriptions and guidelines are summarized from Krueger and Casey (2009).

In focus groups with children or teenagers, smaller groups need to be assembled and about six kids is a good number. The time of the group must be shorter, usually no longer than 1 h. Getting a good moderator that can relate to kids and make them comfortable

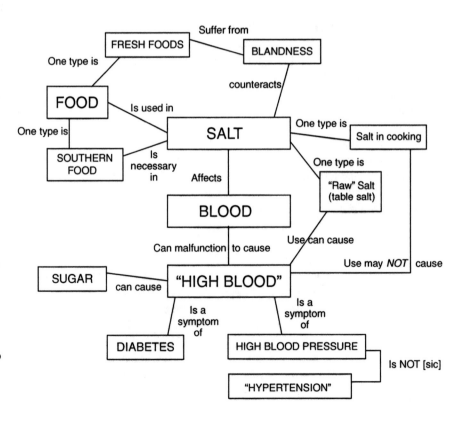

Fig. 16.2 A sample concept map summarizing the key findings from consumers in the salt study (Case Study 2, Smith et al., 2006). Key ideas are represented as boxes (nodes) and are connected by verb phrases to indicate their relationships. When this kind of map is created by individual consumers as part of an interview or focus group exercise it is referred to as "mind-mapping" (Bysted et al., 2003).

talking to new acquaintances is important. Children must be of about the same age, no more than 2 years difference among the group. Avoid recruiting close friends who know each other. If possible, prescreen for kids that are willing to speak up. Food is a good idea. Groups should be conducted in a friendly and possibly familiar location, and one that is not associated with adults in authority such as a school if possible. Questions should be appropriate to the age of the children and yes/no questions should be avoided as usual. It is common to spend a good 15 min in warm-up talking about a popular topic like music or video games.

Telephone interviews allow for more geographic diversity in the group but they also have restrictions and special requirements. Because no travel is involved, incentives can be small or non-monetary. The disadvantage is that any body language or facial expressions are lost. Phone interviews are conducted like a conference call in which the participants call in via a pre-arranged service. Participants should be instructed to call in from a private, comfortable place in which they will not be tempted to be multi-tasking or doing other business during the call. The clarity of the connections should be carefully assessed when people sign in. Groups must be smaller (4–5 participants) and shorter (about an hour). People should state their names before comments, at least at first until it is clear by voice who is speaking. Be prepared to stimulate the discussion when silences occur and intervene when conversations seem to be stuck on one topic.

The Internet offers several options as alternatives to the traditional discussion group. There must be a secure, password-protected system. Like phone interviews, there can be geographic diversity but there are no visual observations of expressions or body language. There may be less spur-of-the-moment synergistic interactions because responses may be delayed. Before or during the group, images, sounds, complex concepts, and such can be posted on another website for viewing. The Internet interaction will avoid some of the issues of social dominance or hierarchies that can surface in a face-to-face group.

One option for the Internet is to set up a chat room with a specified time to enter and interact. Typically the chat room could have six to eight participants and last up to 90 min. A sequence of questions are posted often ahead of time for consideration. Information for inspection and comment before the group can be posted on another website. A chat room will tend to get more top-of-mind comments that may not be well thought-out responses. The format favors people who type fast and do not edit their thoughts, and such persons may dominate the exchange. Moderators should be careful to identify their own comments, questions, and probes, for example, by using all capital letters. Participants should be warned if time is running out on a topic so those who are waiting to respond can do so. An obvious advantage of the chat room is that it can provide a written record of all comments.

Another alternative is an Internet bulletin board. This functions like a chat room extended over time with posted comments. Participants must agree to spend 15–30 min per day reading and posting throughout the duration of the project and they should be informed of the duration. The extended timetable may entail some attrition as personal schedules change or emergencies arise. Bulletin boards can evoke more reflective in-depth comments than a chat room. Participants are urged to post comments on the general topic of the day as well as respond to comments of others. The moderator may summarize a previous day's findings for further commentary or amendments.

16.8.2 Alternatives to Traditional Questioning

Within classical focus groups, there are a number of special activities that can be done that involve more "doing" than "talking." There are no rules that prohibit the use of tasks or consumer interactions with sample products or sensory experiences. Bystedt et al. (2003) list a variety of techniques as alternatives to traditional question and answer. These include free association (what comes immediately to mind when I say X?), mind-mapping (drawing a map of your ideas similar to the concept map discussed above), making collages of the product image from magazine cutouts, responding to a pre-arranged deck of visual images (which of these pictures is associated with the product concept?), various techniques for laddering, and several other methods to stimulate imagery and probe associations.

Sometimes tasting or using a product may help bring to mind advantages and concerns. To facilitate discussion, participants can be asked to bring along a similar product or one they regularly use from the category. Having been presented an idea or prototype, they can be asked to write down three positive and three negative aspects of the product, in their opinion. This can provide some time for reflection and if written, can be collected later to see if they voiced the actual items they wrote down. Such information could be presented to subsequent groups for their commentary. Building from recorded positives and negatives is very much in the tradition of Merton, who originally focused his group discussions on the positive and negative comments recorded during the playing of the radio programs (Stewart and Shamdasani, 1990). Writing down comments on index cards to be passed to the moderator can allow anonymous opinions to be opened for discussion.

Interaction with products can involve other organized tasks. For example, participants can be asked to sort a group of items into clusters or pairs and give the reasons for their categorization. This can be done individually or as a group exercise. They can be given triads of items, as in the repertory grid method (McEwan and Thomson, 1988) in which they describe the reasons why two of the three items are similar to each other and different from a third. Alternatively, they can be given pairs and asked to describe similarities and differences. Another method for exploration is a product placement or mapping task, in which participants place objects on the table to represent their similarities, differences, dimensions of difference, and perhaps categorical relationships (Bystedt et al., 2003; Light et al., 1992; Risvik et al., 1994). The reverse task of dividing up the group sequentially can also be performed. Participants are asked to divide the product first into two piles, to discuss the reasons for this division, and then further divide the piles if possible (see Ellmore and Heymann, 1999 and Olson, 1981 for examples).

Probing information on underlying attitudes and motivations can also benefit from some special techniques. Additional tools have been borrowed from the projective techniques in psychoanalysis (Stewart and Shamdasani, 1990). These are called projective in the Freudian sense in that respondents need not couch their responses in terms of their own feelings, but state them as belonging to someone else. For example,

participants could describe the type of house or household that a person who would use the product would have. "Making up a story" about the product or user is another approach (Bystedt et al., 2003). Seemingly unrelated aspects can give insights into how the product is perceived in relation to attitudes. What kind of car does the person drive who would buy this? What sports do they like? What TV shows do they watch? The line of questioning can include constructing a complete story about the potential user of the product. Participants may be asked to draw a stick figure of the person and label/describe them. Another common projective technique is "filling in bubbles." This describes a partially blank or ambiguous picture, such as a consumer thinking, and a "bubble" appearing above her as in a cartoon or comic strip. In this way, participants can place the attitudes outside themselves onto the person in the picture. Many variations on these techniques are available and more continue to be developed.

16.9 Conclusions

Some people consider observational or qualitative methods "soft" science or not even scientific at all, that is, an unfortunate and unduly pejorative view of this form of research, which can be systematic and verifiable (see discussions of "grounded theory" in Eriksson and Kovalainen, 2008). The dominant model for scientific research over the past 150 years has been that of controlled experimentation and logical hypothesis testing. This "positivist" model has served us well, but it is not the only way to gather useful information. Almost all science begins with observation of some phenomenon in order to develop a framework for later experimentation and tests of theory. Field biology using observational methods to study animal behavior is one example, and in principle, it is not unlike the kinds of methods discussed in this chapter (except perhaps that there is more interaction of the observed and the observer). As always, the key is to find the right tool to address the objectives of the study or research project:

The decision to use a focus group or some other research tool must be based on the appropriateness of the method for obtaining answers for specific research questions. It has been noted before that to a man with a hammer, everything is a nail. . . . . Focus groups are useful for particular purposes and specific situations — for exploring the way particular groups of individuals think and talk about a phenomenon, for generating ideas and for generating diagnostic information. For these purposes, focus groups represent a rigorous scientific method of inquiry (Stewart and Shamdasani, 1990, p. 140).

A number of myths surround the use of focus groups (Krueger and Casey, 1994). One common misconception is that they are inexpensive. If three groups are conducted in each geographic location or three groups of each demographic segment, the budget will grow quite large in a hurry. A second myth is that they give rapid feedback. If one considers all the steps from recruitment, hiring test agencies or moderators, conducting the study and the analysis, it can easily take a minimum of 6 weeks to get the information needed. Another common notion is that a one-way mirror is required. With video transmission this is hardly necessary and the situation may seem more natural to the participants without it. Giving up the one-way mirror allows a greater flexibility for the kinds of locations that can be used for the study. Finally, there is the question of whether one needs an expert facilitator

(or conversely that anyone can run a focus group). Certainly some skills are needed, particularly that of being a good listener.

Qualitative research methods have become important tools for consumer research in the methods available to sensory specialists. For uses such as exploring important sensory attributes and issues related to functional product characteristics (packaging, use directions, convenience, etc.) they are quite valuable. As sensory specialists become more a part of cross-functional teams during product development, they will be exposed to an increasing array of consumer-centric qualitative and quantitative methods, many of which are used for conceptual optimization (Moskowitz et al., 2006). It is important for sensory professionals to understand these tools and how they are used, in order to make the best use of the information. Conducting qualitative research is in some ways like sailing a boat or taking a hot air balloon ride. Some things are under your control, many others are not, and each experience provides some surprises. They are almost always valuable learning experiences.

Appendix: Sample Report Group Report

Boil-in-bag Pasta Project Followup Groups

Abstract

Three discussion groups were conducted following a home use test of the boil-in-bag pasta product. Potential areas for improvement were identified in bag strength, sauce formulation and usage instructions. The convenience aspect of the product was perceived as a major point of attractiveness to the participants. There was some interest in a low calorie version of the product.

Objective

To assess consumer reactions to the boil-in-bag pasta product in an in-depth group interview, to allow further probing of issues identified during the formal in home test and quantitative questionnaire.

Methods

[appropriate methods would be described here or appended]

Results

1. The major consumer-perceived advantage was the product's convenience:

   "I really liked the product since you could just pop the thing in boiling water and pull the bag out five minutes later with a completely cooked dish with sauce and everything. You just throw the pouch away and cleanup was easy."

   "I liked the speed of preparation. When I get home from work, my kids are screaming for their dinner, and well, you know, my husband doesn't lift a finger to help, so I need to get this done to keep the kids from rioting."

2. Problems were seen with the sauce flavor, particularly regarding salt level:

   "The pasta was nice and firm, but I thought the sauce was, you know, way too salty. My husband is on a low salt diet for high blood pressure, and he just went right through the roof."

   "The herb sauce was just too strong. It didn't seem like an authentic Italian dish to me. My mother's version was much more subtle."

3. Problems were seen with the bag strength:

   "With both the products I tried, the bag broke. I would never buy a product like this again if that happened just once. It makes a terrible mess and besides, is just a waste of money since you can't do anything with the food once it gets into a whole pot of hot water."

4. Usage instructions were not clear, especially regarding done-ness:

   "It says to cook until firm. But, uh, you know, like how do I tell its firm when its inside a bag and in a pot of boiling water?"

5. There was some interest in nutritionally-oriented versions of the product:

"I liked the taste, but when I read the nutritional information, I was surprised at the fat level as well as the sodium. I mean, if we can have lite beer and lite everything else these days, we should have a line of these products that's better for you, too."

Conclusions

[appropriate conclusions would appear here]

Recommendations

[appropriate recommendations would appear here]
[in some companies, they will appear at the top of the report]
[abstract may then be replaced with "executive summary"]

Disclaimer

"Qualitative research provides a rich source of information in clarifying existing theories, creating hypotheses and giving directions to future research. This research is based on a limited non-random sample of participants. Such qualitative research is not projectable. No statistical inferences should be drawn from these results. Any results should be viewed as tentative without quantitative corroboration."

References

Ares, G., Gimenez, A. and Gambaro, A. 2008. Understanding consumers' perception of conventional and functional yogurts using word association and hard laddering. Food Quality and Preference, 19, 636–643.

Bisogni, C. A., Ryan, G. J. and Regenstein, J.M. 1986. What is fish quality? Can we incorporate consumer perceptions? In: D. E. Kramer and J. Liston (eds.), Seafood Quality Determination. Elsevier Applied Science, Amsterdam, pp. 547–563.

Bogue, J., Sorenson, D. and O'Keeffe, M. 2009. Cross-category innovativeness as a source of new product ideas: Consumers' perceptions of over-the-counter pharmacological beverages. Food Quality and Preference, 20, 363–371.

Bystedt, J., Lynn, S. and Potts, D. 2003. Moderating to the MAX. Paramount Market, Ithaca, NY.

Casey, M. A. and Krueger, R. A. 1994. Focus group interviewing. In: H. J. H. MacFie and D. M. H. Thomson (eds.), Measurement of Food Preferences. Blackie Academic and Professional, London, pp. 77–96.

Chambers, E., IV and Smith, E. A. 1991. The uses of qualitative research in product research and development. In: H. T. Lawless and B. P. Klein (eds.), Sensory Science Theory and Applications in Foods. Marcel Dekker, New York, pp. 395–412.

Cooper, P. 2007. In search of excellence. The evolution and future of qualitative research. ESOMAR World Research Paper.

Dransfield, E., Morrot, G., Martin, J.-F. and Ngapo, T. M. 2004. The application of a text clustering statistical analysis to aid the interpretation of focus group interviews. Food Quality and Preference, 15, 477–488.

Ellmore, J. R. and Heymann, H. 1999. Perceptual maps of photographs of carbonated beverages created by traditional and free-choice profiling. Food Quality and Preference, 10, 219–227.

Ellmore, J. R., Heymann, H., Johnson, J., and Hewett, J. E. 1999. Preference mapping: Relating acceptance of "creaminess" to a descriptive sensory map of a semi-solid. Food Quality and Preference, 10, 465–475.

Eriksson, P. and Kovalainen, A. 2008. Qualitative Methods in Business Research. Sage, London.

Falk, L. W., Bisogni, C. A. and Sobal, J. 1996. Food choice of older adults: A qualitative investigation. Journal of Nutrition Education, 28, 257–265.

Furst, T., Connors, M., Bisogni, C. A., Sobal, J. and Falk, L. W. 1996. Food choice: A conceptual model of the process. Appetite, 36, 247–266.

Galvez, F. C. F. and Resurreccion, A. N. A. 1992. Reliability of the focus group technique in determining the quality characteristics of mungbean [Vigna Radiata (L.) Wilzec] noodles. Journal of Sensory Studies 7, 315–326.

Goldman, A. E. and McDonald, S. S. 1987. The Group Depth Interview, Principles and Practice. Prentice-Hall, New York.

Grebitus, C. and Bruhn, M. 2008. Analyzing semantic networks of pork quality by means of concept mapping. Food Quality and Preference, 19, 86–96.

Hashim, I. B., Resurreccion, A. V. A. and McWatters, K. H. 1996. Consumer attitudes toward irradiated poultry. Food Technology 50(3), 77–80.

Janas, B. G., Bisogni, C. A. and Sobal, J. 1996. Cardiac patients' mental representations of diet. Journal of Nutrition Education, 28, 223–229.

Kennedy, F. 1976. The focused group interview and moderator bias. Marketing Review, 31, 19–21.

Krippendorf, K. 1980. Content Analysis: An Introduction to Its Methodology. Sage, Beverly Hills.

Krueger, R. A. 1994. Focus Groups: A Practical Guide for Applied Research, Second Edition. Sage, Newbury Park, CA.

Krueger, R. A. and Casey, M. A. 2009. Focus Groups, Fourth Edition. Sage, Thousand Oaks, CA.

Krystallis, A., Maglaras, G. and Mamalis, S. 2008. Motivations and cognitive structures of consumers in their purchasing of functional foods. Food Quality and Preference, 19, 525–538.

Light, A., Heymann, H. and Holt. D. 1992. Hedonic responses to dairy products: Effects of fat levels, label information and risk perception. Food Technology 46(7), 54–57.

Marlowe, P. 1987. Qualitative research as a tool for product development. Food Technology, 41(11), 74, 76, 78.

McEwan, J. A. and Thomson, D. M. H. 1988. An investigation of factors influencing consumer acceptance of chocolate confectionary using the repertory grid method. In: D. M. H. Thomson (ed.), Food Acceptability. Elsevier Applied Science, London, pp. 347–361.

Moskowitz, H. R., Beckley, J. H. and Resurreccion, A. V. 2006. Sensory and Consumer Research in Food Product Design and Development. Blackwell, Ames, IO (IFT).

Novak, J. D. and Gowin, D. B. 1984. Learning How to Learn. University Press, Cambridge.

Olson, J. C. 1981. The importance of cognitive processes and existing knowledge structures for understanding food acceptance. In: J. Solms and R. L. Hall (eds.), Criteria of Food Acceptance. Foster, Zurich, pp. 69–81.

Raz, C., Piper, D., Haller, R., Nicod, H., Dusart, N. and Giboreau, A. 2008. From sensory marketing to sensory design: How to drive formulation using consumers' input. Food Quality and Preference, 19, 719–726.

Resurreccion, A. V. 1998. Consumer Sensory Testing for Product Development. Aspen, Gaithersburg, MD.

Reynolds, F. D. and Johnson, D. K. 1978. Validity of focus group findings. Journal of Advertising Research, 18(3), 21–24.

Risvik, E., McEwan, J. A., Colwill, J. S., Rogers, R. and Lyon, D. H. 1994. Projective mapping: A tool for sensory analysis and consumer research. Food Quality and Preference 5, 263–269.

Schutz, H. G. and Judge, D. S. 1984. Consumer perceptions of food quality. In: J. V. McLoughlin and B. M. McKenna (eds.), Research in Food Science and Nutrition, Vol. 4. Food Science and Human Welfare. Boole, Dublin, pp. 229–242.

Smith, S. L., Quandt, S. A., Arcury, T. A., Wetmore, L. K., Bell, R. A. and Vitolins, M. Z. 2006. Aging and eating in the rural, southern United States: Beliefs about salt and its effect on health. Social Science and Medicine, 62, 189–198.

Stewart, D.W. and Shamdasani, P. N. 1990. Focus Groups: Theory and Practice. Applied Social Research Methods, Series Vol. 20. Sage, Newbury Park, CA.

Stone, H. and Sidel, J. L. 1993. Sensory Evaluation Practices. Academic, New York.

von Hippel, E. 1986. Lead users: A source of novel product concepts. Management Science 32, 791–805.

Wells, W. D. 1974. Group interviewing. In: R. Ferber (ed.), Handbook of Marketing Research. McGraw-Hill, New York.

Chapter 17

Quality Control and Shelf-Life (Stability) Testing

Abstract Two routine functions of a sensory department may be quality control testing and the measurement of product stability or shelf life. These activities may involve any of the three main kinds of sensory testing or modifications of them. However, there are unique constraints for these tests, different types of analyses, and specific models for these data. This chapter discusses different procedures for sensory quality control, presents a recommended procedure, and outlines the programmatic requirements for establishing and maintaining a sensory QC function. The second section of the chapter presents an introduction to shelf-life testing, its special considerations, and some of the models used for stability testing data.

Consumer researchers are well aware of the quality of products. The food industry constantly faces the demand to maintain both quality and profitability simultaneously. Quality, however, is an elusive concept and as such must be operationalized and measured in order for it to be maintained.
—H. R. Moskowitz (1995)

Contents

H.T. Lawless, H. Heymann, *Sensory Evaluation of Food*, Food Science Text Series,
DOI 10.1007/978-1-4419-6488-5_17, © Springer Science+Business Media, LLC 2010

17.1 Introduction: Objectives and Challenges

Product quality has been defined in a variety of different ways (Lawless, 1995). Most sensory researchers focus on issues of consumer satisfaction as a measure of quality (Cardello, 1995; Moskowitz, 1995) although there is an historic tradition of using expert judges, commodity graders, or government inspectors to be the arbiters of product quality (Bodyfelt et al., 1988; York, 1995). This tradition is tied to use of the senses for detection of well-known defects or expected problem areas. The approach was well suited to standard commodities where minimum levels of quality could be insured, but excellence was rarely the issue. Another strong tradition has been the emphasis on conformance to specifications (Muñoz et al., 1992). This approach is useful in the manufacturing of durable goods whose attributes and performance could be measured using instrumental or objective means. Another popular definition of quality has been fitness for use (Lawless, 1995). This definition recognizes that quality does not exist in a vacuum, but only in a context or frame of reference for the consumer. Finally, the reliability or consistency in sensory and performance experiences with a product has been recognized as an important feature of product quality. Consumer expectations arise out of experience, and maintaining the constancy of that experience does a lot to build consumer confidence.

There are a number of challenges and problems that face a sensory evaluation program when trying to provide sensory information for quality control (QC). Difficult situations occur in the manufacturing environment where sensory assessment is needed during the processing itself. Such online sensory quality testing is likely to be done under tight time constraints, for example, while the product is cooling and before a decision is made to bottle or pack a production run. Only a few qualified judges may be available on third shift in the middle of the night when these decisions have to be made. There is little luxury involved in terms of time, and a detailed descriptive evaluation and statistical analysis may not be possible due to time and resource constraints. At the same time, a flexible and comprehensive system may be desired, one that is also applicable to raw materials testing, finished products, packaging materials, and shelf-life

tests (Reece, 1979). Such constraints and demands often entail compromises in sensory practices.

A basic requirement of any sensory QC system is the definition of standards or tolerance limits on a sensory basis for the product. This requires calibration studies. If the sensory QC program is new, management may be surprised to learn that some research needs to be done before the QC panel can be trained and begin to operate. Sometimes the identification of standard products and tolerance limits may incur more expense than the sensory panel operation itself, especially if consumers are used to define the limits of what is acceptable quality. Maintaining reference standards for a standard quality product may also present difficulties. Foods and consumer products may have short shelf lives, and even with optimal storage conditions the standards will need to be replaced sooner or later. It is difficult to prevent some drift in the product over time. Multiple references including both optimally stored and fresh products may be needed (Wolfe, 1979). Some products simply change with age and this is a desirable feature like the proteolysis in ham or in cheese ripening (Dethmers, 1979). Furthermore, the frame of reference of the panel can drift or change seasonally. This makes it difficult to insure that a sensory specification of a standard product is in fact the same as the last standard.

Other barriers to acceptance involve the different ways that sensory evaluation is performed as opposed to traditional quality control. Most sensory tests are designed to look at a few or limited number of products. Sometimes the products are even considered to be identical, as in a homogeneous product evaluated from the same batch, like a well-mixed tank-produced beverage. The major source of variability in this sensory test is in the measuring instruments, the panelists. Statistical tests are designed to look at mean scores against the background of variation among people. This is quite different from the usual operation of quality control, where many samples of the product are taken and measured only once or a very few times on an instrument. The variability measured by traditional QC and pictured in control charts and other plots is across products, not instruments. Sensory QC has to deal with both sources of variation. In the instrumental measures, one can sample hundreds of products and take one measurement on each. In sensory QC, there may be one sample of each product but multiple measurements across panelists.

17.2 A Quick Look at Traditional Quality Control

Traditional quality control involves three major requirements: the establishment of specifications, the establishment of tolerance limits, and a sampling plan appropriate to the product being manufactured or the system being monitored. By specification, we mean the characteristics of the ideal or average product. To set tolerance limits, the liabilities of Type I error, rejecting product that is acceptable and therefore incurring unnecessary cost, must be weighed against the potential for Type II error, letting bad product into the marketplace and offending loyal consumers. This is a management decision that will impact the nature of the tolerance limits that are set. These tolerance limits are the levels of variability and/or the ranges that are deemed acceptable (in specification) versus unacceptable (out of specification).

Historically, such analysis was born in the advent of statistical quality control. W.A. Shewhart, working with Bell Labs in the early 1900s, noticed that there was variability in the functioning of some signal transmission components and that since these were often buried, they were a problem to dig up and repair. Failures or severe problems needed to be differentiated from the normal expected variability in these systems. To address this, he coined the ideas of assignable cause versus chance cause variation. The idea was that some variation was to be expected, but when the observation was outside of some common range, it was likely that some other cause was at work, and the item would need to be replaced, repaired, or otherwise dealt with. Thus the approach was statistical and involved a number of charts or graphs depicting this variation and the limits at which an assignable cause might be suspected. His notions of statistical quality control were later adopted by W.E. Deming in support of the war effort and later in the reconstruction of Japanese industrial practices.

There are several common types of charts used in statistical quality control and the sensory evaluation specialist should be familiar with them as they are part of the common language used by traditional quality control departments. Three kinds of control charts are common: X-bar charts, R charts, and I charts. Various rules exist for warning levels and action levels using these charts. Warning levels generally mean that the process needs to be investigated but no change

is necessary. If action levels are surpassed, then there is good evidence for an out-of-control situation or assignable cause, and the process must be changed.

The X-bar chart plots the mean scores for different test batches over some time period. Typically three to five products are pulled for evaluation (Muñoz et al., 1992). Upper and lower confidence limits (UCL and LCL) are generally set by ± 3 standard errors (sometimes referred to as 3-sigma) as shown in Fig. 17.1. Out-of-control conditions are spotted by a number of criteria, such as a point beyond the 3-sigma UCL or LCL, or some pattern of points such as "nine points in a row all on one side of the historical mean" or "six points in a row all increasing or decreasing" (Nelson, 1984). The R-chart measures the range of observations in any batch of product. Upper and lower limits are set as in the X-bar charts with warning and action levels at 2-sigma and 3-sigma (or 95 and 99% confidence levels). Sometimes the X-bar chart and the R chart may be combined to give a fuller picture. If only one unit is evaluated per batch, then the mean and range cannot be used, only the observed value itself. This is plotted on an I chart (Muñoz et al., 1992). As in the X and R charts, a mean and confidence limits can be set, as well as warning levels and action levels.

17.3 Methods for Sensory QC

17.3.1 Cuttings: A Bad Example

Muñoz and coauthors (1992) give both good and bad examples of applications of sensory QC procedures. Here is an example of a poor implementation of the in/out procedure:

> The panel consists of a small group of company employees (4 to 5) mainly from the management team. The panel evaluates a large amount of production samples (up to 20 to 40) per session without standardized and controlled protocols. Each product is discussed to determine if it is to be considered "in" or "out" of specifications. In this program, no defined specifications or guidelines for product evaluation exist, and no training or product orientation was held. As a result, each panelist makes decisions based on his or her individual experience and familiarity with production, or based on the highest ranking person on the panel. (p. 141)

This scenario highlights some of the pitfalls of a pass/fail procedure. It resembles a common daily

Fig. 17.1 *X*-bar and *R* charts
for a hypothetical analysis of
product thickness ratings over
several batches. The *X*-bar
chart shows the historic mean
and upper and lower control
limits, usually set at three
standard deviations. The
X-bar chart shows one batch
with a mean value below the
lower limits and the batches
on either side show a range
beyond the range limit as
well. This should suggest
action and/or investigation by
process control personnel.
Batch 13 was also above the
range limit suggesting an
out-of-control situation.
Batches 5–10 also show the
alarm pattern of six points on
one side of the historic mean.

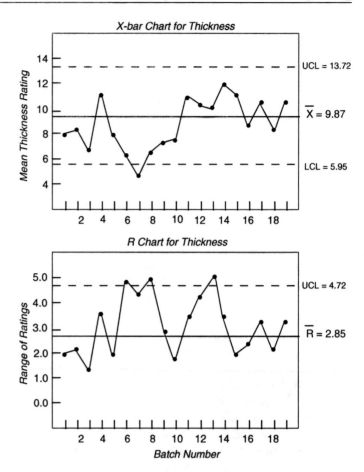

check on production that was often done by a convened
committee of technical personnel and managers, called
"cuttings." Without the guidance of a sensory evaluation specialist, such a method can be put in place with
a number of poor practices, such as having an open
discussion to reach consensus and determining a final
score.

17.3.2 In–Out (Pass/Fail) System

Muñoz et al. (1992) discuss four different approaches
to sensory quality assessment. Their book, *Sensory
Evaluation in Quality Control*, gives a detailed treatment of each. One of the methods is the in/out or pass–
fail procedure. This method differentiates normal production from products that are considered different or
outside specifications. It is a popular procedure at the
plant level and is used in some binary decision-making

scenarios such as Canadian fish inspection (York,
1995).

Panelists are trained to recognize the characteristics that defined "out-of-spec" products as well as the
range of characteristics that are considered "in spec"
(Nakayama and Wessman, 1979). This enhances the
uniformity of criteria among the panelists. As in any
yes/no procedure, the effects of bias and criterion setting can be as influential as the actual sensory experience (see Section 5.8). Different panelists may be more
or less conservative in the degree of sensory difference
they require in order to call something out of spec. In
quality control, the liability of differences in criterion
setting is high, since there are always pressures to pass
poor products to maintain productivity. Obviously, the
presentation of blind control samples is necessary to
estimate a false alarm rate (false positives), and the
introduction of purposely defective samples can be
useful in estimating the false-negative (miss) rate.

Muñoz and coauthors stressed the need for standardized protocols for sample handling and evaluation and the need for independent judgments, rather than discussion and consensus. York (1995) described how government fish inspectors are involved in standards development workshops. Training includes definition of sensory characteristics that define wholesomeness, taint, and decomposition and how these characteristics at different levels contribute to the binary decision for acceptance or rejection.

The major advantages of the in/out procedure are its apparent simplicity and use as a decision-making tool. It is especially suited to simple products or those with a few variable attributes. The disadvantages include the criterion-setting problems described above. Also, the method does not necessarily provide diagnostic reasons for rejection or failure, so there is a lack of direction to be used in fixing problems. It may also be difficult to relate these data to other measures such as microbial or instrumental analyses of food quality. The data necessarily consist of mere frequency counts of the number of panelists judging the product out of spec. Finally, it may also be difficult for some panelists to be analytical and look for specific problems and defects, while at the same time providing an overall integrated judgment of product quality.

17.3.3 Difference from Control Ratings

A second major approach to sensory quality control is to use ratings for an overall degree of difference from a standard or control product. This works well if it is feasible to maintain a constant "gold standard" product for comparison (Muñoz et al., 1992). It is also well suited to products where there is a single sensory characteristic or just a few sensory characteristics that vary in production. The procedure uses a single scale as illustrated in the paper by Aust et al. (1985) such as the following:

extremely different	the same as
from the standard	the standard

Ratings on this scale may be transcribed from zero (rightmost point) to ten (leftmost point). For purposes of rapid analysis, a simple 10-point category scale can be used. Additional points along the scale are sometimes labeled with other verbal descriptions of different degrees of difference.

Training with a range of references and establishing the nature and conditions for reproducing the control sample are critical in this procedure. The panelists must be shown samples in training that represent points along the scale. These can be cross-referenced to consumer opinion or chosen by management tastings (Muñoz et al., 1992). Preferably there is some consumer input for calibration at an early stage of the program development. Muñoz and coauthors also presented a more descriptive version where differences from control on several individual attributes of a flaked breakfast cereal were evaluated. This more detailed procedure can provide more actionable information about the attributes responsible for any differences. If just a single scale is used, panelists may weight attributes differently in determining an overall degree of difference. Specific characteristics may be more or less influential for a given panelist.

Management should choose some level of the difference as a cutoff for action. The scale is useful in that it provides for a range of differences that are acceptable. At some point, regular users of the product will notice and object to differences, and this should be the benchmark for action standards. If at all possible, the panelists should *not* be informed of where the breakpoint in decision making occurs along the scale. If they know where management sets the cutoff, they may become too cautious and tend to give scores that may approach but not surpass the cutoff (Rutenbeck, 1985).

As with the pass/fail method, an important part of this procedure is the introduction of blind control samples. During every test session, a blind-labeled sample of the standard should be inserted into the test set, to be compared against the labeled version of itself. This can help establish the baseline of variation on the scale, since two products are rarely rated as identical. Another way to think about this is that it provides a false alarm rate or an estimate of a placebo effect (Muñoz et al., 1992). In the original paper by Aust et al. (1985), an additional control sample was a product from a different batch of the same production. Thus a test product's variability could be measured against the response bias or variation within the ratings of the standard against itself, as well as the batch-to-batch variation. This approach could also be useful in comparing products from different manufacturing sites. Aust et al. proposed an analysis of variance model for this design. If the control comparison is simply the standard against itself, a paired or dependent *t*-test can compare the mean scores for the test product against the mean score of the standard product rated against itself. This presumes that there are sufficient judges to warrant a statistical test. With small panels, more qualitative criteria for action have to be adopted, e.g., any three out of five panelists below some cutoff score. The false alarm level should also be considered in making decisions. The ratings for the blind standard against itself must be low relative to the test product scores in order to reject a batch.

The major disadvantage of this test, like the yes/no procedure, is that it does not necessarily provide any diagnostic information on the reasons for the difference if only the single scale is used. Of course, open-ended reasons for difference can be given, or additional questions, scales, or checklists can be provided for attributes that are common problems or show common variation.

17.3.4 Quality Ratings with Diagnostics

A third method, similar to the overall difference from control method, is to use quality ratings. This entails an even more complex judgment procedure on the part of the panelists, since it is not only the differences that matter but also how they are weighted in determining product quality. This idea of an integrated quality score is part of the tradition of food commodity judging, as discussed in Section 17.6.

There are three main abilities of the trained or expert judge that are necessary in order to use a quality judging system. The expert judge must maintain a mental standard of what the ideal product is in terms of sensory characteristics. Second, the judge must learn to anticipate and identify common defects that arise as a function of poor ingredients, poor handling or production practices, microbial problems, storage abuse, and so on. Finally, the judge needs to know the weight or influence of each defect at different levels of severity and how they detract from overall quality. This usually takes the form of a point deduction scheme. In the case of seafood, deterioration as a function of aging

or mishandling will go through a sequence of flavor changes and sensory spoilage characteristics. These changes in sensory characteristics can be translated into a scale for fish quality (Regenstein, 1983).

The common characteristics of quality ratings are these (Muñoz et al., 1992): Scales directly represent the quality judgment, rather than just sensory difference, and can use words like poor to excellent. This wording itself can be a motivator, as it gives the impression to panelists that they are directly involved in decision making. Quality grading works best when there is management or industry consensus on what is good. In some cases specific product characteristics can be rated in addition to overall quality, for example, quality of texture, flavor, appearance. In some schemes like wine judging, quality scores for individual attributes are then summed to give the overall score (Amerine and Roessler, 1981). Unfortunately, the quality scoring approach is prone to abuse, where small numbers of poorly trained judges evaluate dozens of product "cuttings," use their own personal criteria, and use consensus (discussion) methods to make decisions. Muñoz and coauthors presented this example of good practice:

> The panel consists of 8 to 12 panelists who are trained in the procedures to assess the quality of a given product type. They learned the company's quality guidelines, which were established using the input from consumers and management. ... These guidelines are shown to panelists by actual products representing various quality levels. The program was designed by a sensory professional using sound methodology and adequate testing controls for the evaluation process. In routine evaluations, panelists rate "overall" quality as well as the quality of selected attributes using a balanced scale (very poor to excellent). The data are treated like interval data and panel means are used to summarize the results of the evaluations. The results are provided to management, which makes decisions on the disposition of the production batches evaluated. (p. 109)

Although there are apparent time and cost advantages in this direct approach, there are also disadvantages. The ability to recognize all the defects and integrate them into a quality score may require a lengthy training process. There is a liability that individual subjectivity in likes and dislikes can creep into the judge's evaluations. The specialized vocabulary of technical defects may seem arcane to non-technical managers. Finally, with small panels, statistical difference tests are rarely applied to such data, so the method is primarily qualitative.

17.3.5 Descriptive Analysis

A fourth approach to sensory quality control is a descriptive analysis method, as described in Chapter 10. The goal is to provide intensity ratings for individual sensory attributes by a trained panel. The focus is on the perceived intensity of single attributes and not quality or overall difference. Intensity rating of single sensory characteristics demands an analytical frame of mind and focused attention on dissecting the sensory experience into its component parts. Muñoz and coauthors called this as a "comprehensive descriptive method," but they do allow for limitation of the scorecard to a small set of critical attributes. For QC purposes, attention to a few critical attributes may be appropriate.

As in the other techniques, calibration must be done. Specifications for the descriptive profile must be set via consumer testing and/or management input. This will consist of a range of allowable intensity scores for the key attributes. Table 17.1 shows an example of a descriptive evaluation of potato chip samples and the range of sensory specifications, as previously determined in a calibration study with consumers and/or management input (from Muñoz et al., 1992). This sample is below the acceptable specification limits in evenness of color and is too high in cardboard flavor, a characteristic of lipid oxidation problems.

Table 17.1 Evaluation of potato chip samples using descriptive specification

	Mean panel score	Acceptable range
Appearance		
Color intensity	4.7	3.5–6.0
Even color	4.8	6.0–12.0
Even size	4.1	4.0–8.5
Flavor		
Fried potato	3.6	3.0–5.0
Cardboard	5.0	0.0–1.5
Painty	0.0	0.0–1.0
Salty	12.3	8.0–12.5
Texture		
Hardness	7.5	6.0–9.5
Crisp/crunch	13.1	10.0–15.0
Denseness	7.4	7.4–10.0

From Muñoz et al. (1992)

Descriptive analysis requires extensive panel training. Panelists should be shown reference standards to learn the meaning of the key attributes. Next they must be shown intensity standards to anchor their quantitative ratings on the intensity scale. They do not have to be shown examples that are labeled as "in specification" or "out of specification," however, since that decision is based on the overall profile of the product and is done by the sensory panel leader or QC management. Defective samples can be used in training intensity ratings, but the actual cutoff points are better kept in confidence by managers making the decisions about product disposition (Muñoz et al., 1992). This will avoid the tendency for panelists to gravitate toward scores that are just within the acceptable range.

Advantages. The detail and quantitative nature of the descriptive specification lends itself well to correlation with other measures such as instrumental analysis. The second advantage is that it presents less of a cognitive burden on panelists, once they have adopted the analytical frame of mind. They are not required to integrate their various sensory experiences into an overall score, but merely report their intensity perceptions of the key attributes. Finally, the reasons for defects and corrective actions are easier to infer since specific characteristics are rated. These can be more closely associated with ingredients and process factors than an overall quality score.

Limitations. Because the method depends on good intensity anchoring, it tends to be more laborious in panel training than some of the other techniques. Due to the need for data handling and statistical analysis, as well as having a sufficient number of trained judges on hand for the method, it is better suited for quality evaluation of finished products. It may be difficult to arrange for descriptive evaluation for ongoing production, particularly on later work shifts if production is around the clock. The training regimen is difficult and time consuming to set up since examples must be found for the range of intensities for each sensory attribute in the evaluation. This can require a lot of technician time in sample preparation. Another liability is that problems may occur in some attributes that were not included on the scorecard and/or outside of the training set. Thus, the method lends itself to situations where the problem areas are well known and the production and ingredient variability can be easily reproduced to make up the training set.

17.3.6 A Hybrid Approach: Quality Ratings with Diagnostics

Gillette and Beckley proposed a reasonable compromise between the quality rating method and a comprehensive descriptive approach at the 1992 meeting of the Institute of Food Technologists (described in Beckley and Kroll, 1996). The centerpiece of this procedure is a scale for overall quality. The quality scale is accompanied by a group of diagnostic scales for individual attributes. These attributes are key sensory components that are known to vary in production. Muñoz et al. (1992, pp. 138–139) describe a similar modification of the overall quality ratings method to include the collection of descriptive information on key attributes. In the method of Gillette and Beckley the main scale takes this form:

procedures, the boundaries for out-of-spec product and the selection of a gold standard must be undertaken before training, preferably in a consumer study, but at least with management input. These defined samples must then be shown to subjects to establish concept boundaries. In other words, tolerance ranges must be shown to panelists (Nakayama and Wessman, 1979).

17.3.7 The Multiple Standards Difference Test

Amerine et al. (1965) mentioned a variation on simple difference tests that would include a non-uniform or variable standard. This has come to be known as the multiple standards difference test. Although there is scant literature on the procedure, it has apparently

1	2	3	4	5	6	7	8	9	10
Reject		Unacceptable			Acceptable			Match	

On this scale, a product that is so clearly deficient to call for immediate disposal gets a score of 1 or 2. Products that are unacceptable to ship but might be reworked or blended get a score in the range of 3–5. If evaluation is online during processing, these batches would not be filled into retail containers or packaged but would be held for rework or blending. If the samples are different from the standard but in an acceptable range, they receive scores of 6–8, and samples that are a near match or considered identical to the standard receive a 9 or 10, respectively. According to Muñoz et al. (1992), the use of the terms "acceptable" and "unacceptable" here is unfortunate, for it gives the panelists an impression of the action standards for products passing or failing and the feeling that they are responsible for decisions about product disposition. This creates a tendency to use the middle to the upper end of the scale, to avoid grading products as unacceptable (Rutenbeck, 1985).

The advantages of this method are its outward simplicity in using an overall rating and the addition of attribute scales to supply reasons for product rejection. The method also recognizes that there are situations where products will not match the gold standard exactly, but still are acceptable to ship. As in the other

enjoyed some popularity. The idea is to give a forced-choice test in which participants pick which one of several alternative products is the most different from the rest of the set. The simplest approach is to have one test product and K alternative versions of the standard product. Rather than representing identical versions of a gold standard, the standards are now chosen to represent the acceptable range of production variability. The choice of standards to represent the range of acceptable variation is critical to the success of this approach. Historically, this method resembles Torgerson's "method of triads" of which the triangle test is a special case (Ennis et al., 1988). Pecore et al. (2006) and Young et al. (2008) used a similar approach, except that overall degree of difference rating was used (discussed below). The choice of products to be included in the set of acceptable standards is critical. If they do not reasonably bracket the range of acceptable variation, then the test will be too sensitive (if the range is small) or too insensitive to detecting bad samples (if their range is too large).

If there are a large number of testers ($N = 25$ or more), as in a discrimination procedure, the z-score approximation to the binomial distribution may be used for hypothesis testing. The approximation is

$$z = \frac{(P - 1/k) - (1/2N)}{\sqrt{(1/k)(1 - 1/k)/N}} \qquad (17.1)$$

where k is the total number of alternatives (test product plus variable standards), P is the proportion selecting the test product as the outlier (the most different sample), and N is the number of judges. This is the same formula as in the triangle and other forced-choice procedures, except that k may be 4 or larger, depending on the number of references.

Although this method appears simple at first, there are a few concerns and potential pitfalls in its application. First, in many QC situations, it may not be feasible to conduct a difference test with sufficient numbers of judges to have a meaningful and sensitive application of the statistical significance test. Second, the failure to reject the null (failure to get a significant result) does not necessarily imply sensory equivalence. From a statistical perspective, it is difficult to have confidence in a "no-difference" result, unless the power of the test is very high. Statistical confidence in the equivalence decision can only be obtained after beta-risk is estimated against a suitable alternative hypothesis (see Appendix E on test power). One approach is to use the analysis for significant similarity, as outlined in Chapter 5. This will necessarily entail a larger number of judges ($N = 80$ or so). Finally, the tests for overall difference such as the triangle procedure are known to have high inherent variability, so the introduction of even more variability by multiple standards makes it very difficult to get a significant difference and reject product. This factor may contribute to a high level of beta-risk, i.e., the chance of missing a true difference.

these three pieces of data, three comparisons can be made: Each mean difference score of the test–control pair is compared to the mean difference score of the control–control pair. Also the *average* test–control rating is compared to the same baseline. Thus the within-control variation is taken into account in the comparisons, which must be significantly exceeded if the test lot is to be found different (and thus actionable). Of course, this kind of test requires a sufficient panel size to get a meaningful and statistically powerful test. Later, Young et al. (2008) extended the model and procedure to include two test lots as well as two control lots and used an incomplete block design to limit the six comparisons to three comparisons per panelist. The critical comparison in this case is between the average of the four means comparing tests to controls versus the average of the mean control–control score and the mean test–test lot score. So the baseline becomes the average difference within control lots and within test lots.

17.4 Recommended Procedure: Difference Scoring with Key Attribute Scales

This method is similar to the hybrid procedure of Gillette and Beckley except that it substitutes the overall difference scale for the quality scale. This avoids the problem that panelists may react to words like "reject" and avoid them. So the method is similar to that of Section 17.3.6. Some category or line version of the difference scale should be used such as the following:

1	2	3	4	5	6	7	8	9	10
Completely different			Very different			Somewhat different		Match	

A similar approach to the multiple standards choice test was described by Pecore et al. (2006) and Young et al. (2008) but using degree of difference ratings, rather than a choice test. This approach was part of the original intent of the degree of difference test proposed by Aust et al. (1985). In this method, a test lot is compared to each of two control lots. The two control lots are also compared to each other. From

The ballot should also include diagnostics on key attributes, those that will vary in production and are likely to cause consumer rejection. For attributes that can be too strong or too weak, using just-right scales is appropriate. Some defects may be a problem at higher levels, and intensity scales are useful for those attributes. Others may warrant product rejection at any level whatsoever, and a checklist can be provided for

Fig. 17.2 A sample ballot for apple juice, using the recommended procedure of degree-of-difference scale plus diagnostic attribute ratings. Note that some attributes use the just-about-right scale while others are better suited to a simple intensity scale. For off-flavors or defects that are objectionable at any level, a simple checklist is useful. Note that this method must be used with a well-trained panel.

Apple Juice QC Ballot

Sample____589_____   Judge_____14 (MK)____
Date/session___12/25/09____   Plant site_Dunkirk_____

Overall difference Rating:

1	2	3	4	5	6	7	8√	9	10
Extremely			Moderately			Slightly		Match	
Different			different			different			

Attributes:

	too low		about right		too high
Sweet	__	_X_	__	__	__
Sour	__	__	__	_X_	__
Color	__	__	_X_	__	__

	too sour		about right		too sweet
Sweet/sour	__	_X__	__	__	__
RATIO					

Strength:

	none / low				very strong
Sweet	__	_X__	__	__	__
Sour	__	__	_X_	__	__
Apple Aroma	__	__	_X_	__	__
Apple Flavor	__	__	_X_	__	__

Off aroma   _X_   __   __   __   __
(list/describe) _____

Off Flavor   _X_   __   __   __   __
(list/describe) _____

Checklist: (circle any defects)

Vinegar-like   Butyric   Lactic acid   Painty/solvent   Fusel Oil

Sauerkraut-like   Other fermented   Bitter   Astringent   Musty

Other (list)_____

Comments_____

those more serious faults. A sample ballot for apple juice is shown in Fig. 17.2.

Screening panelists for sensory acuity and a good training regimen are key here, as in setting up other quality control panels. The screening procedure should use the types of products that people are going to eventually judge and insure that they can discriminate among common levels of ingredients like sugar or acid content and process variables like heating times or processing temperatures. A sample screening procedure for an apple juice panel is shown in Appendix 1 of this chapter. Screening should involve a number of attributes and, if possible, different tasks or tests (Bressan and Behling, 1977). The top performers can

be invited for panel training and others who score well can be kept on file for future replacements as panelist attrition occurs (plan from day 1!). Ideally the screening pool of volunteers should be two to three times the desired panel size. Supervisory approval is a key to good attendance and participation.

After screening, training may take six to ten sessions depending on the complexity of the product. Gross differences are illustrated early in training and smaller differences shown as training progresses. The goal is to solidify the conceptual structure of the panelists so they know the category boundaries for the quality ratings and the expected levels of the sensory attributes. Panelists must also come to recognize how off-flavors, poor texture, or appearance problems factor into their overall score.

With small panels, there is no statistical analysis, but rules of thumb must be established for taking actions. It is difficult to apply mean ratings to less than about eight panelists. Since there are differences in individual sensory ability, poor ratings by just a few individuals may be indicative of potential problems. Thus action criteria should take into account negative minority opinions and weight them more heavily than high outliers or a few panelists who thought the product matched the standard (but who may have missed some important difference). For example, if two panelists rate the sample on scale point two, but the rest give it six, seven, or eight, the mean score could be in the acceptable range in spite of the two panelists who spotted potentially important problems. The panel leader should take note of the two low values and at least call for retesting of this questionable sample. Of course, consistent patterns of disagreement between panelists are a hint that some retraining may be needed.

17.5 The Importance of Good Practice

In all small-panel sensory assessments, the general principles of good testing become especially important since there are shortcuts often taken in these procedures that are not part of standard sensory evaluation practices. Most notably, quality assessment may not entail any statistical analysis, due to the small numbers of panelists. Statistical methods provide some insurance against false alarms due to random variation or errors of missing important differences. Without the aid of statistical analysis, other safeguards for insuring the quality of the information take on an even higher level of importance.

It is worth considering an example concerning pork inspection since it illustrates many of the pitfalls involved in small-panel experiments. This was a study of boar taint or sex aroma from androstenone, a problem odor in the fatty tissues of adult male swine. The goal was to correlate sensory panel scores for this taint with instrumental measures of androstenone content (Thompson and Pearson, 1977). Two sensory analyses were done. In the first, three to five panelists sampled boar taint aroma in the packinghouse

using a hot-iron technique to elicit the aroma and came to a consensus judgment using a 6-point scale for odor intensity. A second evaluation was done after the samples were sent to the laboratory for instrumental analysis. In this case, three panelists were screened for sensitivity to androstenone, and means were calculated from a 9-point scale for odor intensity. Evaluations were performed in a laboratory exhaust hood and preparation procedures were standardized. The correlation with instrumental measures was +0.27 for the first evaluation (not significantly different from zero correlation) and +0.40 (statistically significant) in the second. The increased correlation could be due to a number of methodological factors that were improvements in the second evaluation. These include (1) better evaluation location (fume hood versus packing house), (2) screening of judges, (3) constancy of panel members instead of people dropping in and out, (4) averaging scores versus a consensus procedure, and (5) a more standardized sample preparation method. Each of the shortcuts might have been introduced on some practical grounds, but their combined effect was to increase the error level in the data. This made a difference in the statistical significance and conclusions of the study regarding instrumental-sensory correlations.

Table 17.2 gives a number of guidelines for good sensory practice in quality evaluations and Table 17.3 gives guidelines for judges (adapted from Nelson and Trout, 1964). As in any other sensory test, product samples should be blind coded and presented in different random orders to each panelist. If production personnel are used in the panel who know the identities of some of the products pulled for evaluation, another technical person must blind code them and insert blind controls into the test set. The person who pulls the samples must not evaluate the samples. It is not reasonable to expect that person to be objective and discount any knowledge of the product identity. Serving temperature, volume, and any other details concerning product preparation and the tasting method should be standardized and controlled. Facilities should be odor free and distraction free. Evaluations should be made in a clean sensory testing environment with booths or separators, not on the benchtop of an analytical instrument lab or on the manufacturing floor (Nakayama and Wessman, 1979). Warm-up samples are useful. Blind replicates can be introduced to check judge consistency. Judges should taste a representative

Table 17.2 Ten guidelines for sensory quality testing

1. Establish standards for optimum quality ("gold standard") target plus ranges of acceptable and unacceptable products
2. Standards should be calibrated by consumer testing if possible. Alternatively, experienced personnel may set standards but these should be checked against consumer opinion (users of product)
3. Judges must be trained, i.e., familiarized with standards and limits of acceptable variation
4. Unacceptable product standards should include all types of defects and deviations likely to occur from materials, processing, or packaging
5. Judges may be trained to give diagnostic information on defects, if standards are available typifying these problems. Scaled responses for intensity or checklists may be used
6. Data should always be gathered from at least several panelists. Ideally, statistically meaningful data should be gathered (ten or more observations per sample
7. Test procedures should follow rules of good sensory practice—blind testing, proper environment, test controls, random orders
8. Blind presentation of standards within each test should be used to check for judge's accuracy. It is important to include a (blind) gold standard for reference purposes as well
9. Judge reliability may be tested by blind duplicates
10. Panel agreement is necessary. If unacceptable variation or disagreement occurs, re-training is warranted

Table 17.3 Guidelines for participation in sensory assessments

1. Be in correct physical and mental condition
2. Know the score card
3. Know the defects and the range of probable intensities
4. For some foods and beverages, it is useful to observe aroma immediately after opening the sample container
5. Taste a sufficient volume (Be professional—not timid!)
6. Pay attention to the sequence of flavors
7. Rinse, occasionally, as the situation and product type warrant
8. Concentrate. Think about your sensations and block out all other distractions
9. Do not be too critical. Also, do not gravitate to the middle of the scale
10. Do not change your mind. Often the first impression is valuable, especially for aromas
11. Check your scoring after the evaluation. Get feedback on how you are doing
12. Be honest with yourself. In the face of other opinions, "stick to your guns"
13. Practice. Experience and expertise come slowly. Be patient
14. Be professional. Avoid informal lab banter and ego trips Insist on proper experimental controls—watch out for benchtop "experiments"
15. Do not smoke, drink, or eat for at least 30 min before participation
16. Do not wear perfume, cologne, aftershave, etc. Avoid fragranced soaps and hand lotions

Modified from Nelson and Trout (1964)

portion (not end of batch or other anomalous parts of production).

Other rules of thumb apply to the judges or panelists. Panelists should be screened, qualified, and motivated with suitable incentives. They must not be overtaxed or asked to test too many samples in one day. Rotation of the panel at regular intervals can improve motivation and relieve boredom. Judges should be in good physical condition, i.e., free from ailments like colds or allergies that would detract from their performance. They should not be mentally harried from other problems on the job when arriving for testing, but should be relaxed and able to concentrate on the task at hand. They must be trained to recognize the attributes, scoring levels and, of course, know the scorecard. Judgments should be independent without conferring, jury style. Discussion or feedback can be given later to provide for ongoing calibration. A special liability arises when manufacturing personnel are used who have a lot of pride in the product accompanied by false confidence in the infallibility of the manufacturing process. Such panelists may be unwilling to "rock the boat" and call attention to problem areas. Testing with blind out-of-spec samples, known defects, and other such "catch trials" accompanied by feedback when they pass defective samples can help counteract this overly positive attitude.

The data should consist of interval scale measurements where possible. If large panels are used (ten or more judges), statistical analysis is appropriate and data can be summarized by means and standard errors. If very small panels are used, the data should be treated qualitatively. Frequency counts of individual scores should be reported and considered in action standards. Deletion of outliers can be considered, but a few low scores (i.e., a minority opinion) may be indicative of an important problem, as noted above. Re-tasting may be warranted in situations where there is strong disagreement or high panel variability.

17.6 Historical Footnote: Expert Judges and Quality Scoring

17.6.1 Standardized Commodities

The food industry benefits from standardization of grades for foods that are minimally processed from raw ingredients, from a single source without multiple components, and closely associated with a single agricultural commodity. Such "food commodities" include many dairy products such as milk, cheese, and butter; fruits such as olives; some kinds of meat; and wine. Various industries and governments have established quality grades for food commodities or systems for scoring them based on two main factors: similarity to a product ideal and lack of defects. The value in such quality grading, scoring, or monitoring is the assurance to the consumer that the product will have the sensory properties that they have come to expect.

Sometimes these systems are defined by international organizations in order to provide standards of identity for the food commodity. An example is the International Olive Oil Council (COI). The COI provides written standards for sensory evaluation including definitions for the vocabulary of sensory properties and defects, a standardized scorecard, a point system for assigning grades or classifications, methods for panel training, certification of laboratories evaluating olive oil, and even specifications of the tasting glasses that are to be used in the evaluations. Their website provides all of this information in the various languages of olive oil producing countries (International Olive Oil Council, 2007).

Two further examples of commodity judging systems by trained or expert panels are shown below. The sensory specialist should search for such professional organizations and specifications if they are assigned to develop methods for such a food commodity. The methods are poorly suited to processed engineered foods that do not fall into the category of a standardized commodity, but they can provide a useful starting place for development of a quality monitoring system for a closely related product. The sensory specialist should be careful, however, not to force-fit a standardized grading scheme to a product that is substantially different. For example, the quality evaluation scheme for grading vanilla ice cream would be only poorly suited to sensory testing on a frozen yogurt product made from goat's milk.

17.6.2 Example 1: Dairy Product Judging

A longstanding tradition in the field of dairy products has been the quality grading schemes for assessing product defects and assigning overall quality scores. The American Dairy Science Association continues to hold a decades-old student judging competition, in which students and teams of students attempt to duplicate the quality scores of established experts. Various defective products are supplied, and students must be able to recognize the defect, subtract the appropriate penalty given the type and severity of the problem, and arrive at an overall score (Bodyfelt et al., 1988). The support for quality judging in dairy products is not universal, however. Some countries like New Zealand have replaced the overall quality judging method with ratings on specific key attributes for dairy product analysis.

However, these methods do persist and find some utility in small plant quality control and in government inspections (Bodyfelt et al., 1988; York, 1995). An example of the quality judging scheme for cottage cheese is shown in Table 17.4, listing defects and their point deduction values for slight, definite, and pronounced levels of sensory intensity. An extensive discussion of quality judging for dairy products can be found in "Sensory Evaluation of Dairy Products" by Bodyfelt et al. (1988).

Table 17.4 A point deduction scheme for cottage cheese quality grading. Cottage cheese scoring guide

	Slight	Distinct	Pronounced
Appearance (5 points maximum):			
Lacks cream	4	3	2
Shattered curd	4	3	2
Free cream	4	2	1
Free whey	4	2	1
Texture (5 points maximum):			
Weak/soft	4	3	2
Firm/rubbery	4	2	1
Mealy/grainy	4	2	1
Pasty	3	2	1
Gelatinous	3	2	1
Flavor (10 points maximum):			
High acid	9	7	5
High salt	9	8	7
Flat	9	8	7
Bitter	7	4	1
Diacetyl/coarse	9	7	6
Feed	9	7	5
Acetaldehyde/green	9	7	5
Lacks freshness	8	5	1
Malty	6	3	1
Oxidized	5	3	1
Fruity	5	3	1
Musty	5	3	1
Yeasty	4	2	1
Rancid	4	2	1

Rate the presence of each defect as slight, distinct, or pronounced. Give scores for appearance, texture, and flavor based on the table.

Other problems may include discoloration, matted curd, slimy texture, foreign flavors, unclean flavors (describe), and fermented flavors

Modified from Bodyfelt et al. (1988)

Such methods are poorly suited to food research where the processed or engineered food is not a standard commodity and/or when the sensory changes are not likely to be a set of predictable defects. In new food product development, it is not necessarily clear what consumers or segments of consumers may like, so assignment of quality scores based on some arcane or traditional knowledge of experts is not useful. The dairy judging methods have been repeatedly criticized for lack of applicability to research problems, violations of sensory evaluation principles, and problems in scaling and statistical analysis (Hammond et al., 1986; McBride and Hall, 1979; O'Mahony, 1981; Pangborn and Dunkley, 1964; Sidel et al., 1981). Furthermore, the opinions of expert judges and standard point deduction schemes may not correspond to consumer opinion as shown in Fig. 17.3. The oxidized defects

in milk were viewed less critically on the average by consumers than the suggested ADSA scores would dictate. Of course, having a point deduction scheme that is more severe that the average consumer opinion provides a kind of safety net and insures that the most sensitive consumers will not be offended by poor products. The liability in a stringent "safety net" is that acceptable product batches will be rejected.

17.6.3 Example 2: Wine Scoring

Beyond manufacturing control and government inspection there are other situations in which the consuming public desires information on product quality. Rather than deducting points from some widely accepted standard, there are also products

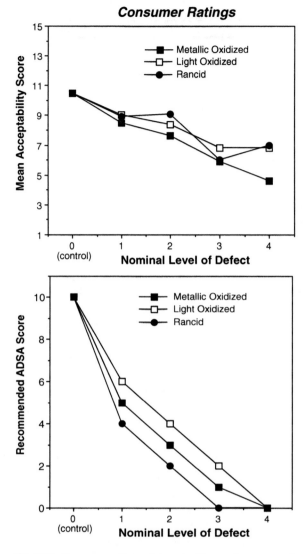

Fig. 17.3 Consumer ratings of abused milk samples compared to the recommended ADSA scores for those products based on the recipes and rating systems shown in Bodyfelt et al. (1988). From Lawless and Claassen (1993) with permission.

where excellence is recognized beyond the merely acceptable. Garvin (1987) remarked that considerations of quality should include the ability to please consumers, not just protect them from annoyances. This idea is further developed in the Kano model in Chapter 19. Some products show a wide range of better-than-acceptable variation. Wines are a good example. What kind of quality measurement system can go beyond point deductions for defects to provide degrees of difference on the positive end of the quality continuum?

An early method for wine quality assessment was the 20-point rating system developed at the University of California at Davis (Amerine and Roessler, 1981; Ough and Baker, 1961). This was an additive scheme for giving overall quality scores. It was based on the analysis of quality for sensory categories such as appearance, body, flavor, and aftertaste, as well as some specific attributes like sweetness, bitterness, and acidity. As shown in Table 17.5, different points are given for different categories, i.e., there is uneven weighting, presumably due to the different contributions of each category to the overall quality. Note that this does not produce a scale value in the psychophysical sense, but a score.

Like other quality grading schemes, this method can be criticized on a number of grounds. First, the weighting system is somewhat arbitrary—different versions can be found—and it was based on the expert opinion of the method's originators, rather than any consumer opinion. Second, whether wine quality can actually be captured by an additive scheme is questionable. Some defects (e.g., bitterness) are simply too serious to provide any good score at all, even though all other attributes might add up to some positive number. Some versions of the technique try to allow for this by providing a few overall quality points to add into the total, a kind of global fudge factor. Also, it is an anecdotal observation that some judges who gain experience with the technique score an overall quality level of wines first and do not bother assigning individual category points to start. Rather, they first decide on an overall score and then allot points into the individual categories, using the method backward.

A simplified alternative procedure is based on hedonic scoring by experienced fine wine drinkers (Goldwyn and Lawless, 1991). The assumption was that a small panel of experienced tasters can provide recognition of good to superior products based on their personal likes and dislikes. The method works to the extent that fine wine drinkers form a cultural and linguistic community (Solomon, 1990) with known and consensual standards of taste at least within a given geographical area. The method used a balanced 14-point hedonic scale (like extremely to dislike extremely) with no neutral center point. Wines were tasted twice and the second final score was recorded. The procedure followed principles of good practice such as randomized orders, blind coding, independent judgments (no conferring), standardized presentation,

Table 17.5 Example of 20-point wine scoring scheme

Characteristic	Scoring guide	Maximum points
Appearance	Cloudy 0, clear 1, brilliant 2	2
Color	Distinctly off 0, slightly off 1, correct 2	2
Aroma and bouquet	Vinous 1, distinct but not varietal 2, varietal 3 subtract 2 for off-odors and 1 for bottle bouquet	4
Vinegary	Obvious 0, slight 1, none 2	2
Total acidity	Distinctly high or low 0, slightly high or low 1, normal 2	2
Sweetness	Too high or low 0, normal 1	1
Body	Too high or low 0, normal 1	1
Flavor	Distinctly abnormal 0, slightly abnormal 1, normal 2	2
Bitterness	Distinctly high 0, slightly high 1, normal 2	2
General quality	Lacking 0, slight 1, impressive 2	2
Total score		20

Modified from Amerine and Roessler (1981)

palate cleansers, and a reasonable pace of tasting. Flights consisted of seven wines and the pace was limited to at least 30 min per flight, allowing time for palate recovery. Blind duplicates were periodically introduced to check on judge reliability.

These methods represent improvements over the types of informal consensus tastings done by juries to award medals at state fair competitions. Such evaluations have almost no scientific merit, i.e., they are of about the same value as a movie critic's review. Analysis of 3 years of data from blind duplicate samples in the California state wine judging has shown that among non-defective wines, any grade or medal may be assigned to any wine in different competitions, and about 90% of judges were unable to reproduce their scores (Hodgson, 2008).

17.7 Program Requirements and Program Development

17.7.1 Desired Features of a Sensory QC System

Rutenbeck (1985) and Mastrian (1985) outlined the program development of a sensory QC system in terms of specific tasks. These included research into availability and expertise of panelists, availability or access to reference materials, and time constraints. Panelist selection, screening, and training on objective terms (such as "high saltiness," as opposed to vague terms like "poor quality") must be undertaken. Sampling

schemes must be developed and agreed upon as well as standard procedures for sample handling and storage. Data handling, report format, historical archiving, and tracking and panelist monitoring are all important tasks. It is extremely important that a sensory evaluation coordinator with a strong technical background in sensory methods should be assigned to carry out these tasks (Mastrian, 1985). Aside from these practical operational concerns, the system should also have certain features that maintain the quality of the evaluation procedures themselves. For example, a method for measuring the overall effectiveness of the system should be identified (Rutenbeck, 1985). External auditing at periodic intervals may be useful (Bauman and Taubert, 1984).

Gillette and Beckley (1992) listed requirements for a good in-plant sensory QC program and ten other desirable features. These concerns are taken from the perspective of ingredient suppliers to a major food manufacturer but can be modified to fit other manufacturing situations. A sensory QC program must involve human evaluation of the products. It must be acceptable to both suppliers and customers. Results must be easily communicated so that reasons for rejection and actions to be taken are both made clear. It should take into account an acceptable range of deviation, recognizing that some products will not match the gold standard but will still be acceptable to consumers. Of course, the program must be able to detect unacceptable production samples.

Additional desirable features include the following: Potential transfer over time to an instrumental measure is a good goal if the evaluations are very repetitive, as instruments do not become fatigued or bored with the testing regimen. This presumes that tight sensory–instrumental correlations can be established. Ideally, a sensory QC program should provide rapid detection for online corrections. Information should be quantitative and interface with other QC methods. As quality control and shelf-life tests are often similar, the methods will be more useful if they are transferable to shelf-life monitoring. Many of the changes over time in a stored product are also quality problems, such as deterioration in texture, browning, oxidation, syneresis, oiling-out, staling, and off-flavor development (Dethmers, 1979). Sensory evaluations may include raw ingredient testing as well as in-process and finished products. A good sensory QC program will produce a track record of actually flagging bad products to prevent further problems down the line or in consumer opinion in the marketplace.

Considering these desired features, there are some traditional test methods from the mainstream of sensory evaluation that simply do not apply well to in-plant quality control work. Problems arise if the tests cannot handle a sufficient volume of production samples. Any test procedure that has a slow turnaround in analysis and reporting of results will not be suitable for online corrective action in the manufacturing environment. For example, it is difficult to implement a descriptive analysis panel for QC work if decisions have to be made in the middle of the night on third shift and the data cannot be statistically analyzed through an automated system. At first glance, finding defective products would seem to suggest that a simple difference test from a standard product would be a good approach. However, most sensory difference tests take the form of forced-choice tasks, like the triangle procedure. The triangle test is useful for detecting any difference at all, but is not suitable when there is a range of acceptable variation. Just because a product is found to be different from the standard does not mean that it is unacceptable.

17.7.2 Program Development and Management Issues

Management may need to be educated as to the cost and practical issues that are involved in sensory QC. Rutenbeck (1985) described the "selling" of a sensory QC program and suggested calculations of measurable results, such as reductions in consumer complaints, cost savings in avoiding rework or scrapping materials, and potential impact on sales volume. Manufacturing executives unfamiliar with sensory testing can easily underestimate the complexity of sensory tests, the need for technician time to setup, the costs of panel startup and panelist screening, and training of technicians and panel leaders as well as panelist incentive programs (Stouffer, 1985). If employees are used as panelists, another stumbling block can be the personnel time away from the person's main job to come to sensory testing (and any associated costs). However, panel participation can be a welcome break for workers, can enhance their sense of participation in corporate quality programs, can expand their job skills and their view of manufacturing, and does not necessarily result in a loss of productivity. There are considerable advantages in using panelists from the processing operation, notably in accessibility and interest (Mastrian, 1985). Arranging for a sensory testing space may also involve some startup costs. An important issue concerns what will be done to insure continuity in the program. Management must be made to see that the sensory instrument will need maintenance, calibration, and eventual replacement. Concerns include panelist attrition and retraining, refreshment, or replacement of reference standards (Wolfe, 1979).

An early issue in program development concerns the definition of standards and cutoffs or specification limits (Stevenson et al., 1984). Management or preferably experienced technical personnel can do the evaluation and set the limits. This approach is fast and simple, but risky, since there is no consumer input (McNutt, 1988). The safest but slowest and most expensive approach is to give a range of products with representative production variation to consumers for evaluation. This calibration set should include known defects that are likely to occur and all ranges of processing and ingredient variables. As a small number of consumers will always be insensitive to any sensory differences, a conservative estimate of problem areas should be set based on rejection or failing scores from a minority of participants.

A third issue concerns the level of thoroughness in sampling that is needed for management comfort versus the cost of overtesting. Ideal quality control programs would sample materials along all stages of production, in every batch and every shift (Stouffer,

1985). This is rarely practical for sensory testing. Sampling multiple products from a batch or production run or performing replicate measurements with a sensory panel will give insurance against missing out-of-spec products, but will increase time and costs of testing.

Additional challenges arise from reporting structures, multiple test sites, and the temptation to substitute instruments for sensory panels. A built-in conflict of interest occurs when a QC department reports directly to manufacturing, since manufacturing is usually rewarded for productivity. A separate reporting structure may be desirable for quality control, so that executives committed to a corporate quality program can insulate the QC department from pressures to pass bad products. Across multiple plants, there is a need to standardize sensory QC procedures and coordinate activities (Carlton, 1985; Stouffer, 1985). This includes maintenance of consistent production samples and reference materials that can be sent to all other plants for comparison. Setting up similar sensory QC systems in different countries and cultures may present difficult challenges to the sensory program coordinator. Considerations in panel setup in other cultures are given in Carlton (1985). Finally, instruments cannot replace sensory evaluation for many important product characteristics (Nakayama and Wessman, 1979). Odor analysis is a good example. In other cases, the instrumental–sensory relationship may be nonlinear or indicate changes that are not perceivable at all (Rutenbeck, 1985; Trant et al., 1981).

17.7.3 The Problem of Low Incidence

A special problem with QC testing as well as shelf-life studies is that the majority of the evaluations result in positive results (favorable decisions for manufacturing). This is in the very nature of the test scenario. Good products are much more often tested than defective ones, and there is a much lower incidence of negative test results than those found in research support testing. This can present a special challenge to the credibility of the sensory testing program.

Table 17.6 shows an incidence diagram for a fairly high rate of problem products, in this case 10% (the credibility problem gets even worse if there is a lower rate of defective products). In over 1,000 tests, then,

100 are objectively defective, while 900 are objectively trouble free. If the tests are properly done, and there is statistical protection against Type I and Type II errors, there will still be some occasions where errors do occur. For the sake of easy calculation, let the long-term alpha- and beta-risks for the testing program be 10%. This means that 10% of the time when a defective product is sent for testing, it will go undetected by the evaluation, and 10% of the time a product which has no defects will be flagged, due to random error. This will lead to 810 correct "pass" decisions and 90 correct detections of sensory problems. Unfortunately, due to the high incidence of good products being tested, the 10% false alarm rate leads to 90 products also being flagged where there is no true sensory problem. Note that this assumes that there is good sensory testing and proper statistical treatment of the results! The problem arises when the sensory QC leader picks up the phone and calls the manufacturing manager and "rings the alarm bell." Given this incidence, the probability of being right about the problem is only 50%, in other words, no better than a coin toss! Even if alpha is reduced to the usual 5%, there is still a one-in-three chance of false alarms.

How can this occur if good sensory testing is done and proper statistics are applied? The answer is that our normal inferential statistics are used to view the outcome chart in Table 17.6 across rows and not down columns. The problem in sensory QC is that given a low incidence of problems, there is simply a high rate of false alarms relative to correct detections. This can hurt the credibility of the program if manufacturing managers develop a feeling that the sensory department is prone to "cry wolf." Thus it is wise to build in a system for additional or repeated testing of product failures to insure that marginal products are in fact defective before action is taken.

17.8 Shelf-Life Testing

17.8.1 Basic Considerations

Shelf-life or stability testing is an important part of quality maintenance for many foods. It is an inherent part of packaging research because one of the primary functions of food packaging is to preserve the integrity of a food in its structural, chemical, microbiological

Table 17.6 Bayesian incidence chart

	Outcome of evaluation		
	Problem reported	No problem reported	Incidence (=total across row)
Problem exists	90	10	100
(description)	("hit rate")	(Type II error)	
No problem exists	90	810	900
	("false alarm")	(correct acceptance)	
(Total)	180	820	1000

Let alpha = 0.10 and beta = 0.10
Assume 1,000 tests are conducted, with a 10% rate of faulty products
Numbers in cells show estimated numbers of problems reported or not, based on alpha and beta rates of 10%
Given that a problem was reported, you stand a 50/50 chance of having made the wrong decision (90/180)!

and sensory properties. A good review of shelf-life testing can be found in the packaging text by Robertson (2006) and the reader is referred there for further information on modeling and accelerated storage tests. For many foods, the microbiological integrity of the food will determine its shelf life, and this can be estimated using standard laboratory practices; no sensory data are required. The sensory aspects of a food are the determining factor for the shelf life of foods that do not tend to suffer from microbiological changes such as baked goods. Sensory tests on foods are almost always destructive tests, so sufficient samples must be stored and available, especially during the period in which the product is expected to deteriorate (Gacula, 1975).

Shelf-life testing may employ any of the three major kinds of sensory tests, discrimination, descriptive, or affective, depending on the goals of the program (Kilcast, 2000). Thus one can view shelf-life tests as no special category of sensory testing, but simply a program of repeated testing using accepted methods. The objectives of the study may dictate what method is most suitable to answer the research questions (Dethmers, 1979). For a designed study to evaluate the effects of a new packaging film, a simple discrimination test might be appropriate to test for changes versus the existing packaging. For purposes of establishing an open dating system, consumer acceptability tests would be appropriate to establish the time that the product is likely to become unacceptable. If the product is new, a descriptive analysis profile is needed to establish the full sensory specification of what a fresh product tastes like. If a product has failed a consumer evaluation, it is often appropriate to submit the samples to descriptive testing to try and understand the reasons for failure and which aspects have deteriorated (Dethmers, 1979). If the purpose is

to establish a suitable degree of stability, i.e., that the failure time exceeds the typical distribution and use time by consumers, a combination of two tests may be appropriate. It is cost efficient in this case to perform discrimination tests against a fresh control or standard product and then perform consumer acceptance tests if any difference is detected.

According to Peryam (1964) and Dethmers (1979) a shelf-life program will involve the following steps: (1) formulating objectives, (2) obtaining representative samples, (3) determining the physical and chemical composition of the test products, (4) setting up a test design, (5) choosing the appropriate sensory method, (6) choosing the storage conditions, (7) establishing the control product or products to which the stored product will be compared, (8) conduct the periodic testing, and (9) determine the shelf life based on the results.

Important strategic choices include the nature of the control product and the storage conditions. Storage conditions should mimic the conditions found in distribution and those in stores, unless some accelerated storage conditions are required. Ideal conditions are generally a poor choice. The control product presents special problems. If a fresh product is available at the different time intervals, how does one know that subsequent batches have not drifted or changed since the initial product was manufactured? If a fresh product is stored from the initial batch under ideal conditions, how does one know it has not changed? There is no perfect solution to this problem. Sometimes a study will involve more than one standard. Reference standards should be clearly identified by date, lot number, production location, etc. A separate program may be instituted to insure the integrity and constancy of reference standards. Descriptive evaluation may be beneficial for this purpose. Options for references include the following: current plant product that has passed QC, current pilot-plant prototype, historical product, optimally stored product, a written descriptive profile, and a mental reference (Wolfe, 1979).

Two main choices are used for criteria for product failure. These include a cutoff point from a critical descriptive attribute (or set of them) and consumer data when the product is rejected as unacceptable. Statistical modeling with equations such as a hazard function or survival analysis is discussed below. Note that product failure is an all-or-none phenomenon, and decreases in sensory measures such as falling acceptability or increasing percents of consumer rejection are more continuous in nature. This opens the opportunity for other kinds of models, such as logistic regression against percent rejecting (Giminez et al., 2007).

17.8.2 Cutoff Point

The choice of a cutoff point has two implications. The first occurs when the cutoff point itself is used as an action standard. When the product gets to this point, we consider it to have reached the end of its useful life. It is no longer salable. The time estimate may be used for some purpose like open dating or "use-by" dates printed on the product package. The second implication is that when a product in a designed study reaches this cutoff point, it defines "failure" and will be used as a data point in some kind of statistical modeling such as survival analysis.

Determination of a cutoff point requires careful consideration. Several options are available including (1) a significant difference in a discrimination test, (2) some degree of difference from control product on a scaled attribute or overall degree of difference scale, and (3) consumer reaction. Consumer data may involve a significant difference in acceptability ratings from control, a cutpoint on an acceptance score, or some percent of consumer rejection (e.g., 50% or 25%). Giminez et al. (2007) found the first significant difference to be too conservative an estimate in the sense that acceptance scores were still above 6 on the 9-point scale. This makes sense because two products may differ but still be acceptable (Kilcast, 2000). Another option is to use any value less than 6 on the 9-point hedonic scale (6 = like slightly, i.e., just

above neutral) (Muñoz et al., 1992). Another option is to use consumer rejection ("I would not buy/eat this product") (Hough et al., 2003). These two measures are not necessarily equivalent. Giminez et al. (2008) found that for certain baked products, consumers might not like the product, but they would answer "yes" when asked if they would consume it at home (having already purchased it). This finding suggests that consumer rejection may not be sufficiently conservative, i.e., that a product may become disliked and even generate consumer complaints before it reaches the point of rejection. Giminez et al. (2007) found that acceptability scores could be related to percent of rejection by logistic regression analysis. The logistic equations for two different countries (Spain and Uruguay) for a baked product were different, a warning about cultural and/or national differences. Logistic regression is a useful general approach to data in the form of proportions that accumulate in an S-shaped curve. The general form is

$$\ln \frac{p}{1-p} = b_0 + b_1 X \qquad (17.2)$$

where p is the proportion rejected and X is the variable that is the predictor, such as time, or in this case acceptability scores (b are constants).

17.8.3 Test Designs

Several options are available for shelf-life tests regarding how samples are stored and test times. The simplest method is to make one large batch of product, store it under normal conditions, and test it at various intervals. However, this is not very efficient in terms of test time and risks the panel drifting its criteria. Another option is to stagger the production times, so all the products of different ages are tested on the same day. A variation on this is to store the product under conditions that essentially stop all aging processes, for example, at very low temperature. This is obviously not possible with all products (you cannot freeze lettuce). Then products are pulled from the optimal storage conditions at different times and allowed to age at normal temperatures. Another variation of this procedure is to allow products to age for different times and then place them into the optimal storage conditions, pulling everything out of storage at the test date.

17.8.4 Survival Analysis and Hazard Functions

The literature on survival analysis is very large, because a number of different fields use these kinds of statistical models, such as actuarial science for the insurance industry. Some of the models are similar to those used in chemical kinetics. These functions are useful when the product has a single process or a group of processes that are occurring at about the same time. However, some products will show a "bathtub" function with two phases of product failure (Robertson, 2006). In the early stages, some product failures occur due to faulty packaging or improper processing (see Fig. 17.4). Then the remaining products from that batch enter a period of product stability. After some time, X, the products begin to fail again, due to deterioration. Gacula and Kubala (1975) suggest that the shelf-life modeling should only consider those failures after time X_2.

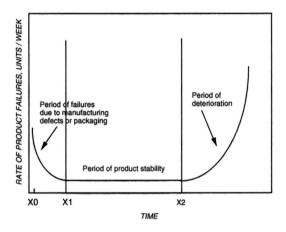

Fig. 17.4 The "bathtub" function showing a common pattern of failure rates changing over the sampling time in a shelf-life study. From time X_0 to X_1, some products will fail due to improper processing or faulty packaging. This is followed by a period of fairly low failure rates when products are stable or within specification limits. At time X_2, failures start to increase markedly. Researchers fitting hazard functions or doing survival analysis for estimation of shelf life should consider using only those times after X_2 in curve fitting as earlier failures are due to causes other than the time-related deterioration.

Survival analysis as applied to sensory data has two main tasks, the fitting of a hazard function to the data and the interpolation of some point used as the criterion for shelf life (such as 25 or 50% consumer rejection). Various functions have been used to fit the function of failure data (or percent of failures) over time. Percent survival (one minus percent of failures) often takes the form of a decaying exponential function.

An important choice of model includes the distribution used to fit the percent of failures. Many distributions have been tried (Gacula and Kubala, 1975) but two useful models are a log–normal distribution (surviving is a positively skewed distribution, few people live to 100) and a Weibull distribution. Weibull functions are useful distributions that can be used to fit a variety of data sets. They include a shape parameter and a scale parameter. When the shape parameter takes a value greater than 2, the distribution is approximately bell shaped and symmetric. These equations for failure take the following forms:

$$F(t) = \Phi\left(\frac{\ln(t) - \mu}{\sigma}\right) \text{(log --normal)} \qquad (17.3)$$

where Φ is the cumulative normal distribution function, t is time, $F(t)$ is the failure proportion at time t, μ is the mean failure time, and σ is the standard deviation.

$$F(t) = 1 - \exp\left[-\exp\left(\frac{\ln(t) - \mu}{\sigma}\right)\right] \quad \text{(Weibull)}$$
$$(17.4)$$

where $\exp(x)$ is the notation for e^x.

If we make two substitutions and determine the mean (μ) and standard deviation (σ) of our failure times, a simple model can help us find the time for a given percentage of failures from the fitted Weibull equation. Let $\rho = \exp(-\mu)$. Then the following relationship holds for any proportion $(F(t))$:

$$t = \frac{-\ln(1 - F(t))^\sigma}{\rho} \qquad (17.5)$$

Using the log–normal model for $F(t)$, a simple graphic method for finding the interpolated 50% failure level is to do the following: For N samples of foods sampled over time that have known failure times T_i, rank all the batches, i as to the time of failure ($i = 1$ to N). Calculate the median ranks, MR values. The median rank can be found in some statistical tables or estimated as

$$MR = (i - 0.3)/(N + 0.4)$$

Plot the median rank on log probability paper versus T_i and interpolate at the 50% point. If a straight line fits the data, the 50% point can be estimated from the linear equation and standard deviations estimated from the probability paper. This is essentially a fit of log MRs to *z*-scores. Other percentages may be interpolated, of course, because 50% may be too high a failure percentage for many products in setting a useful limit. An equivalent mathematical solution is given in Appendix 2 of this chapter.

Hough et al. (2003) point out that the usual sensory experiment produces censored data. That is, for any batch that has failed, we only know that the time of failure was in some interval between the last test and the current test. Similarly, for a batch that has not failed at the final interval, we only know that its failure time is sometime after that final test. So the data are censored and the survival function can be estimated using maximum likelihood techniques.

17.8.5 Accelerated Storage

Sooner or later, product developers may figure out that sensory specialists do not own a time machine and cannot deliver shelf-life estimates without a long study. So they may request some accelerated storage tests to shorten the time. Such tests are based on the idea that at higher temperatures, many chemical reactions will proceed in a predictable manner, according to simple kinetic models, and thus the shelf life at long time intervals can be simulated by shorter intervals at higher temperatures (Mizrahi, 2000). Kinetic models are often based on the Arrhenius equation, and rate constants can be found from experiments conducted at different temperatures. Some of the models are shown in Appendix 3 of this chapter.

Problems in accelerated testing occur when changes in the product due to temperature are not the same as those due to storage time at normal temperatures (Robertson, 2006). Obviously, trying to measure the shelf life of a frozen food at higher temperatures makes little sense. Other foods may not follow simple predictive models as multiple processes occur with different rates. For example, at higher temperatures phase change may occur, solid to liquid. Carbohydrates in the amorphous state may crystallize. The water activity of dry foods may increase with temperature causing an increased reaction rate and overprediction of true shelf life at the normal temperature. If two reactions with

different kinetic constants change at different rates with different temperatures, the one with the higher value may come to predominate. The sensory specialist should be familiar with the logic of this testing and the modeling that is commonly done, as well as the pitfalls.

17.9 Summary and Conclusions

Insuring the quality of products on a sensory basis is an important corporate goal in a competitive environment. Consumers have fixed expectations and will become disloyal to a brand if they experience substandard products. However, in spite of the need for sensory quality control, setting up and maintaining an in-plant QC program is difficult and costly. Commitment to the program is the corporate equivalent of diet and exercise. Everyone admits that it is a good idea. However, maintaining program integrity, avoiding shortcut procedures, and dealing with dwindling panel size can be challenging. The success of any program demands strong management commitment. Without management support, especially from manufacturing, a sensory QC program is bound to fail. In a typical case the program will amount to nothing more than "rubber stamping" of supervisory opinion, thus supporting a management policy that maximizes productivity at the expense of producing unacceptable products. Such programs will blow hot and cold, usually receiving some emergency attention when a truly bad product hits the retail shelves and consumer complaints filter back (Rutenbeck, 1985). After a period of improved production, some complacency may set in and loss of interest in sensory QC efforts (until the next disaster).

Implementation of a sensory QC program will involve four technical tasks. First, a range of products must be prepared for establishing quality specifications and limits. Specifications must be set in a research study with consumers or by management or expert opinion after sampling the various products. This range of products can also be used in panel training. The second step in the program is recruitment, screening, and training, in other words, panel setup. Next, standard protocols for product sampling, handling, storage, serving, blind coding, and maintenance of reference standards must be established. The fourth step is in systematizing the paperflow (Mastrian, 1985). This includes establishing standard reporting formats and processes for data handling, recommendations, and action criteria. This activity should also include mechanisms for archiving results and tracking both products and panelist performance across time.

Appendix 1: Sample Screening Tests for Sensory Quality Judges

Part 1. Paired comparison of sweetness levels
    Adjust samples to three levels, e.g., 10, 11, and 12% sucrose wt/vol.
    Give four pairs in counterbalanced orders, e.g.,

10 versus 11%, 11 versus 10% (hard discrimination)
10 versus 12 %, 12 versus 10% (easier discrimination)

    Use a different order for different panelists. Blind code with random 3-digit labels.

Good performance: All four correct.
Acceptable performance: One error if other test sections perfect.

Part 2. Multiple choice odor identification. Done with blotters in capped jars
    Circle correct answers on sheet. Make up multiple forms with different orders.
    Use random 3-digit codes on bottles. Odors represent common notes in the product.
    Use four alternatives, e.g., fruity, smoky, vinegar, onion

(a) Dilute ethyl hexanoate (or similar ester)
(b) Dilute ethyl 2-methyl butyrate
(c) Dilute vinegar
(d) Dilute phenylethanol
(e) Trans-2-hexenol(dilute until green or leafy smell is obtained)

Good performance: 4/5 correct
Acceptable performance:3/5 correct

Part 3. Odor discrimination test

Run triangle tests with base juice and base juice + 1% vinegar.
Run triangle test with base juice + 0.1% butyric acid.
Subjects should sniff first, then taste.

Provide palate cleansers (water, crackers).
Run duplicates of each test.

Good performance: 3/4 correct.
Acceptable performance: 2/4 correct.

Part 4. Acidity test
    Adjust pH to about 0.5 versus 1% titratable acidity.
    Use four paired comparisons, as in sweetness test above.

Good performance: 3/4 correct.
Acceptable performance: 2/4 correct.

After scoring, rank order candidates from highest to lowest.
Invite the top 50% from each shift for training.
Send thank you notes to all the people who try out.
Keep the rest "on file" for possible replacements if scores are acceptable.

Appendix 2: Survival/Failure Estimates from a Series of Batches with Known Failure Times

This procedure follows the graphic method given in Section 17.8.3 but allows a more exact fit by least squares regression.

1. For N samples of foods sampled over time that have known failure times T_i, rank all the batches, i as to the time of failure ($i = 1$ to N).
2. Calculate the median ranks, MR values. The median rank can be found in some statistical tables or estimated as

$$MR = (i - 0.3)/(N + 0.4)$$

3. Convert each T_i to $\ln(T_i)$, called Y_i. This will permit a fit of MR to the log–normal model.
4. Calculate the z-score for each MR at each T_i. Call this X_i.
5. Regress Y against X using least squares to get the linear equation $Y = a + bX$.

This is equivalent to finding the straight line fit to the log probability plot described in Section 17.8.

6. Then solve for $Y = 0$ (z-score for 50%) which is $X = -a/b$, to get the 50th percentile.
7. Convert back to the original units by exponentiating Time at 50% failure $= e^X = e^{-a/b}$.

Appendix 3: Arrhenius Equation and Q_{10} Modeling

The reaction time for product failures may be linear (zero order) or a decaying exponential (first order). Both allow determination of a rate constant, K. Let us consider a cutoff point rating on some scale, R, as the event to be modeled as function of time, t. R could also be any event that signals product failure.

The zero-order equation is

$$R = R_o - kt \qquad (17.6)$$

And the first-order relationship is

$$R = R_o e^{-kt} \qquad (17.7)$$

where R_o is the rating or failure at $t=0$ and

$$\ln \frac{R}{R_o} = -kt \qquad (17.8)$$

Reaction rates are also dependent on temperature, so in accelerated storage studies, the Arrhenius equation provides a starting point or a generally useful approximation:

$$k = k_o e^{\left(\frac{-E_A}{RT}\right)} \qquad (17.9)$$

and

$$\ln k = \ln k_o - \frac{E_A}{R}\left(\frac{1}{T}\right) \qquad (17.10)$$

where k is the rate constant to be estimated, k_o is a constant independent of temperature (also known as the Arrhenius, pre-exponential, collision, or frequency factor), E_A is the activation energy (J/mol), R is the ideal gas constant, T is temperature (absolute, K).

So a plot of $\ln(k)$ versus $1/T$ can be used to find the activation energy, E_A.

This sometimes takes its derivative form:

$$\frac{d(\ln k)}{dT} = \frac{E_A}{RT^2} \qquad (17.11)$$

The activation energy is fictitious in a way, because there is not a single chemical reaction going on during the aging of a food product, but a large number of simultaneous processes. Nonetheless, we can think of this as useful for two reasons. First, it gives an indication of the fragility of the food (lower activation energy would mean faster deterioration). Second, the E_A value becomes useful in predicting what happens at different temperatures. Of specific interest is predicting what will happen at "normal" temperature given that an accelerated storage study has been conducted at higher temperatures.

Experiments are often performed at varying temperatures, 10°C apart, to generate what is known as the Q_{10} factor.

$$Q_{10} = \frac{k_{T+10}}{k_T} = \frac{S_T}{S_{T+10}} \qquad (17.12)$$

where k_{T+10} and k_T are the rate constants at temperature T and $T + 10$, and S_T and S_{T+10} are the corresponding shelf-life estimates. Note that the ratio of rate constants is the inverse of the ratio of the shelf-life times.

This produces some useful relationships, for example, to use in estimating the activation energy, E_A

$$\ln Q_{10} = \frac{10 E_A}{RT^2} \qquad (17.13)$$

Again, E_A can give some idea of the susceptibility of the product to deterioration. Once the Q_{10} factor has been determined for a product, the time–temperature relationship can be predicted for the accelerated tests. A useful factor to determine is an acceleration factor, AF. This will help us convert from the accelerated temperature back to a usage temperature or normal storage condition temperature like 20°C. Hough (2010) gives the following example, based on an off-flavor rating (OF) as a function of time and temperature:

$$OF_{T,temp} = OF_o + (AF)k_u(T_{temp}) \qquad (17.14)$$

where OF_o is the off-flavor at time zero, k_u is the rate constant at usage temperature, and T_{temp} is the accelerated test temperature. Knowing E_A, we can also estimate the AF from

$$AF = \exp\left[\frac{E_A}{R}\left(\frac{1}{T_u} - \frac{1}{T_{test}}\right)\right] \quad (17.15)$$

where $\exp(X)$ is e^X, T_u is the usage temperature, and T_{test} is the accelerated test temperature.

For example, if we determine that the E_A is 6,500 cal/mol, then we can calculate an acceleration factor based on a test at 40°C and a usage temperature at 20°C:

$$AF = \exp\left[6500\left(\frac{1}{293} - \frac{1}{313}\right)\right] = 4.13$$

Suppose we determine that at accelerated temperature of 40°C, we have a failure time of 35 days. Then we can find the failure time, FT_u, at usage temperature T_u, we merely multiply by the acceleration factor

$$FT_u = FT_{test}(AF) = 35(4.13) = 145$$

Thus our accelerated test predicts a (mean) failure time at 145 days for the product stored at room temperature or about 20°C.

References

Amerine, M. R. and Roessler, E. B. 1981. Wines, Their Sensory Evaluation, Second Edition. W.H. Freeman, San Francisco, CA.

Amerine, M. R., Pangborn, R. M. and Roessler, E. B. 1965. Principles of Sensory Evaluation of Foods. Academic, New York, NY.

Aust, L. B., Gacula, M. C., Beard, S. A. and Washam, R. W. 1985. Degree of difference test method in sensory evaluation of heterogeneous product types. Journal of Food Science, 50, 511–513.

Bauman, H. E. and Taubert, C. 1984. Why quality assurance is necessary and important to plant management. Food Technology, 38(4), 101–102.

Beckley, J. P. and Kroll, D. R. 1996. Searching for sensory research excellence. Food Technology, 50(2), 61–63.

Bodyfelt, F. W., Tobias, J. and Trout, G. M. 1988. Sensory Evaluation of Dairy Products. Van Nostrand/AVI, New York, NY.

Bressan, L. P. and Behling, R. W. 1977. The selection and training of judges for discrimination testing. Food Technology, 31, 62–67.

Cardello, A. V. 1995. Food quality: Relativity, context and consumer expectations. Food Quality and Preference, 6, 163–170.

Carlton, D. K. 1985. Plant sensory evaluation within a multi-plant international organization. Food Technology, 39(11), 130–133, 142.

Dethmers, A. E. 1979. Utilizing sensory evaluation to determine product shelf life. Food Technology, 33(9), 40–43.

Ennis, D. M., Mullen, K. and Frijters, J. E. R. 1988. Variations of the method of triads: Unidimensional Thurstonian models. British Journal of Mathematical and Statistical Psychology, 41, 25–36.

Gacula, M. C. 1975. The design of experiments for shelf life study. Journal of Food Science, 40, 399–403.

Gacula, M. C. and Kubala, J. J. 1975. Statistical models for shelf life failures. Journal of Food Science, 40, 404–409.

Garvin, D. A. 1987. Competing on the eight dimensions of quality. Harvard Business Review, 65(6), 101–109.

Gillette, M. H. and Beckley, J. H. 1992. In-Plant Sensory Quality Assurance. Paper presented at the Annual Meeting, Institute of Food Technologists, New Orleans, LA, June, 1992.

Giminez, A., Ares, G. and Gambaro, A. 2008. Survival analysis to estimate sensory shelf life using acceptability scores. Journal of Sensory Studies, 23, 571–582.

Giminez, A., Varela, P., Salvador, A., Ares, G., Fiszman, S. and Garitta, L. 2007. Shelf life estimation of brown pan bread: A consumer approach. Food Quality and Preference, 18, 196–204.

Goldwyn, C. and Lawless, H. 1991. How to taste wine. ASTM Standardization News, 19(3), 32–27.

Hammond, E., Dunkley, W., Bodyfelt, F., Larmond, E, and Lindsay, R. 1986. Report of the committee on sensory data to the journal management committee. Journal of Dairy Science, 69, 298.

Hodgson, R. T. 2008. An examination of judge reliability at a major U.S. wine competition. Journal of Wine Economics, 3, 105–113.

Hough, G. 2010. Sensory Shelf Life Estimation of Food Products. CRC Press, Boca Raton, FL.

Hough, G., Langohr, K., Gomez, G. and Curia, A. 2003. Survival analysis applied to sensory shelf life of foods. Journal of Food Science, 68, 359–362.

International Olive Oil Council. 2007. Sensory analysis of olive oil. Method for the organoleptic assessment of virgin olive oil. http://www.internationaloliveoil.org/.

Kilcast, D. 2000. Sensory evaluation methods for shelf-life assessment. In: D. Kilcast and P. Subramaniam (eds.), The Stability and Shelf-Life of Food. CRC/Woodhead, Boca Raton, FL, pp. 79–105.

Lawless, H. T. 1995. Dimensions of quality: A critique. Food Quality and Preference, 6, 191–196.

Lawless, H. T. and Claassen, M. R. 1993. Validity of descriptive and defect-oriented terminology systems for sensory analysis of fluid milk. Journal of Food Science, 58, 108–112, 119.

Mastrian, L. K. 1985. The sensory evaluation program within a small processing operation. Food Technology, 39(11), 127–129.

McBride, R. L. and Hall, C. 1979. Cheese grading versus consumer acceptability: An inevitable discrepancy. Australian Journal of Dairy Technology, June, 66–68.

McNutt, K. 1988. Consumer attitudes and the quality control function. Food Technology, 42(12), 97, 98, 108.

Mizrahi, S. 2000. Accelerated shelf-life tests. In: D. Kilcast and P. Subramaniam (eds.), The Stability and Shelf-life of Foods. CRC /Woodhead, Boca Raton, FL, pp. 107–142.

Moskowitz, H. R. 1995.Food Quality: conceptual and sensory aspects. Food Quality and Preference, 6, 157–162.

Muñoz, A. M., Civille, G. V. and Carr, B. T. 1992. Sensory Evaluation in Quality Control. Van Nostrand Reinhold, New York, NY.

Nakayama, M. and Wessman, C. 1979. Application of sensory evaluation to the routine maintenance of product quality. Food Technology, 33(9), 38, 39 ,44.

Nelson, L. 1984. The Shewart control chart-tests for special causes. Journal of Quality Technology, 16, 237–239.

Nelson, J. and Trout, G. M. 1964. Judging Dairy Products. AVI, Westport, CT.

O'Mahony, M. 1981. Our-industry today—psychophysical aspects of sensory analysis of dairy products: A critique. Journal of Dairy Science, 62, 1954–1962.

Ough, C. S. and Baker, G. A. 1961. Small panel sensory evaluations of wines by scoring. Hilgardia, 30, 587–619.

Pangborn, R. M. and Dunkley, W. L. 1964. Laboratory procedures for evaluating the sensory properties of milk. Dairy Science Abstracts, 26, 55–62.

Pecore, S., Stoer, N., Hooge, S., Holschuh, N., Hulting, F. and Case, F. 2006. Degree of difference testing: A new approach incorporating control lot variability. Food Quality and Preference, 17, 552–555.

Peryam, D. R. 1964. Consumer preference evaluation of the storage stability of foods. Food Technology, 18, 214.

Reece, R. N. 1979. A quality assurance perspective on sensory evaluation. Food Technology, 33(9), 37.

Robertson, G. L. 2006. Food Packaging, Principles and Practice, Second Edition. CRC/Taylor and Francis, Boca Raton, FL.

Rutenbeck, S. K. 1985. Initiating an in-plant quality control/sensory evaluation program. Food Technology, 39(11), 124–126.

Regenstein, J. M. 1983. What is fish quality? Infofish, June, 23–28.

Sidel, J. L., Stone, H. and Bloomquist, J. 1981. Use and misuse of sensory evaluation in research and quality control. Journal of Dairy Science, 64, 2292–2302.

Solomon, G. E. A. 1990. The psychology of novice and expert wine talk. American Journal of Psychology, 103, 495–517.

Stevenson, S. G.,Vaisey-Genser, M. and Eskin, N. A. M. 1984. Quality control in the use of deep frying oils. Journal of the American Oil Chemist's Society, 61, 1102–1108.

Stouffer, J. C. 1985. Coordinating sensory evaluation in a multi-plant operation. Food Technology, 39(11), 134–135.

Thompson, R. H. and Pearson, A. M. 1977. Quantitative determination of 5 Androst-16-en-3-one by gas chromatography-mass spectrometry and its relationship to sex odor intensity of pork. Journal of Agricultural and Food Chemistry, 25, 1241–1245.

Trant, A. S., Pangborn, R. M. and Little, A. C. 1981. Potential fallacy of correlating hedonic responses with physical and chemical measurements. Journal of Food Science, 46, 583–588.

Wolfe, K. A. 1979. Use of reference standards for sensory evaluation of product quality. Food Technology, 33(9), 43–44.

York, R. K. 1995. Quality assessment in a regulatory environment. Food Quality and Preference, 6, 137–141.

Young, T. A., Pecore, S., Stoer, N., Hulting, F., Holschuh, N. and Case, F. 2008. Incorporating test and control product variability in degree of difference tests. Food Quality and Preference, 19, 734–736.

Chapter 18

Data Relationships and Multivariate Applications

Abstract Multivariate statistics have found great application in all areas of quantitative sensory science. In this chapter we will briefly describe the two major work horses in the field: principal component analysis (PCA) and canonical variate analysis (CVA). PCA should be used with mean data and CVA with raw data, namely data including replicate observations. We also discuss generalized Procrustes analysis (GPA) which is used with free-choice profiling data as well as in any situation where one may want to compare the data spaces associated with multiple data measurements on the same products. Lastly we discuss (as a preliminary to further in-depth discussion in Chapter 19) internal and external preference mapping. We conclude by stressing that multivariate analyses should always be performed in conjunction with univariate analyses.

The researcher will find that there are certain costs associated with benefits of using multivariate procedures. Benefits from increased flexibility in research design, for instance, are sometimes negated by increased ambiguity in interpretation of results.

—Tabachnik and Fidell (1983) (Italics added)

Contents

18.1 Introduction

Descriptive sensory tests are often performed to determine the effects of changes in raw material, processing, and packaging on the sensory qualities of products. It is also frequently desirable to relate the hedonic results to the sensory and/or instrumental results of the same study. In all of these cases, multiple attributes on a single set of samples were evaluated and must now be analyzed. To do this, one must use a group of analysis tools known as multivariate statistics. During the last 30 years, the widespread access to computers and the increasing sophistication in statistical packages have expanded the use and utility of multivariate statistical analyses. This "easy access" often tempts novices to use these techniques with sometimes surprising (and often suspect) results. Before using any of these statistical techniques, the user must be sure that the method is used appropriately and correctly. Many multivariate techniques require additional statistical assumptions beyond those of the simple univariate tests. This leads to additional

H.T. Lawless, H. Heymann, *Sensory Evaluation of Food*, Food Science Text Series,
DOI 10.1007/978-1-4419-6488-5_18, © Springer Science+Business Media, LLC 2010

liabilities and potential pitfalls when using multivariate methods. Even when used correctly, it is safest to draw conclusions from multivariate results when these results converge with other information. Occasionally one can use these techniques for hypothesis generation, but only rarely are multivariate statistical techniques stand-alone methods that "prove" a point.

In general multivariate statistical analyses aim to extract information from the product–attribute matrix and to present it in understandable form. Their great advantage lies in giving the sensory specialist the ability to detect broader patterns of interrelationships among products and among sensory attributes than given by individual univariate analyses. There are numerous multivariate techniques and prior to analysis the sensory specialist must determine which technique is most appropriate in the specific case. In this chapter we will give an overview of a few of the frequently used multivariate techniques used in sensory studies—these analyses fall within the realm of sensometrics. This overview is a very brief introduction to specifically principal component analysis, multivariate analysis of variance, discriminant/canonical variate analysis, Procrustes analysis, and preference mapping analyses. Some very useful multivariate techniques (STATIS, multi-factor analysis, partial least squares analysis, cluster analysis) are not covered and the reader is encouraged to refer to the large number of available textbooks. Additionally, multi-dimensional scaling (MDS) sometimes classified as a multivariate technique although it does not use multiple dependent measures as input, only some measure of overall similarity, is discussed in Chapter 19. The following texts are good introductions to multivariate statistical methods: Anderson (2003), Hair et al. (2005), Johnson and Wichern (2007), Krzanowski (1988), Stevens (1986), and Tabachnik and Fidell (2006). Useful sensometrics textbooks are Dijksterhuis (1997), Gower and Dijksterhuis (2004), Martens and Martens (2001), and Meullenet et al. (2007).

18.2 Overview of Multivariate Statistical Techniques

18.2.1 Principal Component Analysis

Principal component analysis (PCA) is a multivariate technique that simplifies and describes interrelationships among multiple dependent variables (in sensory data these are usually the descriptors) and among objects (in sensory data these are usually the products) (Anderson, 2003; Tabachnik and Fidell, 2006). The PCA should be performed on the mean data for products averaged across panelists and replications. If one wants to use PCA with raw data one should use the PCA with confidence ellipses as described by Husson et al. (2004, 2006) in SensomineR (Lê and Husson, 2008). PCA transforms the original dependent variables into new uncorrelated dimensions, and this simplifies the data structure and helps one to interpret the data (Johnson and Wichern, 2007). PCA will be discussed in this chapter and in Chapter 19.

The product of a PCA is frequently a graphical representation of the interrelationships among variables and objects. The technique is very useful when several dependent variables are collinear (correlated with one another), a situation that often occurs with sensory descriptive data. From the ANOVA, we may find that many descriptors significantly discriminate among the samples; however, several descriptors may be describing the same characteristic of the product. For example, in descriptive studies, panelists often evaluate both aroma and flavor attributes of the products, yet it is very possible that the aroma and flavor attributes are redundant and measure the same underlying characteristics (Heymann and Noble, 1989). The PCA shows these redundancies by transforming the original data into a new set of variables called principal components; redundant or highly positively correlated attributes will lie close to one another in the new space. Products will have values on these new variables (PCs) just as they did on the original attributes. These are sometimes called factor scores. They allow plotting of the products in the new principal component space.

The principal components are obtained through a linear combination of the dependent variables that maximizes the variance within the sample set. The first principal component (PC) accounts for the maximum possible amount of variance among the samples. Subsequent PCs account for successively smaller amounts of the total variance in the data set and are uncorrelated with (orthogonal to or at 90° angles to) prior PCs. If there are more samples than variables (the ideal situation, see below) then the total number of PCS that can be extracted from a data set is equal to the number of dependent variables. The linear combinations of the PCs can be based on the data correlation matrix (the data are standardized) or the data covariance matrix (the data are not standardized). The correlation matrix should be used when the variables were measured on widely divergent scales since in that case scale range can affect the outcome dramatically. Sensory scientists usually use the covariance matrix since sensory descriptive data are usually measured on the same scale (say a 15 cm unstructured line scale).

If all the PCs are retained then the PCA acts as a method of data transformation without loss of information. This is similar to transforming the temperature from the Fahrenheit to the Celsius scale. However, usually the PCA is performed to simplify and describe interrelationships. In this case, the first few PCs account for the majority of the variance in the data set, and often only these components are retained for further interpretation. Thus, once the PCA has been performed, the analyst must decide how many PCs should be retained. In general, it is recommended that one should use a combination of the criteria listed below when determining the number of PCs to retain (Hatcher and Stepanski, 1994; Stevens, 1986). The usual criteria are:

(1) The Kaiser criterion states that one should retain and interpret PCs with eigenvalues[1] greater than 1. This criterion is based on the assumption that the retained PCs should explain more variance than a single dependent variable, and a PC with an eigenvalue equal to 1 explains the same amount of variance as a single dependent variable (Kaiser, 1960). The Kaiser criterion may be too lenient and retain too many PCs. It is usually fairly accurate when the original data set had more than 20 dependent variables that had high communality (Stevens, 1986). The communality of each variable is the amount of variance associated with that variable that is account for by the retained PCs. If all PCs are retained, then the communality of all dependent variables will be 100%; if fewer PCs are retained, then the communality of each variable

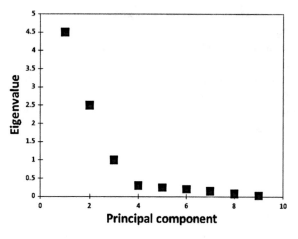

Fig. 18.1   A scree plot of the eigenvalues for a PCA with nine descriptors. The *dotted lines* indicate the elbow in the graph. In this case the scree test would indicate that one should keep three principal components.

depends on how well the retained PCs describe the original data space.

(2) The scree test is a graphical method where the eigenvalues associated with each PC are plotted on a scatter plot (Fig. 18.1). PCs are retained based on the identification of an "elbow" or a break in the graph. PCs appearing before the "elbow" are retained, while those after the break are not (Cattell, 1966). The test derives its name from the scree or talus found at the bottom of vertical cliffs. The scree test tends to retain too few PCs, but the test is reasonably accurate for data sets with more than 250 observations and mean communalities over 0.60 (Cattell and Vogelmann, 1977; Stevens, 1986).

(3) Often the analysts will retain the number of PCs that account for a pre-specified proportion of the variance in the data set. The amount of variance is frequently pre-specified as 70, 80, or 85%.

(4) The last criterion used is that of common sense and interpretability. In other words, retaining the PCs that make sense based on existing knowledge of the subject under investigation. Interpretability is based on several criteria (Hatcher and Stepanski, 1994), namely, variable loading on a given dimension should share some common meaning; variables loading on different dimensions should measure different meanings; and the factor pattern should display a simple structure.

---

[1] An eigenvalues is proportional to the amount of variance collected by the new factor, the PC. Eigenvalues greater than 1 explain more variance than one of the original variables (descriptors).

The result of a PCA is an unrotated factor pattern matrix. The factor pattern may be rotated if it is difficult to interpret the unrotated pattern. The goal of the rotation is to derive a PCA with simple structure. In a PCA with simple structure the variables load highly on a single PC. In a two-dimensional PCA, unlike PCAs with higher dimensions, it is possible to manually rotate the axes. For example, Fig. 18.2 (top)

is the PCA map of hypothetical data. It is clear that rotating the two PCs in the direction of the arrows will lead to Fig. 18.2 (bottom), which has a simpler structure. When one chooses to retain more than two PCs a mathematical rotation must be performed. During mathematical rotation, the PC loadings are transformed either with retention of orthogonality (usually used) or not (rarely used). Orthogonal rotations such as varimax and quartimax retain the orthogonality (uncorrelated aspect) of the PCs, while oblique rotation such a promax and orthoblique do not (Stevens, 1986).

Once a simple structure has been obtained, the PCs must be interpreted. This is done by describing the relationships of the dependent variables (descriptors) to one another and to the retained PCs. Descriptors loaded heavily (either positively or negatively) on a particular PC are used to interpret that dimension. What is meant by an attribute that is loaded heavily? Hatcher and Stepanski (1994) indicated that these are attributes that have loadings larger than absolute 0.40. However, Stevens (1986) is more conservative and suggests that important loadings are those that are twice the size of significant correlation coefficient for the specific sample size. We recommend that the analyst decides on a loading value that he/she feels comfortable with (in our case this is usually 0.75 or so) and that he/she then only uses these loadings (attributes) to describe each PC. Another view is that all loadings are meaningful and should be interpreted. Small loadings (near zero) mean that that PC is not related to those variables, and this can also be useful information.

Finally, the samples (objects or products) are plotted into the PCA space described by the retained PCs. Scores are calculated for each sample to determine its location on the retained PCs. Samples further apart on the PC map are perceptually more different from each other than samples grouped closer together (Coxon, 1982). Husson et al. (2004, 2005) and Monteleone et al. (1998) have developed bootstrapping methods to determine the 95% confidence intervals around the products in the PCA space (Fig. 18.3).

PCA is extensively used with sensory descriptive data. A few examples are Bredie et al. (1997), Guinard et al. (1998), Wortel and Wiechers (2000), Lotong et al. (2002), van Oirschot et al. (2003), and Pickering (2009).

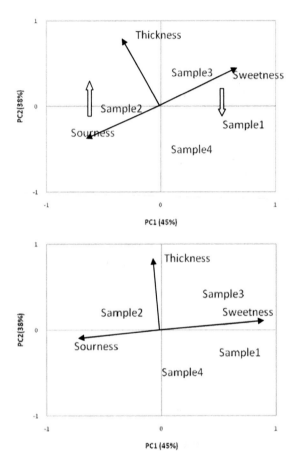

Fig. 18.2 *Top*: Unrotated two-dimensional PCA solution of hypothetical data. *Open arrows* indicate direction of rotation. *Bottom*: Manually rotated PCA solution of hypothetical data. This PCA plot may be interpreted as follows: PC1 explains 45% of the variance in the data set and this PCA is a contrast between sweetness and sourness, PC2 explains an additional 38% of the variance and it is primarily a function of thickness. Sample 1 is sweet and less thick than sample 3; sample 4 is balanced between sour and sweet and is less thick than the other samples; sample 2 is thicker than samples 1 and 4 but less thick than sample 3. It is also more sour than the other samples and less sweet. Sample 3 is somewhat similar in sourness and sweetness to sample 4 but thicker in consistency.

Fig. 18.3 An example of a PCA plot with 95% confidence ellipses generated by bootstrapping. *Circles* that overlap are not significantly different from one another at the 95% level (reprinted with permission from Lê et al., 2008).

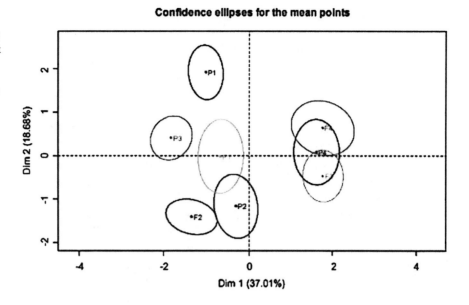

18.2.2 Multivariate Analysis of Variance

Multivariate analysis of variance (MANOVA) is a procedure that allows one to determine whether significant differences exist among treatments when compared on all dependent variables of interest (Ståhle and Wold, 1990; Stevens, 1986). Like univariate analysis of variance (ANOVA), MANOVA tests for differences between two or more treatments, but, in contrast to ANOVA where one evaluates one dependent variable at a time, in MANOVA all dependent variables are evaluated simultaneously.

In sensory descriptive analysis, multiple descriptors are used to describe and evaluate a product set. For example in a descriptive analysis of ice creams, the panelists might evaluate six ice creams for iciness, smoothness, hardness, and melt rate. Usually each attribute (dependent variable) is analyzed by ANOVA, requiring four ANOVAs (one for each attribute) to be performed on the data. Theoretical analysis has revealed that a large number of ANOVAs performed on the same data set may lead to an inflated overall Type I error (Stevens, 1986). For example, suppose a panel evaluated two yogurts by descriptive analysis using eight descriptors. The data are then analyzed using *t*-tests, one *t*-test for each descriptor. Remember that the *F*-value for a two-sample ANOVA is equal to the square of the *t*-test value. The alpha (Type I

error) is fixed at 0.05. If we assume that all eight tests are independent (which is not entirely true since each of the yogurts was evaluated by the same panelists and some of the variables are probably collinear, that is, related to each other), then the overall probability of no Type I error is $(0.95) \times (0.95) \times (0.95) \times (0.95) \times (0.95) \times (0.95) \times (0.95) \times (0.95) \approx 0.66$. Thus the probability of at least one false rejection (given that all null hypotheses are true) is equal to $1–0.66=0.34$. From this simple example, it is easy to see that the overall Type I error quickly becomes very high when multiple tests are performed. It is also not possible to accurately estimate the increase in size of the Type I error. Ideally, the data should be analyzed by MANOVA prior to the individual ANOVAs, since performing a MANOVA prior to individual ANOVAs protects against this situation (Hatcher and Stepanski, 1994; Johnson and Wichern, 2007; Stevens, 1986). There are a minority of statisticians that do not feel that multiple ANOVAs necessarily lead to better control of the Type I error (Huberty and Morris, 1989, Ståhle and Wold, 1990). The MANOVA provides a single *F*-statistic, based on Wilks' Lambda (λ), which assesses the influence of all descriptors simultaneously. A significant MANOVA *F*-statistic (due to a small Wilks' lambda) indicates that the samples differ significantly across the dependent variables. At this point, an ANOVA on each dependent variable should be performed to determine which dependent variables significantly differentiate among the samples. On the other hand, a non-significant MANOVA *F*-statistic (based on a larger Wilks' lambda) indicates that the samples do not differ across the dependent variables and that individual ANOVAs are not warranted.

MANOVA protects the sensory specialist from another problem associated with multiple ANOVAs. Individual ANOVAs do not account for one very important piece of information, namely, the collinearity (correlations) among the descriptive variables. MANOVA includes collinearity (through the covariance matrix) into the test statistic. The effect of correlations among the dependent variables is taken into account within the analysis. In addition, the possibility exists that samples do not differ on any one variable but that some combination of variables significantly discriminate among the samples. MANOVA allows the sensory specialist to explore this possibility whereas performing individual ANOVAs does not. Determining that a combination of variables discriminates among samples when single variables do not is important protection against a Type II error, that is, against missing a true difference. Sensory specialists should use this tool, especially when there is potentially a market impact in making a mistake by declaring products to be equivalent when in fact

they are different. Examples of MANOVA in the sensory literature are Lee et al. (2008), Cano-López et al. (2008), Adhikari et al. (2003), and Montouto-Graña et al. (2002).

18.2.3 Discriminant Analysis (Also Known as Canonical Variate Analysis)

Discriminant analysis has two functions—classification and separation (Huberty, 1984). We prefer to use the name discriminant analysis (DA) for the classification function and canonical variate analysis (CVA) for the separation function. DA is rarely used in pure sensory science studies; but is frequently used in classification of samples based on chemical and instrumental analyses (Luan et al., 2008; Martín et al., 1999; Pillonel et al., 2005; Serrano et al., 2004). On the other hand, CVA is very frequently used with sensory data (Delarue and Sieffermann, 2004; Etaio et al., 2008; Martin et al., 2000) and similarly to PCA provides a two-dimensional or three-dimensional graphic display of the relationships within and between products (Fig. 18.4). CVA is especially useful when one wants to use raw data to get some information of between-product to within-product variation.

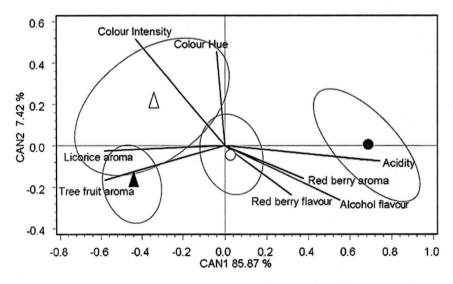

Fig. 18.4 Canonical variate analysis plot of sensory data for significant parameters. Attribute loadings (*lines*) and factor scores (*circles*) for carbonic maceration wines made with Tempranillo (o) or Tempranillo and Viura (•), and for wines made from destemmed grapes (*triangles*) made with Tempranillo (△) or Tempranillo and Viura (▲). Ellipses at 90% of confidence. CAN, canonical variable. *Ellipses* that overlap are not different from one another at the 90% level (reprinted with permission from Etaio et al., 2008).

The easiest way to understand how sensory specialists use canonical variate analysis (CVA) is to refer back to univariate ANOVA. The ANOVA indicates which of the main or interaction effects are significant. However, if the sample mean effect is significant the ANOVA does not indicate which samples differ from one another. To determine this, one of the mean separation techniques, such as Fisher's protected least significant difference (LSD), the honestly significant difference (HSD), Dunnett's test, Duncan's test, has to be applied to the data. The means separation test allows the sensory specialist to determine which samples differ from one another. Fisher's protected LSD requires that the ANOVA for the specific main effect or interaction must be significant before one can examine differences between pairs of means by calculating the LSD (Snedecor and Cochran, 1989). Similarly, the CVA is the multi-dimensional mean separation technique for MANOVA (Ståhle and Wold, 1990).

If the specific main effect or interaction is significant in the MANOVA then one can use CVA to get a graphical map of the sample mean separation (Chatfield and Collins, 1980; Fig. 18.4). This technique has been used extensively in the sensory literature (a few examples are Adhikari et al., 2003; Etaio et al., 2009; Lund et al., 2009; Martin et al., 2000; Wienberg and Martens, 2000). Heymann and Noble (1989) compared CVA and PCA and found that the CVA of the raw sensory descriptive data matrix gave superior results. Brockhoff (2000) showed that CVA is a better choice for sensory descriptive data analysis than PCA because it accounts for uncertainties and error correlations in the raw data.

18.2.4 Generalized Procrustes Analysis

Generalized Procrustes analysis (GPA) is a statistical technique that derives a consensus configuration from two or more data sets (Dijksterhuis, 1997; Gower, 1975; Gower and Dijksterhuis, 2004). The requirement is that all of these data sets have to include the same products. The technique is named after Procrustes, an inn-keeper and highway robber in Greek mythology who had only one bed in his inn. For better or worse, he made all his customers fit the bed by stretching them to fit or by hacking of their limbs to fit (Kravitz, 1975). The GPA in a sense force fits the individual data sets into a single consensus space.

In a GPA, two or more configurations of points in a multi-dimensional space are matched by translation (making the origins equal, i.e., centering), scale change (stretching or shrinking), and rotation or reflection (Gower, 1975). The analysis proceeds through an iterative process that minimizes a value known as the Procrustes statistic, s^{**} (Langron, 1983). The Procrustes statistic is the residual distance between the individual configurations and the consensus configuration at the completion of the GPA, i.e., it is a measure of badness of fit.

When GPA is used with sensory data the individual data sets can come from individual panelists or from different data-collection methods. For example when GPA is used with free-choice profiling data, the individual data sets are the data from each individual panelist (Dijksterhuis, 1997; Heymann, 1994a; Meudic and Cox, 2001). Similarly, it is possible to analyze descriptive data through GPA using the data from each panelist as the individual data sets used to derive the consensus configuration (Dijksterhuis and Punter, 1990, Heymann, 1994b). However, it is also possible to use GPA to integrate data derived by different methods. For example, one can use GPA to compare hedonic and descriptive sensory data (Popper et al., 1997) or to compare descriptive data derived by different panels and methods (Alves and Oliveira, 2005; Aparicio et al., 2007; Delarue and Sieffermann, 2004; Heymann, 1994b; Martin et al., 2000) or to compare data collected instrumentally with data collected by sensory means (Berna et al., 2005; Chung et al., 2003; Dijksterhuis, 1997).

When GPA is performed on data from individual panelists the translation phase standardizes the scores for each panelist by centering at their origin. This is similar to removing the main effect for panelist from the main effect for sample in an ANOVA model. During the scale change phase, the GPA adjusts for the effect of panelists using scales differentially. During the rotation/reflection phase, the GPA minimizes panelist inconsistencies in the use of attributes. This phase is the reason why GPA may be sued to analyze free-choice profiling data, because the analysis takes into account the possibility that panelists may use different terms to describe the same sensations. GPA is also useful in analyzing descriptive profiling data when the sensory specialist is not sure that all panelists consistently used the terms to describe their sensations. In this case the assumption is that the panelists' scores represent different inherent configurations. A Procrustes "ANOVA" can be calculated to determine which of the above transformations were the most important in the formation of the consensus configuration (Dijksterhuis and Punter, 1990).

Like the PCA, GPA provides a simplified configuration based on correlation patterns among variables. The GPA provides a consensus map of the data in a two-dimensional or three-dimensional space. It is possible to have a GPA solution with more than three dimensions but these are frequently very difficult to interpret. The consensus configuration is interpreted similarly to the PCA map (Fig. 18.5). Additionally, it is possible to plot the individual panelists' data spaces and to compare the different panelists with each other (Fig. 18.6). A plot of the panelists' variance explained by dimension (Fig. 18.7) allows the sensory specialist to determine which dimensions were more important to which panelists. It is also possible to plot the descriptors used by the individual panelists into the consensus space (Fig. 18.8). These descriptors are interpreted in the same fashion as the descriptors on a PCA plot.

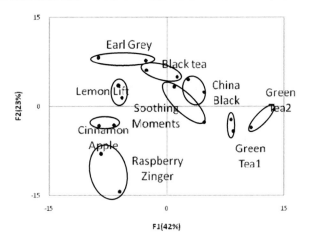

Fig. 18.6 Generalized Procrustes plot of samples evaluated by individual panelists. For clarity only two panelists (#1 and #4) are plotted. The ellipses indicate the positions of these panelists' samples in the consensus space. Samples enclosed in larger ellipses "fit" less well.

18.3 Relating Consumer and Descriptive Data Through Preference Mapping

Multi-dimensional preference mapping is a perceptual mapping method that yields a graphical display of hedonic data (MacFie and Thomson, 1988). Preference mapping is also discussed in Chapter 19. On a single plot, hedonic information for each consumer participating in the study is simultaneously presented in a multi-dimensional space representing and containing the products evaluated (Kuhfield, 1993). The resulting perceptual map provides a clear presentation of the relationships among the products and the individual differences in liking by consumers for these products. In this chapter we will discuss the nuts and bolts of preference mapping and in Chapter 19 the use of preference mapping will be described further.

With this methodology, consumers evaluate six[2] or more products and score their hedonic responses for each product. The data analysis is on an individual

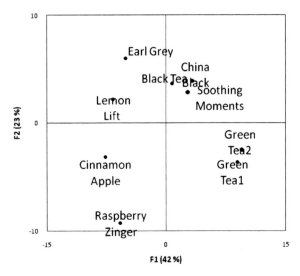

Fig. 18.5 Generalized Procrustes consensus plot. This plot is interpreted as follows: Dimension 1 (F1) accounts for 42% of the variance and dimension 2 (F2) for an additional 23%. F1 is loaded with the green teas on the *right* and the flavored and perfumed teas on the *left*. F2 is loaded with the black teas on the positive side and the raspberry zinger on the negative side. The two green teas are similar in sensory characterics to each other. The China Black and the black teas are also quite similar and they are similar to the Soothing Moments tea.

---

[2] This is based on simulation studies done by Lavine et al. (1988) who showed that one should select no more than $n/3$ PCs, with n=number of samples, in an external preference map (principal regression). Examples of studies with fewer samples are found (Gou et al., 1998) but the results should be viewed with caution.

Fig. 18.7 A plot of the generalized Procrustes variance by panelist (configuration) and dimension (factor). For clarity only the variances associated with two panelists (#1 and #4) are plotted. Panelist 4 had much more varaince explained in dimension 1 than panelist 1. For dimensions 2 through 5 panelist 1 had more variance explained than panelist 4.

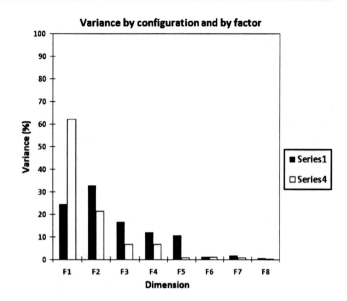

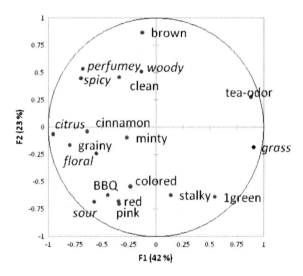

Fig. 18.8 The positions of the descriptors in the consensus space. Descriptors used by panelist 4 is in italics. There does not seem to be much consensus in the use of attributes between these two panelists.

rather than an aggregated (group) level. Preference mapping may be performed by either internal or external analysis. In the simplest form of internal preference mapping, sometimes called MDPREF, the only data used to derive the preference map are the consumers' hedonic data. Thus the entire perceptual map is only based on the acceptance data from the consumers. In the simplest version of external preference mapping, sometimes called PREFMAP, data from an external source are used to derive the preference map, called a product space, and the consumers' hedonic data are then projected into this space through polynomial regression. According to van Kleef et al. (2006), the two main branches of preference mapping "emphasize fundamentally different perspectives on the same data" with internal preference mapping providing "a clear advantage on marketing actionability and new product creativity" and external preference mapping is "more actionable for food technological tasks" (Table 18.1).

Table 18.1 The basic differences between internal and external preference mapping (based on van Kleef et al., 2006)

	Internal preference mapping	External preference mapping
Emphasizes	Preference	Sensory perception
Product positions in map	Account for variation in hedonic or preference data	Account for variation in sensory data (usually descriptive data)
First map dimension	Explains maximum variability in preference directions between products	Explains maximum variability in sensory directions between products
Preference data	Drive orientation of product space	Is supplementary: fitted into the sensory product space
Sensory data	Is supplementary: fitted into the preference-driven product space	Drive orientation of the product space

Please note that 'if analyzed in maximum dimensionality (i.e., number of consumers in internal and number of products [if fewer products than variables] in external preference analysis), the two approaches will show identical results, but with different geographical orientations in space. In practice, however, these data are never analysed in maximum dimensionality, as the purpose is to visualise the most important information in a lower dimensional space' (van Kleef et al., 2006).

18.3.1 Internal Preference Mapping

This analysis is usually a PCA with the products as the samples (rows) and the consumer hedonic scores as the variables (columns). The purpose of the internal preference map is to find a small number of principal components (usually two or three) that explain a large percentage of the variation in the consumer hedonic responses. It is felt that these PCs then indicate underlying perceptual concepts that "explain" the consumer hedonic scores. In this format the internal preference map is a vector model with each consumer represented by an arrow from the zero point intersection pointing in the direction on increased preference for that consumer. Essentially, the arrow indicates that for a specific consumer "more is better" in the direction of the arrow.

It is likely that eventually the consumer will find that more is no longer liked more—the product may become cloying sweet, etc. Thus models such as unfolding models, that are similar to multidimensional scaling models (Busing et al., 2009; DeSarbo et al., 2009; MacKay, 2001, 2006), which indicate ideal points would be more useful. These have been used very infrequently in sensory studies and are more popular in marketing research studies (DeSarbo et al., 2009). Most published studies of sensory hedonic internal preference maps use the vector model (Alves et al., 2008; Ares et al., 2009; Costell et al., 2000; Resano et al., 2009; Rødbotten et al., 2009; Yackinous et al., 1999).

An example of a perceptual map from internal preference mapping is shown in Fig. 18.9. As mentioned before, in order to have a reasonable perceptual map the sensory specialist should have the consumers evaluate at least six products that span the perceptual space (Lavine et al., 1988). The products should differ from one another otherwise the consumers may not have an differentiation in liking scores. There are examples of spaces with fewer products (Gou et al., 1998) but the interpretation of these spaces should be done extremely cautiously since overfitting is a serious problem. For internal preference mapping all consumers should evaluate all the products. It is possible to do imputation, usually mean substitution, if there are a few missing values in the consumer data (Hedderley and Wakeling, 1995). Monteleone et al. (1998) described a boot strapping procedure to determine the 95% confidence ellipses around the products, based on the consumer hedonic scores. These authors also used a permutation test to determine if a specific consumer is significantly fitted into the internal preference map.

18.3.1.1 Extended Internal Preference Mapping

The basic internal preference map is based on *only* the consumer hedonic data. The PCs can be interpreted based on the sensory specialist's product knowledge, as was done in Fig. 18.9. However, the specialist may have access to descriptive data on the same products and could then do an extended internal preference map, where the external information on the products is projected into the internal preference map through regression (Jaeger et al., 1998; Daillant-Spinnler et al., 1996; Martínez et al., 2002; Santa Cruz et al., 2003; van Kleef et al., 2006). This allows one to "name" the underlying perceptual dimensions (Fig. 18.10).

18.3.2 External Preference Mapping

In this case a product space is usually created from sensory profiling data, although the data used to create the product space may be obtained from descriptive analysis methods, from free-choice profiling, from multi-dimensional scaling techniques, from instrumental measurements, etc. These methods differ in their basic principles but they can all be analyzed to yield a spatial representation or a map. For the descriptive data a product space is derived by PCA or CVA (Ares et al., 2009; Lovely and Meullenet, 2009; Schmidt et al.,

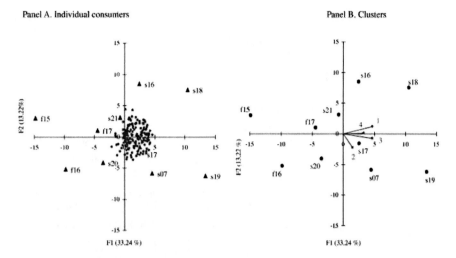

Panel A. Individual consumers   Panel B. Clusters

Fig. 18.9 Internal preference map based on 10 ham samples. PC1 and PC2 account for 33.24 and 13.22% of the variance, respectively. Every consumer is shown on the map as a *black dot*, which corresponds to the endpoint of the fitted vector. A vector line for each consumer can be obtained by drawing a line between the endpoint and the origin. The length of the vector line indicates how well that individual's preference is explained by the dimensions that are plotted. *Panel A* shows the positions of the individual consumers and that in *panel B* the consumer clusters. *Panel A* indicates that most consumers are positioned to the right of the map in the direction of Spanish. Only about 7% of the consumers are located to the left of the map in the direction of the French (f15, f16, f17) and two additional Spanish samples of unspecified origin (S20, S21). *Panel B* shows the four clusters of consumers (based on *k*-means clustering). The clusters are almost superimposed and make interpretation of this internal preference map difficult (reprinted with permission from Resano et al., 2009).

(a)   (b)

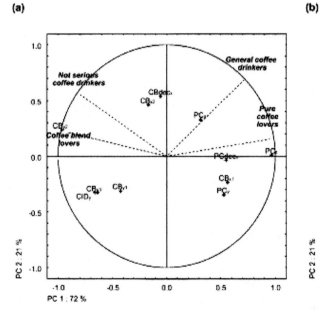 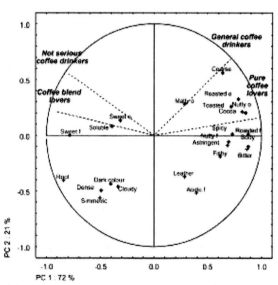

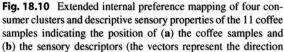

Fig. 18.10 Extended internal preference mapping of four consumer clusters and descriptive sensory properties of the 11 coffee samples indicating the position of (a) the coffee samples and (b) the sensory descriptors (the vectors represent the direction of liking for the consumer clusters: PC, pure coffees; CB, coffee blends; CID, chicory instant drink; dec, decaffeinated; XYZ refers to three coffee manufacturers: f, flavor; o, odor/aroma.) (reprinted with permission from Geel et al., 2005).

2010; Young et al., 2004). For the free-choice profiling a GPA will yield a product space (Gou et al., 1998) and the result of a multi-dimensional scaling of similarity data is also a product space (Faye et al., 2006). A product space can also be obtained from instrumental measurements, for example, Gámbaro et al. (2007) used color measurements to create a product space into which they projected the consumer hedonic scores for honey color. It is important to realize that "for external [preference] analysis to be successful it is essential that the external stimulus [product] space contains dimensions which pertain to preference" (Jaeger et al., 2000).

The individual consumers' hedonic responses (or clusters of consumer responses) are projected into the product space by regressing each consumer's responses onto the spatial dimensions of the products (Fig. 18.11). Each consumer's hedonic scores can be

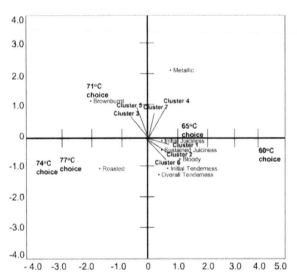

Fig. 18.11 External preference map of combined consumer data with descriptive analysis data for US choice *longissimus* steaks (loin) that were cooked to various endpoint temperatures. Clusters 1, 2, and 6 liked rare steaks that were defined by juiciness, tenderness, and bloody attributes. In addition, clusters 2 and 6 preferred medium rare steaks over steaks cooked to increased endpoint temperatures. Cluster 4 consumers preferred rare, medium rare, and medium steaks over other treatments. It appears that these consumers like all attributes but either do not like roasted and brown/burnt flavors or like steaks that are as juicy and/or tender as possible. Cluster 3 does not like rare steaks, predominantly due to the bloody and metallic attributes (reprinted with permission from Schmidt et al., 2010).

regressed as a series of polynomial preference models: elliptical ideal point with rotation, elliptical ideal point, circular ideal point, and vector models (Coxon, 1982; McEwan, 1996; Schlich, 1995). McEwan (1996) cautions that the elliptical and quadratic models tend to lead to saddle-type ideal points and are thus difficult to interpret, with the result that these models are rarely used. However, Johansen et al. (2009) found a saddle point for one of their clusters and it was relatively easy to interpret (Fig. 18.12). The variance explained by each model is determined, and the most appropriate model is identified for each individual consumer (Callier and Schlich, 1997). If the variance explained by all models for a specific consumer is low, then the behavior of that consumer was not adequately explained by the product space (Callier and Schlich, 1997).

Why does this happen? There are several reasons. It is possible that some consumers do not differentiate among the products at all; they would thus not fit well into the product space. Additionally, some consumers may base their hedonic responses on factors that were not included in the product space derived from the analytical sensory data. The information used by these consumers may have been lost during construction of the product space, or consumers may have used other sensory or non-sensory cues not included in the descriptive analysis of the products. Additionally, some consumers simply yield inconsistent, unreliable responses, possibly because they changed their criteria for acceptance during the test.

Consumer fit ranges from a low of 36% (Helgesen et al., 1997) to less than 50% Tunaley et al. (1988) to nearly 69% (Monteleone et al., 1998). Guinard et al (2001) found that 75% of their beer consumers fit the sensory perceptual map derived for 24 beers. They felt that this occurred due to the widely divergent and large number of beers used in the study. On the other hand, Elmore et al. (1999) found that more than 90% of their consumers fit in to a descriptive sensory space on the perception of creaminess. The major difference between the studies was that the Elmore and coworkers had carefully designed the samples, served to the panelists and the consumers, to be quite different from one another. When samples actually differ perceptually the consumer has a much better chance at determining true like and dislikes—one of use calls this stretching the space to cover all possible responses

Fig. 18.12 Saddle point contour plot for the second consumer segment. Cheese presented to consumer group 1 is marked with a *square* and cheese presented to consumer group 2 is marked with a *circle*. The average scores from the 30 consumers are shown (reprinted with permission from Johansen et al., 2009).

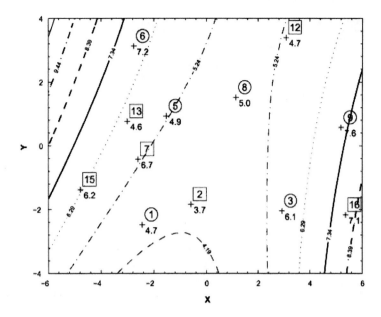

18.4 Conclusions

to the products. In an effort to improve consumer fit in an external preference map, Faber et al. (2003), developed a heuristic or a common sense rule to determine the number of PCs to keep to improve the fit. In their case the fit improved from 51% with two PCs to 80% with five PCs. However, we would caution the sensory specialist to be careful of potentially overfitting the space!

One of the drawbacks of external preference mapping was that all consumers must evaluate all products; however, work by Slama et al. (1998) and Callier and Schlich (1997) have shown that it is possible to have consumers evaluate subsets of products and still get a reasonable external preference map based on the quadratic and the vector models, respectively. Recently Johansen et al. (2009) selected subsets of products for consumer hedonic evaluation from the results of the PCA on the descriptive data. They then analyzed the resultant data using a fuzzy cluster analysis and found that the method worked relatively well for their cheese samples.

Other external preference map techniques are through the use of partial least squares (PLS) where the product space and the consumer space related to the product space are created simultaneously through an iterative process (Martens and Martens, 2001; Meullenet et al., 2002) and logistic regression (Malundo et al., 2001).

In this chapter and in the next chapter a number of multivariate techniques were discussed. In many cases the output from one of these methods is a two-dimensional or three-dimensional map of the loadings (the positions of the attributes) and the scores (the positions of the products). If both the loadings and the scores are plotted in the same map then the map is often called a biplot. The use of multivariate statistics with sensory science data is very useful but there is a danger of over-reliance on the biplot as can be seen by looking at a journal such as *Food Quality and Preference*. This and other journals have a large number of articles dealing with sensory methodology and its related statistical branch, now called sensometrics. A common theme in these articles is that almost no matter what the sensory test, sooner or later the data are visualized (after some multivariate statistical manipulation) as a biplot. This biplotting has become so common that it seems to be obligatory. However, there is a clear and present danger in assuming that we have done good sensory science just because we have a multivariate analysis that produced a biplot.

It is quite possible, for example, to have some attribute from a descriptive analysis that is uncorrelated with other variables, but has a strong influence on

Table 18.2   A summary of the multivariate techniques described in this chapter

Technique	Input	Output	Other information
Principal component analysis			
PCA	Means of scaled[a] data	Product/attribute space	Variance explained
Canonical variate analysis			
CVA	Raw (with replication) scaled data	Product/attribute space	Between sample to within sample variance explained
Generalized Procrustes analysis			
GPA	Individual matrices of scaled data	Consensus product/attribute space	Fit of individuals
Internal preference map	Hedonic data	Product space with consumers as vectors	
External Preference map	Configuration derived from scaled data plus hedonic data		Product/attribute space with consumers as vectors, ideal points, etc.

[a]Scaled data could be from descriptive analysis attributes as well as instrumental and/or chemical measurements

consumer acceptance or rejection. If it is uncorrelated, it will not appear in the first two or three principal components and would not be represented in the biplot at all. It is also possible for two attributes to be highly correlated in the two-dimensional space but not be correlated once one looks at the third or higher dimensions. The sensory scientist should always be mindful that the biplot is a two-dimensional or three-dimensional *representation* of a much larger dimensional space. So the practicing sensory scientist must always be careful to examine all the traditional univariate statistical analyses (one attribute at a time) and resist being seduced by the latest hot statistical technique that results in a two-dimensional or three-dimensional perceptual map. Perceptual maps are only one (arguably minor) tool in the techniques available for data summarization. One of us frequently tells students that biplot are a "virtual" reality and univariate data analyses are reality!

Table 18.2 gives a short overview of the multivariate techniques described in this chapter and should help sensory scientists decided which technique to use where.

References

Adhikari, K., Heymann, H. and Huff, H. E. 2003. Textural characteristics of lowfat, fullfat and smoked cheeses: Sensory and instrumental approaches. Food Quality and Preference, 14, 211–218.

Alves, L. R., Battochio, J. R., Cardosa, J. M. P., De Melo, L. L. M. M., Da Silva, V. S., Siqueira, A. C. P. and Bolini, H. M. A. 2008. Time-intensity profile and internal preference mapping of strawberry jam. Journal of Sensory Studies, 23, 125–135.

Alves, M. R. and Oliveira, M. B. 2005. Monitorization of consumer and naïf panels in the sensory evaluation of two types of potato chips by predictive biplots applied top generalized Procrustes and three-way Tucker-1 analysis. Journal of Chemometrics, 19, 564–574.

Anderson, T. W. 2003. An Introduction to Multivariate Statistical Analysis, Third Edition. Wiley-Interscience, Chichester, UK.

Aparicio, J. P., Medina, M. A. T. and Rosales, V. L. 2007. Descriptive sensory analysis in different classes of orange juice by robust free-choice profile method. Analytica Chimica Acta, 595, 238–247.

Ares, G., Giménez, C. and Gámbaro, A. 2009. Use of an open-ended question to identify drivers of liking of milk desserts. Comparison with preference mapping techniques. Food Quality and Preference, doi:10–1016/j.foodqual. 2009.05.006.

Berna, A. Z., Buysens, S., Di Natale, C., Grün, I. U., Lammertyn, J. and Nicolai, B. M. 2005. Relating sensory analysis with electronic nose and headspace fingerprint MS for tomato aroma profiling. Postharvest Biology and Technology, 36, 143–155.

Bredie, W. L. P., Hassell, G. M., Guy, R. C. E. and Mottram, D. S. 1997. Aroma characteristics of extruded wheat flour and wheat starch containing added cysteine and reducing sugars. Journal of Cereal Science, 25, 57–63.

Brockhoff, P. B. 2000. Multivariate analysis of sensory data: Is CVA better than PCA? Levnedsmiddelkongres 2000, Copenhagen, Denmark.

Busing, F. M. T. A., Heiser, W. J. and Cleaver, G. 2009. Restricted unfolding: Preference analysis with optimal transformations of preferences and attributes. Food Quality and Preference, doi:10–1016/j.foodqual.2009.08.006.

Callier, P. and Schlich, P. 1997. La cartographie des préférences incomplétes–Validation par simulation. Sciences Des Aliments, 17, 155–172.

Cano-López, M., Bautista-Ortín, A. B., Pardo-Mínguez, F., López-Roca, J. M., Gómez-Plaza, E. 2008. Sensory descriptive analysis of red wine aged with oak chips in stainless steel tanks or used barrels: Effect of the contact time and size of the oak chips. Journal of Food Quality, 31, 645–660.

Cattell, R. B. 1966. The scree test for the number of factors. Multivariate Behavioral Research, 1, 245–276.

Cattell, R. B. and Vogelmann, S. 1977. A comprehensive trial of the scree and KG criteria for determining the number of factors. Multivariate Behavioral Research, 12, 289–325.

Chatfield, C. and Collins, A. J. 1980. Introduction to Multivariate Analysis. Chapman and Hall, London.

Chung, S.-J., Heymann, H. and Grün, I. U. 2003. Application of GPA and PLSR in correlating sensory and chemical data sets. Food Quality and Preference, 14, 485–495.

Costell, E., Pastor, E. V., Izquierdo, L. 2000. Relationships between acceptability and sensory attributes of peach nectars using internal preference mapping. European Food Research and Technology, 211, 199–204.

Coxon, A. P. M. 1982. Three-way and further extensions of the basic model. The Users Guide to Multidimensional Scaling. Heinemann Educational Books, London.

Daillant-Spinnler, B., MacFie, H. J. H., Beyts, P. K. and Hedderley, D. 1996. Relationships between perceived sensory properties and major preference directions of 12 varieties of apples from the southern hemisphere. Food Quality and Preference, 7, 113–126.

Delarue, J. and Sieffermann, J.-M. 2004. Sensory mapping using flash profile. Comparison with a conventional descriptive method for the evaluation of the flavor of fruit dairy products. Food Quality and Preference, 15, 383–392.

DeSarbo, W. S., Atalay, A. S. and Blanchard, S. J. 2009. A three-way clusterwise multidimensional unfolding procedure for the spatial representation of context dependent preferences. Computational Statistics and Data Analysis, 53, 3217–3230.

Dijksterhuis, G. B. 1997. Multivariate data analysis in sensory and consumer sciences. Food and Nutrition Press, Trumbull, Connecticut, USA.

Dijksterhuis, G. B. and Punter, P. 1990. Interpreting generalized Procrustes analysis "analysis of variance" tables. Food Quality and Preference, 2, 255–265.

Elmore, J. R., Heymann, H., Johnson, J. and Hewett, J. E. 1999. Preference mapping: Relating acceptance of 'creaminess' to a descriptive sensory map of a semi-solid. Food Quality and Preference, 10, 465–475.

Etaio, I., Elortondo, F. J. P., Albisu, M., Gaston, E., Ojeda, M. and Schlich, P. 2009. Sensory attribute evolution in bottled young red wines from Rioja Alavesa. European Food Research and Technology, 228, 695–705.

Etaio, I., Elortondo, F. J. P., Albisu, M., Gaston, E., Ojeda, M. and Schlich, P. 2008. Effect of winemaking process and addition of white grapes on the sensory and physicochemical characteristics of young red wines. Australian Journal of Grape and Wine Research, 14, 211–222.

Faber, N. M., Mojet, J. and Poelman, A. A. M. 2003. Simple improvement of consumer fit in external preference mapping. Food Quality and Preference, 14, 455–461.

Faye, P., Brémaud, D., Teillet, E., Courcoux, P., Giboreau, A. and Nicod, H. 2006. An alternative to external preference mapping based on consumer perceptive mapping. Food Quality and Preference, 17, 604–614.

Gámbaro, A., Ares, G., Giménez, A. and Pahor, S. 2007. Preference mapping of color of Uruguayan honeys. Journal of Sensory Studies, 22, 507–519.

Geel, L., Kinnear, M. and de Kock, H. L. 2005. Relating consumer preferences to sensory attributes in instant coffee. Food Quality and Preference, 16, 237–244.

Gou, P., Guerrero, L. and Romero, A. 1998. The effect of panel selection and training on external preference mapping using a low number of samples. Food Science and Technology International, 4, 85–90.

Gower, J. C. 1975. Generalized Procrustes analysis. Psychometrika, 40, 33–51.

Gower, J. C. and Dijksterhuis, G. B. 2004. Procrustes Problems. Oxford Statistical Science. Oxford University Press, Oxford, UK.

Guinard, J.-X., Souchard, A., Picot, M., Rogeaux, M. and Sieffermann, J-M. 1998. Sensory determinants of the thirst-quenching character of beer. Appetite, 31, 101–115.

Guinard, J.-X., Uotani, B. and Schlich, P. 2001. Internal and external mapping of preferences for commercial lager beers: Comparison of hedonic ratings by consumers blind versus with knowledge of brand and price. Food Quality and Preference, 12, 243–255.

Hair, J. F., Black, B., Babin, B. and Anderson, R. E. 2005. Multivariate Data Analysis, Sixth Edition. Prentice Hall, New York.

Hatcher, L. and Stepanski, P. J. 1994. A Step-By-Step Approach to Using the SAS System for Univariate and Multivariate Statistics. SAS Institute, Cary, NC.

Hedderley, D. and Wakeling, I. 1995. A comparison of imputation techniques for internal preference mapping, using Monte Carlo simulation. Food Quality and Preference, 6, 281–297.

Helgesen, H., Solheim, R. and Næs, T. 1997. Consumer preference mapping of dry fermented lamb sausages. Food Quality and Preference, 8, 97–109.

Heymann, H. 1994a. A comparison of free choice profiling and multidimensional scaling of vanilla samples. Journal of Sensory Studies, 9, 445–453.

Heymann, H. 1994b. A comparison of descriptive analysis of vanilla by two independently trained panels. Journal of Sensory Studies, 9, 21–32.

Heymann, H. and Noble, A. C. 1989. Comparison of canonical variate and principal component analyses. Journal of Food Science, 54, 1355–1358.

Huberty, A. J. 1984. Issues in the use and interpretation of discriminant analysis. Psychological Bulletin, 95, 156–171.

Huberty, C. J. and Morris, J. D. 1989. Multivariate analysis versus multiple univariate analyses. Psychological Bulletin, 105, 302–308.

Husson, F., Bocquet, V. and Pagès, J. 2004. Use of confidence ellipses in a PCA applied to sensory analysis application to the comparison of monovarietal ciders. Journal of Sensory Studies, 19, 510–518.

Husson, F., Lê, S. and Pagès, J. 2005. Confidence ellipse for the sensory profiles obtained by principal component analysis. Food Quality and Preference, 16, 245–250.

Husson, F., Lê, S. and Pagès, J. 2006. Variability of the representation of the variables resulting from PCA in the case of a conventional sensory profile. Food Quality and Preference, 18, 933–937.

Jaeger, S. R., Andani, Z., Wakeling, I. N. and MacFie, H. J. H. 1998. Consumer preferences for fresh and aged apples: A cross-cultural comparison. Food Quality and Preference, 9, 355–366.

Jaeger, S. R., Wakeling, I. N., MacFie, H. J. H. 2000. Behavioural extensions to preference mapping: The role of synthesis. Food Quality and Preference, 11, 349–359.

Johansen, S. B., Hersleth, M. and Næs, T. 2009. A new approach to product set selection and segmentation in preference mapping. Food Quality and Preference, doi:10.1016/j.foodqual.2009.05.007.

Johnson, R. A. and Wichern, D. W. 2007. Applied Multivariate Statistical Analysis, Sixth Edition. Prentice-Hall, New York.

Kaiser, H. F. 1960. The application of electronic computers in factor analysis. Education and Psychological Measurement, 20, 141–151.

Kravitz, D. 1975. Who's who in Greek and Roman mythology? Clarkson N. Potter, New York.

krzanowski, W. J. 1988. Priciples of Multivariate Analysis: A User's Perspective. Claredon Press, Oxford, pp. 53–85.

Kuhfield, W. F. 1993. Graphical methods for marketing research. In: Marketing Research Methods in the SAS System: A Collection of Papers and Handouts. SAS Institute, Cary NC.

Langron, S. P. 1983. The application of Procrustes statistics to sensory profiling. In: A. A. Willliams and R. K. Atkin (eds.), Sensory Quality in Food and Beverages: Definition, Measurement and Control. Horwood, Chichester, UK, pp. 89–95.

Lavine, B. K., Jurs, P. C. and Henry, D. R. 1988. Chance classifications by non-linear discriminat functions. Journal of Chemometrics, 2, 1–10.

Lê, S. and Husson, F. 2008. SensoMineR: A package for sensory data analysis. Journal of Sensory Studies, 23, 14–25.

Lê, S., Pagès, J. and Husson, F. 2008. Methodology for the comparison of sensory profiles provided by several panels: Application to a cross-cultural study. Food Quality and Preference, 19, 179–184.

Lee, S. M., Chung, S.-J., Lee, O.-H., Lee, H.-S., Kim, Y.-K. and Kim, K.-O. 2008. Development of sample preparation, presentation procedure and sensory descriptive analysis of green tea. Journal of Sensory Studies, 23, 45–467.

Lotong, V., Chambers, D. H., Dus, C., Chambers, E. and Civille, G. V. 2002. Matching results of two independent highly trained sensory panels using different descriptive analysis methods. Journal of Sensory Studies, 17, 429–444.

Lovely, C. and Meullenet, J.-F. 2009. Comparison of preference mapping techniques for the optimization of strawberry yogurt. Journal of Sensory Studies, 24, 457–478.

Luan, F., Liu, H. T., Wen, Y. Y. and Zhang, X. Y. 2008. Classification of the fragrance properties of chemical compounds based on support vector machine and linear discriminant analysis. Flavour and Fragrance Journal, 23, 232–238.

Lund, C. M., Thompson, M. K., Benkwitz, F., Wohler, M. W., Triggs, C. M., Gardner, R., Heymann, H. and Nicolau, L. 2009. New Zealand Sauvignon blanc distinct flavor characteristics: Sensory, chemical and consumer aspects. American Journal of Enology and Viticulture, 60, 1–12.

MacFie, H. J. H. and Thomson, D. M. H. 1988. Preference mapping and multidimensional scaling. In: J. R. Piggott (ed.), Sensory Analysis of Foods. Elsevier Applied Science, New York, pp. 381–409.

MacKay, D. 2006. Chemometrics, econometrics, psychometrics – How best to handle hedonics. Food Quality and Preference, 17, 529–535.

MacKay, D. 2001. Probalistic unfolding models for sensory data. Food Quality and Preference, 12, 427–436.

Malundo, T. M. M., Shewfelt, R. L., Ware, G. O. and Baldwin, E. A. 2001. An alternative method for relating consumer and descriptive data used to identify critical flavor properties of mango (Mangifera, indica L.). Journal of Sensory Studies, 16, 199–214.

Martens, H. and Martens, M. 2001. Multivariate Analysis of Quality: An Introduction. Wiley, Chichester, UK.

Martínez, C., Santa Cruz, M. J., Hough, G. and Vega, M. J. 2002. Preference mapping of cracker type biscuits. Food Quality and Preference, 13, 535–544.

Meullenet, J.-F., Xiong, R. and Findlay, C. 2007. Multivariate and probalistic analyses of sensory science problems. Wiley-Blackwell, New York.

Martin, N., Molimard, P., Spinnler, E. and Schlich, P. 2000. Comparison of odour profiles performed by two independent trained panels following the same disruptive analysis procedures. Food Quality and Preference, 11, 487–495.

Martín, Y. G., Pavón, J. L. P., Cordero, B. M. and Pinto, C. G. 1999. Classification of vegetable oils by linear discriminant analysis of electronic nose data. Analytica Chimica Acta, 384, 83–94.

McEwan, J. A. 1996. Preference mapping for product optimization. In: T. Naes and E. Risvik (eds.), Multivariate Analysis of Sensory Data. Elsevier, London, pp. 71–102.

Meudic, B. and Cox, D. N. 2001. Understanding Malaysian consumers' perception of breakfast cereals using free choice profiling. Food Australia, 53, 303–307.

Meullenet, J.-F., Xiong, R., Monsoor, M. A., Bellman-Homer, T., Dias, P., Zivanovic, S., Fromm, H. and Liu, Z. 2002. Preference mapping of commercial toasted white corn tortillas. Journal of Food Science, 67, 1950–1957.

Monteleone, E., Frewer, L., Wakeling, I. and Mela, D. J. 1998. Individual differences in starchy food consumption: The application of preference mapping. Food Quality and Preference, 9, 211–219.

Montouto-Graña, M., Fernández-Fernández, E., Vázquez-Odériz, M., Romero-Rodríguez, M. 2002. Development of a sensory profile for the specific denomination "Galician potato". Food Quality and Preference, 13, 99–106.

Pickering, G. J. 2009. Optimizing the sensory characteristics and acceptance of canned cat foodL use of a human taste panel. Journal of Animal Physiology and Animal Nutrition, 93, 52–60.

Pillonel, L., Bütikofer, U., Schlichtherle-Cerny, H., Tabacchi, R. and Bosset, J. O. 2005. Geographic origin of European Emmental. Use of discriminant analysis and artificial neural network for classification purposes. International Dairy Journal, 15, 557–562.

Popper, R., Heymann, H. and Rossi, F. 1997. Three multivariate approaches to relating consumer to descriptive data. In: A. M. Muñoz (ed.), Relating Consumer, Descriptive and

Laboratory Data to Better Understand Consumer Responses. ASTM Publication Code Number 28–030097–36. ASTM, West Conshohocken, PA, pp. 39–61.

Resano, H., Sanjuán, A. I. and Albisu, L. M. 2009. Consumers' acceptability and actual choice: An exploratory research on cured ham in Spain. Food Quality and Preference, 20, 391–398.

Rødbotten, M., Martinsen, B. K., Borge, G. I., Mortvedt, H. S., Knutsen, S. H., Lea, P. and Næs, T. 2009. A cross-cultural study of preference for apple juice with different sugar and acid contents. Food Quality and Preference, 20, 277–284.

Santa Cruz, M. J., Garitta, L. V. and Hough, G. 2003. Note: Relationships of consumer acceptability and sensory attributes of Yerba mate (Ilex paraguariensis St. Hilaire) using preference mapping. Food Science and Technology International, 9, 346–347.

Schlich, P. 1995. Preference mapping: Relating consumer preferences to sensory or instrumental measurements. In: P. Etievant and P. Schreier (eds.), Bioflavour: Analysis/Precursor Studies/ Biotechnology. INRA Editions, Versailles, France.

Schmidt, T. B., Schilling, M. W., Behrends, J. M., Battula, V., Jackson, V., Sekhon, R. K. and Lawrence, T. E. 2010. Use of cluster analysis and preference mapping to evaluate consumer acceptability of choice and select bovine M. Longissimus Lumborum steaks cooked to various endpoint temperatures. Meat Science, 84, 46–53.

Slama, M., Heyd, B., Danzart, M. and Ducauze, C. J. 1998. Plans D-optimaux: une stratégie de reduction du nombre de produits en cartographie des préférences. Sciences des Aliments, 18, 471–483.

Serrano, S., Villarejo, M., Espejo, R. and Jodral, M. 2004. Chemical and physical parameters of Andalusian honey: Classification of citrus and eucalyptus honeys by discriminant analysis. Food Chemistry, 87, 619–625.

Snedecor, G. W. and Cochran, W. G. 1989. Statistical Methods, Eighth Edition. Iowa State University, Ames, IA.

Ståhle, L. and Wold, S. 1990. Multivariate analysis of variance (MANOVA). Chemometrics and Intelligent Laboratory Systems, 9, 127–141.

Stevens, J. 1986. Applied Multivariate Statistics for the Social Sciences. Erlbaum Press, New York, NY.

Tabachnik, L. and Fidell, B. 2006. Using Multivariate Statistics, Fifth Edition, Allyn and Bacon.

Tunaley, A. Thomson. D. M. H. and McEwan, J. A. 1988. An investigation of the relationship between preference and sensory characteristics of nine sweeteners. In: D. M. H. Thomson (ed.), Food Acceptability. Elsevier Applied Science, New York.

van Kleef, E., van Trijp, H. C. M. and Luning, P. 2006. Internal versus external preference analysis: An exploratory study on end-user evaluation. Food Quality and Preference, 17, 387–399.

Van Oirschot, Q. E. A., Rees, D. and Aked, J. 2003. Sensory characteristics of five sweet potato cultivars and their changes during storage under tropical conditions. Food Quality and Preference, 14, 673–680.

Wienberg, L. and Martens, M. 2000. Sensory quality criteria for cold versus warm green peas studies by multivariate data analysis. Journal of Food Quality, 23, 565–581.

Wortel, V. A. L. and Wiechers, J. W. 2000. Skin sensory performance of individual personal care ingredients and marketed personal care products. Food Quality and Preference, 11, 121–127.

Yackinous, C., Wee, C. and Guinard, J-X. 1999. Internal preference mapping of hedonic ratings for Ranch salad dressings varying in fat and garlic flavor. Food Quality and Preference, 10, 401–409.

Young, N. D., Drake, M., Lopetcharat, K. and McDaniel, M. R. 2004. Preference mapping of cheddar cheese with varying maturity levels. Journal of Dairy Science, 87, 11–19.

Chapter 19

Strategic Research

Abstract Sensory professionals often assist their companies with strategic research. One common example is the category appraisal, in which competitive products are evaluated relative to one's own. Often the information is summarized by perceptual mapping, using multivariate statistical analyses. An important part of product development is optimization of specific attributes. A third area involves identifying patterns of consumer preferences and groups to whom different versions of a product may be appealing.

Thus, what is of supreme importance in war is to attack the enemy's strategy . . . Therefore I say: 'Know the enemy and know yourself; in a hundred battles you will never be in peril.'
—Sun Tzu, The Art of War (Ch. 3, v. 4, 31)

Contents

19.1 Introduction

19.1.1 Avenues for Strategic Research

A full-service sensory evaluation program is more than a department that merely fulfills test requests. Such technical services are critically important, of course, to provide information about product development and optimization of sensory attributes. Routine testing can also provide support for questions of quality maintenance—in sensory quality control, shelf-life testing, and other common services. An important service arises when advertising claim substantiation requires sensory data, as discussed in Chapter 13. This service and its statistical basis are discussed in the ASTM standard for claim substantiation (ASTM, 2008) and also by Gacula (1993). Many sensory departments, especially those in larger and forward-looking companies, are also providing strategic research and long-term research guidance to their product development and marketing clients.

The distinction between strategic and tactical research is obviously based on a military metaphor. Tactical research concerns all of the focused activities

H.T. Lawless, H. Heymann, *Sensory Evaluation of Food*, Food Science Text Series,
DOI 10.1007/978-1-4419-6488-5_19, © Springer Science+Business Media, LLC 2010

aimed at launching new products and positioning or re-positioning existing brands. This is where the bulk of corporate expenditures on product development and sensory research are aimed. Substantial funds are spent on positioning and pricing studies and on advertising research before and after a new product launch to gain a point or two of market share (Laitin and Klaperman, 1994). However, these funds are not well spent if the bigger picture of the product category and a long-range view of consumer needs and trends are not seen. Research efforts may have resulted in a better apple when consumers really wanted a better orange. At the tactical stage, the company is stuck with the apple and has to do their best with it (Laitin and Klaperman, 1994). To avoid this kind of problem, some companies use innovative research techniques such as perceptual mapping. Strategic research may also identify consumer trends and demographic changes and uncover new product or even whole new business opportunities (Miller and Wise, 1991; Von Arx, 1986).

The most common question in the realm of strategic research is, "How do our products stack up relative to the competition?" The approach is different than the marketing research attack on this question, which would usually be based on market share, profitability, or some other sales-related measure. In contrast, the sensory department can provide information on the perceived performance of products, usually on a blind basis. The relative strengths and/or weaknesses of the company's products in terms of sensory characteristics can be assessed in isolation from the contaminating influences of complex concepts, positioning, brand image, label claims, price, and promotions. Assessment of one's own products and those of the competition is an integral part of such systematic product development schemes as Quality Function Deployment (QFD) and the "house of quality" methods (Benner et al., 2003). These methods seek to connect known consumer "wants" with efficient new product delivery of those characteristics.

A second avenue for strategic research comes from the continuing growth of qualitative research methods as a sensory tool. Many sensory professionals now receive training for moderating group interviews. Their services may be highly valued by product research clients who desire the kind of rich and probing information provided and for the creative ideas that are often generated (Goldman and McDonald, 1987). There is a growing need to communicate effectively with consumers at the early stages of research. Qualitative methods such as group interviews can meet this need (Von Arx, 1986). Consumer interviews not only are tools for answering advertising and positioning questions but can also be used to address more concrete questions about desired sensory attributes and features like convenience and packaging issues. These methods are discussed in detail in Chapter 16. The group depth interview is an important tool in uncovering the reasons for brand preferences, perceived shortcomings, or faults in one's product relative to the competition, points of superiority that should be emphasized or strengthened and opportunities for improvement and new product ideas.

On the borderline of tactical and strategic research is the evaluation of new or alternative versions of a product (Laitin and Klaperman, 1994). Once the relative importance of different product attributes is established, consumer needs and priorities become better defined. Should the package be re-sealable? Do consumers want additional flavor variations? Have they been adding other foods in unusual combinations that suggest a product variation (e.g., granola to yogurt)? Is there a desire for a low-sodium version, a low-fat version, or other nutritional modification? Does the product perform well in a microwave oven, and if not, should that performance feature be improved? Various combinations and profiles can be evaluated both in conceptual stages and in prototypes (Mantei and Teorey, 1989). Redirection of the concept or refinement may be necessary (Von Arx, 1986). At this point, a sensory evaluation department can become involved in exploratory research to determine if consumer perception of a product prototype, in terms of sensory characteristics and performance, matches the target of the conceptual developers and the product research team. The overall goal is to facilitate more successful product development with less waste of work, time, and money (Benner et al., 2003; Von Arx, 1986).

A related area for strategic research, both in marketing and sensory evaluation, is the identification of product profiles that represent undeveloped combinations of characteristics that would have potential consumer appeal. For example, at one time people had radios and alarm clocks, but no clock radios. This new product filled an undeveloped niche in the appliance market. In foods, there are a wealth of items that are oven-ready, and some (but not all) of these have microwavable counterparts or are themselves microwavable. This represents an opportunity for some products that do not yet offer this feature of convenience in preparation. Analysis of the entire scope of products that fulfill a similar purpose can help identify these unfilled niches and bring innovative products to the market. This procedure is sometimes referred to as market gap analysis (Laitin and Klaperman, 1994). Very often the niche will be defined by a set of sensory or performance characteristics as part and parcel of the defining concept and the sensory group can be of assistance in measuring the fit relative to expectations. The sensory and performance characteristics must be discovered, defined, and measured during exploratory research, and conformance of the product prototypes to the desired target must be measured in sensory tests. A full-service sensory evaluation department can assist in all phases of the process.

Another important area for strategic research is the identification of consumer segments. A segmentation study seeks to identify groups of consumers who respond in a similar way and who are definably different from other groups in their perceptions, needs, or response to product attributes. Various multivariate techniques are available such as cluster analysis that can group individuals on the basis of correlated responses across attributes on a questionnaire or survey (Plaehn and Lundahl, 2006; Qannari et al., 1997; Wajrock et al., 2008). Consumer segmentation may then be defined on the basis of usage habits or sensory preferences (Miller and Wise, 1991; Moskowitz and Krieger, 1998).

19.1.2 Consumer Contact

A full-service sensory evaluation department can interact directly with consumers in a number of ways. In fact, the sensory department has unique opportunities to monitor consumer reactions to the company's products. When combined with strategic activities such as category reviews, the sensory department can be a major conduit for consumer input that can affect executive decisions. Opportunities can also arise through interactions with other departments. For example, many companies maintain consumer hotlines or toll-free telephone numbers for comments and complaints. These communications are periodically summarized

in reports and the sensory professionals in charge of certain product lines should monitor these summaries carefully. Complaints usually represent "the tip of the iceberg" of a larger problem and may help identify important issues to be addressed in future optimization or product improvements.

An important avenue for consumer contact arises in home placement tests. If at all possible, the sensory professional should not assign 100% of the interviews to a field service, but should reserve a small percentage of the actual interviews to be conducted in person. A marketing research group typically will delegate all of the interviewing as well as the statistical analysis to subcontracting field agencies. Statistical summaries will reflect majority opinion and can miss important segments and minority opinions. For example, if the vast majority of consumers liked the product, but two people cut their fingers while opening the package, there is an important issue in package design that needs to be addressed. Such infrequent problems and strong negative opinions of those people can get lost in group averages and "top box" scores. Face-to-face contact can facilitate the probing of issues that were missed on the formal quantitative questionnaire.

A third opportunity for direct consumer contact is in focus group moderating, as noted above. In some home placement tests, it may be cost effective to conduct group interviews as a follow-up to the structured questionnaire survey. Issues may be identified in the questionnaire data that require further investigation, and a recall of participants for group interviews may allow those issues to be probed.

19.2 Competitive Surveillance

19.2.1 The Category Review

A category review or category appraisal is a survey of most or all of the products that serve a similar function and are viewed as belonging to the same group by consumers. A "category" is often a group of products that appear in the same part of a food store or in the same aisle or on one section of shelf space. For example, cold breakfast cereals are a category and a distinct category from hot breakfast cereals. The category review is an important strategic search to identify and characterize the company's products and their competitors. The information may include sales and marketing data, physical characteristics, objective sensory specifications (such as descriptive data), and consumer perceptions and opinions. In the sections that follow, we will examine category review research from a sensory and consumer perspective. A full-service sensory department will be capable of conducting such an extensive review at periodic intervals as called for by changes in the market and the appearance of new or innovative products. In many ways the category review is similar to the product evaluations conducted by Consumer's Union for publication in their magazine, Consumer Reports. As noted above, evaluation of competitive products is an important step in methods for systematic new product development (Benner et al., 2003).

A review of a product category might be limited to key brands or could be quite comprehensive in scope. In some product categories the number of producers is quite large, or alternatively a small number of large companies may each have an extensive offering of different products within the category, as in the breakfast cereal industry. It may be advantageous to sample all the products that might be substitutable in the consumer's mind. Inclusion in the study can be based on market share data, such as warehouse case movement information. In a large and diversified category, it is probably wise to include the top 80 or 90% of brands. In the case of a category that is relatively new or in which there are limited data, a store retrieval study can precede the formal sensory and consumer work to see what is out there. A guideline for the size of a store retrieval study in the United States is approximately ten stores in each of ten cities to get a geographical representation from different areas of the country. The stores should represent different types of outlets for those products (e.g., grocery, convenience, food club/warehouse). Field agencies can be hired to purchase the products (usually one of each and every variety they see) and send them back to the originating sensory department. The sensory department can then catalog what was actually found and how often different brands appeared. If seasonal changes are involved, it may be necessary to repeat the retrieval or spread the purchasing over time. The results of the store retrieval can be used to help select competitive products for inclusion in the main study, based on frequency of retrieval as an estimate of market penetration. They are

also a rich source of qualitative information and can be used for idea generation.

If conducted by a sensory evaluation department, the category review will probably involve several phases of analytical descriptive testing and assessment of consumer perceptions. It may be advantageous to coordinate the sensory data collection with brand image questions as determined by marketing concerns, either in a parallel study or as part of the same large research program. Of course, the sensory questions will be focused on sensory attributes and perception of performance (Muñoz et al., 1996) and will be conducted on a blind basis if it is possible to do so. In some cases it may not be possible to have a fully blind study when the competitive products are well known. Sometimes re-packaging can be done to disguise the product identity or brand identity for the sensory study. However, there are limits to this and common sense should be used as a guide. An aerosol air freshener with a distinctive pink cap may be a dead giveaway, but changing the cap might change the dispersal pattern of the product and the resulting consumer perception. In such a case it is probably less damaging to stay with the cap color than risk a change in product performance. Cap color simply becomes part of the product perception and can be analyzed for its potential effects. Similar problems can arise in food products with unique or distinctive packaging features.

Mullet (1988) discussed the use of multivariate techniques in assessing a brand image relative to competitive products. Relative "position" in this case refers to geometric modeling and the perceptual standing in the product set (the spatial imagery underlying the idea of "position" is clear). Relative position is interpreted via attribute scales from questionnaires, and dimensions of the perceptual model are derived from these attributes. The overall goal is to inform management about attributes that might be strengthened or changed through changes in formulation, processing, changes in marketing strategy, or advertising. Mullet illustrated this approach with a consumer study of beer brand perceptions and showed four analyses: factor analysis, multi-dimensional scaling to create a perceptual map, discriminant analysis, and correspondence analysis Hoffman and frank (1986). Multivariate analyses were discussed in Chapter 18, and the use of these tools for perceptual mapping is discussed in the next section.

19.2.2 Perceptual Mapping

Almost all perceptual maps have two important common features. Products are represented as points in the space (if three dimensional) or plane (if two dimensional). First, products that are similar to one another will be positioned close to one another in the map and products that are very different will be far apart. Which positions are similar or dissimilar can be a matter of interpretation, although there are some techniques that will plot confidence intervals around the positions of points in the model. The techniques are not suited to hypothesis testing about product differences—they are best used for comprehending the pattern of relationships among a set of products. A second feature of most perceptual maps is that vectors corresponding to product attributes can be projected through the space to help interpret the positions of different products and the meanings of the axes or other directions through the space. These may be provided by the analysis itself as is the case in factor analysis or PCA or are added in a second step of data collection as in some multidimensional scaling studies (e.g., Lawless et al., 1995; Popper and Heymann, 1996).

The overall goals of perceptual mapping fit well with strategic research. Johnson (1988) stated these goals as (1) learning how the products in a class are perceived with respect to strengths, weaknesses, and similarities, (2) learning what potential buyers want, and (3) learning how to produce or modify a product to optimize its appeal. Ideally, the map will relate to consumer's opinions, acceptance, or desire for a product so that the appeal or "density of demand" through the space can be determined. This is the essential goal of preference mapping (see Sections 18.3 and 19.4).

A variety of multivariate statistical techniques are available to produce pictorial representations that capture the relationships among a set of products (Elmore and Heymann, 1999; Mullet, 1988). Most of the procedures provide a simplified picture or map in two or three dimensions (rarely more). A complex multidimensional set of products is then described by a smaller set of dimensions or factors or derived attributes, sometimes referred to as "latent variables." This simplification of a large data set into a spatial representation that is easily grasped is one of the attractive features of perceptual mapping. However, simplification entails a risk of losing important details

about product differences. Thus the safest uses of these procedures are in exploration and in conjunction with more traditional methods such as univariate analysis of variance (Popper and Heymann, 1996). By "univariate" we mean analyzing each response scale or attribute separately from the others. The analysis of variance then provides information about differences among products for each individual attribute. This can aid in interpretation of the simplified picture as well.

Perceptual maps have other limitations. Perceptual maps represent consumer perceptions at one point in time. The static nature of a single map limits its value as a predictor of future behavior (Johnson, 1988). However, multiple studies can be conducted to compare the perceptions after different changes are made in a product. For example, maps may be constructed before and after consumers are given information about the products and their expectations and points of focus are manipulated. A second limitation is that the map will usually represent the majority opinion, so segments of differing opinions may not be captured in the aggregate summary. Also, the degree to which an individual's likes and dislikes are correlated with the dimensions of the map is necessarily limited by the degree to which the map corresponds to his or her perceptions. If the map is not a good summary for that person, any preference directions will not be clear.

Lawless et al. (1995) suggested criteria for evaluating the utility of a perceptual map. These are shown in Table 19.1. Considerations involve the correspondence of the model to the data or goodness of fit, precision and reliability, the model's validity, and the overall usefulness of the modeling exercise, i.e., what was learned. Reliability may be assessed from analysis of split data sets, by the positions of duplicate pairs, or the positions of near-duplicate pairs from similar production runs or batches. In terms of validity, the map should relate to descriptive attributes and/or to consumer preferences. A useful map can elicit new hypotheses or add confirming evidence to support previous findings. Utility is also a function of visualization—a map that tells a story in few dimensions and is easily interpreted is more useful than a complex, ambiguous model. Finally, the data collection and computation should both be rapid, simple, and cost effective. In the section below, multivariate techniques for mapping will be described and some examples given from the more common and popular techniques (see also Chapter 18).

Table 19.1 Desired qualities in perceptual mapping

Goodness of fit:	High variance accounted for, low badness-of-fit measures (e.g., stress)
Reliability:	Blind duplicates should plot together
Reliability:	Similar pairs (batches) should plot nearby
Dimensionality:	Model has a few dimensions and can be plotted
Interpretation:	Map should be interpretable
Validity:	Map should relate to descriptive attributes
Validity:	Map should relate to consumer preferences
Payoff:	Map should suggest new hypotheses
Payoff:	Map may help confirm previous hypotheses
Cost efficiency:	Data collection is rapid, simple

19.2.3 Multivariate Methods: PCA

In the class of methods commonly referred to as factor analysis, the technique of principal component analysis (PCA) has a long history in sensory and consumer research. The input to PCA usually consists of attribute ratings describing a set of products. Often the mean ratings are used as input, although in some cases raw data from individuals are used (Kohli and Leuthesser, 1993). Given that many attributes have been evaluated, some will be correlated. A product that receives a high value on one attribute will receive a high value on a positively correlated attribute. The PCA finds these patterns of correlation and substitutes a new variable, called a factor, for the group of original attributes that were correlated. The analysis then seeks a second and third group of attributes and derives a factor for each, based on the variance left over. This is analogous to finding a new set of axes in space to replace the N-dimensional space of the original data set with a smaller set of axes or dimensions. The original attributes have a correlation with the new dimensions, called a factor loading, and the products will have values on the new dimensions, called factor scores. The factor loadings are useful in interpreting the dimensions and the factor scores show the relative positions (and therefore similarities and differences) among the products in the map or picture (see Section 18.2 for further discussion). Examples are given below.

Principal component analysis can be applied to any data set where there are attribute ratings for a set of products as in descriptive analysis. An example of the application of PCA to descriptive data can be found in a study of a creamy-textured semi-solid dessert, vanilla puddings (Elmore et al., 1999). The goal was to illuminate those product dimensions that would influence the creaminess of the product, a complex sensory characteristic. Texture variation was induced by modifying the starch type and content, and milk fat, and sodium salts. Puddings differed on 16 sensory attributes. These were reduced to a set of three factors, explaining 81% of the original variance, a considerable simplification. Examination of the three correlated groups of attributes showed that they could be interpreted as being related to thickness, smoothness, and dairy flavor. This result was intuitively appealing since the overall creaminess of semi-solid foods appears to be determined by a combination of such elemental sensory attributes.

Case study: air fresheners. Another example of how PCA can simplify a complex data set, produce insights, and generate hypotheses can be found descriptive data from a strategic category review. Figure 19.1 shows a perceptual map based on descriptive analysis of aerosol air fresheners in the US market circa 1986. This is a sensory space determined from fragrance analysis. At that time the market leader had a large number of different fragrances and several competing companies were also represented in a category review. Fifty-eight aerosol air fresheners were included, based on market share, and were submitted to a trained descriptive panel for characterization. Mean values on fragrance descriptive scales were submitted to PCA. The dimensions of the space, as interpreted from the factor loading are roughly as follows: Products in the upper left corner are high in spicy fragrance notes and represent high-intensity "odor killer" types of products. Items on the far right represent citrus (usually lemon types) so the right-to-left dimension contrasts spicy with citrus types of fragrances. Items in the front and left front quadrant of the map (the

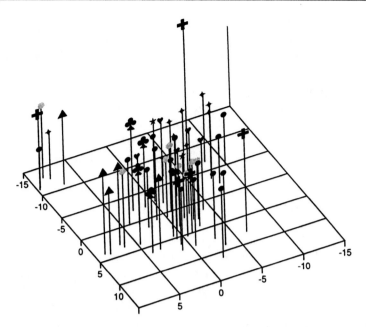

Fig. 19.1 Principal component analysis of fragrance evaluations (trained descriptive panel) of aerosol air fresheners circa 1986. Symbols represent different brands (companies). Three factors were extracted from nine original rating scales, representing spicy versus citrus, floral/sweet, and woody dimensions. Results are further described in the text.

main grouping) are green-floral and sweet-floral types, respectively. Finally, items that are high in the vertical dimension tend to have some woody character (e.g., cedar).

This map can be used to characterize different corporate strategies, from a sensory perspective. First note that the market leader symbolized by the circles has a large number of products and a high concentration in the floral zone. This could arise if new products were launched at regular intervals on the basis of fixed test criteria and older similar smelling products were not "retired." If the new products were selected from a pool of candidates using simple hedonics and consumer acceptance, they might be somewhat similar in odor type, for example, since floral types generally score well. The company symbolized by the triangles is concentrated in the sweet floral zone of the map. This strategy could lead to a kind of "cannibalization"— competition with oneself for market share in similar product types. Avoiding duplication and overlap as well as reaching the maximum number of consumers is one important marketing strategy, the so-called TURF analysis (Miaoulis et al., 1990). The company shown by circles also suffers from this problem of too many similar products. Having too many similar products

could also lead to problems in maintaining the in-store shelf space, a difficulty for the sales force in dealing with retailers.

The company shown by the crosses appears to have a different strategy—a smaller number of products are represented, and they are distributed throughout the space to sample diverse fragrance types, perhaps appealing to different consumer segments. The brand shown by the cloverleaf symbols, a newcomer to the air freshener business, has yet a different strategy. Their four fragrance types are in the same part of the space and surround the best-selling product of the market leader at that time. Their competitive strategy seems to be to attempt to steal market share by copying the most popular and successful fragrance type. Finally, note that there is a section between the floral, fruity, and spicy types that is open for new and different products to fill. This gap suggests new product opportunities that could differentiate a product from the existing market types. This sample map illustrates how a corporate strategy can be described or discovered using sensory data (in this case from a trained descriptive panel). The perceptual map can be helpful in seeing the relative positions and sensory qualities of one's own products relative to competition.

A variety of other techniques will also yield perceptual maps, such as generalized procrustes analysis or GPA, discriminant analysis, and partial least squares (PLS) (Dijksterhuis, 1997; Fox, 1988). Discriminant analysis will produce a map by examining the variance of the means of different products relative to the amount of error or disagreement among people rating the products (Johnson, 1988; Kohli and Leuthesser, 1993). The discriminant analysis will find a weighted combination of all the attributes that would produce the highest F-ratio. It then proceeds to find the best weighted combination of attributes that produces a new dimension uncorrelated with the first combination, and so on (Johnson, 1988). Discriminant analysis may come up with slightly different patterns of factors than a PCA, since it is looking a discrimination of products relative to error or disagreement among people, whereas the PCA simply looks for patterns of correlation (Kohli and Leuthesser, 1993). A number of these multivariate techniques were discussed in Chapter 18.

19.2.4 Multi-dimensional Scaling

An alternative to using attribute ratings and PCA is multi-dimensional scaling (MDS) (Kohli and Leuthesser, 1993). An introduction can be found in Schiffman et al. (1981) and a review by Popper and Heymann (1996). MDS programs use some measure of the similarity of products as input. From these similarity estimates, a map is constructed in as many dimensions as the experimenter requests from the software. Similarity may be found from direct ratings of the similarity of pairs of products or from derived measures of similarity. Derived measures include the following: frequencies that items are sorted together in a sorting task, a correlation coefficient across an attribute profile, and numbers of errors in a set of discrimination tests (so-called confusability measures). So this class of methods is very flexible and can be applied to a variety of situations with a minimum of statistical constraints (Popper and Heymann, 1996). Similarity ratings are considered to be less biasing than rating specific attributes since the participant is not directed to use any particular words or dimensions in assessing similarity—it is up to them. The PCA

depends on the attributes that are selected for description and analysis, but it is not guaranteed that these are the ones that might be important to consumers. MDS methods such as sorting (described below) allow consumers to use whatever criteria they deem appropriate for the product set.

The number of dimensions in the model is a function of the interpretability of the output, its communication value, and the degree to which the model fits the data. For a set of N products, the data can always be fit perfectly by a model with $N–1$ dimensions (two points define a line, three define a plane, and so on). Reducing the model to fewer and fewer dimensions or increasing the number of products in the experiment will increase the difficulty of fitting the model to the data. This badness of fit is measured in MDS programs and is called "stress." Stress reflects the sum of squared deviations between the distances in the model and the (dis)similarities of the input data. As the program proceeds to find its best solution, stress is minimized through an iterative process of moving the points around in the space to achieve the best fit to the data. MDS programs can try to minimize these deviations based on actual measured distances or on the basis of relating the rank orders, called non-metric programs (Kruskal, 1964).

Traditionally, input to MDS was obtained by similarity ratings of all possible product pairs, often rated by marking a line scale. The scale was analogous to an overall degree of difference scale and anchored with suitable terms like "very similar" at one end and "very dissimilar" at the other. The major problem in applying MDS to foods and consumer products has always been that a large number of paired comparisons are required for similarity ratings (Katahira, 1990). For a group of N products, there are $N(N–1)/2$ possible pairs. So for a small set of five products there are ten pairs. Tasting 10 pairs or 20 products is feasible, but using only 5 products does not make a very interesting study. For sets of 10 or 20 products, the numbers of pairs are 45 or 190, respectively. Tasting 90 or 380 foods is simply out of the question unless the participants return for multiple evaluation sessions. This difficulty has led to an emphasis on incomplete statistical designs for multivariate studies in general and MDS in particular (Bijmolt, 1996; Kohli and Leuthesser, 1993; Malhotra et al., 1988). An alternative approach is to use one of the derived measures of similarity such as sorting.

19.2.5 Cost-Efficient Methods for Data Collection: Sorting

A derived measure of similarity that is rapid and easy to obtain comes from the simple task of having consumers categorize or sort the product set into groups of similar products. This method appeared in the early MDS literature in studies of person perception and studies of word meanings such as kinship terms in anthropology (Rosenberg and Kim, 1975, Rosenberg et al., 1968). Similarity can be inferred from the number of times two items are sorted into the same groups, summed across a panel of participants. Items that are like one another should be placed in the same group very often and items that are unlike one another should be placed together rarely or not at all. Another way to treat the data is to transform each individual similarity matrix into a cross product or covariance matrix (Abdi and Valentin, 2007). Individual data matrices are of course a series of zeros and ones, not very informative about distance or similarity. But the covariance matrix looks at the pattern across the entire row and column for each product and compares it to the pattern for each other product. This is a little indirect ("the friend of my enemy is also my enemy") but provides for a more graded or scaled value, rather than a simple binary entry, in each person's data matrix. A program has been developed to analyze individual judge data from sorting, called DISTATIS (Abdi et al., 2007). Sorting with MDS has been applied to consumer product fragrances (Lawless, 1989), cheeses (Lawless et al. 1995), oxidation odors (MacRae et al., 1992), vanilla samples (Heymann, 1994), mouthfeel words (Bertino and Lawless, 1993), ice cream novelties (Wright, 1995), snack bars (King et al., 1998), and grape jellies (Tang and Heymann, 2002), to list just a few.

The sorting technique is simple, rapid, and easy for panelists to perform with sets of about 10–20 items. In the inspection phase, as participants are beginning to make their categories, it is useful to allow them to make notes to aid their memory as they taste the products. The most reasonable application is in product sets that are moderately dissimilar, i.e., a range of differences where sometimes items will be grouped together and sometimes not. The results (MDS configurations) appear to stabilize with about 20 participants. Large groups of consumers are not required in data collection, adding to its overall efficiency. Another advantage is that the consumers can decide for themselves what characteristics are most important to differentiate the groups; there is no imposition of any attributes by the experimenter.

Case study: cheeses. A map from an exploratory study of cheeses is shown in Fig. 19.2 (Lawless et al.,

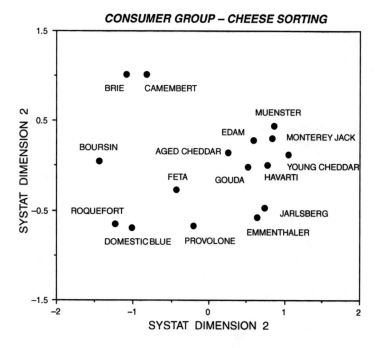

Fig. 19.2 Multi-dimensional analysis of sorting data of cheeses after sorting by a group of 16 cheese consumers. Replotted from Lawless et al. (1995), copyright 1995, used with kind permission of Elsevier Science Ltd, The Boulevard, Langford Lane, Kidlington OX51GB, UK.

1995). Similar pairs are positioned close together in the model—the blue cheeses are close to one another, the white moldy cheeses are close to one another, and the "Swiss"-type cheeses, Jarlsberg and Emmenthaler, are also close. These three different pairs are found in different corners of the map. The unusual cheese, feta (from goat milk), is unlike any other cheese and thus is placed in the center of the map, a compromise position on the part of the program. In other words, feta was an outlier in the data but becomes an "inlier" in the model. Determining whether a centrally positioned item is actually an outlier requires examination of the input data. Intermediate positions between clusters can sometimes represent intermediate or blended sensory character. This pattern for centrally located items was seen in an early study of sorting citrus and woody odors. A blended or ambiguous set of odors having both citrus and woody attributes fell midway between a citrus group and a pine-woody cluster in the output (Lawless, 1989).

There are several limitations to MDS studies in general and to sorting methods in particular. Most MDS programs do not produce any confidence intervals around the points in space, although there are some exceptions (Bijmolt, 1996; Ramsay, 1977; Schiffman et al., 1981). There is usually some subjectivity in interpretation of positions in the map. The reliability of the map can be unclear. There are a few approaches to gain some insight or feeling for the stability of the model. One is to test twice as many participants as are needed and split the data set into two arbitrary halves. If the resulting maps are similar the results may be considered reliable. Another approach is to impose further analyses on the map such as cluster analysis to aid in the interpretation of groupings, clusters, or categories. This approach was taken in the study of terpene aromas (Lawless, 1989) and mouthfeel characteristics (Bertino and Lawless, 1993). Another "trick" is to insert a blind duplicate of one of the products to see whether duplicate items plot close together in the map. In sorting, a duplicate pair should be sorted together most frequently in the set, unless there is a lot of batch-to-batch or sample-to-sample variation.

In comparisons of MDS sorting of terpene aromas by people with different degrees of training, good agreement among groups was observed (Lawless and Glatter, 1990). That is, trained or experienced panelists and untrained consumers all tended to give similar responses, an effect also observed for the cheese data shown earlier. This may be an advantage in that basic perceptual dimensions are uncovered by this procedure—dimensions that are relatively uninfluenced by higher "cognitive" considerations or conceptualizations of the product set. On the other hand, it may also reflect an insensitivity of the sorting method to differences among people. The sorting task may oversimplify relationships. Perhaps this finding is not surprising since the sorting is a group-derived measure of association. Kohli and Leuthesser (1993) recommend the use of MDS when products are not very complex and most participants will extract common underlying dimensions in judging overall similarity.

19.2.6 Vector Projection

In order to interpret an MDS map or model it was common practice to examine the edges and opposing corners of the map to gain insight into the contrasts being made by people during their ratings or sortings. However, a second step of data collection can add more objectivity to the interpretation process (Popper and Heymann, 1996). At the conclusion of the MDS phase, people can be asked about their criteria for similarity or sorting. The most frequently mentioned attributes could then be used on a ballot for profiling in a subsequent session. A simple follow-up session for re-tasting of the products can be held. Each product need only be tasted once by a subject and rated using the attributes. Mean ratings can then be regressed against the coordinates of the product points in space to find a direction through the space that represents that attribute. The regression weights are related to the degree of correlation of the attribute with the dimensions of the model and the overall R^2 indicates whether or not there is a relationship between the attribute ratings and the positions of the products. Discussions of this procedure can be found in Schiffman et al. (1981) and in Kruskal and Wish (1978). Figure 19.3 shows the vector projection for the cheeses shown in Fig. 19.2. We can see a group of flavor-related vectors and a group of texture-related vectors, roughly at right angles to one another.

This method of vector projection is mathematically equivalent to some external PREFMAP procedures (see Sections 18.3 and 19.4). The basic goal is to find a direction through the space such that the coordinates

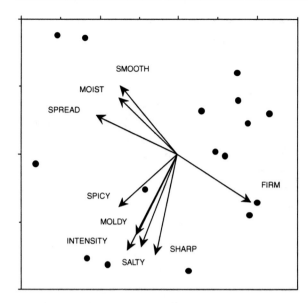

Fig. 19.3 Attribute ratings of the cheese shown in Fig. 19.2 were regressed against the positions in the model to plot vectors to aid in interpretation of the perceptual map. Only attributes with significant differences among cheese in ANOVA and regression p-values less than 0.01 are plotted. Replotted from Lawless et al. (1995), copyright 1995, used with permission of Elsevier Science Ltd, The Boulevard, Langford Lane, Kidlington OX51GB, UK.

along the new vector (think of it as a new axis or ruler) will be maximally correlated with the original scores for each product on that attribute. This is shown in Fig. 19.5. Obviously, if the scores were highly correlated with values on the X-axis of the perceptual map and uncorrelated with the Y-axis, the vector would fall right on the X-axis. If it was equally and positively correlated with X and Y, then it would fall at a 45° angle pointing to the upper right quadrant. If it were equally and negatively correlated with both axes it would point to the lower left. The standardized regression coefficients (beta weights) give the direction for the vector in a unitized space (−1 to +1).

19.2.7 Cost-Efficient Methods for Data Collection: Projective Mapping, aka Napping

Another cost-effective and rapid method for assessing product similarities and producing a map is projective mapping. This technique instructs the consumer

to place each product on a surface such as a large blank sheet of paper. Products are placed so that their positions and distances represent their similarities and dissimilarities. Like sorting and direct similarity scaling, the criterion used by each consumer is up to him/her, and no structure or point of view or attributes are imposed by the experimenter. Thus, we are free to discover what is actually important to that person. This is a potential advantage over PCA, which at least in the beginning weights all the attributes equally (it is only the pattern of correlation that matters). The data are the X and Y coordinates of each product, which can be transformed of course into a distance matrix. Obviously, the data set is a bit richer from that of sorting, because the individual similarity data do not consist of zeros and ones, but actual scaled distance measures.

This method was introduced by Risvik and colleagues in the 1990s, but it did not receive much attention until recently (Risvik et al., 1994, 1997). Pages and colleagues re-introduced it as the "napping" procedure, based on the French word for tablecloth (Pages, 2005; Perrin et al., 2008). An important addition to their version of the procedure was the analysis by multiple factor analysis (MFA), which exists now in the R language as part of a free add-on library (R Development Core Team, 2009). This useful program can uncover more than two dimensions in the data, depending on how individual consumers might pay attention to different attributes. For example, if half of the consumer group maps the products based on taste and texture and the other half uses color and texture, the MFA will arrive at a group configuration with three dimensions, and 50% of the variance will be assigned to texture (the common attribute) and 25% each to color and taste (Nestrud and Lawless, 2010a). Thus the method, when analyzed by MFA, is not limited to two underlying attributes, as one might expect from a planar array produced by each consumer.

Projective mapping or napping has been applied to a variety of products, such as cheeses (Barcenas et al., 2004), wines (Pages, 2005; Perrin et al., 2008), citrus juices (Nestrud and Lawless, 2008), chocolates (Kennedy and Heymann, 2009), and apples (Nestrud and Lawless, 2010b). In some cases it has produced a more informative or richer set of information, as indicated by better correlations with attribute ratings and the vector projections through the group configurations (Nestrud and Lawless, 2010b). Like sorting, the method is fast and easy for consumers and should find wide application in exploratory work as well as category appraisals.

19.3 Attribute Identification and Classification

19.3.1 Drivers of Liking

A popular concept in recent years has been the notion that some attributes may be more critical than others in determining consumer preferences for a particular product. Such a critical attribute can be called a "driver of liking" (van Trijp et al., 2007). There are a number of ways to identify these critical attributes, some qualitative and some quantitative. Qualitatively, you can find out what aspects of a product are important to consumers by interviews like focus groups (see Chapter 16). This assumes that people can articulate what is important to them and they are doing so faithfully. Neither assumption is very solid. A related approach is to ask consumers directly for "importance" ratings. This also relies on people's ability to accurately report their opinions. Quantitatively, you can attempt to relate sensory changes in the product to changes in liking or preference. There are number of ways to go about this, but they all assume that you can make perceivable changes in the product over a range of ingredients or processes and that these changes matter to consumers.

Assume you start by looking at the variables one at a time. The simple correlation between the sensory scores and the liking scores should give us some idea of the strength of the relationship. Another approach is to use just-about-right scales (see Chapter 14) and what is called penalty analysis, i.e., what the cost is in overall liking for not being "just right." A third approach is to use intensity scales, but place a rating for the ideal product along with the actual product rating to get an idea of deviation from ideal (van Trijp et al., 2007). Embedded in these notions is that the slope of the line relating hedonic response to sensory intensity is an indicator of the importance of that attribute. If it has a steep slope, small changes in the sensory attribute cause large changes in liking. Thus it is probably a "driver." One liability of this approach concerns whether you have made a meaningful change

in the sensory intensity. If the range of sensory variation is too small, you may fail to find a correlation due to restricting the range (van Trijp et al., 2007). If the variation is too large, you may have concocted some products that would never be seen on the marketplace, i.e., something so wacky that it by default receives low ratings. So it is hard to know how much to vary a given attribute in this approach. Common sense is called for. Beware, however, that the slope or correlation for different attributes may be a function of how effectively you have spanned the realistic product variation.

As another approach, one could try to build a multiple regression model, a multiple linear model such that the overall liking for a product would be determined by some linear combination of the variables you changed (e.g., Hedderly and Meiselman, 1995). The regression weights (linear coefficients) would give us some idea of the strength of the relationship between the predictors and overall liking for the products. For example, the liking for a fruit beverage might be a function of sweetness and sourness, which in turn is driven by the psychophysical relationships between sugar level and sweetness and between acid level and sourness. However, life is rarely that simple. Acid and sugar interact, perceptually, to partially mask each other through mixture suppression (see Chapter 2). So this approach is somewhat limited by the covariation among the predictor variables in many products. To address the correlation problem, you can perform a PCA or other data reduction procedure and then regress liking against the new factors (PCs or latent variables), but then they become more difficult to interpret.

Another approach useful with more discrete attributes (rather than psychophysical or continuous ones) is conjoint analysis (discussed at length in Moskowitz et al., 2006). In this approach, combinations of attributes or product features are varied, and overall liking is assessed. For example, would you like a jelly or fruit spread that has high sweetness, low fruit solids, and no seeds? One that has medium sweetness, high fruit solids, and seeds? All the combinations can be presented, and then from the pattern of results, the contribution of sweetness, solids, and the variable with/without seeds can be calculated, often by specialized software packages. This approach was historically used with durable goods (washing machines, automobiles, etc.).

The typical conjoint measurement design is similar to a factorial design we would use in analysis of variance: all possible combinations are presented an equal number of times. Other mixture designs have been used (Gacula et al., 2009) for product optimization, in which continuous variables are combined at various levels to find an optimum combination.

19.3.2 The Kano Model

The Kano model proposes that not all attributes are created equal, and they contribute to overall acceptance in different ways (Riviere et al., 2006). The model in its usual English translation looks at satisfaction as the primary consumer response, but one can think of this dimension as a generally positive versus generally negative response to a product (e.g., delighted versus disgusted). Satisfaction, of course, differs from liking in that it includes meeting expectations. The second dimension is the delivery of that attribute, from unfulfilled to fulfilled/delivered. How well is this aspect of the product executed? These two axes and the three attribute classes are shown in Fig. 19.4.

There are three classes of attributes in the Kano model (Matzler and Hinterhuber, 1998). The first attribute is a *performance attribute*. A performance attribute is expected, and if not delivered will leave the consumer dissatisfied. When it is fully delivered or in

high amounts or intensity, the consumer is delighted. So, more of this attribute is better. In some foods, sweetness acts like this when there is no optimum or bliss point. This performance-type attribute is similar to the vector models for product optimization discussed under preference mapping (Section 19.4). The second kind of attribute is one that is expected, and if not delivered, will dissatisfy a consumer. When delivered, satisfaction is only neutral because the attribute is expected. I expect my car door, when opened, not to scrape the curb. I expect my dry breakfast cereal to start out crisp. If my car door scrapes the curb, or my bran flakes are limp or soggy right out of the box, I'm unhappy. These "must-haves" are givens, basic requirements. Consumers may not even articulate them in focus groups or surveys, because they are considered obvious. My hotel room must have toilet paper. But if it has three rolls, I'm not going to brag about it. The third type of attribute is unexpected and not required, so if it is not delivered there is no problem. If it is delivered, it delights the customer, generates excitement or other emotions. These unexpected benefits can drive product innovations.

Other aspects can be added to the model. A common one is a time dimension. Over time, the delightful attributes may come to be expected and then they turn into performance attributes. Further in time, they may become expected and required, so they become must-haves. Pay-at-the-pump gas stations and Internet service in hotel rooms have gone through this transition.

A

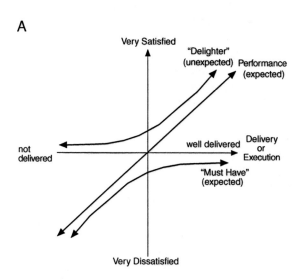

B

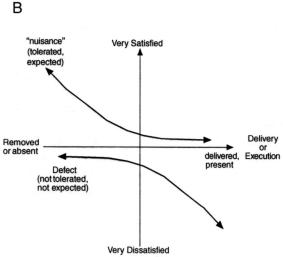

Fig. 19.4 The Kano model (for details, see text).

Many frozen foods are expected to be microwavable, even some baked goods and pizza crusts. In product development, the trick is to prioritize these attributes as consumer needs and figure what the potential payoff may be for delivery versus the penalty for underperformance. Of course, attributes that are unexpected delighters may be points of innovation and differentiation from the competition. For food research, we would add two more, not originally part of the Kano model. First, product defects, which can be viewed as "negative-must-haves." If they are present, they dissatisfy the customer, but they are not expected so when they are not present, there is no added benefit. The second is a nuisance attribute. This is one that is expected and even tolerated, although its removal would increase customer satisfaction. At one time, chewing gum would stick to your dentures. People expected it to. Gummy candy sticks in the spaces between my teeth, but I eat it anyway. The advent of chewing gum that did not stick to your bridgework was a plus for some of us. Olives have pits, grapes have seeds, and lobsters have shells that are tough to break open. These attributes are often tolerated, but a pitted olive, seedless table grape variety, or a lobster tail served without the shell all have consumer appeal.

19.4   Preference Mapping Revisited

19.4.1 Types of Preference Maps

Preference maps are a special class of perceptual maps in which products and consumer preferences are both illustrated (Elmore et al., 1999). In the previous chapter, two kinds of preference maps were described, internal and external. Internal preference maps are basically derived from a PCA of consumer acceptance data. As such, they have limited utility in understanding the product differences and reasons for preference. A more useful tool is the external preference map. In this method, the product space is derived separately, for example, from a PCA on descriptive data. The placement of the products represents their similarities and differences, and the attributes can be projected as vectors through the plot. Then consumer acceptance data can also be projected. Directions can be discovered for the optimal products (for the group as a whole or for

individual preferences) or parts of the space that would represent an optimum point for each consumer. These two approaches can be called vector models and ideal point models, respectively. Both provide the opportunity to look for segments of consumers that may prefer one part of the product space, i.e., one style of product.

19.4.2 Preference Models: Vectors Versus Ideal Points

With vector models, as in the Kano performance attributes, more is better. There is a theoretical direction through the space that lines up best with a consumer's likes and dislikes. As you move on a positive direction along this line, the product's acceptance improves. The direction that best fits the consumer (assuming one does) is the one that provides a maximum correlation between the acceptance ratings and the positions marked off on this vector from perpendicular lines dropped from each product. Think of the vector as a ruler or axis and each product (point in space) has a value on this ruler. Those values have been maximally correlated with the person's acceptance ratings. This is the same notion behind the vector fitting discussed above in Section 19.2.6. In the case of intensity attributes, the products on the outside of the space usually have more of that attribute than products on the opposite side or in the center. So there is a certain sense to this kind of model for descriptive attributes, especially if the space is derived from a PCA, which is after all based on linear correlations among a group of intensity scales.

A good example of this kind of external preference mapping is the classic paper by Greenhof and MacFie (1994), showing individual vector directions for a group of consumers for some meat products. Notably, almost all directions through the space were represented. If one had tried to fit a group average vector, the result may have been meaningless or shown nearly zero correlation. There were some apparent clusters of preference vectors with a high density, suggesting segments of consumers who liked the same kind of product.

An alternative model to the vector approach is a model in which a specific spot in the product space has the highest liking or acceptability. For many attributes, the "more is better" (or "less is better" in the case of a defect or undesired quality) does not fit very well. That is, there is an optimum or "bliss point" as discussed in the just-about-right scaling section. Sweetness in many foods is a good example. We like our fruit juices and some of our wines to have a certain level of sweetness, but more than that is too much. Extending this idea into the product space or perceptual map, there is often some collection of attributes and intensity values that seems to be the best combination. If you come from New York or Vermont, you may prefer your cheddar cheese sharp, firm and crumbly, with a good hit of acid and fecal notes, but if you are from the west coast of the United States you may like it mild, moist, and pasty. These two cheese consumers would have ideal points in different parts of the sensory space. The vector model and the ideal point model are contrasted in Fig. 19.5.

Finding the optimum point for a consumer is mathematically straightforward, as long as there are hedonic (acceptability) ratings for all the products by each consumer. The trick is to find the point that has the maximum inverse correlation of the acceptance ratings with the distance from each product. This makes sense because products that you like should be closer to your ideal point than products you dislike. This is shown in Fig. 19.5, where the length of the rays from the ideal product spot must have the maximum negative correlation with a person's liking ratings. The higher the correlation, the better the fit of the product space/model to that person's favorite type of product. It is possible of course to have a poor fit. The person may not have strong preferences or the combination that person prefers is an unlikely combination that does not reflect the usual pattern of correlation turned up by the PCA. Another desirable feature of an ideal point model is some estimate of the gradient or density around the person's ideal point. For some people, small deviations from the ideal will make a large difference in acceptability. Other people may be more tolerant of sensory changes, and the ideal point position is a little fuzzier. So a kind of contour plot around each point is a useful kind of information. Various types of preference maps and models are discussed extensively in Meullenet et al. (2007).

19.5 Consumer Segmentation

A traditional approach to consumer segmentation has been to look for some patterns of liking and disliking across a set of products and then try to identify the

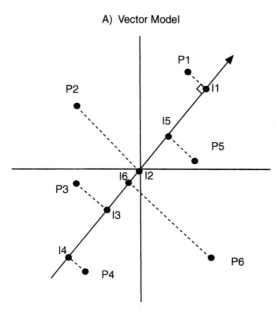

A) Vector Model

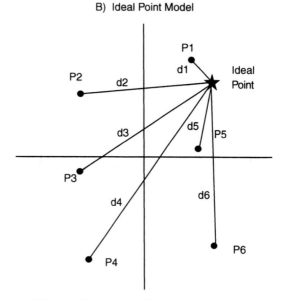

B) Ideal Point Model

Fig. 19.5 Vector versus ideal point models for six hypothetical products P1–P6. For the vector model, the direction through the space is such that the intersection points from the dropped perpendicular segments, I1–I6, are maximally correlated with the original liking ratings. For the ideal point model, the position of the ideal point is such that the distances to the products, d1–d6, are maximally inversely correlated with the original liking ratings.

characteristics of those liking segments. Age, gender, and any number of demographic or lifestyle patterns can then be correlated to try to characterize the kinds of people who like different types of products. The marketing and advertising literature are full of such approaches. An alternative approach is one in which the sensory scientist can contribute, and that is segmentation by sensory properties, and, more specifically, sensory optimization. Even with a Kano-type classification of attributes, it may be important to look for different patterns among different groups of consumers (e.g., Riviere et al., 2006). There are powerful statistical tools for examining patterns of responses and/or demographic data. Perhaps the most common one is some form of cluster analysis, which can group people based on the similarities in the patterns of their responses. A variety of clustering algorithms are available (Plaehn and Lundahl, 2006; Qannari et al., 1997; Wajrock et al., 2008).

Sensory segmentation is a potentially powerful tool for getting the highest ratings for a new product or line extension of existing items. Consider a scenario in which a company wishes to launch a single product with the maximum possible overall liking score. Product developers and sensory scientists team up and they make the single best product based on a number of attributes that they vary and optimize. The combination looks pretty good and the product is launched. But they may have missed an important pattern. If they were aware that the consumer base for this product category had three distinctly different sensory/preference segments, they might have made three products in different styles in order to please those segments. The missed opportunity is that the scores for each of the different styles (scores from their appropriate segment) may have been much higher than the single score for the composite product. Without realizing it, they probably created "airline food" that is acceptable to a wide audience and does not offend anyone, but does not really delight anyone at all.

It is well established that the relationship between sensory intensity and liking is often an inverted U-shaped curve (Moskowitz, 1981) or an inverted V (Conner and Booth, 1992). However, Pangborn (1970) showed that consumers could be grouped based on salt or sweet preferences in products. Some like increasing levels of saltiness while others prefer none at all and yet a third group shows the classic inverted U with an

optimum or bliss point. When you average these three groups you get a somewhat flattened inverted U, but this obscures the two groups that have monotonic relationships (see Fig. 19.6). A classic example of this was uncovered by Moskowitz and Krieger (1998) in their re-analysis of a set of coffee preference data shared among the European Sensory Network members in 1995. The key finding is shown in Fig. 19.7. When segmented by country, five countries showed more or less similar patterns with some intermediate level of bitterness appearing optimal. However, when individual data were considered, there were three clear groups, one favoring no bitterness at all (more is worse), one favoring high bitterness (more is better), and a third group with a very steep optimum. The important result was that all three sensory segments were included in all five countries. So segmentation by sensory preference would be a much better strategy for a coffee company than trying to tailor a universal coffee, and making slightly different styles for each country was probably a waste of time. Note that the optimum ratings for two of the sensory-based groups in Fig.19.7b are higher than the optima from the geographically based groups (Fig. 19.7a). Another example of sensory segmentation is found in Moskowitz et al. (1985).

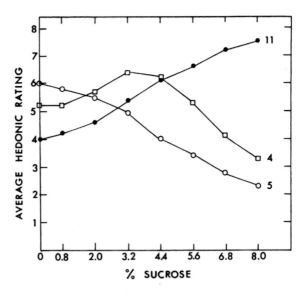

Fig. 19.6 Data from Pangtborn (1970) showing different segments of consumers liking different amounts of sweetness in a product. Three distinct groups were found.

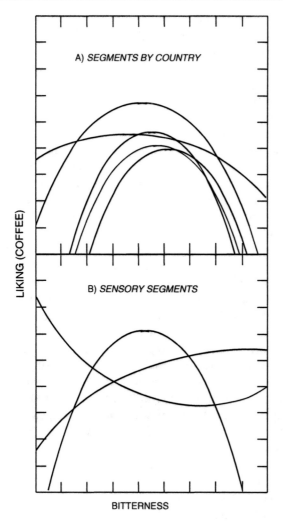

Fig. 19.7 Data replotted from Moskowitz and Krieger (1998). For optimizing coffee bitterness, all five European countries showed similar inverted U functions. However, within each country there were distinct sensory-based segments (Curves are quadratic functions fit to the data).

19.6 Claim Substantiation Revisited

Claim substantiation is another strategic activity that sensory evaluation may participate in using blind consumer tests. Regulatory agencies and networks have certain requirements and expectations for these kinds of tests (NBC, 2009) and the results are often challenged by competitors (ASTM, 2008). Sticky legal battles may ensue, with one side trying to prove that the test was invalid because of some methodological flaw. The sensory department may be called upon to defend

the test methods and results. Here are some guidelines for preference testing that have been extracted from the ASTM and NBC documents, which are very similar.

For claims of superiority as well as equivalence, the consumer central location test (CLT) with a paired presentation is generally required. Home use tests, trained panels, and experts are not favored. The test should be double blind, meaning neither the server nor the consumer knows the identity of the products. This requires an extra person compared to most CLT's for research purposes because the person preparing the samples must not be the server. The products must be tested in several different regions of the country. Products must be representative of what consumers find on the shelves in their area, usually requiring store retrievals of products with the same or similar use-by or expiration dates.

The "no-preference" option is required by NBC but discussed with varying opinions in the ASTM guide. To satisfy the authorities, a no-preference response should be offered. For superiority, the winning product must have a significant difference from the total sample, which implies that no-preference votes are added to the losing proportion.

The simple formula for a binomial preference test is given as follows:

$$z = \frac{P - 0.5}{0.5/\sqrt{N}} \tag{19.1}$$

where P is the proportion for the winner, N is the total sample size, and z must exceed 1.645. Note that this test is one tailed, unlike most preference tests, which are two tailed! This is justified on the basis that you are looking only for a win, not for a loss for your product. NBC explicitly states that no-preference votes must *not* be split among the two products, because "the response is fundamentally different." The sample size required for superiority claims run from 200 to 300 consumers per pair of products. If one wishes to make the claim that it is "America's best" meaning preferred to all competitors, there are requirements for national sampling. Also the products should represent at least 85% of the market by volume (unless there are a large number of small regional brands).

The statistical test may also be conducted between the preference votes only, with the no preference omitted from the analysis. In that case, a qualified claim is made, for example, "Among those with a preference our hot dog beat the leading competitor." This qualification is referred to as a "super" in advertising lingo. The super is required if there were 20% or more no-preference votes. Obviously if there were a very high percentage of no-preference votes (e.g., 50%), then any claim would be difficult to substantiate.

A parity result is treated somewhat differently than the equality discussion of Chapter 5. Some authorities such as NBC recommend that the analysis be done as a two-tailed test at the 90% confidence interval. Then a non-significant result is accepted as evidence of parity, as long as your product does not lose at that level. Other companies set the alpha level even lower (as insurance against missing a difference, Type II error), for example, at 67%. Although this lowers the beta-risk, the use of a two-tailed test actually raises it a bit. For a parity result to be substantiated, a sample size of at least 500 is now required.

19.7 Conclusions

19.7.1 Blind Testing, New Coke, and the Vienna Philharmonic

Sensory-based optimization is only one aspect of product development that helps insure product success in the marketplace. Obviously the concept development, advertising, positioning, promotion, and other marketing strategies are equally, if not more, important. However, most marketers will admit that you cannot make a bad product successful. It must combine a good product idea with the delivery of sensory and performance attributes. But this raises the question, because the product concept is so important, does it make sense to do blind consumer tests at all? Some authors (e.g., Gladwell, 2005) have argued that they are virtually irrelevant and even potentially harmful. Products are not purchased or consumed on a blind basis. The classic debacle from an overreliance on blind testing was the story of New Coke. In the 1980s, Coke had lost some market share to their prime competitor, Pepsi, and was faced with the curious and annoying finding that in blind paired-preference tests, Pepsi won. This led to the reformulation of Coke's flagship product to make it sweeter and more citrusy, i.e., more like Pepsi.

Loyal Coke drinkers rejected this flavor. They rebelled. "Classic Coke" was brought back to the market and New Coke died a quiet death.

In contrast to this story, however, Gladwell later considers what is perhaps the greatest success story for blind testing. This was the advent of the professional musical audition conducted from behind a screen, so that the judges would not know the identity (or gender!) of the performer. He tells the story of Abbie Conant, who was playing trombone (a "man's instrument") for the Royal Opera of Turin, when a chance to audition for the Vienna Philharmonic opened up, and "Herr" Conant was invited to try out. Because one of the applicants was the son of a well-known musician from the Munich orchestra, they decided to conduct the auditions behind a screen. Conant played Ferdinand David's Konzertino for trombone, a "warhorse" piece in the German repertoire. Although she "cracked" one note, the judges were so impressed that they sent the remaining applicants home. To their surprise, Herr Conant turned out to be Frau Conant. Gladwell goes on to describe how the use of the screen led to more and more women being hired for symphony orchestras (a rarity 100 years ago), a classic case of where blind testing worked to create a virtual social revolution in the music world.

So what is the place of blind sensory tests in strategic research and competitive surveillance? One can ask whether the debacle of New Coke could have been avoided, and the answer in our opinion is yes. One point that many sensory practitioners know is that in a paired comparison test, often the sweeter of two products will win. This does not mean that the product will be preferred after an entire portion is consumed or after the product is used for a period of time. Thus there is also the well-known limitation to the predictive value of a central location test (CLT) versus the more extended testing that one can get in a home use test (HUT). Finally, one can ask whether the outcome of a blind CLT should have been given so much weight in the face of a product that appeared to be driven by its image and its extensive history of brand-loyal users. The paired-preference test was just one piece of data in a sea of other information. The loss of market share turned the corporate attention to technical reformulation, but overlooked other factors, such as the highly effective use by Pepsi of the pop star Michael Jackson in its advertising.

19.7.2 The Sensory Contribution

Making a successful innovative product depends on getting a good idea at the beginning. Much of this can be achieved by front-end research, using techniques such as the qualitative methods outlined in Chapter 16. For an extensive discussion of "getting the right idea" to start with, see Moskowitz et al. (2006). This chapter has examined product optimization, consumer segmentation, competitive analysis, and perceptual modeling. These are all areas in which the sensory professional can participate and reach beyond the world of simple product testing. In these arenas, the sensory professional can have a strong impact and influence on the strategy of a company in making successful products. Most importantly, we bring a special expertise to the table and a special point of view that can complement other styles of thinking and perception on the product team.

References

Abdi, H. and Valentin, D. 2007. DISTATIS. How to analyze multiple distance matrices. In: N. Salkind (ed.), Encyclopedia of Measurement and Statistics. Sage, Thousand Oaks, CA, pp. 1–15.

Abdi, H., Valentin, D., Chollet, S. and Chrea, C. 2007. Analyzing assessors and products in sorting tasks: DISTATIS, theory and applications. Food Quality and Preference, 18, 627–640.

ASTM International. 2008. Standard Guide for Sensory Claim Substantiation. Designation E 1958–07. Vol. 15.08 Annual Book of ASTM Standards. ASTM International, Conshohocken, PA, pp. 186–212.

Barcenas, P., Elortondo, F. J. P. and Albisu, M. 2004. Projective mapping in sensory analysis of ewes milk cheeses: A study on consumers and trained panel performance. Food Research International, 37, 723–729.

Benner, M., Linneman, A. R., Jongen, W. M. F. and Folstar, P. 2003. Quality function deployment (QFD) – Can it be used to develop food products? Food Quality and Preference, 14, 327–339.

Bertino, M. and Lawless, H. T. 1993. Understanding mouthfeel attributes: A multidimensional scaling approach. Journal of Sensory Studies, 8, 101–114.

Bijmolt, T. H. A. 1996. Multidimensional Scaling in Marketing: Toward Integrating Data Collection and Analysis. Thesis, Economische Wetenschappen, Rijksuniversiteit Groningen.

Conner, M. T. and Booth, D. A. 1992. Combined measurement of food taste and consumer preference in the individual: Reliability, precision and stability data. Journal of Food Quality, 15, 1–17.

Dijksterhuis, G. 1997. Multivariate Data Analysis in Sensory and Consumer Science. Food and Nutrition, Trumbull, CT.

Elmore, J. R. and Heymann, H. 1999. Perceptual maps of photographs of carbonated beverages created by traditional and free-choice profiling. Food Quality and Preference, 10, 219–227.

Elmore, J. R., Heymann, H., Johnson, J., and Hewett, J. E. 1999. Preference mapping: Relating acceptance of "creaminess" to a descriptive sensory map of a semi-solid. Food Quality and Preference, 10, 465–475.

Fox, R. J. 1988. Perceptual mapping using the basic structure matrix decomposition. Journal of the American Marketing Association, 16, 47–59.

Gacula, M. C., Jr. 1993. Design and Analysis of Sensory Optimization. Food and Nutrition, Trumbull, CT.

Gacula, M., Singh, J., Bi, J. and Altan, S. 2009. Statistical Methods in Food and Consumer Research. Elsevier/Academic, Amsterdam.

Gladwell, M. 2005. Blink. Little, Brown and Co, New York, NY.

Goldman, A. E. and McDonald, S. S. 1987. The Group Depth Interview, Principles and Practice. Prentice-Hall, New York, NY.

Greenhof, K. and MacFie, H. J. H. 1994. Preference mapping in practice. In: H. J. H. MacFie and D. M. H. Thomson (eds.), Measurement of Food Preferences. Chapman & Hall, London, UK.

Hedderly, D. I. and Meiselman, H. L. 1995. Modeling meal acceptability in a free choice environment. Food Quality and Preference, 6, 15–26.

Heymann, H. 1994. A comparison of free choice profiling and multidimensional scaling of vanilla samples. Journal of Sensory Studies, 9, 445–453.

Hoffman, D. L. and Franke, G. R. 1986. Correspondence analysis: Graphical representation of categorical data in marketing research. Journal of Marketing Research, 23, 213–227.

Johnson, R. 1988. Adaptive perceptual mapping. Applied Marketing Research, 28, 8–11.

Katahira, H. 1990. Perceptual mapping using ordered logit analysis. Marketing Science, 9, 1–17.

Kennedy, J. and Heymann, H. 2009. Projective mapping and descriptive analysis of milk and dark chocolates. Journal of Sensory Studies, 24, 220–233.

King, M. J., Cliff, M. A., and Hall, J. W. 1998. Comparison of projective mapping and sorting data collection and multivariate methodologies for identification of similarity-of-use of snack bars. Journal of Sensory Studies, 13, 347–358.

Kohli, C. S. and Leuthesser, L. 1993. Product positioning: A comparison of perceptual mapping techniques. Journal of Product and Brand Management, 2, 10–19.

Kruskal, J. B. 1964. Nonmetric multidimensional scaling: A numerical method. Psychometrika, 29, 1–27.

Kruskal, J. B. and Wish, M. 1978. Multidimensional Scaling. Sage, Beverly Hills, CA, pp. 87–89.

Lawless, H. T. 1989. Exploration of fragrance categories and ambiguous odors using multidimensional scaling and cluster analysis. Chemical Senses, 14, 349–360.

Lawless, H. T. and Glatter, S. 1990. Consistency of multidimensional scaling models derived from odor sorting. Journal of Sensory Studies, 5, 217–230.

Lawless, H. T., Sheng, N. and Knoops, S. S. C. P. 1995. Multidimensional scaling of sorting data applied to cheese perception. Food Quality and Preference, 6, 91–98.

Laitin, J. A. and Klaperman, B. A. 1994. The brave new world of marketing research. Medical Marketing and Media, July, 44–51.

MacRae, A. W., Rawcliffe, T. Howgate, P. and Geelhoed, E. 1992. Patterns of odour similarity among carbonyls and their mixtures. Chemical Senses, 17, 119–125.

Malhotra, N., Jain, A. and Pinson, C. 1988. Robustness of multidimensional scaling in the case of incomplete data. Journal of Marketing Research, 24, 169–173.

Mantei, M. M. and Teorey, T. J. 1989. Incorporating behavioral techniques into the systems development life cycle. MIS Quarterly, 13, 257–274.

Matzler, K. and Hinterhuber, H. H. 1998. How to make product development projects more successful by integrating Kano's model of customer satisfaction into quality function deployment. Technovation, 18, 25–38.

Meullenet, J.-F., Xiong, R. and Findlay, C. 2007. Multivariate and Probabilistic Analyses of Sensory Science Problems. Blackwell, Ames, IA.

Miaoulis, G., Free, V. and Parsons, H. 1990. Turf: A new approach for product line extensions. Marketing Research, March 1990, 28–40.

Moskowitz, H. R. 1981. Relative importance of perceptual factors to consumer acceptance: Linear vs. quadratic analysis. Journal of Food Science, 46, 244–248.

Moskowitz, H. and Krieger, B. 1998. International product optimization: A case history. Food Quality and Preference, 9, 443–454.

Moskowitz, H. R., Beckley, J. H. and Resurreccion, A. V. A. 2006. Sensory and Consumer Research in Food Product Design and Development. Blackwell/IFT, Ames, IA.

Moskowitz, H. R., Jacobs, B. E. and Lazar, N. 1985. Product response segmentation and the analysis of individual differences in liking. Journal of Food Quality, 8, 169–181.

Miller, J. and Wise, T. 1991. Strategic and tactical research. Agricultural Marketing, 29, 38–41.

Mullet, G. M. 1988. Applications of multivariate methods in strategic approaches to product marketing and promotion. Food Technology, 51(11), 145, 152, 153, 155, 156.

Muñoz, A. M., Chambers, E. C. and Hummer, S., 1996. A multifaceted category research study: How to understand a product category and its consumer response. Journal of Sensory Studies, 11, 261–294.

NBC Universal. 2009. Advertising Guidelines. Department of Advertising Standards, NBC Universal, Inc.

Nestrud, M. A. and Lawless, H. T. 2008. Perceptual mapping of citrus juices using projective mapping and profiling data from culinary professionals and consumers. Food Quality and Preference, 19, 431–438.

Nestrud, M. A. and Lawless, H. T. 2010a. Recovery of subsampled dimensions and configurations derived from Napping data by MFA and MDS. Manuscript available from the authors.

Nestrud, M. A. and Lawless, H. T. 2010b. Perceptual mapping of applies and cheeses using projective mapping and sorting. Journal of Sensory Studies, 25, 390–405.

Pages, J. 2005. Collection and analysis of perceived product inter-distances using multiple factor analysis: Application to the study of 10 white wines from the Loire Valley. Food Quality and Preference, 16, 642–649.

Pangborn, R. M. 1970. Individual variation in affective responses to taste stimuli. Psychonomic Science, 21, 125–126.

Perrin, L., Symoneaux, R., Maitre, I., Asselin, C., Jourjon, F. and Pages, J. 2008. Comparison of three sensory methods for use with the Napping® procedure: Case of ten wines from Loire valley. Food Quality and Preference, 19, 1–11.

Plaehn, D. and Lundahl, D. S. 2006. An L-PLS preference cluster analysis on French consumer hedonics to fresh tomatoes. Food Quality and Preference, 17, 243–256.

Popper, R. and Heymann, H. 1996. Analyzing differences among products and panelists by multidimensional scaling. In: T. Naes and E. Risvik (eds.), Multivariate Analysis of Data in Sensory Science, Elsevier Science, Amsterdam, pp. 159–184.

Qannari, E. M., Vigneau, E., Luscan, P., Lefebvre, A. C. and Vey, F. 1997. Clustering of variables, application in consumer and sensory studies. Food Quality and Preference, 8, 423–428.

R Development Core Team. 2009. R: A language and environment for statistical computing. R Foundation for Statistical Computing, Vienna, Austria. http://www.R-project.org.

Ramsay, J. O. 1977. Maximum likelihood estimation in multidimensional scaling. Psychometrika, 42, 241–266.

Riviere, P., Monrozier, R., Rogeaux, M., Pages, J. and Saporta, G. 2006. Adaptive preference target: Contribution of Kano's model of satisfaction for an optimized preference analysis using a sequential consumer test. Food Quality and Preference, 17, 572–581.

Risvik, E., McEwan, J. A., Colwoll, J. S., Rogers, R. and Lyon, D.H. 1994. Projective mapping: A tool for sensory analysis and consumer research. Food Quality and Preference 5, 263–269.

Risvik, E., McEwan, J. A. and Rodbotten, M. 1997. Evaluation of sensory profiling and projective mapping data. Food Quality and Preference, 8, 63–71.

Rosenberg, S. and Kim, M.P. 1975. The method of sorting as a data- gathering procedure in multivariate research. Multivariate Behavioral Research, 10, 489–502.

Rosenberg, S., Nelson, C. and Vivekananthan, P. S. 1968. A multidimensional approach to the structure of personality impressions. Journal of Personality and Social Psychology, 9, 283–294.

Schiffman, S. S., Reynolds, M. L. and Young, F. W. 1981. Introduction to Multidimensional Scaling. Academic, New York, NY.

Sun Tzu (Sun Wu) circa 350 B.C.E. The Art of War. S. B. Griffith, trans. Oxford University, 1963.

Tang, C. and Heymann, H. (2002). Multidimensional sorting, similarity scaling and free choice profiling of grape jellies. Journal of Sensory Studies, 17, 493–509.

Van Trijp, H. C. M., Punter, P. H., Mickartz, F. and Kruithof, L. 2007. The quest for the ideal product: Comparing different methods and approaches. Food Quality and Preference, 18, 729–740.

Von Arx, D. W. 1986. The many faces of market research: A company study. The Journal of Consumer Marketing, 3(2), 87–90.

Wajrock, S., Antille, N., Rytz, A., Pineau, N. and Hager, C. 2008. Partitioning methods outperform hierarchical methods for clustering consumers in preference mapping. Food Quality and Preference, 19, 662–669.

Wright, K. 1995. Attribute Discovery and Perceptual Mapping: A Comparison of Techniques. Master's Thesis, Cornell University.

Similarity, Equivalence Testing, and Discrimination Theory

Harry T. Lawless and Hildegarde Heymann

H.T. Lawless, H. Heymann, *Sensory Evaluation of Food*, Food Science Text Series,
DOI 10.1007/978-1-4419-6488-5, © Springer Science+Business Media, LLC 2010

DOI 10.1007/978-1-4419-6488-5_20

On page 105, Equation 5.2 should read as

$$P_{adjusted} = \frac{P_{observed} - P_{chance}}{1 - P_{chance}}$$

The online version of the original chapter can be found at
http://dx.doi.org/10.1007/978-1-4419-6488-5_5

Measurement of Sensory Thresholds

Harry T. Lawless and Hildegarde Heymann

H.T. Lawless, H. Heymann, *Sensory Evaluation of Food*, Food Science Text Series,
DOI 10.1007/978-1-4419-6488-5,© Springer Science+Business Media, LLC 2010

DOI 10.1007/978-1-4419-6488-5_20

On page 132, Equation 6.2 should read as

$$P_{req} = P_{chance} + P_{corr}(1 - P_{chance})$$

The online version of the original chapter can be found at
http://dx.doi.org/10.1007/978-1-4419-6488-5_6

Descriptive Analysis

Harry T. Lawless and Hildegarde Heymann

H.T. Lawless, H. Heymann, *Sensory Evaluation of Food*, Food Science Text Series,
DOI 10.1007/978-1-4419-6488-5,© Springer Science+Business Media, LLC 2010

DOI 10.1007/978-1-4419-6488-5_20

On page 253, The below reference is incorrect and to be ignored along with citation on page 231.

Caul, J. F. 1967. The profile method of flavor analysis. Advances
  in Food Research, 7, 1–140.

The online version of the original chapter can be found at
http://dx.doi.org/10.1007/978-1-4419-6488-5_10

Appendix A

Basic Statistical Concepts for Sensory Evaluation

Contents

It is important when taking a sample or designing an experiment to remember that no matter how powerful the statistics used, the inferences made from a sample are only as good as the data in that sample. . . . No amount of sophisticated statistical analysis will make good data out of bad data. There are many scientists who try to disguise badly constructed experiments by blinding their readers with a complex statistical analysis.

—O'Mahony (1986, pp. 6, 8)

This chapter provides a quick introduction to statistics used for sensory evaluation data including measures of central tendency and dispersion. The logic of statistical hypothesis testing is introduced. Simple tests on pairs of means (the *t*-tests) are described with worked examples. The meaning of a *p*-value is reviewed.

A.1 Introduction

The main body of this book has been concerned with using good sensory test methods that can generate quality data in well-designed and well-executed studies. Now we turn to summarize the applications of statistics to sensory data analysis. Although statistics are a necessary part of sensory research, the sensory scientist would do well to keep in mind O'Mahony's admonishment: statistical analysis, no matter how clever, cannot be used to save a poor experiment. The techniques of statistical analysis, do however, serve several useful purposes, mainly in the efficient summarization of data and in allowing the sensory scientist to make reasonable conclusions from the information gained in an experiment. One of the most important conclusions is to help rule out the effects of chance variation in producing our results. "Most people, including scientists, are more likely to be convinced by phenomena that cannot readily be explained by a chance hypothesis" (Carver, 1978, p. 387).

Statistics function in three important ways in the analysis and interpretation of sensory data. The first is the simple description of results. Data must be summarized in terms of an estimate of the most likely values to represent the raw numbers. For example, we can describe the data in terms of averages and standard

H.T. Lawless, H. Heymann, *Sensory Evaluation of Food*, Food Science Text Series,
DOI 10.1007/978-1-4419-6488-5, © Springer Science+Business Media, LLC 2010

deviations (a measure of the spread in the data). This is the descriptive function of statistics. The second goal is to provide evidence that our experimental treatment, such as an ingredient or processing variable, actually had an effect on the sensory properties of the product, and that any differences we observe between treatments were not simply due to chance variation. This is the inferential function of statistics and provides a kind of confidence or support for our conclusions about products and variables we are testing. The third goal is to estimate the degree of association between our experimental variables (called independent variables) and the attributes measured as our data (called dependent variables). This is the measurement function of statistics and can be a valuable addition to the normal sensory testing process that is sometimes overlooked. Statistics such as the correlation coefficient and chi-square can be used to estimate the strength of relationship between our variables, the size of experimental effects, and the equations or models we generate from the data.

These statistical appendices are prepared as a general guide to statistics as they are applied in sensory evaluation. Statistics form an important part of the equipment of the sensory scientist. Since most evaluation procedures are conducted along the lines of scientific inquiry, there is error in measurement and a need to separate those outcomes that may have arisen from chance variation from those results that are due to experimental variables (ingredients, processes, packaging, shelf life). In addition, since the sensory scientist uses human beings as measuring instruments, there is increased variability compared to other analytical procedures such as physical or chemical measurements done with instruments. This makes the conduct of sensory testing especially challenging and makes the use of statistical methods a necessity.

The statistical sections are divided into separate topics so that readers who are familiar with some areas of statistical analysis can skip to sections of special interest. Students who desire further explanation or additional worked examples may wish to refer to O'Mahony (1986), *Sensory Evaluation of Foods, Statistical Methods and Procedures*. The books by Gacula et al. (2009), *Statistical Methods in Food and Consumer Research*, and Piggott (1986), *Statistical Procedures in Food Research*, contain information on more complex designs and advanced topics. This appendix is not meant to supplant courses in statistics, which are recommended for every sensory professional.

It is very prudent for sensory scientists to maintain an open dialogue with statistical consultants or other statistical experts who can provide advice and support for sensory research. This advice should be sought early on and continuously throughout the experimental process, analysis, and interpretation of results. R. A. Fisher is reported to have said, "To call the statistician after the experiment is done may be no more than asking him to perform a postmortem examination: he may be able to tell you what the experiment died of" (Fisher, Indian Statistical Congress, 1938). To be fully effective, the sensory professional should use statistical consultants early in the experimental design phase and not as magicians to rescue an experiment gone wrong. Keep in mind that the "best" experimental design for a problem may not be workable from a practical point of view. Human testing can necessarily involve fatigue, adaptation and loss of concentration, difficulties in maintaining attention, and loss of motivation at some point. The negotiation between the sensory scientist and the statistician can yield the best practical result.

A.2 Basic Statistical Concepts

Why are statistics so important in sensory evaluation? The primary reason is that there is variation or error in measurement. In sensory evaluation, different participants in a sensory test simply give different data. We need to find the consistent patterns that are not due to chance variation. It is against this background of uncontrolled variation that we wish to tell whether the experimental variable of interest had a reliable effect on the perceptions of our panelists. Unfortunately, the variance in our measurements introduces an element of risk in making decisions. Statistics are never completely foolproof or airtight. Decisions even under the best conditions of experimentation always run the risk of being wrong. However, statistical methods help us to minimize, control, and estimate that risk.

The methods of statistics give us rules to estimate and minimize the risk in decisions when we generalize from a sample (an experiment or test) to the greater population of interest. They are based on consideration of three factors: the actual measured values, the error or variation around the values, and the number of observations that are made (sometimes referred to as "sample size," not to be confused with the size of a food sample that is served). The interplay of these three factors forms the basis for statistical calculations in all of the major statistical tests used with sensory data, including *t*-tests on means, analysis of variance, and *F*-ratios and comparisons of proportions or frequency counts. In the case of *t*-test on means, the factors are (1) the actual difference between the means, (2) the standard deviation or error inherent in the experimental measurement, and (3) the sample size or number of observations we made.

How can we characterize variability in our data? Variation in the data produces a distribution of values across the available measurement points. These distributions can be represented graphically as histograms. A histogram is a type of graph, a picture of frequency counts of how many times each measurement point is represented in our data set. We often graph these data in a bar graph, the most common kind of histogram. Examples of distributions include sensory thresholds among a population, different ratings by subjects on a sensory panel (as in Fig. A.1), or judgments of product liking on a 9-point scale across a sample of consumers. In doing our experiment, we assume that our measurements are more or less representative of the entire population of people or those who might try our product. The experimental measurements are referred to as a sample and the underlying or parent group as a population. The distribution of our data bears some resemblance to the parent population, but it may differ due to the variability in the experiment and error in our measuring.

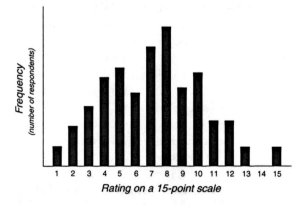

Fig. A.1 A histogram showing a sample distribution of data from a panel's ratings of the perceived intensity of a sensory characteristic on a 15-point category scale.

A.2.1 Data Description

How do we describe our measurements? Consider a sample distribution, as pictured in Fig. A.1. These measurements can be characterized and summarized in a few parameters. There are two important aspects we use for the summary. First, what is the best single estimate of our measurement? Second, what was the variation around this value?

Description of the best or most likely single value involves measures of central tendency. Three are commonly used: the mean is commonly called an average and is the sum of all data values divided by the number of observations. This is a good representation of the central value of data for distributions that are symmetric, i.e., not too heavily weighted in high or low values, but evenly dispersed. Another common measure is the median or 50th percentile, the middle value when the data are ranked. The median is a good representation of the central value even when the data are not symmetrically distributed. When there are some extreme values at the high end, for example, the mean will be unduly influenced by the higher values (they pull the average up). The median is simply the middle value after the measurements are rank ordered from lowest to highest or the average of the two middle values when there is an even number of data points. For some types of categorical data, we need to know the mode. The mode is the most frequent value. This is appropriate when our data are only separated into name-based categories. For example, we could ask for the modal response to the question, when is the product consumed (breakfast, lunch, dinner, or snack)? So a list of items or responses with no particular ordering to the categories can be summarized by the most frequent response.

The second way to describe our data is to look at the variability or spread in our observations. This is usually achieved with a measure called the standard deviation. This specifies the degree to which our measures are dispersed about the central value.

The standard deviation of such an experimental sample of data (*S*) has the following form:

$$S = \sqrt{\frac{\sum_{i=1}^{N} (X_i - M)^2}{N - 1}} \qquad \text{(A.1)}$$

where M = mean of X scores = $(\Sigma X)/N$.

The standard deviation is more easily calculated as

$$S = \sqrt{\frac{\sum_{i=1}^{N} X_i^2 - \left((\Sigma X)^2 / N\right)}{N - 1}} \qquad (A.2)$$

Since the experiment or sample is only a small representation of a much larger population, there is a tendency to underestimate the true degree of variation that is present. To counteract this potential bias, the value of N–1 is used in the denominator, forming what is called an "unbiased estimate" of the standard deviation. In some statistical procedures, we do not use the standard deviation, but its squared value. This is called the sample variance or S^2 in this notation.

Another useful measure of variability in the data is the coefficient of variation. This weights the standard deviation for the size of the mean and can be a good way to compare the variation from different methods, scales, experiments, or situations. In essence the measure becomes dimensionless or a pure measure of the percent of variation in our data. The coefficient of variation (CV) is expressed as a percent in the following formula:

$$CV(\%) = 100 \frac{S}{M} \qquad (A.3)$$

where S is the sample standard deviation and M is the mean value. For some scaling methods such as magnitude estimation, variability tends to increase with increasing mean values, so the standard deviation by itself may not say much about the amount of error in the measurement. The error changes with the level of mean. The coefficient of variation, on the other hand, is a relative measure of error that takes into account the intensity value along the scale of measurement.

The example below shows the calculations of the mean, median, mode, standard deviation, and coefficient of variation for data shown in Table A.1.

$N = 41$

Mean of the scores $= (\Sigma X)/N = (2 + 3 + 3 + 4 + \ldots + 11 + 12 + 13) / 41 = 7.049$

Median = middle score = 7

Mode = most frequent score = 6

Standard deviation = S

Table A.1 First data set, rank ordered

2	5	7	9
3	5	7	9
3	6	7	9
4	6	8	9
4	6	8	10
4	6	8	10
4	6	8	10
5	6	8	11
5	6	8	11
5	7	9	12
			13

$$S = \sqrt{\frac{\sum_{i=1}^{N} X_i^2 - \left((\Sigma X)^2 / N\right)}{N - 1}}$$

$$= \sqrt{\frac{2,303 - (83,521)/41}{40}} = 2.578$$

CV (%) = 100 (S/mean) = 100 (2.578/ 7.049) = 36.6%.

A.2.2 Population Statistics

In making decisions about our data, we like to infer from our experiment to what might happen in the population as a whole. That is, we would like our results from a subsample of the population to apply equally well when projected to other people or other products. By population, we do not necessarily mean the population of the nation or the world. We use this term to mean the group of people (or sometimes products) from which we drew our experimental panel (or samples) and the group to which we would like to apply our conclusions from the study. The laws of statistics tell us how well we can generalize from our experiment (or sensory test) to the rest of the population of interest. The population means and standard deviations are usually denoted by Greek letters, as opposed to standard letters for sample-based statistics.

Many things we measure about a group of people will be normally distributed. That means the values form a bell-shaped curve described by an equation usually attributed to Gauss. The bell curve is symmetric around a mean value—values are more likely to be close to the mean than far from it. The curve is described by its parameters of its mean and its standard deviation as shown in Fig. A.2. The standard deviation

Fig. A.2 The normal distribution curve is described by its parameters of its mean and its standard deviation. Areas under the curve mark off discrete and known percentages of observations.

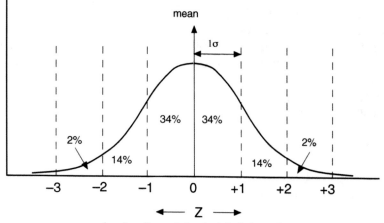

Important properties of the normal distribution curve:

1) areas (under the curve) correspond to proportions of the population.

2) each standard deviation subsumes a known proportion

3) Since proportions are related to probabilities, we know how likely or unlikely certain values are going to be. Extreme scores (away from the mean) are rare or improbable.

mean

1σ

34%   34%

2%   2%

14%   14%

−3   −2   −1   0   +1   +2   +3

← Z →

(marks off equal standard deviation units)

of a population, σ, is similar to our formula for the sample standard deviation as is given by

$$\sigma = \sqrt{\frac{\sum_{i=1}^{N}(X_i - \mu)^2}{N}} \qquad (A.4)$$

where

X = each score (value for each person, product); μ = population mean; N = number of items in population.

How does the standard deviation relate to the normal distribution? This is an important relationship, which forms the basis of statistical risk estimation and inferences from samples to populations. Because we know the exact shape of the normal distribution (given by its equation), standard deviations describe known percentages of observations at certain degrees of difference from the mean. In other words, proportions of observations correspond to areas under the curve. Furthermore, any value, X, can be described in terms of a Z-score, which states how far the value is from the mean in standard deviation units. Thus,

$$Z = \frac{X - \mu}{\sigma} \qquad (A.5)$$

Z-scores represent differences from the mean value but they are also related to areas under the normal curve. When we define the standard deviation as one unit, the Z-score is also related to the area under the curve to the left of right of its value, expressed as a percentage of the total area. In this case the z-score becomes a useful value to know when we want to see how likely a certain observation would be and when we make certain assumptions about what the population may be like. We can tell what percent of observations will lie a given distance (Z-score) from the mean. Because the frequency distribution actually tells us how many times we expect different values to occur, we can convert this z-score to a probability value (sometimes called a p-value), representing the area under the curve to the left or right of the Z-value. In statistical testing, where we look for the rarity of calculated event, we are usually examining the "tail" of the distribution or the smaller area that represents the probability of values more extreme than the z-score. This probability value represents the area under the curve outside our given z-score and is the chance (expected frequency) with which we would see a score of that magnitude or one that is even greater. Tables converting z-values to p-values are found in all statistics texts (see Table A).

A.3 Hypothesis Testing and Statistical Inference

A.3.1 The Confidence Interval

Statistical inference has to do with how we draw conclusions about what populations are like based on samples of data from experiments. This is the logic that is used to determine whether our experimental variables had a real effect or whether our results were likely to be due to chance or unexplained random variation. Before we move on to this notion of statistical decision making, a simpler example of inferences about populations, namely confidence intervals, will be illustrated.

One example of inference is in the estimation of where the true population values are likely to occur based on our sample. In other words, we can examine the certainty with which our sample estimates will fall inside a range of values on the scale of measurement. For example, we might want to know the following information: Given the sample mean and standard deviation, within what interval is the true or population value likely to occur? For small samples, we use the t-statistic to help us (Student, 1908). The t-statistic is like Z, but it describes the distribution of small experiments better than the z-statistic that governs large populations. Since most experiments are much smaller than populations, and sometimes are a very small sample indeed, the t-statistic is useful for much sensory evaluation work. Often we use the 95% confidence interval to describe where the value of the mean is expected to fall 95% of the time, given the information in our sample or experiment.

For a mean value M of N observations, the 95% confidence interval is given by

$$M \pm t \left(S/\sqrt{N} \right) \qquad (A.6)$$

where t is the t-value corresponding to N–1 degrees of freedom (explained below), that includes 2.5% of expected variation in the upper tail outside this value and 2.5% in the lower tail (hence a two-tailed value, also explained below). Suppose we obtain a mean value of 5.0 on a 9-point scale, with a standard deviation of 1.0 in our sample, and there are 15 observations. The t-value for this experiment is based on 14 or $n - 1$ degrees of freedom and is shown in Table B to be

2.145. So our best guess is that the true mean lies in the range of $5 \pm 2.145(1/\sqrt{15})$ or between 4.45 and 5.55. This could be useful, for example, if we wanted to insure that our product had a mean score of at least 4.0 on this scale. We would be fairly confident, given the sample values from our experiment that it would in fact exceed this value.

For continuous and normally distributed data, we can similarly estimate a 95% confidence interval on the median (Smith, 1988), given by

$$\text{Med} \pm 1.253t \left(S/\sqrt{N} \right) \qquad (A.7)$$

For larger samples, say $N > 50$, we can replace the t-value with its Z approximation, using $Z = 1.96$ in these formulas for the 95% confidence interval. As the number of observations increases, the t-distribution becomes closer to the normal distribution.

A.3.2 Hypothesis Testing

How can we tell if our experimental treatment had an effect? First, we need to calculate means and standard deviations. From these values we do further calculations to come up with values called test statistics. These statistics, like the Z-score mentioned above, have known distributions, so we can tell how likely or unlikely the observations will be when chance variation alone is operating. When chance variation alone seems very unlikely (usually one chance in 20 or less), then we reject this notion and conclude that our observations must be due to our actual experimental treatment. This is the logic of statistical hypothesis testing. It is that simple.

Often we need a test to compare means. A useful statistic for small experiments is called Student's t-statistic. Student was the pseudonym of the original publisher of this statistic, a man named Gosset who worked for the Guinness Brewery and did not want other breweries to know that Guinness was using statistical methods (O'Mahony, 1986). By small experiments, we mean experiments with numbers of observations per variable in the range of about 50 or less. Conceptually, the t-statistic is the difference between the means divided by an estimate of the error or uncertainty around those means, called the standard error of the means.

Imagine that we did our experiment many times and each time calculated a mean value. These means themselves, then, could be plotted in a histogram and would have a distribution of values. The standard error of the mean is like the standard deviation of this sampling distribution of means. If you had lots of time and money, you could repeat the experiment over and over and estimate the population values from looking at the distribution of sample mean scores. However, we do not usually do such a series of experiments, so we need a way to estimate this error. Fortunately, the error in our single experiment gives us a hint of how likely it is that our obtained mean is likely to reflect the population mean. That is, we can estimate that the limits of confidence are around the mean value we got. The laws of statistics tell us that the standard error of the mean is simply the sample standard deviation divided by the square root of the number of observations ("N"). This makes sense in that the more observations we make, the more likely it is that our obtained mean actually lies close to the true population mean.

In order to test whether the mean we see in our experiment is different from some other value, there are three things we need to know: the mean itself, the sample standard deviation, and the number of observations. An example of this form of the t-test is given below, but first we need to take a closer look at the logic of statistical testing.

The logical process of statistical inference is similar for the t-tests and all other statistical tests. The only difference is that the t-statistic is computed for testing differences between two means, while other statistics are used to test for differences among other values, like proportions, standard deviations, or variances. In the t-test, we first assume that there is no difference between population means. Another way to think about this is that it implies that the experimental means were drawn from the same parent population. This is called the null hypothesis. Next, we look at our t-value calculated in the experiment and ask how likely this value would be, given our assumption of no difference (i.e., a true null hypothesis). Because we know the shape of the t-distribution, just like a Z-score, we can tell how far out in the tail our calculated t-statistic lies. From the area under the curve out in that tail, we can tell what percent of the time we could expect to see this value. If the t-value we calculate is very high and positive or very low and negative, it is unlikely—a rare event given our assumption. If this rarity passes some arbitrary cutoff

point, usually one chance in 20 (5%) or less, we conclude that our initial assumption was probably wrong. Then we make a conclusion that the population means are in fact different or that the sample means were drawn from different parent populations. In practical terms, this usually implies that our treatment variable (ingredients, processing, packaging, shelf life) did produce a different sensory effect from some comparison level or from our control product. We conclude that the difference was not likely to happen from chance variation alone. This is the logic of null hypothesis testing. It is designed to keep us from making errors of concluding that the experiment had an effect when there really was only a difference due to chance. Furthermore, it limits our likelihood of making this mistake to a maximum value of one chance in 20 in the long run (when certain conditions are met, see postscript at the end of this chapter).

A.3.3 A Worked Example

Here is a worked example of a simple t-test. We do an experiment with the following scale, rating a new ingredient formulation against a control for overall sweetness level:

□	□	□	□	□	□	□

much less sweet   about the same   much more sweet

We convert their box ratings to scores 1 (for the leftmost box) through 7 (for the rightmost). The data from ten panelists are shown in Table A.2.

We now set up our null hypothesis and an alternative hypothesis different from the null. A common notation is to let the symbol H_0 stand for the null hypothesis and

Table A.2 Data for t-test example

Panelist	Rating
1	5
2	5
3	6
4	4
5	3
6	7
7	5
8	5
9	6
10	4

H_a stand for the alternative. Several different alternatives are possible, so it takes some careful thought as to which one to choose. This is discussed further below. The null hypothesis in this case is stated as an equation concerning the population value, not our sample, as follows:

H_0: $\mu = 4.0$. This is the null hypothesis.
H_a: $\mu \neq 4.0$ This is the alternative hypothesis.

Note that the Greek letter "mu" is used since these are statements about population means, not sample means from our data. Also note that the alternative hypothesis is non- directional, since the population mean could be higher or lower than our expected value of 4.0. So the actual t-value after our calculations might be positive or negative. This is called a two-tailed test. If we were only interested in the alternative hypothesis (H_a) with a "greater than" or "less than" prediction, the test would be one tailed (and out critical t-value would change) as we would only examine one end of the t-distribution when checking for the probability and significance of the result.

For our test against a mean or fixed value, the t-test has the following form:

$$t = \frac{M - \mu}{S/\sqrt{N}} \qquad (A.8)$$

where M is the sample mean, S is the standard deviation, N is the number of observations (judges or panelists, usually), and μ is the fixed value or population mean.

Here are the calculations from the data set above:

Mean $= \Sigma X/N = 5.0$
$\Sigma X = 50$
$\Sigma X^2 = 262$
$(\Sigma X)^2 = 2500$

$$S = \frac{\sqrt{(262) - (2500)/10}}{9} = 1.155$$

$$t = \frac{5.0 - 4.0}{1.155/\sqrt{10}} = \frac{1}{0.365} = 2.740$$

So our obtained t-value for this experiment is 2.740. Next we need to know if this value is larger than what

we would expect by chance less than 5% of the time. Statistical tables for the t-distribution tell us that for a sample size of 10 people (so degrees of freedom $= 9$), we expect a t-value of ± 2.262 only 5% of the time. The two-tailed test looks at both high and low tails and adds them together since the test is non-directional, with t high or low. So this critical value of $+2.262$ cuts off 2.5% of the total area under the t-distribution in the upper half and -2.262 cuts off 2.5% in the lower half. Any values higher than 2.262 or lower than -2.262 would be expected less than 5% of the time. In statistical talk, we say that the probability of our obtained result then is less than 0.05, since $2.738 > 2.262$. In other words, we obtained a t-value from our data that is even more extreme than the cutoff value of 2.262.

So far all of this is some simple math, and then a cross-referencing of the obtained t-value to what is predicted from the tabled t-values under the null hypothesis. The next step is the inferential leap of statistical decision making. Since the obtained t-value was bigger in magnitude than the critical t-value, H_0 is rejected and the alternative hypothesis is accepted. In other words, our population mean is likely to be different than the middle of our scale value of 4.0. We do not actually know how likely this is, but we know that the experiment would produce the sort of result we see only about 5% of the time when the null is true. So we infer that it is probably false. Looking back at the data, this does not seem too unreasonable since seven out of ten panelists scored higher than the null hypothesis value of 4.0. When we reject the null hypothesis, we claim that there is a statistically significant result. The use of the term "significance" is unfortunate, for in simple everyday English it means "important." In statistical terms significance only implies that a decision has been made and does not tell us whether the result was important or not. The steps in this chain of reasoning, along with some decisions made early in the process about the alpha-level and power of the test, are shown in Fig. A.3.

A.3.4 A Few More Important Concepts

Before going ahead, there are some important concepts in this process of statistical testing that need further explanation. The first is degrees of freedom. When we look up our critical values for a statistic, the values are

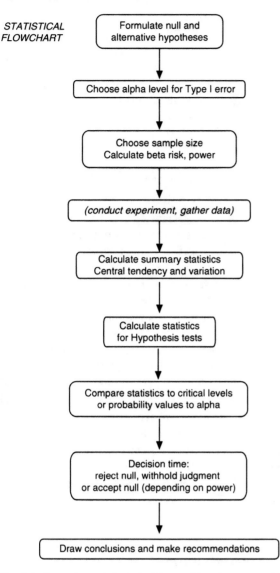

*STATISTICAL
FLOWCHART*

Formulate null and
alternative hypotheses

Choose alpha level for Type I error

Choose sample size
Calculate beta risk, power

(conduct experiment, gather data)

Calculate summary statistics
Central tendency and variation

Calculate statistics
for Hypothesis tests

Compare statistics to critical levels
or probability values to alpha

Decision time:
reject null, withhold judgment
or accept null (depending on power)

Draw conclusions and make recommendations

Fig. A.3 Steps in statistical decision making in an experiment. The items before the collection of the data concern the experimental design and statistical conventions to be used in the study. After the data are analyzed the inferential process begins, first with data description, then computation of the test statistic, and then comparison of the test statistic to the critical value for our predetermined alpha-level and the size of the experiment. If the computed test statistic is greater in magnitude than the critical value, we reject the null hypothesis in favor of the alternative hypothesis. If the computed test statistic has a value smaller in magnitude than the critical value, we can make two choices. We can reserve judgment if the sample size is small or we can accept the null hypothesis if we are sure that the power and sensitivity of the test are high. A test of good power is in part determined by having a substantial number of observations and test sensitivity is determined by having good experimental procedures and controls (see Appendix E).

frequently tabled not in terms of how many observations were in our sample, but how many degrees of freedom we have. Degrees of freedom have to do with how many parameters we are estimating from our data relative to the number of observations. In essence, this notion asks how much the resulting values would be free to move, given the constraints we have from estimating other statistics. For example, when we estimate a mean, we have freedom for that value to move or change until the last data point is collected. Another way to think about this is the following: If we knew all but one data point and already knew the mean, we would not need that last data point. It would be determined by all the other data points and the mean itself, so it has no freedom to change. We could calculate what it would have to be. In general, degrees of freedom are equal to the sample size, minus one for each of the parameters we are estimating. Most statistics are tabled by their degrees of freedom. If we wanted to compare the means from two groups of N_1 and N_2 observations, we would have to calculate some parameters like means for each group. So the total numbers of degrees of freedom are $N_1 - 1 + N_2 - 1$, or $N_1 + N_2 - 2$.

A second important consideration is whether our statistical test is a one- or a two-tailed test. Do we wish to test whether the mean is simply different from some value or whether it is larger or smaller than some value? If the question is simply "different from" then we need to examine the probability that our test statistic will fall into either the low or high tail of its distribution. As stated above in the example of the simple *t*-test, if the question is directional, e.g., "greater than" some value, then we examine only one tail. Most statistical tables have entries for one- and two-tailed tests. It is important, however, to think carefully about our underlying theoretical question. The choice of statistical alternative hypotheses is related to the research hypothesis. In some sensory tests, like paired preference, we do not have any way of predicting which way the preference will go, and so the statistical test is two-tailed. This is in contrast to some discrimination tests like the triangle procedure. In these tests we do not expect performance below chance unless there is something very wrong with the experiment. So the alternative hypothesis is that the true proportion correct is greater than chance. The alternative is looking in one direction and is therefore one-tailed.

A third important statistical concept to keep in mind is what type of distribution you are concerned with.

There are three different kinds of distributions we have discussed. First, there are overall population distributions. They tell us what the world would look like if we measured all possible values. This is usually not known, but we can make inferences about it from our experiments. Second, we have sample distributions derived from our actual data. What does our sample look like? The data distribution can be pictured in a graph such as a histogram. Third, there are distributions of test statistics. If the null hypothesis is true, how is the test statistic distributed over many experiments? How will the test statistic be affected by samples of different sizes? What values would be expected, what variance due to chance alone? It is against these expected values that we examine our calculated value and get some idea of its probability.

A.3.5 Decision Errors

Realizing that statistical decisions are based on probabilities, it is clear that some uncertainty is involved. Our test statistic may only happen 5% of the time under a true null hypothesis, but the null might still be true, even though we rejected it. So there is a chance that our decision was a mistake and that we made an error. It is also possible sometimes that we fail to reject the null, when a true difference exists. These two kinds of mistakes are called Type I and Type II errors. A Type I error is committed when we reject the null hypothesis when it is actually true. In terms of a t-test comparison of means, the Type I error implies that we concluded that two population means are different when they are in fact the same, i.e., our data were in fact sampled from the same parent population. In other words, our treatment did not have an effect, but we mistakenly concluded that it did. The process of statistical testing is valuable, though, because it protects us from committing this kind of error and going down blind alleys in terms of future research decisions, by limiting the proportion of times we could make these decisions. This upper limit on the risk of Type I error (over the long term) is called alpha-risk.

As shown in Table A.3, another kind of error occurs when we miss a difference that is real. This is called a Type II error and is formally defined as a failure to reject the null hypothesis when the alternative hypothesis is actually true. Failures to detect a difference in

Table A.3 Statistical errors in decision making

		Outcome of sensory evaluation	
		Difference reported	No difference reported
True situation	Products are different	Correct decision	Type II error Prob. is beta-risk
	Products are not different	Type I error Prob. is alpha-risk	Correct decision

a t-test or more generally to fail to observe that an experimental treatment had an effect can have important or even devastating business implications. Failing to note that a revised manufacturing process was in fact an improvement would lose the potential benefit if the revision were not adopted as a new standard procedure. Similarly, revised ingredients might be passed over when they in fact produce improvements in the product as perceived by consumers. Alternatively, bad ingredients might be accepted for use if the modified product's flaws are undetected. It is necessary to have a sensitive enough test to protect against this kind of error. The long-term risk or probability of making this kind of mistake is called beta-risk, and one minus the beta-risk is defined as the statistical power of the test. The protection against Type II error by statistical means and by experimental strategy is discussed in Appendix E.

A.4 Variations of the t-Test

There are three kinds of t-tests that are commonly used. One is a test of an experimental mean against a fixed value, like a population mean or a specific point on a scale like the middle of a just-right scale, as in the example above. The second test is when observations are paired, for example, when each panelist evaluates two products and the scores are associated since each pair comes from a single person. This is called the paired t-test or dependent t-test. The third type of t-test is performed when different groups of panelists evaluate the two products. This is called the independent groups t-test. The formulas for each test are similar, in that they take the general form of a difference between means divided by the standard error. However, the actual computations are a bit different. The section below gives examples of the three comparisons of means involving the t-statistic.

One type of t-test is the test against a population mean or another fixed value, as we saw above in our example and Eq. (A.8). The second kind of t-test is the test of paired observations also called the dependent t-test. This is a useful and powerful test design in which each panelist evaluates both products, allowing us to eliminate some of the inter-individual variation. To calculate this value of t, we first arrange the pairs of observations in two columns and subtract each one from the other member of the pair to create a difference score. The difference scores then become the numbers used in further calculations. The null hypothesis is that the mean of the difference scores is zero. We also need to calculate a standard deviation of these difference scores, and a standard error by dividing this standard deviation by the square root of N, the number of panelists

$$t = \frac{M_{\mathrm{diff}}}{S_{\mathrm{diff}}/\sqrt{N}} \qquad (A.9)$$

where M_{diff} is the mean of the difference scores and S_{diff} is the standard deviation of the difference scores. Here is an example of a t-test where each panelist tasted both products and we can perform a paired t-test. Products were rated on a 25-point scale for acceptance. Note that we compute a difference score (D) in this situation, as shown in Table A.4.

Table A.4 Data for paired t-test example

Panelist	Product A	Product B	Difference	(Difference)^2
1	20	22	2	4
2	18	19	1	1
3	19	17	−2	4
4	22	18	−4	16
5	17	21	4	16
6	20	23	3	9
7	19	19	0	0
8	16	20	4	16
9	21	22	1	1
10	19	20	1	1

Calculations:

sum of $D = 10$, mean of $D = 1$
sum of $D^2 = 68$
standard deviation of $D =$

$$S_{\mathrm{diff}} = \sqrt{\frac{\sum_{i=1}^{N} D_i^2 - ((\Sigma D)^2/N)}{N-1}}$$

$$= \sqrt{\frac{68 - (100/10)}{9}} = 2.539,$$

and t comes from

$$t = \frac{M_{\mathrm{diff}}}{S_{\mathrm{diff}}/\sqrt{N}} = \frac{1.0}{2.5390/\sqrt{10}} = 1.25$$

This value does not exceed the tabled value for the 5%, two-tailed limit on t (at 9 df), and so we conclude there is insufficient evidence for a difference. In other words, we do not reject the null hypothesis. The two samples were rather close, compared to the level of error among panelists.

The third type of t-test is conduced when there are different groups of people, often called an independent groups t-test. Sometimes the experimental constraints might dictate situations where we have two groups that taste only one product each. Then a different formula for the t-test applies. Now the data are no longer paired or related in any way and a different calculation is needed to estimate the standard error, since two groups were involved and they have to be combined somehow to get a common estimate of the standard deviations. We also have some different degrees of freedom, now given by the sum of the two group sizes minus 2 or $(N_{\mathrm{Group1}} + N_{\mathrm{Group2}} - 2)$. The t-value is determined by

$$t = \frac{M_1 - M_2}{SE_{\mathrm{pooled}}} \qquad (A.10)$$

where M_1 and M_2 are the means of the two groups and SE_{pooled} is the pooled standard error. For the independent t-test, the pooled error requires some work and gives an estimate of the error combining the error levels of the two groups. The pooled standard error for two groups, X and Y, is given by the following formula:

$$SE_{\mathrm{pooled}} = \sqrt{[1/N_1 + 1/N_2]\frac{\left[\Sigma x^2 - \left((\Sigma x)^2/N_1\right) + \Sigma y^2 - \left((\Sigma y)^2/N_2\right)\right]}{(N_1 + N_2 - 2)}}$$

(A.11)

Here is a worked example of an independent group's t-test. In this case, we have two panels, one from a manufacturing site and one from a research site, both evaluating the perceived pepper heat from an ingredient submitted for use in a highly spiced product.

The product managers have become concerned that the plant QC panel may not be very sensitive to pepper heat due to their dietary consumption or other factors, and that the use of ingredients is getting out of line with what research and development personnel feel is an appropriate level of pepper. So the sample is evaluated by both groups and an independent group's t-test is performed. Our null hypothesis is that there is no difference in the population means and our alternative hypothesis that the QC plant will have lower mean ratings in the long run (one-tailed situation). The data set is comprised of pepper heat ratings on a 15-point category scale as shown in Table A.5.

Table A.5 Data for independent group's t-test

Manufacturing QC panel (X)	R&D test panel (Y)
7	9
12	10
6	8
5	7
8	7
6	9
7	8
4	12
5	9
3	

First, some preliminary calculations:

$N_1 = 10$ $\Sigma x = 63$ Mean $= 6.30$ $\Sigma x^2 = 453$ $(\Sigma x)^2 = 3969$
$N_2 = 9$ $\Sigma y = 79$ Mean $= 8.78$ $\Sigma y^2 = 713$ $(\Sigma y)^2 = 6291$

Now we have all the information we need to calculate the value of

$$SE_{pooled} = \sqrt{(1/10 + 1/9)\frac{\left[453 - \frac{3969}{10} + 713 - \frac{6241}{9}\right]}{(10 + 9 - 2)}} = 0.97$$

$t = [(6.30 - 8.78)]/0.97 = -2.556$.

Degrees of freedom are 17 ($= 10 + 9 - 2$). The critical t-value for a one-tailed test at 17 df is 1.740, so this is a statistically significant result. Our QC panel does seem to be giving lower scores for pepper heat than the R&D panel.

Note that the variability is also a little higher in the QC panel. Our test formula assumes that the variance is about equal. For highly unequal variability (1 SD more than three times that of the other) some adjustments must be made. The problem of unequal variance becomes more serious when the two groups are also very different in size. The t-distribution becomes a poor estimate of what to expect under a true null, so the alpha-level is no longer adequately protected. One approach is to adjust the degrees of freedom and formulas for this are given in advanced statistics books (e.g., Snedecor and Cochran, 1989). The non-pooled estimates of the t-value are provided by some statistics packages and it is usually prudent to examine these adjusted t-values if unequal group size and unequal variances happen to be the situation with your data.

A.4.1 The Sensitivity of the Dependent t-Test for Sensory Data

In sensory testing, it is often valuable to have each panelist try all of the products in our test. For simple paired tests of two products, this enables the use of the dependent t-test. This is especially valuable when the question is simply whether a modified process or ingredient has changed the sensory attributes of a product. The dependent t-test is preferable to the separate-groups approach, where different people try each product. The reason is apparent from the calculations. In the dependent t-test, the statistic is calculated on a difference score. This means that the differences among panelists in overall sensory sensitivity or even in their idiosyncratic scale usage are removed from the situation. It is common to observe that some panelists have a "favorite" part of the scale and may restrict their responses to one section of the allowable responses. However, with the dependent t-test, as long as panelists rank order the products in the same way, there will be a statistically significant result. This is one way to partition the variation due to subject differences from the variation due to other sources of error. In general, partitioning of error adds power to statistical tests, as shown in the section on repeated measures (or complete block) ANOVA (see Appendix C). Of course, there are some potential problems in having people evaluate both products, like sequential order effects and possible fatigue and carry-over effects. However, the advantage gained in the sensitivity of the test usually far outweighs the liabilities of repeated testing.

A.5 Summary: Statistical Hypothesis Testing

Statistical testing is designed to prevent us from concluding that a treatment had an effect when none was really present and our differences were merely due to chance or the experimental error variation. Since the test statistics like Z and t have known distributions, we can tell whether our results would be extreme, i.e., in the tails of these distributions a certain percent of the time when only chance variation was operating. This allows us to reject the notion of chance variation in favor of concluding that there was an actual effect. The steps in statistical testing are summarized in the flowchart shown in Fig. A.3. Sample size or the number of observations to make in an experiment is one important decision. As noted above, this helps determine the power and sensitivity of the test, as the standard errors decrease as a function of the square root of N. Also note that this square root function means that the advantage of increasing sample size becomes less as N gets larger. In other words there is a law of diminishing returns. At some point the cost considerations in doing a large test will outweigh the advantage in reducing uncertainty and lowering risk. An accomplished sensory professional will have a feeling for how well the sensitivity of the test balances against the informational power and uncertainty and risks involved and about how many people are enough to insure a sensitive test. These issues are discussed further in the section on beta-risk and statistical power.

Note that statistical hypothesis testing by itself is a somewhat impoverished manner of performing scientific research. Rather than establishing theorems, laws, or general mathematical relationships about how nature works, we are simply making a binary yes/no decision, either that a given experimental treatment had an effect or that it did not. Statistical tests can be thought of as a starting point or a kind of necessary hurdle that is a part of experimentation in order to help rule out the effects of chance. However, it is not the end of the story, only the beginning. In addition to statistical significance, the sensory scientist must always describe the effects. It is easy for students to forget this point and report significance *but fail to describe what happened*.

A.6 Postscript: What p-Values Signify and What They Do Not

No single statistical concept is probably more often misunderstood and so often abused than the obtained p-value that we find for a statistic after conducting an analysis. It is easy to forget that this p-value is based on a hypothetical curve for the test statistic, like the t-distribution, that is calculated under the assumption that the null hypothesis is true. So the obtained p-value is taken from the very situation that we are trying to reject or eliminate as a possibility. Once this fact is realized, it is easier to put the p-value into proper perspective and give it the due respect it deserves, but no more.

What does the p-value mean? Let us reiterate. It is the probability of observing a value of the test statistic (t, z, r, chi-square, or F-ratio) that is as large or larger than the one we obtain in our experimental analysis, when the null hypothesis is true. That much, no more and no less. In other words, assuming a true null, how likely or unlikely would the obtained value of the t-test be? When it becomes somewhat unlikely, say it is expected less than 5% of the time, we reject this null in favor of the alternative hypothesis and conclude that there is statistical significance. Thus, we have gained some assurance of a relationship between our experimental treatments and the sensory variables we are measuring. Or have we? Here are some common misinterpretations of the p-value:

(1) The p-value (or more specifically, $1-p$) represents the odds-against chance. Absolutely false (Carver, 1978). This puts the cart before the horse. The chance of observing the t-statistic under a true null is not the same as the chance of the null (or alternative) being true given the observations. A $p < 0.05$ does not mean there is only a 5% chance of the null being true. In mathematical logic, the probability of A given B is not necessarily the same as B given A. If I find a dead man on my front lawn, the chance he was shot in the head is quite slim (less than 5%) at least in my neighborhood, but if I find a man shot through the head, he is more than likely dead (95% or more).

(2) A related way of misinterpreting the p-value is to say that the p-value represents the chance of

making a Type I error. This is also not strictly true although it is widely assumed to be true (Pollard and Richardson, 1987). Indeed, that is what our alpha-cutoff is supposed to limit, in the long run. But the actual value of alpha in restricting our liability depends also upon the incidence of true differences versus no differences in our long-term testing program. Only when the chance of true difference is about 50% is the alpha-value an accurate reflection of our liability in rejecting a true null. The incidence is usually not known, but can be estimated. In some cases like quality control or shelf-life testing, there are a lot more "no difference" situations, and then alpha wildly underestimates our chances of being correct, once the null is rejected. The following table shows how this can occur. In this example, we have a 10% incidence of a true difference, alpha is set at 0.05 and the beta-risk (chance of missing a difference) is 10%. For ease of computation, 1000 tests are conducted and the results are shown in Table A.6.

Table A.6 Incidence diagram

(True state)	Incidence	Difference found	Difference not found
Difference exists	100	90	10 (at $\beta = 10\%$)
Difference does not exist	900	45 (at $\alpha = 5\%$)	855

The chance of being correct, having decided there is a significant difference, is 90/(90+45) or 2/3. There is a 1/3 chance (45/135) of being wrong once you have rejected the null (not 5%!), even though we have done all our statistics correctly and are sure we are running a good testing program! The problem in this scenario is the low probability before hand of actually being sent something worth testing. The notion of estimating the chance of being right or wrong given a certain outcome is covered in the branch of statistics known as Bayesian statistics, after Bayes theorem which allows the kind of calculation illustrated in Table A.6 (see Berger and Berry, 1988).

A related mistake is made when we use the words "confident" to describe our level of significance ($1 - p$ or $1 - \alpha$). For example, a well-known introductory statistics text gives the following incorrect information: "The probability of rejecting H_0 erroneously (committing a Type I error) is

known, it is equal to alpha. ... Thus you may be 95% *confident* that your decision to reject H_0 is correct" (Welkowitz et al., 1982, p. 163) [italics inserted]. This is absolutely untrue, as the above example illustrates. With a low true incidence of actual differences, the chance of being wrong once H_0 is rejected may be very high indeed.

(3) One minus the *p*-value gives us the reliability of our data, the faith we should have in the alternative hypothesis, or is an index of the degree of support for our research hypothesis in general. All of these are false (Carver, 1978). Reliability, certainly, is important (getting the same result upon repeated testing) but is not estimated simply as $1-p$. A replicated experiment has much greater scientific value than a low *p*-value, especially if the replication comes from another laboratory and another test panel.

Interpreting *p*-values as evidence for the alternative hypothesis is just as wrong as interpreting them as accurate measures of evidence against the null. Once again, it depends upon the incidence or prior probability. There is a misplaced feeling of elation that students and even some more mature researchers seem to get when we obtain a low *p*-value, as if this was an indication of how good their experimental hypothesis was. It is surprising then that journal editors continue to allow the irrelevant convention of adding extra stars, asterisks, or other symbols to indicate low *p*-values (**0.01, ***0.001, etc.) *beyond the pre-set alpha-level* in experimental reports. The information given by these extra stars is minimal. They only tell you how likely the result is under a true null, which you are deciding is false anyway.

> Overzealous teachers and careless commentators have given the world the impression that our standard statistical measures are inevitable, necessary, optimal and mathematically certain. In truth, statistics is a branch of rhetoric and the use of any particular statistic ... is nothing more than an exhortation to your fellow humans to see meaning in data as you do. (Raskin, 1988, p. 432).

A.7 Statistical Glossary

Alpha-risk. The upper acceptable limit on committing Type I errors (rejecting a true null hypothesis) set by the experimenter before the study, often at 5% or less.

Beta-risk. The upper acceptable limit on committing Type II errors (accepting a false null hypothesis). See Appendix VI.

Degrees of freedom. A value for the number of observations that are unconstrained or free to vary once our statistical observations are calculated from a sample data set. In most cases the degrees of freedom are given by the number of observations in the data minus one.

Dependent variable. The variable that is free to move in a study, what is measured (such as ratings, numbers of correct judgments, preference choices) to form the data set.

Distribution. A collection of values describing a data set, a population, or a test statistic. The distribution plots the values (usually on the horizontal axis) against their frequency or probability of occurrence (on the vertical axis).

Independent variable. The experimental variable or treatment of interest that is manipulated by the experimenter. A set of classes, conditions, or groups that are the subject of study.

Mean. A measure of central tendency. The arithmetic mean or average is the sum of all the observed values divided by the number of observations. The geometric mean is the Nth root of the product of N observations.

Null hypothesis. An assumption about underlying population values. In simple difference testing for scaled data (e.g., where the t-test is used) the null assumes that the population mean values for two treatments are equal. In simple difference testing on proportions (e.g., where the data represent a count of correct judgments, as in the triangle test) the null hypothesis is that the population proportion correct equals the chance probability. This is often misphrased as "there is no difference" (a conclusion from the experiment, not a null hypothesis).

One- and two-tailed tests. Describes the consideration of only one or two ends of a statistic's distribution in determining the obtained p-value. In a one-tailed test, the alternative hypothesis is directional (e.g., the population mean of the test sample is greater than the control sample) while in a two-tailed test, the alternative hypothesis does not state a direction (e.g., the population mean of the test sample is not equal to the population mean of the control sample).

Parameter. A characteristic that is measured about something, such as the mean of a distribution.

P-value. The probability of observing a test statistics as large or larger than the one calculated from an experiment, when the null hypothesis is true. Used as the basis for rejecting the null hypothesis when compared to the pre-set alpha-level. Often mistakenly assumed to be the probability of making an error when the null hypothesis is rejected.

Sample size. The number of observations in our data, usually represented by the letter "N."

Standard deviation. A measure of variability in a data sample or in a population.

Statistic. A value calculated from the data, with a known distribution, based on certain assumptions.

Treatment. A word often used to describe two different levels of an experimental variable. In other words, what has been changed about a product and is the subject of the test. See independent variable.

Type I error. Rejecting the null hypothesis when it is true. In simple difference testing, the treatments are thought to be different when in fact they are the same.

Type II error. Accepting the null hypothesis when it is false. In simple difference testing, the treatments are thought to be equal when in fact the population values for those treatments are different.

References

Berger, J. O. and Berry, D. A. 1988. Statistical analysis and the illusion of objectivity. American Scientist, 76, 159–165.

Carver, R. P. 1978. The case against statistical significance testing. Harvard Educational Review, 48, 378–399.

Gacula, M., Singh, J., Bi, J. and Altan, S. 2009. Statistical Methods in Food and Consumer Research, Second Edition. Elsevier/Academic, Amsterdam.

O'Mahony, M. 1986. Sensory Evaluation of Food. Statistical Methods and Procedures. Marcel Dekker, New York.

Piggott, J. R. 1986. Statistical Procedures in Food Research. Elsevier Applied Science, London.

Pollard, P. and Richardson, J. T. E. 1987. On the probability of making Type I errors. Psychological Bulletin, 102, 159–163.

Raskin, J. 1988. Letter to the editor. American Scientist, 76, 432.

Smith, G. L. 1988. Statistical analysis of sensory data. In: J. R. Piggott (ed.), Sensory Analysis of Foods. Elsevier, London.

Snedecor, G. W. and Cochran, W. G. 1989. Statistical Methods, Eighth Edition. Iowa State University, Ames, IA.

Student. 1908. The probable error of a mean. Biometrika, 6, 1–25.

Welkowitz, J., Ewen, R. B. and Cohen, J. 1982. Introductory Statistics for the Behavioral Sciences. Academic, New York.

Appendix B

Nonparametric and Binomial-Based Statistical Methods

Contents

Although statistical tests provide the right tools for basic psychophysical research, they are not ideally suited for some of the tasks encountered in sensory analysis.

—M. O'Mahony (1986, p. 401)

Frequently, sensory evaluation data do not consist of measurements on continuous variables, but rather are frequency counts or proportions. The branch of statistics that deals with proportions and ranked data is called nonparametric statistics. This chapter illustrates statistics used on proportions and ranks, with worked examples.

B.1 Introduction to Nonparametric Tests

The t-test and other "parametric" statistics work well for situations in which the data are continuous as with some rating scales. In other situations, however, we categorize performance into right and wrong answers or we count the numbers who make a choice of one product over another. Common examples of this kind of testing include the triangle test and the paired preference tests. In these situations, we want to use a kind of distribution for statistical testing that is based on discrete, categorical data. One example is the binomial distribution and is described in this section. The binomial distribution is useful for tests based on proportions, where we have counted people in different categories. The binomial distribution is a special case where there are only two outcomes (e.g., right and wrong answers in a triangle test). Sometimes we may have more than two alternatives for classifying responses, in which case multinomial distribution statistics apply. A commonly used statistic for comparing frequencies when there are two or more response categories is the chi-square statistic. For example, we might want to know if the meals during which a food product is consumed (say, breakfast, lunch, dinner, snacks) differed among teenagers and adults. If we asked consumers about the situation in which they most commonly consumed the product, the data would consist of counts of the frequencies for each group of consumers. The chi-square statistic could then be used to compare these two frequency distributions. It will also indicate whether there is any association between the response categories (meals) and the age group or whether these two variables are independent.

Some response alternatives have more than categorical or nominal properties and represent rankings of responses. For example, we might ask for ranked preference of three or more variations of flavors for a new product. Consumers would rank them from most appealing to least appealing, and we might want to know whether there is any consistent trend, or whether all flavors are about equally preferred across the group. For rank order data, there are a number of statistical techniques within the nonparametric toolbox.

Since it has been argued that many sensory measurements, even rating scales, do not have interval-level properties (see Chapter 7), it often makes sense to apply a nonparametric test, especially those based on ranks, if the researcher has any doubts about the level of measurement inherent in the scaling data. The nonparametric tests can also be used as a check on conclusions from the traditional tests. Since the non-parametric tests involve fewer assumptions than their parametric alternatives, they are more "robust" and less likely to lead to erroneous conclusions or mis-estimation of the true alpha-risk when assumptions have been violated. Furthermore, they are often quick and easy to calculate, so re-examination of the data does not entail a lot of extra work. Nonparametric methods are also appropriate when the data deviate from a normal distribution, for example, with a pattern of high or low outliers, marked asymmetry or skew.

When data are ranked or have ordinal-level properties, a good measure of central tendency is the median. For data that are purely categorical (nominal level), the measure of central tendency to report is the mode, the most frequent value. Various measures of dispersion can be used as alternatives to the standard deviation. When the distribution of the data is not normally distributed, the 95% confidence interval for the median of N scores can be approximated by

$$\frac{N+1}{2} \pm 0.98\sqrt{N} \qquad (B.1)$$

When the data are reasonably normal, the confidence interval for the median is given by

$$\text{Med} \pm 1.253t(S/\sqrt{N}) \qquad (B.2)$$

where t is the two-tailed t-value for N–1 degrees of freedom (Smith, 1988). Another simple alternative is to state the semi-interquartile range or one-half the difference between the data values from the 75th and 25th percentiles.

There are several nonparametric versions of the correlation coefficient. One commonly used is the rank order correlation attributed to Spearman. Nonparametric statistical tests with worked examples are given in Siegel (1956), Hollander and Wolfe (1973), Conover (1980), and a book tailored for sensory evaluation by Rayner et al. (2005). It is advisable that the sensory professional have some familiarity with the common nonparametric tests so they can be used when the assumptions of the parametric tests seem doubtful. Many statistical computing packages will also offer nonparametric modules and various choices of these tests. The sections below illustrate some of the common binomial, chi-square, and rank order statistics, with some worked examples from sensory applications.

B.2 Binomial-Based Tests on Proportions

The binomial distribution describes the frequencies of events with discrete or categorical outcomes. Examples of such data in product testing would be the proportion of people preferring one product over another in a test or the proportion answering correctly in a triangle test. The distribution is based on the binomial expansion, $(p+q)^n$, where p is the probability of one outcome, q is the probability of the other outcome ($q = 1-p$), and n is the number of samples or events. Under the null hypothesis in most discrimination tests, the value of p is determined by the number of alternatives and so equals one-third in the triangle test and one-half in the duo–trio or paired comparison tests.

A classic and familiar example of binomial-based outcomes is in tossing a coin. Assuming it is a fair coin puts the expected probability at one-half for each outcome (heads or tails). Over many tosses (analogous to many observations in a sensory test) we can predict the likely or expected numbers of heads and tails and how often these various possibilities are likely to occur. To predict the number of each possibility, we "expand" the combinations of $(p+q)^n$ letting p represent the numbers of heads and q the numbers of tails in n total throws as follows:

For one throw, the values are $(p+q)^1$ or $p + q = 1$. One head or one tail can occur and they will occur with probability $p = q = 1/2$. The coefficients (multipliers) of each term divided by the total number of outcomes gives us the probability of each combination. For two throws, the values are $(p+q)^2$ so the expansion is $p^2 + 2pq + q^2 = 1$ (note that the probabilities total 1). p^2 is associated with the outcome of two heads, and the multiplicative rule of probabilities tells us that the chances are $(1/2) (1/2) = 1/4$. $2pq$ represents one head and one tail (this can occur two ways) and the probability is $1/2$. In other words, there are four possible combinations and 2 of them include one head and one tail, so the chance of this is 2/4 or 1/2. Similarly, q^2 is associated with two tails and the probability is 1/4. Three throws will yield the following outcomes: $(p+q)^3 = p^3 + 3p^2q + 3pq^2 + q^3$. This expansion tells us that there is a 1/8 chance of three heads or three tails, but there are three ways to get two heads and one tail (HHT, HTH, THH) and similarly three ways to get one head and two tails, so the probability of these two outcomes is 3/8 for each. Note that there are eight possible outcomes for three throws or more generally 2^n outcomes of n observations (Fig. B.1).

As such an expansion continues with more events, the distribution of events, in terms of the possible numbers of one outcome, will form a bell-shaped distribution (when $p = q = 1/2$), much like the normal distribution bell curve. The coefficient for each term in the expansion is given by the formula for combinations, where an outcome appears A times out of N tosses (that is A heads and $N–A$ tails), as follows:

$$\text{Coefficient} = \frac{N!}{(N - A)!A!}$$

This is the number of times the particular outcome can occur. When the coefficient is multiplied by $p^A q^{N-A}$ we get the probability of that outcome. Thus we can find an exact probability for any sample based on the expansion. This is manageable for small samples, but as the number of observations becomes large,

the binomial distribution begins to resemble the normal distribution reasonably well and we can use a z-score approximation to simplify our calculations. For small samples, we can actually do these calculations, but reference to a table can save time.

Here is an example of a small experiment (see O'Mahony, 1986 for similar example). Ten people are polled in a pilot test for an alternative formula of a food product. Eight prefer the new product over the old: two prefer the old product. What is the chance that we would see a preference split of 8/10 or more, if the true probabilities were 1/2 (i.e., a 50/50 split in the population)?

The binomial expansion for 10 observations, $p = q = 1/2$ is

$$p^{10} + 10p^9q + 45\,p^8q^2 + 120\,p^7q^3\ldots \text{ etc.}$$

In order to see if there is an 8-to-2 split or larger we need to calculate the proportions of times these outcomes can occur. Note that this includes the values in the "tail" of the distribution, which includes the outcomes of a 9-to-1 and a 10-to-none preference split. So we only need the first three terms or $(1/2)^{10} + 10 (1/2)^9(1/2) + 45 (1/2)^8 (1/2)^2$.

This sums to about 0.055 or 5.5%. Thus, if the true split in the population as a whole was 50/50, we would see a result this extreme (or more extreme) only about 5% of the time. Note that this is *about* where we reject the null hypothesis. But we have only looked at one tail of the distribution for this computation. In a preference test we would normally not predict at the outset that one item is preferred over another. This requires a two-tailed test and so we need to double this value, giving a total probability of 11%. Remember, this is the exact probability of seeing an 8–2 split or something more extreme (i.e., 9 to 1 or 10 to zero).

For small experiments, we can sometimes go directly to tables of the cumulative binomial distribution, which gives us exact probabilities for our outcomes based on the ends of the expansion equation. For larger experiments ($N > 25$ or so), we can use the normal distribution approximation. The extremeness of a proportion can be represented by a z-score, rather than figuring all the probabilities and expansion terms. The disparity from what is expected by chance can be expressed as the probability value associated with that z-score. The formula for a binomial based z-score is

Fig. B.1 The binomial expansion is shown graphically for tossing a coin with outcomes of heads and tails for various numbers of throws. (**a**) frequencies (probabilities) associated with a single toss. (**b**) frequencies expected from two tosses, (**c**) from three tosses, (**d**) from four tosses, and (**e**) from ten tosses. Note that as the number of events (or observations) increases, the distribution begins to take on the bell-shaped appearance of the normal distribution.

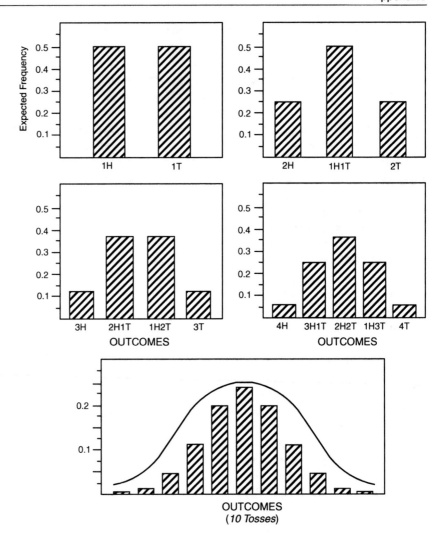

$$z = \frac{(P_{\mathrm{obs}} - p) - (1/2N)}{\sqrt{pq/N}} = \frac{(x - Np) - (0.5)}{\sqrt{Npq}} \quad \text{(B.3)}$$

where P_{obs} is the proportion observed, p is the chance probability $q = 1-p$, N is the number of observations, and x is the number of those outcomes observed ($P_{\mathrm{obs}} = x/N$).

The continuity correction accounts for the fact that we cannot have fractional observations and the distribution of the binomial outcomes is not really a continuous measurement variable. In other words, there are a limited number of whole number outcomes since we are counting discrete events (you cannot have half a person prefer product A over product B). The continuity correction accounts for this approximation by adjusting by the maximum amount of deviation from a continuous variable in the counting process or one-half of one observation.

The standard error of the proportion is estimated to be the square root of p times q divided by the number of observations (N). Note that as with the t-value, our standard error, or the uncertainly around our observations, decreases as the reciprocal of the square root of N. Our certainty that the observed proportion lies near to the true population proportion increases as N gets large.

Tables for minimum numbers correct, commonly used for triangle tests, paired preference tests, etc., solve this equation for X as a function of N and z (see Roessler et al., 1978). The tables show the minimum number of people who have to get the test correct to reject the null hypothesis and then conclude that a difference exists. For one-tailed discrimination tests at an alpha-risk of 0.05, the z-value is 1.645. In this case the minimum value can be solved from the inequalities where Eq. (B.3) is solved for X, and the equal sign is changed to "greater than" (Z must exceed 1.645), and rounded up to the nearest whole number since you cannot have a fraction of a person.

Given the value of 1.645 for Z (at $p = 0.05$), and 1/3 for p and 2/3 for q as in a triangle test, the inequality can be solved for X and N as follows:

$$X \geq \frac{2N+3}{6} + 0.775\sqrt{N} \qquad \text{(B.4)}$$

and for tests in which p is $\frac{1}{2}$, the corresponding equation is

$$X \geq \frac{N+1}{2} + 0.8225\sqrt{N} \qquad \text{(B.5)}$$

We can also use these relationships to determine confidence intervals on proportions. The 95% confidence interval on an observed proportion, P_{obs} ($= X/N$, where X is the number correct in a choice test) is equal to

$$P_{obs} \pm Z\sqrt{pq/N} \qquad \text{(B.6)}$$

where Z will take on the value of 1.96 for the two-tailed 95% intervals for the normal distribution. This equation would be useful for estimating the interval within which a true proportion is likely to occur. The two-tailed situation is applicable to a paired preference test as shown in the following example. Suppose we test 100 consumers and 60% show a preference for product A over product B. What is the confidence interval around the level of 60% preference for product A and does this interval overlap the null hypothesis value of 50%? Using Eq. (B.6),

$$P_{obs} \pm Z\sqrt{pq/N} = 0.60 \pm 1.96\sqrt{0.5(0.5/100)} = 0.60 \pm 0.098$$

In this case the lower limit is above 50%, so there is just enough evidence to conclude that the true population proportion would not fall at 50%, given this result, 95% of the time.

B.3 Chi-Square

B.3.1 A Measure of Relatedness of Two Variables

The chi-square statistic is a useful statistic for comparing frequencies of events classified in a table of categories. If each observation can be classified by two or more variables, it enters into the frequency count for a part of a matrix or classification table, where rows and columns represent the levels of each variable. For example, we might want to know whether there is any relationship between gender and consumption of a new reduced-fat product. Each person could be classified as high- versus low-frequency users of the product and also as male or female. This would create a two-way table with four cells representing the counts of people who fall into one of the four groups. For the sake of example, let us assume we had a 50/50 split in sampling the two sexes and also an even proportion of our high- and low-frequency groups. Intuitively, we would expect 25% of observations to fall in each cell of our table, assuming no difference between men and women in frequency of use. To the extent that one or more cells in the table is disproportionally filled or lacking in observations, we would find evidence of an association or lack of independence of gender and product use. Table B.1 shows two examples, one with no association between the variables and the other with a clear association (numbers represent counts of 200 total participants).

Table B.1 Examples of different levels of association

| | No association | | Clear association | | |
| | Usage group | | Usage group | | |
	Low	High	Low	High	(Total)
Males	50	50	75	25	(100)
Females	50	50	20	80	(100)
(Totals)	(100)	(100)	(95)	(105)	(200)

In the left example, the within-cell entries for frequency counts are exactly what we would expect based on the marginal totals, with one-half of the groups classified according to each variable, we expect one-fourth of the total in each of the four cells (Of course, a result of exactly 25% in each cell would rarely be found in real life. In the right example, we see that females are more inclined to fall into the high-usage group and that the reverse is true for males. So knowing gender helps us predict something about the usage group, and conversely, knowing the usage, we can make a prediction about gender. So we conclude that there is a relationship between these two variables.

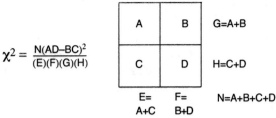

$$\chi^2 = \frac{N(AD-BC)^2}{(E)(F)(G)(H)}$$

B.3.2 Calculations

More generally, the chi-square statistic is useful for comparing distributions of data across two or more variables. The general form of the statistic is to (1) compute the expected frequency minus the observed frequency, (2) square this value, (3) divide by the expected frequency, and (4) sum these values across all cells (Eq. (B.7)). The expected frequency is what would be predicted from random chance or from some knowledge or theory of what is a likely outcome based on previous or current observations.

$$\chi^2 = \sum \frac{(\text{observed} - \text{expected})^2}{\text{expected}} \qquad (B.7)$$

The statistic has degrees of freedom equal to the number of rows minus one, times the number of columns minus one. For a 2×2 table, the function is mathematically equivalent to the z-formula from the binomial probability in Eq. (B.3) (i.e., $\chi^2 = z^2$, see postscript to this chapter for a proof). For small samples, $N < 50$, we can also use a continuity correction, as with the Z-formula, where we subtract 1/2, so the Yates correction for continuity gives us this equation:

$$\chi^2 \text{ Yates} = \sum \frac{(|\text{observed} - \text{expected}| - 0.5)^2}{\text{expected}} \qquad (B.8)$$

Note that the absolute value must be taken before the continuity correction is subtracted and that the subtraction is before squaring (this is incorrect in some texts).

A simple form of the test for 2×2 matrices and a computational formula are shown in Fig. B.2.

Some care is needed in applying chi-square tests, as they are temptingly easy to perform and so widely applicable to questions of cross-classification and association between variables. Tests using chi-square usually assume that each observation is independent, e.g., that each tally is a different person. It is not appropriate

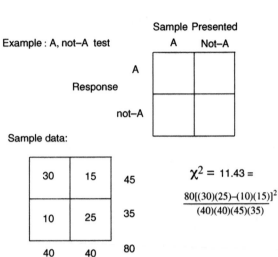

Fig. B.2 Some uses of the chi-square test for 2×2 contingency tables. The example shows the short cut formula applied to the A, not-A test situation. The same analysis applies to the same/different test. However, this is only appropriate if there are different individuals in each cell, i.e., each tester only sees one product. If the testers see both versions, then the McNemar test is appropriate instead of the simple chi-square.

for related-samples data such as repeated observations on the same person. The chi-square test is not robust if the frequency counts in any cells are too small, usually defined as minimum count of five observations (expected) as rule of thumb. Many statistical tests are based upon the chi-square distributions, as we will see in the section on rank order statistics.

B.3.3 Related Samples: The McNemar Test

The chi-square statistic is most often applied to independent observations classified on categorical variables. However, many other statistical tests follow a chi-square distribution as a test statistic. Repeated observations of a group on a simple dichotomous variable (two classes) can be tested for change or difference using the McNemar test for the significance of changes. This is a simple test well suited to before-and-after experiments such as the effect of information on attitude change. It can be applied to any situation where test panelists view two products and their responses are categorized into two classes. For example, we might want to see if the number of people who report liking a product changes after the presentation of some information such as nutritional content. Stone and Sidel (1993) give an example of using the McNemar test for changes to assess whether just-right scales show a difference between two products.

The general form of the test classifies responses in a two-by-two matrix, with the same response categories as rows and columns. Since the test is designed to examine changes or differences, the two cells with the same values of row and column variables are ignored. It is only the other two corners of the table where the classification differs that we are interested in. Table B.2 gives example, with the frequency counts represented by the letters "a" through "d."

Table B.2 Example for McNemar calculations

	Before information is presented	
	Number liking the product	Number disliking or neutral
After information is presented		
Number liking	a	b
Number disliking or neutral	c	d

The McNemar test calculates the following statistic:

$$\chi^2 = \frac{(|b - c| - 1)^2}{b + c} \tag{B.9}$$

Note that the absolute value of the difference is taken in the two cells where change occurs and that the other two cells (a and d) are ignored. The obtained value must exceed the critical value of chi-square for $df = 1$, which is 3.84 for a two-tailed test and 2.71 for a one-tailed test with a directional alternative hypothesis. It is important that the expected cell frequencies be larger than 5. Expected frequencies are given by the sum of the two cells of interest, divided by two. Table B.3 gives an example testing for a change in preference response following a taste test among 60 consumers.

Table B.3 Sample data for McNemar test

	Before tasting	
	Prefer product A	Prefer product B
After tasting		
Prefer product A	12	33
Prefer product B	8	7

And so our calculated value becomes:

$$\chi^2 = \frac{(|33 - 8| - 1)^2}{33 + 8} = 576/41 = 14.05$$

This is larger than the critical value of 3.84, so we can reject the null hypothesis (of no change in preference) and conclude that there was a change in preference favoring Product A, as suggested by the frequency counts. Although there was a 2-to-1 preference for B before tasting (marginal totals of 40 versus 20), 33 of those 40 people switched to product A while less than half of those preferring product A beforehand switched in the other direction. This disparity in changes drives the significance of the McNemar calculations and result. This test is applicable to a variety of situations, such as the balanced A, not-A, and same/different tests in which each panelist judges both kinds of trials and thus the data are related observations. The generalization of the McNemar test to a situation with related observations and multiple rows and columns is the Stuart test for two products or the Cochran-Mantel-Haenzel test for more than two products.

B.3.4 The Stuart–Maxwell Test

A useful test for 3×3 matrices such as those generated from just-about-right (JAR) scale data is the Stuart–Maxwell test discussed in Chapter 14. For example, we might have two products rated on JAR scales and want to see if there is any difference in the distribution of ratings. The data are collapsed into three categories, those above just-right, those at or near just-right, and those below the just-right point. Then the frequencies in the off-diagonal cells are used to calculate a chi-square variable. The cells with identical classifications (the diagonal) are not used. The critical value is compared to a chi-square value for two degrees of freedom, which is 5.99. The calculations are shown in Fig. B.3 and a worked example in Fig. B.4.

Fig. B.3 The calculations involved in the Stuart–Maxwell test as applied to just-about-right (JAR) scale data.

Stuart Maxwell Calculations

1. Entries A thorugh I are the cell totals.

2. Average the off-diagonal pairs, P1 - P3

3. Find differences of row and column totals:

D1=C1–R1   D2=C2–R2   D3=C3–R3

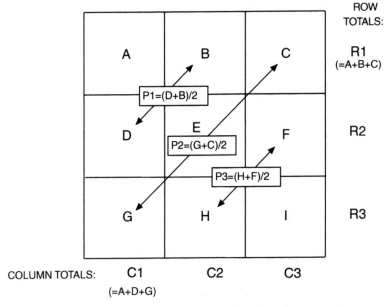

4. Chi-square is calculated. Note that the cell averages (P1, P2, P3) are multiplied by the squared differences (D1, D2, D3) of row and column totals in which they DO NOT participate.

$$\text{Chi-square} = \frac{[(P1)(D3)^2 + (P2)(D2)^2 + (P3)(D1)^2]}{2[(P1)(P2) + (P2)(P3) + (P1)(P3)]}$$

B.3.5 Beta-Binomial, Chance-Corrected Beta-Binomial, and Dirichlet Multinomial Analyses

These three models can be used for replicated data from choice tasks. The beta-binomial is applicable to replicated tests where there are two outcomes (e.g., right and wrong answers) or two choices, as in a preference test. The Dirichlet multinomial is applicable to tests where there are more than two choices, such as a preference test with the no preference option. The equations below describe how to conduct tests for overdispersion, when panelists or consumers are not acting like random events (like

flipping coins) but rather show consistent patterns of response. Worked examples are not shown here but can be found in Gacula et al. (2009) and Bi (2006). Students are urged to look at those worked examples before attempting these tests. Maximum likelihood solutions are also given in those texts, using S-plus programs.

In all the examples below, the letters n, r, and m are used to refer to the number of panelists, replicates, and choices, respectively. We think these are easier to remember, but they are different than the notations of Bi and Gacula who use n for replicates and k for panelists (be forewarned). Lowercase x with subscripts will refer to a single observation or count of choices for a given panelist and/or replicate.

Fig. B.4 A worked example of the Stuart–Maxwell test.

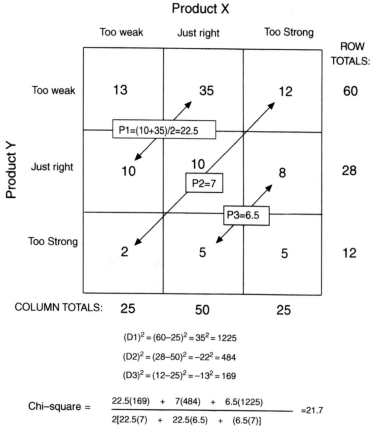

Example for Just–About–Right Scales:

$(D1)^2 = (60-25)^2 = 35^2 = 1225$

$(D2)^2 = (28-50)^2 = -22^2 = 484$

$(D3)^2 = (12-25)^2 = -13^2 = 169$

$$\text{Chi–square} = \frac{22.5(169) \quad + \quad 7(484) \quad + \quad 6.5(1225)}{2[22.5(7) \quad + \quad 22.5(6.5) \quad + \quad (6.5(7)]} = 21.7$$

This value is then compared to the critical value of chi–square for 2 df (=5.99). As 21.7>5.99, there was a significant difference in the ratings of the two products. Inspection of the 3X3 matrix suggests that Product Y is too weak relative to Product X.

B.3.5.1 Beta-Binomial

The beta-binomial model assumes that the performance of panelists is distributed like a beta distribution (Bi, 2006). This distribution has two parameters, but they can be summarized in a statistic called gamma. Gamma varies from zero to one and is a measure of the degree to which there are systematic patterns of response versus apparent random variation across replications.

First we calculate a mean proportion and a variance parameter, μ and S, respectively:

$$\mu = \frac{\sum_{i=1}^{n} x_i/r}{n} \qquad (B.10)$$

where each x_i is the number of correct judgments summed across replicates for that panelist. So μ is the mean of the number of correct replicates. S is defined as:

$$S = \sum_{i=1}^{n} (x_i - \mu)^2 \qquad (B.11)$$

and then we can calculate our gamma:

$$\gamma = \frac{1}{r-1} \left[\frac{rS}{\mu(1-\mu)n} - 1 \right] \qquad (B.12)$$

where r is the number of replicates, S is a measure of dispersion, μ is the mean proportion correct for the group (looking at each person's individual proportions as shown below), and n is the number of judges.

To test whether the beta-binomial or binomial is a better fit, we use the following Z-test, sometimes called Tarone's Z-test (Bi, 2006, p. 114):

$$Z = \frac{E - nr}{\sqrt{2nr(r-1)}} \qquad (B.13)$$

where E is another measure of dispersion,

$$E = \sum_{i=1}^{n} \frac{(x_i - rm)^2}{m(1-m)} \qquad (B.14)$$

and m is the mean proportion correct

$$m = \sum_{i=1}^{n} x_i / nr \qquad (B.15)$$

If this Z is not significant, you have some justification for combining replicates and looking at the total proportion correct over $n \times r$ trials.

If we wish to test our obtained value of μ against some null hypothesis value of μ_o, we can use another simple Z-test:

$$Z = \frac{|\mu - \mu_o|}{\sqrt{\text{Var}(\mu)}} \qquad (B.16)$$

where $\text{Var}(\mu)$ is

$$\text{Var}(\mu) = \frac{\mu(1-\mu)}{nr}[(r-1)\gamma + 1] \qquad (B.17)$$

Using this general equation, most significance tests can be done on any two-choice format such as forced choice or preference tests. The appealing factor is that the above equation has taken into account the overdispersion in the data. That is, the issue of whether there are segments of panelists performing consistently versus apparent random performance from replicate to replicate has been addressed.

B.3.5.2 The Chance-Corrected Beta-Binomial

Some authors have argued that the beta-binomial is unrealistic, because there is a lower limit on the population mean performance that is dictated by the chance performance level (see Bi, 2006, for a whole chapter on this approach). This is intuitively appealing, although some published comparisons of the two models show only modest differences with real data sets.

Let p be the mean proportion correct, now we can defined a chance-corrected mean proportion as

$$\hat{\mu} = \frac{p - C}{1 - C} \qquad (B.18)$$

where C is the chance proportion correct, e.g., 1/3 for the triangle test or the chance-expected preference in a two product test of $\frac{1}{2}$. Now we need our variance parameter, S,

$$S = \sum_{i=1}^{n} (p_i - \mu)^2 \qquad (B.19)$$

where p_i is the proportion correct for panelist i. Now we need a new and slightly more complex estimate of gamma as

$$\gamma = \frac{1}{(r-1)(p-C)}\left[\frac{rS}{n(1-p)} - p\right] \qquad (B.20)$$

The same Z-test still applies for testing again a null proportion, μ_o, but now we need a new variance calculation, actually two of them:

$$\text{Var}(\hat{\mu}) = \frac{\text{Var}(p)}{(1-C)^2} \qquad (B.21)$$

and

$$\text{Var}(p) = (1-C)^2(1-\hat{\mu})\left[(r-1)\hat{\mu}\gamma + \frac{C}{1-C} + \hat{\mu}\right]/nr \qquad (B.22)$$

(whew! But now we have worked gamma back into the picture).

B.3.5.3 The Dirichlet-Multinomial Model

This model extends the reasoning of the beta-binomial approach to the situation where there are more than two alternatives (Gacula et al., 2009). In its simplest form, it can be used to test against some fixed proportions like an equal one-third split in a preferences test or a 35/30/35% split if one uses the commonly observed no preference rate from identical samples of 30% (see Chapter 13).

Suppose we have three options: "prefer product A", "no preference," and "prefer product B." Let X_1 be the sum for product A over all choices, X_2 be the sum of no preference, and X_3 the sum for product B. Let there be n panelists, r replicates, and m choices (in this case three). We have N total observations ($= n \times r$). The first thing we can do is try to see if there is a pattern of responding analogous to a nonzero gamma in Tarone's Z-test. This is yet another $\underline{Z}$-statistic, given by the following formula:

$$Z = \frac{N \sum_{j=1}^{m} 1/X_j \sum_{i=1}^{n} x_{ij}(x_{ij} - 1) - [nr(r - 1)]}{\sqrt{2(m - 1)[nr(r - 1)]}} \quad \text{(B.23)}$$

where x_{ij} is the total number of that choice, j, for panelist i, multiplied by $x_{ij}-1$, then summed across all panelists, then weighted by $1/X_j$. Repeat for each choice, j.

We can also do a simple test against expected proportions, based on a weighted chi-square with 2df. But first we need the heterogeneity parameter, C, which is analogous to 1-gamma in the beta-binomial model. Let $p_j = X_j/N$, where we just convert the total for each choice to the corresponding proportion:

$$C = \frac{r}{(n-1)(m-1)} \sum_{j=1}^{m} 1/p_j \sum_{j=1}^{n} (\frac{x_{ij}}{m} - p_j)^2 \quad \text{(B.24)}$$

Once we have our correction factor, C, for panelist "patterns" or overdispersion, we can perform a simple χ^2 test as follows:

$$\chi^2 = \frac{nr}{C} \sum_{j=1}^{m} \frac{(p_j - p_{\text{exp}})^2}{p_{\text{exp}}} \quad \text{(B.25)}$$

where p_j is again our observed proportion for each choice, and p_{exp} is the proportion we expect based on our theory. This is tested against a χ^2 distribution with $m-1$ degrees of freedom. A significant χ^2 would indicate a deviation from our expected proportions. Such a test could also be applied to just-about-right data if we have some basis for assuming some reasonable or predicted distribution of results in the JAR categories.

B.4 Useful Rank Order Tests

B.4.1 The Sign Test

A simple nonparametric test of difference with paired data is the sign test. The simplest case of ranking is the paired comparison, when only two items are to be ranked as to which is stronger or which is preferred. The sign test based on comparisons between two samples is based on binomial statistics. The sign test can also be used with any data such as scaled responses that have at least ordinal properties. Obviously, in cases of no difference, we expect the number of rankings in one direction (for example, product A over B) to equal the number of rankings in the opposite direction (product B over A) so the null probability of 1/2 can be used.

In a two-sample case, when every panelist scores both products, the scores can be paired. Probabilities can be examined from the binomial tables, from the critical value tables used for discrimination tests (for one-tailed hypotheses) with $p = 1/2$ or from the paired preference tables (for two tailed) (Roessler et al., 1978). The sign test is the nonparametric parallel of the dependent groups or paired t-test. Unlike the t-test, we do not need to fulfill the assumption of normally distributed data. With skewed data, the t-test can be misleading since high outliers will exert undue leverage on the value of the mean. Since the sign test only looks for consistency in the direction of comparisons, the skew or outliers are not so influential. There are also several nonparametric counterparts to the independent groups t-test. One of these, the Mann–Whitney U-test, is shown below.

Table B.4 gives an example of the sign test. We simply count the direction of paired scores and assume a

Table B.4 Data for sign test example

Panelist	Score, Product A	Score, Product B	Sign, for $B > A$
1	3	5	+
2	7	9	+
3	4	6	+
4	5	3	−
5	6	6	O
6	8	7	−
7	4	6	+
8	3	7	+
9	7	9	+
10	6	9	+

50/50 split under the null hypothesis. In this example, panelists scored two products on a rating scale (every panelist tasted both products) so the data are paired. Plus or minus "signs" are given to each pairing for whether A is greater than B or B is greater than A, respectively, hence the name of the test. Ties are omitted, losing some statistical power, so the test works best at detecting differences when there are not too many ties.

Count the number of +'s (= 7), and omit ties. We can then find the probability of (at least) 7/9 in a two-tailed binomial probability table, which is 0.09. Although this is not enough evidence to reject the null, it might warrant further testing, as there seems to be a consistent trend.

B.4.2 The Mann–Whitney U-Test

A parallel test to the independent groups t-test is the Mann–Whitney U-test. It is almost as easy to calculate as the sign test and thus stands as a good alternative to the independent groups t-test when the assumptions of normal distributions and equal variance are doubtful. The test can be used for any situation in which two groups of data are to be compared and the level of measurement is at least ordinal. For example, two manufacturing sites or production lines might send representative samples of soup to a sensory group for evaluation. Mean intensity scores for saltiness might be generated for each sample and then the two sets of scores would be compared. If no difference were present between the two sites, then rankings of the combined scores would find the two sites to be interspersed. On the other hand, if one site was producing consistently more salty soup than another, then that site should move toward higher rankings and the other site toward lower rankings. The U-test is sensitive to just such patterns of overlap versus separation in a set of *combined* ranks.

The first step is to rank the combined data and then find the sum of the ranks for the smaller of the two groups. For a small experiment, with the larger of the two groups having less than 20 observations, the following formula should be used:

$$U = n_1 n_2 + [n_1(n_1 + 1)/2] + R_1 \qquad (B.26)$$

where n_1 is the smaller of the two samples, n_2 is the larger of the two samples, and R_1 is the sum of the ranks assigned to the smaller group. The next step is to test whether U has the correct form, since it may be high or low depending upon the trends for the two groups. The smaller of the two forms is desired. If U is larger than $n_1 n_2/2$, it is actually a value called U' and must be transformed to U by the formula, $U = n_1 n_2 - U'$.

Critical values for U are shown in Table E. Note that the obtained value for U must be equal to *or smaller than* the tabled value in order to reject the null, as opposed to other tabled statistics where the obtained value must exceed a tabled value . If the sample size is very large, with n_2 greater than 20, the U statistic can be converted to a z-score by the following formula, analogous to the difference between means divided by a standard deviation:

$$z = \frac{[U - (n_1 n_2)/2]}{\sqrt{[n_1 n_2 (n_1 + n_2 + 1)/12]}} \qquad (B.27)$$

If there are ties in the data, the standard deviation(denominator) in the above formula needs adjustment as follows:

$$SD = \sqrt{[n_1 n_2/(N(N-1))][((N^3 - N)/12) - \Sigma T]} \qquad (B.28)$$

where $N = n_1 + n_2$ and $T = (t^3 - t)/12$ where t is the number of observations tied for a given rank. This demands an extra housekeeping step where ties must be counted and the value for T computed and summed before Z can be found.

A worked example for a small sample is shown next. In our example of salty scores for soups, let us assume we have the following panel means (Table B.5).

So there were 6 samples (= n_2) taken from site A and 5 (= n_1) from site D. The ranking of the 11 scores would look like this (Table B.6).

Table B.5 Data for Mann–Whitney U-test

Site A	Site D
4.7	8.2
3.5	6.6
4.3	4.1
5.2	5.5
4.2	4.4
2.7	

Table B.6 Ranked data for Mann–Whitney U-test

Score	Rank	Site
8.2	1	D
6.6	2	D
5.5	3	D
5.2	4	A
4.7	5	A
4.4	6	D
4.3	7	A
4.2	8	A
4.1	9	D
3.5	10	A
2.7	11	A

R_1 is then the sum of the ranks for site D ($= 1 + 2 + 3 + 6 + 9 = 21$). Plugging into the formula we find that $U = 30 + 15 - 21 = 24$.

Next, we check to make sure we have U and not U' (the smaller of the two is needed). Since U is larger than $n_1 n_2 / 2 = 15$, we did in fact obtain U', so we subtract U from 30 giving a value of 6. This is then compared to the maximum critical value in Table E. For these sample sizes, the U value must be three or smaller to reject the null at a two-tailed probability of 0.05, so there is not enough evidence to reject the null in this comparison. Inspection of the rankings shows a lot of overlap in the two sites, in spite of the generally higher scores at site D. The independent groups t-test on these data also give a p value higher than 0.05, so there is agreement in this case. Siegel (1956) states that the Mann–Whitney test is about 95% as powerful as the corresponding t-test. There are many other nonparametric tests for independent samples, but the Mann–Whitney U-test is commonly used and simple to calculate.

B.4.3 Ranked Data with More Than Two Samples, Friedman and Kramer Tests

Two tests are commonly used in sensory evaluation for ranked products where there are three or more items being compared. The Friedman "analysis of variance" on ranked data is a relatively powerful test that can be applied to any data set where all products are viewed by all panelists, that is, there is a complete ranking by each participant. The data set for the Friedman test thus takes the same form as a one-way analysis of variance with products as columns and panelists as rows, except

that ranks are used instead of raw scores. It is also applicable to any data set where the rows form a set of matched observations that can be converted to ranks. The Friedman test is very sensitive to a pattern of consistent rank orders. The calculated statistic is compared to a chi-square value that depends upon the number of products and the number of panelists. The second test that is common in sensory work is Kramer's rank sum test. Critical values for significance for this test were recalculated and published by Basker (1988) and by Newell and MacFarlane (1987) (see Table J). A variation of the Friedman test is the rank test of Page (1963), which is a little more powerful than the Friedman test, but is only used when you are testing against one specific predicted ranking order. Each of these methods is illustrated with an example below.

Example of the Friedman test: Twenty consumers are asked to rank three flavor submissions for their appropriateness in a chocolate/malted milk drink. We would like to know if there is a significant overall difference among the candidates as ranked. The Friedman test constructs a chi-square statistic based on column totals, T_j, in each of the J columns. For a matrix of K rows and J columns, we compared the obtained value to a chi-square value of $J-1$ degrees of freedom. Here is the general formula:

$$\chi^2 = \left\{ \frac{12}{[K(J)(J+1)]} \left[\sum_{j=1}^{J} T_j^2 \right] \right\} - 3K(J+1)$$

(B.29)

Table B.7 shows the data and column totals. So the calculations proceed as follows:

$$\chi^2 = \left\{ \frac{12}{[20(3)(4)]} [(43.5)^2 + (46.5)^2 + (30)^2] \right\} - 3(20)(4) = 7.725$$

In the chi-square table for $J-1$ degrees of freedom, in this case df $= 2$, the critical value is 5.99. Because our obtained value of 7.7 exceeds this, we can reject the null. This makes sense since product C had a predominance of first rankings. Note that in order to compare individual samples, we require another test. The sign test is appropriate, although if many pairs of samples are compared, then the alpha level needs to be reduced to compensate for the experiment-wise increase in risk. Another approach is to use the least-significant-difference (LSD) test for ranked data, as follows:

$$LSD = 1.96 \sqrt{\frac{K(J)(J+1)}{6}}$$

(B.30)

Table B.7 Data for Friedman test on ranks

Panelist	Ranks Product A	Product B	Product C
1	1	3	2
2	2	3	1
3	1	3	2
4	1	2	3
5	3	1	2
6	2	3	1
7	3	2	1
8	1	3	2
9	3	1	2
10	3	1	2
11	2	3	1
12	2	3	1
13	3	2	1
14	2	3	1
15	2.5	2.5	1
16	3	2	1
17	3	2	1
18	2	3	1
19	3	2	1
20	1	2	3
Sum (column totals, T_j)	43.5	46.5	30

for J items ranked by K panelists. Items whose rank sums differ by more than this amount may be considered significantly different.

An example of the (Kramer) Rank Sum test. We can also use the rank sum test directly on the previous data set to compare products. We merely need the differences in the rank sums (column totals):

Differences: A versus B = 3.0
                       B versus C = 16.5
                       A versus C = 13.5

Comparing to the minimum critical differences at $p < 0.05$, (= 14.8, see J), there is a significant difference between B and C, but not any other pair. What about the comparison of A versus C, where the difference was close to the critical value? A simple sign test between A and C would have yielded a 15 to 5 split, which is statistically significant (two-tailed $p = 0.042$). In this case when the rank sum test was so close to the cutoff value, it would be wise to examine the data with an additional test.

B.4.4 Rank Order Correlation

The common correlation coefficient, r, is also known as the Pearson product–moment correlation coefficient. It is a useful tool for estimating the degree

of linear association between two variables. However, it is very sensitive to outliers in the data. If the data do not achieve an interval scale of measurement, or have a high degree of skew or outliers, the nonparametric alternative given by Spearman's formula should be considered. The Spearman rank order correlation was one of the first to be developed (Siegel, 1956) and is commonly signified by the Greek letter, ρ (rho). The statistic asks whether the two variables line up in similar rankings. Tables of significance indicate whether an association exists based on these rankings.

The data must first be converted to ranks, and a difference score calculated for each pair of ranks, similar to the way differences are computed in the paired t-test. These differences scores, d, are then squared and summed. The formula for rho is as follows:

$$\rho = \frac{6 \sum d^2}{(N^3 - N)} \tag{B.31}$$

Thus the value for rho is easy to calculate unless there are a high proportion of ties. If greater than one-fourth of the data are tied, an adjustment should be made. The formula is very robust in the case of a few ties, with changes in rho usually only in the third decimal place. If there are many ties, a correction must be calculated for each tied case based on $(t^3-t)/12$ where t is the number of items tied at a given rank. These values are then summed for all the ties for each variable x and y, to give values T_x and T_y. rho is then calculated as follows:

$$\rho = \frac{\sum x^2 + \sum y^2 - \sum d^2}{2\sqrt{\sum x^2 \sum y^2}} \tag{B.32}$$

and

$$\sum x^2 = [(N^3 - N)/12] - \sum T_x \text{ and similarly}$$

$$\sum y^2 = [(N^3 - N)/12] - \sum T_y \tag{B.33}$$

For example, if there are two cases for X in which two items are tied and one case in which three are tied, the T_x becomes the sum:

$$\sum T_x = (2^3 - 2)/12 + (2^3 - 2)/12 + (3^3 - 3)/12 = 3$$

and this quantity is then used as ΣT_x in Eq. (B.17).

Suppose we wished to examine whether there was a relationship between mean chewiness scores for a set of products evaluated by a texture panel and mean scores on scale for hardness. Perhaps we suspect that the same underlying process variable gives rise to textural problems observable in both mastication and initial bite. Mean panel scores over ten products might look like Table B.8, with the calculation of rho following:

Table B.8 Data and calculations for rank order correlation

Product	Chewiness	Rank	Hardness	Rank	Difference	D^2
A	4.3	7	5.0	6	1	1
B	5.6	8	6.1	8	0	0
C	5.8	9	6.4	9	0	0
D	3.2	4	4.4	4	0	0
E	1.1	1	2.2	1	0	0
F	8.2	10	9.5	10	0	0
G	3.4	5	4.7	5	0	0
H	2.2	3	3.4	2	1	1
I	2.1	2	5.5	7	5	25
J	3.7	6	4.3	3	3	9

The sum of the D^2 values is 36, so rho computes to the following:

$$1 - 6(36)/(1,000 - 10) = 1 - 0.218 = 0.782$$

This is a moderately high degree of association, significant at the 0.01 level. This is obvious from the good agreement in rankings, with the exception of product I and J. Note that product F is a high outlier on both scales. This inflates the Pearson correlation to 0.839, as it is sensitive to the leverage exerted by this point that lies away from the rest of the data set.

B.5 Conclusions

Some of the nonparametric parallels to common statistical tests are shown in Table B.1. Further examples can be found in statistical texts such as Siegel (1956). The nonparametric statistical tests are valuable to the sensory scientist for several reasons and it should be part of a complete sensory training program to become familiar with the most commonly used tests. Also, the binomial distribution forms the basis for the choice tests commonly used in discrimination testing, so it is important to know how this distribution is derived and when it approximates normality. The chi-square statistics are useful for a wide range of problems involving categorical variables and as a nonparametric measure of association. They also form the basis for other statistical tests such as the Friedman and McNemar tests. Nonparametric tests may be useful for scaled data

where the interval-level assumptions are in doubt or for any data set when assumptions about normality of the data are questionable. In the case of deviations from the assumptions of a parametric test, confirmation with a nonparametric test may lend more credence to the significance of a result (Table B.9).

Table B.9 Parametric and nonparametric statistical tests

Purpose	Parametric test	Nonparametric parallel
Compare two products (matched data)	Paired (dependent) t-test on means	Sign test
Compare two products (separate groups)	Independent groups t-test on means	Mann–Whitney U-test
Compare multiple products (complete block design)	One-way analysis of variance with repeated measures	Friedman test or rank sum test
Test association of two variables	Pearson (product–moment) correlation coefficient	Spearman rank order correlation

Nonparametric tests are performed on ranked data instead of raw numbers.

Other nonparametric tests are available for each purpose, the listed ones are common.

B.6 Postscript

B.6.1 Proof showing equivalence of binomial approximation Z-test and χ^2 test for difference of proportions

Recall that

$$\chi^2 = \sum \frac{(\text{observed} - \text{expected})^2}{\text{expected}} \quad (B.34)$$

and

$$z = \frac{x/N - p}{\sqrt{pq/N}} \quad (B.35)$$

where

X = number correct,
N = total judgments or panelists,
p = chance proportion,
q = 1–p.

Note that continuity corrections have been omitted for simplicity.

Alternative Z-formula (multiply Eq. (B.35) by N/N)

$$z = \frac{x - Np}{\sqrt{pqN}} \qquad (B.36)$$

Although the χ^2 distribution changes shape with different df, the general relationship of the χ^2 distribution to the Z-distribution is that χ^2 at 1 df is a square of Z. Note that critical χ^2 at 1 df $= 3.84 = 1.96^2 = Z_{0.95}^2$:

$$z^2 = \frac{(x - Np)^2}{pqN} \qquad (B.37)$$

and

$$z^2 = \frac{x^2 - 2xNp + N^2p^2}{pqN} \qquad (B.38)$$

The proof will now proceed to show the equivalence of Eq. (B.38) to χ^2.

Looking at any forced choice test, the χ^2 approach requires these frequency counts:

	Correct judgments	Incorrect
Observed	X	$N–X$
Expected	Np	Nq

$$\chi^2 = \frac{(x - Np)^2}{Np} + \frac{[(N - x) - Nq]^2}{Nq} \qquad (B.39)$$

Simplifying $(N–X)–Nq$ to $N(1–q)–X$
then since $p = 1–q$
$(N–X)–Nq = Np–X$

Thus we can recast Eq. (B.39) as

$$\chi^2 = \frac{(x - Np)^2}{Np} + \frac{(Np - X)^2}{Nq} \qquad (B.40)$$

and expanding the squared terms

$$\chi^2 = \frac{(x^2 - 2xNp + N^2p^2)}{Np} + \frac{(x^2 - 2xNp + N^2p^2)}{Nq} \qquad (B.41)$$

To place them over a common denominator of Npq, we will multiple the left expression by q/q and the right expression by p/p giving

$$\chi^2 = \frac{(qx^2 - 2xNpq + qN^2p^2)}{Npq} + \frac{(px^2 - 2xNpp + pN^2p^2)}{Npq} \qquad (B.42)$$

Collecting common terms

$$\chi^2 = [(q + p)x^2 - (q + p)2xNp + (q + p)N^2p^2]/Npq \qquad (B.43)$$

Recall that $q + p = 1$, so Eq. (B.43) simplifies to

$$\chi^2 = [(1)x^2 - (1)2xNp + (1)N^2p^2]/Npq \qquad (B.44)$$

and dropping the value 1 in each of the three terms in the numerator gives Eq. (B.38), the formula for Z^2:

$$z^2 = \frac{x^2 - 2xNp + N^2p^2}{pqN} = \chi2$$

Recall that the continuity correction was omitted for simplicity of the calculations. The equivalence holds *if and only if* the continuity correction is either *omitted from both* analyses or *included in both* analyses. If it is omitted from one analysis but not the other, the one from which it is omitted will stand a better chance of attaining significance.

References

Basker, D. 1988. Critical values of differences among rank sums for multiple comparisons. Food Technology, 42(2), 79, 80–84.

Bi, J. 2006. Sensory Discrimination Tests and Measurements. Blackwell, Ames, IA.

Conover, W. J. 1980. Practical Nonparametric Statistics, Second Edition. Wiley, New York.

Gacula, M., Singh, J., Bi, J. and Altan, S. 2009. Statistical Methods in Food and Consumer Research, Second Edition. Elsevier/Academic, Amsterdam.

Hollander, M. and Wolfe, D. A. 1973. Nonparametric Statistical Methods. Wiley, New York.

Newell, G. J. and MacFarlane, J. D. 1987. Expanded tables for multiple comparison procedures in the analysis of ranked data. Journal of Food Science, 52, 1721–1725.

O'Mahony, M. 1986. Sensory Evaluation of Food. Statistical Methods and Procedures. Marcel Dekker, New York.

Page, E. B. 1963. Ordered hypotheses for multiple treatments: A significance test for linear ranks. Journal of the American Statistical Association, 58, 216–230.

Rayner, J. C. W., Best, D. J., Brockhoff, P. B. and Rayner, G. D. 2005. Nonparametric s for Sensory Science: A more Informative Approach. Blackwell, Ames IA.

Roessler, E. B., Pangborn, R. M., Sidel, J. L. and Stone, H. 1978. Expanded statistical tables estimating significance in paired-preference, paired-difference, duo-trio and triangle tests. Journal of Food Science, 43, 940–943.

Siegel, S. 1956. Nonparametric Statistics for the Behavioral Sciences. McGraw Hill, New York.

Smith, G. L. 1988. Statistical analysis of sensory data. In: J. R. Piggott (ed.), Sensory Analysis of Foods. Elsevier Applied Science, London.

Stone, H. and Sidel, J. L. 1993. Sensory Evaluation Practices, Second Edition. Academic, San Diego.

Appendix C

Analysis of Variance

Contents

For tests with more than two products and data that consist of attribute scale values, analysis of variance followed by planned comparisons of means is a common and useful statistical method. Analysis of variance and related tests are illustrated in this chapter, with worked examples.

C.1 Introduction

C.1.1 Overview

Analysis of variance is the most common statistical test performed in descriptive analysis and many other sensory tests where more than two products are compared using scaled responses. It provides a very sensitive tool for seeing whether treatment variables such as changes in ingredients, processes, or packaging had an effect on the sensory properties of products. It is a method for finding variation that can be attributed to some specific cause, against the background of existing variation due to other perhaps unknown or uncontrolled causes. These other unexplained causes produce the experimental error or noise in the data.

The following sections illustrate some of the basic ideas in analysis of variance and provide some worked examples. As this guide is meant for students and practitioners, some theory and development of models has been left out. However, the reader can refer to the statistics texts such as Winer (1971), Hays (1973), O'Mahony (1986), and Gacula et al. (2009). A particularly useful book is the *Analysis of Variance for Sensory Data*, by Lea et al. (1998), Lundahl and McDaniel (1988). We have tried to use the same nomenclature as O'Mahony (1986) since that work is already familiar to many workers in sensory evaluation and of Winer (1971), a classic treatise on ANOVA for behavioral data.

C.1.2 Basic Analysis of Variance

Analysis of variance is a way to examine differences among multiple treatments or levels and to compare several means at the same time. Some experiments have many levels of an ingredient or process variable. Factors in ANOVA terminology mean independent variables, the variables that you manipulate, i.e., the variables under your direct control in an experiment. Analysis of variance estimates the variance (squared deviations) attributable to each factor. This can be thought of as the degree to which each factor or variable moves the data away from the grand or overall mean of the data set. It also estimates the variance due to error. Error can be thought of as other remaining variation not attributable to the factors we manipulate.

In an analysis of variance we construct a ratio of the factor variance to the error variance. This ratio follows the distribution of an F-statistic. A significant F-ratio for a given factor implies that at least one of the individual comparisons among means is significant for that factor. We use a model, in which there is some overall mean for the data and then variation around that value. The means from each of our treatment levels and their differences from this grand mean represent a way to measure the effect of those treatments. However, we have to view those differences in the light of the random variation that is present in our experiment. So, like the t- or z-statistic, the F-ratio is a ratio of signal-to-noise. In a simple two product experiment with one group of people testing each product, the F statistic is simply the square of the t-value, so there is an obvious relationship between the F- and t-statistics.

The statistical distributions for F indicate whether the ratio we obtain in the experiment is one we would expect only rarely by the operation of chance. Thus we apply the usual statistical reasoning when deciding to accept or reject a null hypothesis. The null hypothesis for ANOVA is usually that the means for the treatment levels would all be equal in the parent population. Analysis of variance is thus based on a model, a linear model, that says that any single data point or observation is result of several influences—the grand mean, plus (or minus) whatever deviations are caused by

each treatment factor, plus the interactions of treatment factors, plus error.

C.1.3 Rationale

The worked example below will examine this in more detail, but first a look at some of the rationale and derivation. The rationale proceeds as follows:

(a) We wish to know whether there are any significant differences among multiple means, relative to the error in our experimental measures.
(b) To do this, we examine variance (squared standard deviations).
(c) We look at the variance of our sample means from the overall ("grand") mean of all of our data. This is sometimes called the variance due to "treatments." Treatments are just the particular levels of our independent variable.
(d) This variance is examined relative to the variance within treatments, i.e. the unexplained error or variability not attributed to the treatments themselves.

The test is done by calculating a ratio. When the null is true (no difference among product means) it is distributed as an F-statistic. The F-distribution looks like a t-distribution squared (and is in the same family as the chi square distribution). Its exact shape changes and depends upon the number of degrees of freedom associated with our treatments or products (the numerator of the ratio) and the degrees of freedom associated with our error (the denominator of the ratio).

Here is a mathematical derivation and a similar but more detailed explanation can be found in O'Mahony (1986). Variance (the square of a standard deviation) is noted by S^2, x represents each score, and M is the mean of x scores or $(\Sigma x)/N$. Variance is the mean difference of each score from the mean, given by

$$S^2 = \frac{\sum_{i=1}^{N}(X_i - M)^2}{N - 1} \qquad (C.1)$$

and computationally by

$$S^2 = \frac{\sum_{i=1}^{N} X_i^2 - \frac{(\Sigma X)^2}{N}}{N - 1} \qquad (C.2)$$

This expression can be thought of as the "mean squared deviation." For our experimental treatments, we can speak of the means squared due to treatments, and for error, we can speak of the "mean squared error." The ratios of these two quantities give us the F-ratio, which we compare to the expected distribution of the F-statistic(under a true null). Note that in the computational formula for S^2, we accumulate sums of squared observations. Sums of squares form the basis of the calculations in ANOVA.

To calculate the sums of squares, it is helpful to think about partitioning the total variation.

Total variance is partitioned into variance between treatments and variance within treatments (or error). This can also be done for the sums of squares (SS):

$$SS_{total} = SS_{between} + SS_{within} \qquad (C.3)$$

This is useful since SS_{within} ("error") is tough to calculate—it is like a pooled standard deviation over many treatments. However, SS_{total} is easy! It is simply the numerator of our overall variance or

$$SS_{total} = \sum_{i=1}^{N} X_i^2 - \frac{(\Sigma X)^2}{N} \text{ over all } x \text{ data points.}$$
$$(C.4a)$$

So we usually estimate SS_{within} (error) as SS_{total} minus $SS_{between}$. A mathematical proof of how the SS can be partitioned like this is found in O'Mahony (1986), appendix C, p. 379.

C.1.4 Calculations

Based on these ideas, here is the calculation in a simple one-way ANOVA. "One-way" merely signifies that there is only one treatment variable or factor of interest. Remember, each factor may have multiple levels, which are usually the different versions of the product to be compared. In the following examples, we will talk in terms of products and sensory judges or panelists.

Let $T =$ a total (It is useful to work in sums)
let $a =$ number of products (or treatments)
let $b =$ number of panelists per treatment.
The product, $ab = N$.

$$SS_{total} = \sum_{i=1}^{N} X_i^2 - \frac{T^2}{N} \qquad (C.4b)$$

T without subscript is the grand total of all data or simply Σx, over all data points.

O'Mahony calls T^2/N a "correction factor" or "C," a useful convention:

$$SS_{between} = (1/b) \sum T_a^2 - T^2/N \qquad (C.5)$$

where the "a" subscript refers to different products. Now we need the error sums of squares, which is simply from

$$SS_{within} = SS_{total} - SS_{between} \qquad (C.6)$$

The next step is to divide each SS by its associated degrees of freedom to get our mean squares. We have mean squares associated with products and mean squares associated with error. In the final step, we use the ratio of these two estimates of variance to form our F-ratio.

C.1.5 A Worked Example

Our experimental question is did the treatment we used on the products make any difference? In other words, are these means likely to represent real differences, or just the effects of chance variation? The ANOVA will help address these questions. The sample data set is shown in Table C.1.

Table C.1 Data set for simple one-way ANOVA

Panelist	Product A	Product B	Product C
1	6	8	9
2	6	7	8
3	7	10	12
4	5	5	5
5	6	5	7
6	5	6	9
7	7	7	8
8	4	6	8
9	7	6	5
10	8	8	8
Totals	61	68	79
Means	6.1	6.8	7.9

First, column totals are calculated and a grand total, as well as the correction factor ("C") and the sum of the squared data points.

Sums	$T_a = 61$	$T_b = 68$	$T_c = 79$
	(Product A)	(Product B)	(Product C)

Grand T (sum of all data) $= 208$
$T^2/N = (208)^2/30 = 1442.13$ (O'Mahony's "C" factor)
$\Sigma(x^2) = 1530$ (sum of all squared scores)

Given this information, the sums of squares can be calculated as follows:

$\text{SS}_{\text{total}} = 1530 - 1442.13 = 87.87$
SS due to treatments ("between") $= (T_a^2 + T_b^2 + T_c^2)/b - T^2/N$
(remember b is the number of panelists)
$= (61^2 + 68^2 + 79^2)/10 - 1442.13 = 16.47$

Next, we need to find the degrees of freedom. The total degrees of freedom in the simple one-way ANOVA are the number of observation minus one (30–1=29). The degrees of freedom for the treatment factor are the number of levels minus one. The degrees of freedom for error are the total degrees of freedom minus the treatment ("between") df.

df total $= N-1 = 29$
df for treatments $= 3-1 = 2$
df for error $= \text{df}_{\text{total}} - \text{df}_{\text{between}} = 29-2 = 27$

Finally, a "*source table*" is constructed to show the calculations of the mean squares (our variance estimates) for each factor, and then to construct the *F*-ratio. The mean squares are the SS divided by the appropriate degrees of freedom (MS=SS/df). Table C.2 is the source table.

Table C.2 Source table for first ANOVA

Source of variance	SS	df	Mean squares	F
Total	87.867	(29)		
Between	16.467	2	8.233	3.113
Within (error)	71.4	27	2.644	

A value of F = 3.119 at 2 and 27 degrees of freedom is just short of significance at $p = 0.06$. Most statistical software programs will now give an exact *p*-value for the *F*-ratio and degrees of freedom. If the ANOVA is done "by hand" then the *F*-ratio should be compared to the critical value found in a table such as Table D. We see from this table that the critical value for 2 and 27 df is about 3.35 (we are interpolating here between 2, 26 and 2, 28 df), and our obtained value did not exceed this critical value.

C.2 Analysis of Variance from Complete Block Designs

C.2.1 Concepts and Partitioning Panelist Variance from Error

The complete block analysis of variance for sensory data occurs when all panelists view all products, or all levels of our treatment variable (Gacula and Singh, 1984). This type of design is also called the "repeated measures" analysis of variance in the behavioral sciences, when the experimental subject participates in all conditions (O'Mahony, 1986; Winer, 1971). Do not confuse the statistical term, "repeated measures" with replication. The design is analogous to the dependent or paired observations *t*-test, but considers multiple levels of a variable, not just two. Like the dependent *t*-test, it has added sensitivity since the variation due to panelist differences can be partitioned from the analysis, in this case taken out of the error term. When the error term is reduced, the *F*-ratio due to the treatment or variable of interest will be larger, so it is "easier" to find statistical significance. This is especially useful in sensory evaluation, where panelists, even well trained ones, may use different parts of the scale or they may simply have different sensitivities to the attribute being evaluated. When all panelists rank order products the same, the complete block ANOVA will usually produce a significant difference between products, in spite of panelists using different ranges of the scale.

The example below shows the kind of situation where a complete block analysis, like the dependent *t*-test, will have value in finding significant differences. In this example, two ratings by two subjects are shown in Fig. C.1. The differences between products, also called "within subject differences" are in the same direction and of the same magnitude. The "within-subject" effects in repeated measures terminology corresponds to between-treatment effects in simple ANOVA terminology. (This can be confusing.)

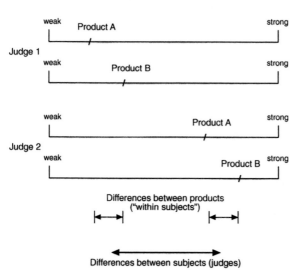

Panelist vs. product variation

Line scale ratings for two products, A and B,
by two panelists, Judge 1 and Judge 2.

(Ratings indicated by this mark: **/**)

Fig. C.1 Two hypothetical panelists rating two products. They agree on the rank order of the products and the approximate sensory difference, but use different parts of the scale. The differences can be separated into the between-products (within panelist) differences and the difference between the panelists in the overall part of the scale they have used. The dependent t-test separates these two sources of difference by converting the raw scores to difference scores (between products) in the analysis. In a complete block design when panelist variation can be partitioned, the ANOVA provides a more sensitive comparison than leaving the inter-individual difference in the error term.

In this example, the difference between panelists in the part of the scale they use is quite large. In any conventional analysis, such variation between people would swamp the product effect by creating a large error term. However, the panelist differences can be pulled out of the error term in a complete block design, i.e. when every panelist evaluated all of the products in the experiment.

To see the advantage of this analysis, we will show examples with and without the partitioning of panelist variance. Here is a worked example, first without partitioning panelist effects or as if there were three independent groups evaluating each product. This is a simple one-way ANOVA as shown above. Three products are rated. They might differ in having three levels of an ingredient. The sample data set is shown in Table C.3, with one small change from the first example of simple ANOVA. Note that panelist #10 has

Table C.3 Data set for the complete block design

Panelist	Product A	Product B	Product C
1	6	8	9
2	6	7	8
3	7	10	12
4	5	5	5
5	6	5	7
6	5	6	9
7	7	7	8
8	4	6	8
9	7	6	5
10	1	2	3
Totals	54	62	74
Means	5.4	6.2	7.4

now produced the values 1, 2, and 3 instead of 8, 8, and 8. The panelist is no longer a non-discriminator, but is probably insensitive.

Here is how the one-way ANOVA would look:

$$\text{Sums} \qquad T_a = 54 \qquad \underline{T_b} = 62 \qquad T_c = 74$$

Grand T (sum of all data points) $= 190$
$T^2/N = (190)^2/30 = 1203.3$ (O'Mahony's "C" factor)
$\Sigma(x^2) = 1352$
$SS_{total} = 1352 - 1203.3 = 148.7$
SS due to products $= (T_a^2 + T_b^2 + T_c^2)/b - T^2/N$
(remember b here refers to the number of panelists)
$= (54^2 + 62^2 + 74^2)/10 - 1203.3 = 20.3$
$SS_{error} = SS_{Total} - SS_{Products} = 148.7 - 20.3 = 128.4$

Table C.4 shows the source table.

Table C.4 Source table for complete block ANOVA

Source of variance	SS	df	Mean squares	F
Total	148.67	(29)		
Between	20.26	2	10.13	2.13
Within (error)	128.4	27	4.76	

For 2 and 27 degrees of freedom, this F gives us a $p = 0.14$ ($p > 0.05$, not significant). The critical F-ratio for 2 and 27 degrees of freedom is about 3.35 (interpolated from values in Table D).

Now, here is the difference in the complete block ANOVA. An additional computation requires row sums and sums of squares for the row variable, which is our panelist effect as shown in Table C.3. In the one-way analysis, the data set was analyzed as if there were 30 different people contributing the ratings. Actually, there were ten panelists who viewed all products. This

Table C.5 Data set for the complete block design showing panelist calculations (rows)

Panelist	Product A	Product B	Product C	Σpanelist	(Σpanelist)^2
1	6	8	9	23	529
2	6	7	8	21	441
3	7	10	12	19	841
4	5	5	5	15	225
5	6	5	7	18	324
6	5	6	9	20	400
7	7	7	8	22	484
8	4	6	8	18	324
9	7	6	5	18	324
10	8	8	8	6	36
Totals	61	68	79		3,928

fits the requirement for a complete block design. We can thus further partition the error term into an effect due to panelists ("between-subjects" effect) and residual error. To do this, we need to estimate the effect due to inter-panelist differences. Take the sum across rows (to get panelist sums), then square them. Sum again down the new column as shown in Table C.5. The panelist sum of squares is analogous to the product sum of squares, but now we are working across rows instead of down the columns:

$$SS_{panelists} = \sum \left(\sum panelist \right)^2 /3 - C = 3928/3 - 1203.3 = 106$$

"C", once again is the "correction factor" or the grand total squared, divided by the number of observations. In making this calculation, we have used nine more degrees of freedom from the total, so these are no longer available to our estimate of error df below.

A new sum of squares for residual error can now be calculated:

$$SS_{error} = SS_{total} - SS_{products} - SS_{panelists} = 148.7 - 20.3 - 106 = 22.36$$

and the mean square for error (MS error) is $SS_{error}/18 = 22.36/18 = 1.24$

Note that there are now only 18 degrees of freedom left for the error since we took another nine to estimate the panelists' variance. However, the mean square error has shrunk from 4.76 to 1.24. Finally, a new F-ratio for the product effect ("within subjects") having removed the between-subjects effect from the error term as shown in our source table, Table C.6.

So the new $F = MS_{products}/MS_{error} = 10.15/1.24 = 8.17$

At 2 and 18 degrees of freedom, this is significant at $p = 0.003$, and it is now bigger than the critical F for 2 and 18 degrees of freedom ($F_{crit} = 3.55$)

Table C.6 Source table for two-way ANOVA

Source of variance	SS	df	Mean squares	F
Total	148.7	(29)		
Products	20.3	2	10.13	8.14
Panelists	106	9		
Error	22.4	18	1.24	

Why was this significant when panelist variance was partitioned, but not in the usual one-way ANOVA? The answer lies in the systematic variation due to panelists' scale use and the ability of the two-way ANOVA to remove this effect from the error term. Making error smaller is a general goal of just about every sensory study, and here we see a powerful way to do this mathematically, by using a specific experimental design.

C.2.2 The Value of Using Panelists As Their Own Controls

The data set in the complete block example was quite similar to the data set used in the one-way ANOVA illustrated first. The only change was in panelist #10, who rated the products all as an 8 in the first example. In the second example, this non-discriminating panelist was removed and data were substituted from an insensitive panelist, but one with correct rank ordering. This panelist rated the products 1, 2, and 3 following the general trend of the rest of the panel, but on an overall lower level of the scale.

Notice the effect of substituting a person who is an outlier on the scale but who discriminates the products in the proper rank order. Because these values are quite low, they add more to the overall variance than to the product differences, so the one-way ANOVA goes from nearly significant ($p = 0.06$) to much less evidence against the null ($p = 0.14$). In other words, the panelist who did not differentiate the products, but who sat in the middle of the data set was not very harmful to the one-way ANOVA, but the panelist with overall low values contributes to error, even though he or she discriminated among the products. Since the complete block design allows us to partition out overall panelist differences, and focus just on product differences, the fact that he or she was a low rater does not hurt this type of analysis. The F-ratio for products is now significant ($p = 0.003$). In general, the panelists are monotonically increasing, with the exceptions of #4, 5, and 9 (dotted lines) as shown in Fig. C.2. The panelist with low ratings follows the majority trend and thus helps the situation.

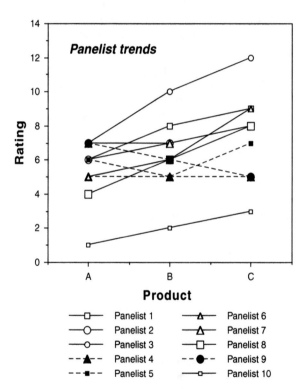

Fig. C.2 Panelist trends in the complete block example. Note that panelist #10 has rank ordered the products in same way as the panel means, but is a low outlier. This panelist is problematic for the one-way ANOVA but less so for when the panelist effects are partitioned as in the repeated measures models.

The same statistically significant result is obtained in a Friedman "analysis of variance" on ranks (see Appendix B). A potential insight here is the following: having a complete design allows repeated measures ANOVA. This allows us to "get rid of" panelist differences in scale usage, sensory sensitivity, anosmia, etc., and focus on product trends. Since humans are notoriously hard to calibrate, this is highly valuable in sensory work.

C.3 Planned Comparisons Between Means Following ANOVA

Finding a significant F-ratio in ANOVA is only one step in statistical analysis of experiments with more than two products. It is also necessary to compare treatment means to see which pairs were different. A number of techniques are available to do this, most based on variations on the t-test. The rationale is to avoid inflated risk of Type I error that would be inherent in making comparisons just by repeating t-tests. For example, the Duncan test attempts to maintain "experiment-wise" alpha at 0.05. In other words, across the entire set of paired comparisons of the product means, we would like to keep alpha-risk at a maximum of 5%. Since risk is a function of number of tests, the critical value of the t-statistic is *adjusted* to maintain risk at an acceptable level.

Different approaches exist, differing in assumptions and degree of "liberality" in amount of evidence needed to reject the null. Common types include the tests called Scheffé, Tukey, or HSD (honestly-significant-difference), Newman-Keuls, Duncans, LSD ("least-significant-difference"). The Scheffé test is most conservative and the LSD test the least (for examples see Winer, 1971, pp. 200–201). The Duncan procedure guards against Type I error among a set of comparisons, as long as there is already a significant F-ratio found in the ANOVA. This is a good compromise test to use for sensory data. The LSD test and the Duncan test are illustrated below.

The least significant difference, or LSD test is quite popular, since you simply compute the difference between means required for significance, based on your error term from the ANOVA. The error term is a pooled estimate of error considering all your treatments together. However, the LSD test does little to protect you from making too many comparisons, since the critical values do not increase with the numbers of comparisons you make, as is the case with some of the other statistics such as Duncan and Tukey (HSD) tests:

$$LSD = t\sqrt{\frac{2MS_{error}}{N}} \qquad (C.7)$$

where n is the number of panelists in a one-way ANOVA or one factor repeated measures, and t is the t-value for a two-tailed test with the degrees of freedom for the error term. The difference between the means must be *larger than the LSD*.

Calculations for the Duncan multiple range test use a "studentized range statistic", usually abbreviated with a lower case "q". The general formula for comparing pairs of individual means is to find the quantity to the right of this inequality and compare it to q:

$$q_p \leq \frac{Mean_1 - Mean_2}{\sqrt{\frac{2MS_{error}}{N}}} \qquad (C.8)$$

The calculated value must exceed a tabled value of q_p, which is based on the number of means separating the two we wish to compare, when all the means are rank ordered. MS_{error} is the error term associated with that factor in the ANOVA from which the means originate, n is the number of observations contributing to each mean, and q_p is the studentized range statistic from Duncan's tables (see table G). The subscript, p, indicates the number of means between the two we are comparing (including themselves), when they are rank ordered. If we had three means, we would use the value for p of 2 for comparing adjacent means when ranked and for p of 3 for comparing the highest and lowest means. The degrees of freedom are $n-1$. Note that the values for q are similar to but slightly greater than the corresponding t-values.

The general steps proceed as follows:

1. Conduct ANOVA and find MS error term
2. Rank order the means
3. Find q values for each p (number of means between, plus 2) and $n-1$ df.
4. Compare q to the formula in Eq. C.8 or
5. Find critical differences than must be exceeded by the values of

$$Difference \geq q_p\sqrt{\frac{2MS_{error}}{N}}$$

Note that this is just like the LSD test, but uses q instead of t. These critical differences are useful when you have lots of means to compare.

Here is a sample problem. From a simple one-way ANOVA on four observations, the means were as follows: Treatment $A = 9$, Treatment $B = 8$, and Treatment $C = 5.75$. The $MS_{error} = 0.375$. If we compare treatments A and C, the quantity to exceed q becomes

$$\frac{(9 - 5.75)}{\sqrt{2(0.375)/4}} = 3.25/0.433 = 7.5$$

The critical value of q, for $p = 3$, alpha $= 0.05$ is 4.516, so we can conclude that treatments A and C were significantly different.

An alternative computation is to find a critical difference by multiplying our value of q times the denominator (our error term) to find the difference between the means that must be exceeded for significance. This is sometimes easier to tabulate if you are comparing a number of means "by hand". In the above example, using the steps above for finding a critical difference, we multiply q (or 4.516) by the denominator term for the pooled standard error (0.433), giving a critical difference of 1.955. Since 9–5.75 (= 3.25) exceeds the critical difference of 1.955, we can conclude that these two samples were different.

C.4 Multiple Factor Analysis of Variance

C.4.1 An Example

In many experiments, we will have more than one variable of interest, for example, two or more ingredients or two or more processing changes. The applicable statistical tool for analysis of scaled data where we have two or more independent variables (called factors) is the multiple factor analysis of variance. These are called two-way ANOVAs for two variables, three-way for three variables, and so on.

Here is a simple sample problem and the data set is shown in Table C.7. We have two sweeteners, sucrose and high-fructose corn syrup (HFCS), being blended in a food (say a breakfast cereal), and we would like to understand the impact of each on the sweetness of the product. We vary the amount of each sweetener added

Table C.7 Data set for a two-factor analysis of variance (entries such as 1,1,2,4 represent four data points)

	Factor 1: Level of sucrose		
	Level 1	Level 2	Level 3
Level of HFCS	2%	4%	6%
Level A (2%)	1,1,2,4	3,5,5,5	6,4,6,7
Level B (4%)	2,3,4,5	4,6,7,5	6,8,8,9
Level C (6%)	5,6,7,8	7,8,8,6	8,8,9,7

to the product (2, 4 and 6% of each) and have a panel of four individuals rate the product for sweetness. (Four panelists are probably too few for most experiments but this example is simplified for the sake of clarity.) We use three levels of each sweetener, in a factorial design. A factorial design means that each level of one factor is combined with every level of the other factor.

We would like to know whether these levels of sucrose had any effect, whether the levels of HFCS had any effect and whether the two sweeteners in combination produced any result that would not be predicted from the average response to each sweetener. This last item we call an interaction (more on this below).

First, let us look at the cell means, and marginal means, shown in Table C.8:

Table C.8 Means for two factor experiment

	Factor (variable) 1			
Factor 2	Level 1	Level 2	Level 3	Row mean
Level A	2.0	4.5	5.75	4.08
Level B	3.5	5.5	7.75	5.58
Level C	6.5	7.25	8.0	7.25
Column mean	4.0	5.75	7.17	5.63 (Grand mean)

Next let us look at some graphs of these means to see what happened. Figure C.3 shows the trends in the data.

C.4.2 Concept: A Linear Model

Here is what happened in the analysis of the previous data set: The ANOVA will test hypotheses from a general linear model. This model states that any score in the data set is determined by a number of factors:

Score = Grand mean + Factor 1 effect + Factor 2 effect + Interaction effect + Error.

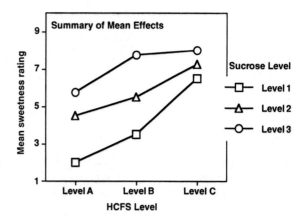

Fig. C.3 Means from the two-factor sweetener experiment.

In plain English, there is an overall tendency for these products to be sweet that is estimated by the grand mean. For each data point, there is some perturbation from that mean due to the first factor, some due to the second factor, some due to the particular ways in which the two factors interact or combine, and some random error process. The means for the null hypothesis are the population means we would expect from each treatment, averaged across all the other factors that are present. These can be thought of as "marginal means" since they are estimated by the row and column totals (we would often see them in the margins of a data matrix as calculations proceeded).

For the effects of our two sweeteners, we are testing whether the marginal means are likely to be equal (in the underlying population), or whether there was some systematic differences among them, and whether this variance was large relative to error, in fact so large that we would rarely expect such variation under a true null hypothesis.

The ANOVA uses an F-ratio to compare the effect variance to our sample error variance. The exact calculations for this ANOVA are not presented here, but they are performed in the same way as the two-factor complete design ANOVA (sometimes called "repeated measures" (Winer, 1971)) that is illustrated in a later section.

The output of our ANOVA will be presented in a table like Table C.9.

We then determine significance by looking up the critical F-ratio for our numerator and denominator degrees of freedom. If the obtained F is greater than the tabulated F, we reject the null hypothesis and conclude

Table C.9 Source table for two-way ANOVA with interaction

Effect	Sums of squares	df	MS	F
Factor 1	60.22	2	30.11	25.46
Error	7.11	6	1.19	
Factor 2	60.39	2	30.20	16.55
Error	10.94	6	1.82	
Interaction	9.44	4	2.36	5.24
Error	5.22	12	0.43	

that our factor had some effect. The critical F-values for these comparisons are 5.14 for the sweetener factors (2 and 6 df, both significant) and 3.26 (for 4 and 12 df) for the interaction effect (see Table D). The interaction arises since the higher level of sweetener 2 in Fig. C.3 shows a flatter slope than the slopes at the lower levels. So there is some saturation or flattening out of response, a common finding at high sensory intensities.

C.4.3 A Note About Interactions

What is an interaction? Unfortunately, the word has both a common meaning and a statistical meaning. The common meaning is when two things act upon or influence one another. The statistical meaning is similar, but it does not imply that a physical interaction, say between two food chemicals occurred. Instead, the term "interaction" means that the effect of one variable *changed* depending upon the level of the other variable. Here are two examples of interaction. For the sake of simplicity, only means are given to represent two variables at two points each.

In the first example, two panels evaluated the firmness of texture of two food products. One panel saw

a big difference between the two products while the second found only a small difference. This is visible as a difference in the slope of the lines connecting the product means. Such a difference in slope is called a magnitude interaction when both slopes have the same sign, and it is fairly common in sensory research. For example, panelists may all evaluate a set of products in the same rank order, but some may be more fatigued across replications. Decrements in scores will occur for some panelists more than others, creating a panelist by replicate interaction.

The second example of an interaction is a little less common. In this case the relative scores for the two products change position from one panel to the other. One panel sees product 1 as deserving a higher rating than product 2, while the other panel finds product 2 to be superior. This sort of interaction can happen with consumer acceptance ratings when there are market segments or in descriptive analysis if one panel misunderstands the scale direction or it is misprinted on the ballot (e.g., with end-anchor words reversed). This is commonly called a crossover interaction. Figure C.4 shows these interaction effects. A crossover interaction is much more serious and can be a big problem when the interaction effect is part of the error term as in some ANOVAs (see Sections C.5 below and C.6.1).

C.5 Panelist by Product by Replicate Designs

A common design in sensory analysis is the two-way ANOVA with all panelists rating all products (complete block) and replicated ratings. This design

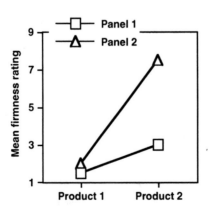
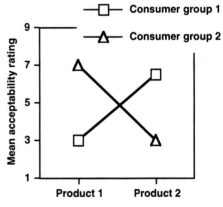

Fig. C.4 Interaction effect. *Upper panel*: Magnitude interaction. *Lower panel*: Crossover interaction.

would be useful with a descriptive panel, for example, where panelists commonly evaluate all the products. An example of the two factors is when there is one set of products and replications. Each score is a function of the panelist effect, treatment effect, replication effect, interactions, and error.

Error terms for treatment and replication are the interaction effects of each with panelists, which form the denominator of the F-ratio. This is done because panelist effects are random effects (see Section C.6.1), and the panelist by treatment interaction is embedded in the variance estimate for treatments. For purposes of this example, treatments and replications are considered fixed effects (this is a mixed model or TYPE III in some statistical programs like SAS). The sample data set is shown in Table C.10. Once again there are a small number of panelists so that the calculations will be a little simpler. Of course, in most real sensory studies the panel size would be considerably larger, e.g., 10–12 for descriptive data and 50–100 for consumer studies.

Table C.10 Data set for a two-factor ANOVA with partitioning of panelist variation

Product	Replicate 1			Replicate 2		
	A	B	C	A	B	C
Panelist 1	6	8	9	4	5	10
Panelist 2	6	7	8	5	8	8
Panelist 3	7	10	12	6	7	9

The underlying model says that the total variance is a function of the product effect, replicate effect, panelist effect, the three two-way interactions, the three-way interaction, and random error. We have no estimate of the smallest within-cell error term other than the three-way interaction. Another way to think about this is that each score deviates from the grand mean as a function of that particular product mean, that particular panelist mean, that particular replication mean, plus (or minus) any other influences from the interactions.

Here are the calculations, step by step. This is a little more involved than our examples so far. We will call the effect of each factor a "main effect" as opposed to the interaction effects and error.

Step 1. First, we calculate sums of squares and main effects.

As in the one way ANOVA with repeated measures, there are certain values we need to accumulate:

Grand total $= 135$
(Grand total)$^2/N = T^2/N = 18{,}255/18 = 1012.5$
(O'Mahony's "correction factor", C)
Sum of squared data $= 1083$

There are three "marginal sums" we need to calculate, in order to estimate main effects.

The product marginal sums (across panelists and reps):

$\Sigma A = 34 \quad (\Sigma A)^2 = 1156$
$\Sigma B = 45 \quad (\Sigma B)^2 = 2025$
$\Sigma C = 56 \quad (\Sigma C)^2 = 3136$

The sum of squares for products then becomes

$$SS_{products} = [(1156 + 2025 + 3136)/6] - \text{correction factor, } C$$
$$= 052.83 - 1012.5 = 40.33$$

(We will need the value, 1052.83, later. Let us call it PSS1, for "partial sum of squares" #1)

Similarly, we calculate replicate and panelist sums of squares.

The replicate marginal sums (across panelists and products):

$\Sigma rep1 = 73 \quad (\Sigma rep1)^2 = 5{,}329$
$\Sigma rep2 = 62 \quad (\Sigma rep2)^2 = 3{,}844$

The sum of squares for replicates then becomes

$$SS_{reps} = [(5329 + 3844)/9] - \text{correction factor}$$
$$= 1019.2 - 1012.5 = 6.72$$

Note: the divisor, 9, is not the number of reps (2), but the number of panelists times the number of products ($3 \times 3 = 9$). Think of this as the number of observations contributing to each marginal total. (We need the value, 1019.2, later in calculations. Let us call it PSS2).

As in other repeated measures designs, we need the panelist sums (across products and reps):

$\Sigma pan1 = 42 \quad (\Sigma pan1)^2 = 1764$
$\Sigma pan2 = 42 \quad (\Sigma pan2)^2 = 1764$
$\Sigma pan3 = 51 \quad (\Sigma pan3)^2 = 2601$

The sum of squares for panelists then becomes

$$SS_{pan} = [(1764 + 1764 + 2601)/6] - \text{correction factor, } C$$
$$= 1021.5 - 1012.5 = 9.00$$

(We will need the value, 1021.5, later in calculations. Let us call it PSS3, for partial sum of squares #3)

Step 2. Next, we need to construct summary tables of interaction sums.

Here are the rep-by-product interaction calculations. We obtain a sum for each replicate by product combination and then square them. The three interaction tables are shown in Table C.11.

Table C.11 Interaction calculations

Product	Rep 1	Rep 2	Squared values	
A	19	15	361	225
B	25	20	625	400
C	29	27	841	729

Product	Panelist 1	Panelist 2	Panelist 3	Squared values		
A	10	11	13	100	121	169
B	13	15	17	169	225	289
C	19	16	21	361	256	441

Panelist	Rep 1	Rep 2	Squared values	
1	23	19	529	361
2	21	21	441	441
3	29	22	841	484

First for the product by replicate table, we obtain the following information:

Sum of squared values = 3181; 3181/3 = 1060.3 (= PSS4, needed later)

To calculate the sum of squares, we need to subtract the PSS values for each main effect and then add back the correction term (this is dictated by the underlying variance model):

$$SS_{rep \times prod} = (3181/3) - PSS1 - PSS2 + C$$
$$= 1060.3 - 1052.83 - 1019.2 + 1012.5$$
$$= 0.77$$

Next we look at the panelist by product interaction information. The panelist by product interaction calculations are based on the center of Table C.11. Once again, we accumulate the sums for each combination and then square them giving these values:

Sum of squared values = 2131; 2131/2 = 1065.5 (= PSS5, needed later)

$$SS_{pan \times prod} = (2131/2) - PSS1 - PSS3 + C$$
$$= 1065.5 - 1052.83 - 1021.5 + 1012.5$$
$$= 3.67$$

Here are the replicate by panelist interaction calculations, based on the lower part of Table C.11:

Sum of squared values = 3097; 3097/3 = 1032.3 (= PSS6, needed later)

$$SS_{R \times pan} = (3097/3) - PSS2 - PSS3 + C$$
$$= 1032.3 - 1021.5 - 1019.2 + 1012.5$$
$$= 4.13$$

The final estimate is for the sum of squares for the three-way interaction. This is all we have left in this design, since we are running out of degrees of freedom. This is found by the sum of the squared (data) values, minus each PSS from the interactions, plus each PSS from the main effects, minus the correction factor. Do not worry too much about where this comes from, you would need to dissect the variance component model to fully understand it:

$$SS \, (3 \, way) = \sum x^2 - PSS4 - PSS5 - PSS6 + PSS1$$
$$+ PSS2 + PSS3 - C = 1083 - 1060.3$$
$$- 1065.5 - 1032.3 + 1052.83$$
$$+ 1019.2 + 1021.5 - 1012.5 = 5.93$$

Step 3. Using the above values, we can calculate the final results as shown in the source Table C.12.

Table C.12 Source table for panelist by product by replicate ANOVA

Effect	Sum of squares	df	Mean square	F
Products	40.33	2	20.17	21.97
Prod × panelist	3.67	4	0.92	
Replicates	6.72	1	6.72	3.27
Rep × panelist	4.13	2	2.06	
Product × rep	0.77	2	0.39	0.26
Prod × rep × panelist	5.93	4	1.47	

Note that the error terms for each effect are the interaction with panelists. This is dictated by the fact that panelists are a random effect and this is a "mixed model" analysis.

So only the product effect was significant. The critical F-ratios were 6.94 for the product effect (2,4 df), 19.00 for the replicate effect (1,2 df), and 6.94 for the interaction (2,4 df).

Degrees of freedom are calculated as follows:

For the main effects, df = levels−1, e.g., three products gives 2 df.

For interactions, df = product of df for individual factors, (e.g., prod × pan df = (3−1) × (3−1) = 4).

C.6 Issues and Concerns

C.6.1 Sensory Panelists: Fixed or Random Effects?

In a fixed effects model, specific levels are chosen for a treatment variable, levels that may be replicated in other experiments. Common examples of fixed effect variables might be ingredient concentrations, processing temperatures, or times of evaluation in a shelf life study. In a random effects model, the values of a variable are chosen by being randomly selected from the population of all possible levels. Future replications of the experiment might or might not select this exact same level, person, or item. The implication is that future similar experiments would also seek another random sampling rather than targeting specific levels or spacing of a variable. In this ANOVA model, the particular level chosen is thought to exert a systematic influence on scores for other variables in the experiment. In other words, interaction is assumed.

Examples of random effects in experimental design are common in the behavioral sciences. Words chosen for a memory study or odors sampled from all available odor materials for a recognition screening test are random, not fixed, stimulus effects. Such words or odors represent random choices from among the entire set of such possible words or odors and do not represent specific levels of a variable that we have chosen for study. Furthermore, we wish to generalize to all such possible stimuli and make conclusions about the parent set as a whole and not just the words or odors that we happened to pick. An persistent issue is whether sensory panelists are ever fixed effects. The fixed effects model is simpler and is the one most people learn in a beginning statistics course, so it has unfortunately persisted in the literature even though behavioral science dictates that human subjects or panelists are a random effect, even as they are used in sensory work.

Although they are never truly randomly sampled, panelists meet the criteria of being a sample of a larger population of potential panelists and of not being available for subsequent replications (for example, in another lab). Each panel has variance associated with its composition, that is, it is a sample of a larger population. Also, each product effect includes not only the differences among products and random error, but also the interaction of each panelist with the product variable. For example, panelists might have steeper or shallower slopes for responding to increasing levels of the ingredient that forms the product variable. This common type of panelist interaction necessitates the construction of F-ratios with interaction terms in the denominator. Using the wrong error term (i.e., from simple fixed effects ANOVAs) can lead to erroneous rejection of the null hypothesis.

Fixed effects are specific levels of a variable that experimenters are interested in, whereas random effects are samples of a larger population to which they wish to generalize the other results of the experiment. Sokal and Rohlf (1981) make the following useful distinction: [Fixed versus random effects models depend] "on whether the different levels of that factor can be considered a random sample of more such levels, or are fixed treatments whose differences the investigator wishes to contrast." (p. 206).

This view has not been universally applied to panelist classification within the sensory evaluation community. Here are some common rejoinders to this position.

When panelists get trained, are they no longer a random sample and therefore a fixed effect. This is irrelevant. We wish to generalize these results to any such panel of different people, similarly screened and trained, from the population of qualifying individuals. Hays (1973) puts this in perspective by stating that even though the sample has certain characteristics, that does not invalidate its status as a sample of a larger group: "Granted that only subjects about whom an inference is to be made are those of a certain age, sex, ability to understand instructions, and so forth, the experimenter would, nevertheless, like to extend his inference to all possible such subjects." (p. 552).

A second problem arises about the use of the interaction term in mixed model ANOVAs. We can assume no interaction in the model or even test for the existence of a significant interaction. The answer is that you can, but why choose a riskier model, inflating your chance of Type I error? If you test for no significant interaction, you depend upon a failure to reject the null, which is an ambiguous result, since it can happen from a sloppy experiment with high error variance, just as well as from a situation where there is truly no effect. So it is safer to use a mixed model where panelists are considered random. Most statistical packages will select the interaction effect as the error term when you specify panelists as a random effect, and some even assume it as the default.

Further discussion of this issue can be found in the book by Lea et al. (1998), Lundahl and McDaniel (1988) and in Lawless (1998) and the articles in the same issue of that journal.

C.6.2 A Note on Blocking

In factorial designs with two or more variables, the sensory specialist will often have to make some decisions about how to group the products that will be viewed in a single session. Sessions or days of testing often form one of the blocks of an experimental design. The examples considered previously are fairly common in that the two variables in a simple factorial design are often products and replicates. Of course judges, are a third factor, but a special one. Let us put judges aside for the moment and look at a complete block design in which each judge participates in the evaluation of all products and all of the replicates. This is a common design in descriptive analysis using trained panels.

Consider the following scenario: Two sensory technicians (students in a food company on summer internships) are given a sensory test to design. The test will involve comparisons of four different processed soft cheese spreads and a trained descriptive panel is available to evaluate key attributes such as cheese flavor intensity, smoothness, and mouthcoating. Due to the tendency of this product to coat the mouth, only four products can be presented in each session. The panel is available for testing on four separate sessions on different days. There are then two factors to be assigned to blocks of sessions in this experiment, the products and the replicates.

Technician "A" decides to present one version of the cheese spread on each day, but replicate it four times within a session. Technician "B" on the other hand presents all four products on each day, so that the sessions (days) become blocked as replicates. Both technicians use counterbalanced orders of presentation, random codes, and other reasonable good practices of sensory testing. The blocking schemes are illustrated in Fig. C.5.

Which design seems better? A virtually unanimous opinion among sensory specialists we asked is that assigning all four products within the same session is better than presenting four replicates of the same product within a session. The panelists will have a more stable frame of reference within a session than

across sessions, and this will improve the sensitivity of the product comparison. There may be day-to-day variations in uncontrolled factors that may confound the product comparisons across days (changes in conditions, changes in the products while aging, or in the panelists themselves) and add to random error. Having the four products present in the same session lends a certain directness to the comparison without any burden of memory load. There is less likelihood of drift in scale usage within a session as opposed to testing across days.

Why then assign products within a block and replicates across sessions? Simply stated, the product comparison is most often the more critical comparison of the two. Product differences are likely to be the critical question in a study. A general principle for assignment of variables to blocks in sensory studies where the experimental blocks are test sessions: Assign the variable of greatest interest within a block so that all levels of that factor are evaluated together. Conversely, assign the variable of secondary interest across the blocks if there are limitations in the number of products that can be presented.

C.6.3 Split-Plot or Between-Groups (Nested) Designs

It is not always possible to have all panelists or consumers rate all products. A common design uses different groups of people to evaluate different levels of a variable. In some cases, we might simply want to compare two panels, having presented them with the same levels of a test variable. For example, we might have a set of products evaluated in two sites in order to see if panelists are in agreement in two manufacturing plants or between a QC panel and an R and D panel. In this case, there will be repeated measures on one variable (the products) since all panelists see all products. But we also have a between-groups variable that we wish to compare. We call this a "split plot" design in keeping the nomenclature of Stone and Sidel (1993). It originates from agricultural field experiments in which plots were divided to accommodate different treatments. Bear in mind that we have one group variable and one repeated measures variable. In behavioral research, these are sometimes called "between-subjects" and "within-subjects" effects. Examples of these designs can be found in Stone and Sidel (1993) and Gacula et al. (2009).

Fig. C.5 Examples of blocking strategy for the hypothetical example of the processed cheese spread.

A Blocking Problem

Four product are available on four days. Four replicates are desired.
The product causes carry-over and fatigue and tends to coat the mouth,
so only four products can be tested in any one day.
How are the two factors to be assigned to the blocks of days (sessions)?

Approach "A" : assign replicates within a block, products across days
(sessions):

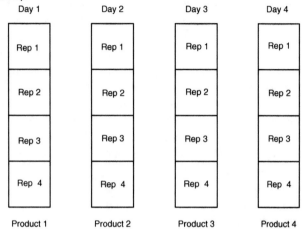

Approach "B": Test all four products in each day, once:

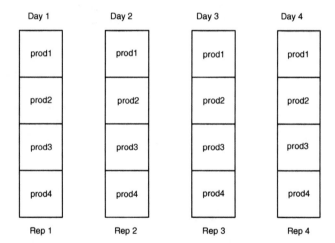

Which blocking scheme is better and why?

C.6.4 Statistical Assumptions and the Repeated Measures ANOVA

The model underlying the repeated measures analysis of variance from complete block designs has more assumptions than the simple one-way ANOVA. One of these is an assumption that the covariance (or degree of relationship) among all pairs of treatment levels is the same. Unfortunately, this is rarely the case in observations of human judgment.

Consider the following experiment: We have sensory judges examine an ice cream batch for shelf life in a heat shock experiment on successive days. We are lucky enough to get the entire panel to sit for every experimental session, so we can use a complete block or repeated measures analysis. However, their frame of reference is changing slightly, a trend that seems to be affecting the data as time progresses. Their data from adjacent days are more highly correlated than their data from the first and last test. Such time-dependencies violate the assumption of "homogeneity of covariance."

But all is not lost. A few statisticians have suggested some solutions, if the violations were not too bad (Greenhouse and Geisser, 1959; Huynh and Feldt, 1976). Both of these techniques adjust your degrees of freedom in a conservative manner to try and account for a violation of the assumptions and still protect you from that terrible deed of making a Type I error. The corrections are via an "epsilon" value that you will sometimes see in ANOVA package printouts, and adjusted p-values, often abbreviated G-G or H-F. Another solution is to use a multivariate analysis of variance approach or MANOVA, which does not labor under the covariance assumptions of repeated measures. Since most packaged printouts give you MANOVA statistics nowadays anyway, it does not hurt to give them a look and see if your conclusions about significance would be any different.

C.6.5 Other Options

Analysis of variance remains the bread-and-butter everyday statistical tool for the vast majority of sensory experiments and multi-product comparison. However, it is not without its shortcomings. One concern is that we end up examining each scale individually, even when our descriptive profile many contain many scales for flavor, texture, and appearance. Furthermore, many of these scales are intercorrelated. They may be providing redundant information or they may be driven by the same underlying of latent causes. A more comprehensive approach would include multivariate techniques such as principal components analysis, to assess these patterns of correlation. Analysis of variance can also be done following principal components, using the factor scores rather than the raw data. This has the advantage of simplifying the analysis and reporting of results although the loss of detail about individual scales may not always be desired. See Chapter 18 for descriptions of multivariate techniques and their applications.

A second concern is the restrictive assumptions of ANOVA that are often violated. Normal distributions,

equal variance, and in the case of repeated measures, homogeneity of covariance, are not always the case in human judgments. So our violations lead to unknown changes in the risk levels. Risk may be underestimated as the statistical probabilities of our analysis are based on distributions and assumptions that are not always descriptive of our experimental data. For such reasons, it has recently become popular to use MANOVA. Many current statistical analysis software packages offer both types of analysis and some even give them automatically or as defaults. The sensory scientist can then compare the outcomes of the two types of analyses. If they are the same, the conclusions are straightforward. If they differ, some caution is warranted in drawing conclusions about statistical significance.

The analyses shown in this section are relatively simple ones. It is obvious that more complex experimental designs are likely to come the way of a sensory testing group. In particular, incomplete designs in which people evaluate only some of the products in the design are common. Product developers often have many variables at several different levels each to screen. We have stressed the complete block designs here because of the efficient and powerful partitioning of panelist variance that is possible. Discussions of incomplete designs are their efficiency can be found in various statistics texts.

We recognize that it is unlikely at this point in the history of computing that many sensory professionals or statistical analysis services will spend much time doing ANOVAs by hand. If the experimental designs are complex or if many dependent measures have been collected, software packages are likely to be used. In these cases the authors of the program have taken over the burden of computing and partitioning variance. However, the sensory scientist can still make decisions on a theoretical level. For example, we have included all interaction terms in the above analyses, but the linear models upon which ANOVAs are based need not include such terms if there are theoretical or practical reasons to omit them. Many current statistical analysis packages allow the specification of the linear model, giving discretionary modeling power to the scientist. In some cases it may be advantageous to pool effects or omit interactions from the model if their variance contribution is small. This will increase the degrees of freedom for the remaining factors and increase the chances of finding a significant effect.

References

Gacula, M. C., Jr. and Singh, J. 1984. Statistical Methods in Food and Consumer Research. Academic, Orlando, FL.

Gacula, M., Singh, J., Bi, J. and Altan, S. 2009. Statistical Methods in Food and Consumer Research. Elsevier/Academic, Amsterdam.

Greenhouse, S. W. and Geisser, S. 1959. On methods in the analysis of profile data. Psychometrika, 24, 95–112.

Hays, W. L. 1973. Statistics for the Social Sciences, Second Edition. Holt, Rinehart and Winston, New York.

Huynh, H. and Feldt, L. S. 1976. Estimation of the Box correction for degrees of freedom in the randomized block and split plot designs. Journal of Educational Statistics, 1, 69–82.

Lawless, H. 1998. Commentary on random vs. fixed effects for panelists. Food Quality and Preference, 9, 163–164.

Lea, P., Naes, T. and Rodbotten, M. 1998. Analysis of Variance for Sensory Data. Wiley, Chichester, UK.

Lundahl, D. S. and McDaniel, M. R. 1988. The panelist effect—fixed or random? Journal of Sensory Studies. 3, 113–121.

O'Mahony, M. 1986. Sensory Evaluation of Food. Statistical Methods and Procedures. Marcel Dekker, New York.

Sokal, R. R. and Rohlf, F. J. 1981. Biometry, Second Edition. W. H. Freeman, New York.

Stone, H. and Sidel, J. L. 1993. Sensory Evaluation Practices, Second Edition. Academic, San Diego.

Winer, B. J. 1971. Statistical Principles in Experimental Design, Second Edition. McGraw-Hill, New York.

Appendix D

Correlation, Regression, and Measures of Association

Contents

The correlation coefficient is frequently abused. First, correlation is often improperly interpreted as evidence of causation. . . . Second; correlation is often improperly used as a substitute for agreement.

—Diamond (1989)

This chapter is a short introduction to correlation and regression. Pearson's correlation coefficient and the coefficient of determination for interval data are discussed followed by a section on linear regression. There is an example on how to calculate a linear regression. An extremely brief discussion of multiple linear regression is followed by a discussion of other measures of association. These are Spearman's rank correlation coefficient for ordinal data and Cramér's measure for nominal data.

D.1 Introduction

Sensory scientists are frequently confronted with the situation where they would like to know if there is a significant association between two sets of data. For example, the sensory specialist may want to know if the perceived brown color intensity (dependent variable) of a series of cocoa powder–icing sugar mixtures increased as the amount of cocoa (independent variable) in the mixture increased. Another example, the sensory scientist may want to know if the perceived sweetness of grape juice (dependent variable) is related to the total concentration of fructose and glucose (independent variable) in the juice, as determined by high-pressure liquid chromatography.

In these cases we need to determine whether there is evidence for an association between independent and dependent variables. In some cases, we may also be able to infer a cause and effect relationship between independent and dependent variables. The measures of association between two sets of data are called correlation coefficients and if the size of the calculated correlation coefficient leads us to reject the null hypothesis of no association then we know that the change in the independent variables—our treatments, ingredients, or processes—were associated with a change in the dependent variables—the variables we measured, such as sensory-based responses or consumer acceptance. However, from one point of view, this hypothesis testing approach is scientifically impoverished. It is not a very informative method of scientific research, because the statistical hypothesis decision is a binary or yes/no relationship. We conclude that either there is evidence for an association between our treatments and our observations or there is not. Having rejected the null hypothesis, we still do not know the degree of relationship or the tightness of the association between our variables. Even worse, we have not yet specified any type of mathematical model or equation that might characterize the association.

Associational measures also known as correlation coefficients can address the first of these two questions. In other words, the correlation coefficient will allow us to decide whether there is an association between the two data series. Modeling, in which we attempt to fit a mathematical function to the relationship, addresses the second question. The most widespread measure of association is the simple correlation coefficient (Pearson's correlation coefficient). The most common approach to model fitting is the simple linear regression. These two approaches have similar underlying calculations and are related to one another. The first sections of this chapter will show how correlations and regressions are computed and give some simple examples. The later sections will deal with related topics: how to build some more complex models between several variables, how measures of association are derived from other statistical methods, such as analysis of variance, and finally how to compute measures of association when the data do not have interval scale properties (Spearman rank correlation coefficient and Cramér's $\hat{\phi}'^2$).

Correlation is of great importance in sensory science because it functions as a building block for other statistical procedures. Thus, methods like principal components analysis draw part of their calculations from measures of correlation among variables. In modeling relationships among data sets, regression and multiple regression are standard tools. A common application is the predictive modeling of consumer acceptability of a product based upon other variables. Those variables may be descriptive attributes

as characterized by a trained (non-consumer) panel or they may be ingredient or processing variables or even instrumental measures. The procedure of multiple regression allows one to build a predictive model for consumer likes based on a number of such other variables. A common application of regression and correlation is in sensory–instrumental relationships. Finally, measures of correlation are valuable tools in specifying how reliable our measurements are—by comparing panelist scores over multiple observations and to assess agreement among panelists and panels.

When the sensory scientist is investigating possible relationships between data series the first step is to plot the data in a scatter diagram with the X-series on the horizontal axis and the Y-series on the vertical axis. Blind application of correlation and regression analyses may lead to wrong conclusions about the relationship between the two variables. As we will see the most common correlation and regression methods estimate the parameters of the "best" straight line through the data (regression line) and the closeness of the points to the line (simple correlation coefficient). However, the relationship may not necessarily be described well by a straight line, in other words the relationship may not necessarily be linear. Plotting the data in scatter diagrams will alert the specialist to problems in fitting linear models to data that are not linearly related (Anscombe, 1973). For example, the four data sets listed in Table D.1 and plotted in Fig. D.1 clearly are not all accurately described by a linear model. In all four cases the 11 observation pairs have mean x-values equal to 9.0, mean y-values equal to 7.5, a correlation

Table D.1 The Anscombe quartet—four data sets illustrating principles associated with linear correlation[1]

a		b		c		d	
x	y	x	y	x	y	x	y
4	4.26	4	3.10	4	5.39	8	6.58
5	5.68	5	4.74	5	5.73	8	5.76
6	7.24	6	6.13	6	6.08	8	7.71
7	4.82	7	7.26	7	6.42	8	8.84
8	6.95	8	8.14	8	6.77	8	8.47
9	8.81	9	8.77	9	7.11	8	7.04
10	8.04	10	9.14	10	7.46	8	5.25
11	8.33	11	9.26	11	7.81	8	5.56
12	10.84	12	9.13	12	8.15	8	7.91
13	7.58	13	8.74	13	12.74	8	6.89
14	9.96	14	8.10	14	8.84	19	12.50

[1] Anscombe (1973)

Fig. D.1 Scatter plots of the Anscombe quartet data (Table D.1) (redrawn from Anscombe).

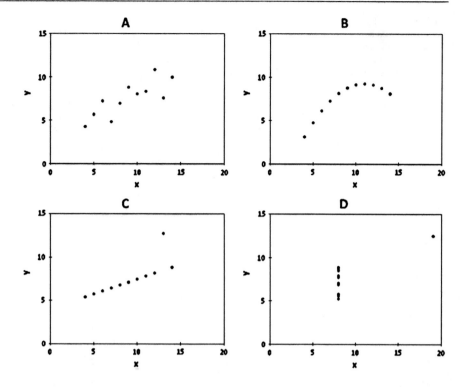

coefficient equal to 0.82, and a regression line equation of $y = 3 + 0.5x$.

However as Fig. D.1 clearly indicates the data sets are very different. These data sets are the so-called Anscombe quartet created by their author to highlight the perils of blindly calculating linear regression models and Pearson correlation coefficients without first determining through scatter plots whether a simple linear model is appropriate (Anscombe, 1973).

D.2 Correlation

When two variables are related, a change in one is usually accompanied by a change in the other. However, the changes in the second variable may not be linearly related to changes in the first. In these cases, the correlation coefficient is not a good measure of association between the variables (Fig. D.2).

The concomitant change between the variables may occur because the two variables are causally related. In other words, the change in the one variable causes the change in the other variable. In the cocoa–icing sugar example the increased concentration of cocoa powder in the mixture caused the increase in perceived brown color. However, the two variables may not be necessarily causally related because a third factor may drive the changes in both variables or there may be several intervening variables in the causal chain between the two variables (Freund and Simon, 1992). An anecdotal example often used in statistics texts is the following: Some years after World War II statisticians found a correlation between the number of storks and the number of babies born in England. This did not mean that the storks "caused" the babies. Both variables were related to the re-building of England (increase in families and in the roofs which storks use as nesting places) after the war. One immediate cautionary note in associational statistics is that the causal inference is not often clear. The usual warning to students of statistics is that "correlation does not imply causation."

The rate of change or dependence of one variable on another can be measured by the slope of a line relating the two variables. However, that slope will numerically depend upon the units of measurement. For example, the relationship between perceived sweetness and molarity of a sugar will have a different slope if concentration is measured in percent-by-weight or if units should change to milli-molar instead of molar concentration. In order to have a measure of association that is free from the particular units that are chosen, we must

Fig. D.2 The correlation coefficient could be used as a summary measure for plots A and B, but not for plots C and D (in analogy to Meilgaard et al., 1991).

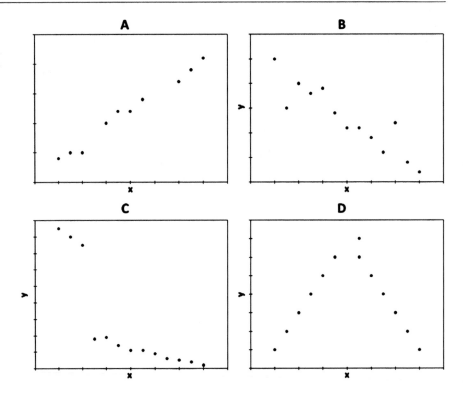

standardize both variables so that they have a common system of measurement. The statistical approach to this problem is to replace each score with its standard score or difference from the mean in standard deviation units. Once this is done, we can measure the association by a measure known as the Pearson correlation coefficient (Blalock, 1979). When we have not standardized the variables, we still can measure the association, but now it is simply called covariance (the two measures "vary together") rather than correlation. The measure of correlation is the extremely useful and very widespread in its applications.

The simple or Pearson correlation coefficient is calculated using following computational equation (Snedecor and Cochran, 1980):

$$r = \frac{\sum xy - \frac{\sum x \sum y}{n}}{\sqrt{\left[\sum x^2 - \frac{(\sum x)^2}{n}\right]\left[\sum y^2 - \frac{(\sum y)^2}{n}\right]}} \quad (D.1)$$

where the series of x data points = the independent variable and the series of y data points = the dependent variable.

Each data set has n data points and the degrees of freedom associated with the simple correlation coefficient are $n-2$. If the calculated r-value is larger than

the r-value listed in the correlation table (Table F2) for the appropriate alpha, then the correlation between the variables is significant.

The value of Pearson's correlation coefficient always lies between -1 and $+1$. Values of r close to absolute 1 indicate that a very strong linear relationship exists between the two variables. When r is equal to zero, then there is no linear relationship between the two variables. Positive values of r indicate a tendency for the variables to increase together. Negative values of r indicate a tendency of large values of one variable to be associated with small values of the other variable.

D.2.1 Pearson's Correlation Coefficient Example

In this study a series of 14 cocoa–icing sugar mixtures were rated by 20 panelists for brown color intensity on a nine-point category scale. Was there a significant correlation between the percentage of cocoa powder added to the icing sugar mixture and the perceived brown color intensity?

Cocoa added, %	Mean Brown color intensity	X^2	Y^2	XY
30	1.73	900	2.98[a]	51.82
35	2.09	1225	4.37	73.18
40	3.18	1600	10.12	127.27
45	3.23	2025	10.42	145.23
50	4.36	2500	19.04	218.18
55	4.09	3025	16.74	225.00
60	4.68	3600	21.92	280.91
65	5.77	4225	33.32	375.23
70	6.91	4900	47.74	483.64
75	6.73	5625	45.26	504.54
80	7.05	6400	49.64	563.64
85	7.77	7225	60.42	660.68
90	7.18	8100	51.58	646.36
95	8.54	9025	73.02	811.82
$\Sigma X = 875$	$\Sigma Y = 73.32$	$\Sigma X^2 = 60375$	$\Sigma Y^2 = 446.56$	$\Sigma XY = 5167.50$[a]

[a] Values rounded to decimals after calculation of squares and products

$$r = \frac{5167.50 - \frac{875 \times 72.32}{14}}{\sqrt{\left[60.375 - \frac{(875)^2}{14}\right]\left[446.56 - \frac{(73.32)^2}{14}\right]}} = 0.9806$$

At an alpha-level of 5% the tabular value of the correlation coefficient for 12 degrees of freedom is 0.4575 (Table F2). The calculated value $= 0.9806$ exceeds the table value and we can conclude that there is a significant association between the perceived brown color intensity of the cocoa–icing sugar mixture and the percentage of cocoa added to the mixture. Since the only ingredient changing in the mixture is the amount of brown color we can also conclude that the increased amount of cocoa causes the increased perception of brown color.

D.2.2 Coefficient of Determination

The coefficient of determination is the square of Pearson's correlation coefficient (r^2) and it is the estimated proportion of the variance of the data set Y that can be attributed to its linear correlation with X, while $1-r^2$ (the coefficient of non-determination) is the proportion of the variance of Y that is free from effect of X (Freund and Simon, 1992). The coefficient of determination can range between 0 and 1 and the closer the r^2 value is to 1 the better the straight line fits.

D.3 Linear Regression

Regression is a general term for fitting a function, usually a linear one, to describe the relationship among variables. Various methods are available for fitting lines to data, some based on mathematical solutions and others based on iterative or step-by-step trials to minimize some residual error or badness-of-fit measure. In all of these methods there must be some measurement of how good the fit of the model or equation is to the data. The least-squares criterion is the most common measure of fit for linear relationships (Snedecor and Cochran, 1980, Afifi and Clark, 1984). In this approach, the best fitting straight line is found by minimizing the squared deviations of every data point from the line in the y-direction (Fig. D.3).

The simple linear regression equation is

$$y = a + bx \tag{D.2}$$

where a is the value of the estimated intercept; b the value of the estimated slope.

The estimated least-squares regression is calculated using the following equations:

$$b = \frac{\Sigma xy - \frac{(\Sigma x)(\Sigma y)}{n}}{\Sigma x^2 - \frac{(\Sigma x)^2}{n}} \tag{D.3}$$

$$a = \sum y/n - b = \sum x/n \tag{D.4}$$

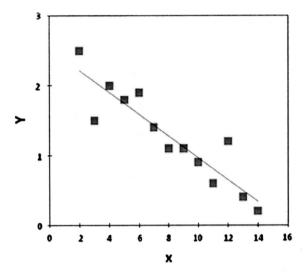

Fig. D.3 Least-squares regression line. The least-squares criterion minimizes the squared residuals for each point. The *arrow* indicates the residual for an example point. The *dashed line* indicates the mean X and Y values. The equation for the line is $Y = bx + a$.

It is possible to assess the goodness of fit of the equation in different ways. The usual methods used are the coefficient of determination and the analysis of variance (Piggott, 1986).

D.3.1 Analysis of Variance

In a linear regression it is possible to partition the total variation into the variation explained by the regression analysis and the residual or unexplained variation (Neter and Wasserman, 1974). The F-test is calculated by the ratio of the regression mean square to the residual mean square with 1 and $(n–2)$ degrees of freedom. This tests whether the fitted regression line has a non-zero slope. The equations used to calculate the total sums of squares and the sums of squares associated with the regression are as follows:

$$SS_{total} = \sum (Y_i - \bar{Y})^2 \qquad (D.5)$$

$$SS_{regression} = \sum (\hat{Y} - \bar{Y}) \qquad (D.6)$$

$$SS_{residual} = \sum (Y_i - \hat{Y})^2 = SS_{total} - SS_{regression} \qquad (D.7)$$

where Y_i is the value of a specific observation, $\bar{Y}$ is the mean for all the observations, and $\hat{Y}$ is the predicted value for the specific observation.

D.3.2 Analysis of Variance for Linear Regression

Source of variation	Degrees of freedom	Sum of squares	Mean squares	F-value
Regression	1	$SS_{regression}$	$SS_{regression}/1$	$MS_{regression}/MS_{residual}$
Residual	$n–2$	$SS_{residual}$	$SS_{residual}/(n–2)$	
Total	$n–1$	SS_{total}		

D.3.3 Prediction of the Regression Line

It is possible to calculate confidence intervals for the slope of the fitted regression line (Neter and Wasserman, 1974). These confidence intervals lie on smooth curves (the branches of a hyperbola) on either side of the regression line (Fig. D.4). The confidence intervals are at their smallest at the mean value of

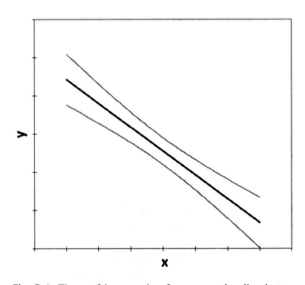

Fig. D.4 The confidence region for a regression line is two curved lines on either side of the linear regression line. These *curved lines* are closest to the regression line at the mean values for X and Y.

the X-series and the intervals become progressively larger as one moves away from the mean value of the X-series. All predictions using the regression line should fall within the range used to calculate the fitted regression line. No predictions should be made outside this range. The equations to calculate the confidence intervals are as follows:

$$\hat{Y}_0 \pm t \left\{ 1 + \frac{1}{n} + \frac{(x_0 - \bar{x})^2}{\sum (x_i - \bar{x})^2} \right\}^{\frac{1}{2}} S \qquad (D.8)$$

where x_0 is the point you are estimating around; x_i are the measured points; $\bar{x}$ is the mean value for the x-series; n is the number of observations; and S is the residual mean square of regression (see below). The t-value is determined by the degrees of freedom $(n-1)$ since this is a two-tailed test at an $\alpha/2$ level (Rawlings et al., 1998).

D.3.4 Linear Regression Example

We will return to the cocoa powder–icing sugar mixture example used as an example in Pearson's correlation coefficient section. A linear regression line was fitted to the data using the equations given above:

$$b = \frac{5167.50 - \frac{875 \times 72.32}{14}}{60,375 - \frac{(875)^2}{14}} = 0.102857$$

$$a = 5.237 - (0.102857)(62.5) = -1.19156$$

The fitted linear regression equation is $y = 0.1028 X - 1.1916$ and the coefficient of determination is 0.9616 with 12 degrees of freedom. However, we should *round* this equation to $y = 0.10 X - 1.19$ since our level of certainty does not justify four decimal places!

D.4 Multiple Linear Regression

Multiple linear regression (MLR) calculates the linear combination of independent variables (more than one X) that is maximally correlated with the dependent variable (a single Y). The regression is performed in a least-squares fashion by minimizing the sum of squares of the residual (Afifi and Clark, 1984; Stevens, 1986). The regression equation must be cross-validated by applying the equation to an independent sample from the same population—if predictive power drops

sharply then the equation has only limited utility. In general, one needs about 15 subjects (or observations) *per independent variable* to have some hope of successful cross-validation, although some scientists will do multiple linear regression with as few as five times as many observations as independent variables. When the independent variables are intercorrelated or multicollinear it spells real trouble for the potential use of MLR. Multicollinearity limits the possible magnitude of the multiple correlation coefficient (R) and makes determining the importance of a given independent variable in the regression equation difficult as the effects of the independent variables are confounded due to high inter-correlations. Also, when variables are multicollinear then the order that independent variables enter the regression equation makes a difference with respect to the amount of variance on Y that each variable accounts for. It is only with totally uncorrelated independent variables that the order has no effect. As seen in Chapter 18, principal component analysis (PCA) creates orthogonal (non-correlated) principal components (PCs). It is possible to MLR using these new variables when one started with highly multicollinear data (as is often found in sensory descriptive data). The biggest problem with doing MLR on PCs is that it may be very difficult to interpret the resultant output. Multiple linear regression analyses can be performed using any reputable statistical analysis software package (Piggott, 1986).

D.5 Other Measures of Association

D.5.1 Spearman Rank Correlation

When the data are not derived from an interval scale but from an ordinal scale the simple correlation coefficient is not appropriate. However, the Spearman rank correlation coefficient is appropriate. This correlation coefficient is a measure of the association between the *ranks* of independent and dependent variables. This measure is also useful when the data are not normally distributed (Blalock, 1979).

The Spearman coefficient is often indicated by the symbol ρ (rho); however, sometimes r_s is also used. Similar to the simple correlation coefficient the Spearman coefficient ranges between −1 and 1. Values of r_s close to absolute 1 indicate that a very strong relationship exists between the *ranks* of the two variables. When ρ is equal to zero, then there is no relationship between the *ranks* of the two variables. Positive values of ρ indicate a tendency for the ranks of the variables to increase together. Negative values of ρ indicate a tendency of large rank values of one variable to be associated with small rank values of the other variable. The Spearman correlation coefficient for data with a few ties is calculated using the following equation; the equation to use for data with many ties see Appendix B:

$$\rho = 1 - \frac{6\Sigma d^2}{n(n^2 - 1)} \qquad (D.9)$$

where n is the number of ranked products and d the differences in ranks for each product between the two data series.

Critical values of ρ are found in Spearman rank correlation tables (Table F1). When the value for n is more than 60 then the Pearson tabular values can be used to determine the tabular values for the Spearman correlation coefficient.

D.5.2 Spearman Correlation Coefficient Example

We are returning to the cocoa powder example used for Pearson's correlation coefficient. In this case the two panelists were asked to rank the perceived brown intensities of the 14 cocoa powder–icing sugar mixtures. Was there a significant correlation between the ranks assigned by the two panelists to the cocoa mixtures?

$$r_s = 1 - \frac{6 \times 42}{14(156 - 1)} = 0.8839$$

The tabular value for the Spearman correlation coefficient with 14 ranks at an alpha-value of 5% is equal to 0.464 (see Table F1). We can conclude that the two-panelist rank ordering of the brown color intensities of the cocoa powder–icing sugar mixtures were significantly similar. However, there is no direct causal relationship between the rank order of panelist A and that of panelist B.

D.5.3 Cramér's V Measure

When the data are not derived from an interval or an ordinal scale but from a nominal scale then the appropriate measure of association is the Cramér measure, $\hat{\phi}'^2$ (phi–hat prime squared) (Herzberg, 1989). This association coefficient is a squared measure and can range from 0 to 1. The closer the value of Cramér's V measure is to 1 the greater the association between the two *nominal* variables, the closer Cramér's V measure is to zero the smaller the association between the two *nominal* variables. In practice, you may find that a Cramer's V of 0.10 provides a good minimum threshold for suggesting there is a substantive relationship between two variables. The Cramér coefficient of association is calculated using the following equation:

$$V = \sqrt{\frac{x^2}{n(q - 1)}} \qquad (D.10)$$

Cocoa in mixture, %	Panelist A(X)	Panelist B(Y)	d	d^2
30	1	1	0	0
35	2	2	0	0
40	4	3	1	1
45	6	4	2	4
50	7	7	0	0
55	5	8	−3	9
60	9	6	3	9
65	3	5	−2	4
70	10	9	1	1
75	8	10	−2	4
80	11	12	1	1
85	14	13	1	1
90	13	11	2	4
95	12	14	−2	4
				$\Sigma d^2 = 42$

where n is equal to the sample size; q is the smaller of the category variables represented by the rows (r) and the columns (c); and χ^2 is chi-square observed for the data (see Table C).

D.5.4 Cramér Coefficient Example

The following data set is hypothetical. One hundred and ninety consumers of chewing gum indicated which flavor of chewing gum they usually used. The sensory scientist was interested in determining whether there was an association between gender and gum flavors.

Observed values

	Fruit flavor gum	Mint flavor gum	Bubble gum	
Men	35	15	50	100
Women	12	60	18	90
	47	75	68	190

Expected value for women using fruit flavored gum is $47 \times (90/190) = 22.263$ and for men using bubble gum the expected value is $68 \times (100/190) = 35.789$.

Expected values

	Fruit flavor gum	Mint flavor gum	Bubble gum	
Men	24.737	39.474	35.789	100.000
Women	22.263	35.526	32.210	89.999
	47.000	75.000	67.999	189.998

$$\chi^2 = \sum \frac{(O_{ij} - E_{ij})^2}{E_{ij}}$$

$$\chi^2 = \frac{(35 - 24.737)^2}{24.737} + \frac{(15 - 39.474)^2}{39.474}$$

$$+ \frac{(50 - 35.789)^2}{35.789} + \frac{(12 - 22.263)^2}{22.263}$$

$$+ \frac{(60 - 35.526)}{35.526} + \frac{(18 - 32.210)^2}{32.210} = 52.935$$

Thus the calculated $\chi^2 = 52.935$ and the degrees of freedom are equal to $(r-1)\times(c-1)$.

In this case with two rows and three columns df $= (2-1)\times(3-1) = 2$. The tabular value for $\chi^2_{0.05,\, df=2}$ is 5.991 (see Table C). The χ^2 value is significant. However, to determine whether there is an association between the genders and their use of gum flavors we use the following equation:

$$V = \sqrt{\frac{\chi^2}{n(q-1)}} = \sqrt{\frac{52.935}{190(2-1)}} = 0.572830714$$

The Cramér value of association is 0.5278. There is some association between gender and the use of gum flavors.

References

Afifi, A. A. and Clark, V. 1984. Computer-aided multivariate analysis. Lifetime Learning, Belmont, CA, pp. 80–119.

Anscombe, F. J. 1973. Graphs in statistical analysis. American Statistician, 27, 17–21.

Blalock, H. M. 1979. Social Statistics, Second Edition. McGraw-Hill Book, New York.

Diamond, G. A. 1989. Correlation, causation and agreement. The American Journal of Cardiology, 63, 392.

Freund, J. E. and Simon, G. A. 1992. Modern Elementary Statistics, Eighth Edition. Prentice Hall, Englewood Cliffs, NJ, p. 474.

Herzberg, P. A. 1989. Principles of Statistics. Robert E. Krieger, Malabar, FL, pp. 378–380.

Neter, J. and Wasserman, W. 1974. Applied Linear Models. Richard D. Irwin, Homewood, IL, pp. 53–96.

Piggott, J. R. 1986. Statistical Procedures in Food Research. Elsevier Applied Science, New York, NY, pp. 61–100.

Rawlings, J. O., Pantula, S. G. and Dickey, D. A. 1998. Applied Regression Analysis: A Research Tool. Springer, New York.

Snedecor, G. W. and Cochran, W. G. 1980. Statistical Methods, Seventh Edition. Iowa State University, Ames, IA, pp. 175–193.

Stevens, J. 1986. Applied Multivariate Statistics for the Social Sciences. Lawrence Erlbaum Associates, Hillsdale, NJ.

Appendix E

Statistical Power and Test Sensitivity

Contents

Research reports in the literature are frequently flawed by conclusions that state or imply that the null hypothesis is true. For example, following the finding that the difference between two sample means is not statistically significant, instead of properly concluding from this failure to reject the null hypothesis that the data do not warrant the conclusion that the population means differ, the writer concludes, at least implicitly that there is no difference. The latter conclusion is always strictly invalid, and it is functionally invalid unless power is high.

—J. Cohen (1988)

The power of a statistical test is the probability that if a true difference or effect exists, the difference or effect will be detected. The power of a test becomes important, especially in sensory evaluation, when a no-difference decision has important implications, such as the sensory equivalence of two formulas or products. Concluding that two products are sensorially similar or equivalent is meaningless unless the test has sufficient power. Factors that affect test power include the sample size, alpha level, variability, and the chosen size of a difference that must be detected. These factors are discussed and worked examples given.

E.1 Introduction

Sensory evaluation requires experimental designs and statistical procedures that are sensitive enough to find differences. We need to know when treatments of interest are having an effect. In food product development, these treatments usually involve changes in food constituents, the methods of processing, or types of packaging. A purchasing department may change suppliers of an ingredient. Product development may test for the stability of a product during its shelf life. In each of these cases, it is desirable to know when a product has become perceivably different from some comparison or control product, and sensory tests are conducted.

In normal science, most statistical tests are done to insure that a true null hypothesis is not rejected without cause. When enough evidence is gathered to show that our data would be rare occurrences given the null assumption, we conclude that a difference did occur. This process keeps us from making the Type I error discussed in Appendix A. In practical terms, this keeps a research program focused on real effects and insures that business decisions about changes are made with some confidence.

However, another kind of error in statistical decision making is also important. This is the error associated with accepting the null when a difference did occur. Missing a true difference can be as dangerous as finding a spurious one, especially in product research. In order to provide tests of good sensitivity, then, the sensory evaluation specialist conducts tests using good design principles and sufficient numbers of judges and replicates. The principles of good practice are discussed in Chapter 3. Most of these practices are aimed at reducing unwanted error variance. Panel screening, orientation, and training are some of the tools at the disposal of the sensory specialist that can help minimize unwanted variability. Another example is in the use of reference standards, both for sensory terms and for intensity levels in descriptive judgments.

Considering the general form of the t-test, we discover that two of the three variables in the statistical formula are under some control of the sensory scientist. Remember that the t-test takes this form:

$$t = \text{difference between means/standard error}$$

and the standard error is the sample standard deviation divided by the square root of the sample size (N). The denominator items can be controlled or at least influenced by the sensory specialist. The standard deviation or error variance can be minimized by good experimental controls, panel training, and so on. Another tool for reducing error is partitioning, for example in the removal of panelist effects in the complete block ANOVA ("repeated measures") designs or in the paired t-test. As the denominator of a test statistic (like a F-ratio or a t-value) becomes smaller, the value of the test statistic becomes larger and it is easier to reject the null. The probability of observing the results (under the assumption of a true null) shrinks. The second factor under the control of the sensory professional is the sample size. The sample size usually refers to the number of judges or observations. In some ANOVA models additional degrees of freedom can also be gained by replication.

It is sometimes necessary to base business decisions on acceptance of the null hypothesis. Sometimes we conclude that two products are sensorially similar, or

that they are a good enough match that no systematic difference is likely to be observed by regular users of the product. In this scenario, it is critically important that a sensitive and powerful test be conducted so that a true difference is not missed, otherwise the conclusion of "no difference" could be spurious. Such decisions are common in statistical quality control, ingredient substitution, cost reductions, other reformulations, supplier changes, shelf life and packaging studies, and a range of associated research questions. The goal of such tests is to match an existing product or provide a new process or cost reduction that does not change or harm the sensory quality of the item. In some cases, the goal may be to match a competitor's successful product. An equivalence conclusion may also be important in advertising claims, as discussed in Chapter 5.

In these practical scenarios, it is necessary to estimate the power of the test, which is the probability that a true difference would be detected. In statistical terms, this is usually described in an inverse way, first by defining the quantity beta as the long-term probability of missing a true difference or the probability that a Type II error is committed. Then one minus beta is defined as the power of the test. Power depends upon several interacting factors, namely the amount of error variation, the sample size, and the size of the difference one wants to be sure to detect in the test. This last item must be defined and set using the professional judgment of the sensory specialist or by management. In much applied research with existing food products, there is a knowledge base to help decide how much a change is important or meaningful.

This chapter will discuss the factors contributing to test power and give some worked examples and practical scenarios where power is important in sensory testing. Discussions of statistical power and worked examples can also be found in Amerine et al. (1965), Gacula and Singh (1984), and Gacula (1991, 1993). Gacula's writings include considerations of test power in substantiating claims for sensory equivalence of products. Examples specific to discrimination tests can be found in Schlich (1993) and Ennis (1993). General references on statistical power include the classic text by Cohen (1988), his overview article written for behavioral scientists (Cohen, 1992) and the introductory statistics text by Welkowitz et al. (1982). Equivalency testing is also discussed at length by Wellek (2003), Bi (2006), and ASTM (2008). Let the reader note that many scientific bodies have rejected the idea of using test power as justification for accepting the null, and prefer an approach that proves that any difference lies within a specified or acceptable interval. This idea is most applicable to proving the equivalence of measured variables (like the bioequivalence of drug delivery into the bloodstream). However, this equivalence interval approach has also been taken for simple sensory discrimination testing (see Ennis, 2008; Ennis and Ennis, 2009).

E.2 Factors Affecting the Power of Statistical Tests

E.2.1 Sample Size and Alpha Level

Mathematically, the power of a statistical test is a function of four interacting variables. Each of these entails choices on the part of the experimenter. They may seem arbitrary, but in the words of Cohen, "all conventions are arbitrary. One can only demand of them that they not be unreasonable" (1988, p. 12). Two choices are made in the routine process of experimental design, namely the sample size and the alpha level. The sample size is usually the number of judges in the sensory test. This is commonly represented by the letter "N" in statistical equations. In more complex designs like multi-factor ANOVA, "N" can reflect both the number of judges and replications, or the total number of degrees of freedom contributing to the error terms for treatments that are being compared. Often this value is strongly influenced by company traditions or lab "folklore" about panel size. It may also be influenced by cost considerations or the time needed to recruit, screen, and/or train and test a sufficiently large number of participants. However, this variable is the one most often considered in determinations of test power, as it can easily be modified in the experimental planning phase.

Many experimenters will choose the number of panelists using considerations of desired test power. Gacula (1993) gives the following example. For a moderate to large consumer test, we might want to know whether the products differ one half a point on the 9-point scale at most in their mean values. Suppose we had prior knowledge that for this product, the standard deviation is about 1 scale point ($S = 1$), we can find the

required number of people for an experiment with 5% alpha and 10% beta (or 90% power). This is given by the following relationship:

$$N = \frac{(Z_\alpha + Z_\beta)^2 S^2}{(M_1 - M_2)^2}$$

$$\frac{(1.96 + 1.65)^2 1^2}{(0.5)^2} \cong 52 \qquad (E.1)$$

where M_1–M_2 is the minimal difference we must be sure to detect and Z_α and Z_β are the Z-scores associated with the desired Type I and Type II error limits. In other words, there are 52 observers required to insure that a one-half point difference in means can be ruled out at 90% power when a non-significant result is obtained. Note that for any fractional N, you must round up to the next whole person.

The second variable affecting power is the alpha level, or the choice of an upper limit on the probability of rejecting a true null hypothesis (making a Type I error). Usually we set this value at the traditional level of 0.05, but there are no hard and fast rules about this magical number. In many cases in exploratory testing or industrial practice, the concern over Type II error—missing a true difference—are of sufficient concern that the alpha level for reporting statistical significance will float up to 0.10 or even higher. This strategy shows us intuitively that there is a direct relationship between the size of the alpha level and power, or in other words, an inverse relationship between alpha-risk and beta-risk. Consider the following outcome: we allow alpha to float up to 0.10 or 0.20 (or even higher) and still fail to find a significant p-value for our statistical test. Now we have an inflated risk of finding a spurious difference, but an enhanced ability to reject the null. If we still fail to reject the null, even at such relaxed levels, then there probably is no true difference among our products. This assumes no sloppy experiment, good laboratory practices, and sufficient sample size, i.e., meeting all the usual concerns about reasonable methodology. The inverse relationship between alpha and beta will be illustrated in a simple example below.

Because of the fact that power increases as alpha is allowed to rise, some researchers would be tempted to raise alpha as a general way of guarding against Type II error. However, there is a risk involved in this, and that is the chance of finding false positives or spurious random differences. In any program of repeated testing, the strategy of letting alpha float up as a cheap way to increase test power should not be used. We have seen cases in which suppliers of food ingredients were asked to investigate quality control failures of their ingredient submissions, only to find that the client company had been doing discrimination tests with a lax alpha level. This resulted in spurious rejections of many batches that were probably within acceptable limits.

E.2.2 Effect Size

The third factor in the determination of power concerns the effect size one is testing against as an alternative hypothesis. This is usually a stumbling block for scientists who do not realize that they have already made two important decisions in setting up the test—the sample size and alpha level. However, this third decision seems much more subjective to most people. One can think of this as the distance between the mean of a control product and the mean of a test product under an alternative hypothesis, in standard deviation units. For example, let us assume that our control product has a mean of 6.0 on some scale and the sample has a standard deviation of 2.0 scale units. We could test whether the comparison product had a value of less than 4.0 or greater than 8.0, or one standard deviation from the mean in a two-tailed test. In plain language, this is the size of a difference that one wants to be sure to detect in the experiment.

If the means of the treatments were two standard deviations apart, most scientists would call this a relatively strong effect, one that a good experiment would not want to miss after the statistical test is conducted. If the means were one standard deviation apart, this is an effect size that is common in many experiments. If the means were less than one half of one standard deviation apart, that would a smaller effect, but one that still might have important business implications. Various authors have reviewed the effect sizes seen in behavioral research and have come up with some guidelines for small, medium, and large effect sizes based on what is seen during the course of experimentation with humans (Cohen, 1988; Welkowitz et al., 1982).

Several problems arise. First, this idea of effect size seems arbitrary and an experimenter may not have any knowledge to aid in this decision. The sensory professional may simply not know how much of a consumer impact a given difference in the data is likely to produce. It is much easier to "let the statistics make the decision" by setting an alpha level according to tradition and concluding that no significant difference means that two products are sensorially equal. As shown above, this is bad logic and poor experimental testing. Experienced sensory scientists may have information at their disposal that makes this decision less arbitrary. They may know the levels of variability or the levels important to consumer rejection or complaints. Trained panels will show standard deviations around 10% of scale range (Lawless, 1988). The value will be slightly higher for difficult sensory attributes like aroma or odor intensity, and lower for "easier" attributes like visual and some textural attributes. Consumers, on the other hand will have intensity attributes with variation in the realm of 25% of scale range and sometimes even higher values for hedonics (acceptability). Another problem with effect size is that clients or managers are often unaware of it and do not understand why some apparently arbitrary decision has to enter into scientific experimentation.

The "sensitivity" of a test to differences involves both power and the overall quality of the test. Sensitivity entails low error, high power, sufficient sample size, good testing conditions, good design, and so on The term "power" refers to the formal statistical concept describing the probability of accepting a true alternative hypothesis (e.g. finding a true difference). In a parallel fashion, Cohen (1988) drew an important distinction between effect size and "operative effect size" and showed how a good design can increase the effective sensitivity of an experiment. He used the example of a paired t-test as opposed to an independent groups t-test. In the paired design subjects function as their own controls since they evaluate both products. The between-person variation is "partitioned" out of the picture by the computation of difference scores. This effectively takes judge variation out of the picture.

In mathematical terms, this effect size can be stated for the t-test as the number of standard deviations separating means, usually signified by the letter "d". In the case of choice data, the common estimate is our old friend d' (d-prime) from signal detection theory, sometimes signified as a population estimate by the

Greek letter delta (Ennis, 1993). For analyses based on correlation, the simple Pearson's *r* is a common and direct measure of association. Various measures of effect size (such as variance accounted for by a factor) in ANOVAs have been used. Further discussion of effect sizes and how to measure them can be found in Cohen (1988) and Welkowitz et al. (1982).

E.2.3 How Alpha, Beta, Effect Size, and N Interact

Diagrams below illustrate how effect size, alpha, and beta interact. As an example, we perform a test with a rating scale, e.g., a just-about-right scale, and we want to test whether the mean rating for the product is higher than the midpoint of the scale. This is the simple *t*-test against a fixed value, and our hypothesis is one tailed. For the simple one-tailed *t*-test, alpha represents the area under the *t*-distribution to the right of the cutoff determined by the limiting *p*-value (usually 5%). It also represents the upper tail of the sampling distribution of the mean as shown in Fig. E.1. The value of beta is shown by the area underneath the alternative

hypothesis curve to the left of the cutoff as shaded in Fig. E.1. We have shown the sampling distribution for the mean value under the null as the bell-shaped curve on the left. The dashed line indicates the cutoff value for the upper 5% of the tail of this distribution. This would be the common value set for statistical significance, so that for a give sample size (*N*), the *t*-value at the cutoff would keep us from making a Type I error more than 5% of the time (when the null is true). The right-hand curve represents the sampling distribution for the mean under a chosen alternative hypothesis. We know the mean from our choice of effect size (or how much of a difference we have decided is important) and we can base the variance on our estimate from the sample standard error. When we choose the value for mean score for our test product, the *d*-value becomes determined by the difference of this mean from the control, divided by the standard deviation. Useful examples are drawn in Gacula's (1991, 1993) discussion and in the section on hypothesis testing in Sokal and Rohlf (1981).

In this diagram, we can see how the three interacting variables work to determine the size of the shaded area for beta-risk. As the cutoff is changed by changing the alpha level, the shaded area would become larger or

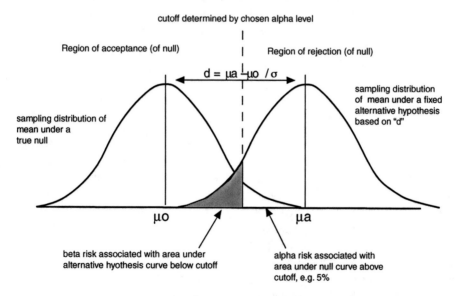

Fig. E.1 Power shown as the tail of the alternative hypothesis, relative to the cutoff determined by the null hypothesis distribution. The diagram is most easily interpreted as a one-tailed *t*-test. A test against a fixed value of a mean would be done against a population value or a chosen scale point such the midpoint of a just-right scale. The value of the mean for the alternative

hypothesis can be based on research, prior knowledge, or the effect size, *d*, the difference between the means under the null and alternative hypotheses, expressed in standard deviation units. Beta is given by the shaded area underneath the sampling distribution for the alternative hypothesis, below the cutoff determined by alpha. Power is one minus beta.

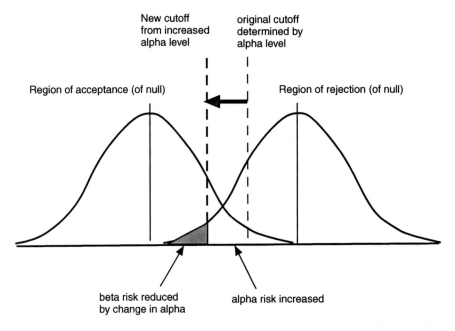

Effect of increasing alpha level
to decrease beta risk

New cutoff
from increased
alpha level

original cutoff
determined by
alpha level

Region of acceptance (of null)

Region of rejection (of null)

beta risk reduced
by change in alpha

alpha risk increased

Fig. E.2 Increasing the alpha level decreases the area associated with beta, improving power (all other variables held equal).

smaller (see Fig. E.2). As the alpha-risk is increased, the beta-risk is decreased, all other factors being held constant. This is shown by shifting the critical value for a significant *t*-statistic to the left, increasing the alpha "area," and decreasing the area associated with beta.

A second influence comes from changing the effect size or alternative hypothesis. If we test against a larger *d*-value, the distributions would be separated, and the area of overlap is decreased. Beta-risk decreases when we choose a bigger effect size for the alternative hypothesis (see Fig. E.3). Conversely, testing for a small difference in the alternative hypothesis would pull the two distributions closer together, and if alpha is maintained at 5%, the beta-risk associated with the shaded area would have to get larger. The chances of missing a true difference are very high if the alternative hypothesis states that the difference is very small. It is easier to detect a bigger difference than a smaller one, all other things in the experiment being equal.

The third effect comes from changing the sample size or the number of observations. The effect of increasing "*N*" is to shrink the effective standard deviation of the sampling distributions, decreasing the standard error of the mean. This makes the distributions

taller and thinner so there is less overlap and less area associated with beta. The *t*-value for the cutoff moves to the left in absolute terms.

In summary, we have four interacting variables and knowing any three, we can determine the fourth. These are alpha, beta, "*N*," and effect size. If we wish to specify the power of the test up front, we have to make at least two other decisions and then the remaining parameter will be determined for us. For example, if we want 80% test power (beta = 0.20), and alpha equal to 0.05, and we can test only 50 subjects, then the effect size we are able to detect at this level of power is fixed. If we desire 80% test power, want to detect 0.5 standard deviations of difference, and set alpha at 0.05, then we can calculate the number of panelists that must be tested (i.e., "*N*" has been determined by the specification of the other three variables). In many cases, experiments are conducted only with initial concern for alpha and sample size. In that case there is a monotonic relationship between the other two variables that can be viewed after the experiment to tell us what power can be expected for different effect sizes. These relationships are illustrated below. Various freeware programs are available for estimating power and

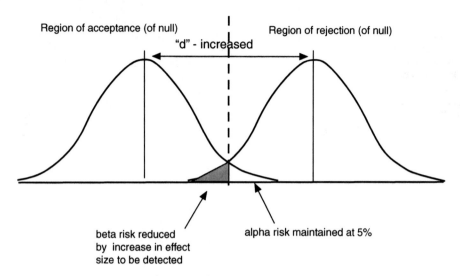

Effect of increasing alternative
hypothesis effect size ("d")
to decrease beta risk

Region of acceptance (of null)   Region of rejection (of null)

"d" - increased

beta risk reduced   alpha risk maintained at 5%
by increase in effect
size to be detected

Fig. E.3 Increasing the effect size that must be detected increases the power, reducing beta. Larger effects (larger d, difference between the means of the alternative and null hypotheses) are easier to detect.

sample size (e.g., Erdfelder et al., 1996). Tables for the power of various statistical tests can also be found in Cohen (1988). The R library "pwr" package specifically implements power analyses outlined in Cohen (1988).

E.3 Worked Examples

E.3.1 The t-Test

For a specific illustration, let us examine the independent groups t-test to look at the relationship between alpha, beta, effect size, and "N." In this situation, we want to compare two means generated from independent groups, and the alternative hypothesis predicts that the means are not equal (i.e., no direction is predicted). Figure E.4 shows the power of the two-tailed independent groups t-test as a function of different sample sizes (N) and different alternative hypothesis effect sizes (d). (Note that N here refers to the total sample, not N for each group. For very different sample sizes per group, further calculations must be done.) If we set the lower limit of acceptable power at 50%,

we can see from these curves that using 200 panelists would allow us to detect a small difference of about 0.3 standard deviations. With 100 subjects this difference must be about 0.4 standard deviations, and for small sensory tests of 50 or 20 panelists (25 or 10 per group, respectively) we can only detect differences of about 0.6 or 0.95 standard deviations, respectively, with 50/50 chance of missing a true difference. This indicates the liabilities in using a small sensory test to justify a "parity" decision about products.

Often, a sensory scientist wants to know the required sample size for a test, so they can recruit the appropriate number of consumers or panelists for a study. Figure E.5 shows the sample size required for different experiments for a between-groups t-test and a decision that is two tailed. An example of such a design would be a consumer test for product acceptability, with scaled data and each of the products placed with a different consumer group (a so-called monadic design). Note that the scale is log transformed, since the group size becomes very large if we are looking for small effects. For a very small effect of only 0.2 standard deviations, we need 388 consumers to have a minimal power level of 0.5. If we want to increase power to 90%, the number exceeds

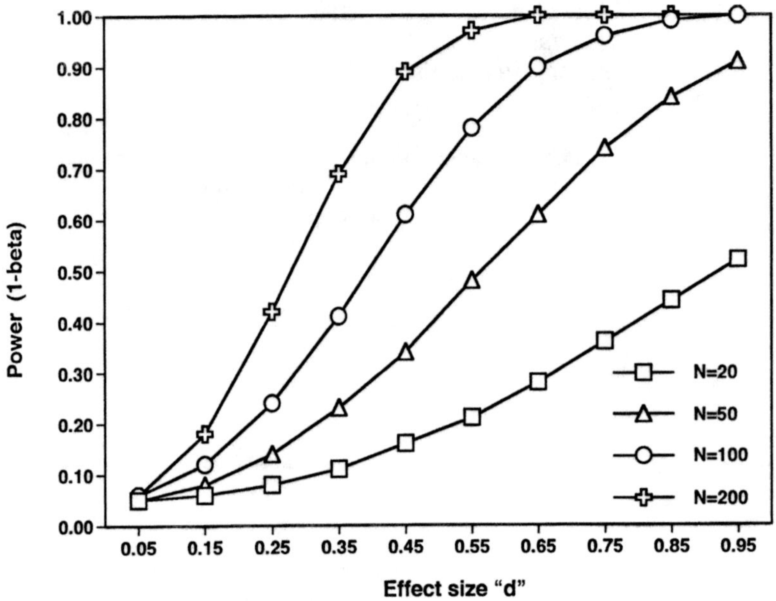

Fig. E.4 Power of the
two-tailed independent groups
t-test as a function of different
sample sizes (*N*) and different
alternative hypothesis effect
sizes (*d*); the decision is
two-tailed at alpha = 0.05.
The effect size "*d*" represents
the difference between the
means in standard deviations.
Computed from the
GPOWER program of
Erdfelder et al. (1996).

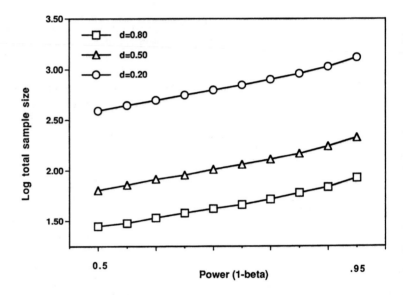

Fig. E.5 Number of judges
required for independent
groups *t*-test at different levels
of power; the decision is two
tailed at alpha = 0.05. Note
that the sample size is plotted
on a log scale. Computed
from the GPOWER program
of Erdfelder et al. (1996).

1,000. On the other hand, for a big difference of 0.8 standard deviations (about 1 scale point on the 9-point hedonic scale) we only need 28 consumers for 50% power and 68 consumers for 90% power. This illustrates why some sensory tests done for research and product development purposes are smaller than the corresponding marketing research tests. Market research tests may be aimed at finding small advantages in a mature or optimized product system, and this requires a test of high power to keep both alpha- and beta-risks low.

E.3.2 An Equivalence Issue with Scaled Data

Gacula (1991, 1993) gives examples of calculations of test power using several scenarios devoted to substantiating claims of product equivalence. These are mostly based on larger scale consumer tests, where the sample size justifies the use of the normal distribution (Z) rather than the small sample *t*-test. In such an experiment, the calculation of power is straightforward, once the mean difference associated with the alternative hypothesis is stated. The calculation for power follows this relationship:

$$\text{Power} = 1 - \beta = 1 - \Phi\left[\frac{X_c - \mu_D}{\text{SE}}\right] \quad (E.2)$$

where X_c represents the cutoff value for a significantly higher mean score, determined by the alpha level. For a one-tailed test, the cutoff is equal to the mean plus 1.645 times the standard error (or 1.96 standard errors for a two-tailed situation). The Greek letter Φ represents the value of the cumulative normal distribution; in other words we are converting the Z-score to a proportion or probability value. Since many tables of the cumulative normal distribution are given in the larger proportion, rather than the tail (as is true in Gacula's tables), it is sometimes necessary to subtract the tabled value from 1 to get the other tail. The parameter μ_D represents the mean difference as determined by the alternative hypothesis. This equation simply finds the area underneath the alternative hypothesis Z-distribution, beyond the cutoff value X_c. A diagram of this is shown below.

Here is a scenario similar to one from Gacula (1991). A consumer group of 92 panelists evaluates two products and gives them mean scores of 5.9 and 6.1 on a 9-point hedonic scale. This is not a significant difference, and the sensory professional is tempted to conclude that the products are equivalent. Is this conclusion justified?

The standard deviation for this study was 1.1, giving a standard error of 0.11. The cutoff values for the 95% confidence interval are then 1.96 standard errors, or the mean plus or minus 0.22. We see that the two means lie within the 95% confidence interval so the statistical conclusion of no difference seems to be justified. A two-tailed test is used to see whether the new product is higher than the standard product receiving a 5.9. The two-tailed test requires a cutoff that is 1.96 standard errors above, or 0.22 units above the mean. This sets our upper cutoff value for X_c at 5.9 + 0.22 or 6.12. Once this boundary has been determined, it can be used to split the distributions expected on the basis of the alternative hypotheses into two sections. This is shown in Fig. E.6. The section of the distribution that is higher than this cutoff represents the detection of a difference or power (null rejected) while the section that

is lower represents the chance of missing the difference or beta (null accepted).

In this example, Gacula originally used the actual mean difference of 0.20 as the alternative hypothesis. This would place the alternative hypothesis mean at 5.9 + 0.2 or 6.1. To estimate beta, we need to know the area in the tail of the alternative hypothesis distribution to the left of the cutoff. This can be found once we know the distance of the cutoff from out alternative mean of 6.1. In this example, there is a small difference from the cutoff of only 6.1–6.12 or 0.02 units on the original scale, or 0.02 divided by the standard error to give about 0.2 Z-score units from the mean of the alternative to the cutoff. Essentially, this mean lies very near to the cutoff and we have split the alternative sampling distribution about in half. The area in the tail associated with beta is large, about 0.57, so power is about 43% (1 minus beta). Thus the conclusion of no difference is not strongly supported by the power under the assumptions that the true mean lies so close to 5.9. However, we have tested against a small difference as the basis for our alternative hypothesis. There is still a good chance that such a small difference does exist.

Suppose we relax the alternative hypothesis. Let us presume that we determined before the experiment that a difference of one-half of one standard deviation on our scale is the lower limit for any practical importance. We could then set the mean for the alternative hypothesis at 5.9 plus one half of the standard deviation (1.1/2 or 0.55). The mean for the alternative now becomes 5.9 + 0.55 or 6.45. Our cutoff is now 6.12–6.45 units away (0.33) or 0.33 divided by the standard error of 0.11 to convert to Z-score units, giving a value of 3. This has effectively shifted the expected distribution to the right while our decision cutoff remains the same at 6.12. The area in the tail associated with beta would now be less than 1% and power would be about 99%. The choice of an alternative hypothesis can greatly affect the confidence of our decisions. If the business decision justifies a choice of one half of a scale unit as a practical cutoff (based on one-half of one standard deviation) then we can see that our difference of only 0.2 units between mean scores is fairly "safe" when concluding no difference. The power calculations tell us exactly how safe this would be. There is only a very small chance of seeing this result or one more extreme, if our true mean score was 0.55 units higher. The observed events are

Fig. E.6 Power first depends upon setting a cutoff value based upon the sample mean, standard error, and the alpha level. In the example shown, this value is 6.12. The cutoff value can then be used to determine power and beta-risk for various expected distributions of means under alternative hypotheses. In Gacula's first example, the actual second product mean of 6.1 was used. The power calculation gives only 43%, which does not provide a great deal of confidence in a conclusion about product equivalence. The lower example shows the power for testing against an alternative hypothesis that states that the true mean is 6.45 or higher. Our sample and experiment would detect this larger difference with greater power.

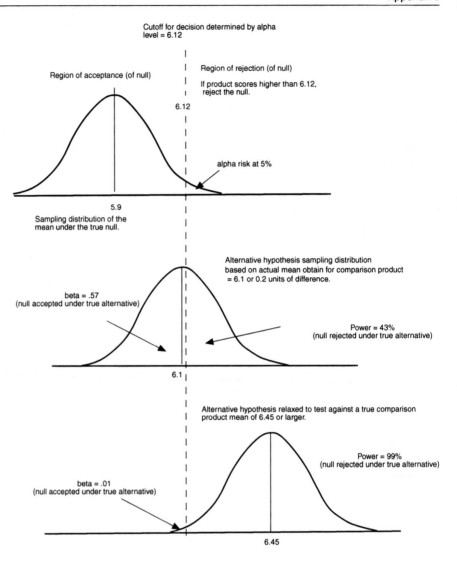

Cutoff for decision determined by alpha level = 6.12

Region of acceptance (of null)

Region of rejection (of null)

If product scores higher than 6.12, reject the null.

6.12

alpha risk at 5%

5.9

Sampling distribution of the mean under the true null.

Alternative hypothesis sampling distribution based on actual mean obtain for comparison product = 6.1 or 0.2 units of difference.

beta = .57 (null accepted under true alternative)

Power = 43% (null rejected under true alternative)

6.1

Alternative hypothesis relaxed to test against a true comparison product mean of 6.45 or larger.

Power = 99% (null rejected under true alternative)

beta = .01 (null accepted under true alternative)

6.45

fairly unlikely given this alternative, so we reject the alternative in favor of the null.

E.3.3 Sample Size for a Difference Test

Amerine et al. (1965) gave a useful general formula for computing the necessary numbers of judges in a discrimination test, based on beta-risk, alpha-risk, and the critical difference that must be detected. This last item is conceived of as the difference between the chance probability, p_o and the probability associated with an alternative hypothesis, p_a. Different models for this are discussed in Chapter 5 [see also Schlich (1993) and Ennis (1993)]. For the sake of example, we will take

the chance-adjusted probability for 50% correct, which is halfway between the chance probability and 100% detection (i.e., 66.7% for the triangle test).

$$N = \left[\frac{Z_\alpha \sqrt{p_o(1 - p_o)} + Z_\beta \sqrt{p_a(1 - p_a)}}{|p_o - p_a|} \right]^2 \quad \text{(E.3)}$$

for a one-tailed test (at $a = 0.05$), $Z_\alpha = 1.645$, and if beta is kept to 10% (90% power) then $Z_\beta = 1.28$. The critical difference, p_o-p_a, has a strong influence on the equation. In the case where it is set to 33.3% for the triangle test (a threshold of sorts) we then require 18 respondents as shown in the following calculation:

$$N = \left[\frac{1.645\sqrt{0.333(.667)} + 1.28\sqrt{0.667(0.333)}}{|0.333 - 0.667|} \right]^2 = 17.03$$

So you would need a panel of about 18 persons to protect against missing a difference this big and limit your risk to 10%. Note that this is a fairly gross test, as the difference we are trying to detect is large. If half of a consumer population notices a difference, the product could be in trouble.

Now, suppose we do not wish to be this lenient, but prefer a test that will be sure to catch a difference about half this big at the 95% power level instead of 90%. Let us change one variable at a time to see which has a bigger effect. First the beta-risk goes from 90 to 95% and if we remain one tailed, $Z\beta$ now equals 1.645. So the numbers become

$$N = \left[\frac{1.645\sqrt{0.333(0.667)} + 1.645\sqrt{0.667(0.333)}}{|0.333 - 0.667|}\right]^2 = 21.55$$

So with the increase only in power, we need four additional people for our panel. However, if we decrease the effect size we want to detect by half the numbers become

$$N = \left[\frac{1.645\sqrt{0.333(0.667)} + 1.28\sqrt{0.667(0.333)}}{|0.167|}\right]^2 = 68.14$$

Now the required panel size has quadrupled. The combined effect of changing both beta and testing for a smaller effect is

$$N = \left[\frac{1.645\sqrt{0.333(0.667)} + 1.645\sqrt{0.667(0.333)}}{|0.167|}\right]^2 = 86.20$$

Note that in this example, the effect of halving the effect size (critical difference) was greater than the effect of halving the beta-risk. Choosing a reasonable alternative hypothesis is a decision that deserves some care. If the goal is to insure that almost no person will see a difference, or only a small proportion of consumers (or only a small proportion of the time) a large test may be necessary to have confidence in a "no-difference" conclusion. A panel size of 87 testers is probably a larger panel size than many people would consider for a triangle test. Yet it is not unreasonable to have this extra size in the panel when the research question and important consequences concern a parity or equivalence decision. Similar "large" sample sizes can be found in the test for similarity as outlined by Meilgaard et al. (1991).

E.4 Power in Simple Difference and Preference Tests

The scenarios in which we test for the existence of a difference or the existence of a preference often involve important conclusions when no significant effect is found. These are testing situations where acceptance of the null and therefore establishing the power of the test are of great importance. Perhaps for this reason, power and beta-risk in these situations have been addressed by several theorists. The difference testing approaches of Schlich and Ennis are discussed below and a general approach to statistical power is shown in the introductory text by Welkowitz et al. (1982).

Schlich (1993) published risk tables for discrimination tests and offered a SAS routine to calculate beta-risk based on exact binomial probabilities. His article also contains tables of alpha-risk and beta-risk for small discrimination tests and minimum numbers of testers, and correct responses associated with different levels of alpha and beta. Separate tables are computed for the triangle test and for the duo–trio test. The duo–trio table is also used for the directional paired comparison as the tests are both one tailed with chance probability of one-half. The tables showing minimum numbers of testers and correct responses for different levels of beta and alpha in the triangle test are abridged and shown for the triangle test and for the duo–trio test as Tables N1 and N2.

The effect size parameter is stated as the chance-adjusted percent correct. This is based on Abbott's formula, where the chance adjusted proportion, p_d, is based on the difference between the alternative hypothesis percent correct, p_a, and the chance percent correct, p_o, by the following formula:

$$p_d = \frac{p_a - p_o}{1 - p_o} \tag{E.4}$$

This is the so-called discriminator or guessing model discussed in Chapter 5. Schlich suggests the following guidelines for effect size, that 50% above chance is a large effect (50% discriminators), 37.5% above chance is a medium effect, and 25% above chance is a small effect.

Schlich also gave some examples of useful scenarios in which the interplay of alpha and effect size are driven by competing business interests. For example, manufacturing might wish to make a cost reduction by changing an ingredient, but if a spurious difference is found they will not recommend the switch and will not save any money. Therefore the manufacturing decision is to keep alpha low. A marketing manager, on the other hand, might want to insure that very few if any consumers can see a difference. Thus they wish to test against a small effect size, or be sure to detect even a small number of discriminators. Keeping the test power higher (beta low) under both of these conditions will drive the required sample size (N) to a very high level, perhaps hundreds of subjects, as seen in the examples below.

Schlich's tables provide a crossover point for a situation in which both alpha and beta will be low given a sufficient number of testers and a certain effect size. If fewer than the tabulated number ("x") answer correctly, the chance of Type I error will increase should you decide that there is a difference, but the chance of Type II error will decrease should you decide that there is no difference. Conversely, if the number of correct judges exceeds that in the table, the chance of finding a spurious difference will decrease should you reject the null, but the chance of missing a true difference will increase if you accept the null. So it is possible to use these minimal values for a decision rule. Specifically, if the number is less than x, accept the null and there will be lower beta-risk than that listed in the column heading. If the number correct is greater than X, reject the null and alpha-risk will be lower than that listed. It is also possible to interpolate to find other values using various routines that can be found on the web.

Another set of tabulations for power in discrimination tests has been given by Ennis (1993). Instead of basing the alternative hypothesis on the proportions of discriminators, he has computed a measure of sensory difference or effect size based on Thurstonian modeling. These models take into account the fact that different tests may have the same chance probability level, but some discrimination methods are more difficult than others. The concept of "more difficult" shows up in the signal detection models as higher variability in the perceptual comparisons. The more difficult test requires a bigger sensory difference to obtain the same number of correct judges. The triangle test is more difficult than the three-alternative forced-choice test (3-AFC). In the 3-AFC, the panelist's attention is usually directed to a specific attribute rather than choosing on the odd sample. However, the chance

percent correct for both triangle and 3-AFC test is 1/3. The correction for guessing, being based on the chance level, does not take into account the difficulty of the triangle procedure. The "difficulty" arises due to the inherent variability in judging three pairs of differences as opposed to judging simply how strong or weak a given attribute is.

Thurstonian or signal detection models (see Chapter 5) are an improvement over the "proportion of discriminators" model since they do account for the difference in inherent variability. Ennis's tables use the Thurstonian sensory differences symbolized by the lower case Greek letter delta, δ. Delta represents the sensory difference in standard deviations. The standard deviations are theoretical variability estimates of the sensory impressions created by the different products. The delta values have the advantage that they are comparable across methods, unlike the percent correct or the chance-adjusted percent correct. Table E.1 shows the numbers of judges required for different levels of power (80, 90%) and different delta values in the duo–trio, triangle, 2AFC (paired comparison), and 3AFC tests. The lower numbers of judges required for the 2AFC and 3AFC tests arise from their higher sensitivity to differences, i.e., lower inherent variability under the Thurstonian models.

In terms of delta values, we can see that the usual discrimination tests done with 25 or 50 panelists will

Table E.1 Numbers of judges required for different levels of power and sensory difference for paired comparison (2-AFC), duo–trio, 3-AFC, and triangle tests with alpha = 0.05

δ	2-AFC	Duo–trio	3-AFC	Triangle
80% power				
0.50	78	3092	64	2742
0.75	35	652	27	576
1.00	20	225	15	197
1.25	13	102	9	88
1.50	9	55	6	47
1.75	7	34	5	28
2.00	6	23	3	19
90% power				
0.50	108	4283	89	3810
0.75	48	902	39	802
1.00	27	310	21	276
1.25	18	141	13	124
1.50	12	76	9	66
1.75	9	46	6	40
2.00	7	31	5	26

Abstracted from Ennis (1993)

only detect gross differences ($\delta > 1.5$) if the triangle or duo–trio procedures are used. This fact offers some warning that the "non-specific" tests for overall difference (triangle, duo–trio) are really only gross tools that are better suited to giving confidence when a difference is detected. The AFC tests, on the other hand, like a paired comparison test where the attribute of difference is specified (e.g., "pick which one is sweeter") are safer when a no-difference decision is the result.

Useful tables for the power of a triangle test can be found in Chambers and Wolf (1996). A more generally useful table for various simple tests was given by Welkowitz et al. (1982) where the effect size and sample size are considered jointly to produce a power table as a function of alpha. This produces a value we will tabulate as the capital Greek letter delta (Δ, to distinguish it from the lowercase delta in Ennis's tables), while the raw effect size is given by the letter "d." Δ can be thought of as the d-value corrected for sample size. The Δ and d-values take the forms shown in Table E.2 for simple statistical tests. Computing these delta values, which take into account the sample size, allows the referencing of power questions to one simple table, (Table E.3). In other words, all of these simple tests have power calculations via the same table.

Here is worked example, using a two-tailed test on proportions (Welkowitz et al., 1982). Suppose a marketing researcher thinks that a product improvement will produce a preference difference of about 8% against the standard product. In other words, he expects that in a preference test, the split among consumers of this product would be something like 46% preferring the standard product and 54% preferring the new product. He conducts a preference test with 400 people, considered by "intuition" to be a hefty sample size and finds no difference. What is the power of the test and what is the certainty that he did miss a true difference of that size?

Table E.3 Effect size adjusted for sample size to show power as a function of alpha

Two-tailed alpha	0.05	0.025	0.01	0.005
One-tailed alpha	0.10	0.05	0.02	0.01
Δ	Power			
0.2	0.11	0.05	0.02	0.01
0.4	0.13	0.07	0.03	0.01
0.6	0.16	0.09	0.03	0.01
0.8	0.21	0.13	0.06	0.04
1.0	0.26	0.17	0.09	0.06
1.2	0.33	0.22	0.13	0.08
1.4	0.40	0.29	0.18	0.12
1.6	0.48	0.36	0.23	0.16
1.8	0.56	0.44	0.30	0.22
2.0	0.64	0.52	0.37	0.28
2.2	0.71	0.59	0.45	0.36
2.4	0.77	0.67	0.53	0.43
2.6	0.83	0.74	0.61	0.51
2.8	0.88	0.80	0.68	0.59
3.0	0.91	0.85	0.75	0.66
3.2	0.94	0.89	0.78	0.70
3.4	0.96	0.93	0.86	0.80
3.6	0.97	0.95	0.90	0.85
3.8	0.98	0.97	0.94	0.91
4.0	0.99	0.98	0.95	0.92

Reprinted with permission from Welkowitz et al. (1982), Table H

The d-value becomes 0.08 and the delta value is 1.60. Referring to Table E.3, we find that with alpha set at the traditional 5% level, the power is 48% so there is still a 52% chance of Type II error (missing a difference) even with this "hefty" sample size. The problem in this example is that the alternative hypothesis predicts a close race. If the researcher wants to distinguish advantages that are this small, even larger tests must be conducted to be conclusive.

We can then turn the situation around and ask how many consumers should be in the test given the small win that is expected, and the importance of a "no-preference" conclusion? We can use the following relationship for proportions:

$$N = 2\left(\frac{\Delta}{d}\right)^2 \tag{E.5}$$

For a required power of 80% and keeping alpha at the traditional 5% level, we find that a delta value of 2.80 is required. Substituting in our example, we get

$$N = 2\left(\frac{2.80}{0.08}\right)^2 = 2450$$

Table E.2 Conversion of effect size (d) to delta (Δ) value, considering sample size

Test	d-value	Δ
One-sample t-test	$d = (\mu_1 - \mu_2)/\sigma$	$\Delta = d\sqrt{N}$
Dependent t-test	$d = (\mu_1 - \mu_2)/\sigma$	$\Delta = d\sqrt{N}$
Independent t-test	$d = (\mu_1 - \mu_2)/\sigma$	$\Delta = d\sqrt{\frac{2N_1N_2}{N_1+N_2}}$
Correlation	r	$\Delta = d\sqrt{N-1}$
Proportions	$\frac{p_0 - p_a}{\sqrt{p_0(1-p_0)}}$	$\Delta = d\sqrt{N}$

This might not seem like a common consumer test for sensory scientists, who are more concerned with alpha-risk, but in marketing research or political polling of close races, these larger samples are sometimes justified, as our example shows.

E.5 Summary and Conclusions

Equations for the required sample sizes for scaled data and for discrimination tests were given by Eqs. (E.1) and (E.3), respectively. The equation for power for scaled data was given in Eq. (E.2). The corresponding equation for choice data from discrimination tests is

$$\text{Power} = 1 - \beta = 1 - \Phi\left[\frac{Z_\alpha\sqrt{p_0(1-p_0)/N} - (p_0 - p_a)}{\sqrt{p_a(1-p_a)/N}}\right]$$

(E.6)

Table E.4 summarizes these formulae.

A finding of "no difference" is often of importance in sensory evaluation and in support of product research. Many business decisions in foods and consumer products are made on the basis of small product changes for cost reduction, a change of process variables in manufacturing, a change of ingredients or suppliers. Whether or not consumers will notice the change is the inference made from sensory research. In many cases, insurance is provided by performing a sensitive test under controlled conditions. This is the philosophy of the "safety net" approach, paraphrased as follows: "If we do not see a difference under controlled conditions using trained (or selected, screened, oriented, etc.) panelists, then consumers are unlikely to notice this change under the more casual and variable conditions of natural observation." This logic depends upon the assumption that the laboratory test is in fact more sensitive to sensory differences than the consumer's normal experience. Remember that the consumer has extended opportunities to observe the product under a variety of conditions, while the laboratory-based sensory analysis is often limited in time, scope, and the conditions of evaluation.

As stated above, a conclusion of "no difference" based only on a failure to reject the null hypothesis is not logically airtight. If we fail to reject the null, at least three possibilities arise: First, there may have been too much error or random noise in the experiment, so the statistical significance was lost or swamped by large standard deviations. It is a simple matter to do a sloppy experiment. Second, we may not have tested a sufficient number of panelists. If the sample size is too small, we may miss statistical significance because the confidence intervals around our observations are simply too wide to rule out potentially important sensory differences. Third, there may truly be no difference (or no practical difference) between our products. So a failure to reject the null hypothesis is ambiguous and it is simply not proper to conclude that two products are sensorially equivalent simply based on a failure to reject the null. More information is needed.

One approach to this is experimental. If the sensory test is sensitive enough to show a difference in some other condition or comparison, it is difficult to argue that the test was simply not sensitive enough to find any difference in a similar study. Consideration of a track record or demonstrated history of detecting differences with the test method is helpful. In a particular laboratory and with a known panel, it is reasonable to conclude that a tool, which has often shown differences in the past, is operating well and is sufficiently discriminative. Given the history of the sensory procedure under known conditions, it should be possible to use this sort of common sense approach to minimize risk in decision making. In an ongoing sensory testing program for discrimination, it would be reasonable to use a panel of good size (say 50 screened testers), perform a replicated test, and know whether the panel had shown reliable differences in the past.

Another approach is to "bracket" the test comparison with known levels of difference. In other words, include extra products in the test that one would expect to be different. Baseline or positive and negative control comparisons can be tested and if the panel finds significant differences between those benchmark

Table E.4 Sample size and power formulas (see text for details)

Form of data	Sample size	Power
Proportion or frequency	$N = \left[\dfrac{Z_\alpha\sqrt{p_0(1-p_0)}+Z_\beta\sqrt{p_a(1-p_a)}}{|p_0-p_a|}\right]^2$	$1 - \Phi\left[\dfrac{Z_\alpha\sqrt{p_0(1-p_0)/N}-(p_0-p_a)}{\sqrt{p_a(1-p_a)/N}}\right]$
Scaled or continuous	$N = \dfrac{(Z_\alpha+Z_\beta)^2 S^2}{(M_1-M_2)^2}$	$1 - \Phi\left[\dfrac{X_c-\mu_D}{\text{SE}}\right] = 1 - \Phi\left[\dfrac{Z_\alpha(\text{SE})-\mu_D}{\text{SE}}\right]$

products, we have evidence that the tool is working. If the experimental comparison was not significant, then the difference is probably smaller than the sensory difference in the benchmark or bracketing comparisons that did reach significance. Using meta-analytic comparisons (Rosenthal, 1987), a conclusion about relative effect size may be mathematically tested. A related approach is to turn the significance test around, as in the test for significant similarity discussed in Chapter 5. In that approach, the performance in a discrimination test must be at or above chance, but significantly below some chosen cutoff for concluding that products are different, practically (not statistically) speaking.

The third approach is to do a formal analysis of the test power. When a failure to reject the null is accompanied by evidence that the test was of sufficient power, reasonable scientific conclusions may be stated and business decisions can be made with reduced risk. Sensory scientists would do well to make some estimates of test power before conducting any experiment where a null result will generate important actions. It is easy to overestimate the statistical power of a test. On the other hand, it is possible to design an overly sensitive test that finds small significant differences of no practical import. As in other statistical areas, considerations of test power and sensitivity must also be based on the larger framework of practical experience and consumer and/or marketplace validation of the sensory procedure.

Finally, it should be noted that many fields have rejected the approach that sufficient test power allows one to accept the null for purposes of equivalence. For example, in bioequivalence of drugs, one must demonstrate that the test drug falls within a certain range of the control or comparison (USFDA, 2001). This has led to an interval testing approach to equivalence, also discussed by Wellek (2003) and more specifically for sensory testing by Bi (2006), Ennis (2008), and Ennis and Ennis (2009).

References

Amerine, M. A., Pangborn, R. M. and Roessler, E. B. 1965. Principles of Sensory Evaluation of Food. New York: Academic Press.

ASTM. 2008. Standard guide for sensory claim substantiation. Designation E-1958-07. Annual Book of Standards, Vol. 15.08. ASTM International, West Conshohocken, PA, pp. 186–212.

Bi, J. 2006. Sensory Discrimination Tests and Measurements. Blackwell, Ames, IA.

Chambers, E. C. IV and Wolf, M. B. 1996. Sensory Testing Methods, Second Edition. ASTM Manual Series MNL 26. ASTM International, West Conshohocken, PA.

Cohen, J. 1988. Statistical Power Analysis for the Behavioral Sciences, Second Edition. Lawrence Erlbaum Associates, Hillsdale, NJ.

Cohen, J. 1992. A power primer. Psychological Bulletin, 112, 155–159.

Ennis, D. M. 1993. The power of sensory discrimination methods. Journal of Sensory Studies, 8, 353–370.

Ennis, D. M. 2008. Tables for parity testing. Journal of Sensory Studies, 32, 80–91.

Ennis, D. M. and Ennis, J. M. 2010. Equivalence hypothesis testing. Food Quality and Preference, 21, 253–256.

Erdfelder, E., Faul, F. and Buchner, A. 1996. Gpower: A general power analysis program. Behavior Research Methods, Instrumentation and Computers, 28, 1–11.

Gacula, M. C., Jr. 1991. Claim substantiation for sensory equivalence and superiority. In: H. T. Lawless and B. P. Klein (eds.), Sensory Science Theory and Applications in Foods. Marcel Dekker, New York, pp. 413–436.

Gacula, M. C. Jr. 1993. Design and Analysis of Sensory Optimization. Food and Nutrition, Trumbull, CT.

Gacula, M. C, Jr. and Singh, J. 1984. Statistical Methods in Food and Consumer Research. Academic, Orlando, FL.

Lawless, H. T. 1988. Odour description and odour classification revisited. In: D. M. H. Thomson (ed.), Food Acceptability. Elsevier Applied Science, London, pp. 27–40.

Meilgaard, M., Civille, G. V. and Carr, B. T. 1991. Sensory Evaluation Techniques, Second Edition. CRC, Boca Raton.

Rosenthal, R. 1987. Judgment Studies: Design, Analysis and Meta-Analysis. University Press, Cambridge.

Schlich, P. 1993. Risk tables for discrimination tests. Food Quality and Preference, 4, 141–151.

Sokal, R. R. and Rohlf, F. J. 1981. Biometry. Second Edition. W. H. Freeman, New York.

U. S. F. D. A. 2001. Guidance for Industry. Statistical Approaches to Bioequivalence. U. S. Dept. of Health and Human Services, Food and Drug Administration, Center for Drug Evaluation and Research (CDER). http://www.fda.gov/cder/guidance/index.htm.

Welkowitz, J., Ewen, R. B. and Cohen, J. 1982. Introductory Statistics for the Behavioral Sciences. Academic, New York.

Wellek, S. 2003. Testing Statistical Hypothesis of Equivalence. Chapman and Hall, CRC, Boca Raton, FL.

Appendix F

Statistical Tables

Contents

Table F.A Cumulative probabilities of the standard normal distribution. Entry area $1-\alpha$ under the standard normal curve from $-\infty$ to $z(1-\alpha)$

z	0	0.01	0.02	0.03	0.04	0.05	0.06	0.07	0.08	0.09
0	0.5000	0.5040	0.5080	0.5120	0.5160	0.5199	0.5239	0.5279	0.5319	0.5359
0.1	0.5398	0.5438	0.5478	0.5517	0.5557	0.5596	0.5636	0.5675	0.5714	0.5753
0.2	0.5793	0.5832	0.5871	0.5910	0.5948	0.5987	0.6026	0.6064	0.6103	0.6141
0.3	0.6179	0.6217	0.6255	0.6293	0.6331	0.6368	0.6406	0.6443	0.6480	0.6517
0.4	0.6554	0.6591	0.6628	0.6664	0.6700	0.6736	0.6772	0.6808	0.6844	0.6879
0.5	0.6915	0.6950	0.6985	0.7019	0.7054	0.7088	0.7123	0.7157	0.7190	0.7224
0.6	0.7257	0.7291	0.7324	0.7357	0.7389	0.7422	0.7454	0.7486	0.7517	0.7549
0.7	0.7580	0.7611	0.7642	0.7673	0.7704	0.7734	0.7764	0.7794	0.7823	0.7852
0.8	0.7881	0.7910	0.7939	0.7967	0.7995	0.8023	0.8051	0.8078	0.8106	0.8133
0.9	0.8159	0.8186	0.8212	0.8238	0.8264	0.8289	0.8315	0.8340	0.8365	0.8389
1	0.8413	0.8438	0.8461	0.8485	0.8508	0.8531	0.8554	0.8577	0.8599	0.8621
1.1	0.8643	0.8665	0.8686	0.8708	0.8729	0.8749	0.8770	0.8790	0.8810	0.8830
1.2	0.8849	0.8869	0.8888	0.8907	0.8925	0.8944	0.8962	0.8980	0.8997	0.9015
1.3	0.9032	0.9049	0.9066	0.9082	0.9099	0.9115	0.9131	0.9147	0.9162	0.9177
1.4	0.9192	0.9207	0.9222	0.9236	0.9251	0.9265	0.9279	0.9292	0.9306	0.9319
1.5	0.9332	0.9345	0.9357	0.9370	0.9382	0.9394	0.9406	0.9418	0.9429	0.9441
1.6	0.9452	0.9463	0.9474	0.9484	0.9495	0.9505	0.9515	0.9525	0.9535	0.9545
1.7	0.9554	0.9564	0.9573	0.9582	0.9591	0.9599	0.9608	0.9616	0.9625	0.9633
1.8	0.9641	0.9649	0.9656	0.9664	0.9671	0.9678	0.9686	0.9693	0.9699	0.9706
1.9	0.9713	0.9719	0.9726	0.9732	0.9738	0.9744	0.9750	0.9756	0.9761	0.9767
2	0.9772	0.9778	0.9783	0.9788	0.9793	0.9798	0.9803	0.9808	0.9812	0.9817
2.1	0.9821	0.9826	0.9830	0.9834	0.9838	0.9842	0.9846	0.9850	0.9854	0.9857
2.2	0.9861	0.9864	0.9868	0.9871	0.9875	0.9878	0.9881	0.9884	0.9887	0.9890
2.3	0.9893	0.9896	0.9898	0.9901	0.9904	0.9906	0.9909	0.9911	0.9913	0.9916
2.4	0.9918	0.9920	0.9922	0.9925	0.9927	0.9929	0.9931	0.9932	0.9934	0.9936
2.5	0.9938	0.9940	0.9941	0.9943	0.9945	0.9946	0.9948	0.9949	0.9951	0.9952
2.6	0.9953	0.9955	0.9956	0.9957	0.9959	0.9960	0.9961	0.9962	0.9963	0.9964
2.7	0.9965	0.9966	0.9967	0.9968	0.9969	0.9970	0.9971	0.9972	0.9973	0.9974
2.8	0.9974	0.9975	0.9976	0.9977	0.9977	0.9978	0.9979	0.9979	0.9980	0.9981
2.9	0.9981	0.9982	0.9982	0.9983	0.9984	0.9984	0.9985	0.9985	0.9986	0.9986
3	0.9987	0.9987	0.9987	0.9988	0.9988	0.9989	0.9989	0.9989	0.9990	0.9990

Table F.B Table of critical values for the t-distribution

df	Level of significance for one-tailed test						
	0.01	0.05	0.025	0.01	0.005	0.001	0.0005
	Level of significance for two-tailed test						
	0.02	0.1	0.05	0.02	0.01	0.002	0.001
1	3.078	6.314	12.706	31.821	63.656	318.289	636.578
2	1.886	2.92	4.303	6.965	9.925	22.328	31.600
3	1.638	2.353	3.182	4.541	5.841	10.214	12.924
4	1.533	2.132	2.776	3.747	4.604	7.173	8.610
5	1.476	2.015	2.571	3.365	4.032	5.894	6.869
6	1.440	1.943	2.447	3.143	3.707	5.208	5.959
7	1.415	1.895	2.365	2.998	3.499	4.785	5.408
8	1.397	1.860	2.306	2.896	3.355	4.501	5.041
9	1.383	1.833	2.262	2.821	3.250	4.297	4.781
10	1.372	1.812	2.228	2.764	3.169	4.144	4.587
11	1.363	1.796	2.201	2.718	3.106	4.025	4.437
12	1.356	1.782	2.179	2.681	3.055	3.930	4.318
13	1.350	1.771	2.160	2.650	3.012	3.852	4.221
14	1.345	1.761	2.145	2.624	2.977	3.787	4.140
15	1.341	1.753	2.131	2.602	2.947	3.733	4.073
16	1.337	1.746	2.120	2.583	2.921	3.686	4.015
17	1.333	1.740	2.110	2.567	2.898	3.646	3.965
18	1.330	1.734	2.101	2.552	2.878	3.610	3.922
19	1.328	1.729	2.093	2.539	2.861	3.579	3.883
20	1.325	1.725	2.086	2.528	2.845	3.552	3.850
21	1.323	1.721	2.080	2.518	2.831	3.527	3.819
22	1.321	1.717	2.074	2.508	2.819	3.505	3.792
23	1.319	1.714	2.069	2.500	2.807	3.485	3.768
24	1.318	1.711	2.064	2.492	2.797	3.467	3.745
25	1.316	1.708	2.060	2.485	2.787	3.450	3.725
26	1.315	1.706	2.056	2.479	2.779	3.435	3.707
27	1.314	1.703	2.052	2.473	2.771	3.421	3.689
28	1.313	1.701	2.048	2.467	2.763	3.408	3.674
29	1.311	1.699	2.045	2.462	2.756	3.396	3.660
30	1.310	1.697	2.042	2.457	2.750	3.385	3.646
60	1.296	1.671	2.000	2.390	2.660	3.232	3.460
120	1.289	1.658	1.980	2.358	2.617	3.160	3.373
∞	1.282	1.645	1.960	2.326	2.576	3.091	3.291

Table F.C Table of critical values of the chi-square (χ^2) distribution

Alpha	0.1	0.05	0.025	0.01	0.005
df					
1	2.71	3.84	5.02	6.64	7.88
2	4.61	5.99	7.38	9.21	10.60
3	6.25	7.82	9.35	11.35	12.84
4	7.78	9.49	11.14	13.28	14.86
5	9.24	11.07	12.83	15.09	16.75
6	10.65	12.59	14.45	16.81	18.55
7	12.02	14.07	16.01	18.48	20.28
8	13.36	15.51	17.54	20.09	21.96
9	14.68	16.92	19.02	21.67	23.59
10	15.99	18.31	20.48	23.21	25.19
11	17.28	19.68	21.92	24.73	26.76
12	18.55	21.03	23.34	26.22	28.30
13	19.81	22.36	24.74	27.69	29.82
14	21.06	23.69	26.12	29.14	31.32
15	22.31	25.00	27.49	30.58	32.80
16	23.54	26.30	28.85	32.00	34.27
17	24.77	27.59	30.19	33.41	35.72
18	25.99	28.87	31.53	34.81	37.16
19	27.20	30.14	32.85	36.19	38.58
20	28.41	31.41	34.17	37.57	40.00
21	29.62	32.67	35.48	38.93	41.40
22	30.81	33.92	36.78	40.29	42.80
23	32.01	35.17	38.08	41.64	44.18
24	33.20	36.42	39.36	42.98	45.56
25	34.38	37.65	40.65	44.31	46.93
26	35.56	38.89	41.92	45.64	48.29
27	36.74	40.11	43.20	46.96	49.65
28	37.92	41.34	44.46	48.28	50.99
29	39.09	42.56	45.72	49.59	52.34
30	40.26	43.77	46.98	50.89	53.67
40	51.81	55.76	59.34	63.69	66.77
50	63.17	67.51	71.42	76.15	79.49
60	74.40	79.08	83.30	88.38	91.95
70	85.53	90.53	95.02	100.43	104.22
80	96.58	101.88	106.63	112.33	116.32
90	107.57	113.15	118.14	124.12	128.30
100	118.50	124.34	129.56	135.81	140.17

Table F.D1 Critical values of the F-distribution at $\alpha = 0.05$

df1 df2	1	2	3	4	5	10	20	30	40	50	60	70	80	100	∞
5	6.61	5.79	5.41	5.19	5.05	4.74	4.56	4.50	4.46	4.44	4.43	4.42	4.42	4.41	4.37
6	5.99	5.14	4.76	4.53	4.39	4.06	3.87	3.81	3.77	3.75	3.74	3.73	3.72	3.71	3.68
7	5.59	4.74	4.35	4.12	3.97	3.64	3.44	3.38	3.34	3.32	3.30	3.29	3.29	3.27	3.24
8	5.32	4.46	4.07	3.84	3.69	3.35	3.15	3.08	3.04	3.02	3.01	2.99	2.99	2.97	2.94
9	5.12	4.26	3.86	3.63	3.48	3.14	2.94	2.86	2.83	2.80	2.79	2.78	2.77	2.76	2.72
10	4.96	4.10	3.71	3.48	3.33	2.98	2.77	2.70	2.66	2.64	2.62	2.61	2.60	2.59	2.55
11	4.84	3.98	3.59	3.36	3.20	2.85	2.65	2.57	2.53	2.51	2.49	2.48	2.47	2.46	2.42
12	4.75	3.89	3.49	3.26	3.11	2.75	2.54	2.47	2.43	2.40	2.38	2.37	2.36	2.35	2.31
13	4.67	3.81	3.41	3.18	3.03	2.67	2.46	2.38	2.34	2.31	2.30	2.28	2.27	2.26	2.22
14	4.60	3.74	3.34	3.11	2.96	2.60	2.39	2.31	2.27	2.24	2.22	2.21	2.20	2.19	2.14
15	4.54	3.68	3.29	3.06	2.90	2.54	2.33	2.25	2.20	2.18	2.16	2.15	2.14	2.12	2.08
16	4.49	3.63	3.24	3.01	2.85	2.49	2.28	2.19	2.15	2.12	2.11	2.09	2.08	2.07	2.02
17	4.45	3.59	3.20	2.96	2.81	2.45	2.23	2.15	2.10	2.08	2.06	2.05	2.03	2.02	1.97
18	4.41	3.55	3.16	2.93	2.77	2.41	2.19	2.11	2.06	2.04	2.02	2.00	1.99	1.98	1.93
19	4.38	3.52	3.13	2.90	2.74	2.38	2.16	2.07	2.03	2.00	1.98	1.97	1.96	1.94	1.89
20	4.35	3.49	3.10	2.87	2.71	2.35	2.12	2.04	1.99	1.97	1.95	1.93	1.92	1.91	1.86
22	4.30	3.44	3.05	2.82	2.66	2.30	2.07	1.98	1.94	1.91	1.89	1.88	1.86	1.85	1.80
23	4.26	3.40	3.01	2.78	2.62	2.25	2.03	1.94	1.89	1.86	1.84	1.83	1.82	1.80	1.75
26	4.23	3.37	2.98	2.74	2.59	2.22	1.99	1.90	1.85	1.82	1.80	1.79	1.78	1.76	1.71
28	4.20	3.34	2.95	2.71	2.56	2.19	1.96	1.87	1.82	1.79	1.77	1.75	1.74	1.73	1.67
30	4.17	3.32	2.92	2.69	2.53	2.16	1.93	1.84	1.79	1.76	1.74	1.72	1.71	1.70	1.64
35	4.12	3.27	2.87	2.64	2.49	2.11	1.88	1.79	1.74	1.70	1.68	1.66	1.65	1.63	1.57
40	4.08	3.23	2.84	2.61	2.45	2.08	1.84	1.74	1.69	1.66	1.64	1.62	1.61	1.59	1.53
45	4.06	3.20	2.81	2.58	2.42	2.05	1.81	1.71	1.66	1.63	1.60	1.59	1.57	1.55	1.49
50	4.03	3.18	2.79	2.56	2.40	2.03	1.78	1.69	1.63	1.60	1.58	1.56	1.54	1.52	1.46
60	4.00	3.15	2.76	2.53	2.37	1.99	1.75	1.65	1.59	1.56	1.53	1.52	1.50	1.48	1.41
70	3.98	3.13	2.74	2.50	2.35	1.97	1.72	1.62	1.57	1.53	1.50	1.49	1.47	1.45	1.37
80	3.96	3.11	2.72	2.49	2.33	1.95	1.70	1.60	1.54	1.51	1.48	1.46	1.45	1.43	1.35
100	3.94	3.09	2.70	2.46	2.31	1.93	1.68	1.57	1.52	1.48	1.45	1.43	1.41	1.39	1.31
∞	3.86	3.01	2.62	2.39	2.23	1.85	1.59	1.48	1.42	1.38	1.35	1.32	1.30	1.28	1.16

Table F.D2 Critical values of the F-distribution at $\alpha = 0.01$

df1	1	2	3	4	5	10	20	30	40	50	60	70	80	100	∞
df2															
3	34.12	30.82	29.46	28.71	28.24	27.23	26.69	26.50	26.41	26.35	26.32	26.29	26.27	26.24	26.15
4	21.20	18.00	16.69	15.98	15.52	14.55	14.02	13.84	13.75	13.69	13.65	13.63	13.61	13.58	13.49
5	16.26	13.27	12.06	11.39	10.97	10.05	9.55	9.38	9.29	9.24	9.20	9.18	9.16	9.13	9.04
6	13.75	10.92	9.78	9.15	8.75	7.87	7.40	7.23	7.14	7.09	7.06	7.03	7.01	6.99	6.90
7	12.25	9.55	8.45	7.85	7.46	6.62	6.16	5.99	5.91	5.86	5.82	5.80	5.78	5.75	5.67
8	11.26	8.65	7.59	7.01	6.63	5.81	5.36	5.20	5.12	5.07	5.03	5.01	4.99	4.96	4.88
9	10.56	8.02	6.99	6.42	6.06	5.26	4.81	4.65	4.57	4.52	4.48	4.46	4.44	4.42	4.33
10	10.04	7.56	6.55	5.99	5.64	4.85	4.41	4.25	4.17	4.12	4.08	4.06	4.04	4.01	3.93
11	9.65	7.21	6.22	5.67	5.32	4.54	4.10	3.94	3.86	3.81	3.78	3.75	3.73	3.71	3.62
12	9.33	6.93	5.95	5.41	5.06	4.30	3.86	3.70	3.62	3.57	3.54	3.51	3.49	3.47	3.38
13	9.07	6.70	5.74	5.21	4.86	4.10	3.66	3.51	3.43	3.38	3.34	3.32	3.30	3.27	3.19
14	8.86	6.51	5.56	5.04	4.70	3.94	3.51	3.35	3.27	3.22	3.18	3.16	3.14	3.11	3.03
15	8.68	6.36	5.42	4.89	4.56	3.80	3.37	3.21	3.13	3.08	3.05	3.02	3.00	2.98	2.89
16	8.53	6.23	5.29	4.77	4.44	3.69	3.26	3.10	3.02	2.97	2.93	2.91	2.89	2.86	2.78
17	8.40	6.11	5.19	4.67	4.34	3.59	3.16	3.00	2.92	2.87	2.83	2.81	2.79	2.76	2.68
18	8.29	6.01	5.09	4.58	4.25	3.51	3.08	2.92	2.84	2.78	2.75	2.72	2.71	2.68	2.59
19	8.19	5.93	5.01	4.50	4.17	3.43	3.00	2.84	2.76	2.71	2.67	2.65	2.63	2.60	2.51
20	8.10	5.85	4.94	4.43	4.10	3.37	2.94	2.78	2.69	2.64	2.61	2.58	2.56	2.54	2.44
30	7.56	5.39	4.51	4.02	3.70	2.98	2.55	2.39	2.30	2.25	2.21	2.18	2.16	2.13	2.03
40	7.31	5.18	4.31	3.83	3.51	2.80	2.37	2.20	2.11	2.06	2.02	1.99	1.97	1.94	1.83
50	7.17	5.06	4.20	3.72	3.41	2.70	2.27	2.10	2.01	1.95	1.91	1.88	1.86	1.82	1.71
60	7.08	4.98	4.13	3.65	3.34	2.63	2.20	2.03	1.94	1.88	1.84	1.81	1.78	1.75	1.63
70	7.01	4.92	4.07	3.60	3.29	2.59	2.15	1.98	1.89	1.83	1.78	1.75	1.73	1.70	1.57
80	6.96	4.88	4.04	3.56	3.26	2.55	2.12	1.94	1.85	1.79	1.75	1.71	1.69	1.65	1.53
100	6.90	4.82	3.98	3.51	3.21	2.50	2.07	1.89	1.80	1.74	1.69	1.66	1.63	1.60	1.47
∞	6.69	4.65	3.82	3.36	3.05	2.36	1.92	1.74	1.63	1.57	1.52	1.48	1.45	1.41	1.23

Table F.E Critical values of U for a one-tailed alpha at 0.025 or a two-tailed alpha at 0.05

n_1	5	6	7	8	9	10	11	12	13	14	15	16	17	18	19	20
n_2																
5	2	3	5	6	7	8	9	11	12	13	14	15	17	18	19	20
6	3	5	6	8	10	11	13	14	16	17	19	21	22	24	25	27
7	5	6	8	10	12	14	16	18	20	22	24	26	28	30	32	34
8	6	8	10	13	15	17	19	22	24	26	29	31	34	36	38	41
9	7	10	12	15	17	21	23	26	28	31	34	37	39	42	45	48
10	8	11	14	17	20	23	26	29	33	36	39	42	45	48	52	55
11	9	13	16	19	23	26	30	33	37	40	44	47	51	55	58	62
12	11	14	18	22	26	29	33	37	41	45	49	53	57	61	65	69
13	12	16	20	24	28	33	37	41	45	50	54	59	63	67	72	76
14	13	17	22	26	31	36	40	45	50	55	59	64	67	74	78	83
15	14	19	24	29	34	39	44	49	54	59	64	70	75	80	85	90
16	15	21	26	31	37	42	47	53	59	64	70	75	81	86	92	98
17	17	22	28	34	39	45	51	57	63	67	75	81	87	93	99	105
18	18	24	30	36	42	48	55	61	67	74	80	86	93	99	106	112
19	19	25	32	38	45	52	58	65	72	78	85	92	99	106	113	119
20	20	27	34	41	48	55	62	69	76	83	90	98	105	112	119	127

Reworked from Auble, D. 1953. Extended tables for the Mann–Whitney U-statistic. Bulletin of the Institute of Educational Research, 1(2). Indiana University

Table F.F1 Table of critical values of ρ (Spearman Rank correlation coefficient)

One-tailed alpha values			
0.05	0.025	0.01	0.005
Two-tailed alpha values			
0.10	0.05	0.02	0.01
n			
4 0.807			
4 1.000			
5 0.900	1.000	1.000	
6 0.829	0.886	0.943	1.000
7 0.714	0.786	0.893	0.929
8 0.643	0.738	0.833	0.881
9 0.600	0.700	0.783	0.833
10 0.564	0.648	0.745	0.794
11 0.536	0.618	0.709	0.755
12 0.503	0.587	0.678	0.727
13 0.484	0.560	0.648	0.703
14 0.464	0.538	0.626	0.679
15 0.446	0.521	0.604	0.654
16 0.429	0.503	0.582	0.635
17 0.414	0.488	0.566	0.618
18 0.401	0.472	0.550	0.600
19 0.391	0.460	0.535	0.584
20 0.380	0.447	0.522	0.570
21 0.370	0.436	0.509	0.556
22 0.361	0.425	0.497	0.544
23 0.353	0.416	0.486	0.532
24 0.344	0.407	0.476	0.521
25 0.337	0.398	0.466	0.511
26 0.331	0.390	0.457	0.501
27 0.324	0.383	0.449	0.492
28 0.318	0.375	0.441	0.483
29 0.312	0.368	0.433	0.475
30 0.306	0.362	0.425	0.467
35 0.283	0.335	0.394	0.433
40 0.264	0.313	0.368	0.405
45 0.248	0.294	0.347	0.382
50 0.235	0.279	0.329	0.363

Reworked from Ramsey, P. H. 1989. Critical values for Spearman's rank order correlation. Journal of Educational and Behavioral Statistics, 14, 245–253

Table F.F2 Table of critical values of r (Pearson's correlation coefficient)

df (n–2)	One-tailed alpha values			
	0.05	0.025	0.01	0.005
	Two-tailed alpha values			
	0.1	0.05	0.02	0.01
1	0.988	0.997	0.999	0.999
2	0.900	0.950	0.980	0.990
3	0.805	0.878	0.934	0.959
4	0.729	0.811	0.882	0.917
5	0.669	0.754	0.833	0.875
6	0.622	0.707	0.789	0.834
7	0.582	0.666	0.750	0.798
8	0.549	0.632	0.716	0.765
9	0.521	0.602	0.685	0.735
10	0.497	0.576	0.658	0.708
11	0.476	0.553	0.634	0.684
12	0.458	0.532	0.612	0.661
13	0.441	0.514	0.592	0.641
14	0.426	0.497	0.574	0.623
15	0.412	0.482	0.558	0.606
16	0.400	0.468	0.542	0.590
17	0.389	0.456	0.528	0.575
18	0.378	0.444	0.516	0.561
19	0.369	0.433	0.503	0.549
20	0.360	0.423	0.492	0.537
21	0.352	0.413	0.482	0.526
22	0.344	0.404	0.472	0.515
23	0.337	0.396	0.462	0.505
24	0.330	0.388	0.453	0.496
25	0.323	0.381	0.445	0.487
26	0.317	0.374	0.437	0.479
27	0.311	0.367	0.43	0.471
28	0.306	0.361	0.423	0.463
29	0.301	0.355	0.416	0.456
30	0.296	0.349	0.409	0.449
35	0.275	0.325	0.381	0.418
40	0.257	0.304	0.358	0.393
45	0.243	0.288	0.338	0.372
50	0.231	0.273	0.322	0.354

Table F.G Critical values for Duncan's multiple range test (p, df, $\alpha = 0.05$)

	Number of means bracketing comparison (p)[a]						
	2	3	4	5	10	15	20
df							
1	17.969	17.969	17.969	17.969	17.969	17.969	17.969
2	6.085	6.085	6.085	6.085	6.085	6.085	6.085
3	4.501	4.516	4.516	4.516	4.516	4.516	4.516
4	3.926	4.013	4.033	4.033	4.033	4.033	4.033
5	3.635	3.749	3.796	3.814	3.814	3.814	3.814
6	3.461	3.586	3.649	3.68	3.697	3.697	3.697
7	3.344	3.477	3.548	3.588	3.625	3.625	3.625
8	3.261	3.398	3.475	3.521	3.579	3.579	3.579
9	3.199	3.339	3.42	3.47	3.547	3.547	3.547
10	3.151	3.293	3.376	3.43	3.522	3.525	3.525
11	3.113	3.256	3.341	3.397	3.501	3.510	3.510
12	3.081	3.225	3.312	3.37	3.484	3.498	3.498
13	3.055	3.200	3.288	3.348	3.470	3.49	3.490
14	3.033	3.178	3.268	3.328	3.457	3.484	3.484
15	3.014	3.16	3.25	3.312	3.446	3.478	3.480
16	2.998	3.144	3.235	3.297	3.437	3.473	3.477
17	2.984	3.130	3.222	3.285	3.429	3.469	3.475
18	2.971	3.117	3.21	3.274	3.421	3.465	3.474
19	2.96	3.106	3.199	3.264	3.415	3.462	3.474
20	2.95	3.097	3.190	3.255	3.409	3.459	3.473
22	2.933	3.080	3.173	3.239	3.398	3.453	3.472
24	2.919	3.066	3.160	3.226	3.390	3.449	3.472
26	2.907	3.054	3.149	3.216	3.382	3.445	3.471
28	2.897	3.044	3.139	3.206	3.376	3.442	3.470
30	2.888	3.035	3.131	3.199	3.371	3.439	3.470
35	2.871	3.018	3.114	3.183	3.360	3.433	3.469
40	2.858	3.005	3.102	3.171	3.352	3.429	3.469
60	2.829	2.976	3.073	3.143	3.333	3.419	3.468
80	2.814	2.961	3.059	3.130	3.323	3.414	3.467
120	2.800	2.947	3.045	3.116	3.313	3.409	3.466
∞	2.772	2.918	3.017	3.089	3.294	3.399	3.466

Reworked from Harter, H. L. 1960. Critical values for Duncan's new multiple range test. Biometrics, 16, 671–685.

[a]Number of means, when rank ordered, between the pair being compared and including the pair itself

Table F.H1 Critical values[a] of the triangle test for similarity (maximum number correct as a function of the number of observations (*N*), beta, and proportion discriminating)

N	Beta	Proportion discriminating		
		10%	20%	30%
30	0.05			11
	0.1		10	11
36	0.05		11	13
	0.1	10	12	14
42	0.05	11	13	16
	0.1	12	14	17
48	0.05	13	16	19
	0.1	14	17	20
54	0.05	15	18	22
	0.1	16	20	23
60	0.05	17	21	25
	0.1	18	22	26
66	0.05	19	23	28
	0.1	20	25	29
72	0.05	21	26	30
	0.1	22	27	32
78	0.05	23	28	33
	0.1	25	30	34
84	0.05	25	31	35
	0.1	27	32	38
90	0.05	27	33	38
	0.1	29	35	38
96	0.05	30	36	42
	0.1	31	38	44

Created in analogy to Meilgaard, M., Civille, G. V., Carr B. T. 1991. Sensory Evaluation Techniques. CRC Boca Raton, FL. Using B.T. Carr's Discrimination Test Analysis Tool EXCEL program
[a]Accept the null hypothesis with 100(1–beta) confidence if the number of correct choices does not exceed the tabled value for the allowable proportion of discriminators

Table F.H2 Critical values[a] of the duo–trio and paired comparison tests for similarity (maximum number correct as a function of the number of observations (N), beta, and proportion discriminating

N	Beta	Proportion discriminating		
		10%	20%	30%
32	0.05	12	14	15
	0.1	13	15	16
36	0.05	14	16	18
	0.1	15	17	19
40	0.05	16	18	20
	0.1	17	19	21
44	0.05	18	20	22
	0.1	19	21	24
48	0.05	20	22	25
	0.1	21	23	26
52	0.05	22	24	27
	0.1	23	26	28
56	0.05	24	27	29
	0.1	25	28	31
60	0.05	26	29	32
	0.1	27	30	33
64	0.05	28	31	34
	0.1	29	32	36
68	0.05	30	33	37
	0.1	31	35	38
72	0.05	32	35	39
	0.1	33	37	41
76	0.05	34	38	41
	0.1	35	39	43
80	0.05	36	40	44
	0.1	37	41	46
84	0.05	38	42	46
	0.1	39	44	48

Created in analogy to Meilgaard, M., Civille, G. V., Carr B. T. 1991. Sensory Evaluation Techniques. CRC Boca Raton, FL. Using B.T. Carr's Discrimination Test Analysis Tool EXCEL program

[a]Accept the null hypothesis with 100(1–beta) confidence if the number of correct choices does not exceed the tabled value for the allowable proportion of discriminators

Table F.I   Table of probabilities for values as small as observed values of x associated with the binomial test $(p=0.50)$[a,b]

x	0	1	2	3	4	5	6	7	8	9	10	11	12	13	14	15
N																
5	0.031	0.188	0.500	0.813	0.969											
6	0.016	0.109	0.344	0.656	0.891	0.984										
7	0.008	0.063	0.227	0.500	0.773	0.938	0.992									
8	0.004	0.035	0.145	0.363	0.637	0.855	0.965	0.996								
9	0.002	0.020	0.090	0.254	0.500	0.746	0.910	0.980	0.998							
10	0.001	0.011	0.055	0.172	0.377	0.623	0.828	0.945	0.989	0.999						
11	0.000	0.006	0.033	0.113	0.274	0.500	0.726	0.887	0.967	0.994						
12	0.000	0.003	0.019	0.073	0.194	0.387	0.613	0.806	0.927	0.981	0.997					
13	0.000	0.002	0.011	0.046	0.133	0.291	0.500	0.709	0.867	0.954	0.989	0.998				
14		0.001	0.006	0.029	0.090	0.212	0.395	0.605	0.788	0.910	0.971	0.994	0.999			
15		0.000	0.004	0.018	0.059	0.151	0.304	0.500	0.696	0.849	0.941	0.982	0.996			
16		0.000	0.002	0.011	0.038	0.105	0.227	0.402	0.598	0.773	0.895	0.962	0.989	0.998		
17		0.000	0.001	0.006	0.025	0.072	0.166	0.315	0.500	0.685	0.834	0.928	0.975	0.994	0.999	
18			0.001	0.004	0.015	0.048	0.119	0.240	0.407	0.593	0.760	0.881	0.952	0.985	0.996	0.999
19			0.000	0.002	0.010	0.032	0.084	0.180	0.324	0.500	0.676	0.820	0.916	0.968	0.990	0.998
20			0.000	0.001	0.006	0.021	0.058	0.132	0.252	0.412	0.588	0.748	0.868	0.942	0.979	0.994
21			0.000	0.001	0.004	0.013	0.039	0.095	0.192	0.332	0.500	0.668	0.808	0.905	0.961	0.987
22				0.000	0.002	0.008	0.026	0.067	0.143	0.262	0.416	0.584	0.738	0.857	0.933	0.974
23				0.000	0.001	0.005	0.017	0.047	0.105	0.202	0.339	0.500	0.661	0.798	0.895	0.953
24				0.000	0.001	0.003	0.011	0.032	0.076	0.154	0.271	0.419	0.581	0.729	0.846	0.924
25					0.000	0.002	0.007	0.022	0.054	0.115	0.212	0.345	0.500	0.655	0.788	0.885
26					0.000	0.001	0.005	0.014	0.038	0.084	0.163	0.279	0.423	0.577	0.721	0.837
27					0.000	0.001	0.003	0.010	0.026	0.061	0.124	0.221	0.351	0.500	0.649	0.779
28						0.000	0.002	0.006	0.018	0.044	0.092	0.172	0.286	0.425	0.575	0.714
29						0.000	0.001	0.004	0.012	0.031	0.068	0.132	0.229	0.356	0.500	0.644
30						0.000	0.001	0.003	0.008	0.021	0.049	0.100	0.181	0.292	0.428	0.572
35								0.000	0.001	0.003	0.008	0.020	0.045	0.088	0.155	0.250
40										0.000	0.001	0.003	0.008	0.019	0.040	0.077

[a]These values are one tailed. For a two-tailed test double the value.
[b]The alpha level is equal to (1–probability)

Table F.J Critical values for the differences between rank sums ($\alpha = 0.05$)

	Number of samples									
	3	4	5	6	7	8	9	10	11	12
Number of panelists										
3	6	8	11	13	15	18	20	23	25	28
4	7	10	13	15	18	21	24	27	30	33
5	8	11	14	17	21	24	27	30	34	37
6	9	12	15	19	22	26	30	34	37	42
7	10	13	17	20	24	28	32	36	40	44
8	10	14	18	22	26	30	34	39	43	47
9	10	15	19	23	27	32	36	41	46	50
10	1	15	20	24	29	34	38	43	48	53
11	11	16	21	26	30	35	40	45	51	56
12	12	17	22	27	32	37	42	48	53	58
13	12	18	23	28	33	39	44	50	55	61
14	13	18	24	29	34	40	46	52	57	63
15	13	19	24	30	36	42	47	53	59	66
16	14	19	25	31	37	42	49	55	61	67
17	14	20	26	32	38	44	50	56	63	69
18	15	20	26	32	39	45	51	58	65	71
19	15	21	27	33	40	46	53	60	66	73
20	15	21	28	34	41	47	54	61	68	75
21	16	22	28	35	42	49	56	63	70	77
22	16	22	29	36	43	50	57	64	71	79
23	16	23	30	37	44	51	58	65	73	80
24	17	23	30	37	45	52	59	67	74	82
25	17	24	31	38	46	53	61	68	76	84
26	17	24	32	39	46	54	62	70	77	85
27	18	25	32	40	47	55	63	71	79	87
28	18	25	33	40	48	56	64	72	80	89
29	18	26	33	41	49	57	65	73	82	90
30	19	26	34	42	50	58	66	75	83	92
35	20	28	37	45	54	63	72	81	90	99
40	21	30	39	48	57	67	76	86	96	106
45	23	32	41	51	61	71	81	91	102	112
50	24	34	44	54	64	75	85	96	107	118
55	25	34	46	56	67	78	90	101	112	124
60	26	37	48	59	70	82	94	105	117	130
65	27	38	50	61	73	85	97	110	122	135
70	28	40	52	64	76	88	101	114	127	140
75	29	41	53	66	79	91	105	118	131	145
80	30	42	55	68	81	94	108	122	136	150
85	31	44	57	70	84	97	111	125	140	154
90	32	45	58	72	86	100	114	129	144	159
95	33	46	60	74	88	103	118	133	148	163
100	34	47	61	76	91	105	121	136	151	167

Reworked from Newell, G. and MacFarlane, J. 1988. Expanded tables for multiple comparison procedures in the analysis of ranked data. Journal of Food Science, 52, 1721–1725

Table F.K Critical values[a] of the beta binomial distribution

	Gamma							
	0	0.1	0.2	0.3	0.4	0.5	0.6	0.8
$p = 1/3$, one sided[b]								
N								
20	19	19	19	19	19	19	19	20
25	22	23	23	23	23	24	24	24
30	26	27	27	27	27	28	28	28
35	30	30	31	31	31	32	32	32
40	34	34	34	34	35	35	36	36
45	38	38	38	39	39	39	39	40
50	41	42	42	42	43	43	43	44
55	45	45	46	46	46	47	47	48
60	49	49	49	50	50	50	51	51
70	56	56	57	57	58	58	58	59
80	63	64	64	64	65	65	66	66
90	70	71	71	72	72	72	73	74
100	77	78	79	79	79	80	80	81
125	95	96	96	97	97	98	98	99
150	113	114	114	115	115	116	116	117
200	148	149	149	150	151	151	152	153
$p = 1/2$, one sided[b]								
20	26	26	26	26	27	27	27	27
25	31	32	32	32	32	33	33	33
30	37	37	37	38	38	38	39	39
35	42	43	43	43	44	44	44	45
40	48	48	49	49	49	50	50	50
45	53	54	54	54	55	55	55	56
50	59	59	60	60	60	61	61	61
55	64	65	65	65	66	66	66	67
60	70	70	70	71	71	72	72	73
70	80	81	81	82	82	82	83	84
80	91	91	92	92	93	93	94	94
90	101	102	103	103	104	104	104	105
100	112	113	113	114	114	115	115	116
125	138	139	140	140	141	141	142	143
150	165	165	166	167	167	168	169	170
200	217	218	218	219	220	221	221	223
$p = 1/2$, two sided[c]								
20	27	27	27	28	28	28	28	29
25	32	33	33	33	34	34	34	35
30	38	38	39	39	39	40	40	41
35	44	44	44	45	45	46	46	46
40	49	50	50	50	51	51	52	52
45	55	55	56	56	56	57	57	58
50	60	61	61	62	62	62	63	64
55	66	66	67	67	68	68	68	69
60	71	72	72	73	73	74	74	75
70	82	83	83	84	84	85	85	86
80	93	93	94	95	95	96	96	97
90	104	104	105	105	106	107	107	108
100	114	115	116	116	117	118	118	119
125	141	142	142	143	144	144	145	146
150	167	168	169	170	171	171	172	173
200	220	221	222	223	224	224	225	227

[a]Values are rounded up to 1 except where the exact value was less than 0.05 higher than the integer

[b]When used for discrimination tests, the total number of correct choices must equal or exceed the tabled value

[c]For this test, when used for preference tests, the total number of preference choices for the larger proportion (more preferred item) must equal or exceed the tabled value

Table F.L Minimum numbers of correct judgments[a] to establish significance at probability levels of 5 and 1% for paired difference and duo–trio tests (one tailed, $p = 1/2$) and the triangle test (one tailed, $p = 1/3$)

Paired difference and duo–trio tests				Triangle test	
	Probability levels			Probability levels	
Number of trials (n)	0.05	0.01	Number of trails (n)	0.05	0.01
7	7	7	5	4	5
8	7	8	6	5	6
9	8	9	7	5	6
10	9	10	8	6	7
11	9	10	9	6	7
12	10	11	10	7	8
13	10	12	11	7	8
14	11	12	12	8	9
15	12	13	13	8	9
16	12	14	14	9	10
17	13	14	15	9	10
18	13	15	16	9	11
19	14	15	17	10	11
20	15	16	18	10	12
21	15	17	19	11	12
22	16	17	20	11	13
23	16	18	21	12	13
24	17	19	22	12	14
25	18	19	23	12	14
26	18	20	24	13	15
27	19	20	25	13	15
28	19	21	26	14	15
29	20	22	27	14	16
30	20	22	28	15	16
31	21	23	29	15	17
32	22	24	30	15	17
33	22	24	31	16	18
34	23	25	32	16	18
35	23	25	33	17	18
36	24	26	34	17	19
37	24	26	35	17	19
38	25	27	36	18	20
39	26	28	37	18	20
40	26	28	38	19	21
41	27	29	39	19	21
42	27	29	40	19	21
43	28	30	41	20	22
44	28	31	42	20	22
45	29	31	43	20	23
46	30	32	44	21	23
47	30	32	45	21	24
48	31	33	46	22	24
49	31	34	47	22	24
50	32	34	48	22	25
60	37	40	49	23	25
70	43	46	50	23	26
80	48	51	60	27	30
90	54	57	70	31	34
100	59	63	80	35	38
			90	38	42
			100	42	45

[a]Created in EXCEL 2007 using B. T. Carr's Discrimination Test Analysis Tool EXCEL program (used with permission)

Table F.M Minimum numbers of correct judgments[a] to establish significance at probability levels of 5 and 1% for paired preference test (two tailed, $p = 1/2$)

Trials (n)	0.05	0.01	Trails (n)	0.05	0.01
7	7	7	45	30	32
8	8	8	46	31	33
9	8	9	47	31	33
10	9	10	48	32	34
11	10	11	49	32	34
12	10	11	50	33	35
13	11	12	60	39	41
14	12	13	70	44	47
15	12	13	80	50	52
16	13	14	90	55	58
17	13	15	100	61	64
18	14	15	110	66	69
19	15	16	120	72	75
20	15	17	130	77	81
21	16	17	140	83	86
22	17	18	150	88	92
23	17	19	160	93	97
24	18	19	170	99	103
25	18	20	180	104	108
26	19	20	190	109	114
27	20	21	200	115	119
28	22	22	250	141	146
29	21	22	300	168	173
30	21	23	350	194	200
31	22	24	400	221	227
32	23	24	450	247	253
33	23	25	500	273	280
34	24	25	550	299	306
35	24	26	600	325	332
36	25	27	650	351	359
37	25	27	700	377	385
38	26	28	750	403	411
39	27	28	800	429	437
40	27	29	850	455	463
41	28	30	900	480	490
42	28	30	950	506	516
43	29	31	1,000	532	542
44	29	31			

[a]Created in EXCEL 2007 using B. T. Carr's Discrimination Test Analysis Tool EXCEL program (used with permission)

Table F.N1 Minimum number of responses (n) and correct responses (x) to obtain a level of Type I and Type II risks in the triangle test. P_d is the chance-adjusted percent correct or proportion of discriminators

Type I risk	Type II risk					
	0.20		0.10		0.05	
	N	X	N	X	N	X
$P_d = 0.50$						
0.10	12	7	15	8	20	10
0.05	16	9	20	11	23	12
0.01	25	15	30	17	35	19
$P_d = 0.40$						
0.10	17	9	25	12	39	14
0.05	23	12	30	15	40	19
0.01	35	19	47	24	56	28
$P_d = 0.30$						
0.10	30	14	43	19	54	23
0.05	40	19	53	24	66	29
0.01	62	30	82	38	97	44
$P_d = 0.20$						
0.10	62	26	89	36	119	47
0.05	87	37	117	48	147	59
0.01	136	59	176	74	211	87

Abstracted from Schlich, P. 1993. Risk tables for discrimination tests. Food Quality and Preference, 4, 141–151.

Table F.N2 Minimum number of responses (n) and correct responses (x) to obtain a level of Type I and Type II risks in the duo–trio test. P_c is the chance-adjusted percent correct or proportion of discriminators

Type I risk	Type II risk					
	0.20		0.10		0.05	
	N	X	N	X	N	X
$P_d = 0.50$						
0.10	19	13	26	17	33	21
0.05	23	16	33	22	42	27
0.01	40	28	50	34	59	39
$P_d = 0.40$						
0.10	28	18	39	24	53	32
0.05	37	24	53	33	67	41
0.01	64	42	80	51	96	60
$P_d = 0.30$						
0.10	53	32	72	42	96	55
0.05	69	42	93	55	119	69
0.01	112	69	143	86	174	103
$P_d = 0.20$						
0.10	115	65	168	93	214	117
0.05	158	90	213	119	268	148
0.01	252	145	325	184	391	219

Abstracted from Schlich, P. 1993. Risk tables for discrimination tests. Food Quality and Preference, 4, 141–151

Table F.01 d' and B (variance factor) values for the duo–trio and 2-AFC (paired comparison) difference tests

	Duo–trio		2-AFC	
P_C	d'	B	d'	B
0.51	0.312	70.53	0.036	3.14
0.52	0.472	36.57	0.071	3.15
0.53	0.582	25.28	0.107	3.15
0.54	0.677	19.66	0.142	3.15
0.55	0.761	16.32	0.178	3.16
0.56	0.840	14.11	0.214	3.17
0.57	0.913	12.55	0.250	3.17
0.58	0.983	11.40	0.286	3.18
0.59	1.050	10.52	0.322	3.20
0.60	1.115	9.83	0.358	3.22
0.61	1.178	9.29	0.395	3.23
0.62	1.240	8.85	0.432	3.25
0.63	1.301	8.49	0.469	3.27
0.64	1.361	8.21	0.507	3.29
0.65	1.421	7.97	0.545	3.32
0.66	1.480	7.79	0.583	3.34
0.67	1.569	7.64	0.622	3.37
0.68	1.597	7.53	0.661	3.40
0.69	1.565	7.45	0.701	3.43
0.70	1.715	7.39	0.742	3.47
0.71	1.775	7.36	0.783	3.51
0.72	1.835	7.36	0.824	3.56
0.73	1.896	7.38	0.867	3.61
0.74	1.957	7.42	0.910	3.66
0.75	2.020	7.49	0.954	3.71
0.76	2.084	7.58	0.999	3.77
0.77	2.149	7.70	1.045	3.84
0.78	2.216	7.84	1.092	3.91
0.79	2.284	8.01	1.141	3.99
0.80	2.355	8.21	1.190	4.08
0.81	2.428	8.45	1.242	4.18
0.82	2.503	8.73	1.295	4.29
0.83	2.582	9.05	1.349	4.41
0.84	2.664	9.42	1.406	4.54
0.85	2.749	9.86	1.466	4.69
0.86	2.840	10.36	1.528	4.86
0.87	2.935	10.96	1.593	5.05
0.88	3.037	11.65	1.662	5.28
0.89	3.146	12.48	1.735	5.54
0.90	3.263	13.47	1.812	5.84
0.91	3.390	14.67	1.896	6.21
0.92	3.530	16.16	1.987	6.66
0.93	3.689	18.02	2.087	7.22
0.94	3.867	20.45	2.199	7.95

Table F.01 (continued)

P_C	Duo–trio		2-AFC	
	d'	B	d'	B
0.95	4.072	23.71	2.326	8.93
0.96	4.318	28.34	82.476	10.34
0.97	3.625	35.52	2.660	12.57
0.98	5.040	48.59	2.900	16.72
0.99	5.701	82.78	3.290	27.88

B-factors are used to compute variance of the d' values, where
$\mathrm{Var}(d') = B/N$, where N is the sample size
Reprinted with permission from "Tables for Sensory Methods,
The Institute for Perception, February, 2002"

Table F.02 d' and B (variance factor) values for the triangle
and 3-AFC difference tests

P_C	Triangle		3-AFC	
	d'	B	d'	B
0.34	0.270	93.24	0.024	2.78
0.35	0.429	38.88	0.059	2.76
0.36	0.545	25.31	0.093	2.74
0.37	0.643	19.17	0.128	2.72
0.38	0.728	15.67	0.162	2.71
0.39	0.807	13.42	0.195	2.69
0.40	0.879	11.86	0.229	2.68
0.41	0.948	10.71	0.262	2.67
0.42	1.013	9.85	0.295	2.66
0.43	1.075	9.17	0.328	2.65
0.44	1.135	8.62	0.361	2.65
0.45	1.193	8.18	0.394	2.64
0.46	1.250	7.82	0.427	2.64
0.47	1.306	7.52	0.459	2.64
0.48	1.360	7.27	0.492	2.63
0.49	1.414	7.06	0.524	2.63
0.50	1.466	6.88	0.557	2.64
0.51	1.518	6.73	0.589	2.64
0.52	1.570	6.60	0.622	2.64
0.53	1.621	6.50	0.654	2.65
0.54	1.672	6.41	0.687	2.65
0.55	1.723	6.34	0.719	2.66
0.56	1.774	6.28	0.752	2.67
0.57	1.824	6.24	0.785	2.68
0.58	1.874	6.21	0.818	2.69
0.59	1.925	6.19	0.852	2.70
0.60	1.976	6.18	0.885	2.71
0.61	2.027	6.18	0.919	2.73
0.62	2.078	6.19	0.953	2.75
0.63	2.129	6.21	0.987	2.77
0.64	2.181	6.28	1.022	2.79
0.65	2.233	6.29	1.057	2.81
0.66	2.286	6.32	1.092	2.83

Table F.02 (continued)

P_C	Triangle		3-AFC	
	d'	B	d'	B
0.67	2.339	6.38	0.128	2.86
0.68	2.393	6.44	1.164	2.89
0.69	2.448	6.52	1.201	2.92
0.70	2.504	6.60	1.238	2.95
0.71	2.560	6.69	1.276	2.99
0.72	2.618	6.80	1.314	3.03
0.73	2.676	6.91	1.353	3.07
0.74	2.736	7.04	1.393	3.12
0.75	2.780	7.18	1.434	3.17
0.76	2.860	7.34	1.475	3.22
0.77	2.924	7.51	1.518	3.28
0.78	2.990	7.70	1.562	3.35
0.79	3.058	7.91	1.606	3.42
0.80	3.129	8.14	1.652	3.50
0.81	3.201	8.40	1.700	3.59
0.82	3.276	8.68	1.749	3.68
0.83	3.355	8.99	1.800	3.79
0.84	3.436	9.34	1.853	3.91
0.85	3.522	9.74	1.908	4.04
0.86	3.611	10.19	1.965	4.19
0.87	3.706	10.70	2.026	4.37
0.88	3.806	11.29	2.090	4.57
0.89	3.913	11.97	2.158	4.80
0.90	4.028	12.78	2.230	5.07
0.91	4.152	13.75	2.308	5.40
0.92	4.288	14.92	2.393	5.81
0.93	4.438	16.40	2.487	6.30
0.94	4.607	18.31	2.591	6.95
0.95	4.801	20.88	2.710	7.83
0.96	5.031	24.58	2.850	9.10
0.97	5.316	30.45	3.023	11.10
0.98	5.698	41.39	3.253	14.85
0.99	6.310	71.03	3.618	25.00

B-factors are used to compute variance of the d' values, where $Var(d') = B/N$, where N is the sample size

Reprinted with permission from "Tables for Sensory Methods, The Institute for Perception, February, 2002"

Table F.P Random permutations of nine

6	4	9	3	8	7	2	5	1	2	1	6	7	5	8	4	3	9
4	2	1	9	3	8	7	6	5	9	8	3	7	6	4	5	2	1
3	5	4	1	6	8	7	9	2	3	6	2	4	9	7	1	8	5
5	3	4	2	1	6	8	9	7	4	9	5	7	1	3	8	6	2
8	7	1	9	2	5	6	4	3	1	7	2	6	9	3	5	4	8
3	6	9	7	2	8	5	1	4	6	7	5	9	8	3	1	4	2
3	1	7	6	5	2	4	9	8	4	8	7	3	5	6	9	1	2
3	1	2	9	4	5	6	8	7	8	3	9	6	7	1	4	5	2
1	3	5	7	2	6	8	9	4	4	3	5	9	8	2	1	7	6
6	3	8	9	7	4	2	5	1	6	8	7	9	5	2	1	4	3
1	7	5	3	6	8	4	2	9	8	5	1	7	9	3	6	4	2
6	3	9	7	5	8	1	4	2	8	2	1	4	6	9	5	3	7
7	5	1	2	8	4	9	3	6	3	5	1	4	2	7	9	8	6
1	2	4	8	9	3	6	5	7	2	6	3	9	7	5	8	4	1
4	6	3	9	5	7	2	8	1	9	6	8	5	2	4	7	1	3
7	6	1	5	4	8	2	9	3	8	3	2	5	9	6	4	1	7
3	9	7	5	4	6	8	1	2	7	3	4	2	1	9	5	8	6
1	3	5	7	6	8	2	4	9	6	5	4	3	2	1	7	9	8
2	9	4	7	1	3	5	8	6	1	5	4	2	6	7	9	3	8
5	2	8	3	4	7	1	9	6	6	5	1	4	9	7	2	3	8
2	1	8	7	3	5	9	4	6	7	8	1	2	3	4	5	9	6
5	7	2	8	6	3	4	9	1	3	9	1	4	6	5	8	2	7
4	1	6	2	5	3	7	9	8	8	6	5	7	4	3	9	2	1
1	6	7	9	4	8	2	5	3	8	9	2	5	4	3	7	1	6
9	8	5	1	6	2	3	7	4	5	4	3	6	9	8	1	7	2
5	3	1	6	7	8	2	9	4	1	9	7	2	3	8	4	5	6
1	3	2	7	8	5	4	6	9	4	1	2	6	3	5	7	8	9
3	4	9	7	5	8	1	6	2	5	2	3	7	4	6	8	9	1
5	4	6	8	2	1	7	9	3	4	6	8	9	2	3	1	7	5
1	3	7	9	4	8	6	2	5	4	2	9	3	1	7	6	8	5
6	2	5	1	9	8	4	7	3	2	5	6	9	4	7	3	1	8
5	2	9	8	3	1	4	6	7	4	9	2	6	1	5	7	3	8
8	5	1	3	6	2	9	7	4	6	3	2	4	9	1	5	8	7
1	7	4	3	2	9	5	6	8	2	3	6	4	5	8	7	1	9
9	3	4	5	6	7	1	8	2	6	1	4	5	8	7	2	3	9
1	6	4	3	5	9	7	8	2	7	8	9	4	2	5	3	6	1
4	5	9	8	1	2	3	6	7	7	3	8	1	9	2	6	5	4
9	8	5	4	2	7	3	1	6	7	2	1	9	5	4	6	3	8
9	8	2	6	4	5	7	1	3	9	6	3	8	7	2	5	4	1
9	3	1	5	6	2	4	8	7	7	1	8	2	3	9	5	4	6
4	7	6	9	3	2	1	8	5	7	3	4	9	1	5	2	6	8
7	1	8	5	6	9	4	2	3	2	3	7	9	4	8	5	6	1

Each row with a column has the number 1–9 in random order. Start with any row (do not always start with the first or last rows) and read either from right to left or from left to right

Table F.Q Random numbers

8	2	0	3	1	4	5	8	2	1	7	2	7	3	8	5	5	2	9	0	6	3	1	8	4
0	8	7	3	3	1	9	7	5	2	5	7	8	9	8	0	3	8	2	5	1	2	7	5	2
2	3	3	8	8	1	4	2	4	0	2	6	1	8	9	5	2	8	9	8	3	4	0	1	0
4	7	5	5	8	3	0	7	7	1	9	1	8	1	7	4	1	7	1	3	7	9	3	3	7
1	9	3	9	5	3	4	9	5	5	2	7	5	8	0	3	4	8	8	1	2	7	5	3	4
2	8	7	8	1	4	1	4	9	4	2	4	1	5	2	9	4	8	2	1	5	2	8	1	9
8	4	8	5	1	3	9	8	6	0	7	2	1	9	0	2	0	8	7	0	8	0	1	3	0
0	3	8	8	4	7	5	1	5	1	7	3	4	5	2	0	7	4	7	9	8	6	7	7	4
3	5	3	1	9	3	7	4	9	5	0	2	0	1	4	6	2	5	4	5	8	5	0	9	2
3	4	5	9	5	2	7	9	8	9	0	5	5	8	5	1	7	7	3	5	5	4	7	7	2
4	1	5	3	0	9	1	3	7	2	5	8	7	7	1	3	6	3	9	7	8	7	9	1	7
7	2	9	5	6	7	8	5	4	5	3	4	5	4	1	9	8	8	7	5	7	9	3	1	8
5	9	2	8	9	8	6	4	4	1	5	3	7	7	0	8	0	2	5	6	0	8	1	2	0
1	3	3	3	9	0	5	2	8	7	4	0	9	0	3	7	3	1	7	9	4	5	5	2	8
4	8	0	1	0	8	6	2	1	0	0	5	0	3	1	5	4	9	0	3	7	4	7	0	1
7	7	0	8	6	3	2	8	8	5	8	9	5	8	4	0	5	9	1	8	0	5	4	9	4
3	3	8	5	7	5	7	4	3	4	5	7	9	8	9	5	0	7	7	6	8	8	8	5	9
9	1	7	1	3	6	9	2	9	1	9	4	2	3	3	0	8	1	8	7	7	6	4	7	2
6	2	2	8	0	9	4	5	3	7	2	5	4	8	8	5	6	6	5	0	4	6	5	6	8
1	7	5	9	0	0	2	0	5	8	5	8	5	1	9	5	3	3	7	4	0	5	8	2	4
0	3	9	6	9	4	7	3	5	7	0	8	5	4	7	1	1	8	5	3	2	8	0	9	8
3	0	8	2	8	1	4	4	1	8	7	8	6	9	9	9	7	5	8	9	8	4	5	9	0
9	4	9	1	2	2	0	1	3	2	4	8	7	9	1	8	8	2	9	8	3	2	8	2	9
7	2	5	1	4	4	9	8	5	2	8	5	5	1	0	8	2	6	2	0	8	9	2	2	3
9	9	2	5	7	4	3	1	2	3	8	4	1	5	2	4	0	4	2	2	8	7	1	8	2
2	0	9	1	8	9	4	4	8	1	4	8	8	7	9	2	5	0	8	9	3	3	0	1	2
8	5	2	8	1	2	1	7	7	1	4	7	8	1	4	2	7	3	7	4	0	0	1	2	9
1	2	9	9	8	4	2	5	3	2	7	4	3	2	3	3	8	5	3	3	8	5	5	3	2
3	2	8	3	7	9	6	0	4	8	8	0	5	4	1	1	4	9	0	5	0	9	4	4	1
0	9	3	4	1	1	9	5	8	3	2	4	6	7	3	4	4	9	2	3	7	2	5	7	8
8	7	5	3	4	2	1	5	5	0	1	2	4	7	5	5	2	8	8	7	8	2	8	0	3
9	6	0	1	3	0	5	3	8	6	2	9	6	0	3	4	7	8	1	1	9	1	6	5	3

Start on any column or row and read from right to left or left to right or up and down to create random numbers of three digits to label your sample cups

Author Index

Subject Index

CPSIA information can be obtained at www.ICGtesting.com
Printed in the USA
LVOW09*1440070816

499405LV00004B/15/P

9 781441 964878